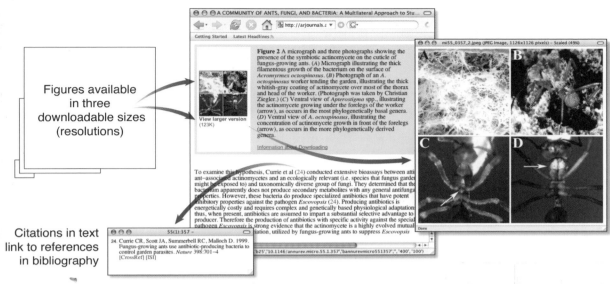

Figures available
in three
downloadable sizes
(resolutions)

Citations in text
link to references
in bibliography

References in Annual Reviews
article bibliography link out to
sources of cited articles online

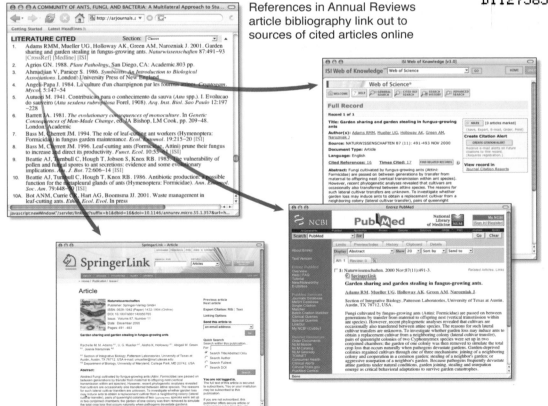

Annual Review of
Microbiology

Annual Review of Microbiology

Volume 62, 2008

Susan Gottesman, *Editor*
National Cancer Institute, Bethesda

Caroline S. Harwood, *Associate Editor*
University of Washington

www.annualreviews.org • science@annualreviews.org • 650-493-4400

Annual Reviews
4139 El Camino Way • P.O. Box 10139 • Palo Alto, California 94303-0139

Annual Reviews
Palo Alto, California, USA

International Standard Serial Number: 0066-4227
International Standard Book Number: 978-0-8243-1162-9
Library of Congress Catalog Card Number: 49-432

TYPESET BY APTARA
PRINTED AND BOUND BY FRIESENS CORPORATION, ALTONA, MANITOBA, CANADA

Preface

The process of putting together an Annual Review volume is long but enjoyable—a chance to wander through the most exciting science in your field, try to envision what will be ripe for reviewing a year hence, and then wait to see what appears. For the *Annual Review of Microbiology* (as for the other volumes), the Editorial Committee meets once a year, having for the previous year jotted down thoughts and notes from meetings, seminars, and reading; we present our favorite ideas to the group, look at proposals that have been submitted, and try to put together a balanced list of topics and authors. Then we send out invitations, and wait. The reviews are due about 18 months later, for a volume that appears more than two years later. Thus, I am writing this preface in a kind of time warp, with the reality of who was involved in planning this volume a somewhat distant memory.

Some major changes to *ARM* have occurred between the spring of 2006, when this volume was planned, and now. While my name may appear on the spine of this volume as Editor, don't let that mislead any readers. This volume, as well as next year's, owes its form and existence entirely to Nick Ornston's enthusiastic and energetic leadership. After 25 years under the very able oversight of Nick Ornston, I have been handed the somewhat overwhelming job of trying to fill his shoes as Editor of *ARM*. It seems unlikely I will be able to conjure up the literary references, from Kermit the Frog to Dionysus, that Nick did in his prefaces through the years. Nick was only the third Editor of *ARM* in its history, and it is humbling to know that most of what I learned from *ARM* during much of my career was thanks to his abilities to pick just the right Editorial Committee members and authors. I plan to lean on his advice heavily in the coming years, as I've done since I joined the Editorial Committee.

A major loss for the *ARM* family was the death of Albert Balows in the fall of 2006. Al, a renowned clinical microbiologist, brought his broad interests, his ability to come up with the perfect entertaining title for a proposed article, as well as his knowledge about the more clinical aspects of our field to *ARM* as a member of the Editorial Committee and, since 1980, as Associate Editor. When travel became difficult, he was still fully engaged, calling in his suggestions for this volume and others. We will miss his enthusiasm and his insights.

This year we welcome Peggy Cotter and Stephen Hajduk as Editorial Committee members. Nancy A. Moran joined us to help plan this volume and since has become a member of our Editorial Committee. William Burkholder and John Boothroyd also helped in the planning, and we thank them all for their contributions.

As I begin my term as Editor, I am extremely grateful for the expertise and support of Cleo X. Ray as production editor, the generous advice and encouragement of Nick Ornston, and the enthusiasm of Sam Gubins, Editor-in-Chief of Annual Reviews.

Susan Gottesman

Editor

**Annual Review of
Microbiology**

Volume 62, 2008

Contents

Indexes

Errata

An online log of corrections to *Annual Review of Microbiology* articles may be found at
http://micro.annualreviews.org/

Related Articles

The APOBEC3 Cytidine Deaminases: An Innate Defensive Network Opposing
Exogenous Retroviruses and Endogenous Retroelements
Ya-Lin Chiu and Warner C. Greene

From the *Annual Review of Medicine*, Volume 59 (2008)

Bacterial and Fungal Biofilm Infections
A. Simon Lynch and Gregory T. Robertson

The Challenge of Hepatitis C in the HIV-Infected Person
David L. Thomas

Hide-and-Seek: The Challenge of Viral Persistence in HIV-1 Infection
Luc Geeraert, Günter Kraus, and Roger J. Pomerantz

Herpes Simplex: Insights on Pathogenesis and Possible Vaccines
David M. Koelle and Lawrence Corey

From the *Annual Review of Phytopathology*, Volume 46 (2008)

Plants as a Habitat for Beneficial and/or Human Pathogenic Bacteria
Heather L.L. Tyler and Eric W. Triplett

Breaking the Barriers: Microbial Effector Molecules Subvert Plant Immunity
Silke Robatzek and Vera Göhre

Yeast as a Model Host to Explore Plant Virus-Host Interactions
Peter D. Nagy

Living in Two Worlds: The Plant and Insect Lifestyles of *Xyella fastidiosa*
Subhadeep Chatterjee, Rodrigo P.P. Almeida, and Steven Lindow

Plant Viruses as Biotemplates for Materials and Their Use in Nanotechnology
Trevor Douglas, Mark Young, Debbie Willits, and Masaki Uchida

From the *Annual Review of Plant Biology*, Volume 59 (2008)

Roots, Nitrogen Transformations, and Ecosystem Services
Louise E. Jackson, Martin Burger, and Timothy R. Cavagnaro

Coordinating Nodule Morphogenesis with Rhizobial Infection in Legumes
Giles E.D. Oldroyd and J. Allan Downie

The Fortunate Professor

Stanley Falkow

Department of Microbiology and Immunology and Medicine (Infectious Diseases and Geographic Medicine), Stanford University School of Medicine, Stanford, California 94305; email: falkow@stanford.edu

Annu. Rev. Microbiol. 2008. 62:1–18

First published online as a Review in Advance on March 17, 2008

The *Annual Review of Microbiology* is online at micro.annualreviews.org

This article's doi: 10.1146/annurev.micro.62.081307.162931

0066-4227/08/1013-0001$20.00

Key Words

plasmid, bacterial pathogenicity, R factors

Abstract

My professional life can be summarized by a quote from the Talmud.

Much have I learned from my teachers,

More from my colleagues,

But most from my students.

It is the fortunate professor who learns from the student.

Contents

INTRODUCTION

I am honored to be asked to contribute the Prefatory chapter for Volume 62 of the *Annual Review of Microbiology*. The mandate was to provide a retrospective view of my career in microbiology, and I was encouraged to include personal reminiscences and events that were significant to me. This is not an easy task under any circumstance. In my case, it took on added meaning because three years ago I was diagnosed with a bone marrow disease, which led me to begin to bring my career to an end. I feel quite well and my disease, like a number of microbial diseases I studied in my career, has an imprecise natural history. The state of my health was known to some of the Committee Members when I was invited to submit this article. I thought, "Well, I guess they wanted to invite me before it's too late."

I would like to take this opportunity to share some of the things I learned in my professional life and to outline the help and guidance I received from so many people along the way. During the four decades I worked as an active scientist, I helped train 35 graduate students and 61 postdoctoral students and infectious diseases fellows, and I collaborated with and published scientific reports with 75 other scientists. I dedicate this article to them, especially my students. I must apologize; I could not possibly mention them all or thank them all sufficiently in the space provided for this short memoir.

MY DECISION TO BECOME A BACTERIOLOGIST

My father, Jacob Falkowitzki (?), was born in Kiev before World War I and immigrated to Albany, New York; my mother, Mollie Gingold, was born in the United States after her family arrived from Bialystock, Poland. I was born in 1934, at the peak of the Great Depression. My first language was Yiddish, and my early years were spent in a noisy, colorful environment of tenement row houses filled with a mélange of languages, smells, and customs.

In 1943 my family moved to Newport, Rhode Island. I was taken from an intense urban neighborhood of Jews, Italians, and Poles into a conservative New England seaside community. I attended a public school and was a terrible student by both objective and subjective criteria, but I loved to read, although I didn't want to read what was assigned to me in school. Somehow I discovered the public library and happened on a book called *Microbe Hunters* by Paul de Kruif. De Kruif describes in colorful detail the microbe hunters Louis Pasteur, Robert Koch, Paul Ehrlich, Elie Metchnikoff, and others, who showed that microbes could cause disease. He described their work to understand the human systems that oppose infection and their search for cures. These microbe hunters became, and remain, my heroes. Their search to understand microbes was to me the most extraordinary adventure that I could imagine. It still is.

Thus, at age 11, I decided to become a bacteriologist. My first view of bacteria was in spoiled milk, barely discernable in my Gilbert Company Hall of Science microscope, but no doubt about it, here were Antony van Leeuwenhoek's tumbling animalcules. Protozoa in a grass infusoria were even more wondrous. I wish I could report that these revelations had a major impact on my endeavors as a student. They did not. I continued my record of failure in mathematics, and my science grades in chemistry and biology were average, at best. Fortunately, I was put on the right path through the intervention of Lottie Brindle, a remarkable English teacher, who told me, among other things, "If you stop being such a wise guy and listen, you just might make something of yourself." I cannot say that I underwent a complete transformation, but I became a better student.

THE UNIVERSITY OF MAINE (1951–1955)

I left Newport to attend the University of Maine in 1951. After years of declaring that I wanted to be a bacteriologist, I suddenly realized I had fallen in love with a fantasy I had conjured up from a description in a book. At Maine I finally applied myself and became immersed in serious laboratory exercises in chemistry and biology. Toward the end of my freshman year, I wrote a letter to the Newport Hospital offering my services during my summer break with the stated hope to be taught something about bacteriology. They replied that they would be happy to employ me, without pay, to work in the clinical laboratory. My duties would include helping the laboratory technicians, assisting at autopsies and helping the hospital bacteriologist, Alice Schaeffer Sauzette, who had developed a well-known endospore stain (22). In the summers that followed, Alice showed me how to identify the microbes appearing on the plates of growth media on which we spread bits of effluvia emanating from within or on the human body during health and disease. I gram-stained all of it and looked at the slides using a real microscope.

This experience exceeded my childhood fantasy! During that first summer, I saw most of the common infectious diseases of that time and often isolated the microbes from the sick patients. If there were a case of something unusual like diphtheria, meningitis, or whooping cough, I would arrange with a physician to be present when he examined that individual, and sadly, sometimes I saw the patient again on the autopsy table. Many of the physicians at the hospital knew of my interests and would talk to me

about infectious disease epidemics of the past and reminisce about the days before antibiotics. I was fortunate to have been exposed to the discipline of medical microbiology and infectious disease so early in my career; it helped me understand host-pathogen interactions before I even knew that this was the path of research I would follow.

In my junior year I finally took a university course in general bacteriology. There was only one other bacteriology major, Peter Pattee, who subsequently became a distinguished professor at Iowa State University. I showed off my skill at streaking plates and performing the routine chores of culturing and identifying bacteria. I am happy to report that my professors in the bacteriology department, particularly my advisor, E. R. Hitchner, took my self-importance and misplaced pride in stride and then firmly, but gently, pointed out my ignorance.

My undergraduate bacteriology training at Maine included fermentation mechanisms, where we prepared sauerkraut and studied how each group of microbes interacted and sequentially cooperated to transform a crock of alternating layers of shredded cabbage, salt, and water into a wonderful accompaniment for bratwursts. I learned dairy bacteriology. I soaked up virology, parasitology, and, at last, a formal course in medical bacteriology. Hitchner presented soil microbiology lectures with all the reverence of a parson preaching to his congregation. He made me see that microorganisms live not as pure cultures, but as members of complex communities that often depended on one another for their survival. To get the baccalaureate degree in bacteriology, it was also necessary to do a research project and to write a report of the findings. I studied the L-forms of bacteria and this little study eventually became my first scientific paper (5).

GRADUATE SCHOOL (1955–1960)

I started graduate school at the University of Michigan in the fall of 1955. Three events followed that had a major impact on my life. First, I met Alan Campbell, a new assistant professor. I was Alan's teaching assistant, and he gave me a copy of the new book, *The Microbial World*, written by Roger Y. Stanier, Michael Douderoff, and Edward A. Adelberg, which was to be the foundation of his course. He also introduced me to bacterial genetics.

Second, I began to suffer panic attacks. Today, this biochemical anomaly can be treated with a variety of drugs and behavior modification; in 1955, there was nothing available. I've written about this before (7), and there is no doubt it influenced my personal and professional life immensely. The immediate effect was to stop my graduate education. I returned to my job at Newport Hospital. In the evenings I did research, which led to the publication of two papers describing biochemical tests for distinguishing between different species of enteric bacteria, one of which, I'm told, is still used occasionally (6).

Third, these papers put me in contact with C. A. Stuart at Brown University. I asked him how bacteria became pathogenic and how the enteric species had evolved. Rather than answer, he introduced me to Herman Chase, a mouse geneticist, who handed me a copy of a new book, *The Chemical Basis of Heredity* (20). Chase suggested I read as much as I could and come back weekly and we'd discuss any questions I had. I belatedly learned about the discovery of DNA and had the privilege of reading papers by Seymour Benzer, Francois Jacob, and other greats of that era.

Thanks to Chase and Stuart, I was given a graduate fellowship and started graduate school at Brown in the fall of 1957 working in "Doc" Stuart's laboratory. Seymour Lederberg joined the Brown faculty that year. Seymour had worked with Salvatore Luria and planned to duplicate the storied Cold Spring Harbor phage course for Brown biology students. I was Seymour's teaching assistant, and he tutored me as I performed each of the experiments prior to lab class. It was a marvelous opportunity that served me well thereafter.

Then, and today, a first-year graduate student is introduced to the intricacies and uncertainties of scientific papers and is taught to understand the strength of the work that is reported and its possible flaws. In addition, I took courses in genetics taught by Chase. I remember especially Elizabeth LeDuc's cell biology lectures seemed like poetry. All my professors were scholars who asked us innocent questions that had no answers. They led my classmates and me to the cutting edge of thinking about the biology of that era. They posed questions I had never contemplated, and now forced to consider them, I was humbled and yet challenged. Graduate school was one of the happiest times of my professional life.

WALTER REED (1960–1967)

Doc Stuart gave me a paper to read written by Louis S. Baron on the transfer of *Escherichia coli* genes to *Salmonella typhi*. Norton Zinder published similar results using *E. coli* and *Salmonella typhimurium*. I duplicated Baron and Zinder's results and spent most of the summer of 1959 reading everything I could lay my hands on about *E. coli* genetics and performing genetic crosses between different *Salmonella* and *E. coli* Hfr donors. The time had come to do my own genetic experiments and not just read about the work of others. Stuart was near retirement and had severe heart disease so he sent me in September 1960 to perform my thesis research in Lou Baron's laboratory at Walter Reed Army Institute of Research (WRAIR).

Lou was a wonderful man and a gifted bacterial geneticist. He asked me prior to my arrival to work on a naturally occurring *S. typhi* clinical isolate that was notable for its ability to ferment lactose and transfer the *lac*+ genotype at high frequency to other bacteria. My experiments indicated that this strain was nature's version of the F-*lac* episome, which had recently been described for *E. coli* K-12 by Ed Adelberg and Francois Jacob. Moreover, the lactose gene was transferred from the *S. typhi* donor to many different kinds of lactose-negative bacteria. To my astonishment, the experiment even worked for *Serratia marcescens* and *Vibrio cholerae*.

I performed an experiment violating the conventional wisdom that gene transfer only occurred between closely related bacteria. Like every other biology student of that time, I knew about the 1958 Meselson-Stahl experiment on the nature of DNA replication based on showing that DNA of different density, including differences in base composition, could be separated in concentrated solutions of cesium chloride (CsCl). *S. marcescens* had a G+C content of 58%, while *E. coli*, *Salmonella typhi*, and *Shigella flexneri* DNA were all 50%–52% G+C. In a CsCl gradient, molecules with an overall composition of 50% would band at a totally distinct location compared with a 58% G+C molecule. I wondered if the DNA from a culture of *S. marcescens* carrying the F-*lac* genetic factor might show two distinct bands in the CsCl gradient, one corresponding to the chromosome of *Serratia* and the other to a purported extrachromosomal element like the F_0-*lac* factor transferred from the *S. typhi* donor. It was a simple experiment. That is, it was simple if I could make the DNA and if I had a Model E ultracentrifuge and knew how to use it. Within a month, Baron and Sam Formal, who has remained a wonderful mentor to me, arranged an American Society for Microbiology (ASM) travel fellowship and the Army's blessing to work with Julius Marmur at Brandeis University to test this idea.

I spent two weeks at Brandeis under Marmur's guidance. I learned how to purify bacterial DNA. It sounds so trivial now, but in 1960 few could do so. I learned how to perform DNA melting curves, and of course, there was the experiment to determine if we could actually "see" F-DNA. We did (11)! Julius remained a strong force throughout my career until his death. He would send me notes and journal articles inscribed with "You should look at this!"

I used my newfound molecular tools to study *E. coli*, *S. typhi*, and their hybrids. I hoped to understand how pathogenic enteric bacteria differed from their nonpathogenic brethren. I thought pathogens must have genes that defined their ability to be invasive; however, I could not establish this experimentally in *Salmonella* or in experiments I did with Sam Formal in *Shigella*. Thus, in 1962

when I presented my data on *Salmonella* and *Shigella* virulence to an audience at Cold Spring Harbor, their response was outright skepticism. Pathogens were seen as degenerate forms of bacteria that grew at the expense of the host and caused damage by doing so. I argued that pathogenicity was a kind of genetic adaptation for survival; pathogens were not all retrograde microbes or some kind of vicious throwback. I lacked the experimental data to make the point, and moreover, I didn't know what experiments to do to unequivocally demonstrate my view or to convince others that pathogenicity was not equivalent to disease.

I was counseled by my elders at Cold Spring Harbor to forget pathogens and to concentrate more on episomes, like F, and to continue the kind of work I had done with Marmur. By a fortunate accident (such an important ingredient in research), I noted that the newly described R factors, extrachromosomal elements mediating antibiotic resistance, could be seen as a faint band in a CsCl gradient of *Serratia* that carried these genetic elements. Therefore, I decided to focus my research on F and especially the R factors.

I established a close working relationship with two investigators, Naomi Datta at the Royal Postgraduate Medical School in London and Tsutomu (Tom) Watanabe of Keio University School of Medicine in Tokyo, who were working on the genetics of R factors. Our contact was by handwritten letters that detailed our respective progress, thoughts, suggestions, and ideas. It is pleasing to occasionally reread our correspondence and think about how earnest we were, how ignorant we were in some cases, and actually how insightful we were occasionally without realizing it. The problem then and now is most often we don't know which thoughts are ignorant and which thoughts are insightful until some time later.

In contemporary science, most PhD recipients spend three or more years as postdoctoral fellows in another laboratory. This is a time to put into practice the principles learned in pursuing the doctorate and permits a young investigator to become independent and to decide what facet of science to focus on in the decades to follow. I did not have a formal postdoctoral education. Rather, I had been put into a relatively independent research position while I was still a doctoral student. Fortunately, I was invited to join a group at the National Institutes of Health (NIH) that became known as the Lambda Lunch Group. (It still exists today.) It began as an informal group, organized by Gordon Tompkins in the Laboratory of Molecular Biology (LMC), then part of the National Institute of Arthritis and Metabolic Diseases.

The world of molecular biology was making a transition from the bacterial genetics of the Pasteur school to the detailed molecular analysis of more-defined genetic elements with small chromosomes such as the bacteriophage lambda (λ) and, thank goodness, R factors. Thus, on Wednesdays, Lambda lunchniks, which included Gary Felsenfeld, Martin Gellert, Bob Martin, and Bruce Ames, met to listen to a talk or discuss a recent paper. A. L. (Larry) Taylor attended, as well as Gerry Fink and John Roth. All of them, like me, were just beginning in the field. The interactions with the attendees at the Lambda Lunch stood me in good stead for many years thereafter. Also, at NIH I met people like Wally Rowe, Maxine Singer, Malcolm Martin, Bill Hoyer, and Marshall Nirenberg, who answered my questions, explained things, or simply encouraged me with a gentle push and, on occasion, shove to explore new facets of science. Thus, although I didn't have a formal postdoctoral experience, I always thought myself lucky because, rather than having one postdoctoral mentor, I had many.

At one Lambda Lunch I heard about experiments performed by Ellis Bolton and Brian McCarthy at the Carnegie Institution of Washington's Department of Terrestrial Magnetism (DTM) using DNA immobilized in agar gels to quantitatively measure DNA-DNA and DNA-RNA hybrids. With their help I began to measure the relatedness between different bacteria, their viruses, and other extrachromosomal elements with reasonable quantization. I

say this now, when the entire sequence of DNA for almost all pathogenic bacteria is known and stored in a public database and a simple computer command will compare two organisms, base by base, for their similarity. But in 1963, I was ecstatic. This simple experimental tool, together with the other methods I had learned, formed a tangible research plan to study pathogenic bacteria and their episomes and to determine how (if) they differed significantly from nonpathogens. Soon DNA agar columns became passé as we learned that certain kinds of filter paper sufficed, and the Southern blot revolutionized the way scientists looked at DNA homology. Now, of course, we have DNA and RNA microarrays to look at global relationships over entire genomes.

My WRAIR colleagues, John Wohlheiter and Ron Citarella, and I put our collective energy into examining the nucleic acid of F factor, R factors, and other extrachromosomal elements. In 1965 we characterized F-factor DNA (9), and a year later we published a broad study of the molecular nature of R factors (10). From 1965 to 1967, I collaborated with a number of different investigators at WRAIR, NIH, and Carnegie. Gary Felsenfeld and Shalom Hirschman at NIH introduced me to the fine mathematical details of analyzing the melting curves of DNA to gain insight into the evolution of bacteriophage genomes. This was complemented by work I did with Dean Cowie at Carnegie on the same subject. David Kohne, David Kingsbury, and Don J. Brenner were all postdoctoral fellows at Carnegie. Brenner and I worked together for the next seven years, measuring the molecular relationships between members of the *Enterobacteriaceae* using nucleic acid hybridization (2). Don eventually went to the Centers for Disease Control and Prevention (CDC) and was a major force in bacterial taxonomy and medical bacteriology.

GEORGETOWN UNIVERSITY (1967–1972)

Arthur Saz asked me to teach a few lectures to the medical students at Georgetown University.

I discovered teaching was as much fun as research. Saz persuaded me to move across town to join the Georgetown faculty in 1967. The National Science Foundation awarded me my first grant to study "bacterial extrachromosomal elements." I found people who were willing to work with me. Richard Silver, Pat Guerry, Lucy Tompkins, and Vickers Hirschfeld were my first graduate students. My first postdoctoral student was Don LeBlanc. We worked in a laboratory that was about as wide and half as long as a bowling alley lane into which we crammed the ultracentrifuges, liquid scintillation counters, and other equipment necessary in those days to do molecular biology. We began to look at the molecular events associated with the transfer of R-factor DNA into recipient cells using a method developed by David and Dorothy Freifelder. The university environment and the constant questioning by students validated my decision to follow a career of scientific research that had as its foundation teaching others.

I made more than my fair share of mistakes as a young professor. I was still fortunate to have a new set of mentors. Saz talked with me daily and managed to insert Talmudic-like wisdom and subtle and not so subtle messages into our conversations. Another person who influenced me was the dean of the medical school, John C. Rose, who seemed to have a perpetual little smile (not common for deans then and certainly not now). John helped me with my first steps in being a mentor.

The mentor's role is that of an advisor or counselor. It cannot be a friendship in the usual sense of the word nor can it be paternalistic. In so far as is possible, the advice should not be formulated into what would I (the mentor) do, but rather what would be in the best interest of the person I counsel. It requires absolute honesty and trust. It is a learned role and often the learning is through understanding the hopes, aspirations, and pain of the people who are your students. You must listen! Alas, I learned through mistakes made with my students. Errors are forgivable, especially if you recognize the error quickly and

apologize. Most of all, try not to repeat the error again.

The most profound change for me as a professor was the realization that the research work would no longer be done with my own hands but by others. I had difficulty following the directions of others when it came to my own experiments. I did not feel that I could tell others what to do. My students could not be an extension of me. They had to follow their own ideas. I could be a dedicated participant and guide, but the decisions had to be theirs. This is not to say I didn't argue, cajole, whine, and even beg a student to do it my way. In the end, it was their decision. In the paragraphs that follow, it should be explicitly understood that it is really their story to tell and not mine. It has been a journey of shared hopes and ideas.

Saz and Rose helped me organize a symposium on R factors at Georgetown. I invited and finally met Tom Watanabe and Naomi Datta, as well as Pete Guinée from the Rijksinstituut De Volksgezondheid in Utrecht, The Netherlands. I also invited Bob Rownd and several U.S. infectious diseases specialists including David Smith and Vernon Knight to present their work and their ideas on this new research area. This was the first international meeting on the subject of transferable antibiotic resistance, and as a consequence I received a number of invitations from around the world to speak about my work.

The invitations were a major problem for me because for many years I had a dread of flying and travel in general; however, I managed to overcome these phobias and attended the Ciba symposium on extrachromosomal elements in London in the fall of 1968. This meeting was memorable because I had the chance to meet such legendary figures as William Hayes, Eli Wollman, and Werner Arber. I met my own contemporaries like Richard Novick, who was just beginning his pioneering studies on staphylococci. I became reacquainted with Alan Campbell and spent several hours receiving helpful guidance from Salve Luria. This meeting was notable because it officially blessed the term plasmid to replace the term episome (25).

I suppose these memories of my first important scientific meeting are of little consequence to anyone other than me. In reading my story, it may be useful for you to understand that because of my panic attacks, I had been living my life both scientifically and personally in a kind of cocoon, always half afraid and ready at a moment's notice to run. A newfound freedom crystallized at this meeting. As much as this meeting was an important personal milestone, it also had an important scientific impact as well.

Naomi Datta asked me to join her and H. Williams (Willie) Smith at a pub for a drink. Willie, a veterinarian, was interested in the work I presented on the molecular nature of R plasmids and asked if it could be extended to other plasmids. I was unaware of his work and asked Willie what plasmids he had in mind. In the minutes that followed, I was treated to an unaffected explanation of one of the most exciting things I had ever heard in medical microbiology. In his soft Welsh accent, he told me that *E. coli* were the cause of an infectious diarrhea in pigs and other farm animals. He described marvelous experiments that established that the epidemic *E. coli* strains carried a transferable plasmid that encoded one or more enterotoxins (23). He further astounded me by saying a second plasmid was necessary for the bacteria to adhere to the pig's intestinal cells. He offered to send me any or all of his strains and said simply, "Just promise me you'll let me know what you find." I was at the time drinking my first pint of English bitter, and this gracious offer caused me to try to swallow the stuff and say yes simultaneously with somewhat embarrassing consequences for my shirt and innocent bystanders. Willie was a great scientist and we enjoyed a long relationship exchanging a number of letters about our research and slanderous gossip.

I returned to my laboratory at Georgetown resolved to extend our studies to these novel elements, which, wondrously, wedded plasmids and pathogenicity. Shortly thereafter I met Sherwood (Sherry) Gorbach, who had human *E. coli* isolates that produced enterotoxins like those described by Smith. The human strains came from travelers, and Gorbach regaled me with stories of the history of Montezuma's Revenge, guppy tummy, and traveler's quickstep. Yet, this was no laughing matter where the same disease was a leading cause of death in infants and toddlers of developing countries. We showed that one *E. coli* strain harbored three plasmids, one of which encoded an enterotoxin very much like those encoded in the pig strains. However, working with enterotoxin plasmids and the plasmids encoding adherence factors was not as straightforward as working with R factors. There was no selectable marker.

Thanks to seminal work done by E. S. Anderson at the Enteric Reference Laboratory of the Central Public Health Laboratory in London, and Roy Clowes at the University of Texas, Dallas, it was recognized that not all R plasmids were transferable. Rather, quite small molecular species encoding resistance could be mobilized from bacterial cell to bacterial cell by a transmissible plasmid such as F, or in our case Ent, which did not itself confer a resistance trait. This concept of plasmid mobilization was an important concept for understanding plasmid epidemiology. Small plasmids like pSC101 and RSF1010 carrying only resistance genes and a simple replication apparatus became the foundation of the cloning vectors of the future. Naomi Datta and her associate, Bob Hedges, began to classify R plasmids on the basis of whether or not two plasmids could coexist in the same cell (4). If the plasmids could not be coresidents (compatible) in a bacterium, they considered the plasmids to be incompatible. Bob Hedges asked us to take representative examples of each incompatibility class to tell him about their molecular properties. Pat Guerry and I established in 1970 that a human enterotoxin plasmid and Willie Smith's pig Ent plasmid were related to the classic F factor. We included in this study R plasmids from two other incompatibility groups called I and N. These plasmids exhibited virtually no homology to each other, even though they mediated resistance to the same antibiotics. Clearly the bulk of the plasmid genes, those probably used for replication and transfer, were not related. But

Julian Davies and his coworkers showed that different kinds of aminoglycoside antibiotics were neutralized by specific R-factor-encoded enzymes and the enzymatic mechanism of resistance for streptomycin for the N plasmid was the same as it was for the I incompatibility group R plasmid we studied.

SEATTLE (1972–1981)

While I was in the midst of juggling R factors and Ent plasmids and wondering about their evolution, Brian McCarthy called to say he was at the University of Washington, Seattle, and asked if I was interested in applying for a job there. The Department of Microbiology in Seattle was, and is, among the very best in the United States. The faculty was legendary and included the Chair, John Sherris, as well as Helen Whiteley, Neal Groman, Erling Ordell, Howard Douglas, Charles Evans, Russell Weiser, and Eugene Nestor. The negotiations were brief, and on the evening of the Watergate break-in, June 17, 1972, my wife, two daughters, and I boarded an airplane at Dulles and, with two very angry cats tucked away in the baggage hold, moved to Seattle.

The move to Seattle provided me with more resources plus another bonus. The University of Washington environment permitted me to work again with the hospital microbiology laboratory. Every day the clinical microbiology laboratory held an hour-long session called plate rounds. Fritz Schoenknecht and the eminent physician William (Bill) Kirby presided over these meetings. They had a stunning breadth of knowledge about all microorganisms that might be cultured from the human body. I always thought that plate rounds provided me with a unique view of the natural history of microorganisms that inhabit humans. I don't suppose it is fair to say it is akin to what Darwin did while he wandered the Galapagos looking at finches. But, in a way, I learned that even things that seem to be the same, like *E. coli* from a stool sample or urine culture, are often different if you look closely enough, just as the beaks of the Galapagos finches were differ-

ent in length and revealed something of their biology.

Thanks to King Holmes, Marvin Turck, and the late Bob Petersdorf, I was given a joint appointment in the Department of Medicine and became a participant in the Infectious Diseases (ID) Training Program. The young ID faculty included Larry Cory and Walt Stamm, who have remained close friends. This relationship led to a steady stream of infectious diseases physicians in training in my laboratory over the next decade. Among the first of these were Dennis Schaberg, Peter Piot, Lucy Tompkins, and Michael Lovett. Infectious diseases physicians also arrived as sabbatical visitors and included Mitchell Cohen from the CDC and Rainer Laufs from the Hygiene-Institut der Universität Göttingen, Germany (now at the University of Hamburg). They worked beautifully and productively with the graduate students and postdoctoral fellows who joined me in Seattle. This blend forever after became an important feature of my laboratory.

My first order of research in Seattle was to focus on the enterotoxin plasmids. We became more involved with learning the assays to detect enterotoxins. Fortunately, Carlton Gyles, a veterinarian who had worked with Willie Smith, arrived in Seattle as a sabbatical visitor, and he played a major role in our effort. In the beginning, my Seattle laboratory included Jorge Crosa; graduate students Magdalene (Maggie) So, Fred Heffron, and Leonard Mayer; and Pat Guerry, who came with me from Georgetown to work as a research associate. In November of 1972 I traveled to Honolulu to participate in the U.S.-Japan Cooperative Conference on Bacterial Plasmids. I have written about an informal evening snack at a delicatessen where Herb Boyer, Stan Cohen, Charlie Brinton, and I discussed the possibility of splicing together DNA from different sources (8). Several months after the meeting, Herb Boyer called to tell me that the initial splicing experiments linking together pSC101 and RSF1010 worked. We agreed that Maggie So would travel to Herb's lab to try to "clone" the *E. coli* Ent plasmid enterotoxin genes.

It might seem that the discovery of gene splicing would have had an immediate and profound impact on our work. Eventually it did, of course, but it was still necessary to characterize the DNA! We couldn't yet characterize large chromosomes, so we focused on our beloved plasmids. Characterization was elementary but laborious. We first had to separate the few percent of plasmid DNA from the total bacterial DNA by density centrifugation. We measured the molecular mass of the plasmid DNA of interest by sucrose gradient sedimentation. My new colleagues and friends at Seattle, Jimmie Lara and his wife, Stephanie, taught me how to use the electron microscope to photograph circular DNA and to measure its contour length to estimate the molecular mass of plasmid DNA. It was this tedious approach that permitted Lynn Elwell and Hans de Graaf to characterize the R plasmids in *Haemophilus influenzae* and Leonard Mayer to catch a glimpse of the indigenous plasmid of the gonococcus.

Guerry, Crosa, and I were in danger of drowning in a sea of different sized plasmids; however, we were also refining our skills at performing DNA-DNA reassociation experiments, and the smaller plasmids we found that encoded one or two antibiotic-resistant traits began to reveal something about the distinction between the replication machinery of a plasmid and the mysterious origin of the carried genes encoding antibiotic resistance and enterotoxins. Using hybridization experiments, an undergraduate, Richard Sublett, showed that the ampicillin-resistant genes found on diverse plasmids were often identical. I sent these results to Datta and Hedges by airmail, and a packet from them crossed in the mail containing the draft of a manuscript by Alan Jacob and Hedges describing DNA transposition of ampicillin resistance (14).

Fred Heffron became excited by these findings, and we proposed to look at their plasmid carrying the translocated ampicillin by using the DNA-DNA heteroduplex method in the electron microscope as had been done by Phil Sharp in Norman Davidson's lab. One Saturday morning in 1975, it fell to me to sit down at the microscope to search for the telltale molecule—two identical plasmids except for a single region that would appear as a kinky insertion loop (15). My feeling when that first molecule came into view remains one of the most exciting moments in my scientific life. Fred and I danced around the lab to quizzical stares. Fred, Ron Gill, Craig Rubens, and Pat Guerry went on to publish a wonderful set of papers on transposon biology. At virtually the same time, Dennis Kopecko and Stan Cohen, Nancy Kleckner and David Botstein, as well as Doug Berg and Julian Davies, described similar findings using other antibiotic-resistant genes.

The field of recombinant DNA became a hotbed of political activity. The publication of an open letter to the scientific community stating that there should be a moratorium on several forms of experiments brought about a meeting at Asilomar, California, in 1975. A group of us who originally had been asked to help bring some formality to naming plasmids (how many R1 plasmids could the literature tolerate?) (21) were pressed into action to provide guidelines for recombinant DNA experiments involving bacteria, their plasmids, and their phages. Much has been written of the Asilomar meeting (18), and I can add little to that extensive history.

Roy Curtiss, Don Helinski, and I were appointed to the first NIH Recombinant DNA Committee and charged with helping to prepare formal experimental guidelines. This too has been the subject of many historical books and papers (13). I have nothing to add to this except to pay homage to Bill Gartland and Dewitt (Hans) Stetten, Jr., who were the NIH's stewards for the public good. I resigned from the Committee once the first guidelines were adopted to turn my attention to service on a Food and Drug Administration (FDA) committee concerned with antibiotics and animal feed. I was a strong proponent for the removal of antimicrobials in animal feed. Donald Kennedy, who led the FDA in the late 1970s, proposed to remove or strictly control many antimicrobial agents in animal feed and fell into a buzz saw of controversy.

These nonlaboratory facets of my professional life were time consuming. I never minded serving on study sections or editorial boards but this was different. In retrospect I learned a good deal about the interconnection of our political system with science. I will never forget the times I testified before Congress and I even testified before the city council of Cambridge, MA. I also learned that I didn't want to do this if I could avoid it. Yet, it is also a scientist's responsibility to serve the public's interest.

During this same period, I also witnessed the birth of biotechnology. Stan Cohen and Herb Boyer applied for patents on the gene-splicing technology through their respective universities. Herb took the first steps to launch what became Genentech, while Stan Cohen was involved with another company called Cetus-Palo Alto. Over the years, I served as a scientific advisor for a number of biotechnology companies, and I sometimes laugh at how naive we were about turning a scientific idea into an actual product. In the years that followed, I have been privileged to watch the evolution of an extraordinary industry worldwide and to see the application of those primitive ideas become products that have a major impact on people's lives and well-being. I have always thought that biotechnology was the direct result of the research funding philosophy of 1950–1980, which was to encourage individual creativity and to invest in the best people no matter what their precise area of scientific expertise might be.

The recombinant DNA controversy underlined the need for safe cloning vectors and host bacteria. Crosa and I worked closely with the Boyer lab and especially Francisco "Paco" Bolívar to produce vectors that became known as pBR313 and pBR322 (1). An unexpected fallout from this work was the serendipitous discovery that agarose gels could be employed to analyze the plasmids of bacteria (19). We could abandon density gradient centrifugation, sedimentation in sucrose gradients, and examination of circular DNA in the electron microscope. We could look at simple lysates electrophoresed through agarose to instantly tell the number of different plasmid classes in a

microbe and their molecular mass from the distance the molecules migrated into the gel. This simple procedure led to the rapid discovery of the nature of R plasmids in the gonococcus by Marilyn Roberts and Lynn Elwell, to the first plasmid epidemiology studies by Dennis Schaberg, Lucy Tompkins, and Jim Plorde, and to the use of restriction polymorphisms by Jim Kaper in his studies of cholera vibrios from Chesapeake Bay and Louisiana.

We continued to make progress on characterizing the enterotoxin plasmids of E. coli. Maggie So cloned the E. coli ST enterotoxin. Walter Dallas cloned the heat-stable enterotoxin, LT, and introduced DNA sequencing into the laboratory. Steve Moseley did seminal work on the use of DNA hybridization to detect pathogens directly in clinical material using enterotoxin genes as probes. However, I had determined, almost from the moment that I understood the power of recombinant DNA methods, to begin to search once more for the determinants of bacterial pathogens that distinguished them from the more numerous harmless bacteria that inhabit humans.

I took a sabbatical leave in 1978 at the University of Bristol in Mark Richmond's laboratory. I stopped work on R factors in the laboratory. I had been a member of the R-plasmid research community for 18 years, and we continued to interact even as our paths took divergent turns in the years that followed. I remain indebted to my friend, Julian Davies, for his steady good counsel, for teaching me about the ecology and biochemical basis of drug resistance and that "life is too short to drink bad wine."

When I returned from England there was an assembly of new graduate and postdoctoral students waiting who were encouraged to explore aspects of bacterial pathogenicity. Dan Portnoy, with help from Moseley, began to examine the plasmids of Yersinia. Michael Koomey cloned the IgA protease gene from the gonococcus. I accidentally met the wonderful Margaret Pittman at a meeting and her first words to me were, "If you're so smart, why aren't you working on pertussis?" I made the same challenge to Alison Weiss and she surpassed Margaret's dreams in the years to come. Richard and Sheila Hull translated the phenotypes observed by Barbara Minshew and Catherina Svanborg in uropathogenic *E. coli* into cloned virulence genes.

STANFORD UNIVERSITY (1981–PRESENT)

In 1981 I moved to Stanford University as the Chair of the renovated Department of Microbiology and Immunology and was charged with recruiting young faculty and revamping the teaching program. I was able to recruit John Boothroyd, Ed Mocarski, and Mark Davis as the first faculty members of our young department. I taught most of the medical bacteriology lectures during those first years at Stanford, with an emphasis on syndromes rather than using taxonomic groups of organisms. Subsequently, Gary Schoolnik, Ann Arvin, and Harry Greenberg, physicians in clinical departments who also had active basic research interests, accepted a joint appointment in the department as an adjunct to our teaching effort. They were, and continue to be, active in teaching and as graduate student mentors, as well as collaborators in research efforts of common interest.

The move to Stanford had a profound impact on the structure of my laboratory. At both Georgetown and the University of Washington most of the members of my laboratory were graduate students. The graduate program at Stanford was quite small by comparison, and there was a greater emphasis on postdoctoral students. The original cadre of graduate students working on pathogenic determinants traveled with me to Stanford. They were joined by Paul Orndorff, Rod Welch, and David Low as postdoctoral fellows. They selected their research projects from the panel of the cloned genes that the Hulls had isolated from uropathogenic *E. coli*. Orndorff took the clone with the type 1 pilus genes, while Welch and Low selected a cloned sequence that mediated the *E. coli* α-hemolysin (24).

I was especially lucky that first year to have Staffan Normark as a sabbatical visitor. He, like Gordon Dougan and Mark Achtman, who were visitors in Seattle, had decided to change his focus from antibiotic resistance to the study of pathogenicity. Staffan's presence was a constant stimulus not only for his intellect and humor but because he provided a steady stream of visitors like Hans Wolf-Watz from his department at Umeå University in Sweden. It is noteworthy, I think, that when Staffan returned to Umeå, he was so excited about bacterial pathogenicity that he organized an EMBO course on the genetics and molecular biology of pathogens in the summer of 1983. The course was the first of its kind in Europe and attracted a number of participants, many of whom were established investigators in other aspects of microbiology.

The course faculty reflected the youth of the field and included Staffan, Dan Portnoy, Michael Koomey, Alison Weiss, Maggie So, and John Mekalanos. John, when he was a student, had worked for a few weeks in our lab in Seattle on enterotoxins and had now embarked on his career to study *Vibrio cholerae*, its toxin, and its regulation. I joined the course faculty for a five-day period at the end of the course and got to know a number of the participants, especially Rino Rappuoli and Jörg Hacker. I have been told that Staffan's dream course was the launching pad into the bacterial pathogenesis arena for many European scientists.

Alison Weiss departed for her postdoctoral studies, and *Bordetella pertussis* attracted the next generation of students. Scott Stibitz began work on virulence regulation. Bill Black worked on the contribution of pertussis toxin to virulence. David Relman traveled to Siena, Italy, to work with Rino Rappuoli's group to characterize the filamentous hemagglutinin. Subsequently, Jeff Miller and Craig Roy did seminal work on the control of *Bordetella* virulence regulation by a gene they named *bvg*.

I have never been a politic man. At university meetings, it always seemed I had half the people smiling and the other half scowling. I resigned as the Chair in 1985 to return to a life in the laboratory where, all agreed, I was better suited.

The era of research on uropathogenic *E. coli* ended by 1988 in the laboratory with the studies by Agnès Labigne on afimbrial bacterial adhesins and Carleen Collins' work on the role of bacterial urease in bladder infection. Also while Mike Lovett, Lindy Palmer, and Patrick Bavoil had worked diligently on aspects of *Chlamydia trachomatis* pathogenesis, I banned further work on it and *Neisseria gonorrhoeae* despite Jay Shaw's work on entry into epithelial cells by invoking the "it will break your heart" rule. If we were to study a microbe in the laboratory, it had to be reasonably easy to grow, it had to have a genetic transfer system available (or likely available), and there had to be a suitable animal model of infection. The "it will break your heart" rule was undoubtedly shortsighted, not necessarily strictly followed, but, all in all, it worked to save me at least a good deal of frustration.

The era of bacterial invasion began in the laboratory as the work on pertussis and uropathogenicity waned. Pamela Small, a graduate student, established invasion assays for *Yersinia* and *Salmonella* in epithelial cells and introduced us to the use of gentamicin to kill extracellular bacteria, while sparing their intracellular brethren. Ralph Isberg came to the laboratory with the express idea of cloning genes important for bacterial invasion. He chose *Yersinia pseudotuberculosis*. With the simplicity that so often accompanies a fundamental scientific discovery, Isberg cloned a single genetic determinant he called invasin that transformed ordinary *E. coli* K-12 into a microbe that could easily breach the epithelial barrier, even though this extraordinary trait is not itself sufficient to make the organism pathogenic (17). Virginia Miller found other such genes she called *ail* that encoded not only invasion but also serum resistance. These discoveries by Isberg and Miller pointed out our lack of expertise in studying even the most rudimentary aspects of host cells.

The NIH asked me to help review the Rocky Mountain Laboratory (RML) in Hamilton, Montana. There I encountered Claude Garon from my Georgetown days. At the time,

I had just experienced a divorce, and at age 50, I was thinking about what had been missing in my life outside of science. Fly fishing was one such passion I had ignored, and I rediscovered its joys in the Bitterroot River that flowed behind RML. Claude, an expert electron microscopist, encouraged me to come back to RML the following summer to learn how to use microscopy to study infected animal cells. So in the summer of 1986, I took my first micrographs of *Yersinia* invading animal cells at RML. Virginia Miller looked at my first attempts and remarked, "Oh, Stanley, these will never do!" I did get better, and every year since then I have returned to the Bitterroot Valley to explore science and, on occasion, the joys of fly fishing and even took a few EM photographs in focus.

Brett Finlay initiated work on the entry of *S. typhimurium* into polarized epithelial cells. I had given up on *Salmonella* in 1967. Twenty years later he brought it back into my scientific life. Brett used this approach to isolate bacterial mutants defective in entry, and in doing so, with his characteristic enthusiasm, changed the flavor of the laboratory forever by introducing us to many of the fundamental methods used by cell biologists (12). I began to seek out cell biologists like Ari Helenius, Ira Melman, Suzanne Pfeffer, and Kai Simons, who were uniformly enthusiastic and helpful. The work on *Salmonella* quickly expanded. Cathy Lee pursued the regulation of *Salmonella* invasion. Bradley Jones extended Lee's work and also established, with Nafisa Ghori, a model to look at the entry of *Salmonella* into the Peyer's patch through M cells. Carol ("Connie") Francis, a graduate student, and I spent some time together as "students" in adjacent Stanford laboratories run by Stephen Smith and James Nelson where we learned in vivo imaging, confocal microscopy, and the rudiments of epithelial cell biology. Carol, with help from Michael Starnbach, was thrust into the world of cell ruffling and the intricacies of the host cell cytoskeleton. Although epithelial cells were fine experimental tools, it was clear from the histopathology we began to perform on infected animals that our current favorite organisms, *Yersinia* and *Salmonella*, were much more likely to be found in the host macrophages than in epithelial cells. Fortunately, RML purchased a confocal microscope, and in 1990 I began to apply my newly learned skills to help Michelle Rathman follow the trafficking of internalized *Salmonella* in macrophages.

Of course we were not alone in our pursuit of examining the precise interactions of host cells and microorganisms. Frequently I ran into the always enthusiastic Jorge Galán, my good friend Philippe Sansonetti (we shared Sam Formal as our mentor and are recent fellow aviators), and the irrepressible Pascale Cossart at meetings, where we compared notes on our burgeoning interests in cell biology.

Pascale, together with Rino Rappuoli, Pierre Bouquet, and Staffan Normark, organized a meeting that was jointly sponsored by the American Society for Cell Biology and EMBO on epithelial cell cross-talk between microbes and host cells in August 1991 in Arolla, Switzerland. This meeting was an attempt to get the two cultures of cell biology and those interested in microbial pathogenesis to talk with one another. I think it worked well and had a seminal effect on the field. It helped launch the journal *Cellular Microbiology* and a plethora of subsequent meetings on host cell–pathogen interactions (3).

From 1988 to 1995 there were new experimental systems coming into play in the laboratory seemingly from all directions. Or, perhaps it was my lack of direction. It was during this time I also had a glimpse into the future that in many ways paralleled the experience I had when I understood what the impact of gene cloning would have on the future of science. Bill Haseltine invited me to visit his company, Human Genome Science, in 1992. Bill suggested we stop by the nonprofit arm of the company, the Institute for Genomic Research, directed by Craig Venter. We arrived after 6:00 p.m. on a warm summer evening. There was no one there, but I gasped when I saw the rows of DNA sequencing machines, all connected by an umbilicus of cables that disappeared into the ceiling. We followed the cables to a dimly lit room that contained computers and a row of printers spewing ATGC in a rapid staccato. My actual response cannot be printed here, but as Bill explained to me the scope of the project and what Venter envisioned was possible, I was stunned. When I called home that evening, I told my wife, Lucy, I had just seen the future and science would be practiced differently in the very near future.

However, the impact was not immediate in my laboratory. There was a second wave of research on the *Yersinia* pursued by Kathleen McDonough, who began to haunt the ground squirrel dens of Stanford to harvest fleas to infect with the plague bacillus. Dorothy Pierson began to explore the function of the *ail* gene. Jim Bliska, working closely with Jim Dixon's laboratory, soon began to characterize the effect of YopH, tyrosine phosphatase, on macrophages, while Joan Mecsas and Bärbel Raupach compared the different strategies used by *Yersinia* to exploit the host cell cytoskeleton. At the same time Denise Monack began to work on the cellular basis of pathogen-induced cytotoxicity induced by *Yersinia* and *Salmonella*.

New organisms somehow kept appearing! Joe St. Geme finished his training in pediatrics and began to explore the interaction of *H. influenzae* with epithelial cells. David Schauer, a veterinarian, earning his PhD, taught me about the *Citrobacter* species in mice that caused colonic hyperplasia and showed it was a variation on a theme of enteropathogenic *E. coli*. David Relman and I attended an infectious diseases conference on a day when we had discussed a paper by Norm Pace on exploring bacterial communities using 16S RNA probes. We heard about a bacterial disease of patients infected with HIV called bacillary angiomatosis, and he, together with Lucy Tompkins, used the 16S approach to show that the agent was a *Bartonella*.

Lalita Ramakrishnan, another ID fellow, was determined to work on *Mycobacterium tuberculosis*. I told her I was too old to wait a month for a colony to grow. She convinced me to let her focus on the faster-growing *Mycobacterium marinum* instead. I kept getting talked into things! Evi Strauss worked on *Edwardsiella tarda*; Brendan Cormack wanted to work on *Candida* pathogenesis. I told him I knew nothing about fungi; he told me he knew nothing about bacterial pathogenesis. We'd teach each other. He did far better than I. Creg Darby looked at *Caenorhabditis elegans* interacting with *Y. pestis* and found a mechanism that probably accounts for how the plague bacillus blocks the foregut of fleas.

The serendipity of scientific discovery was always our partner. Raphael Valdivia planned to do a thesis project on intracellular trafficking in *Mycobacteria* but became enchanted with the green fluorescent protein (GFP) of a jellyfish. He and Cormack, almost as a lark, planned an experiment that provided a marvelous probe for many investigators in different scientific disciplines. Valdivia instead used GFP to isolate genes expressed only inside of macrophages.

The inevitable departure of people to establish their own laboratories led to other changes. There was a decided increase in individuals who had initially worked in eukaryotic systems and who now wished to study bacterial pathogenesis. Most of these people felt that the microbe could be used as a tool to study the biology of the host. I felt that one had the best of both worlds. You learned as much about the pathogen as you did about the host.

In the last decade my lab was active, our work focused primarily on two pathogens. One was *Salmonella*, my first love. The *Salmonella* work was fueled by Valdivia's discovery of specific genes expressed intracellularly. This work appeared shortly after David Holden's description of signature-tagged mutagenesis and his discovery of the Spi2 pathogenicity island (16). Also, Denise Monack established a persistent infection model in *Salmonella*, which more closely resembled the natural disease seen in typhoid with its hallmark asymptomatic bacterial shedding. Anthea Lee, Corrie Detweiler, and Igor Brodsky looked specifically at some of the Spi2 genes and how they were regulated and functioned in the phagosome.

The second focus of the laboratory became *Helicobacter pylori*. The reasons for this are rather more personal. Lucy Tompkins and I were married in 1983. She had gone from her PhD work in my laboratory in 1967 to Dartmouth Medical School, and subsequently she did her infectious diseases residency at the University of Washington. After our marriage, she moved to the Division of Infectious Diseases at Stanford and established her own laboratory. A few years ago, Lucy was asked to become the Chief of the Division of Infectious Diseases and Geographic Medicine, as well as an associate dean. This inevitably took away the time she could spend in the laboratory. She convinced me to become involved with her and her research associate, Ellyn Segal, to look at *H. pylori* pathogenicity. Ellyn had discovered that the CagA protein caused human cells to elongate and this was accompanied by a phosphorylation event. By going to meetings with Lucy, I knew many of the major investigators in this arena, including Barry Marshall, Marty Blaser, Jeffrey Gordon, and Antonello Covacci. The first time I heard Barry Marshall talk about the isolation of *H. pylori* and his view of its role in gastritis and gastric cancer, I stated unequivocally that I thought it was nonsense based on two misguided physicians drinking a bacterial culture and becoming ill. I confessed my error in public when Barry asked me to deliver one of the major talks at a symposium celebrating the twentieth anniversary of his discovery. *H. pylori* has charisma and many of the students with prior experience in eukaryotic biology were drawn to study it because of its association with gastric cancer.

The genomic revolution touched the laboratory owing to the urging of my Stanford colleague, David Botstein, to forge ahead and use Pat Brown's DNA array technology to look at global sequence homology and global gene expression in host-pathogen interactions. We turned first to *H. pylori* because it possessed a relatively small genome. Nina Salama and Tim McDaniel bravely set out to construct a DNA array of the *H. pylori* genome. Nina studied the phylogeny of *H. pylori* and used the array to define the organism's essential genes. Lucy Thompson, an Australian graduate student, used the array to explore *H. pylori* physiology. On the other side of the coin, Karen Guillemin was involved with the construction of a mouse and a human DNA array. Karen examined the effect of *H. pylori* infection on eukaryotic cells using a human DNA array. Her findings unexpectedly revealed that a large number of genes associated with the tight junctions were upregulated when *H. pylori cagA* gene product was injected into cells. Manuel Amieva, Roger Vogelmann, and James Nelson went on to show how the insertion of CagA into host cells caused cytoskeletal rearrangements and triggered a number of cell signaling pathways. These findings were extended by Scotty Merrell and Sahar El-Etr using human polarized cells. Finally, Anne Mueller, working with our Australian collaborators Adrian Lee and Jani O'Rourke, did an in-depth analysis of the natural history of mucosa-associated lymphoid tissue (MALT) formation and its transition to lymphoma in mice over an 18-month period. Anne and Scotty went on to examine the dynamics of the infection process and introduced the laboratory to the use of laser capture microscopy coupled with DNA microarray analysis.

Despite my protestations about remaining focused these past years, it is true that I could always be convinced by someone to bring another microbe into the fold. Erin Gaynor initiated a study of *Campylobacter jejuni* with a graduate student, Joanna MacKichan. They somehow managed alone to construct a genome array and looked at virulence regulation in the somewhat ignored important enteric pathogen. Wanda Songy tried to discover why certain *Streptococcus bovis* are found in sepsis in colon cancer patients. Last, but not least, Elizabeth Joyce, even today, is trying to understand how the pneumococcus colonizes the nasopharynx.

Kaman Chan and Charlie Kim, my last graduate students, constructed a complete DNA array of *Salmonella* genes. Charlie published several bioinformatics tools while working on this project, as well as investigating nitrite-inducible *Salmonella* genes. Kaman looked at the molecular phylogeny of *Salmonella* and then developed a global method to look at *Salmonella* genes essential for survival in a macrophage during infection.

So, during these past few years, those working in the lab have moved more and more into the host arena and led me to appreciate more fully the nuances and intricacies of the host and pathogen interplay. It seems clear to me that the term bacterial pathogenesis places far too much emphasis on the microbe and downplays the impact of the host on the evolution of the pathogen. Indeed, it is now clearer than ever that the host-pathogen interaction has been one of the driving forces of eukaryotic evolution including ours; our total cellular burden is, after all, 90% microbial.

FINAL THANK YOU

I was a student during the first era of bacterial genetics. I witnessed the emergence of molecular biology. I witnessed and was a beneficiary of the discovery of messenger RNA, the genetic code, recombinant DNA technology, DNA sequencing, PCR, and most recently, genomic sequencing and now array technology and systems biology. I look back with fondness on the days, many years ago now, when we sought to discover the nature of plasmid resistance genes and the role of plasmids like Ent in the evolution of bacteria. I thought then it was the most important problem in science, which is how it should be when one is young. In the years that followed there grew an ever greater passion to understand the biology of pathogens, and it is fitting that pathogenicity is intricately enmeshed with mobile genetic elements not unlike R factors and transposons. Each day still brings up questions more important and more challenging than those from the day before, which is how it should be even as I grow older. My boyhood dream of a life of scientific adventure was surpassed by reality and human creativity, a human characteristic that has not been lost despite the poor record of human history.

I was sad to close my laboratory in 2005. Fortunately Manuel Amieva, Denise Monack, and David Relman are now Stanford faculty members and still study fascinating organisms. They and their students keep me abreast of their latest findings and the newest developments in the host-pathogen research field. I was fortunate to have worked with so many talented and genuinely nice people in my career. They constitute a reservoir of wonderful memories, some of which I've tried to share with you. I still remain in contact with most of them. One never truly loses a former student. There is, at the least, one more letter to write no matter how old they or I become.

Finally, I realized a number of years ago that in life you never get to have as many dogs as you want or to find out how it all turns out. As you get older and are asked to be a keynote speaker or get an award of some kind, you also realize you are being given credit for things you never did and for attributes you never possessed.

ACKNOWLEDGMENTS

I should note that a number of my former students wrote articles about the early years in my laboratory in a volume celebrating my sixtieth birthday. I deliberately did not cite these because this book is not readily available to many people (7). However, it provides another perspective of some of the events I describe here.

Anyone who knows me at all has met my assistant, Sara Fisher. We have worked together for almost 20 years, and she has been the surrogate mother to most of the people who worked with me these past years at Stanford. She has been a mainstay of my professional life as well as my friend.

Lucy Tompkins, my wife, has been my best collaborator and introduced me to what love and joy can mean in life.

I thank Carleen Collins for reading this article and providing me with corrections and cautions that I treasure but did not always follow.

As noted above, after the manuscript was completed, I felt there was only one of my former students, Carleen Collins, who would tell me the unvarnished truth about what I had written. Tragically, just a few weeks later, Carleen was diagnosed with a very aggressive malignant disease and she passed away on February 12, 2008. I treasured her as a person, as a scientist, and as an advisor. I shall miss her as will all of her friends and colleagues.

DISCLOSURE STATEMENT

The author is not aware of any biases that might be perceived as affecting the objectivity of this review.

LITERATURE CITED

1. Bolivar F, Rodriguez RL, Greene PJ, Betlach MC, Heyneker HL, et al. 1977. Construction and characterization of new cloning vehicles. II. Multipurpose cloning system. *Gene* 2:95–113
2. Brenner DJ, Fanning GR, Johnson KE, Citarella RV, Falkow S. 1969. Polynucleotide sequence relationships among members of the Enterobacteriaceae. *J. Bacteriol.* 98:637–50
3. Cossart P, Boquet P, Normark S, Rappuoli R. 1996. Cellular microbiology emerging. *Science* 271:315–16
4. Datta N, Hedges RW. 1971. Compatibility groups among fi-R factors. *Nature* 234:222–23
5. Falkow S. 1957. L-Forms of proteus induced by filtrates of antagonistic strains. *Bacteriology* 73:443–44
6. Falkow S. 1958. Activity of lysine decarboxylase as an aid in the identification of Salmonellae and Shigellae. *Am. J. Clin. Pathol.* 29:598–600
7. Falkow S. 1994. A look through the retrospectoscope. In *Molecular Genetics of Bacterial Pathogenesis*, ed. VL Miller, JB Kaper, DA Portnoy, RR Isberg, pp. xxiii–xxxix. Washington, DC: ASM Press
8. Falkow S. 2001. I'll have the chopped liver please, or how I learned to love the clone. *ASM News* 67:555–59

9. Falkow S, Citarella RV. 1965. Molecular homology of F-merogenote DNA. *J. Mol. Biol.* 12:138–151

10. Falkow S, Citarella RV, Wohlhieter JA, Watanabe T. 1966. The molecular nature of R-factors. *J. Mol. Biol.* 17:110–16

11. Falkow S, Marmur J, Carey WF, Spilman W, Baron LS. 1961. Episomic transfer between *Salmonella typhosa* and *Serratia marcescens*. *Genetics* 46:703–6

12. Finlay BB, Gumbiner B, Falkow S. 1988. Penetration of *Salmonella* through a polarized Madin-Darby canine kidney epithelial-cell monolayer. *J. Cell Biol.* 107:221–30

13. Fredrickson DS. 2001. *The Recombinant DNA Controversy: A Memoir. Science, Politics, and the Public Interest 1974–1981*. Washington, DC: ASM Press. 388 pp.

14. Hedges RW, Jacob A. 1974. Transposition of ampicillin resistance from RP4 to other replicons. *Mol. Gen. Genet.* 132:31–40

15. Heffron F, Sublett R, Hedges RW, Jacob A, Falkow S. 1975. Origin of the TEM beta-lactamase gene found on plasmids. *J. Bacteriol.* 122:250–56

16. Hensel M, Shea JE, Gleeson C, Jones MD, Dalton E, et al. 1995. Simultaneous identification of bacterial virulence genes by negative selection. *Science* 269:400–3

17. Isberg RR, Falkow S. 1985. A single genetic locus encoded by *Yersinia pseudotuberculosis* permits invasion of cultured animal cells by *Escherichia coli* K12. *Nature* 317:262–64

18. Lear J. 1978. *Recombinant DNA: The Untold Story*. New York: Crown. 280 pp.

19. Meyers JA, Sanchez D, Elwell LP, Falkow S. 1976. Simple agarose gel electrophoretic method for the identification and characterization of plasmid deoxyribonucleic acid. *J. Bacteriol.* 127:1529–37

20. McElroy WD, Glass B. 1957. *The Chemical Basis of Heredity*. Baltimore: Johns Hopkins Press. 848 pp.

21. Novick R, Clowes RC, Cohen SN, Curtiss R III, Datta N, et al. 1976. Uniform nomenclature for bacterial plasmids: a proposal. *Bacteriol. Rev.* 40:168–89

22. Schaeffer AB, Fulton D. 1933. A simplified method of staining endospores. *Science* 77:194

23. Smith HW, Halls S. 1967. Studies on *Escherichia coli* enterotoxin. *J. Pathol. Bacteriol.* 93:531–43

24. Welch RA, Hull R, Falkow S. 1983. Molecular cloning and physical characterization of a chromosomal hemolysin from *Escherichia coli*. *Infect. Immun.* 42:178–86

25. Wolstenholme GEW, O'Connor M, eds. 1963. *Ciba Foundation Symposium on Bacterial Episomes and Plasmids*. London: Churchill

Evolution of Intracellular Pathogens

Arturo Casadevall

Department of Microbiology and Immunology, Albert Einstein College of Medicine of
Yeshiva University, Bronx, New York 10461; email: casadeva@aecom.yu.edu

Annu. Rev. Microbiol. 2008. 62:19–33

The *Annual Review of Microbiology* is online at
micro.annualreviews.org

This article's doi:
10.1146/annurev.micro.61.080706.093305

Key Words

host-microbe, facultative, obligate, genome reduction, endosymbiotic,
virulence, pathogenicity

Abstract

The evolution of intracellular pathogens is considered in the context of
ambiguities in basic definitions and the diversity of host-microbe interactions. Intracellular pathogenesis is a subset of a larger world of host-
microbe interactions that includes amoeboid predation and endosymbiotic existence. Intracellular pathogens often reveal genome reduction.
Despite the uniqueness of each host-microbe interaction, there are only
a few general solutions to the problem of intracellular survival, especially
in phagocytic cells. Similarities in intracellular pathogenic strategies between phylogenetically distant microbes suggest convergent evolution.
For discerning such patterns, it is useful to consider whether the microbe is acquired from another host or directly from the environment.
For environmentally acquired microbes, biotic pressures, such as amoeboid predators, may select for the capacity for virulence. Although often
viewed as a specialized adaptation, the capacity for intracellular survival
may be widespread among microbes, thus questioning whether the intracellular lifestyle warrants a category of special distinctiveness.

Contents

INTRODUCTION

The topic of evolution of intracellular pathogens cannot be considered without first grappling with the uncertainties and ambiguities in the meaning of the terms evolution, intracellular, and pathogen. Unfortunately, none of these terms lends itself to a straightforward definition. In combination, they present a vexing problem in semantics that must be addressed before delving into this complex subject. The word evolution connotes change, yet in the context of this essay the word is used to refer to an origin, as the goal is to explore how certain microbes that reside inside cells came to acquire that lifestyle. Because intracellular pathogens are extremely varied, and because most if not all have adopted the strategy of intracellular life as part of their evolutionary trajectory, it may not be possible to propose an overarching notion of evolution for this microbial set. Consequently, any approach

to the subject must be an attempt to discern evolutionary themes in an area where there is no significant fossil record and only a fraction of intracellular pathogens have been studied in detail. Each pathogenic microbe is different and consequently the pathogenic strategy of each intracellular pathogen has unique aspects. This introduces a conundrum because uniqueness argues against generalities in the evolutionary process. However, it is possible to identify generalities by considering the subject from the larger perspective of the mechanisms that are responsible for the virulence and pathogenicity of intracellular pathogens.

The second hurdle in approaching the topic is the designation of a microbe as an intracellular pathogen. There is no good definition for an intracellular pathogen because the term intracellular is vague when considered in the context of microbial life. Intracellular means inside a cell, but when this adjective is applied to pathogenic microbes, one is immediately confronted with many ambiguities. For example, most, if not all, intracellular pathogens must spend some of their life in the extracellular space prior to entry or after cellular exit. An illustrative case of this is the obligate intracellular pathogen *Ehrlichia chaffeensis*, which can be found in extracellular spaces (25). Hence, except for those pathogens that are transferred vertically during host cell replication, the term intracellular refers to one phase of the microbial cycle. Classifying pathogenic microbes as intracellular or extracellular is further complicated because there is often no clear dividing line. By all criteria, viruses are obligate intracellular pathogens, yet viruses are often considered a different class of microbes and are not usually grouped with other intracellular pathogens. In general, microbes are considered intracellular pathogens when their cycle in the host includes residence and/or replication inside host cells. However, this distinction can be blurry. For example, encapsulated bacteria such as *Streptococcus pneumoniae* are generally considered extracellular pathogens, yet these bacteria are often found inside neutrophils. Similarly, *Staphylococcus aureus* is not classically considered

---

Intracellular pathogen: a microbe capable of causing host damage whereby the lifestyle in the host is associated with intracellular residence, survival, or replication

Pathogen: a microbe capable of causing host damage

Obligate intracellular pathogen: a microbe capable of causing host damage that is completely dependent on a host cell for survival and replication

---

an intracellular pathogen but this bacterium can survive and replicate inside several types of cells, and intracellular residence is now considered important for persistence and pathogenesis (41). Even more vexing is the example of *Aspergillus fumigatus*, a fungus that is a common cause of life-threatening disease in immunocompromised patients. This organism grows as hyphae in tissue yet is thought to be inhaled as conidia that germinate and replicate in alveolar macrophages (22). *A. fumigatus* does meet some criteria as an intracellular pathogen, yet most authorities do not consider this fungus as such. Furthermore, replication is not required for the designation of intracellular pathogen, as illustrated by *Trichinella spiralis*, the causative agent of trichinosis, which survives but does not replicate in human skeletal muscle cells. The topic is made even more confusing by many reviews on intracellular pathogens that tend to focus on microbes that survive ingestion by phagocytic cells, a quality that is viewed as an ability to subvert professional phagocytes of the innate immune system. Nevertheless, many intracellular pathogens such as *Toxoplasma gondii* and *T. spiralis* can reside in nonphagocytic cells. Consequently, focusing on the host cell type inhabited is not particularly enlightening in resolving the definitional difficulties. Perhaps the greatest problem with the term intracellular is that it connotes a distinction for a set of microbes that may deserve no such distinction. In other words, having an intracellular pathogenic strategy is currently viewed as a special property, yet there is increasing evidence that the capacity for intracellular residence is a common attribute among pathogenic and nonpathogenic microbes. This notion is explored in this review with the proposal that the entire concept of intracellular pathogenesis requires redefinition.

Finally, the word pathogen confers upon microbes a quality that is not their own, as virulence is one outcome of the interaction between a microbe and a host, and requires a susceptible host (10). For the purposes of this review, pathogen is defined as a microbe capable of damaging a host (9). When considering the topic of the evolution of intracellu-lar pathogens, it is worthwhile to remember that no microbe can be a pathogen without a host, and consequently, it is impossible to discuss the subject of evolution from a microbe-centric perspective. Because this review is focused on pathogenic microbes, it does not consider the endosymbiotic bacteria, which are common in many invertebrate and protozoal species (34, 35). Ancient endosymbiotic bacteria are thought to have been critically important in the evolution of eukaryotes and may have been the progenitors of such organelles as mitochondria and plastids (28). The exclusion of endosymbiotic bacteria is done with the acknowledgment that there is often a thin line between symbiotic and pathogenic intracellular microbes, and that these microbes often employ similar mechanisms for intracellular survival. For example, the obligate intracellular bacterium *Wolbachia* spp. is associated with numerous arthropod species where it passed vertically; the outcome of the host-microbe interaction can be symbiotic, neutral, or pathogenic depending on the tissue and the host (32). In fact, it has been argued that the major difference between pathogenic and nonpathogenic (endosymbiotic bacteria) intracellular microbes is their effect on the host, since both must have similar biological requirements for inhabiting intracellular spaces (17).

Considering the diversity of microbes capable of residing in host cells, the ambiguities in the definitions, and the uncertainty in the boundaries of classification, there is a quixotic element to the idea that the topic of evolution of intracellular pathogens can be approached in a short review. In this regard, despite the ambiguities evident in the term intracellular pathogen, this concept has been enormously influential in other fields, such as immunology (11) (see sidebar, Does the Intracellular Lifestyle Warrant a Special Distinctiveness?). On the other hand, an explosion in knowledge of microbial pathogenesis in recent years allows one to begin to discern the outlines of the landscape over which certain microbes evolved to survive and replicate in host cells as part of their pathogenic mechanisms. In particular, progress

DOES THE INTRACELLULAR LIFESTYLE WARRANT A SPECIAL DISTINCTIVENESS?

Distinguishing pathogenic microbes as intracellular or extracellular has been a highly influential division in both microbiology and immunology. For example, immunologists have often viewed humoral and cellular immunity as the immune arms responsible for defense against extracellular and intracellular pathogens, a concept that is now considered too simplistic. Consequently, the designation itself has the potential to affect how we think in entire disciplines. Because concepts often drive research and serve as a basis for understanding, such sweeping generalizations should be made carefully. As discussed in this review there does not seem to be a clear dividing line between extracellular and intracellular pathogens, except possibly for obligate intracellular pathogens. However, some obligate intracellular pathogens, such as *Ehrlichia* spp., have extracellular phases where they are susceptible to antibody-mediated immunity. Hence, classifying organisms as intracellular and extracellular pathogens may be ultimately a futile exercise fraught with error and misconception.

Genome reduction: a phenomenon observed in many microbes that adopt intracellular lifestyles that is associated with gene loss and significantly smaller genomes than free-living phylogenetic relatives

in genomics has yielded such far-reaching concepts as the association of genome reduction with intracellular lifestyle, which provides optimism for the notion that there may be some general rules that apply to the evolution of intracellular life. Furthermore, the recognition that for microbes such as *Legionella pneumophila* (20) and *Cryptococcus neoformans* (48) virulence is likely to be an outcome of selection pressures in the environment brought on by amoeboid predators opens a new conceptual approach to consider how some intracellular pathogens emerged. Consequently, this review tries to identify common themes among the uniqueness of individual host-microbe interactions. The goal is to look at evolution from a distance without getting lost in the intricacies of individual pathogenic microbes. Although generalizations in biology can be treacherous, especially in light of the variability inherent in host-microbe interactions, they also have the potential to provide useful concepts and reveal important principles.

THE INTRACELLULAR LIFESTYLE IS ANCIENT

When approaching the topic of evolution, it is important to ask when the species, characteristic, or phenomenon under study first became apparent in biological history. Establishing a timeline is a critical step for understanding the origin of a phenomenon in the context of evolutionary and geologic time. Unfortunately, there is no fossil record to provide an estimate of when some organisms acquired the capacity to survive inside other microbes. If we accept an endosymbiotic origin for mitochondria and other eukaryotic organelles (28, 29), then we can conclude that the capacity for intracellular residence is ancient and antedated the emergence of eukaryotic organisms as we know them. The emergence of the intracellular lifestyle in ancient microbes appears to have at least three major requirements: (*a*) size differences between microbes such that one can ingest another; (*b*) a mechanism for particle ingestion on the part of the host and/or host invasion on the part of the smaller entity, and (*c*) a capacity for the ingested microbe to survive within the larger host. Such early interactions could have had varied outcomes including survival of both microbes (symbiosis and mutualism), survival of the host (predation), damage to the host (intracellular pathogenesis), or damage to both microbes (incompatibility and antagonism). Presumably, each outcome was subject to selection pressures that directed the emergence of different types of microbe-microbe or host-microbe relationships.

Today, one can discern outcomes that mirror putative ancient outcomes in extant host-microbe relationships. Endosymbiotic bacteria represent mutualistic interactions of a type that may have once been responsible for the origin of the eukaryotic organelles. Bacterial grazing by amoebae represents predatory-type interactions (30). The ingestion by macrophages of pathogenic bacteria with subsequent bacterial survival and damage to the host cell is the type of interaction that falls into the traditional view of intracellular pathogenesis. It has been suggested that the pre- and postingestional survival strategies of bacterial prey in response to amoeboid predator grazing were the precursors of extra- and intracellular pathogenic strategies, respectively (30). Similarly, postingestional adaptations, such as digestional resistance, intracellular toxin production, and intracellular replication may be linear antecedents of similar phenomena now associated with intracellular pathogens (30). Nevertheless, infection of a host cell with a nonreplication-permissive microbe capable of host damage would represent a deleterious interaction for both entities. An example would be *Toxoplasma gondii* infection of cells in a nondefinitive host that is a biological dead end for the parasite that also damages the host cell. When viewed from the vantage point of evolution, it becomes apparent that what we call intracellular pathogenesis is a part of a continuum of intracellular lifestyle strategies that is both ancient and constantly coevolving.

OBLIGATE AND FACULTATIVE INTRACELLULAR PATHOGENS

Intracellular pathogenic microbes fall into two groups: obligate or facultative. Obligate intracellular pathogens have lost their capacity for living outside of their hosts and these include all viruses, bacteria such as *Rickettsia* and *Chlamydia* spp., and protozoa such as *Plasmodium* spp. In contrast, facultative intracellular pathogens retain the capacity for replication outside their hosts and these include a large number of pathogenic bacteria and fungi. Hence, the designations of obligate and facultative would appear at first glance to represent a clear dividing line for approaching the topic of evolution of intracellular pathogens. However, on closer examination, this distinction is blurred by the phenomenon of genome reduction (see below), suggesting that the difference between obligate and facultative pathogens is simply one of the extent of gene loss that often accompanies host association and dependence. Obligate and facultative intracellular pathogens can also be distinguished by their means of acquisition. Obligate intracellular pathogens are

necessarily acquired from other hosts, whereas facultative intracellular pathogens are acquired from other hosts (e.g., *Mycobacterium tuberculosis*) or directly from the environment (e.g., *C. neoformans*). Obligate intracellular pathogens are dependent on their hosts, and consequently these microbes always have intimate and dependent relationships with their hosts. However, these microbes have mastered escaping from host response mechanisms, and, when it occurs, microbial-mediated disease is usually caused by disruption of the host-microbe relationship. In contrast, facultative intracellular pathogens have more varied relationships with their hosts. Some, like *M. tuberculosis*, cause disease only in a minority of infected hosts and this is associated with transmission of infection. Others, like the bacterium *L. pneumophila* and the fungus *C. neoformans*, are acquired from the environment and have no obvious need for mammalian virulence in their replication or survival. Hence, the outcome of the interaction of these microbes with mammalian hosts is usually a function of the immunological status of the host, and both legionellosis and cryptococcosis are often associated with immune impaired hosts.

THE PHENOMENON OF GENOME REDUCTION

Despite the diversity among intracellular pathogenic strategies, evidence from the available sequenced genomes suggests what may be a general rule, namely, that obligate and facultative intracellular pathogens acquired from other hosts often exhibit genome reduction. In other words, the transition from the status of free-living to intracellular life is associated with the loss of large segments of DNA (reviewed in References 15 and 17). These gene deletions often include DNA segments that encode entire metabolic pathways providing nutrients that can then be acquired from the host. For example, *Rickettsia* spp. have lost numerous genes needed for many metabolic pathways including sugar, purine, and amino acid metabolism (42). *Candidatus* Blochmannia, an obligate endosymbiont of ants, has a severely reduced genome compared with free-living relatives consistent with loss of DNA for host adaptation (53). Another example of genes that are lost in the adoption of the intracellular lifestyle are bacterial toxin-antitoxin loci, which encode for proteins that allow survival in stressful conditions (39).

For some intracellular microbes, genome deletions may have been selected because they make the microbe more fit in its interaction with the host. This phenomenon has been called pathoadaptation, and here the likely mechanism is gene loss that is positively selected because the genes' absence increases microbial fitness in the host (38). An excellent example of this phenomenon is the finding that the gene coding for lysine decarboxylase is lost in *Shigella* spp.; presumably the product of this enzyme inhibits bacterial virulence factors (31). Another example of pathoadaptation is the loss of flagella in many intracellular pathogens, possibly reflecting convergent evolution to shedding a motility mechanism as a trade-off for greater fitness in the host (38). The phenomenon of gene reduction also applies to eukaryotic intracellular pathogens, as revealed from genomic studies of microsporidia, which are highly specialized fungi (23). However, gene gain from eukaryotes can also occur in certain cases, as demonstrated by the finding of numerous eukaryotic-like proteins in *L. pneumophila*, a bacterium that has a long and close evolutionary association with environmental amoebae (12).

Many obligate intracellular pathogenic microbes have also lost mobile elements from their genome, a phenomenon that may translate into a lack of opportunity for gene transfer that in turn reduces the capacity for genetic diversity. An exception to this generalization is *Coxiella burnetii*, which retains mobile elements and has a genome with numerous pseudogenes, suggesting that this organism is a recent convert to obligate intracellular life and may be in the midst of genome reduction (44). In light of these findings with bacterial genomes, it is possible to view viruses as extreme examples of genome reduction, with viral genomes containing only those genes essential for replication and escape of host immune mechanisms.

If the progression from a free-living state to obligate intracellular life is accompanied by irreversible genome loss, then one can envision this evolutionary pathway to be a journey toward niche specialization with no return. In fact, Andersson & Kurland (6) have suggested that the process of genome reduction associated with intracellular pathogenesis may eventually lead to a Muller's ratchet, a phenomenon whereby small asexual genomes accumulate deleterious mutations that are ultimately associated with extinction. This raises the interesting possibility that, for some microbes capable of intracellular survival, the temptation of obligate intracellular life with its bountiful access to host resources and protection may carry with it seeds for their own destruction as distinct biological entities. Even in situations in which the host-microbe relationship evolves into that of mutualistic endosymbionts, such as mitochondria, the obligate intracellular style is associated with a loss of self in the form of genome deletions. Hence, for some pathogenic microbes, the obligate lifestyle may be an evolutionary dead end associated with loss of genome, a narrow ecological niche, dependence on host survival for microbial survival, and the grim possibility of a Muller's ratchet. However, a recent analysis of the *Mycoplasma* genome revealed that it contained a significant proportion of DNA acquired by horizontal gene transfer, suggesting that this outcome may be avoided by certain intracellular pathogens (45). In contradistinction to intracellular pathogens, environmentally ubiquitous microbes, such as *Pseudomonas aeruginosa*, that are capable of occupying diverse ecological niches have large genomes (55).

THE ECOLOGY AT THE SOURCE OF INFECTION

Another distinction that is enormously important when considering the evolution of intracellular pathogens is the environment at the source of infection. In general, microbial infection is acquired either from other hosts or directly from the environment. In this regard, a host can be considered an ecological niche. Hosts provide microbes with protection, the potential of access to nutrients, and transportation. In contrast to environmental residences where a microbe is subject to numerous conditions and complexities, hosts tend to provide microbes with a more constant and predictable environment, with the caveat that survival in a host usually requires adaptation and specialization, which in turn reduce the number of potential sites that the microbe can inhabit. Hosts that serve as sources of infection can be self or nonself. For example, *M. tuberculosis* is usually acquired from other humans, although the initial strains encountered by human populations may have come from cattle or other vertebrate hosts. An example of infection with intracellular pathogens from nonself hosts would be the acquisition of *Yersinia pestis* from fleas and *L. pneumophila* from environmental amoebae. On the other hand, some intracellular pathogens are acquired directly from environment without having to implicate recent residence in other hosts. Examples of this category include the fungi *Cryptococcus neoformans* and *Histoplasma capsulatum*, which cause human infection after the inhalation of aerosolized spores.

The ecology at the source of infection confers tremendous selection pressures that shape the behavior of intracellular pathogens and the consequences of infection. For intracellular pathogens acquired from other hosts, there must be a premium on pathogenic strategies that are permissive to growth without damaging the host too quickly to insure replication and transmission and, in metazoal hosts, to prevent and/or impair an immune response. For some microbes such as *M. tuberculosis*, damage to the host in the form of cavitary tuberculosis is essential for host-to-host transmission, with the caveat that pulmonary disease occurs in only a minority of infected individuals. For host-acquired microbes, the selection pressures include host immune mechanisms and the

optimization of their replicative strategies in their respective hosts. Given that, residence in the intracellular space necessarily reduces the likelihood that such microbes would encounter other microbes to exchange DNA; the intracellular environment is associated with loss of mobile elements from bacterial DNA and possesses the threat of accumulation of deleterious mutations. Some of the facultative intracellular microbes acquired from other hosts, such as *M. tuberculosis*, may be on evolutionary paths to becoming obligate intracellular pathogens, as evidenced by genome reduction relative to other related mycobacteria (21). In this regard, *Mycobacterium leprae* may already have a minimal mycobacterial genome, which precludes growth outside only a few vertebrate hosts (13, 52).

In contrast, intracellular pathogens acquired from the environment are under different types of selection pressures than those acquired from other hosts. For example, microbes acquired directly from the environment would be under biotic and nonbiotic selection pressures in the environment that would not be experienced by microbes that jump from vertebrate to vertebrate host. Soil organisms are exposed to extremes of temperature, light, and humidity, depending on diurnal and weather conditions. Competition for nutrients in soils is fierce given the abundance and diversity of microbial life. Soil-dwelling microbes are under the constant threat of predation from amoebae and small animals. Soil-dwelling microbes need a full array of metabolic machinery to survive in their ecological site and consequently are not under the type of selection pressure that would lead to genome reduction. Hence, one should not anticipate that genome reduction will apply to all intracellular hosts, but rather it should be expected only when the microbe is associated with a host that can potentially supply its metabolic needs. Consistent with this thesis, facultative intracellular pathogens acquired directly from the environment, such as *C. neoformans* and *H. capsulatum*, are autotrophs with minimal nutritional needs that can replicate across a broad range of environmental temperatures. Hence, it appears that whether a microbe is host- or environmentally acquired provides a clear dividing line for comparing and contrasting the evolution of intracellular pathogens.

The dividing line between host and environmental acquisition is blurred by the observation that enteric human pathogenic bacteria can interact with soil protozoa to emerge in forms that may enhance their survival. For example, coincubation of the human enteric pathogen *Salmonella enterica* with a ciliated protozoan of *Tetrahymena* spp. resulted in the ingestion of the bacteria but they were resistant to digestion and were released in defecation vacuoles containing as many as 50 bacteria per vesicle (8). Bacteria in *Tetrahymena*-derived vesicles were significantly less susceptible to microbicidal chemicals or dehydration (8). This experiment provides an example of how bacterial resistance to intracellular digestion by a predatory protozoan can lead to packaging in vesicles that in turn is likely to affect the persistence and infectivity of this pathogen.

THE INTRACELLULAR ENVIRONMENT

By definition, intracellular pathogens reside inside cells during some part of the pathogenic process. In analyzing the sites of intracellular residence, there are two major types of intracellular locations: vesicular and nonvesicular compartments. Vesicular compartments include phagosomes resulting from host-cell- or microbe-induced ingestion and microbial created compartments such as the parasitophorous vacuole of *Toxoplasma gondii*. Nonvesicular compartments include all intracellular locations where the microbe is not enclosed by a membrane, such as free cytoplasmic residence. The majority of intracellular bacteria occupy vesicular compartments and those that are able to access the cytoplasm have specialized mechanisms that mediate phagosomal escape. Phagosomal compartments are generally viewed as harsh environments where in-

gested microbes are confronted with acidification, free-radical fluxes, nutrient deprivation, and a battery of antimicrobial proteins. Consequently, intracellular pathogens that inhabit phagosomal compartments interfere with their maturation and/or are resistant to host killing mechanisms. The cytoplasmic environment is generally considered nutrient rich and largely devoid of antimicrobial defenses (37), but two studies that have addressed this issue have provided inconsistent results. Nonpathogenic bacteria expressing *L. monocytogenes* listeriolysin O were found to escape to the cytoplasm after phagocytosis and replicate in there, consistent with the notion that the cytoplasm is a nutrient-rich and relatively unprotected cellular space (reviewed in Reference 37). However, there was no growth when bacteria that do not normally grow in the cytoplasm as part of their life cycle were placed in host cell cytoplasm by microinjection (18). This observation was interpreted to suggest that the ability of certain microbes to grow in the cytoplasm is an acquired trait in evolution (18).

There is no clear correlation between the type of intracellular residence and whether the microbe is an obligate or facultative intracellular pathogen or whether the microbe is acquired from another host or directly from the environment. Because host and/or environmental pressures are likely to have played a critical part in the emergence of the pathogenic strategy, the inability to correlate intracellular location with either category suggests that this distinction is not a dominant factor in the emergence of an intracellular pathogenic strategy. Instead, intracellular location is likely to be a distal outcome of adaptation to a host rather than a dominant engine driving evolution.

THE FOUR PHASES OF INTRACELLULAR PATHOGENESIS

Taking a bird's eye view of intracellular pathogenesis, it is possible to divide this process into four phases: (*a*) entry, (*b*) survival,

(c) replication, and (d) exit from the host cell. Entry, survival, and exit would appear to be essential for all intracellular pathogens, whereas most pathogens, but not all, also replicate in their host cells. Entry involves the steps necessary for the transition from the extracellular to the intracellular space. Each of these phases is primarily microbe-active or host-active depending on the specific host-microbe interaction. Entry can be achieved by the microbe by active or passive mechanisms. Examples of microbe-active entry are the invasion of host cells by *T. gondii* or the inducement of ingestion by cytoskeletal rearrangement in *Salmonella* spp. Microbe-passive entry mechanisms include opsonin-mediated phagocytosis, which delivers the microbe to a phagosome where it is not killed, as exemplified by the interaction of *C. neoformans* with macrophages. Microbes such as *H. capsulatum* manifest both microbe-active and microbe-passive mechanisms, whereby the fungal cell binds to the complement receptor through a direct fungal adhesion-receptor interaction that then leads to ingestion.

In contrast to entry, which can be a microbe- or host-active process, the type of survival strategy is determined largely by the microbe. Some microbes survive by subverting intracellular antimicrobial mechanisms such as phagosome-lysosome fusion, modulating phagosomal pH, damaging phagosomal membranes, and/or quenching microbicidal oxidative bursts. Others escape from the vesicular compartment, such as a phagosome to the nonvesicular compartment of the cytoplasm. Other microbes, like *T. gondii*, survive by creating their own compartments that are impervious to host microbicidal mechanisms. For all obligate intracellular pathogens and most facultative intracellular pathogens, survival in a permissive cell leads to replication with an increase in the intracellular microbial burden. Again, numerous exit strategies are dictated largely by the specific microbe in question. Some bacteria such as *Listeria monocytogenes* spread from cell to cell by actin tails. Others exit by inducing host cell lysis with release of microbial progeny. Yet others,

like *C. neoformans*, exit the host cell by inducing phagosomal extrusion (5). Because intracellular residence could trap a microbe in its host, it is intriguing to consider that those microbes with specialized exit strategies represent that subset which is fully adapted to the intracellular lifestyle. Again, like the situation with the location of intracellular residence, no overriding theme emerges in the comparison of the various pathogenic strategies. Hence, apart from the overarching themes of entry, survival, replication, and cellular exit, the details associated with these phases of intracellular life for specific pathogens appear to be adaptations suitable for specific host-microbe interactions.

Despite the menagerie of variations used to achieve entry, survival, replication, and exit by different intracellular pathogens, there are only a few general solutions to these problems. Entry requires attachment to cellular receptors, and several facultative intracellular pathogens, including *M. leprae* (43), *H. capsulatum* (26), and *C. neoformans* (50), exploit the complement receptor. Others, including several gram-negative bacteria, induce their own uptake by manipulating host cytoskeletal functions (3). Once inside the cell, many intracellular pathogens ensure their own survival by interfering with the process of phagosome maturation. This strategy is used by such different intracellular pathogens as *M. tuberculosis* (facultative intracellular, mycobacteria, self host-acquired), *L. pneumophila* (facultative intracellular, bacteria, nonself host-acquired), *Chlamydia* spp. (obligate intracellular, bacteria, self host-acquired), *T. gondii* (facultative intracellular, protozoa, nonself host-acquired), and *H. capsulatum* (facultative intracellular, fungus, environmentally acquired). Given that this group includes diverse bacteria, fungi, and protozoa, a likely explanation for the commonalities in the process is convergent evolution.

ORIGIN OF VIRULENCE IN INTRACELLULAR PATHOGENS

The interaction of an intracellular microbe with a host can have positive, neutral, and deleterious outcomes for each of the participants. Positive interactions include mutualism and are illustrated by the existence of endosymbiotic bacteria in invertebrate hosts. Neutral interactions are harder to define without more knowledge of the systems involved, because neutrality implies absence of damage for either party, which is akin to proving a negative. Deleterious interactions include the death of the microbe and damage to the host. Because this review is about intracellular pathogens, the discussion is focused on those interactions that are deleterious to the host.

Virulence is one outcome of the host-microbe interaction whereby the result is host damage, and this damage can come from the microbe, the host immune response, or both (9). Because host-acquired intracellular pathogens are host dependent, there is no obvious requirement for the host-microbe interaction to be deleterious to the host unless host damage is required for microbial replication or transmission. In general, the virulence of host-acquired microbes is a result of an unbalanced interaction whereby microbe- and/or host-mediated damage affects homeostasis. Two well-studied facultative intracellular pathogens acquired from other hosts include *M. tuberculosis* and *L. pneumophila*, which occupy the ecological sites of humans and amoebae, respectively. These organisms manifest similar intracellular survival strategies, and, for both, lung disease has strong components of host-mediated damage. For *M. tuberculosis* pulmonary cavitary formation with cough and the formation of infective aerosols is a critically important mechanism for host-to-host spread. For *L. pneumophila*, human infection is a dead-end event that is unlikely to be a significant selection force for this protozoal associated microbe.

For environment-acquired intracellular pathogens, the encounter with a potential host can have different implications. Using *C. neoformans* and *H. capsulatum* as examples of environmentally acquired facultative intracellular pathogens, one can discern significant differences from host-acquired microbes. Neither of these fungi requires other hosts for completion of its life cycle. For environmentally acquired microbes their interactions with animal hosts are a potential dead end that results in the death of the microbe. Because these interactions are likely to be rare and to involve only a minute fraction of the individuals in the populations, they cannot be expected to provide a major selection pressure for the evolution of intracellular pathogenic strategies. Instead, it is more likely that intracellular pathogenic strategies of environmentally acquired microbes are the result of adaptations to their ecological niche. In this regard, protozoal grazers such as amoebae have emerged as an important biotic force that could select for bacterial characteristics that translate into fitness in vertebrate hosts (34). For example, *Escherichia coli* strains harboring the Shiga toxin genes are more resistant to predation by the protozoan *Tetrahymena pyriformis*, raising the intriguing possibility that this important virulence factor was initially selected for environmental survival and only accidentally functions to harm animal hosts (49). Similarly, the interactions of several pathogenic fungi are similar to those with macrophages, and passage of an avirulent strain of *H. capsulatum* in amoebae restored its virulence (27, 47, 48). For *C. neoformans*, the interactions with amoeboid cells of *Dictyostelium discoideum* also enhanced virulence for mice (46).

HOW COMMON IS THE CAPACITY FOR INTRACELLULAR LIFE?

The phenomenon of intracellular pathogenesis is often viewed as a specialized host-microbe interaction despite considerable evidence that most, if not all, microbes have some capacity for intracellular survival. Host-associated microbes, whether endosymbionts, commensals, or pathogens, all require some capacity for intracellular life. Many human

Table 1 Human pathogenic microbes that can survive ingestion by amoebae[a]

Type	Organism	Reference
Bacteria	*Chlamydia pneumonia*	(14)
	Legionella pneumophila	(34)
	Listeria monocytogenes	(56)
	Escherichia coli	(4)
	Mycobacterium avium	(36)
	Coxiella burnetii	(24)
	Francisella tularensis	(1)
	Vibrio cholerae	(2)
	Helicobacter pylori	(54)
	Pseudomonas aeruginosa	(19)
Fungi	*Cryptococcus neoformans*	(48)
	Histoplasma capsulatum	(47)
	Sporothrix schenckii	(47)
	Blastomyces dermatidis	(47)

[a]Not a complete list. For a more extensive reference list see Reference 19.

intracellular pathogens are capable of survival in amoebae (**Table 1**). Hence, it should not be surprising that investigators often discover that common pathogens not considered intracellular pathogens can survive and often replicate in host cells. Examples of such organisms classified as extracellular that have been recently found to survive and replicate inside cells include *Staphylococcus aureus* (41), *Streptococcus pyogenes* (33), *Enterococcus* spp. (51) and *Helicobacter pylori* (40). In fact, analysis of gene expression patterns of *S. aureus* upon entry into human epithelial cells suggests a response that is adapted to the intracellular environment (16). Environmental microbes such as bacteria and fungi are subject to grazing by protista such as amoebae. Once ingested, their survival is a function of their ability to escape from the host cell and/or inhibit the host microbicidal mechanisms that are presumably designed for nutrient acquisition (30).

Because all soil microbes are in an ecological niche inhabited by innumerable species of amoeboid predators, survival in such an environment would almost certainly require the selection of effective mechanisms for intracellular

survival. If this is the case, then the capacity for intracellular life must be a common and necessary attribute for survival in both environmental and host ecological niches. Hence, the ability of microbes to survive inside other cells might be the rule rather than the exception, with intracellular pathogens representing the subset of microbes that are capable of both intracellular survival and mediating damage to the host, either inadvertently or as a necessary condition for replication and survival. Such a view would redefine the outcome of evolution of intracellular pathogens and shift it from a specialized survival strategy to a common microbial survival mechanism that is sometimes associated with deleterious effects on the host.

CONCLUSION

In the introduction, we considered some of the problems inherent in the terminology of the phrase evolution of intracellular pathogens. However, in approaching this subject it is apparent that there is also a problem of perspective, whereby intracellular pathogenic strategies are seen as specialized microbial adaptations rather than common interactions that sometimes are deleterious to the host. Such viewpoints are critically important to how one might view the evolution of intracellular pathogens. If the capacity for intracellular lifestyle is a common microbial attribute, then similarities in the host subversion mechanisms observed for intracellular pathogens are likely a consequence of divergent evolution whereby the outcome of pathogenicity can arise by chance or as part of microbial specialization to a particular host. However, if the capacity for intracellular lifestyle is relatively rare among the microbiota, then the similarities among numerous unique intracellular survival strategies described for intracellular pathogens represent convergent evolution to solve the problem of intracellular survival. At this time, we do not have sufficient information to choose between these possibilities, and both may apply to different sets of intracellular pathogens. For example, similarities in the intracellular pathogenic strategy of phylogenetically distant microbes such as fungi and bacteria may represent convergent evolution, whereas the variations in intracellular survival mechanisms among gram-negative bacteria may represent divergent evolution from ancient ancestor strategies to survive phagocytic predators. To investigate these possibilities, one would need additional information on the innate capacity of pathogenic and nonpathogenic microbes to survive in both vertebrate and nonvertebrate hosts.

More definitive conclusions about the evolution of intracellular pathogens will require a better delineation of the diversity of microbial life on earth combined with an assessment of potential intracellular living opportunities and threats at various ecological sites. For example, current views on the association of amoebae with the emergence of virulence in such organisms as *L. pneumophila* and *C. neoformans* have been inferred from the interactions of these microbes with only a few amoeboid species. Because the biota contains vast numbers of amoeboid species, it would seem logical to ascertain the generality of observations made with such common laboratory species as *Acanthamoeba castellani* with wild amoebae. Given the immense number of host-microbe interactions, such a project appears to involve a staggering amount of work. However, it is possible that the outlines of the problem will emerge after studying only a few more interactions in the same manner that the completion of a few microbial genomes provided fundamental insights into the relationship of facultative and obligate intracellular pathogens and illustrated the phenomenon of genome reduction.

SUMMARY POINTS

1. Each intracellular pathogen likely adopts a unique intracellular survival strategy. The uniqueness of each microbial strategy follows from the uniqueness of each microbial species and its niche.

2. Despite the uniqueness of each host-microbe interaction, there are relatively few solutions to the problem of intracellular entry, survival, and escape. For phylogenetically distant microbial species, similarities in intracellular pathogenic strategies are probably most easily explained by convergent evolution.

3. For a microbial species, the benefits of intracellular life are balanced by the loss of potential free-living habitats and by genome reduction. Organisms such as *Mycobacterium leprae* have lost their capacity for living independently of their hosts, and increased host specialization severely limits the number of host species available for infection and survival. Loss of identity is exemplified by mitochondria and may be one outcome of the endosymbiotic relationship. Hence, the attraction of intracellular life may ultimately doom such microbes through dependence on vulnerable hosts, Muller's ratchet phenomena, or loss of identity.

4. The ecological site from which a microbe is acquired during infection is an important consideration when analyzing the selection forces responsible for the evolution of intracellular pathogens. Specifically, it is important to consider whether the microbe is under active predation by larger microbes capable of ingesting it, and to identify those phagocytic predators. Amoeboid predators can be found within animal hosts, where they could conceivably provide selection pressures on host-associated flora.

5. Whereas intracellular pathogenesis is often viewed as a specialized lifestyle, there is emerging evidence that most pathogenic microbes are capable of intracellular survival, at least during some stages of the infection and disease cycle. For example, organisms such as *S. aureus*, *S. pyogenes*, *H. pylori*, *B. anthracis* and *Aspergillus fumigatus* are found inside host cells and most can replicate in that environment. Even ciliated organisms that are almost always found in extracellular spaces like *Trypanosoma* spp. have sophisticated mechanisms for invading vertebrate cells (7). Hence, the capacity for intracellular life may be the rule rather than the exception.

DISCLOSURE STATEMENT

The author is not aware of any biases that might be perceived as affecting the objectivity of this review.

ACKNOWLEDGMENTS

Dr. Casadevall is supported in part by NIH awards HL059842, AI033774, AI052733, and AI033142. The author thanks Dr. Liise-anne Pirofski for critical reading of this manuscript and many helpful suggestions.

LITERATURE CITED

1. Abd H, Johansson T, Golovliov I, Sandstrom G, Forsman M. 2003. Survival and growth of *Francisella tularensis* in *Acanthamoeba castellanii*. *Appl. Environ. Microbiol.* 69:600–6
2. Abd H, Saeed A, Weintraub A, Nair GB, Sandstrom G. 2007. *Vibrio cholerae* O1 strains are facultative intracellular bacteria, able to survive and multiply symbiotically inside the aquatic free-living amoeba *Acanthamoeba castellanii*. *FEMS Microbiol. Ecol.* 60:33–39
3. Alonso A, Garcia-del PF. 2004. Hijacking of eukaryotic functions by intracellular bacterial pathogens. *Int. Microbiol.* 7:181–91
4. Alsam S, Jeong SR, Sissons J, Dudley R, Kim KS, Khan NA. 2006. *Escherichia coli* interactions with *Acanthamoeba*: a symbiosis with environmental and clinical implications. *J. Med. Microbiol.* 55:689–94
5. Alvarez M, Casadevall A. 2006. Phagosome extrusion and host-cell survival after *Cryptococcus neoformans* phagocytosis by macrophages. *Curr. Biol.* 16:2161–65
6. Andersson SG, Kurland CG. 1998. Reductive evolution of resident genomes. *Trends Microbiol.* 6:263–68
7. Andrews NW. 1993. Living dangerously: how *Trypanosoma cruzi* uses lysosomes to get inside host cells, and then escapes into the cytoplasm. *Biol. Res.* 26:65–67
8. Brandl MT, Rosenthal BM, Haxo AF, Berk SG. 2005. Enhanced survival of *Salmonella enterica* in vesicles released by a soilborne *Tetrahymena* species. *Appl. Environ. Microbiol.* 71:1562–69
9. Casadevall A, Pirofski L. 1999. Host-pathogen interactions: redefining the basic concepts of virulence and pathogenicity. *Infect. Immun.* 67:3703–13
10. Casadevall A, Pirofski L. 2001. Host-pathogen interactions: the attributes of virulence. *J. Infect. Dis.* 184:337–44
11. Casadevall A, Pirofski LA. 2006. A reappraisal of humoral immunity based on mechanisms of antibody-mediated protection against intracellular pathogens. *Adv. Immunol.* 91:1–44
12. Cazalet C, Rusniok C, Bruggemann H, Zidane N, Magnier A, et al. 2004. Evidence in the *Legionella pneumophila* genome for exploitation of host cell functions and high genome plasticity. *Nat. Genet.* 36:1165–73
13. Cole ST, Eiglmeier K, Parkhill J, James KD, Thomson NR, et al. 2001. Massive gene decay in the leprosy bacillus. *Nature* 409:1007–11

14. Essig A, Heinemann M, Simnacher U, Marre R. 1997. Infection of *Acanthamoeba castellanii* by *Chlamydia pneumoniae*. *Appl. Environ. Microbiol.* 63:1396–99
15. Fraser-Liggett CM. 2005. Insights on biology and evolution from microbial genome sequencing. *Genome Res.* 15:1603–10
16. Garzoni C, Francois P, Huyghe A, Couzinet S, Tapparel C, et al. 2007. A global view of *Staphylococcus aureus* whole genome expression upon internalization in human epithelial cells. *BMC Genomics* 8:171
17. Gil R, Latorre A, Moya A. 2004. Bacterial endosymbionts of insects: insights from comparative genomics. *Environ. Microbiol.* 6:1109–22
18. Goetz M, Bubert A, Wang G, Chico-Calero I, Vazquez-Boland JA, et al. 2001. Microinjection and growth of bacteria in the cytosol of mammalian host cells. *Proc. Natl. Acad. Sci. USA* 98:12221–26
19. Greub G, Raoult D. 2004. Microorganisms resistant to free-living amoebae. *Clin. Microbiol. Rev.* 17:413–33
20. Harb OS, Gao LY, Abu Kwaik Y. 2000. From protozoa to mammalian cells: a new paradigm in the life cycle of intracellular bacterial pathogens. *Environ. Microbiol.* 2:251–65
21. Huard RC, Fabre M, de HP, Lazzarini LC, van SD, et al. 2006. Novel genetic polymorphisms that further delineate the phylogeny of the *Mycobacterium tuberculosis* complex. *J. Bacteriol.* 188:4271–87
22. Ibrahim-Granet O, Philippe B, Boleti H, Boisvieux-Ulrich E, Grenet D, et al. 2003. Phagocytosis and intracellular fate of *Aspergillus fumigatus* conidia in alveolar macrophages. *Infect. Immun.* 71:891–903
23. Keeling PJ, Fast NM. 2002. Microsporidia: biology and evolution of highly reduced intracellular parasites. *Annu. Rev. Microbiol.* 56:93–116
24. La SB, Raoult D. 2001. Survival of *Coxiella burnetii* within free-living amoeba *Acanthamoeba castellanii*. *Clin. Microbiol. Infect.* 7:75–79
25. Li JS, Winslow GM. 2003. Survival, replication, and antibody susceptibility of *Ehrlichia chaffeensis* outside of host cells. *Infect. Immun.* 71:4229–37
26. Long KH, Gomez FJ, Morris RE, Newman SL. 2003. Identification of heat shock protein 60 as the ligand on *Histoplasma capsulatum* that mediates binding to CD18 receptors on human macrophages. *J. Immunol.* 170:487–94
27. Malliaris SD, Steenbergen JN, Casadevall A. 2004. *Cryptococcus neoformans* var. *gattii* can exploit *Acanthamoeba castellanii* for growth. *Med. Mycol.* 42:149–58
28. Margulis L. 1971. Symbiosis and evolution. *Sci. Am.* 225:48–57
29. Margulis L. 1971. The origin of plant and animal cells. *Am. Sci.* 59:230–35
30. Matz C, Kjelleberg S. 2005. Off the hook—how bacteria survive protozoan grazing. *Trends Microbiol.* 13:302–7
31. Maurelli AT, Fernandez RE, Bloch CA, Rode CK, Fasano A. 1998. "Black holes" and bacterial pathogenicity: a large genomic deletion that enhances the virulence of *Shigella* spp. and enteroinvasive *Escherichia coli*. *Proc. Natl. Acad. Sci. USA* 95:3943–48
32. McGraw EA, O'Neill SL. 2004. *Wolbachia pipientis*: intracellular infection and pathogenesis in *Drosophila*. *Curr. Opin. Microbiol.* 7:67–70
33. Medina E, Goldmann O, Toppel AW, Chhatwal GS. 2003. Survival of *Streptococcus pyogenes* within host phagocytic cells: a pathogenic mechanism for persistence and systemic invasion. *J. Infect. Dis.* 187:597–603
34. Molmeret M, Horn M, Wagner M, Santic M, Abu KY. 2005. Amoebae as training grounds for intracellular bacterial pathogens. *Appl. Environ. Microbiol.* 71:20–28
35. Moran NA, Baumann P. 2000. Bacterial endosymbionts in animals. *Curr. Opin. Microbiol.* 3:270–75
36. Mura M, Bull TJ, Evans H, Sidi-Boumedine K, McMinn L, et al. 2006. Replication and long-term persistence of bovine and human strains of *Mycobacterium avium* subsp. *paratuberculosis* within *Acanthamoeba polyphaga*. *Appl. Environ. Microbiol.* 72:854–59
37. O'Riordan M, Portnoy DA. 2002. The host cytosol: front-line or home front? *Trends Microbiol.* 10:361–64
38. Pallen MJ, Wren BW. 2007. Bacterial pathogenomics. *Nature* 449:835–42
39. Pandey DP, Gerdes K. 2005. Toxin-antitoxin loci are highly abundant in free-living but lost from host-associated prokaryotes. *Nucleic Acids Res.* 33:966–76
40. Petersen AM, Krogfelt KA. 2003. *Helicobacter pylori*: an invading microorganism? A review. *FEMS Immunol. Med. Microbiol.* 36:117–26
41. Qazi SN, Harrison SE, Self T, Williams P, Hill PJ. 2004. Real-time monitoring of intracellular *Staphylococcus aureus* replication. *J. Bacteriol.* 186:1065–77

42. Renesto P, Ogata H, Audic S, Claverie JM, Raoult D. 2005. Some lessons from *Rickettsia* genomics. *FEMS Microbiol. Rev.* 29:99–117

43. Schlesinger LS, Horwitz MA. 1991. Phagocytosis of *Mycobacterium leprae* by human monocyte-derived macrophages is mediated by complement receptors CR1 (CD35), CR3 (CD11b/CD18), and CR4 (CD11c/CD18) and IFN-gamma activation inhibits complement receptor function and phagocytosis of this bacterium. *J. Immunol.* 147:1983–94

44. Seshadri R, Paulsen IT, Eisen JA, Read TD, Nelson KE, et al. 2003. Complete genome sequence of the Q-fever pathogen *Coxiella burnetii. Proc. Natl. Acad. Sci. USA* 100:5455–60

45. Sirand-Pugnet P, Lartigue C, Marenda M, Jacob D, Barre A, et al. 2007. Being pathogenic, plastic, and sexual while living with a nearly minimal bacterial genome. *PLoS Genet.* 3:e75

46. Steenbergen JN, Nosanchuk JD, Malliaris SD, Casadevall A. 2003. *Cryptococcus neoformans* virulence is enhanced after intracellular growth in the genetically malleable host *Dictyostelium discoideum. Infect. Immun.* 71:4862–72

47. Steenbergen JN, Nosanchuk JD, Malliaris SD, Casadevall A. 2004. Interaction of *Blastomyces dermatitidis, Sporothrix schenckii*, and *Histoplasma capsulatum* with *Acanthamoeba castellanii. Infect. Immun.* 72:3478–88

48. Steenbergen JN, Shuman HA, Casadevall A. 2001. *Cryptococcus neoformans* interactions with amoebae suggest an explanation for its virulence and intracellular pathogenic strategy in macrophages. *Proc. Natl. Acad. Sci. USA* 18:15245–50

49. Steinberg KM, Levin BR. 2007. Grazing protozoa and the evolution of the *Escherichia coli* O157:H7 *Shiga* toxin-encoding prophage. *Proc. Biol. Sci.* 274:1921–29

50. Taborda CP, Casadevall A. 2002. CR3 (CD11b/CD18) and CR4 (CD11c/CD18) are involved in complement-independent antibody-mediated phagocytosis of *Cryptococcus neoformans. Immunity* 16:791–802

51. Verneuil N, Sanguinetti M, Le BY, Posteraro B, Fadda G, et al. 2004. Effects of the *Enterococcus faecalis* hypR gene encoding a new transcriptional regulator on oxidative stress response and intracellular survival within macrophages. *Infect. Immun.* 72:4424–31

52. Vissa VD, Brennan PJ. 2001. The genome of *Mycobacterium leprae*: a minimal mycobacterial gene set. *Genome Biol.* 2:REVIEWS1023

53. Wernegreen JJ, Lazarus AB, Degnan PH. 2002. Small genome of *Candidatus Blochmannia*, the bacterial endosymbiont of *Camponotus*, implies irreversible specialization to an intracellular lifestyle. *Microbiology* 148:2551–56

54. Winiecka-Krusnell J, Wreiber K, von Euler A, Engstrand L, Linder E. 2002. Free-living amoebae promote growth and survival of *Helicobacter pylori. Scand. J. Infect. Dis.* 34:253–56

55. Wren BW. 2000. Microbial genome analysis: insights into virulence, host adaptation and evolution. *Nat. Rev. Genet.* 1:30–39

56. Zhou X, Elmose J, Call DR. 2007. Interactions between the environmental pathogen *Listeria monocytogenes* and a free-living protozoan (*Acanthamoeba castellanii*). *Environ. Microbiol.* 9:913–22

(p)ppGpp: Still Magical?*

Katarzyna Potrykus and Michael Cashel

Laboratory of Molecular Genetics, National Institute of Child Health and Human Development, National Institutes of Health, Bethesda, Maryland 20892-2785; email: potrykuk@mail.nih.gov; mcashel@nih.gov

Annu. Rev. Microbiol. 2008. 62:35–51

First published online as a Review in Advance on May 2, 2008

The *Annual Review of Microbiology* is online at micro.annualreviews.org

This article's doi: 10.1146/annurev.micro.62.081307.162903

0066-4227/08/1013-0035$20.00

Key Words

stringent response, transcription, stress, pathogenesis, Rel/Spo homologs

Abstract

The fundamental details of how nutritional stress leads to elevating (p)ppGpp are questionable. By common usage, the meaning of the stringent response has evolved from the specific response to (p)ppGpp provoked by amino acid starvation to all responses caused by elevating (p)ppGpp by any means. Different responses have similar as well as dissimilar positive and negative effects on gene expression and metabolism. The different ways that different bacteria seem to exploit their capacities to form and respond to (p)ppGpp are already impressive despite an early stage of discovery. Apparently, (p)ppGpp can contribute to regulation of many aspects of microbial cell biology that are sensitive to changing nutrient availability: growth, adaptation, secondary metabolism, survival, persistence, cell division, motility, biofilms, development, competence, and virulence. Many basic questions still exist. This review tries to focus on some issues that linger even for the most widely characterized bacterial strains.

Contents

INTRODUCTION

Nearly 40 years ago two spots appeared on autoradiograms, as if by magic, from extracts of *Escherichia coli* responding to the stress of amino acid starvation. This response provokes stringent inhibition of stable RNA (rRNA and tRNA) synthesis that is greatly relaxed in *relA* mutants. These spots, first called magic spots, were derivatives of GTP and GDP that differed only by the presence of a pyrophosphate esterified to the ribose 3′ carbon, abbreviated as pppGpp and ppGpp, respectively. Currently we know that (p)ppGpp signals nutritional stress, leading to adjustments of gene expression in most bacteria and plants. If magic can be defined as ignorance of how something happens and how it works, then much of the magic of (p)ppGpp is not lost. This is because fundamental details regarding (p)ppGpp remain uncertain in the best-studied bacterial strains, let alone the diverse bacteria that exploit this regulator in different ways. This is too broad a topic to review here; recent reviews are highly recommended (11, 40, 56, 72).

ALMOST A SINGLE SUPERFAMILY OF ENZYMES EXISTS FOR (p)ppGpp SYNTHESIS AND BREAKDOWN

The sequenced genomes of free-living eubacteria and plants contain one or more variants of *rsh* (Rel Spo homolog) genes. These genes encode large (~750 amino acid) RSH proteins (**Figure 1**). The namesakes for RSH are the RelA and SpoT proteins of *E. coli*; two apparently similar RSH proteins exist among other beta- and gamma-proteobacteria, whereas most other bacteria have a single RSH protein, designated Rel with species names, such as Rel*Mtb*. RSH variants can have end extensions as well as insertions. Small fragments with weakly active synthase have been discovered in *Streptococcus mutans* and *Bacillus subtilis*, but their functions are unknown (39, 50). Similar sequences coexist generally in genomes of the class *Firmicutes* (e.g., bacilli, streptococci, staphylococci, *Listeria*, clostridia) together with a full-length RSH protein. Found among the first sequenced *Rickettsia* genomes are multiple *rsh* fragments whose activities are untested.

There is a small, secreted enzyme from *Streptomyces morookaensis* with no obvious homology to RSH proteins. Under special conditions this enzyme, once a commercial source for (p)ppGpp and (p)ppApp, transfers pyrophosphate residues indiscriminately to ribonucleoside 5′ mono-, di-, and triphosphates as well as synthesizes nucleotides with a 5′

(p)ppGpp: guanosine 5′-triphosphate, 3′-diphosphate; guanosine 5′-diphosphate, 3′-diphosphate

RSH proteins: proteins with Rel and Spo homology

polyphosphate, $2',3'$-cyclic monophosphate (46 and references therein). Such compounds unexpectedly are found in protein crystals and probably are biologically important.

(p)ppGpp Hydrolases and Synthases

The N-terminal half of generic RSH proteins contains catalytic activity domains for hydrolase and synthase. The RelA protein has only synthase activity; its hydrolase is inactive (**Figure 1b**, left). RSH synthases in general have similarities to polymerases, such as DNA polymerase beta (30). RSH (p)ppGpp hydrolases are Mn^{2+}-dependent pyrophosphohydrolases with a conserved His-Asp (HD) motif (1, 30). For the SpoT protein, sequence variants limit its synthase activity, but not its hydrolase (**Figure 1b**, right). Thus, RelA is viewed as specialized for synthesis because of an inactive HD domain sequence and SpoT is viewed as specialized for hydrolysis with a weak synthase. Separate engineered peptides for hydrolase and synthase are active, at least for RelMtb and RelSeq (3, 30), despite an early report of overlapping functions that were deduced from behavior of progressive deletions in SpoT (24). Point mutants that define each domain can help predict activities of new RSH enzymes from their sequence (30).

Regulating the balance of the opposing activities of RSH enzymes is crucial. Equally active, unregulated hydrolase and synthase activities would catalyze a futile cycle of (p)ppGpp synthesis and hydrolysis (4, 43). Too much synthase elevates (p)ppGpp, which provokes a stringent response, inhibits growth and, in *E. coli*, adjusts gene expression to curtail unnecessary activities in nongrowing cells. Too little (p)ppGpp from excess hydrolase makes cells less able to respond appropriately to nutritional stress.

How Are RSH Activities Regulated?

Results from experiments with various RSH proteins indicate that both the N-terminal

domain (NTD) and the C-terminal domain (CTD) can contribute to regulation. Synthase activation (RelA and bifunctional RSH enzymes) (**Figure 1b,c**) seems to occur by a common signal. This involves sensing the inability of tRNA aminoacylation to keep up with the demands of protein synthesis, typically provoked in vivo by amino acid starvation or by adding inhibitors of aminoacyl tRNA synthases. Early in vitro experiments elegantly defined the ribosome idling reaction during elongation (27); this was verified for the RelMtb enzyme (4) with ribosomal activation components (RAC) by using puromycin-treated ribosomes, poly U, and uncharged Phe-tRNA.

The synthase catalytic sites of monofunctional (RelA-like) enzymes have a conserved acidic triad of residues (ExDD) that differs from the conserved basic (RxKD) triad found for bifunctional RSH proteins (64). The authors report that three crucial properties of the NTD synthases sort with the two sequences, even in chimerical enzymes: a mono/dual metal mechanism, a broad/sharp Mg^{2+} optimum for substrate binding, and a major helicity change. Accordingly, one must wonder whether the presence of an active hydrolase constrains synthase catalysis or vice versa. If so, then substituting the acidic triad of monofunctional enzymes for bifunctional RSH proteins (and the reverse) might alter cellular hydrolase or synthase regulatory properties.

Despite the availability of detailed ribosomal structures, little is known of the interactions between RelA or RSH and ribosomes, except that ribosomal mutants of the L11 protein (termed RelC) abolish activation. For RelA and RelMtb, point mutants in the CTD as well as CTD deletions abolish activation under RAC conditions, hinting an activation pathway from ribosome to CTD to NTD. An interesting regulatory role proposed for the CTD of RelA and RelMtb involves oligomerization (3, 25).

The conserved TGS region of the CTD (**Figure 1a**) has now been implicated in the regulation of the strong hydrolase with a weak synthase of SpoT. The ability of SpoT to sense many sources of nutrient stress other

RelA: *E. coli* protein that activates (p)ppGpp synthesis during amino acid starvation

RelMtb: RSH enzyme from *M. tuberculosis*

SpoT: *E. coli* protein that mediates (p)ppGpp elevation during other nutrient stress

Stringent response: positive and negative effects on cells by elevated (p)ppGpp

RAC: ribosomal activation components

TGS: conserved domain on RSH CTD for uncharged ACP binding

than amino acid starvation and to respond by limiting hydrolase has long been puzzling (**Figure 1b**, middle). An exciting mechanism allowing SpoT to sense fatty acid synthesis limitation has been discovered (6). The acyl carrier protein (ACP) binds to the TGS domain of SpoT and this binding is probably influenced by the ratio of unacylated ACP to acylated ACP in the cell. Fatty acid starvation thus leads to a shift in the balance of the two SpoT activities in favor of synthesis. The authors point to parallels between SpoT and RelA sensing. They also raise the possibility that sensing uncharged (unacylated) ACP might explain SpoT-mediated (p)ppGpp accumulation during carbon source starvation because the expected metabolic

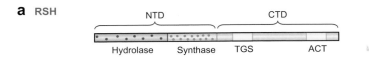

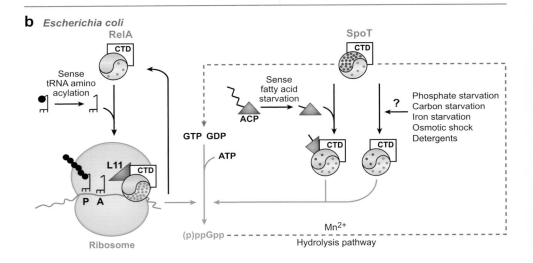

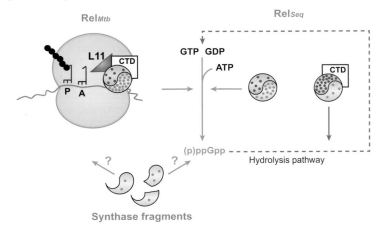

consequences are to limit fatty acid synthesis as well as provoke RelA synthase activation (24). Sensing other nutritional stress may be complex. Phosphate starvation is sensed by SpoT hydrolase to elevate (p)ppGpp, which induces IraP, a RssB antiadaptor that antagonizes RssB activation of RpoS turnover, thereby inducing RpoS (10).

In vitro assays of RelSeq individual activities reveal that a CTD deletion inhibits hydrolase and activates synthase (43) (**Figure 1c**, right). However, a similar CTD deletion affects RelMtb without inhibiting hydrolase, although RAC-dependent activation is lost (3). In the absence of structures for a full-length RSH enzyme, notions of how the CTD alters the balance of hydrolase/synthase activity are speculative. The NTD-CTD boundary for both SpoT and RelSeq is a solvent-accessible region that might be a hinge (43). Because the RelSeq NTD structure shows the head of hydrolase neighbors the tail of synthase, this hinge could allow physical contact between the TGS region and hydrolase and/or synthase sequences. The structures suggest that avoidance of a (p)ppGpp futile cycle may be an intrinsic feature of the catalytic half of the protein. These crystals resolve two mutually exclusive active site conformers (hydrolase-OFF/synthase-ON versus hydrolase-ON/synthase-OFF). Substrate binding to either site is argued to induce the switch between the two conformations to affect catalytic sites 30 Å apart (30). The existence of hydrolase domain point mutants that reverse the synthase defect of some but not all synthase-defective alleles underscores the notion that there is cross-talk between sites (U. Mechold & M. Cashel, unpublished data). Altered CTD structure by ACP or RAC effectors might trigger an allosteric switch between the two NTD conformers either by physical contacts or by inducing a conformational cascade over the full length of the RSH protein. The net effect is an enzyme activity state that favors hydrolase or synthase, not both.

ACP: acyl carrier protein

EFFECTS OF (p)ppGpp ON BACTERIAL PHYSIOLOGY

The many effects of (p)ppGpp on metabolism and physiology are complex and seem to differ greatly among different organisms. Profiling and proteomic studies in different organisms consistent with this trend are beginning to appear but do not yet involve comparing effects of a complete absence of ppGpp with wild type (20, 51).

Figure 1

Cartoon of (p)ppGpp regulation. (*a*) Conserved domains of the N-terminal domain (NTD) and C-terminal domain (CTD) halves of a RSH protein. (*b*) Roles in *Escherichia coli* of RelA (*left*) and SpoT (*right*) displaying the NTD hydrolase/synthase (*yin and yang symbol*) and CTD (*square*), with the balance of hydrolase/synthase shown as the ratio of green to orange dots. The CTD contacts hydrolase and synthase, reflecting possible CTD regulation of each activity. (*Left*) Activation of RelA synthase requires cognate uncharged tRNA, a translating ribosome (*yellow*) with an empty A site paused for lack of cognate charged tRNA, and r-protein L11. Synthesis of (p)ppGpp from GTP (or GDP) involves pyrophosphoryl transfer from ATP and is accompanied by release of RelA. (*Right*) SpoT regulation is depicted in two ways. Acyl carrier protein (ACP) lacking acyl fatty acids (*purple triangle*) binds to the SpoT TGS region of the CTD, which shifts the activity balance to synthesis; this effect also requires other CTD functions. Other stress conditions provoke a similar shift of the activity balance by unknown mechanisms. Hydrolysis of (p)ppGpp regenerates GTP or GDP by an Mn^{2+}-dependent reaction releasing pyrophosphate. (*c*) Regulation of RelMtb activities from *Mycobacterium tuberculosis*; synthase is activated similar to RelA. RelMtb differs from RelA by the added presence of a hydrolase, and RelMtb also differs from SpoT in that the hydrolase is modestly inhibited when sensing stress. A related RelSeq enzyme has strong hydrolase and weak synthase activities without ribosomal activation; removal of its CTD reverses the balance of activities, and structures have been resolved that reflect both activity states. Depicted at the bottom of panel *c* are small fragments with homology to RSH synthase recently discovered to coexist in members of the class *Firmicutes* with a full-length RSH protein. Physiological roles for these proteins are unknown.

Rapid Induction of (p)ppGpp Inhibits Growth, but How?

Induction of (p)ppGpp to high levels without starvation quickly inhibits growth and protein synthesis of exponentially growing *E. coli* (69). How might protein synthesis inhibition occur? It seems unlikely that inhibition of de novo rRNA and tRNA synthesis (via the stringent response) would so quickly block the activity of the pre-existing protein synthesis apparatus. Nor should rapid inhibition occur via (p)ppGpp induction of RpoS-dependent proteins that slow protein synthesis. Substrates for protein synthesis should not be limiting: Amino acid starvation does not occur under these conditions, and *E. coli* GTP levels drop only by half owing to inhibition of IMP dehydrogenase (GuaB). In contrast, more complete inhibition of GuaB in *B. subtilis* severely depletes GTP, leading to rRNA inhibition (36). Depleted GTP also leads to transcriptional regulation of about 200 genes by CodY, a GTP-binding protein (26). Although (p)ppGpp effects are largely indirect in *B. subtilis*, they contribute to the regulation of sporulation, competence, enzyme secretion, antibiotic production, and stress survival. The behavior of a (p)ppGpp-resistant GuaB protein would be interesting in both organisms.

Interactions of (p)ppGpp with protein synthesis elongation factors are generally reversed by equimolar (GTP)GDP (15). A recent report argues that ppGpp inhibits IF2-mediated fMet-Phe initiation dipeptide formation even at equimolar concentrations of (p)ppGpp and GTP, probably by interfering with 30S and 50S subunit interactions (44). This is intriguing because equimolar GTP and (p)ppGpp levels are reached physiologically during a stringent response. Inhibiting translation initiation would be an efficient pathway to limit excessive protein synthesis during nutritional impoverishment. Still, how (p)ppGpp might inhibit protein synthesis has been elusive historically, and independent verifications are in order.

The Extent of Growth Inhibition Differs for ppGpp and pppGpp

E. coli accumulates more ppGpp than pppGpp during amino acid starvation (15). Gratuitous (p)ppGpp induction inhibits growth about eightfold more severely with ppGpp than with pppGpp. This estimate comes from measuring growth rates while inducing only ppGpp or only pppGpp. This is accomplished using P*BAD* promoters and altering the abundance of the enzyme that converts pppGpp to ppGpp (GppA) as well as using (p)ppGpp synthases with different GTP or GDP affinities (U. Mechold & M. Cashel, unpublished data).

Basal Levels Control Growth by Regulating Ribosomal Number

Growth rate control is defined as the systematic variation of cellular RNA, DNA, and protein content as a function of rates of balanced growth. Basal level changes of (p)ppGpp over a 10- to 12-fold range are inversely correlated with growth rate and the number of ribosomes per cell (12). There is now a consensus that (p)ppGpp is a determinant of growth rate control rather than nucleoside triphosphate (NTP) substrate concentrations (65). Nevertheless, the existing literature remains confusing regarding the abolition of growth rate control in (p)ppGpp⁰ strains. Different views on how (p)ppGpp curtails transcription of rRNA are discussed below. We argue that the growth-rate-determining role of basal levels of (p)ppGpp involves rRNA control and differs from the growth inhibitory effects of inducing large amounts of (p)ppGpp.

Inhibition of DNA Replication

Classical studies with *E. coli* concluded that amino acid starvation inhibited DNA replication at the initiation stage at *oriC*, most probably owing to the lack of the DnaA replication initiation protein. It was believed that replication arrest due to (p)ppGpp accumulation in

B. subtilis was different and caused by the binding of an Rtp protein to specific sites about 100–200 kb away from oriC in both directions (for review see Reference 80). This view of B. subtilis behavior changed dramatically when DNA replication was studied with microarrays, revealing that inhibition of elongation in the presence of (p)ppGpp can take place throughout the chromosome, independent of Rtp and the proposed specific arrest sites (79). DNA primase (DnaG) was directly inhibited by (p)ppGpp. Unlike E. coli, B. subtilis accumulates more pppGpp than ppGpp; the more abundant nucleotide is a more-potent DnaG inhibitor. Replication forks were not disrupted, as they did not recruit RecA, thus maintaining genomic integrity. It is unknown whether E. coli behaves like B. subtilis in this respect.

ppGpp cocrystallizes with the B. subtilis Obg protein, which belongs to the conserved, small GTPase protein family (14). Obg interacts with several regulators (RsbT, RsbW, RsbX) necessary for the stress activation of σ^B, the global controller of a general stress regulon in B. subtilis (66). The E. coli ObgE protein (also known as CgtA) stabilizes arrested replication forks, and an obgE depletion causes disruption of cell cycle events, leading to filamentation and polyploidy (22, 23). CgtA is also associated with ribosomes and SpoT and is argued to alter SpoT basal activities (31). However, in a different study, SpoT was not detected when high-salt-washed ribosomes immobilized with a HA-tagged L1 protein were employed (H. Murphy & M. Cashel, unpublished data). In the same study, RelA binding to ribosomes is stoichiometric. It remains possible that SpoT could bind to ribosomes but that L1-tag interferes.

Effects on Phage Replication and Development

A truncated form of IF2 (IF2-2) was recently identified as the E. coli factor necessary to promote assembly of the E. coli replication restart proteins, PriA, PriC, DnaT, and DnaB-DnaC complex, at the phage Mu replication fork (52). This allows the DNA polymerase III complex to

be recruited. Ordinarily, PriA, PriC, and DnaT promote the assembly of a replisome without the initiator protein DnaA and the oriC and play an essential role in restarting replication after stalling of the replication fork (52). The authors mention their unpublished results, in which premixing IF2-2 with high levels of GTP diminished Mu replication in vitro by the PriA-PriC pathway, whereas premixing with ppGpp stimulated the reaction. They further speculate that ppGpp might activate the PriA-PriC restart pathway to ensure that chromosomal replication is completed when ppGpp shuts down the initiation at oriC. Extensive studies have shown that resolution of arrested replication forks has requirements for (p)ppGpp that can be satisfied by RNA polymerase (RNAP) M+ (p)ppGpp^0 suppressor mutants (discussed below) (75).

RNAP: RNA polymerase

The (p)ppGpp levels of the host seem to act as a sensor for phage lambda development, primarily affecting transcription. Modest ppGpp levels inhibit pR and activate pE, pI, and paQ promoters in vivo (67) and have effects in vitro (61, 62) that seem to favor lysogeny. In contrast, absent or high concentrations of (p)ppGpp favor lysis. This unusual concentration dependence similarly affects the switch between lytic and lysogenic growth through regulation of HflB (alias FtsH), a protease responsible for degradation of CII, a lysogeny-promoting phage protein. Again, modest ppGpp levels favor lysogeny by leading to low HflB levels. When ppGpp is either absent or high, HflB protease levels are high; this leads to lower CII levels and favors lysis (67).

(p)ppGpp^0 Physiology

Apart from phage growth, a complete absence of (p)ppGpp generates its own unique phenotypic features in E. coli. These include multiple amino acid requirements, poor survival of aged cultures, aberrant cell division, morphology, and immotility, as well as being locked in a growth mode during entry into starvation (41, 82). The multiple amino acid requirements are of special interest for two reasons: (a) They

M+ mutant: suppresses (p)ppGpp⁰ phenotypes; *E. coli* M+ grows without added amino acids

DnaK suppressor (DksA): protein that reverses thermolability of a *dnaK* mutant

reflect positive regulation by (p)ppGpp at the transcriptional level, and (*b*) they allow isolation of spontaneous mutants growing on minimal glucose medium. These (p)ppGpp^0 phenotypic suppressors are called M+ mutants, which so far map exclusively within RNAP *rpoB*, *rpoC*, and *rpoD* subunit genes (15, 47).

(p)ppGpp ACTS AT THE LEVEL OF TRANSCRIPTION

Inhibition of rRNA synthesis is the classical feature of the stringent response; numerous hypotheses have been made to explain this event (see recommended reviews above). Understanding (p)ppGpp regulation of transcription currently seems based on three key features: (*a*) shared characteristics of promoters affected by (p)ppGpp, (*b*) genetic and structural evidence that RNAP is the target of (p)ppGpp, and (*c*) the DksA protein augments (p)ppGpp regulation. Because early in vitro studies showed no differences between the effect of ppGpp and pppGpp on transcription, most studies use ppGpp exclusively.

Characteristics of Affected Promoters

One of the key elements of promoters inhibited by (p)ppGpp is the presence of a GC-rich discriminator, defined as a region between TATA-box (−10 box) and +1 nt (where +1 is the transcription start site) (76). In addition, the discriminator's activity depends on −35 and −10 sequences, as well as the length of the linker, i.e., the region between them (53). Promoters negatively regulated by ppGpp have a 16-bp linker, in contrast with the 17-bp consensus. Promoters activated by ppGpp seem to have an AT-rich discriminator and longer linkers (for example, the *his* promoter linker is 18 bp). There is also evidence that sensitivity to supercoiling influences ppGpp responses (21).

RNAP Is the Target

Although it is plausible that transcription of rRNA should be regulated by (p)ppGpp during the stringent response, for many years attempts to verify this hypothesis in vitro with pure RNAP were plagued by irreproducibility. Genetic evidence suggesting that RNAP was the target of (p)ppGpp came from the discovery that M+ mutants (also called stringent RNAP mutants) display in vitro and in vivo mimicry of physiology and transcription regulation conferred by (p)ppGpp, even in its absence. Cross-linking ppGpp to RNAP reinforced this notion, although different contacts were deduced (16, 74). Structural details of an association between ppGpp and RNAP came from the analysis of cocrystals that positioned ppGpp in the secondary channel of RNAP near the catalytic center (2). This channel provides access to the catalytic center for NTP substrates during polymerization as well as an entry point for derailed backtracked, nascent RNA in RNAP arrested during elongation. The ppGpp target could be defined by direct contacts with appropriate RpoB and RpoC residues. However, sequence changes in M+ mutants, chemical cross-linking, and cocrystallization provide a different target locations.

DksA Augments Regulation

DksA has many regulatory functions in addition to its ability to restore thermotolerance to a *dnaK* mutant when overexpressed (34). Among these functions was a need for DksA and (p)ppGpp to stimulate the accumulation of RpoS during early stationary phase of growth (13). The regulatory interrelationship between (p)ppGpp and DksA was clarified when DksA was found to be necessary for the stringent response. This finding was followed by discoveries that DksA potentiated (p)ppGpp regulation generally in vitro and in vivo, based on studies of inhibition of a rRNA promoter or activation of selected amino acid biosynthetic promoters (54, 55).

Determining the structure of DksA was a major contribution toward understanding its regulatory properties—the 17-kDa protein is structurally similar to GreA and GreB, which is not evident from sequences (57). Both GreA and GreB are well-characterized transcriptional elongation factors (9). They bind directly to RNAP rather than DNA and act by inserting their N-terminal coiled-coil finger domain through the RNAP secondary channel. Two conserved acidic residues at the tip of the finger domain are necessary to induce RNAP's intrinsic ability to cleave backtracked RNA, whose 3′ end then comes near the catalytic center and is available for polymerization, functionally rescuing the arrested enzyme. Binding of the GreA/B factors to RNAP is thought to occur by contacts between the C-terminal globular domain of Gre and RpoC residues 645–703 at the entrance of the secondary channel that form a coiled-coil (77). DksA also possesses two acidic residues at its finger tip, but it does not induce nucleolytic cleavage activity. Instead, these residues are proposed to stabilize ppGpp binding to RNAP by mutual coordination of an Mg^{2+} ion that is crucial for polymerization (57).

Evidently, GreA/B and DksA are structural homologs with different activities. This diversity has been extended by showing that GreA exhibits antagonistic effects to DksA at the *rrnB* P1 promoter, independent of ppGpp, and acts at an earlier initiation step than DksA does (60). Another example is that GreB, when overproduced, might mimic DksA (63). Two additional factors predicted to have shapes similar to GreB and GreA have been investigated: Gfh1 (37, 70) and TraR (8). It is exciting that these proteins possess different functions, none of which involves specific DNA-binding properties of more common regulators of specific promoters. However, *Pseudomonas aeruginosa* DksA is reported to bind to DNA (58).

Because DksA enhances ppGpp's effect, whether inhibition or activation, it was termed a ppGpp cofactor. Yet, in ppGpp^0 strains DksA overproduction can completely compensate for positive and negative regulation with respect to amino acid auxotrophy, cell-cell aggregation, motility, filamentation, stationary-phase morphology, and stimulating RpoS accumulation (41). DksA and (p)ppGpp can also have opposing roles on cellular adhesion (41). This implies that DksA is not only present to stabilize ppGpp's interaction with RNAP but that each regulator can have different modes of action. These epistatic relationships hint that DksA might function downstream of (p)ppGpp.

The above RNAP crystal studies were performed with RNAP from *Thermus thermophilus*. This organism produces ppGpp upon amino acid starvation, but like *B. subtilis* its rRNA levels in vivo respond to the availability of GTP rather than ppGpp directly (35). On the other hand, ppGpp inhibition of *Tth* rRNA promoters in vitro with *Tth* RNAP was observed but required higher ppGpp concentrations than the measured intracellular pool (35). Perhaps this is because no DksA homolog has been identified in *T. thermophilus* to date. A yet unidentified protein might be required for ppGpp to exert its full effect in vitro. However, the same statement raises concerns over the predicted DksA-RNAP interaction model, in which *E. coli* DksA was docked into the structure of *Tth* RNAP bound with ppGpp. This issue might be resolved by constructing RNAP mutants in the residues predicted to be involved in ppGpp and DksA interactions. However, because these residues are in proximity to the catalytic center, such mutations might alter RNAP properties as well as a mode of action other than the one intended.

Transcription Inhibition

A consensus is building that ppGpp directly inhibits transcription from ribosomal promoters. There are several models of how this might occur (**Figure 2a**). One model relies on the fact that ppGpp and DksA together and independently decrease the stability of the open complexes formed on DNA by RNAP (5, 54, 55). If decreasing open complex stability is the major role of ppGpp and DksA, the model suggests that only intrinsically unstable promoters, such as *rrn* promoters, would be inhibited, whereas those that form relatively stable complexes would be activated. Although appealing in its simplicity, this model does not explain all the instances of inhibition, such as the lambda pR promoter that forms stable open complexes yet is inhibited by ppGpp (61). The lambda pR study suggested that although ppGpp affects many steps of transcription initiation, the first phosphodiester bond formation might be the key target for pR. Similar proposals were made for *rrnB* P1 and *rrnD* P1 promoters in which substrate competition between ppGpp and NTPs was implied (33).

A different model was proposed for an M+ mutant with deletion of four residues of the RNAP rudder sequence (*rpoC* Δ312–315). Transcription of the P*argT* tRNA promoter with RNAP from these mutants formed

a Direct inhibition: **ribosomal promoters**

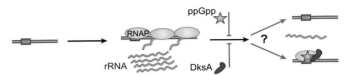

b Direct activation: **amino acid biosynthesis promoters**

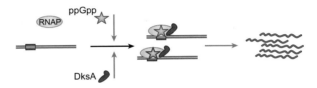

c Models of indirect activation: **alternative σ factors (σ^H, σ^S, σ^N)**

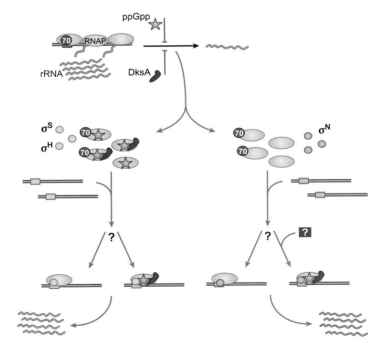

dead-end promoter complexes with features similar to those of stable closed complexes (42). These observations are in agreement with a trapping mechanism, previously proposed to explain ppGpp action (28). In this model, RNAP is trapped by ppGpp in closed complexes and is unable to initiate transcription. Studies with *rrnB* P1 promoters employing ppGpp and DksA also seem to indicate formation of such dead-end complexes (60). Thus, ppGpp seems to act at many levels, and the mechanism of its action is a complex outcome of several factors, intrinsic promoter properties not being the least of them.

Activation of Transcription

Many models calling for direct, passive, or indirect mechanisms had been proposed for activation of transcription by ppGpp; many of the mechanisms are not mutually exclusive (for comprehensive reviews, see References 40 and 72). This is further complicated by the difficulty to distinguish between direct and indirect effects of interconnected cellular processes. To simplify definitions, we propose to use the following criteria. Direct activation occurs when RNAP interacts with effectors, such as ppGpp, DksA, or both, to increase transcription from a given promoter (**Figure 2b**). Activation of a promoter in a pure in vitro system is proof of direct activation. Indirect activation by these effectors of one promoter relies on inhibition of other (strong) promoters, leading to increased availability of RNAP that indirectly activates transcription initiation (**Figure 2c**).

Direct Activation

Historically, models in which ppGpp bound to RNAP would directly activate transcription have not been favored simply because such activation was not demonstrable in vitro. The first promising report implying such a possibility came from coupled transcription-translation system with the use of cellular extracts, demonstrating activation of *hisG* promoter (17). Later, such an effect was demonstrated in a defined in vitro transcription system for lambda paQ promoter (62). However, the in vitro demonstration that certain amino acid biosynthesis promoters (P*argI*, P*thrABC*, P*livJ*, and P*hisG*) are activated by combining ppGpp and purified DksA provides convincing evidence of direct activation according to our definition (36). The precise mechanism is still unknown, although it has been suggested that ppGpp and DksA stimulate the rate of an isomerization step on the pathway to open complex formation (55). It is probable that the other σ^{70}-dependent promoters activated by ppGpp are also affected through a direct mechanism.

Indirect Activation

Transcription directed by the alternative sigma factors σ^S, σ^H, σ^N (40), and σ^E (18) also requires (p)ppGpp for activation in vivo. Attempts to verify this dependence in vitro have failed, with the exception of σ^E, which was recently demonstrated (19). Therefore, a sigma competition model was proposed whereby ppGpp alters the affinity of the housekeeping σ^{70} to

---

Figure 2

Models of how ppGpp inhibits and activates RNA polymerase (RNAP). (*a*) ppGpp and DksA inhibit transcription from ribosomal promoters by interacting directly with RNAP; this may occur either by destabilizing open complexes formed by RNAP on DNA or by trapping RNAP at the promoter site. (*b*) ppGpp and DksA directly activate transcription from certain amino acid biosynthesis promoters and possibly other σ^{70} promoters. (*c*) Inhibition of transcription from ribosomal promoters by ppGpp and DksA liberates core RNAP that can now be engaged in transcription dependent on alternative sigma factors. (*Left*) In the case of σ^H and σ^S it is proposed that ppGpp, and possibly DksA, aids these factors in competing for core RNAP; it is uncertain whether ppGpp together with DksA might further promote transcription at the initiation stage. (*Right*) Neither ppGpp nor DksA aids σ^N in competing for core RNAP or enhances transcription in in vitro assays. It is possible that a yet unidentified factor is necessary to observe a direct activation in vitro.

SPI: *Salmonella*
pathogenicity island

core RNAP, allowing other sigma factors to bind (32, 40, 72). This requires increased core RNAP availability, attained through inhibition of strong promoters, such as *rrn* (**Figure 2c**), that account for most transcripts occurring under normal growth conditions. However, not all RNAPs in the cell are involved in active transcription, but they reside at chromosomal sites; the dynamics of their release and how they can contribute to the competition is unclear (78).

The sigma competition model was mainly based on in vivo observations indicating that σ^S and σ^H promoters lose their dependence on ppGpp for activity when σ^{70} levels are depleted or when σ^{70} is mutated, lowering its affinity for core RNAP (32). ppGpp could reduce σ^{70} competitiveness for core RNAP in vitro when assayed together with σ^H (32). Moreover, some of the ppGpp^0 *rpoD* (σ^{70}) M+ suppressor mutants behaved as if the mutant σ^{70} had less affinity to core RNAP (29, 71). Thus, ppGpp and/or DksA would be expected not only to increase the availability of free RNAP by inhibiting strong σ^{70}-dependent promoters, but also to make core RNAP more available by disturbing σ^{70}-core RNAP interactions. Whether this disturbance happens at the level of association or dissociation with core RNAP is uncertain. Also, in vitro transcription studies employing DksA together with ppGpp have not been reported with σ^S and σ^H factors to date, so the direct activation model cannot be completely excluded.

Similar in vivo observations were made for σ^N (38). Recently, DksA was also found to be required for σ^N-dependent transcription of the *Pseudomonas* Po promoter in vivo (using *E. coli* system), but not in vitro because neither ppGpp nor DksA, added together or separately, could mimic this dependence (7). Competition between σ^{70} and σ^N by in vitro assay was also unaffected. Certain *rpoB* or *rpoC* M+ mutants can stimulate the same σ^N-dependent Po promoter in vivo. However, purified M+ RNAP did not alter transcription initiation from the Po promoter enough to account for the observed high activity in vivo. Nevertheless, it could be verified that the M+ mutant RNAP

bound σ^{70} more poorly than σ^N (71). Either the ability of the M+ mutant RNAP to bypass requirements for ppGpp and DksA for rRNA promoters is not the same for the Po promoter or this promoter has a requirement for an additional, so far unidentified factor present in vivo (**Figure 2c**).

PATHOGENESIS AND (p)ppGpp

A growing number of studies report involvement of (p)ppGpp in processes related to growth, stress, starvation, and survival that affect pathogenicity. A frequent scenario is that when (p)ppGpp is absent, pathogenicity is compromised for reasons that vary with the organism studied. Inhibitory effects can also occur on host interactions that enhance pathogen survival, invasiveness, or persistence. Examples include *Vibrio cholerae*, *Salmonella typhimurium*, *Legionella pneumophila*, *Pseudomonas aeruginosa*, *Campylobacter jejuni*, *Brucella abortus*, *Mycobacterium tuberculosis*, *Listeria monocytogenes*, and *Borrelia burgdorferi* (see Reference 11).

The enterobacterial pathogen *S. typhimurium* accumulates (p)ppGpp in stationary phase to induce *hilA*, a master regulator of *Salmonella* pathogenicity island 1 (SPI 1) and SPI 2 virulence genes (73). The transcriptional basis of regulation is unexplored, although the SPI sequences are AT rich. The SPI 1 genes are involved in host cell invasion and SPI 2 genes are required for replication within the host cell. Deleting *relA* and *spoT* genes, but not *relA* alone, gave a (p)ppGpp^0 state that resulted in strong attenuation in mice and noninvasiveness in vitro (59, 68, 73). Vaccine tests reveal that 30 days after single immunization with the (p)ppGpp^0 strain, mice were protected from challenge with wild-type *Salmonella* at a dose 10^6-fold above the established LD$_{50}$ (48). It is also intriguing that this requirement for (p)ppGpp depends on SpoT function rather than RelA. For another enteric bacterium, enterohemorrhagic *E. coli* (EHEC), adherence capacity, as well as expression in the enterocyte effacement pathogenicity island locus, depends on *relA*, *spoT*, and *dksA* (49).

Another example of the diversity of stress inducing (p)ppGpp is *Helicobacter pylori*, a pathogen with a single RSH enzyme. Although originally found to be unresponsive to amino acid starvation, accumulation of ppGpp seems required for survival, specifically during aerobic shock and acid exposure (45, 81). These conditions are likely to be encountered during the course of infection and transmission.

SUMMARY POINTS

1. Virtually all bacteria and plants synthesize ppGpp and pppGpp, which are regulatory analogs of GDP and GTP.

2. *E. coli* (and all beta- and gamma-proteobacteria) synthesizes (p)ppGpp with RelA and SpoT. Other bacteria contain a single Rel/Spo homolog gene. Members of the class *Firmicutes* have additional small fragments with synthetic activity.

3. (p)ppGpp signals different kinds of environmental stress and leads to adjustments of gene expression and physiology. In *E. coli*, complete elimination of (p)ppGpp seems to lock cells in a growth mode unperturbed by environmental changes.

4. Diverse bacteria exploit (p)ppGpp in fundamentally different ways that can alter many aspects of cell biology.

5. Many basic questions as to how (p)ppGpp is made and how it works remain unanswered.

DISCLOSURE STATEMENT

The authors are not aware of any biases that might be perceived as affecting the objectivity of this review.

ACKNOWLEDGMENTS

We thank Daniel Vinella, Agnieszka Szalewska-Palasz, Rajendran Harinarayanan, Helen Murphy, and the Friday Seminar group for many discussions. We apologize to coauthors and colleagues whose important citations are absent owing to space limitations. This work was supported by the NICHD intramural program of the NIH.

LITERATURE CITED

1. Aravind L, Koonin EV. 1998. The HD domain defines a new superfamily of metal-dependent phospho-hydrolases. *Trends Biochem. Sci.* 23:469–72

2. Artsimovitch I, Patlan V, Sekine S, Vassylyeva MN, Hosaka T, et al. 2004. Structural basis for transcription regulation by alarmone ppGpp. *Cell* 117:299–310

3. Avarbock A, Avarbock D, Teh JS, Buckstein M, Wang ZM, Rubin H. 2005. Functional regulation of the opposing (p)ppGpp synthetase/hydrolase activities of Rel*Mtb* from *Mycobacterium tuberculosis*. *Biochemistry* 44:9913–23

4. Avarbock D, Avarbock A, Rubin H. 2000. Differential regulation of opposing Rel*Mtb* activities by the aminoacylation state of a tRNA.ribosome.mRNA.Rel*Mtb* complex. *Biochemistry* 39:11640–48

5. Barker MM, Gaal T, Josaitis CA, Gourse RL. 2001. Mechanism of regulation of transcription initiation by ppGpp. I. Effects of ppGpp on transcription initiation in vivo and in vitro. *J. Mol. Biol.* 305:673–88

6. First mechanism for sensing stress by SpoT.

6. **Battesti A, Bouveret E. 2006. Acyl carrier protein/Spot interaction, the switch linking SpoT-dependent stress response to fatty acid metabolism.** *Mol. Microbiol.* **62:1048–63**

7. Bernardo LM, Johansson LU, Solera D, Skarfstad E, Shingler V. 2006. The guanosine tetraphosphate (ppGpp) alarmone, DksA and promoter affinity for RNA polymerase in regulation of sigma-dependent transcription. *Mol. Microbiol.* 60:749–64

8. Blankschein MD, Potrykus K, Grace E, Choudhary A, Vinella D, et al. 2008. TraR, a DksA homolog, modulates transcription in the absence of ppGpp. Manuscript submitted

9. Borukhov S, Lee J, Laptenko O. 2005. Bacterial transcription elongation factors: new insights into molecular mechanism of action. *Mol. Microbiol.* 55:1315–24

10. Bougdour A, Gottesman S. 2007. ppGpp regulation of RpoS degradation via antiadaptor protein IraP. *Proc. Natl. Acad. Sci. USA* 104:12896–901

11. Braeken K, Moris M, Daniels R, Vanderleyden J, Michiels J. 2006. New horizons for (p)ppGpp in bacterial and plant physiology. *Trends Microbiol.* 14:45–54

12. Bremer H, Dennis PP. 1996. Modulation of chemical composition and other parameters of the cell by growth rate. In Escherichia coli *and* Salmonella: *Cellular and Molecular Biology*, ed. FC Neidhardt, 1:1553–69. Washington, DC: ASM Press

13. Brown L, Gentry D, Elliott T, Cashel M. 2002. DksA affects ppGpp induction of RpoS at a translational level. *J. Bacteriol.* 184:4455–65

14. Buglino J, Shen V, Hakimian P, Lima CD. 2002. Structural and biochemical analysis of the Obg GTP binding protein. *Structure* 10:1561–92

15. Cashel M, Gentry D, Hernandez VJ, Vinella D. 1996. The stringent response. In Escherichia coli *and* Salmonella: *Cellular and Molecular Biology*, ed. FC Neidhardt, 1:1458–96. Washington, DC: ASM Press

16. Chatterji D, Fujita N, Ishihama A. 1998. The mediator for stringent control, ppGpp, binds to the beta-subunit of *Escherichia coli* RNA polymerase. *Genes Cells* 15:279–87

17. **Choy HE. 2000. The study of guanosine 5′-diphosphate 3′-diphosphate-mediated transcription regulation in vitro using a coupled transcription-translation system.** *J. Biol. Chem.* **275:6783–89**

18. Costanzo A, Ades SE. 2006. Growth phase-dependent regulation of the extracytoplasmic stress factor, σ^E, by guanosine 3′,5′-bispyrophosphate (ppGpp). *J. Bacteriol.* 188:4627–34

19. Costanzo A, Nicoloff H, Barchinger SE, Banta AB, Gourse RL, Ades SE. 2008. ppGpp and DksA likely regulate the activity of the extracytoplasmic stress factor sigma(E) in *Escherichia coli* by both direct and indirect mechanisms. *Mol. Microbiol.* 67:619–32

20. Durfee T, Hansen AM, Zhi H, Blattner FR, Jin DJ. 2008. Transcription profiling of the stringent response in *Escherichia coli*. *J. Bacteriol.* 190:1084–96

21. Figueroa-Bossi N, Guerin M, Rahmouni R, Leng M, Bossi L. 1998. The supercoiling sensitivity of a bacterial tRNA promoter parallels its responsiveness to stringent control. *EMBO J.* 17:2359–67

22. Foti JJ, Persky NS, Ferullo DJ, Lovett ST. 2007. Chromosome segregation control by *Escherichia coli* ObgE GTPase. *Mol. Microbiol.* 65:569–81

23. Foti JJ, Schienda J, Sutera VA Jr, Lovett ST. 2005. A bacterial G protein-mediated response to replication arrest. *Mol. Cell* 17:549–60

24. Gentry DR, Cashel M. 1996. Mutational analysis of *Escherichia coli spoT* gene identifies distinct but overlapping regions involved in ppGpp synthesis and degradation. *Mol. Microbiol.* 19:1373–84

25. Gropp M, Strausz Y, Gross M, Glaser G. 2001. Regulation of *Escherichia coli* RelA requires oligomerization of the C-terminal domain. *J. Bacteriol.* 183:570–79

26. Handke LD, Shivers RP, Sonenshein AL. 2008. Interaction of *Bacillus subtilis* CodY with GTP. *J. Bacteriol.* 190:798–806

27. **Haseltine WA, Block R. 1973. Synthesis of guanosine tetra- and pentaphosphate requires the presence of a codon-specific, uncharged transfer ribonucleic acid in the acceptor site of ribosomes.** *Proc. Natl. Acad. Sci. USA* **70:1564–68**

28. Heinemann M, Wagner R. 1997. Guanosine 3′,5′-bis(diphosphate) (ppGpp)-dependent inhibition of transcription from stringently controlled *Escherichia coli* promoters can be explained by an altered initiation pathway that traps RNA polymerase. *Eur. J. Biochem.* 247:990–99

29. Hernandez VJ, Cashel M. 1995. Changes in conserved region 3 of *Escherichia coli* sigma 70 mediate ppGpp-dependent functions in vivo. *J. Mol. Biol.* 252:536–49

17. First in vitro demonstration of positive control by ppGpp.

27. The first and still the best demonstration of RelA activation.

30. Hogg T, Mechold U, Malke H, Cashel M, Hilgenfeld R. 2004. Conformational antagonism between opposing active sites in a bifunctional RelA/SpoT homolog modulates (p)ppGpp metabolism during the stringent response. *Cell* 117:57–68

31. Jiang M, Sullivan SM, Wout PK, Maddock JR. 2007. G-protein control of the ribosome-associated stress response protein SpoT. *J. Bacteriol.* 189:6140–47

32. Jishage M, Kvint K, Shingler V, Nystrom T. 2002. Regulation of σ-factor competition by the alarmone ppGpp. *Genes Dev.* 16:1260–70

33. Jores L, Wagner R. 2003. Essential steps in the ppGpp-dependent regulation of bacterial ribosomal RNA promoters can be explained by substrate competition. *J. Biol. Chem.* 278:16834–43

34. Kang PJ, Craig EA. 1990. Identification and characterization of a new *Escherichia coli* gene that is a dosage-dependent suppressor of a *dnaK* deletion mutant. *J. Bacteriol.* 172:2055–64

35. Kasai K, Nishizawa T, Takahashi K, Hosaka T, Aoki H, Ochi K. 2006. Physiological analysis of the stringent response elicited in an extreme thermophilic bacterium, *Thermus thermophilus*. *J. Bacteriol.* 188:7111–22

36. Krasny L, Gourse RL. 2004. An alternative strategy for bacterial ribosome synthesis: *Bacillus subtilis* rRNA transcription regulation. *EMBO J.* 23:4473–83

37. Lamour V, Hogan BP, Erie DA, Darst SA. 2006. Crystal structure of *Thermus aquaticus* Gfh1, a Gre-factor paralog that inhibits rather than stimulates transcript cleavage. *J. Mol. Biol.* 356:179–88

38. Laurie AD, Bernardo LM, Sze CC, Skarfstad E, Szalewska-Palasz A, et al. 2003. The role of the alarmone (p)ppGpp in sigma N competition for core RNA polymerase. *J. Biol. Chem.* 278:1494–503

39. Lemos JA, Nascimento MM, Abranches J, Burne RA. 2007. Three gene products govern (p)ppGpp production in *Streptococcus mutans*. *Mol. Microbiol.* 65:1568–81

40. Magnusson LU, Farewell A, Nystrom T. 2005. ppGpp: a global regulator in *Escherichia coli*. *Trends Microbiol.* 13:236–42

41. Magnusson LU, Gummesson B, Joksimovic P, Farewell A, Nystrom T. 2007. Identical, independent, and opposing roles of ppGpp and DksA in *Escherichia coli*. *J. Bacteriol.* 189:5193–202

42. Maitra A, Shulgina I, Hernandez VJ. 2005. Conversion of active promoter-RNA polymerase complexes into inactive promoter bound complexes in *E. coli* by the transcription effector, ppGpp. *Mol. Cell* 17:817–29

43. Mechold U, Murphy M, Brown L, Cashel M. 2002. Intramolecular regulation of opposing (p)ppGpp catalytic activities of Rel*Seq*, the Rel/Spo enzyme from *Streptococcus equisimilis*. *J. Bacteriol.* 184:2878–88

44. Milon P, Tischenko E, Tomsic J, Caserta E, Folkers G, et al. 2006. The nucleotide-binding site of bacterial translation initiation factor 2 (IF2) as a metabolic sensor. *Proc. Natl. Acad. Sci. USA* 103:13962–67

45. Mouery K, Rader BA, Gaynor EC, Guillemin K. 2006. The stringent response is required for *Helicobacter pylori* survival of stationary phase, exposure to acid, and aerobic shock. *J. Bacteriol.* 188:5494–500

46. Mukai J, Hirashima A, Mikuniya T. 1980. Nucleotide 2′,3′-cyclic monophosphokinase from actinomycetes. *Nucleic Acids Symp. Ser.* 22:89–90

47. Murphy H, Cashel M. 2003. Isolation of RNA polymerase suppressors of a (p)ppGpp deficiency. *Methods Enzymol.* 44:596–601

48. Na HS, Kim HJ, Lee HC, Hong Y, Rhee JH, Choy HE. 2006. Immune response induced by *Salmonella typhimurium* defective in ppGpp synthesis. *Vaccine* 24:2027–34

49. Nakanishi N, Abe H, Ogura Y, Hayashi T, Tashiro K, et al. 2006. ppGpp with DksA controls gene expression in the locus of enterocyte effacement (LEE) pathogenicity island of enterohaemorrhagic *Escherichia coli* through activation of two virulence regulatory genes. *Mol. Microbiol.* 61:194–205

50. Nanamiya H, Kasai K, Nosawa H, Yun C-S, Narisawa T, et al. 2008. Identification and functional analysis of novel (p)ppGpp synthetase genes in *Bacillus subtilis*. *Mol. Microbiol.* 67:291–304

51. Nascimento MM, Lemos JA, Abranches J, Lin VK, Burne RA. 2008. Role of RelA of *Streptococcus mutans* in global control of gene expression. *J. Bacteriol.* 190:28–36

52. North SH, Kirtland SE, Nakai H. 2007. Translation factor IF2 at the interface of transposition and replication by the PriA-PriC pathway. *Mol. Microbiol.* 66:1566–78

53. Park JW, Jung Y, Lee SJ, Jin DJ, Lee Y. 2002. Alteration of stringent response of the *Escherichia coli* rnpB promoter by mutations in the −35 region. *Biochem. Biophys. Res. Commun.* 290:1183–87

30. First crystal structures of RSH catalytic region.

39. Discovery of RSH gene synthase fragments.

41. First demonstration that DksA and (p)ppGpp might be more than cofactors.

50. Co-discovery of RSH gene synthase fragments.

54. Paul BJ, Barker MM, Ross W, Schneider DA, Webb C, et al. 2004. DksA: a critical component of the transcription initiation machinery that potentiates the regulation of rRNA promoters by ppGpp and the initiating NTP. *Cell* 118:311–22

55. Paul BJ, Berkmen MB, Gourse RL. 2005. DksA potentiates direct activation of amino acid promoters by ppGpp. *Proc. Natl. Acad. Sci. USA* 102:7823–28

56. Paul BJ, Ross W, Gaal T, Gourse RL. 2004. rRNA transcription in *Escherichia coli*. *Annu. Rev. Genet.* 38:749–70

57. Perederina A, Svetlov V, Vassylyeva MN, Tahirov TH, Yokoyama S, et al. 2004. Regulation through the secondary channel—structural framework for ppGpp-DksA synergism during transcription. *Cell* 118:297–309

58. Perron K, Comte R, van Delden C. 2005. DksA represses ribosomal gene transcription in *Pseudomonas aeruginosa* by interacting with RNA polymerase on ribosomal promoters. *Mol. Microbiol.* 56:1087–102

59. Pizarro-Cerda J, Tedin K. 2004. The bacterial signal molecule, ppGpp, regulates *Salmonella* virulence gene expression. *Mol. Microbiol.* 52:1827–44

60. Potrykus K, Vinella D, Murphy H, Szalewska-Palasz A, D'Ari R, Cashel M. 2006. Antagonistic regulation of *Escherichia coli* ribosomal RNA *rrnB* P1 promoter activity by GreA and DksA. *J. Biol. Chem.* 281:15238–48

61. Potrykus K, Wegrzyn G, Hernandez VJ. 2002. Multiple mechanisms of transcription inhibition by ppGpp at the lambda pR promoter. *J. Biol. Chem.* 277:43785–91

62. Potrykus K, Wegrzyn G, Hernandez VJ. 2004. Direct stimulation of the paQ promoter by the transcription effector guanosine-3′,5′-(bis)pyrophosphate in a defined in vitro system. *J. Biol. Chem.* 279:19860–66

63. Rutherford ST, Lemke JJ, Vrentas CE, Gaal T, Ross W, Gourse RL. 2007. Effects of DksA, GreA, and GreB on transcription initiation: insights into the mechanisms of factors that bind in the secondary channel of RNA polymerase. *J. Mol. Biol.* 366:1243–57

64. Sajish M, Tiwari D, Rananaware D, Nandicoori VK, Prakash B. 2007. A charge reversal differentiates (p)ppGpp synthesis by monofunctional and bifunctional Rel proteins. *J. Biol. Chem.* 282:34977–83

65. Schneider DA, Gourse RL. 2004. Relationship between growth rate and ATP concentration in *Escherichia coli*: a bioassay for available cellular ATP. *J. Biol. Chem.* 279:8262–68

66. Scott JM, Haldenwang WG. 1999. Obg, an essential GTP binding protein of *Bacillus subtilis*, is necessary for stress activation of transcription factor σ^B. *J. Bacteriol.* 181:4653–60

67. Slominska M, Neubauer P, Wegrzyn G. 1999. Regulation of bacteriophage λ development by 5′-diphosphate-3′-diphosphate. *Virology* 262:431–41

68. Song M, Kim HJ, Kim EY, Shin M, Lee HC, et al. 2004. ppGpp-dependent stationary phase induction of genes on *Salmonella* pathogenicity island 1. *J. Biol. Chem.* 279:34183–90

69. Svitil AL, Cashel M, Zyskind JW. 1993. Guanosine tetraphosphate inhibits protein synthesis in vivo. A possible protective mechanism for starvation stress in *Escherichia coli*. *J. Biol. Chem.* 268:2307–11

70. Symersky J, Perederina A, Vassylyeva MN, Svetlov V, Artsimovitch I, Vassylyev DG. 2006. Regulation through the RNA polymerase secondary channel. Structural and functional variability of the coiled-coil transcription factors. *J. Biol. Chem.* 281:1309–12

71. Szalewska-Palasz A, Johansson LUM, Bernardo LMD, Skarfstad E, Stec E, et al. 2007. Properties of RNA polymerase bypass mutants: implications for the role of ppGpp and its cofactor DksA in controlling transcription dependent on σ^{54}. *J. Biol. Chem.* 282:18046–56

72. Szalewska-Palasz A, Wegrzyn G, Wegrzyn A. 2007. Mechanisms of physiological regulation of RNA synthesis in bacteria: new discoveries breaking old schemes. *J. Appl. Genet.* 48:281–94

73. Thompson A, Rolfe MD, Lucchini S, Schwerk P, Hinton JC, Tedin K. 2006. The bacterial signal molecule, ppGpp, mediates the environmental regulation of both the invasion and intracellular virulence gene programs of *Salmonella*. *J. Biol. Chem.* 281:30112–121

74. Toulokhonov II, Shulgina I, Hernandez VJ. 2001. Binding of the transcription effector ppGpp to *Escherichia coli* RNA polymerase is allosteric, modular, and occurs near the N terminus of the β′-subunit. *J. Biol. Chem.* 276:1220–25

75. Trautinger BW, Jaktaji RP, Rusakova E, Lloyd RG. 2005. RNA polymerase modulators and DNA repair activities resolve conflicts between DNA replication and transcription. *Mol. Cell* 19:247–58

76. Travers A. 1980. Promoter sequence for stringent control of bacterial ribonucleic acid synthesis. *J. Bacteriol.* 141:973–76

77. Vassylyeva MN, Svetlov V, Dearborn AD, Klyuyev S, Artsimovitch I, Vassylyev DG. 2007. The carboxy-terminal coiled-coil of the RNAP β′-subunit is the main binding site for Gre factors. *EMBO Rep.* 8:1038–43

78. Wade JT, Struhl K, Busby SJ, Grainger DC. 2007. Genomic analysis of protein-DNA interactions in bacteria: insights into transcription and chromosome organization. *Mol. Microbiol.* 65:21–26

79. Wang JD, Sanders GM, Grossman AD. 2007. Nutritional control of elongation of DNA replication by (p)ppGpp. *Cell* 128:865–75

80. Wegrzyn G. 1999. Replication of plasmids during bacterial response to amino acid starvation. *Plasmid* 41:1–16

81. Wells DH, Gaynor EC. 2006. *Helicobacter pylori* initiates the stringent response upon nutrient and pH shift. *J. Bacteriol.* 188:3726–29

82. Xiao H, Kalman M, Ikehara K, Zemel S, Glaser G, Cashel M. 1991. Residual guanosine 3′,5′-bispyrophosphate synthetic activity of *relA* null mutants can be eliminated by *spoT* null mutations. *J. Biol. Chem.* 266:5980–90

RELATED RESOURCES

Borukhov S, Nudler E. 2008. RNA polymerase: the vehicle of transcription. *Trends Microbiol.* 16:126–34

Claverys J-P, Prudhomme M, Martin B. 2006. Induction of competence regulons as a general response to stress in gram-positive bacteria. *Annu. Rev. Microbiol.* 60:451–75

Foster PL. 2007. Stress-induced mutagenesis in bacteria. *Crit. Rev. Biochem. Mol. Biol.* 42:373–97

Kaiser D. 2006. A microbial genetic journey. *Annu. Rev. Microbiol.* 60:1–25

76. A discriminator characteristic that has withstood the test of time.

79. (p)ppGpp inhibits DnaG.

Evolution, Population Structure, and Phylogeography of Genetically Monomorphic Bacterial Pathogens

Mark Achtman

Environmental Research Institute, University College Cork, Cork, Ireland;
email: m.achtman@ucc.i.e

Annu. Rev. Microbiol. 2008. 62:53–70

The *Annual Review of Microbiology* is online at
micro.annualreviews.org

This article's doi:
10.1146/annurev.micro.62.081307.162832

Key Words

single nucleotide polymorphism, mutation discovery strategy,
genomic comparison, biogeography

Abstract

Genetically monomorphic bacteria contain so little sequence diversity
that sequencing a few gene fragments yields little or no information.
As a result, our understanding of their evolutionary patterns presents
greater technical challenges than exist for genetically diverse microbes.
These challenges are now being met by analyses at the genomic level
for diverse types of genetic variation, the most promising of which are
single nucleotide polymorphisms. Many of the most virulent bacterial
pathogens are genetically monomorphic, and understanding their evolutionary and phylogeographic patterns will help our understanding of
the effects of infectious disease on human history.

Contents

INTRODUCTION

Genetically Monomorphic Pathogens

Levels of genetic diversity are sufficiently high in most microbial taxa that the sequences of several housekeeping gene fragments can provide a medium-resolution overview of their population genetic structure (51). As an extreme example, almost all unrelated isolates of *Helicobacter pylori* possess distinct sequences (48). It is therefore all the more striking that still other taxa have such low levels of sequence diversity that only few polymorphisms, or even none at all, are found upon sequencing a few genes. I refer to such organisms as genetically monomorphic.

Currently, the most prominent examples of genetically monomorphic microbes are pathogenic bacteria, including two of the four Category A bacterial taxa that are thought to represent potential biological agents for bioterrorism (68) [*Bacillus anthracis* (84) and *Yersinia pestis* (3)] and two of the Category B biological

agents [*Burkholderia mallei* (32) and *Escherichia coli* O157:H7 (89)].[1] Other prominent examples are the *Mycobacterium tuberculosis* complex (76), *Mycobacterium leprae* (55), *Salmonella enterica* serovar Typhi (43), *Shigella sonnei* (65), *Chlamydia pneumoniae* (67), and *Treponema pallidum* (37).

Genetically monomorphic pathogens present an intellectual challenge, namely, What evolutionary history has resulted in so little diversity? How, when, and where did they arise? They also present a technical challenge consisting of how to deduce evolutionary histories in the absence of appreciable diversity. I anticipate that these challenges will be successfully met over the next few years owing to rapid developments in genomic sequencing and resequencing, but our knowledge in this area is currently somewhat fragmentary and incomplete. This review article compares and contrasts the population genetic and phylogenetic aspects of selected genetically monomorphic pathogens, identifies existing problems and gaps in our current knowledge, and enumerates approaches that may be particularly rewarding for future research.

MOLECULAR TOOLS FOR RESOLVING POPULATION STRUCTURES

Microbiologists have long searched for discriminatory markers that can subdivide pathogens. Discriminatory markers are needed to test whether outbreaks have a common source, e.g., for forensic purposes. Such markers can enable tracking the epidemic and pandemic spread of particular strains, yielding a long-term overview of epidemiological patterns. They could also potentially be used to reconstruct evolutionary history. However, many molecular methods are not particularly

---

[1]It is not obvious why so many potential agents of bioterrorism should be genetically monomorphic, particularly because humans are only secondary hosts for all these potential bioterrorism agents and therefore irrelevant for their evolutionary trajectories.

suitable for these purposes but rather represent yet another typing method (YATMs) that are only temporarily in fashion and will be difficult to interpret in several decades (1). Multilocus sequence typing (MLST) (52) was designed to provide an enduring alternative approach for the subdivision of microbes on the basis of neutral sequence diversity. MLST has now been developed for >48 microbial taxa, for which publicly available databases facilitate global communication via a universal language (**http://pubmlst.org/databases3.shtml**). However, the level of resolution by MLST is too low for the subdivision of genetically monomorphic bacteria.

Alternative techniques have been used for fine typing of genetically monomorphic pathogens, including pulsed-field gel electrophoresis (PFGE) (78) and scoring the presence or absence of insertion (IS) elements (2, 56, 83) or indels (14, 39, 82, 90). Some of the databases based on these methods are huge, an order of magnitude larger than has yet been achieved by MLST: More than 100,000 *E. coli* and *S. enterica* strains have been analyzed by PFGE (78) and tens of thousands of *M. tuberculosis* strains have been tested by IS*6110* fingerprinting (86). However, all three methods face serious problems. First, any method based on fingerprinting, such as PFGE, is difficult to standardize, both between laboratories and over time within a single laboratory, and is particularly subject to faulty calls. As a result, putative similarity between fingerprint patterns according to database comparisons often requires independent experimental confirmation. Second, the polymorphisms detected by PFGE and IS elements are among the most variable within microbial genomes and do not necessarily accurately reflect the phylogenetic history of the remainder of the genome. For example, differences in PFGE patterns within Typhi reflect major rearrangements in chromosomal order via recombination between rRNA loci (50) and did not correlate with phylogenetic relationships based on genomic single nucleotide polymorphisms (SNPs) (69). Similarly,

the most parsimonious evolutionary history in *Y. pestis* deduced on the basis of acquisition of IS elements was incorrect due to distortion by frequent homoplasies (2). Indel analysis is more conservative than PFGE or IS elements and yields phylogenetic patterns that are consistent with other techniques (28, 39). However, in general, the resolution of indel analysis is relatively low.

Two newer techniques are being used for forensic purposes and outbreak investigations: MLVA (42, 45, 49) and CRISPR analysis (33, 63). MLVA has been used to deduce patterns of spread within local outbreaks of plague in prairie dogs (31), and CRISPR analyses can resolve strains within classical subgroupings of *Y. pestis* at a global scale (85). MLVA and CRISPR analysis can be performed without expensive equipment and are eminently suitable for both clinical microbiology as well as reference centers. Furthermore, large publicly available databases (**http://minisatellites.u-psud.fr/**) are available for implementing these tools in different organisms (33), and 40,000 strains of *M. tuberculosis* have been tested by CRISPR analysis (spoligotyping) (15). Despite these impressive achievements, I argue that neither method can yield a definitive overview of the evolution, population structure, and phylogeography of genetically monomorphic pathogens. MLVA is not suitable for broad phylogenetic analyses due to its high mutation rate and correspondingly high rate of homoplasies. For example, MLVA of *Y. pestis* yielded an incorrect phylogenetic branching order and it was difficult to determine the boundaries of natural populations according to this method (2). MLVA also poses some problems with consistency between different laboratories (49). Phylogenetic patterns based on CRISPR analysis do seem to correlate to some extent with conclusions from other methods, similar to indels. However, the resolution offered by CRISPRs may be insufficient for forensic purposes or outbreak analysis because the number of variable spacer sequences per genome is limited.

YATM: yet another typing method

Multilocus sequence typing (MLST): consists of sequencing several housekeeping gene fragments for a total concatenated length of 3–4 kb

PFGE: pulsed-field gel electrophoresis

Indel: insertion or deletion

SNP: single nucleotide polymorphism

Homoplasy: a genetic change that has occurred independently on multiple occasions

Multilocus VNTR analysis (MLVA): each VNTR corresponds to a microsatellite or minisatellite in eukaryotes; called MIRU in *M. tuberculosis*

CRISPR: clustered regularly interspaced short palindromic repeats; called spoligotypes in *M. tuberculosis*

Single Nucleotide Polymorphisms

sSNP: synonymous SNP

My impression is that phylogenetic analysis of genetically monomorphic pathogens is in the process of moving away from all the tools summarized above, and detailed analyses will increasingly focus on polymorphic SNPs over the next few years. The first description of SNPs as a typing tool for a genetically monomorphic taxon was within the *M. tuberculosis* complex, in which sequencing 31 kb of coding sequences from multiple strains revealed 30 synonymous SNPs (sSNPs) (76). *M. tuberculosis* probably represents the upper limit of what might be considered genetically monomorphic because a subsequent, more systematic analysis identified 36 sSNPs in 8 kb (8) (one sSNP per 200 bp). Approximately 1000 SNPs were found by comparing five complete genomes of *B. anthracis* (62). Still lower levels of sequence diversity are present in other monomorphic pathogens: Eighty-two SNPs were identified within 80 kb in a globally representative collection of Typhi (1 SNP per kilobase) (69) and 76 sSNPS in 3250 nonrepetitive coding sequences (777,520 potential sSNPs; 1 sSNP per 10 kb) in three diverse strains of *Y. pestis* (2).

Even though SNP frequencies are low in monomorphic pathogens, their numbers will increase dramatically in the near future owing to resequencing (11) of the genomes of multiple strains. As with the HapMap project for mapping SNP diversity in the human genome (81), which has facilitated many analyses of various aspects of human diversity, the availability of larger numbers of SNPs will greatly facilitate new approaches to the evolution and typing of genetically monomorphic pathogens. Once larger numbers of SNPs have been defined, various methodologies are available for SNP typing (44), including methods that are appropriate for laboratories without access to high-throughput equipment (12). However, because of phylogenetic discovery bias, the information provided by SNP typing is crucially dependent on the strategy used to identify SNPs (see Appendix 1: Phylogenetic Discovery Bias, below).

CASE STUDIES WITH INDIVIDUAL TAXA

S. enterica Typhi

Typhoid fever represents a serious global public health problem in humans (61), with an estimated 16–33 million cases and 500,000–600,000 deaths annually (**http://www.who.int/vaccine_research/diseases/diarrhoeal/en/index7.html**). Extensive investigations at the beginning of the twentieth century (46) showed that up to 6% of infected individuals became long-lasting healthy carriers, in whom infection of the gall bladder by Typhi persists over long periods in the absence of clinical symptoms. The frequency of typhoid fever (and healthy carriers) has continually dropped since the late nineteenth century in many developed countries (6, 47),[2] but typhoid fever continues to remain a major health problem in southern Asia (17), Africa, and South America (61). Unfortunately, no reliable estimates of the current numbers of healthy carriers are available, although such estimates would be needed to develop prognoses on the future burden of disease. Information on the long-term importance of typhoid fever for human health is also lacking, due to historical difficulties in distinguishing typhoid fever from rickettsial typhus fever and other maladies.

Typhi is a human-adapted serovar of *S. enterica*, with no known close relatives among other serovars (23), except that in the distant past extensive homologous recombination occurred with the ancestors of Paratyphi A, a second human-adapted serovar (20). Typhi is genetically monomorphic (43). It is therefore striking just how much information was gained by a single, recently published, intensive analysis

---

[2]Almost all infectious diseases are much rarer in developed countries today than they were in the nineteenth century. This is often attributed to improved hygiene, clean water, and better waste disposal. Other potential factors are better nutrition and greater social isolation. I remain skeptical as to the true causes of this improvement in human health, except that it is clear that most of the improvement had already happened prior to the introduction of vaccines and antibiotics.

(69) that may represent a paradigm for investigating the evolutionary history of a monomorphic pathogen. Roumagnac et al. (69) assembled a collection of hundreds of strains from southern Asia, Africa, and South America that were isolated between 1958 and 1967 and between 1984 and 2004. One hundred five strains were subjected to mutation discovery using denaturing high-pressure liquid chromatography (dHPLC) of 200 gene fragments (88.7 kb), which identified a total of 82 SNPs. Mutation discovery also revealed one deletion of 4 bp, one apparent import from Paratyphi A, and two 24- to 25-kb imports. [The 24- to 25-kb imports reflect laboratory manipulations during the genetic attenuation of vaccine strains (K. Holt, personal communication).] The 55 polymorphic gene fragments were then subjected to further mutation discovery by dHPLC in 350 additional strains, most of which were isolated in southern Asia, yielding 97 biallelic polymorphisms and 85 haplotypes. The Typhi-specific SNPs define a unique, fully parsimonious, phylogenetic path without any homoplasies, wherein individual haplotypes document the sequential accumulation of mutations from a common root node, H45 (**Figure 1b**). Of the 85 haplotypes, eight were isolated from multiple continents. Because the phylogenetic path indicates that each haplotype has arisen only once, this observation immediately demonstrates that Typhi has spread globally on at least eight occasions. The most recent example of such global spread is haplotype H58. H58 is the most common haplotype in southern Asia and until recently, all isolates of H58 were from southern Asia. H58 has now also been isolated in Africa, suggesting recent intercontinental spread.

The ratio of synonymous to nonsynonymous SNPs in Typhi indicates that the population structure has not been subjected to strong selection, i.e., it is neutral (69). It is therefore interesting to compare this phylogeny to the patterns of SNPs in the *gyrA* gene that encodes a DNA gyrase, within which any of six mutations in two codons

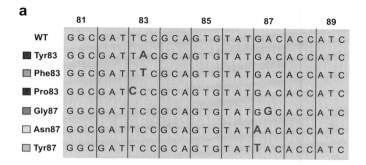

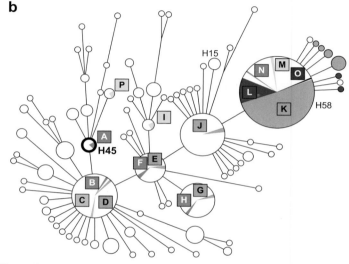

Figure 1

Neutral phylogeny and sources of resistance to nalidixic acid in *Salmonella enterica* Typhi. (*a*) Mutations (*red*) in codons 83 and 87 of *gyrA* that result in resistance to nalidixic acid. Each mutation is indicated by a distinct color code and by a name that specifies the amino acid resulting from that mutation plus the codon position. (*b*) Minimal spanning tree of 85 haplotypes of Typhi based on denaturing high-pressure liquid chromatography of 55 gene fragments among 483 strains. Haplotypes are indicated by circles, whose diameter increases as the number of strains increases. The ancestral haplotype, H45, is indicated by a bold circle. H58 is the common haplotype in southern Asia. Within haplotype circles, white arcs represent strains that are sensitive to nalidixic acid, whereas colored arcs represent strains carrying the *gyrA* mutation with the corresponding color in panel *a*. Each letter represents 1 of 16 unique mutations. Modified from figure 2 in Reference 69 with permission from the author.

can result in resistance to nalidixic acid and lessened susceptibility to fluoroquinolones (**Figure 1a**). Over the past few decades, an increase in the frequency of multiple-drug resistance in Typhi has resulted in the preferential use of fluoroquinolones for disease therapy, with a concomitant increase in the numbers of nalidixic acid–resistant strains. In some areas, such as southern Vietnam, most isolates are now nalidixic acid resistant (17). It was conceivable that this increase in resistance reflected the clonal expansion of cells derived from a single nalidixic acid–resistant mutant, but this was not the case. Five independent mutations were detected within H58 and at least 16 distinct homoplastic mutational events were identified among 119 strains of Typhi that were resistant to nalidixic acid (**Figure 1***b*). These observations provide a striking demonstration of the strength of the selection pressures exerted by antibiotic therapy for acute disease.

The genetic patterns just described are unusual. First, even within such a small sample, many of the ancestral nodes in the tree are represented by extant strains, including the ancestral root node. Similar patterns have been observed in *M. tuberculosis* (8), but in other microbes ancestral haplotypes are lost owing to overgrowth by fitter descendants or during genetic drift. Second, despite the strong selective pressures exerted by antibiotic treatment, strains of Typhi that are sensitive to nalidixic acid continue to be isolated. These considerations stimulated the proposal that the healthy carrier state represents a protected reservoir for Typhi within which ancestral genotypes can continue to proliferate despite independent evolutionary adaptations in other strains during the acute phase of disease (69). The epidemiological implications of this proposal for vaccine development remain to be elucidated by mathematical simulations.

M. tuberculosis

In the early part of the twentieth century, every fourth European suffered from clinical tuberculosis. Although the disease frequency has decreased dramatically in Europe and North America since then, one-third of the human population is infected, with an estimated 9 million new infections and 2 million deaths each year (**http://www.who.int/tb/ publications/tb_global_facts_sep05_en.pdf**). There are two forms of infection, the acute phase associated with clinical disease and an encapsulated lung granuloma in which bacteria can survive for decades in the absence of clinical symptoms (53). Similar to the healthy carrier state for Typhi, granulomas represent a reservoir for genotypes that were being transmitted decades ago. Thus, the reactivation of quiescent granulomas results in the secondary transmission of genotypes that have not necessarily been subjected to the same selective pressures as those from initial acute phase disease. Mathematical simulations indicate that the existence of the quiescent granulomas has marked effects on the dynamics of epidemic transmission, resulting in waves of disease that peak 100 years after the waves started (13).

Genetic diversity in *M. tuberculosis* has been investigated by a multitude of molecular techniques, including RFLP of IS6110 (83), CRISPRs (spoligotyping) (15), MLVA (MIRU) (74), and indel analyses (27). Unfortunately, although 98% of the cases of disease are in the developing world, most analyses have focused on the comparatively few cases in Europe and North America. As a result, where phylogeographic conclusions have been drawn, they have often relied on analyses of the strains isolated from infected immigrants that migrated from developing countries. This is unfortunate because due to the existence of granulomas, such strains represent genotypes that were circulating decades ago and may not necessarily be common today. Furthermore, the focus of many analyses has largely been on tracing outbreaks of closely related strains, and the tools that have been used are not as informative as would be desirable for a global overview of the genetic diversity and phylogeography of *M. tuberculosis*. However, a biogeographic consensus is beginning to emerge because the phylogenetic groupings recognized in five different analyses of global diversity in *M. tuberculosis* that were based on SNPs, indels, and CRISPRs were relatively concordant (28).

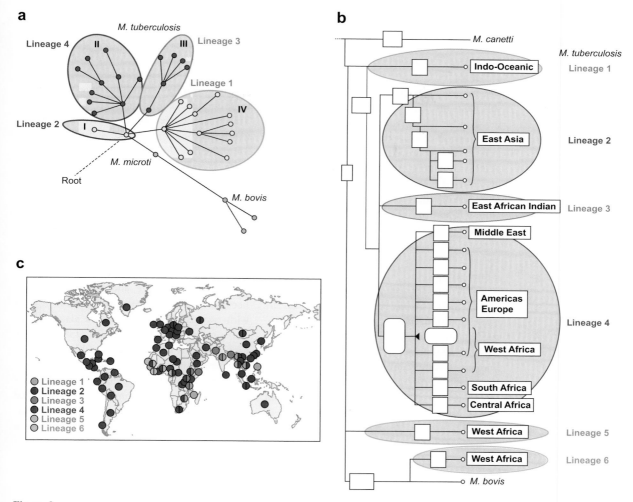

Figure 2

Phylogeographic patterns in the *Mycobacterium tuberculosis* complex. (*a*) Rooted tree based on synonymous single nucleotide polymorphisms discovered by sequencing (8). Original lineage assignments (Lineage I, II, etc.) are indicated by Roman designations within shaded areas. Current lineage assignments according to Gagneux & Small (28) are indicated by Lineage 1, 2, etc. and colored lines around shaded areas. (*b*) Rooted tree based on indels (large sequence polymorphisms) discovered by comparative genomic hybridization (27). Each open box reflects one or more indels, whereas boxes containing text indicate geographic specificities. (*c*) Geographic specificities of lineages indicated by colored segments. Modified from figures 1a, 1b, and 2 in Reference 28 with permission from the author.

M. tuberculosis consists of six phylogenetic groupings, designated Lineage 1 through Lineage 6 by Gagneux & Small (28), two of which were previously designated *Mycobacterium africanum* (**Figure 2a,b**). The six lineages of *M. tuberculosis* show biogeographic specificities in that individual lineages seem to

be associated with particular geographic locations (28) (**Figure 2c**). Some of these associations probably reflect ancient human migrations, as has been inferred for biogeographic patterns in *H. pylori* (24, 48) and *Streptococcus mutans* (16), which are thought to have accompanied humans when they migrated out of

Africa 50,000 years ago, and *M. leprae* (55). The distribution of Lineage 4 in Europe, the Americas, parts of Africa, and Australia probably reflects global colonization by Europeans, as it does for *H. pylori*. However, unlike *H. pylori*, for which biogeography is thought to result from historical isolation by distance (48), the data for *M. tuberculosis* points to increased transmission of individual lineages within certain ethnic groups. Lineage 2 is transmitted efficiently among Americans who were born in China, Lineage 1 among Americans born in the Philippines or Vietnam, and Lineage 4 among Americans born in North America (27). Because transmission is linked to virulence in *M. tuberculosis*, these data also hint at lineage-specific differences in virulence for particular ethnic groups.

On a broader scale, *M. tuberculosis* is part of the *M. tuberculosis* complex, which also includes additional closely related, monomorphic species (*M. pinnipedia*, *M. microti*, *M. caprae*, *M. bovis*) that primarily infect hosts other than humans. Until recently, the phylogenetic history of the *M. tuberculosis* complex seemed simple. On the basis of its genetic monomorphism and an estimated date for the split between *E. coli* and *S. enterica*, this group of organisms was thought to have evolved in the past 15,000–20,000 years (76) (but see the critique of molecular dating below). However, tubercle bacilli from Djibouti, East Africa, some of which were previously designated *M. canetti*, show considerably greater diversity (36), in part because they have acquired mosaic blocks by horizontal genetic exchange (72). Owing to the existence of these bacteria, it has been suggested that the *M. tuberculosis* complex is part of a protospecies that has infected hominids since their origins and initially consisted of a highly diverse group of organisms (36). According to this interpretation, the currently common isolates of the *M. tuberculosis* complex are the progeny of a lineage that underwent an extreme bottleneck in the relatively recent past and much of the original diversity has disappeared, except in Djibouti. A similar interpretation invoking an extreme bottleneck has also been invoked to explain the extreme monomorphism within *M. bovis* in the United Kingdom (73).

Y. pestis

Y. pestis is an endemic cause of septicemia in a variety of rodents and other mammals in diverse countries and continents (77) (**Figure 3a**). Transmission from host to host is predominantly vector-borne, via fleas that are specific for individual host species and geographical areas. Human disease, manifested as bubonic, septicemic, or pneumonic plague, is usually a rare zoonosis, even in endemic regions such as the United States, because it requires close contact with infected animals and the transmission by infected fleas to a largely unsuitable host. However, local outbreaks of plague in humans do occur, sometimes accompanied by considerable media coverage, and can result in panic because of the historical associations of *Y. pestis* with two pandemics, Justinian's plague (541–767 a.d.) and the Black Death (1346–1800s), each of which killed a considerable portion of the European population (**Figure 3c**).

A third pandemic began in 1894 when marine shipping from Hong Kong transmitted rats and fleas infected with *Y. pestis* to the entire globe, resulting in new endemic foci in the United States (4), Madagascar (54), India (18), and elsewhere (**Figure 3c**). However, despite occasional recent and dramatic outbreaks of human-to-human transmission of pneumonic plague (**http://www.who.int/csr/don/archive/disease/plague/en/**), the basic epidemiological features of the third pandemic differ dramatically from those of the Black Death (18), and epidemiologists (70) and historians (19) have proposed that historical plague was caused by a pathogen other than *Y. pestis*. These proposals were largely based on comparisons of endemic disease in India at the beginning of the twentieth century with historical records of epidemic disease, raising the question whether the epidemiological patterns of human plague caused by *Y. pestis* are always similar. Alternative explanations that account for differences between modern and historical plague include different patterns of transmission associated

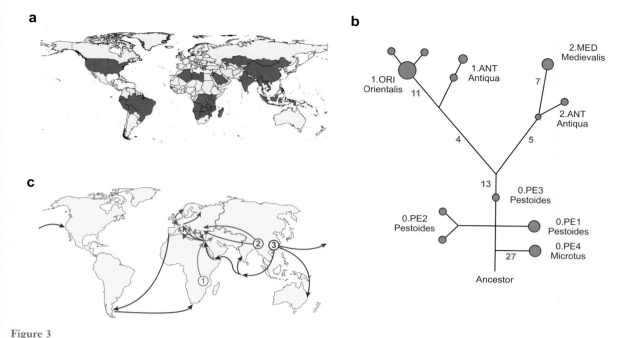

Figure 3

Phylogeographic patterns and genetic diversity in *Yersinia pestis*. (*a*) Endemic areas of plague. Modified from figure 1 in Reference 77 with permission from the author. (*b*) Phylogenetic tree of ancestry based on a combination of MLVA (multilocus VNTR analysis), IS*100* insertions and genotyping with 44 synonymous single nucleotide polymorphisms. Gray numbers along branches indicate numbers of sSNPs. Modified from Reference 2 with permission from the author. (*c*) Routes of historical pandemics. 1. Justinian's plague; 2. the Black Death; 3. the third pandemic. Modified from Reference 3 with permission from the author.

with different species of host and fleas (10) and/or major differences in the frequency of infestation of humans by fleas. Still another possibility that has not been addressed is that different lineages of *Y. pestis* might well be associated with differing epidemiological patterns, especially considering that at least one lineage of *Y. pestis* in China, called biovar Microtus, is relatively avirulent for large mammals (91). To address this last question, we need a deeper understanding of the phylogeographic patterns of global diversity within *Y. pestis* than is currently available.

Y. pestis is a genetically monomorphic clone of *Y. pseudotuberculosis* that changed its ecological niche coincidentally with the acquisition of two plasmids (3).[3] Western scientists have fo-

cused on strains related to the third pandemic, the so-called Orientalis biovar, and endemic strains from Africa (biovar Antiqua) and the Near East (biovar Medievalis). *Y. pestis* from endemic areas in Central and East Asia has not been analyzed in detail by Western scientists, except for a few strains that were designated pestoides. Analyses of endemic disease in Central Asia by Russian scientists (7) and in eastern Asia by Chinese scientists (90) have used different and mutually incomprehensible designations that have not been fully reconciled with the Western designations, except for

---

[3]For most genetically monomorphic pathogens, there is no obvious immediate ancestor. However, *Y. pestis* is a clone of

Y. pseudotuberculosis (3) and *B. mallei* is a clone of *B. pseudomallei* (32). In contrast, although most bacteria that cause anthrax correspond to a clone of the *B. cereus* complex that is called *B. anthracis*, the ability to cause anthrax is not restricted to that clone and *Bacillus* strains from invasive disease are polyphyletic (59).

biovar Microtus described above, but projects directed toward this goal are in progress (85).

Possibly the currently most comprehensive global overview of diversity in *Y. pestis* was provided by a comparison of the lineages detected by MLVA, IS elements, and genotyping with 44 sSNPs in a global sample of biovars Orientalis, Medievalis, Antiqua, Microtus, and pestoides (2). All three methods consistently grouped the strains into eight populations, two within Antiqua, four within pestoides plus Microtus, and one each for most strains within Orientalis and Medievalis. These populations mapped onto a three-branch phylogeny (**Figure 3b**) consisting of an ancestral branch, branch 0, containing pestoides plus biovar Microtus, and two descendant branches, branches 1 and 2. Because of several inconsistencies between population assignments and biovar, each population was assigned a novel designation that included the branch and a mnemonic abbreviation of the biovar, e.g., 1.ORI for Orientalis and 1.ANT1 for one of the two major groups of biovar Antiqua (**Figure 3b**). Branch 0 was found only in Central Asia, with the exception of 0.PE3. 2.ANT populations were all isolated in East Asia and 1.ANT populations were isolated in Africa. The sources of 1.ORI mark the global transmission associated with the third pandemic, and 2.MED is present in the Near East (2) and China (91). These phylogeographic associations are worth further investigation with higher-resolution methods because they may shed light on the detailed history of plague pandemics.

AGE OF MONOMORPHIC PATHOGENS

The simplest explanation for a genetically monomorphic pathogen is that the population size of the ancestors of all extant organisms was so strongly reduced during a recent bottleneck that genetic diversity was abolished. One possibility that can result in such a bottleneck is a crucial genetic event that only happened once, such as a change in ecological niche due to the acquisition of two plasmids by the progenitor of *Y. pestis* (3). Alternatively, bottlenecks also occur within small populations owing to random genetic drift, for example, the genetic monomorphism that is associated with *M. bovis* in the United Kingdom (73). Because most genetically monomorphic pathogens seem to have clonal population structures in which homoplasies and homologous recombination are rare, it would be possible to calculate the time since they arose by a coalescent approach if an appropriate molecular clock rate were available. Such calculations have indicated that the time to the most recent common ancestor (the age) is >17,000 years for *B. anthracis* (84), 10,000–71,000 years for Typhi (69), ≥13,000 years for *Y. pestis* (2), 15,000–20,000 years for the *M. tuberculosis* complex (41, 75), and 40,000 years for *E. coli* O157:H7 (89).

All but one of these calculations were based on a molecular clock rate that was calculated by dividing the synonymous genetic diversity between *E. coli* and *S. enterica* by the 140 million years that were estimated to have elapsed since they separated (57). This clock rate is similar (about twice as high) to the clock rate for *Buchnera aphidicola*, which was calibrated on the fossil record of aphids with whom these bacteria have cospeciated (57). However, even though I am a coauthor on two of these publications, it is inappropriate to use these clock rates to calculate the age of genetically monomorphic pathogens for a number of reasons. (*a*) A universal clock rate for all microbes does not exist (57), probably because their lifestyles and population sizes are so different. (*b*) The clock rates for *E. coli* and *Buchnera* correspond to rates of substitution that are based on the fixed nucleotide polymorphisms that differentiate distinct species. The rates of accumulation of diversity within a species are rates of divergence, which have different dynamics than substitution rates (72). (*c*) A clock rate calibrated at 140 million years is inappropriate for events that may have happened over a timescale that is three orders of magnitude smaller because clock rates accelerate for recent events (40). (*d*) The estimate of 140 million years since *E. coli* separated from *S. enterica* was made in 1987 (58) on the basis of calibrations that would be suspect today (unpublished analyses that are too complicated for this venue). It is therefore time for a reality check and to cast doubt on all these time estimates.

The ideal solution for determining the age of genetically monomorphic pathogens would be to calibrate molecular clock rates against a fossil record. A molecular variant of this approach is to examine the sequence diversity within ancient DNA from samples that can be dated accurately. However, until now this approach has not yielded much new information and claims based on ancient DNA analyses remain highly controversial (see Appendix 2: Ancient DNA, below).

FUTURE PROSPECTS AND OPEN QUESTIONS

We have an initial overview of the dimensions of the global diversity within *Y. pestis*, *B. anthracis*, Typhi, and *M. tuberculosis*. This overview will become more definitive in the near future because I know of genomic resequencing projects for 10–30 strains from each of these taxa whose analyses will probably be published within the next two to three years. SNP-typing is starting to be used for epidemiological surveys, and multiple phylogeographic analyses at the global scale are currently in progress. These analyses will lead to additional genomic resequencing and even better sets of SNPs for forensic and epidemiological surveys. Genetically monomorphic pathogens are particularly suited for such approaches because a definitive set of SNPs can be defined owing to their limited global diversity.

In contrast to the recent dramatic advances in resequencing technology and the amount of available data that will soon be available, our understanding of the evolutionary causes of genetic monomorphism is still primitive and population genetic and ecological analyses are in their infancy. Sophisticated algorithms exist that can calculate believable ages and reconstruct patterns of ancestry. But these algorithms can only first be applied after reliable molecular clock rates have been deduced, which

is not yet the case. We will need to correlate biogeographic patterns with historical and geographic events. And ecological and theoretical approaches will need to be correlated with epidemiological data, which will probably require stronger interdisciplinary collaborations. The same goals apply to other microbes, except that they are easier to attain with genetically monomorphic pathogens than within taxa that are highly diverse. These goals present major challenges but I am convinced that they are attainable and will provide fascinating insights in the next few years.

APPENDIX 1: PHYLOGENETIC DISCOVERY BIAS

A comparison of the genomic sequences from two related strains reveals all the sequence diversity that exists between that pair, including insertions, deletions, rearrangements, and sequence polymorphisms. Therefore, it might only be necessary to compare publicly available genomic sequences to elucidate a set of SNPs appropriate for phylogenetic purposes. A comparison of two genomes yields a linear tree on which all other strains can be mapped (**Figure 4a–d**). However, a linear tree condenses the true diversity of all strains other than the genomic reference strains, and can group strains together that are divergent, particularly if the two genomic sequences are from closely related strains (5). This phenomenon is called phylogenetic discovery bias. Even a comparison of multiple genomic sequences is subject to phylogenetic discovery bias (62), and the discernible genetic diversity is again limited by the diversity of the strains chosen for sequencing (**Figure 4e–g**).

These considerations are not purely theoretical. Polymorphic SNPs that were identified by a comparison of the genomes of *M. tuberculosis* strains H37Rv and CDC1551 with the genome of a relatively distant outgroup (*Mycobacterium bovis*) were used for analyses of the diversity and relationships among large numbers of strains (25, 34, 35). However, H37Rv and CDC1551 are in fact closely related (8, 28) despite initial

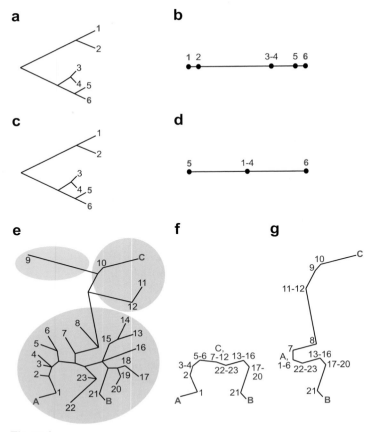

a

1
2
3
4 5
6

b

1 2   3-4   5 6

c

1
2
3
4 5
6

d

5   1-4   6

e

9   10   C
              11
              12
          14
6 7 8   15 13
5   16
4   18
3   23   19 17
2   20
1   21 B
      22

f

                C,
            5-6 7-12 13-16
3-4   22-23   17-
2   20
    A   1   21 B

g

                10 C
              9
          11-12
          7 8
      A,   13-16
      1-6 22-23 17-20
          21 B

Figure 4

Discovery bias in single nucleotide polymorphism discovery based on the
choice of genomic sequences. (*a, c*) A phylogenetic tree of six strains, two of
which (*red*) have been used for genomic comparisons. (*b, d*) A linear phylogeny
results from the comparisons between two reference genomic sequences,
in which the two reference strains are always at the terminus and the genetic
distances to the other strains reflect the positions of the branch points of
those strains along the path linking the reference genome sequences. When
the two reference genomes are as distant as possible (*a*), the linear phylogeny
(*b*) accurately represents the relative distances of the branch points of the other
four individual strains. In contrast, when the reference genomes are closely
related (*c*), the branch positions of the other four strains collapse to a single
point that does not reflect their true distances from the reference genomes.
(*e*) Phylogenetic tree of 26 strains of *Bacillus anthracis* based on MLVA (multilocus
VNTR analysis), with arbitrary designations from 1 to 23 plus designations A,
B, and C for strains whose genomes were sequenced. Colored ellipses indicate
distinct phylogenetic groupings. (*f*) Linear phylogeny based on comparisons
of sequences A and B. (*g*) Linear phylogeny based on comparisons of sequences
B and C. The positions of strains in panels *f* and *g* accurately reflect the
genetic distances of the branch points along the linear phylogeny for individual
strains in panel *e* but collapse additional diversity beyond those branch points
and can assign strains from different groupings to a common location (strains
C, 7–12 in panel *f*) or suggest that they are closely related (9 and 10 in panel *g*).
Modified from figures 1 and 4 in Reference 62 with permission from the author.

claims to the contrary (5). Because of phylogenetic discovery bias, the SNPs identified by
the genomic comparisons did not provide reliable phylogenetic distances and artificially split
M. tuberculosis lineage II into multiple groups
(28). Similarly, because of discovery bias, genotyping based on 232 sSNPs was unable to resolve most isolates within *C. pneumoniae* (67).

Discovery bias is particularly likely in the
absence of information on the global and temporal genetic diversity of a taxon, particularly because some phylogenetic subgroupings
are rare or restricted to particular geographic
locations. For example, only two isolates of
group C were found among >1000 *B. anthracis* isolates from diverse global sources (84).
Similarly, an unusual group of *E. coli* was represented by only two strains in a global collection of >400 strains (88). And the hpAfrica2
group of *H. pylori* was only identified because
biopsies were obtained from healthy individuals
in South Africa (24). Mutation discovery needs
to be performed on large collections of strains
from global sources to ensure that subsequent
high-throughput tests do not exclude important
groupings. The development of SNP-based
genotyping should therefore be preceded, or
at least accompanied, by an overview of the
global genetic diversity of the taxon to be investigated, a goal for which the classical molecular
tools such as MLVA and CRISPR are eminently
suited (**Figure 5**). The composition of panels
of SNPs that seem appropriate for genotyping large strain collections, such as SNPs based
on seven genomes of *B. anthracis* (84), need to
be reconsidered or augmented once additional
groups of strains that were not included in the
initial analyses are discovered (49).

Unfortunately, many studies that are technically impressive seem to have ignored these tru-
isms. For example, mutation discovery within
E. coli O157:H7 bacteria was performed by
Nimblegen array resequencing of one-fourth of
the chromosomal protein-coding regions (1.17
MB) of 11 strains, resulting in the recognition of
almost 1000 SNPs. However, of the 11 strains
that were resequenced, one was from Japan,
one was from Germany, and nine were from

the United States. It seems likely that important additional diversity will be revealed once a more representative global sample is subjected to mutation discovery.

APPENDIX 2: ANCIENT DNA

For over a decade, scientists have attempted to reconstruct the evolutionary history of several monomorphic pathogens through the PCR amplification of specific DNA from skeletons associated with mass graves or specific bone lesions (ancient DNA, aDNA). Some of these reports are patently ludicrous. For example, it has been claimed that Typhi caused the Plague of Athens in 430 B.C. because a PCR product with 93% homology to the *narG* gene of Typhi could be amplified from three teeth of skeletons in a mass grave (60). However, 93% homology indicates immediately that the organism from which the PCR product was obtained is unrelated to the genetically monomorphic Typhi. Indeed the amplified *narG* sequences were only distantly related to all *narG* sequences from *S. enterica*, suggesting that they were amplified from nonpathogenic soil organisms (71). Other reports seem much more convincing, including the amplification of spoligotypes that are characteristic of *M. africanum* from 4000-year-old Egyptian mummies and of spoligotypes characteristic of *M. tuberculosis* from 2500- to 3000-year-old Egyptian mummies, from medieval Germans and Hungarians (92), and from a 2200-year-old skeleton in Britain (80). aDNA from 2000-year-old skeletons in Tuva (Siberia) has been genotyped as *M. bovis* (79), and extensive genotyping with multiple markers has also been performed in other analyses (26, 80). Similarly convincing publications have described the amplification of aDNA specific to *Y. pestis* from dental pulp from Justinian's plague, the Black Death, and more recent samples (21, 22, 66). In one of these analyses, genotyping with MLVA markers was used to claim that the cause of the first and second plague pandemics was biovar Orientalis (22). These experiments were initially only performed in a single laboratory, but recently, successful PCR ampli-

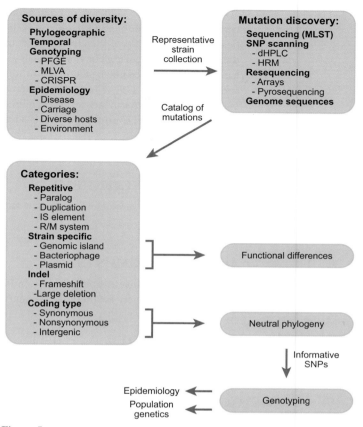

Figure 5

Avoiding phylogenetic discovery bias for single nucleotide polymorphism (SNP)-based epidemiology and population genetics. It is crucial to begin with a representative strain collection that has been chosen on the basis of multiple criteria that are associated with population subdivision, and that contains examples of the diversity that has already been revealed by molecular tests. This strain collection should then be subjected to mutation discovery using methods that will identify unknown SNPs. The resulting catalog of mutations then needs to be assigned to various categories. Strain-specific features are of potential interest for functional analyses that might shed light on different disease syndromes or host adaptation, but neutral phylogenies based on SNPs are preferred for developing SNP-based surveys for epidemiological purposes and to elucidate population genetic patterns.

fication of *Y. pestis* aDNA has been reported on other samples dated to Justinian's plague by independent workers using the same techniques (87).

The number of convincing reports on genotyping aDNA in *M. tuberculosis* is growing rapidly. It is worth noting, however, that experts in aDNA analyses consider all microbial aDNA work to be at high risk of contamination from the environment (38) and are skeptical about the validity of such experiments in the absence of particularly convincing analyses based on sophisticated phylogenetic algorithms and common sense (29). None of the microbial reports on aDNA can satisfy these requirements. Furthermore, control experiments have failed to amplify any aDNA for the *M. tuberculosis* complex from relatively recent tuberculous skeletons in museums (9) and for *Y. pestis* with teeth from mass graves in cold climates (30). I conclude that the question whether aDNA has really been amplified successfully from *M. tuberculosis* and *Y. pestis* has not been definitely resolved.

DISCLOSURE STATEMENT

The author is not aware of any biases that might be perceived as affecting the objectivity of this review.

ACKNOWLEDGMENTS

M.A. is a Principal Investigator of the Scientific Foundation of Ireland (grant 05/FE1/B882) and the Max-Planck Society for the Advancement of Science.

LITERATURE CITED

1. Achtman M. 1996. A surfeit of YATMs? *J. Clin. Microbiol.* 34:1870
2. Achtman M, Morelli G, Zhu P, Wirth T, Diehl I, et al. 2004. Microevolution and history of the plague bacillus, *Yersinia pestis*. *Proc. Natl. Acad. Sci. USA* 101:17837–42
3. Achtman M, Zurth K, Morelli G, Torrea G, Guiyoule A, Carniel E. 1999. *Yersinia pestis*, the cause of plague, is a recently emerged clone of *Yersinia pseudotuberculosis*. *Proc. Natl. Acad. Sci. USA* 96:14043–48
4. Adjemian JZ, Foley P, Gage KL, Foley JE. 2007. Initiation and spread of traveling waves of plague, *Yersinia pestis*, in the western United States. *Am. J. Trop. Med. Hyg.* 76:365–75
5. Alland D, Whittam TS, Murray MB, Cave MD, Hazbon MH, et al. 2003. Modeling bacterial evolution with comparative-genome-based marker systems: application to *Mycobacterium tuberculosis* evolution and pathogenesis. *J. Bacteriol.* 185:3392–99
6. Anderson GW, Hamblen AD, Smith HM. 1936. Typhoid carriers. A study of their disease producing potentialities over a series of years as indicated by a study of cases. *Am. J. Public Health* 26:396–405
7. Anisimov AP, Lindler LE, Pier GB. 2004. Intraspecific diversity of *Yersinia pestis*. *Clin. Microbiol. Rev.* 17:434–64
8. Baker L, Brown T, Maiden MC, Drobniewski F. 2004. Silent nucleotide polymorphisms and a phylogeny for *Mycobacterium tuberculosis*. *Emerg. Infect. Dis.* 10:1568–77
9. Barnes I, Thomas MG. 2006. Evaluating bacterial pathogen DNA preservation in museum osteological collections. *Proc. Biol. Sci.* 273:645–53
10. Beaucournu JC. 1999. Diversité des puces vectrices en fonction des foyers pesteux. *Bull. Soc. Pathol. Exot.* 92:419–21
11. Bentley DR. 2006. Whole-genome resequencing. *Curr. Opin. Genet. Dev.* 16:545–52
12. Black WC 4th, Vontas JG. 2007. Affordable assays for genotyping single nucleotide polymorphisms in insects. *Insect Mol. Biol.* 16:377–87
13. Blower SM, McLean AR, Porco TC, Small PM, Hopewell PC, et al. 1995. The intrinsic transmission dynamics of tuberculosis epidemics. *Nat. Med.* 1:815–21
14. Brosch R, Gordon SV, Marmiesse M, Brodin P, Buchrieser C, et al. 2002. A new evolutionary scenario for the *Mycobacterium tuberculosis* complex. *Proc. Natl. Acad. Sci. USA* 99:3684–89

15. Brudey K, Driscoll JR, Rigouts L, Prodinger WM, Gori A, et al. 2006. *Mycobacterium tuberculosis* complex genetic diversity: mining the fourth international spoligotyping database (SpolDB4) for classification, population genetics and epidemiology. *BMC Microbiol.* 6:23

16. Caufield PW, Saxena D, Fitch D, Li Y. 2007. Population structure of plasmid-containing strains of *Streptococcus mutans*, a member of the human indigenous biota. *J. Bacteriol.* 189:1238–43

17. Chau TT, Campbell JI, Galindo CM, Van Minh HN, Diep TS, et al. 2007. Antimicrobial drug resistance of *Salmonella enterica* serovar Typhi in Asia and molecular mechanism of reduced susceptibility to the fluoroquinolones. *Antimicrob. Agents Chemother.* 51:4315–23

18. Christakos G, Olea RA, Yu HL. 2007. Recent results on the spatiotemporal modelling and comparative analysis of Black Death and bubonic plague epidemics. *Public Health* 121:700–20

19. Cohn SK Jr. 2002. *The Black Death Transformed: Disease and Culture in Early Renaissance Europe.* London: Arnold. 318 pp.

20. Didelot X, Achtman M, Parkhill J, Thomson NR, Falush D. 2007. A bimodal pattern of relatedness between the *Salmonella* Paratyphi A and Typhi genomes: convergence or divergence by homologous recombination? *Genome Res.* 17:61–68

21. Drancourt M, Aboudharam G, Signoli M, Dutour O, Raoult D. 1998. Detection of 400-year-old *Yersinia pestis* DNA in human dental pulp: an approach to the diagnosis of ancient septicemia. *Proc. Natl. Acad. Sci. USA* 95:12637–40

22. Drancourt M, Roux V, Dang LV, Tran-Hung L, Castex D, et al. 2004. Genotyping, Orientalis-like *Yersinia pestis*, and plague pandemics. *Emerg. Infect. Dis.* 10:1585–92

23. Falush D, Torpdahl M, Didelot X, Conrad DF, Wilson DJ, Achtman M. 2006. Mismatch induced speciation in *Salmonella*: model and data. *Philos. Trans. R. Soc. London B Biol. Sci.* 361:2045–53

24. Falush D, Wirth T, Linz B, Pritchard JK, Stephens M, et al. 2003. Traces of human migrations in *Helicobacter pylori* populations. *Science* 299:1582–85

25. Filliol I, Motiwala AS, Cavatore M, Qi W, Hazbon MH, et al. 2006. Global phylogeny of *Mycobacterium tuberculosis* based on single nucleotide polymorphism (SNP) analysis: insights into tuberculosis evolution, phylogenetic accuracy of other DNA fingerprinting systems, and recommendations for a minimal standard SNP set. *J. Bacteriol.* 188:759–72

26. Fletcher HA, Donoghue HD, Taylor GM, van der Zanden AG, Spigelman M. 2003. Molecular analysis of *Mycobacterium tuberculosis* DNA from a family of 18th century Hungarians. *Microbiology* 149:143–51

27. Gagneux S, DeRiemer K, Van T, Kato-Maeda M, de Jong BC, et al. 2006. Variable host-pathogen compatibility in *Mycobacterium tuberculosis*. *Proc. Natl. Acad. Sci. USA* 103:2869–73

28. Gagneux S, Small PM. 2007. Global phylogeography of *Mycobacterium tuberculosis* and implications for tuberculosis product development. *Lancet Infect. Dis.* 7:328–37

29. Gilbert MT, Bandelt HJ, Hofreiter M, Barnes I. 2005. Assessing ancient DNA studies. *Trends Ecol. Evol.* 20:541–44

30. Gilbert MT, Cuccui J, White W, Lynnerup N, Titball RW, et al. 2004. Absence of *Yersinia pestis*–specific DNA in human teeth from five European excavations of putative plague victims. *Microbiology* 150:341–54

31. Girard JM, Wagner DM, Vogler AJ, Keys C, Allender CJ, et al. 2004. Differential plague transmission dynamics determine *Yersinia pestis* population genetic structure on local, regional and global scales. *Proc. Natl. Acad. Sci. USA* 101:8408–13

32. Godoy D, Randle G, Simpson AJ, Aanensen DM, Pitt TL, et al. 2003. Multilocus sequence typing and evolutionary relationships among the causative agents of melioidosis and glanders, *Burkholderia pseudomallei* and *Burkholderia mallei*. *J. Clin. Microbiol.* 41:2068–79

33. Grissa I, Vergnaud G, Pourcel C. 2007. The CRISPRdb database and tools to display CRISPRs and to generate dictionaries of spacers and repeats. *BMC. Bioinform.* 8:172

34. Gutacker MM, Mathema B, Soini H, Shashkina E, Kreiswirth BN, et al. 2006. Single-nucleotide polymorphism-based population genetic analysis of *Mycobacterium tuberculosis* strains from four geographic sites. *J. Infect. Dis.* 193:121–28

35. Gutacker MM, Smoot JC, Migliaccio CA, Ricklefs SM, Hua S, et al. 2002. Genome-wide analysis of synonymous single nucleotide polymorphisms in *Mycobacterium tuberculosis* complex organisms. Resolution of genetic relationships among closely related microbial strains. *Genetics* 162:1533–43

36. Gutierrez MC, Brisse S, Brosch R, Fabre M, Omais B, et al. 2005. Ancient origin and gene mosaicism of the progenitor of *Mycobacterium tuberculosis*. *PLoS Pathog.* 1:e5

37. Harper KN, Ocampo PS, Steiner BM, George RW, Silverman MS, et al. 2008. On the origin of the treponematoses: a phylogenetic approach. *PLoS Negl. Trop. Dis.* 2:e148

38. Hebsgaard MB, Phillips MJ, Willerslev E. 2005. Geologically ancient DNA: fact or artefact? *Trends Microbiol.* 13:212–20

39. Hinchliffe SJ, Isherwood KE, Stabler RA, Prentice MB, Rakin A, et al. 2003. Application of DNA microarrays to study the evolutionary genomics of *Yersinia pestis* and *Yersinia pseudotuberculosis*. *Genome Res.* 13:2018–29

40. Ho SY, Larson G. 2006. Molecular clocks: when times are a-changin. *Trends Genet.* 22:79–83

41. Kapur V, Whittam TS, Musser JM. 1994. Is *Mycobacterium tuberculosis* 15,000 years old? *J. Infect. Dis.* 170:1348–49

42. Keim P, Price LB, Klevytska AM, Smith KL, Schupp JM, et al. 2000. Multiple-locus variable-number tandem repeat analysis reveals genetic relationships within *Bacillus anthracis*. *J. Bacteriol.* 182:2928–36

43. Kidgell C, Reichard U, Wain J, Linz B, Torpdahl M, et al. 2002. *Salmonella typhi*, the causative agent of typhoid fever, is approximately 50,000 years old. *Infect. Genet. Evol.* 2:39–45

44. Kim S, Misra A. 2007. SNP genotyping: technologies and biomedical applications. *Annu. Rev. Biomed. Eng.* 9:289–320

45. Klevytska AM, Price LB, Schupp JM, Worsham PL, Wong J, Keim P. 2001. Identification and characterization of variable-number tandem repeats in the *Yersinia pestis* genome. *J. Clin. Microbiol.* 39:3179–85

46. Ledingham JCG, Arkwright JA. 1912. *The Carrier Problem in Infectious Diseases*. London: Arnold. 319 pp.

47. Levine MM, Black RE, Lanata C. 1982. Precise estimation of the numbers of chronic carriers of *Salmonella typhi* in Santiago, Chile, an endemic area. *J. Infect. Dis.* 146:724–26

48. Linz B, Balloux F, Moodley Y, Manica A, Liu H, et al. 2007. An African origin for the intimate association between humans and *Helicobacter pylori*. *Nature* 445:915–18

49. Lista F, Faggioni G, Valjevac S, Ciammaruconi A, Vaissaire J, et al. 2006. Genotyping of *Bacillus anthracis* strains based on automated capillary 25-loci multiple locus variable-number tandem repeats analysis. *BMC Microbiol.* 6:33

50. Liu SL, Sanderson KE. 1996. Highly plastic chromosomal organization in *Salmonella typhi*. *Proc. Natl. Acad. Sci. USA* 93:10303–8

51. Maiden MC. 2006. Multilocus sequence typing of bacteria. *Annu. Rev. Microbiol.* 60:561–88

52. Maiden MCJ, Bygraves JA, Feil E, Morelli G, Russell JE, et al. 1998. Multilocus sequence typing: a portable approach to the identification of clones within populations of pathogenic microorganisms. *Proc. Natl. Acad. Sci. USA* 95:3140–45

53. McKinney JD, Jacobs WR Jr, Bloom BR. 1998. Persisting problems in tuberculosis. In *Emerging Infections*, ed. RM Krause, pp. 51–146. New York: Academic

54. Migliani R, Chanteau S, Rahalison L, Ratsitorahina M, Boutin JP, et al. 2006. Epidemiological trends for human plague in Madagascar during the second half of the 20th century: a survey of 20,900 notified cases. *Trop. Med. Int. Health* 11:1228–37

55. Monot M, Honore N, Garnier T, Araoz R, Coppee JY, et al. 2005. On the origin of leprosy. *Science* 308:1040–42

56. Motin VL, Georgescu AM, Elliott JM, Hu P, Worsham PL, et al. 2002. Genetic variability of *Yersinia pestis* isolates as predicted by PCR-based IS100 genotyping and analysis of structural genes encoding glycerol-3-phosphate dehydrogenase (*glpD*). *J. Bacteriol.* 184:1019–27

57. Ochman H, Elwyn S, Moran NA. 1999. Calibrating bacterial evolution. *Proc. Natl. Acad. Sci. USA* 96:12638–43

58. Ochman H, Wilson AC. 1987. Evolution in bacteria: evidence for a universal substitution rate in cellular genomes. *J. Mol. Evol.* 26:74–86

59. Okinaka R, Pearson T, Keim P. 2006. Anthrax, but not *Bacillus anthracis*? *PLoS Pathog.* 2:e122

60. Papagrigorakis MJ, Yapijakis C, Synodinos PN, Baziotopoulou-Valavani E. 2006. DNA examination of ancient dental pulp incriminates typhoid fever as a probable cause of the Plague of Athens. *Int. J. Infect. Dis.* 10:206–14

61. Parry CM, Hien TT, Dougan G, White NJ, Farrar JJ. 2002. Typhoid fever. *N. Engl. J. Med.* 347:1770–82

62. Pearson T, Busch JD, Ravel J, Read TD, Rhoton SD, et al. 2004. Phylogenetic discovery bias in *Bacillus anthracis* using single-nucleotide polymorphisms from whole-genome sequencing. *Proc. Natl. Acad. Sci. USA* 101:13536–41

63. Pourcel C, Salvignol G, Vergnaud G. 2005. CRISPR elements in *Yersinia pestis* acquire new repeats by preferential uptake of bacteriophage DNA, and provide additional tools for evolutionary studies. *Microbiology* 151:653–63

64. Prentice MB, Rahalison L. 2007. Plague. *Lancet* 369:1196–207

65. Pupo GM, Lan R, Reeves PR. 2000. Multiple independent origins of *Shigella* clones of *Escherichia coli* and convergent evolution of many of their characteristics. *Proc. Natl. Acad. Sci. USA* 97:10567–72

66. Raoult D, Aboudharam G, Crubezy E, Larrouy G, Ludes B, Drancourt M. 2000. Molecular identification by "suicide PCR" of *Yersinia pestis* as the agent of Medieval Black Death. *Proc. Natl. Acad. Sci. USA* 97:12800–3

67. Rattei T, Ott S, Gutacker M, Rupp J, Maass M, et al. 2007. Genetic diversity of the obligate intracellular bacterium *Chlamydophila pneumoniae* by genome-wide analysis of single nucleotide polymorphisms: evidence for highly clonal population structure. *BMC Genomics* 8:355

68. Rotz LD, Khan AS, Lillibridge SR, Ostroff SM, Hughes JM. 2002. Public health assessment of potential biological terrorism agents. *Emerg. Infect. Dis.* 8:225–30

69. Roumagnac P, Weill F-X, Dolecek C, Baker S, Brisse S, et al. 2006. Evolutionary history of *Salmonella* Typhi. *Science* 314:1301–4

70. Scott S, Duncan CJ. 2001. *Biology of Plagues: Evidence from Historical Populations*. Cambridge, UK: Cambridge Univ. Press. 420 pp.

71. Shapiro B, Rambaut A, Gilbert MT. 2006. No proof that typhoid caused the Plague of Athens (a reply to Papagrigorakis et al.). *Int. J. Infect. Dis.* 10:334–35

72. Smith NH. 2006. A re-evaluation of *M. prototuberculosis*. *PLoS Pathog.* 2:e98

73. Smith NH, Gordon SV, Rua-Domenech R, Clifton-Hadley RS, Hewinson RG. 2006. Bottlenecks and broomsticks: the molecular evolution of *Mycobacterium bovis*. *Nat. Rev. Microbiol.* 4:670–81

74. Sola C, Filliol I, Legrand E, Lesjean S, Locht C, et al. 2003. Genotyping of the *Mycobacterium tuberculosis* complex using MIRUs: association with VNTR and spoligotyping for molecular epidemiology and evolutionary genetics. *Infect. Genet. Evol.* 3:125–33

75. Sreevatsan S, Escalante P, Pan X, Gillies DA II, Siddiqui S, et al. 1996. Identification of a polymorphic nucleotide in *oxyR* specific for *Mycobacterium bovis*. *J. Clin. Microbiol.* 34:2007–10

76. Sreevatsan S, Pan X, Stockbauer K, Connell ND, Kreiswirth BN, et al. 1997. Restricted structural gene polymorphism in the *Mycobacterium tuberculosis* complex indicates evolutionarily recent global dissemination. *Proc. Natl. Acad. Sci. USA* 94:9869–74

77. Stenseth NC, Atshabar BB, Begon M, Belmain SR, Bertherat E, et al. 2008. Plague: past, present, and future. *PLoS Med.* 5:e3

78. Swaminathan B, Barrett TJ, Hunter SB, Tauxe RV. 2001. PulseNet: the molecular subtyping network for foodborne bacterial disease surveillance, United States. *Emerg. Infect. Dis.* 7:382–89

79. Taylor GM, Murphy E, Hopkins R, Rutland P, Chistov Y. 2007. First report of *Mycobacterium bovis* DNA in human remains from the Iron Age. *Microbiology* 153:1243–49

80. Taylor GM, Young DB, Mays SA. 2005. Genotypic analysis of the earliest known prehistoric case of tuberculosis in Britain. *J. Clin. Microbiol.* 43:2236–40

81. The International Hapmap Consortium. 2005. A haplotype map of the human genome. *Nature* 437:1299–320

82. Tsolaki AG, Hirsh AE, DeRiemer K, Enciso JA, Wong MZ, et al. 2004. Functional and evolutionary genomics of *Mycobacterium tuberculosis*: insights from genomic deletions in 100 strains. *Proc. Natl. Acad. Sci. USA* 101:4865–70

83. Van Embden JDA, Cave MD, Crawford JT, Dale JW, Eisenach KD, et al. 1993. Strain identification of *Mycobacterium tuberculosis* by DNA fingerprinting: recommendation for a standardized methodology. *J. Clin. Microbiol.* 31:406–9

84. Van Ert MN, Easterday WR, Huynh LY, Okinaka RT, Hugh-Jones ME, et al. 2007. Global genetic population structure of *Bacillus anthracis*. *PLoS ONE* 2:e461

85. Vergnaud G, Li Y, Gorge O, Cui Y, Song Y, et al. 2007. Analysis of the three *Yersinia pestis* CRISPR loci provides new tools for phylogenetic studies and possibly for the investigation of ancient DNA. *Adv. Exp. Med. Biol.* 603:327–38

86. Victor TC, de Haas PE, Jordaan AM, van der Spuy GD, Richardson M, et al. 2004. Molecular characteristics and global spread of *Mycobacterium tuberculosis* with a western cape F11 genotype. *J. Clin. Microbiol.* 42:769–72

87. Wiechmann I, Grupe G. 2004. Detection of *Yersinia pestis* DNA in two early medieval skeletal finds from Aschheim (Upper Bavaria, 6th century A.D.). *Am. J. Phys. Anthropol.* 126:48–55

88. Wirth T, Falush D, Lan R, Colles F, Mensa P, et al. 2006. Sex and virulence in *Escherichia coli*: an evolutionary perspective. *Mol. Microbiol.* 60:1136–51

89. Zhang W, Qi W, Albert TJ, Motiwala AS, Alland D, et al. 2006. Probing genomic diversity and evolution of *Escherichia coli* O157 by single-nucleotide polymorphisms. *Genome Res.* 16:757–67

90. Zhou D, Han Y, Song Y, Tong Z, Wang J, et al. 2004. DNA microarray analysis of genome dynamics in *Yersinia pestis*: insights into bacterial genome microevolution and niche adaptation. *J. Bacteriol.* 186:5138–46

91. Zhou D, Tong Z, Song Y, Han Y, Pei D, et al. 2004. Genetics of metabolic variations between *Yersinia pestis* biovars and the proposal of a new biovar, microtus. *J. Bacteriol.* 186:5147–52

92. Zink AR, Molnar E, Motamedi N, Palfy G, Marcsik A, Nerlich AG. 2007. Molecular history of tuberculosis from ancient mummies and skeletons. *Int. J. Osteoarch.* 17:380–91

Global Spread and Persistence of Dengue

Jennifer L. Kyle and Eva Harris

Division of Infectious Diseases, School of Public Health, and Graduate Group in Microbiology, University of California, Berkeley, California 94720-7354; email: jennkyle@berkeley.edu, eharris@berkeley.edu

Annu. Rev. Microbiol. 2008. 62:71–92

First published online as a Review in Advance on April 22, 2008

The *Annual Review of Microbiology* is online at micro.annualreviews.org

This article's doi: 10.1146/annurev.micro.62.081307.163005

Key Words

flavivirus, mosquito-borne, *Aedes*, emergence, genotype, RNA virus

Abstract

Dengue is a spectrum of disease caused by four serotypes of the most prevalent arthropod-borne virus affecting humans today, and its incidence has increased dramatically in the past 50 years. Due in part to population growth and uncontrolled urbanization in tropical and subtropical countries, breeding sites for the mosquitoes that transmit dengue virus have proliferated, and successful vector control has proven problematic. Dengue viruses have evolved rapidly as they have spread worldwide, and genotypes associated with increased virulence have expanded from South and Southeast Asia into the Pacific and the Americas. This review explores the human, mosquito, and viral factors that contribute to the global spread and persistence of dengue, as well as the interaction between the three spheres, in the context of ecological and climate changes. What is known, as well as gaps in knowledge, is emphasized in light of future prospects for control and prevention of this pandemic disease.

Contents

INTRODUCTION

Globally, as many as 1 in 100 people are infected each year by one or more of four serotypes of dengue virus (DENV1–4), a mosquito-borne, positive-strand RNA virus in the genus *Flavivirus*, family *Flaviviridae*. Tens of millions of cases of dengue fever (DF) are estimated to occur annually, including up to 500,000 cases of the life-threatening dengue hemorrhagic fever/dengue shock syndrome (DHF/DSS) (156). Epidemic DHF/DSS emerged 50 years ago in Southeast Asia (60) but was first seen in the Americas only in 1981 (75) and in South Asia in 1989 (88). Since the 1950s, the incidence of DHF/DSS has increased over 500-fold, with more than 100 countries affected by outbreaks of dengue (156). DHF/DSS has become one of the top ten causes of pediatric hospitalization in Southeast Asia (155), and the number of cases of DHF/DSS in the Americas alone has grown dramatically (98).

Dengue is associated with explosive urban epidemics and has become a major public health problem, with significant economic, political, and social impact (46). Some of the reasons for the dramatic increase in the geographic spread of dengue, including its more severe forms, are known with some certainty, whereas other reasons remain a subject of debate and speculation. This review highlights the epidemiological history of dengue and explores the various reasons for its dramatic spread and persistence during the past 50 years. To this end, aspects of human, mosquito, and virus biology, ecology, and evolution are discussed, along with the interaction between these spheres (**Figure 1**).

THE HUMAN SPHERE

Introduction

DF is a self-limited though debilitating febrile illness characterized by headache, retro-orbital pain, myalgia, arthralgia, and rash. DHF is marked by increased vascular permeability (plasma leakage), thrombocytopenia, and hemorrhagic manifestations; DSS occurs when fluid leakage into the interstitial spaces results in

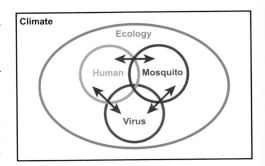

Figure 1

The interplay of human, mosquito, and virus biology contributes to the clinical spectrum and geographic distribution of dengue. Each sphere influences and affects the others, all in the context of ecology and against the backdrop of climate and climate change.

DENV: dengue virus

DF: dengue fever

DHF/DSS: dengue hemorrhagic fever/dengue shock syndrome

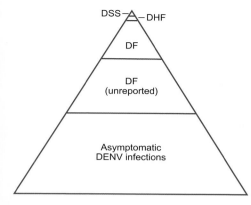

Figure 2

DENV infections and disease are represented by a pyramid. An estimated 100 million infections occur each year, 10%–50% of which are symptomatic, but only a fraction of DF cases are reported. Finally, severe disease due to DENV infection represents the tip of the pyramid. The most severe forms of dengue are DHF and DSS, which present with higher frequency in secondary than in primary DENV infections.

hypovolemic shock, which without appropriate treatment may lead to death (43, 56). Apparent disease due to dengue has been described as the tip of the iceberg (78), as less than 10% of symptomatic dengue cases are reported (156) and 50%–90% of all DENV infections are asymptomatic (10, 17, 29, 140) (**Figure 2**). The most severe disease, DHF/DSS, is found at the very tip of the pyramid, and its incidence varies significantly between primary and secondary DENV infections. A secondary DENV infection results when a person previously infected with one serotype is exposed to a different serotype, and it has been documented as the single most important risk factor for severe dengue (17, 29, 39, 53, 58, 125, 140). For example, Thai data of DENV infections in children under 15 years of age demonstrated that 0.18% of primary infections and 2% of secondary infections manifest as DHF/DSS (78). In this section, we describe dengue and review the reasons for the spread and persistence of epidemic DF and DHF/DSS and what is known about the risk for infection and severe disease.

History and Spread of Dengue

Clinical descriptions of a dengue-like syndrome were recorded as far back as A.D. 992 in China, although the first epidemics of well-documented cases of what are believed to be dengue occurred in 1779–1780 (42). The viral etiology of dengue was suggested experimentally a century ago (7), but it was not until World War II that technical advances enabled Japanese (67) and American (122) investigators to isolate DENV. The first two DENV serotypes were identified at this time, followed by the third and fourth serotypes when DHF/DSS emerged in urban centers in the Philippines and Thailand in 1954 (60, 103). It has been hypothesized that the movement of troops during World War II, together with destruction of the environment and human settlements, contributed to the spread of the viruses and, to a certain extent, their mosquito vectors throughout Southeast Asia and the Western Pacific (42, 114). Since then, Southeast Asia has remained hyperendemic for all four DENV serotypes. In the Americas, the decline and reemergence of epidemic dengue since the 1980s has been even more closely linked to the presence of its mosquito vectors, *Aedes aegypti* and *A. albopictus* (**Figure 3**).

Increases in human population, uncontrolled urbanization, and international travel can explain much of the spread and persistence of dengue in the twentieth and early twenty-first centuries (42, 114). It has been estimated that the minimum population size required to sustain dengue transmission is 10,000–1,000,000 (76), and early epidemics of dengue in Southeast Asia were linked to towns with populations over 10,000 (133). A mathematical model examining the spatiotemporal incidence of DHF over a 15-year period in Thailand estimated that epidemics of DHF originate in the capital city of Bangkok every 3 years and travel outward to the rest of the country at a rate of 148 km per month (23), stressing the role of cities in dengue transmission dynamics. In the tropical and subtropical areas where *A. aegypti* and *A. albopictus*

Secondary infection: a subsequent infection with a heterotypic serotype of DENV, occurring months to years after the primary infection

Hyperendemic: the presence of numerous serotypes of dengue virus cocirculating in one location

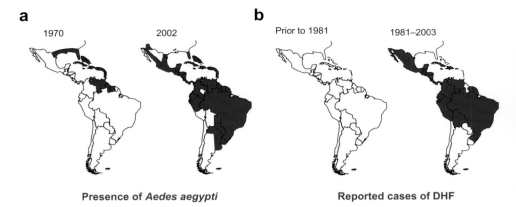

a

1970   2002

Presence of *Aedes aegypti*

b

Prior to 1981   1981–2003

Reported cases of DHF

Figure 3

The spread of *Aedes aegypti* and DHF in the Americas. (*a*) The shaded areas represent the regions in the Americas where *A. aegypti* was present in 1970 (*left*) and in 2002 (*right*). (*b*) Shaded areas represent the countries that reported cases of DHF prior to 1981 (*left*) and between 1981 and 2003 (*right*). The increased distribution of DHF mirrors the dissemination of *A. aegypti*. Reprinted with permission from the U.S. Centers for Disease Control and Prevention.

mosquitoes are present, urban growth has not been accompanied by well-organized water and waste management programs because of the sheer numbers of people migrating from the countryside and because of the lack of resources available for infrastructure and public health measures in these regions (72, 119). In settings where the availability of water is intermittent and piped water supplies may be nonexistent, both indoor and outdoor containers used for water storage comprise key *Aedes* breeding sites (112). The indestructibility of discarded plastics and the increased numbers of unused tires, combined with poor garbage disposal systems, have led to the accumulation of numerous additional breeding sites. Commercial shipping has been linked to the spread of both *A. aegypti* and *A. albopictus* between regions (63, 133, 135), and plane travel has greatly increased the dissemination of dengue viruses via rapid transit of viremic individuals around the world (42). In other words, much of the increase in DENV infections in recent decades can be explained by an increase in both human population and mosquito breeding habitats, combined with the dissemination of both mosquitoes and dengue viruses to new geographic regions.

Risk for DENV Infection

An estimated 3.5 billion people, or half the world's population, are at risk for DENV infection in tropical and subtropical countries (12). Fundamentally, exposure to an infected *A. aegypti* mosquito determines an individual's risk for acquiring dengue. By avoiding the mosquito and eliminating breeding sites around the home and workplace, an individual can mitigate that risk to some extent, although there are factors beyond an individual's immediate control. In the past, epidemics of dengue have occurred in the United States as far north as Philadelphia (42), and *A. aegypti* as well as *A. albopictus* are present in the southern and central United States today. Yet, most dengue cases in the United States are reported in travelers returning from endemic countries (110). This can be attributed to improved infrastructure, such as reliable sources of piped water that remove the need for *Aedes*-friendly water storage containers, air conditioning powered by an electric grid with few interruptions, and screens covering windows and enclosing patios. These variables greatly reduce the exposure of people to mosquitoes, so that if a viremic individual returns to the United States, the possibility of that person contacting an appropriate vector is fairly low, and thus the potential of DENV transmission to the community is low. Even in the context of a region highly endemic for dengue, such as Puerto Rico, a higher incidence of disease is correlated with lower socioeconomic status and lack of window and door screens (149).

The risk of a dengue epidemic can be modeled mathematically by the basic reproductive rate of the virus (R_0), which corresponds to the number of subsequent infections that would be produced in a group of susceptible individuals given the introduction of one infected person (4). For vector-borne diseases, R_0 takes into account measures of transmission capacity, specifically the number of female mosquitoes per human host, the human-biting rate of the mosquito species, the proportion of bites that produce an infection, the average duration of infection in humans, the proportion of bites of viremic humans that result in infected mosquitoes, mosquito mortality rates, and latent periods of incubation in both hosts (4). When $R_0 < 1$, transmission is interrupted; $R_0 = 1$ results in endemicity; and $R_0 > 1$ results in an increase in cases (i.e., a potential epidemic). R_0 values for dengue in endemic regions are estimated to range between 1.33 and 11.6 (57). An additional factor to consider is herd immunity, or the number of immune individuals in the population, which can be represented as "p," such that $(1-p)R_0$ determines the effective R_0. This equation is useful in vaccination programs in terms of estimating the number of vaccinated individuals (p) needed to interrupt transmission ($R_0 < 1$) for a given disease (4, 37).

Risk Factors for DHF/DSS

Risk factors for developing DHF/DSS include pre-existing immunity from a previous DENV infection, time between infections, age, ethnicity and host genetic background, sequence of infecting serotypes, and viral genotype (43, 55). In response to a primary infection with DENV, protective immunity to the infecting serotype is believed to last a lifetime. As evidence, serotype-specific immunity was protective for up to 18 months in human volunteers (121), and neutralizing antibodies and serotype-specific T cells have been found in patients in Cuba, Greece, and Japan 20–40 years after an isolated dengue epidemic (51, 68, 129). However, complete cross-protective immunity from a secondary infection was present in human volunteers for only 1–2 months after a primary DENV infection, with partial immunity present up to 9 months, resulting in a milder disease of shorter duration upon reinfection (121). After the emergence of DHF/DSS, studies seeking to explain the cause of this new and more severe manifestation of dengue identified a second, heterologous DENV infection as a risk factor for DHF/DSS (58). Prospective studies in Thailand, Burma, and Indonesia (17, 29, 39, 125, 140), as well as studies of sequential epidemics of dengue in Cuba (53), confirmed the association of secondary infection with more severe disease. Evidence from Cuba has suggested that increased time between infections may also increase disease severity (53). After an isolated DENV1 epidemic in Cuba in 1977, two separate DENV2 epidemics caused by closely related Southeast Asian DENV2 strains occurred in 1981 and 1997 on the island. Strikingly, death rates were almost 40 times greater when the interval between infections was 20 years, compared with 4 years.

The pathogenic mechanism of DHF/DSS is still poorly understood. A predominant theory regarding DHF/DSS pathogenesis attributes the higher incidence of DHF/DSS among secondary infections to the phenomenon of antibody-dependent enhancement (ADE) (56). The ADE theory postulates that after an initial period of cross-reactive protection, antibodies from a primary infection remain cross-reactive with other DENV serotypes but have waned to nonneutralizing levels. These nonneutralizing antibodies could then mediate an increased uptake of virus into monocyte/macrophage cells via the Fc receptor, leading to increased viral replication and immune activation

R_0: basic reproductive rate

Herd immunity: the threshold level of collective immunity in a population, above which transmission of a particular pathogen will be disrupted and not be maintained

Neutralizing antibodies: antibodies capable of preventing infection of a cell/host by a pathogen

ADE: antibody-dependent enhancement

accompanied by cytokine release (56). Field studies have found evidence of higher levels of viremia in DHF patients, which supports the assertion that increased viral replication is associated with more severe disease (8, 83, 93, 145). A different but complementary theory of immunopathology involves reactivation of cross-reactive memory T cells specific for the previous rather than the current DENV infection, resulting in delayed viral clearance and/or increased cytokine secretion along with increased apoptosis of both infected and uninfected bystander cells (118). There is immunological evidence that this phenomenon of "original antigenic sin" may occur during secondary DENV infections (91). In both theories, cytokines are believed to play a direct role in the immunopathogenesis of DENV, owing to their proinflammatory effects on vascular endothelial cells that presumably lead to leaky junctions and increased capillary permeability (118).

Most epidemiologic studies find that children under age 15 are at increased risk for DHF/DSS, independent of other risk factors (17, 52, 61, 125, 140), which may be related to increased capillary fragility and decreased tolerance for insult to microvascular integrity in this age group (38). A few studies have indicated that Africans and people of African descent may have genetic polymorphisms that confer partial protection against severe dengue (59, 130). Recent work has identified DENV epitopes that, in the context of specific human leukocyte antigen (HLA) types, may be associated with immune enhancement (92, 131, 159). Other studies have more broadly correlated certain HLA types with disease severity and/or protection from severe disease (128, 136). Defining the link between disease risk and HLA type, race, or DENV cellular receptor (123) and cytokine polymorphisms (32) has the potential not only to provide important information regarding the pathogenesis of secondary DENV infection, but also to serve as a potential prognostic tool to identify individuals at increased risk for severe disease.

THE MOSQUITO SPHERE

Introduction

The principal vector of DENV is the *A. aegypti* mosquito, an anthropophilic species that has adapted extremely well to the urban environment, which is found both indoors and outdoors in close proximity to human dwellings (112). *A. aegypti* is believed to have originated in the jungles of Africa and was most likely spread throughout the rest of the world via slave and trading ships during the seventeenth to nineteenth centuries (112, 133). It was noted some time ago that epidemics of dengue seemed to correlate with the spread of *A. aegypti* in South and Southeast Asia, appearing first in port towns and moving inland over time along waterways (133). Now a fully domesticated mosquito, *A. aegypti* is an efficient vector of DENV because of its preference for laying its eggs in artificial containers, biting humans, and remaining indoors, where it has access to its favorite host (112).

A. albopictus is a secondary vector of DENV in Southeast Asia, the Western Pacific, and increasingly in Central and South America (40), but it has also been documented as the sole vector during certain dengue epidemics (3, 28). Prior to 1979, this species was found only in Asia and in the Western Pacific, but it has spread to much of the rest of the world in recent decades (40, 133). The invasion of North America by *A. albopictus* was first confirmed with its discovery in Houston, Texas, in 1985 (18), probably arriving in shipments of used tires from Japan (63, 70). The range of *A. albopictus* stretches farther north than that of *A. aegypti*, and its eggs are somewhat resistant to subfreezing temperatures (63), raising the possibility that *A. albopictus* could mediate a re-emergence of dengue in the United States or Europe. For example, *A. albopictus* can survive the winters in northern Italy (113) and was recently implicated in an outbreak of Chikungunya virus in Italy (117). In this section, we describe the mosquitoes that transmit dengue, the hypothesis of DENV emergence from the

HLA: human leukocyte antigen

jungles of Southeast Asia, the potential for re-emergence, variations in vector competence between *Aedes* strains, mechanisms of viral persistence in mosquitoes between epidemics, past vertical control programs for *A. aegypti*, and current community-based strategies for vector control.

The Sylvatic Cycle and Emergence/Re-Emergence of DENV

A number of forest-dwelling *Aedes* mosquitoes, known as tree-hole mosquitoes, have been identified as vectors of a sylvatic cycle of DENV in the jungles of Africa (DENV2) and Southeast Asia (DENV1–4), involving mainly nonhuman primates. The viruses in this sylvatic cycle are phylogenetically distinct from those in the urban cycle of dengue involving *A. aegypti* (147), though sylvatic strains of DENV may occasionally cause disease in humans (111, 124). It is thought that DENV emerged from the jungles of Southeast Asia, with *A. albopictus* or perhaps other *Aedes* species maintaining the virus in a sylvatic cycle involving nonhuman primates and humans living in the countryside. This hypothesis was reached in part on the basis of studies in the 1950s that documented high levels of anti-DENV antibodies in both nonhuman primates and rural inhabitants in the apparent absence of disease (120, 133, 134). Neutralizing antibodies against DENV1 and DENV2, the only serotypes known at the time, were present in about 50% of children up to 15 years of age in diverse rural communities in Malaysia, in contrast with much lower levels (3%–9%) in the cities of Singapore and Kuala Lumpur (134). In Southeast Asia in the 1950s, *A. aegypti* was still primarily found only in towns and cities, whereas *A. albopictus* was common in coastal and inland rural areas (133). Thus, the combined evidence argues for a rural source for the dengue viruses in Southeast Asia, possibly with *A. albopictus* as the primary vector.

In Africa, a nondomesticated, forest-dwelling subtype of *A. aegypti*, *A. aegypti formosus*, is present that demonstrates a low biting rate for humans; however, sylvatic DENV2 strains have been recovered primarily from other *Aedes* species in the jungles of West Africa, including *A. luteocephalus*, *A. furcifer*, *A. taylori*, and *A. vittatus* (25, 111). The potential for epidemic DENV strains to re-emerge has been addressed experimentally in several studies that suggested that viral adaptation to the vector was required for efficient transmission by *A. aegypti* and *A. albopictus* (90), but that adaptation to vertebrate hosts was not required for the emergence of DENV from a sylvatic cycle (144). Two sylvatic *Aedes* species from West Africa, *A. furcifer* and *A. luteocephalus*, were highly susceptible to both sylvatic and endemic DENV2, raising the possibility that adaptation of DENV to peridomestic mosquitoes does not necessarily result in loss of infectivity for some sylvatic *Aedes* species (26). However, both domestic and forest-dwelling *A. aegypti* from Senegal were poorly infected by sylvatic and endemic DENV strains, and another investigation found that populations of *A. aegypti formosus* from different parts of Africa were less susceptible to DENV2 than were *A. aegypti* from Southeast Asia, South America, and the South Pacific (30). These studies illustrate the complexity of the coevolution of DENV with its mosquito vectors.

Vector Competence

Variations in vector competence among strains of *A. aegypti* and between *A. aegypti* and *A. albopictus* have received a fair amount of attention. Early studies had shown that *A. aegypti*, though clearly correlated with epidemic dengue, was not as easily infected with DENV as *A. albopictus* (69, 115), leading to the hypothesis that the adaptation of DENV to *A. aegypti* had selected for viruses that caused higher viremia (112). However, other studies using field-caught mosquitoes, as opposed to laboratory strains, demonstrated comparable susceptibility between the two species (153) or a higher susceptibility by *A. aegypti* (89, 146). Although differences between the use of laboratory strains versus field-caught mosquitoes may explain some of these discrepant results

Vector competence: the capacity of a vector to transmit a pathogen by virtue of being susceptible to infection and dissemination and subsequently capable of transmission to an appropriate host

(146), the question remains what might account for different vector competence among mosquitoes, and whether coevolution between dengue viruses and *A. aegypti* has occurred since DENV emerged from the jungle. Substantial variation in susceptibility exists between different strains of both *A. aegypti* (14, 47, 138) and *A. albopictus* (48, 85) mosquitoes from different locations. This suggests that genetic differences in the vector may be responsible for varying susceptibilities to DENV, and specific quantitative trait loci (QTLs) in *A. aegypti* have been linked to vector competence (13). Recent sequencing of the *A. aegypti* genome (94) will facilitate identification of genes linked to previously described QTLs associated, for example, with midgut and other barriers to infection. Comparative genomics analyses are also now possible with sequenced *Drosophila melanogaster* and *Anopheles gambiae* genomes (148).

Vertical Transmission of DENV in Mosquitoes

In most endemic countries, dengue displays a seasonal pattern related to temperature and rainfall (33, 57, 112). This has led many investigators over the years to question how the virus overwinters, or persists during dry or cold seasons. One possibility is that a population of infected mosquitoes could survive throughout the interim and introduce the virus during the next season. *Aedes* mosquitoes remain infected with DENV for life, and the longest lifespan recorded to date is 174 days, although a more typical survival rate is 1–2 weeks (112). A second possibility is passage of the virus to the next generation of mosquitoes via survival in an infected egg. Early studies had shown no evidence of vertical transmission of DENV in *Aedes* mosquitoes (112), but more recent studies have demonstrated that vertical transmission is possible both in the laboratory and in the wild (49). Some evidence exists that *A. albopictus* mosquitoes are more efficient at vertical transmission than *A. aegypti*, which would make them a candidate for maintaining DENV during interepidemic periods

(116). Thus, vertical transmission of DENV in mosquitoes is possible, whether or not the mechanism is truly transovarial or mediated by infection of the mature egg at the time of oviposition (112). Finally, given the high number of asymptomatic cases of dengue (10, 17, 29, 140), it is also possible that silent transmission in humans by a reduced number of vectors maintains DENV transmission between epidemics.

Past Experience with Vertical Vector Control Measures

Because *A. aegypti* facilitated the emergence of epidemic dengue in urban centers around the world and is still the primary vector of dengue today, most control efforts have focused on this species. Mosquito control measures are particularly important given the current lack of dengue-specific vaccines or therapeutics (66, 71, 154), and they play a central role worldwide. A fundamental distinction in the design of a vector control program is whether it takes a government-led, vertical (top-down) approach or a community-led, horizontal (bottom-up) approach (41).

A vertical, Pan-American Health Organization–led campaign focused on controlling urban yellow fever in the mid-twentieth century succeeded in eliminating *A. aegypti* from most countries in the Americas by 1965 (135) and had the additional benefit of reducing the incidence of dengue in the region. Nonetheless, the mosquito remained in the northernmost countries of South America and in some locations in the Caribbean, as well as in the United States (135), which discontinued its program in 1969 without having achieved the goal of eradication (132). This campaign established the use of larval source reduction as a means of controlling mosquito populations, as well as three indices used to monitor larval density that are still in use today (35, 155), in particular the house index (HI). The control program also included the use of outdoor insecticidal sprays (DDT and malathion) in and around all breeding sites (98). However, since the program was discontinued in the early 1970s, *A. aegypti* has returned to almost all countries in the Americas (43).

Two other successful vertical control programs were undertaken by the governments of Singapore and Cuba. DHF was first reported in Singapore in 1960 (19), and beginning in 1968, the Vector Control Unit of the Ministry of Health established a program of entomologic surveillance, larval source reduction, public education, and law enforcement targeted to control both *A. aegypti* and *A. albopictus* (97). This program succeeded in bringing the HI down from almost 50% to approximately 2% by 1973, where it has remained until the present time. The seroprevalence of DENV infection in the general population declined to 43% in 1996, with only 6.7% of primary school children positive for anti-DENV antibodies, compared with 71% in some locations in Thailand (141), 69% in Yogyakarta, Indonesia (39), and 90% in urban Nicaragua (10). However, after 15 years of low incidence, Singapore has recently experienced a resurgence in dengue, without a concurrent change in HI values (97). This increase has been attributed in part to lowered herd immunity, increased virus transmission outside the home, and a shift in policy from vector surveillance to case detection. Another element contributing to this resurgence likely includes tens of millions of people who visit, transit through, and commute to work in Singapore every year, leading to a high potential for reintroduction by viremic individuals.

In the case of Cuba, a devastating epidemic of DHF/DSS in 1981, the first in the Americas, resulted in over 10,000 cases of severe illness and 158 deaths. The Cuban government initiated a vertical, systematic campaign aimed at eradicating the *A. aegypti* vector from the island, and *A. aegypti* was eliminated from 13 of Cuba's 14 provinces (75). Some 10,000 health workers remained committed to the control program, and for 15 years no dengue cases were reported in Cuba (74, 75). However, dengue (due to DENV2) re-emerged in Cuba in 1997, though it was detected early and confined to Santiago de Cuba. In 2000, DENV3 was isolated in Cuba for the first time, and it caused an epidemic of DF/DHF in the city of Havana during 2001–2002. Once again, the Cuban government mobilized a major vector control campaign, and every house in Havana was inspected 10 times. Starting with an HI of 0.49% at the beginning of the epidemic, within three months the house index had been reduced to 0.01% (50). In 2006, another outbreak of DF/DHF was reported from 4 of 14 provinces (105); however, few details about this epidemic are publicly known. Unfortunately, even these vector control programs that maintained an HI of less than one percent were not able to prevent the recurrence of epidemic dengue, probably due to low herd immunity combined with constant reintroduction of DENV from international visitors and Cuban medical workers returning from endemic countries.

With their past successes, Singapore and Cuba had long been considered to have model dengue control programs, owing in part to their unique political and geographical situations. These two countries implemented consistent programs and policies that made possible the long-term control of dengue, rather than relying only on emergency responses to manage epidemics. However, both locales have faced reintroductions of dengue in spite of low reported vector indices, likely due in large part to the continued influx of people from endemic regions either as tourists, migrant workers, or recipients of cultural exchange, combined with a highly susceptible native population that, ironically, resulted from the success of vector control programs in these countries.

Community-Based Vector Control Programs

In Southeast Asia, the World Health Organization established an Aedes Research Unit in Bangkok, Thailand, in 1966 to investigate control measures for *A. aegypti* after it was identified as a vector of the newly emerging epidemics of hemorrhagic fever in the region (60, 84). These measures included a new method of pesticide application, ultralow-volume spraying, to reduce adult mosquito populations (84), as well as scaling up the more labor-intensive, but highly effective, targeting of breeding sites in residences by health workers (11, 99). Originally intended as a measure to control or halt an ongoing epidemic by drastically reducing densities of adult mosquitoes in a short period of time (84), ultralow-volume spraying became widely used as a "preventive" measure in many parts of the world during the next two decades (41), despite accumulating evidence of its having little impact on reducing incidence of disease (95).

Various factors eroded the effectiveness of overreliance on both mass pesticide spraying and government-led vertical models for vector control, including increases in pesticide resistance, an awareness of the detrimental side effects of pesticide use, decreased government funding for public health services, and a push from the global health community to move toward horizontal programs integrating education and community participation (72). However, no alternative community-based models were available at the time for vector control programs. Early attempts to establish such a program in Thailand were unsustainable because the community was not involved in the program design and had no stake or understanding of the program and thus did not continue it in the absence of government support (44). Early community-based programs were designed with strong educational components; however, many of them were not successful in motivating community participation. The lesson of early control programs throughout the 1980s and early 1990s was that community-based programs need to incorporate a sense of ownership to be sustainable (44).

The key to dengue control is to close the motivational gap between community knowledge and sustainable practice in reducing mosquito breeding sites. New evidence-based methodologies focus on furnishing community members with key concepts and training so they can gather their own data, evaluate control programs, and generate and implement their own improved interventions based on successes and challenges encountered in their specific geographical and cultural settings. Thus, theoretical education about dengue and its vectors is not enough; people are motivated to change behavior by informed dialogue based on their own evidence that forms the basis for their own decisions. Evidence-based approaches, such as Communication-for-Behavioural-Impact (COMBI) (100) and the SEPA (Socializing Evidence for Participatory Action) program based on CIET methods (5) (**http://www.ciet.org/en/**), are proving more successful in effecting behavioral change and reduction of entomological indices and hold promise as community-based vector control programs, especially in conjunction with some degree of institutional support, though their long-term sustainability and impact on dengue incidence are still under evaluation (L. Lloyd, personal communication; E. Harris, J. Arostegui, J. Coloma & N. Andersson, unpublished results).

Equally important to a successful control program is the ability to effectively target and monitor *A. aegypti* populations as part of a source reduction strategy. A method of identifying highly productive breeding sites that facilitates targeted source reduction efforts has been developed to replace the more traditional larval indices, thus maximizing the effect of control measures (34, 35). This pupal/demographic survey method involves counting the numbers of pupae (the stage between larvae and adult mosquitoes) per container, thus identifying which container types are responsible for the largest output of adult mosquitoes, as well as relating the results to the density of local human populations. Input from the pupal/demographic survey can be combined with temperature and herd immunity values to create mathematical models of transmission thresholds [the container-inhabiting mosquito simulations model (CIMSiM) and the dengue simulation model (DENSIM)] that can provide target values of pupal densities to interrupt transmission for control programs (33, 34, 36). The hope is that these indices will provide a more precise ability to monitor and predict the potential for dengue epidemics than has the traditional HI, which does not necessarily correlate with dengue transmission (35). Another promising avenue has been the development of biological control methods as alternatives to pesticides, including larvivorous fish, larvae-eating copepods (*Mesocyclops*), toxins, insect growth hormones, and viruses targeted to mosquito larvae (reviewed in Reference 86). The principal goal of all vector control programs is to minimize the populations of adult mosquitoes to interrupt or at least minimize the transmission threshold, R_0. Even reducing the vector population without eliminating it can mitigate the impact of an epidemic (33).

THE VIRUS SPHERE

Introduction

The four dengue viruses fall into a distinct serogroup among the mosquito-borne flaviviruses, showing a phylogenetic relationship with the Japanese encephalitis virus group and more distantly with yellow fever virus (77, 82). The current DENV serogroup progenitor is estimated to have arisen approximately 1000 years ago using molecular clock techniques (142), and most phylogenies show that DENV4 is the most divergent serotype, followed by DENV2, and then DENV1 and DENV3 as the most closely related serotypes (82, 142, 157). Phylogenetic analysis of sylvatic and endemic/epidemic strains suggests that each serotype emerged separately from a sylvatic ancestor (147), and this emergence is estimated to have occurred about 125–320 years ago, varying by serotype (142).

Based on sequences of the complete envelope (E) gene or the E-NS1 boundary using a cutoff of 6% divergence (106), DENV1 is currently divided into four to five genotypes, including a sylvatic clade (27, 158). DENV2 is divided into six subtypes, designated as Sylvatic, American, Cosmopolitan, Asian 1, Asian 2, and Asian-American (27, 64, 82, 147), although the two Asian subtypes have on occasion been collapsed into a single Asian genotype (108). DENV3 has been divided into four genotypes (I–IV) (64, 80, 87), sometimes including a genotype V (27). Finally, DENV4 is divided into two endemic genotypes (I–II) and one sylvatic genotype and shows the least genetic diversity among the serotypes, at least among available strains (27, 64, 79, 147). Overall, as further sequences become available, these genotypic structures are likely to be revised, possibly with the appearance of a new genotype or the collapse of two or more genotypes into one. In this section, we discuss the association of certain DENV genotypes or subtypes with disease severity or increased fitness in the context of host immunity.

Viral Genotypes and Virulence

With only 62%–67% homology based on amino acid sequences (152), the four dengue viruses could have been classified as separate viral groups but instead are treated as four DENV serotypes pertaining to a single group. Nonetheless, differences in severity associated with individual serotypes or particular sequences of serotypes in sequential infection have been observed, and it remains an open question whether some serotypes are inherently more pathogenic than others. DENV2 viruses have most commonly been associated with DHF/DSS (9, 17, 53, 96, 125, 140), along with DENV1 and DENV3 viruses (9, 39, 50, 62, 88); DENV4 appears to be the most clinically mild, although it too can cause severe disease (96). DENV2 and DENV4 have been associated with increased disease severity as a secondary infection, whereas DENV1 and DENV3 seem to cause more severe disease in primary infection than do the other two serotypes (9, 62, 96, 145).

In most studies, secondary infection by any of the four DENV serotypes remains the greatest risk factor for severe disease (56). Nonetheless, the association of some DENV genotypes with increased disease severity, whether or not in the context of secondary infection, has now been well documented, in particular involving certain genotypes of DENV2 and DENV3. In general, Southeast Asia appears to serve as a source for viral diversity, generating

UTR: untranslated region

a multitude of strains, some of which are inherently more virulent and perhaps more successful than others, as evidenced by their worldwide spread and possible displacement of earlier DENV strains. Compelling evidence from phylogenetic studies suggests that only DENV2 strains that originated in Southeast Asia are associated with DHF/DSS in the Americas, and not the native American strains that were originally imported from the South Pacific (27, 109). Subsequent functional analysis revealed that Thai DENV strains (Asian genotype) replicated to higher titers than American genotype DENV2 strains in human monocyte-derived macrophages and dendritic cells (21, 102). Full-length sequencing of Asian and American genotypes revealed several key nucleotide differences, particularly at position 390 in the E protein and in the 5′ and 3′ untranslated regions (UTRs) (81). Substitution of N390 found in the Asian genotype by the American genotype D390 reduced virus output from both human monocyte-derived macrophage and dendritic cell cultures (21, 102), and this reduction was enhanced by replacing the Asian genotype 5′ and 3′ UTRs with those from the American genotype (21). The Asian DENV2 strains also disseminated in a larger percentage of field-caught *A. aegypti* mosquitoes compared with American DENV2 strains (6), and when the mosquitoes were coinoculated with equal titers of Asian and American strains, the Asian strains were consistently recovered from a larger percentage of mosquitoes than were the American strains (20). Thus, it is possible that the success of the Southeast Asian DENV2 strains is due in part to more efficient replication in human target cells as well increased transmission by vector mosquitoes. Only one exception to this paradigm has been reported; Shurtleff et al. (127) described the association of DHF with an American genotype DENV2 from Venezuela, as determined by analysis of the 3′ UTR.

Another recent example involves a clade replacement within the DENV2 Asian-American genotype identified by phylogenetic analysis of full-length genomes from Nicaraguan patients; interestingly, this correlates temporally with a large increase in disease severity, and the new clade is significantly associated with severe disease (A. Balmaseda, T. Gomez, M. Henn, C. Rocha & E. Harris, unpublished results). The mechanism(s) responsible for the increased fitness and/or virulence of the new DENV2 clade is currently under investigation. Although an increase in viral virulence must be considered in the context of host immunity, the possibility exists that more virulent dengue viruses will continue to evolve in Southeast Asia and spread worldwide, displacing more benign genotypes in the years to come (107, 157).

The DENV3 serotype provides another convincing example of how increased viral diversity has led to the emergence or evolution of a clade of viruses strongly associated with DHF/DSS. DENV3 genotype III includes isolates from East Africa, South Asia, and Latin America and has been associated with an increase in DHF/DSS in these regions (27, 87). Emergence of epidemic DHF/DSS in Sri Lanka in 1989 led investigators to question the reasons for this occurrence. After eliminating the possibility of a general increase in virus circulation or a change in serotype prevalence on the island (88), the decisive factor was identified as a clade replacement event (87). Both DENV3 III subtypes A and B were present in Sri Lanka in 1989, but only one subtype (B) persisted after 1989 and was involved in all subsequent cases of DHF/DSS on the island. Additionally, DENV3 III subtype B has since spread to the Americas, where it has also been associated with epidemics of DHF/DSS (27, 50, 87). Other genotypes of DENV3 may also be associated with increased severity of disease; genotype I viruses reintroduced into islands of the Western Pacific have been associated with DHF/DSS, in contrast to past epidemics of DF associated with genotype IV (80). As viral strains with increased virulence are identified via the marriage of phylogenetic and epidemiologic analyses, the next challenge will be to define the molecular basis of this increased pathogenesis. With this information in hand, a combination of active surveillance for and rapid detection of viral genotypes with potential for increased virulence could help identify at-risk populations and individuals, respectively.

Virus Evolution and Host Immunity

Intriguing evidence to explain differences in viral virulence in relation to pre-existing host immunity of native and introduced DENV2 genotypes derives from observations in Iquitos, Peru. After years of DENV1 circulation, a large number of secondary infections with DENV2 were documented in Iquitos in the complete absence of severe dengue (150), which was unexpected given the increased risk of DHF/DSS typically observed in secondary DENV infections. The DENV2 genotype present in Iquitos belonged to the native American genotype (150), in contrast to the DENV2 strains that have caused epidemics of DHF/DSS in the Caribbean and throughout Latin America (27, 61, 75, 109). The question arose whether the lack of DHF/DSS due to these secondary DENV2 infections with native American strains was caused by an inherent lack of virulence compared with Southeast Asian DENV2 strains, and/or whether anti-DENV1 antibodies present among the population of Iquitos neutralized or at least mitigated secondary "American" DENV2 infection by virtue of cross-reacting, neutralizing antibodies. In fact, sera from Iquitos residents characterized by a monotypic, anti-DENV1 response have higher cross-reactive neutralizing titers against American DENV2 strains than against Asian DENV2 strains (73). Antibodies arising from a DENV1 infection in Cuban patients also demonstrated higher neutralizing ability against the American DENV2 genotype than against the Asian DENV2 genotype (51), raising the possibility that the Asian DENV2 strains have epitopes divergent from those that may be shared between DENV1 and American DENV2 strains (73).

Several investigators have taken advantage of detailed information available from Bangkok, Thailand, which has remained hyperendemic for all four DENV serotypes since at least 1958

(60), to tease out the correlations between the periodicity of dengue epidemics and potential cross-protection between serotypes. Serotype- and severity-specific data collected between 1973 and 1999 showed that each serotype displays a somewhat different pattern of oscillation across this time period, and that together the four serotypes exhibit rather complex dynamics (96). Mathematical models have been designed to test the theory that the interaction between the periodicity of alternating epidemics due to different serotypes and host immunity can explain the patterns seen in Bangkok (2, 151). One model describes a scenario in which temporary cross-immunity between serotypes and seasonal fluctuations in vector populations explains serotype dynamics in Bangkok (151), and posits that ADE and differences in viral virulence are less important in shaping patterns of transmission (although it does not exclude both playing a key role in disease). Another model postulates that moderate cross-immunity alone can explain the oscillations and periodicity of individual serotypes (2), and that clade replacement events seen within each serotype are also associated with serotype-specific periodicity in combination with cross-reactive protection (2, 158). In particular, the authors propose that clade replacements within DENV1 serotypes in Thailand are best explained by a combination of mutations fixed by stochastic events plus cross-protective immunity to an incipient increase in DENV4 (158), as these two serotypes show a striking out-of-phase periodicity with one another.

The studies discussed above would suggest that viral evolution must then be considered in the context of cocirculation of different serotypes and the presence of cross-protective immunity. Alternatively, other theories have been proposed suggesting that the phenomenon of ADE could explain the periodicity and alternating epidemics caused by multiple serotypes in Mexico and Thailand (24, 31). Although most studies of DENV evolution have shown strong evidence for negative or purifying selection (157, 158), support for adaptive evolution has been reported

Extrinsic incubation period: the latent period in a vector mosquito before the virus has disseminated to the salivary glands, from where it can be transmitted to a vertebrate host as the mosquito takes a blood meal

(15, 16, 143), as well as some examples of recombination (1, 65). Even so, most studies suggest that positive selection and recombination have not played a decisive role in the overall evolution of DENV. To date, it appears that a combination of random genetic drift, rapid evolution, an ever-increasing number of infections, and perhaps the sporadic selectively driven replacement of viral clades characterizes DENV evolution, along with a complex interaction with serotype-specific host immunity that is only now beginning to be unraveled (2, 79, 114, 157). A better understanding of this last interaction is crucial in the face of imminent large-scale tetravalent dengue vaccine trials, because selection pressures due to host immunity will be greatly increased by trials and the eventual implementation of dengue vaccines.

OTHER FACTORS

Climate

The term climate change refers to multiyear, large-scale changes in climate patterns, including fluctuations in both rainfall and temperature; global warming refers to an increase in the average global temperature related to the greenhouse effect, whereby solar radiation is trapped beneath the earth's atmosphere. Changes in the composition of the atmosphere have been predicted to lead to a 2.0°C–4.5°C rise in global temperatures by the year 2100 (126), which could have an impact on vector-borne diseases (137). There has been a great deal of debate on the implications of global warming for human health (22, 139). Models that discuss the specific impact on dengue focus particularly on humidity (54) and temperature (101) in an attempt to predict the impact on the geographic range of mosquito vectors. Other perspectives highlight the overriding importance of infrastructure and socioeconomic differences that exist today and already prevent the transmission of vector-borne diseases, including dengue, even in the continued presence of their vectors (104). At the moment, there is

no consensus, but in the case of dengue it is important to note that even if global warming does not cause the mosquito vectors to expand their geographic range, there could still be a significant impact on transmission in endemic regions. For instance, a 2°C increase in temperature would simultaneously lengthen the lifespan of the mosquito and shorten the extrinsic incubation period of DENV, resulting in more infected mosquitoes for a longer period of time (33).

Public Policy

A great deal of work is currently directed toward the development of tetravalent dengue vaccines and specific antivirals, which will hopefully provide additional tools for reducing the health impact of dengue in the near to mid-term future (66, 71, 154). At present, sustainable vector control programs that can maintain low mosquito densities, as well as good surveillance programs that can quickly identify incipient epidemics and thus trigger mobilization of emergency control measures, will continue to be our most important tools for controlling dengue for some time. Inevitably, these measures will face the challenge that much of public health faces—how to convince both policymakers and residents that only their vigilance now can prevent the need to cope with large epidemics in the future. Dengue will continue to be a challenge for public health officials and policymakers for the reasons discussed in this review, namely, increases in human population, urbanization, and international travel; the plethora of mosquito habitats due to daunting challenges in vector control; and the increasing occurrence of DF and DHF/DSS epidemics related in part to changes in viral virulence and to host immune status. Although we understand the general principles behind the spread and persistence of dengue, and further research questions remain to be explored, knowledge alone is not enough. The overriding question is, Can we take this knowledge and use it to contain or reverse the trend, or will the prevalence of dengue continue to increase in the years to come?

SUMMARY POINTS

1. The incidence of DF and of DHF/DSS has increased dramatically in the past 50 years, and key reasons for this increase include population growth, uncontrolled urbanization, spread of the mosquito vectors, and movement of the virus in conjunction with the rapid transit of people around the globe.

2. The risk for acquiring dengue relates foremost to the host's immune status and exposure to an infected mosquito. Risk factors for DHF/DSS include most importantly previous exposure to a heterotypic dengue virus, as well as the time between infections, age, ethnicity and genetic background, and the genotype and serotype of the infecting virus.

3. Originally found in the jungles and rural areas of Southeast Asia, dengue virus is now maintained primarily in an urban cycle involving humans and *A. aegypti* and *A. albopictus* mosquitoes, and the challenge of controlling urban breeding sites for these vectors has hindered progress in containing the dengue pandemic.

4. Dengue virus may be maintained between epidemic cycles by silent transmission in humans (asymptomatic infections) and/or vertical transmission or overwintering in the mosquito vectors.

5. Past vertical or government-led vector control programs have been somewhat successful using intensive source reduction techniques combined with targeted insecticide use. Ultimately, these programs are either unsustainable and/or unable to prevent dengue transmission. New approaches that encompass both community participation and targeting of highly productive breeding containers via pupal/demographic surveys hold promise to minimize dengue epidemics in the future.

6. Some genotypes and subtypes of dengue virus appear to cause more severe disease; functional analysis in vitro confirms that certain strains may be inherently more virulent in both mammalian cells and mosquitoes. However, there is also clearly a role for host immunity in determining the fitness of dengue virus strains and thus influencing viral evolution.

7. Although development and evaluation of dengue-specific vaccines and therapeutics are currently underway, these tools will not be available for general use in the immediate future. Therefore, our best hope for confronting the continued spread of dengue at the moment is to use the knowledge we already have to design more effective control measures, while pursuing remaining research questions that will allow the design of more effective measures in the future through a better understanding of the complex interaction of human, mosquito, and viral biology.

DISCLOSURE STATEMENT

The authors are not aware of any biases that might be perceived as affecting the objectivity of this review.

ACKNOWLEDGMENTS

The authors wish to thank Josefina Coloma, Scott Balsitis, and Eddie Holmes for critical reading of the manuscript.

LITERATURE CITED

1. Aaskov J, Buzacott K, Field E, Lowry K, Berlioz-Arthaud A, Holmes EC. 2007. Multiple recombinant dengue type 1 viruses in an isolate from a dengue patient. *J. Gen. Virol.* 88:3334–40

2. **Adams B, Holmes EC, Zhang C, Mammen MP Jr, Nimmannitya S, et al. 2006. Cross-protective immunity can account for the alternating epidemic pattern of dengue virus serotypes circulating in Bangkok. *Proc. Natl. Acad. Sci. USA* 103:14234–39**

3. Ali M, Wagatsuma Y, Emch M, Breiman RF. 2003. Use of a geographic information system for defining spatial risk for dengue transmission in Bangladesh: role for *Aedes albopictus* in an urban outbreak. *Am. J. Trop. Med. Hyg.* 69:634–40

4. Anderson R, May R. 1991. *Infectious Diseases of Humans: Dynamics and Control.* Oxford: Oxford Univ. Press

5. Andersson N. 1996. *Evidence-based planning: the philosophy and methods of sentinel community surveillance.* Washington, DC: World Bank Econ. Dev. Inst.

6. Armstrong PM, Rico-Hesse R. 2003. Efficiency of dengue serotype 2 virus strains to infect and disseminate in *Aedes aegypti. Am. J. Trop. Med. Hyg.* 68:539–44

7. Ashburn P, Craig C. 1907. Experimental investigations regarding the etiology of dengue fever. *J. Infect. Dis.* 4:440–75

8. Avirutnan P, Punyadee N, Noisakran S, Komoltri C, Thiemmeca S, et al. 2006. Vascular leakage in severe dengue virus infections: a potential role for the nonstructural viral protein NS1 and complement. *J. Infect. Dis.* 193:1078–88

9. Balmaseda A, Hammond SN, Perez L, Tellez Y, Saborio SI, et al. 2006. Serotype-specific differences in clinical manifestations of dengue. *Am. J. Trop. Med. Hyg.* 74:449–56

10. Balmaseda A, Hammond SN, Tellez Y, Imhoff L, Rodriguez Y, et al. 2006. High seroprevalence of antibodies against dengue virus in a prospective study of schoolchildren in Managua, Nicaragua. *Trop. Med. Int. Health* 11:935–42

11. Bang YH, Pant CP. 1972. A field trial of Abate larvicide for the control of *Aedes aegypti* in Bangkok, Thailand. *Bull. WHO* 46:416–25

12. Beatty M, Letson W, Edgil D, Margolis H. 2007. *Estimating the total world population at risk for locally acquired dengue infection.* Abstract presented at Annu. Meet. Am. Soc. Trop. Med. Hyg., 56th, Philadelphia

13. Bennett KE, Flick D, Fleming KH, Jochim R, Beaty BJ, Black WC. 2005. Quantitative trait loci that control dengue-2 virus dissemination in the mosquito *Aedes aegypti. Genetics* 170:185–94

14. Bennett KE, Olson KE, Muñoz Mde L, Fernandez-Salas I, Farfan-Ale JA, et al. 2002. Variation in vector competence for dengue 2 virus among 24 collections of *Aedes aegypti* from Mexico and the United States. *Am. J. Trop. Med. Hyg.* 67:85–92

15. Bennett SN, Holmes EC, Chirivella M, Rodriguez DM, Beltran M, et al. 2003. Selection-driven evolution of emergent dengue virus. *Mol. Biol. Evol.* 20:1650–58

16. Bennett SN, Holmes EC, Chirivella M, Rodriguez DM, Beltran M, et al. 2006. Molecular evolution of dengue 2 virus in Puerto Rico: positive selection in the viral envelope accompanies clade reintroduction. *J. Gen. Virol.* 87:885–93

17. Burke DS, Nisalak A, Johnson DE, Scott RM. 1988. A prospective study of dengue infections in Bangkok. *Am. J. Trop. Med. Hyg.* 38:172–80

18. CDC. 1986. Epidemiologic notes and reports *Aedes albopictus* introduction—Texas. *MMWR* 35:141–42

19. Chew A, Leng GA, Yuen H, Teik KO, Kiat LY, et al. 1961. A haemorrhagic fever in Singapore. *Lancet* 1:307–10

20. Cologna R, Armstrong PM, Rico-Hesse R. 2005. Selection for virulent dengue viruses occurs in humans and mosquitoes. *J. Virol.* 79:853–59

21. Cologna R, Rico-Hesse R. 2003. American genotype structures decrease dengue virus output from human monocytes and dendritic cells. *J. Virol.* 77:3929–38

22. Colwell RR, Epstein PR, Gubler D, Maynard N, McMichael AJ, et al. 1998. Climate change and human health. *Science* 279:968–69

23. Cummings DA, Irizarry RA, Huang NE, Endy TP, Nisalak A, et al. 2004. Travelling waves in the occurrence of dengue haemorrhagic fever in Thailand. *Nature.* 427:344–47

2. Showed that temporary cross-immunity can explain oscillations and periodicity of individual DENV serotypes in Thailand and linked clade replacement events to potential cross-reactive protection between serotypes.

24. Cummings DA, Schwartz IB, Billings L, Shaw LB, Burke DS. 2005. Dynamic effects of antibody-dependent enhancement on the fitness of viruses. *Proc. Natl. Acad. Sci. USA* 102:15259–64

25. Diallo M, Ba Y, Sall AA, Diop OM, Ndione JA, et al. 2003. Amplification of the sylvatic cycle of dengue virus type 2, Senegal, 1999–2000: entomologic findings and epidemiologic considerations. *Emerg. Infect. Dis.* 9:362–67

26. Diallo M, Sall AA, Moncayo AC, Ba Y, Fernandez Z, et al. 2005. Potential role of sylvatic and domestic African mosquito species in dengue emergence. *Am. J. Trop. Med. Hyg.* 73:445–49

27. Diaz FJ, Black WC, Farfan-Ale JA, Lorono-Pino MA, Olson KE, Beaty BJ. 2006. Dengue virus circulation and evolution in Mexico: a phylogenetic perspective. *Arch. Med. Res.* 37:760–73

28. Effler PV, Pang L, Kitsutani P, Vorndam V, Nakata M, et al. 2005. Dengue fever, Hawaii, 2001–2002. *Emerg. Infect. Dis.* 11:742–49

29. Endy TP, Chunsuttiwat S, Nisalak A, Libraty DH, Green S, et al. 2002. Epidemiology of inapparent and symptomatic acute dengue virus infection: a prospective study of primary school children in Kamphaeng Phet, Thailand. *Am. J. Epidemiol.* 156:40–51

30. Failloux AB, Vazeille M, Rodhain F. 2002. Geographic genetic variation in populations of the dengue virus vector *Aedes aegypti*. *J. Mol. Evol.* 55:653–63

31. Ferguson N, Anderson R, Gupta S. 1999. The effect of antibody-dependent enhancement on the transmission dynamics and persistence of multiple-strain pathogens. *Proc. Natl. Acad. Sci. USA* 96:790–94

32. Fernandez-Mestre MT, Gendzekhadze K, Rivas-Vetencourt P, Layrisse Z. 2004. TNF-alpha-308A allele, a possible severity risk factor of hemorrhagic manifestation in dengue fever patients. *Tissue Antigens* 64:469–72

33. Focks D, Barrera R. 2007. Dengue transmission dynamics: assessment and implications for control. In *Report of the Scientific Working Group Meeting on Dengue*, pp. 92–108. Geneva: WHO

34. Focks DA, Brenner RJ, Hayes J, Daniels E. 2000. Transmission thresholds for dengue in terms of *Aedes aegypti* pupae per person with discussion of their utility in source reduction efforts. *Am. J. Trop. Med. Hyg.* 62:11–18

35. Focks DA, Chadee DD. 1997. Pupal survey: an epidemiologically significant surveillance method for *Aedes aegypti*: an example using data from Trinidad. *Am. J. Trop. Med. Hyg.* 56:159–67

36. **Focks DA, Daniels E, Haile DG, Keesling JE. 1995. A simulation model of the epidemiology of urban dengue fever: literature analysis, model development, preliminary validation, and samples of simulation results. *Am. J. Trop. Med. Hyg.* 53:489–506**

37. Fox J, Elveback L, Scott W, Gatewood L, Ackerman E. 1971. Herd immunity: basic concept and relevance to public health immunization practices. *Am. J. Hyg.* 94:179–89

38. Gamble J, Bethell D, Day NP, Loc PP, Phu NH, et al. 2000. Age-related changes in microvascular permeability: a significant factor in the susceptibility of children to shock? *Clin. Sci.* 98:211–16

39. Graham RR, Juffrie M, Tan R, Hayes CG, Laksono I, et al. 1999. A prospective seroepidemiologic study on dengue in children four to nine years of age in Yogyakarta, Indonesia. I. Studies in 1995–1996. *Am. J. Trop. Med. Hyg.* 61:412–19

40. Gratz NG. 2004. Critical review of the vector status of *Aedes albopictus*. *Med. Vet. Entomol.* 18:215–27

41. Gubler DJ. 1989. *Aedes aegypti* and *Aedes aegypti*-borne disease control in the 1990s: top down or bottom up. Charles Franklin Craig Lecture. *Am. J. Trop. Med. Hyg.* 40:571–78

42. **Gubler DJ. 1997. Dengue and dengue hemorrhagic fever: its history and resurgence as a global public health problem. See Ref. 45, pp. 1–22**

43. Gubler DJ. 1998. Dengue and dengue hemorrhagic fever. *Clin. Microbiol. Rev.* 11:480–96

44. Gubler DJ, Clark GG. 1996. Community involvement in the control of *Aedes aegypti*. *Acta Trop.* 61:169–79

45. Gubler DJ, Kuno G, eds. 1997. *Dengue and Dengue Hemorrhagic Fever*. Wallingford, UK: CAB International

46. Gubler DJ, Meltzer M. 1999. Impact of dengue/dengue hemorrhagic fever on the developing world. *Adv. Virus Res.* 53:35–70

47. Gubler DJ, Nalim S, Tan R, Saipan H, Sulianti Saroso J. 1979. Variation in susceptibility to oral infection with dengue viruses among geographic strains of *Aedes aegypti*. *Am. J. Trop. Med. Hyg.* 28:1045–52

48. Gubler DJ, Rosen L. 1976. Variation among geographic strains of *Aedes albopictus* in susceptibility to infection with dengue viruses. *Am. J. Trop. Med. Hyg.* 25:318–25

36. Developed weather-driven simulation model of urban dengue transmission that allows customization by input of location-specific parameters such as human demographics, herd immunity, and mosquito density.

42. Comprehensive review of the historical incidence of dengue and reasons for its dramatic expansion shortly after World War II.

49. Gunther J, Martinez-Munoz JP, Perez-Ishiwara DG, Salas-Benito J. 2007. Evidence of vertical transmission of dengue virus in two endemic localities in the state of Oaxaca, Mexico. *Intervirology* 50:347–52
50. Guzman M, Pelaez O, Kouri G, Quintana I, Vazquez S, et al. 2006. Caracterizacion final y lecciones de la epidemia de dengue 3 en Cuba, 2001–2002. *Rev. Panam. Salud Publica* 19:282–89
51. Guzman MG, Alvarez M, Rodriguez-Roche R, Bernardo L, Montes T, et al. 2007. Neutralizing antibodies after infection with dengue 1 virus. *Emerg. Infect. Dis.* 13:282–86
52. Guzman MG, Kouri G, Bravo J, Valdes L, Vazquez S, Halstead SB. 2002. Effect of age on outcome of secondary dengue 2 infections. *Int. J. Infect. Dis.* 6:118–24
53. Guzman MG, Kouri G, Valdes L, Bravo J, Vazquez S, Halstead SB. 2002. Enhanced severity of secondary dengue-2 infections: death rates in 1981 and 1997 Cuban outbreaks. *Rev. Panam. Salud Publica* 11:223–27
54. Hales S, de Wet N, Maindonald J, Woodward A. 2002. Potential effect of population and climate changes on global distribution of dengue fever: an empirical model. *Lancet* 360:830–34
55. Halstead S. 1997. Epidemiology of dengue and dengue hemorrhagic fever. See Ref. 45, pp. 23–44
56. Halstead SB. 2007. Dengue. *Lancet* 370:1644–52
57. Halstead SB. 2008. Dengue virus-mosquito interactions. *Annu. Rev. Entomol.* 53:273–91
58. **Halstead SB, Nimmannitya S, Cohen SN. 1970. Observations related to pathogenesis of dengue hemorrhagic fever. IV. Relation of disease severity to antibody response and virus recovered. *Yale J. Biol. Med.* 42:311–28**
59. Halstead SB, Streit TG, Lafontant JG, Putvatana R, Russell K, et al. 2001. Haiti: absence of dengue hemorrhagic fever despite hyperendemic dengue virus transmission. *Am. J. Trop. Med. Hyg.* 65:180–83
60. Hammon WM, Rudnick A, Sather GE. 1960. Viruses associated with epidemic hemorrhagic fevers of the Philippines and Thailand. *Science* 131:1102–3
61. Hammond SN, Balmaseda A, Perez L, Tellez Y, Saborio SI, et al. 2005. Differences in dengue severity in infants, children, and adults in a 3-year hospital-based study in Nicaragua. *Am. J. Trop. Med. Hyg.* 73:1063–70
62. Harris E, Videa E, Perez L, Sandoval E, Tellez Y, et al. 2000. Clinical, epidemiologic, and virologic features of dengue in the 1998 epidemic in Nicaragua. *Am. J. Trop. Med. Hyg.* 63:5–11
63. **Hawley WA, Reiter P, Copeland RS, Pumpuni CB, Craig GB Jr. 1987. *Aedes albopictus* in North America: probable introduction in used tires from northern Asia. *Science* 236:1114–16**
64. **Holmes EC, Twiddy SS. 2003. The origin, emergence and evolutionary genetics of dengue virus. *Infect. Genet. Evol.* 3:19–28**
65. Holmes EC, Worobey M, Rambaut A. 1999. Phylogenetic evidence for recombination in dengue virus. *Mol. Biol. Evol.* 16:405–9
66. Hombach J. 2007. Vaccines against dengue: a review of current candidate vaccines at advanced development stages. *Rev. Panam. Salud Publica* 21:254–60
67. Hotta S. 1952. Experimental studies on dengue. I. Isolation, identification and modification of the virus. *J. Infect. Dis.* 90:1–9
68. Innis B. 1997. Antibody responses to dengue virus infection. See Ref. 45, pp. 221–44
69. Jumali, Sunarto, Gubler DJ, Nalim S, Eram S, Sulianti Saroso J. 1979. Epidemic dengue hemorrhagic fever in rural Indonesia. III. Entomological studies. *Am. J. Trop. Med. Hyg.* 28:717–24
70. Kambhampati S, Black WC, Rai KS. 1991. Geographic origin of the US and Brazilian *Aedes albopictus* inferred from allozyme analysis. *Heredity* 67(Pt. 1):85–93
71. Keller TH, Chen YL, Knox JE, Lim SP, Ma NL, et al. 2006. Finding new medicines for flaviviral targets. *Novartis Found. Symp.* 277:102–14
72. Knudsen AB, Slooff R. 1992. Vector-borne disease problems in rapid urbanization: new approaches to vector control. *Bull. WHO* 70:1–6
73. Kochel TJ, Watts DM, Halstead SB, Hayes CG, Espinoza A, et al. 2002. Effect of dengue-1 antibodies on American dengue-2 viral infection and dengue haemorrhagic fever. *Lancet* 360:310–12
74. Kouri G, Guzman MG, Valdes L, Carbonel I, del Rosario D, et al. 1998. Reemergence of dengue in Cuba: a 1997 epidemic in Santiago de Cuba. *Emerg. Infect. Dis.* 4:89–92
75. Kouri GP, Guzman MG, Bravo JR, Triana C. 1989. Dengue haemorrhagic fever/dengue shock syndrome: lessons from the Cuban epidemic, 1981. *Bull. WHO* 67:375–80

58. One of the first studies to document enhanced disease during secondary DENV infection.

63. First description of the means of transport and origin of *A. albopictus* in the United States.

64. Comprehensive discussion of DENV evolution.

76. Kuno G. 1997. Factors influencing the transmission of dengue viruses. See Ref. 45, pp. 61–88

77. Kuno G, Chang GJ, Tsuchiya KR, Karabatsos N, Cropp CB. 1998. Phylogeny of the genus Flavivirus. *J. Virol.* 72:73–83

78. Kurane I, Ennis FE. 1992. Immunity and immunopathology in dengue virus infections. *Semin. Immunol.* 4:121–27

79. Lanciotti RS, Gubler DJ, Trent DW. 1997. Molecular evolution and phylogeny of dengue-4 viruses. *J. Gen. Virol.* 78(Pt. 9):2279–84

80. Lanciotti RS, Lewis JG, Gubler DJ, Trent DW. 1994. Molecular evolution and epidemiology of dengue-3 viruses. *J. Gen. Virol.* 75(Pt. 1):65–75

81. **Leitmeyer KC, Vaughn DW, Watts DM, Salas R, Villalobos I, et al. 1999. Dengue virus structural differences that correlate with pathogenesis. *J. Virol.* 73:4738–47**

81. First demonstration of the molecular basis of enhanced virulence of Southeast Asian DENV2 viruses.

82. Lewis JA, Chang GJ, Lanciotti RS, Kinney RM, Mayer LW, et al. 1993. Phylogenetic relationships of dengue-2 viruses. *Virology* 197:216–24

83. Libraty DH, Endy TP, Houng HS, Green S, Kalayanarooj S, et al. 2002. Differing influences of virus burden and immune activation on disease severity in secondary dengue-3 virus infections. *J. Infect. Dis.* 185:1213–21

84. Lofgren CS, Ford HR, Tonn RJ, Jatanasen S. 1970. The effectiveness of ultralow-volume applications of malathion at a rate of 6 US fluid ounces per acre in controlling *Aedes aegypti* in a large-scale test at Nakhon Sawan, Thailand. *Bull. WHO* 42:15–25

85. Lourenco de Oliveira R, Vazeille M, de Filippis AM, Failloux AB. 2003. Large genetic differentiation and low variation in vector competence for dengue and yellow fever viruses of *Aedes albopictus* from Brazil, the United States, and the Cayman Islands. *Am. J. Trop. Med. Hyg.* 69:105–14

86. McCall P, Kittayapong P. 2007. Control of dengue vectors: tools and strategies. In *Report of the Scientific Working Group Meeting on Dengue*, pp. 110–19. Geneva: WHO

87. **Messer WB, Gubler DJ, Harris E, Sivananthan K, de Silva AM. 2003. Emergence and global spread of a dengue serotype 3, subtype III virus. *Emerg. Infect. Dis.* 9:800–9**

87. First demonstration of a clade replacement event in DENV3 linked to emergence of DHF/DSS.

88. Messer WB, Vitarana UT, Sivananthan K, Elvtigala J, Preethimala LD, et al. 2002. Epidemiology of dengue in Sri Lanka before and after the emergence of epidemic dengue hemorrhagic fever. *Am. J. Trop. Med. Hyg.* 66:765–73

89. Mitchell CJ, Miller BR, Gubler DJ. 1987. Vector competence of *Aedes albopictus* from Houston, Texas, for dengue serotypes 1 to 4, yellow fever and Ross River viruses. *J. Am. Mosq. Control Assoc.* 3:460–65

90. Moncayo AC, Fernandez Z, Ortiz D, Diallo M, Sall A, et al. 2004. Dengue emergence and adaptation to peridomestic mosquitoes. *Emerg. Infect. Dis.* 10:1790–96

91. Mongkolsapaya J, Dejnirattisai W, Xu XN, Vasanawathana S, Tangthawornchaikul N, et al. 2003. Original antigenic sin and apoptosis in the pathogenesis of dengue hemorrhagic fever. *Nat. Med.* 9:921–27

92. Mongkolsapaya J, Duangchinda T, Dejnirattisai W, Vasanawathana S, Avirutnan P, et al. 2006. T cell responses in dengue hemorrhagic fever: Are cross-reactive T cells suboptimal? *J. Immunol.* 176:3821–29

93. Murgue B, Roche C, Chungue E, Deparis X. 2000. Prospective study of the duration and magnitude of viraemia in children hospitalised during the 1996–1997 dengue-2 outbreak in French Polynesia. *J. Med. Virol.* 60:432–38

94. Nene V, Wortman JR, Lawson D, Haas B, Kodira C, et al. 2007. Genome sequence of *Aedes aegypti*, a major arbovirus vector. *Science* 316:1718–23

95. Newton EA, Reiter P. 1992. A model of the transmission of dengue fever with an evaluation of the impact of ultralow volume (ULV) insecticide applications on dengue epidemics. *Am. J. Trop. Med. Hyg.* 47:709–20

96. Nisalak A, Endy TP, Nimmannitya S, Kalayanarooj S, Thisayakorn U, et al. 2003. Serotype-specific dengue virus circulation and dengue disease in Bangkok, Thailand from 1973 to 1999. *Am. J. Trop. Med. Hyg.* 68:191–202

97. Ooi EE, Goh KT, Gubler DJ. 2006. Dengue prevention and 35 years of vector control in Singapore. *Emerg. Infect. Dis.* 12:887–93

98. PAHO. 1994. *Dengue and Dengue Hemorrhagic Fever in the Americas: Guidelines for Prevention and Control.* Washington, DC: Pan American Health Organization

99. Pant CP, Mount GA, Jatanasen S, Mathis HL. 1971. Ultra-low-volume ground aerosols of technical malathion for the control of *Aedes aegypti* L. *Bull. WHO* 45:805–17

100. Parks W, Lloyd L. 2004. *Planning social mobilization and communication for dengue fever prevention and control: a step-by-step guide*. Geneva: WHO. **http://www.who.int/tdr/publications/publications/planning_dengue.htm**

101. Patz JA, Martens WJ, Focks DA, Jetten TH. 1998. Dengue fever epidemic potential as projected by general circulation models of global climate change. *Environ. Health Perspect.* 106:147–53

102. Pryor MJ, Carr JM, Hocking H, Davidson AD, Li P, Wright PJ. 2001. Replication of dengue virus type 2 in human monocyte-derived macrophages: comparisons of isolates and recombinant viruses with substitutions at amino acid 390 in the envelope glycoprotein. *Am. J. Trop. Med. Hyg.* 65:427–34

103. Quintos F, Lim L, Juliano L, Reyes A, Lacson P. 1954. Hemorrhagic fever observed among children in the Philippines. *Philipp. J. Pediatr.* 3:1–9

104. Reiter P. 2001. Climate change and mosquito-borne disease. *Environ. Health Perspect.* 109(Suppl. 1):141–61

105. Reuters, "Cuba says dengue outbreak caused deaths, no figures," *Reuters AlertNet*, October 27, 2006, **http://www.alertnet.org/thenews/newsdesk/N27171520.htm**

106. Rico-Hesse R. 1990. Molecular evolution and distribution of dengue viruses type 1 and 2 in nature. *Virology* 174:479–93

107. Rico-Hesse R. 2003. Microevolution and virulence of dengue viruses. *Adv. Virus Res.* 59:315–41

108. Rico-Hesse R, Harrison LM, Nisalak A, Vaughn DW, Kalayanarooj S, et al. 1998. Molecular evolution of dengue type 2 virus in Thailand. *Am. J. Trop. Med. Hyg.* 58:96–101

109. **Rico-Hesse R, Harrison LM, Salas RA, Tovar D, Nisalak A, et al. 1997. Origins of dengue type 2 viruses associated with increased pathogenicity in the Americas. *Virology* 230:244–51**

110. Rigau-Perez JG, Gubler DJ, Vorndam AV, Clark GG. 1997. Dengue: a literature review and case study of travelers from the United States, 1986–1994. *J. Travel Med.* 4:65–71

111. Robin Y, Cornet M, Heme G, Gonidec G. 1980. Isolement du virus de la dengue au Sénégal. *Ann. Virol.* 131E:149–54

112. Rodhain F, Rosen L. 1997. Mosquito vectors and dengue virus-vector relationships. See Ref. 45, pp. 45–60

113. Romi R. 1995. History and updating on the spread of *Aedes albopictus* in Italy. *Parassitologia* 37:99–103

114. Rosen L. 1977. The Emperor's New Clothes revisited, or reflections on the pathogenesis of dengue hemorrhagic fever. *Am. J. Trop. Med. Hyg.* 26:337–43

115. Rosen L, Roseboom LE, Gubler DJ, Lien JC, Chaniotis BN. 1985. Comparative susceptibility of mosquito species and strains to oral and parenteral infection with dengue and Japanese encephalitis viruses. *Am. J. Trop. Med. Hyg.* 34:603–15

116. Rosen L, Shroyer DA, Tesh RB, Freier JE, Lien JC. 1983. Transovarial transmission of dengue viruses by mosquitoes: *Aedes albopictus* and *Aedes aegypti*. *Am. J. Trop. Med. Hyg.* 32:1108–19

117. Rosenthal E. 2007. As earth warms up, tropical virus moves to Italy. *New York Times*, December 23, 2007

118. Rothman AL, Ennis FA. 1999. Immunopathogenesis of dengue hemorrhagic fever. *Virology* 257:1–6

119. Rudnick A. 1978. Ecology of dengue virus. *Asian J. Infect. Dis.* 2:156–60

120. Rudnick A, Marchette NJ, Garcia R. 1967. Possible jungle dengue—recent studies and hypotheses. *Jpn. J. Med. Sci. Biol.* 20(Suppl.):69–74

121. Sabin AB. 1950. The dengue group of viruses and its family relationships. *Bacteriol. Rev.* 14:225–32

122. Sabin AB, Schlesinger RW. 1945. Production of immunity to dengue with virus modified by propagation in mice. *Science* 101:640–42

123. Sakuntabhai A, Turbpaiboon C, Casademont I, Chuansumrit A, Lowhnoo T, et al. 2005. A variant in the CD209 promoter is associated with severity of dengue disease. *Nat. Genet.* 37:507–13

124. Saluzzo JF, Cornet M, Castagnet P, Rey C, Digoutte JP. 1986. Isolation of dengue 2 and dengue 4 viruses from patients in Senegal. *Trans. R. Soc. Trop. Med. Hyg.* 80:5

125. Sangkawibha N, Rojanasuphot S, Ahandrik S, Viriyapongse S, Jatanasen S, et al. 1984. Risk factors in dengue shock syndrome: a prospective epidemiologic study in Rayong, Thailand. I. The 1980 outbreak. *Am. J. Epidemiol.* 120:653–69

109. First linkage of Southeast Asian DENV2 genotype to enhanced disease in the Americas by phylogenetic analysis.

126. Schnoor JL. 2007. The IPCC fourth assessment. *Environ. Sci. Technol.* 41:1503

127. Shurtleff AC, Beasley DW, Chen JJ, Ni H, Suderman MT, et al. 2001. Genetic variation in the 3′ noncoding region of dengue viruses. *Virology* 281:75–87

128. Sierra B, Alegre R, Perez AB, Garcia G, Sturn-Ramirez K, et al. 2007. HLA-A, -B, -C, and -DRB1 allele frequencies in Cuban individuals with antecedents of dengue 2 disease: advantages of the Cuban population for HLA studies of dengue virus infection. *Hum. Immunol.* 68:531–40

129. Sierra B, Garcia G, Perez AB, Morier L, Rodriguez R, et al. 2002. Long-term memory cellular immune response to dengue virus after a natural primary infection. *Int. J. Infect. Dis.* 6:125–28

130. Sierra B, Kouri G, Guzman MG. 2007. Race: a risk factor for dengue hemorrhagic fever. *Arch. Virol.* 152:533–42

131. Simmons CP, Dong T, Chau NV, Dung NT, Chau TN, et al. 2005. Early T-cell responses to dengue virus epitopes in Vietnamese adults with secondary dengue virus infections. *J. Virol.* 79:5665–75

132. Slosek J. 1986. *Aedes aegypti* mosquitoes in the Americas: a review of their interactions with the human population. *Soc. Sci. Med.* 23:249–57

133. Smith CE. 1956. The history of dengue in tropical Asia and its probable relationship to the mosquito *Aedes aegypti*. *J. Trop. Med. Hyg.* 59:243–51

134. Smith CE. 1958. The distribution of antibodies to Japanese encephalitis, dengue, and yellow fever viruses in five rural communities in Malaya. *Trans. R. Soc. Trop. Med. Hyg.* 52:237–52

135. Soper FL. 1967. *Aedes aegypti* and yellow fever. *Bull. WHO* 36:521–27

136. Stephens HA, Klaythong R, Sirikong M, Vaughn DW, Green S, et al. 2002. HLA-A and -B allele associations with secondary dengue virus infections correlate with disease severity and the infecting viral serotype in ethnic Thais. *Tissue Antigens* 60:309–18

137. Sutherst RW. 2004. Global change and human vulnerability to vector-borne diseases. *Clin. Microbiol. Rev.* 17:136–73

138. Tardieux I, Poupel O, Lapchin L, Rodhain F. 1990. Variation among strains of *Aedes aegypti* in susceptibility to oral infection with dengue virus type 2. *Am. J. Trop. Med. Hyg.* 43:308–13

139. Taubes G. 1997. Apocalypse not. *Science* 278:1004–6

140. Thein S, Aung MM, Shwe TN, Aye M, Zaw A, et al. 1997. Risk factors in dengue shock syndrome. *Am. J. Trop. Med. Hyg.* 56:566–72

141. Tuntaprasart W, Barbazan P, Nitatpattana N, Rongsriyam Y, Yoksan S, Gonzalez JP. 2003. Seroepidemiological survey among schoolchildren during the 2000–2001 dengue outbreak of Ratchaburi Province, Thailand. *Southeast Asian J. Trop. Med. Public Health* 34:564–68

142. Twiddy SS, Holmes EC, Rambaut A. 2003. Inferring the rate and time-scale of dengue virus evolution. *Mol. Biol. Evol.* 20:122–29

143. Twiddy SS, Woelk CH, Holmes EC. 2002. Phylogenetic evidence for adaptive evolution of dengue viruses in nature. *J. Gen. Virol.* 83:1679–89

144. Vasilakis N, Shell EJ, Fokam EB, Mason PW, Hanley KA, et al. 2007. Potential of ancestral sylvatic dengue-2 viruses to re-emerge. *Virology* 358:402–12

145. Vaughn DW, Green S, Kalayanarooj S, Innis BL, Nimmannitya S, et al. 2000. Dengue viremia titer, antibody response pattern, and virus serotype correlate with disease severity. *J. Infect. Dis.* 181:2–9

146. Vazeille M, Rosen L, Mousson L, Failloux AB. 2003. Low oral receptivity for dengue type 2 viruses of *Aedes albopictus* from Southeast Asia compared with that of *Aedes aegypti*. *Am. J. Trop. Med. Hyg.* 68:203–8

147. Wang E, Ni H, Xu R, Barrett AD, Watowich SJ, et al. 2000. Evolutionary relationships of endemic/epidemic and sylvatic dengue viruses. *J. Virol.* 74:3227–34

148. Waterhouse RM, Kriventseva EV, Meister S, Xi Z, Alvarez KS, et al. 2007. Evolutionary dynamics of immune-related genes and pathways in disease-vector mosquitoes. *Science* 316:1738–43

149. Waterman SH, Novak RJ, Sather GE, Bailey RE, Rios I, Gubler DJ. 1985. Dengue transmission in two Puerto Rican communities in 1982. *Am. J. Trop. Med. Hyg.* 34:625–32

150. Watts DM, Porter KR, Putvatana P, Vasquez B, Calampa C, et al. 1999. Failure of secondary infection with American genotype dengue 2 to cause dengue haemorrhagic fever. *Lancet* 354:1431–34

151. Wearing HJ, Rohani P. 2006. Ecological and immunological determinants of dengue epidemics. *Proc. Natl. Acad. Sci. USA* 103:11802–7

133. Early documentation of the distribution of *A. aegypti* and dengue in Southeast Asia just before the emergence of DHF/DSS.

152. Westaway E, Blok J. 1997. Taxonomy and evolutionary relationships of flaviviruses. See Ref. 45, pp. 147–73
153. Whitehead RH, Yuill TM, Gould DJ, Simasathien P. 1971. Experimental infection of *Aedes aegypti* and *Aedes albopictus* with dengue viruses. *Trans. R. Soc. Trop. Med. Hyg.* 65:661–67
154. Whitehead SS, Blaney JE, Durbin AP, Murphy BR. 2007. Prospects for a dengue virus vaccine. *Nat. Rev. Microbiol.* 5:518–28
155. WHO. 1997. *Dengue Haemorrhagic Fever: Diagnosis, Treatment, Prevention and Control.* Geneva: WHO. 2nd ed.
156. WHO. 2000. *Strengthening Implementation of the Global Strategy for Dengue Fever/Dengue Haemorrhagic Fever Prevention and Control.* Presented at Report of the Informal Consultation, Geneva, Switzerland
157. Zanotto PM, Gould EA, Gao GF, Harvey PH, Holmes EC. 1996. Population dynamics of flaviviruses revealed by molecular phylogenies. *Proc. Natl. Acad. Sci. USA* 93:548–53
158. Zhang C, Mammen MP Jr, Chinnawirotpisan P, Klungthong C, Rodpradit P, et al. 2005. Clade replacements in dengue virus serotypes 1 and 3 are associated with changing serotype prevalence. *J. Virol.* 79:15123–30
159. Zivna I, Green S, Vaughn DW, Kalayanarooj S, Stephens HA, et al. 2002. T cell responses to an HLA-B*07-restricted epitope on the dengue NS3 protein correlate with disease severity. *J. Immunol.* 168:5959–65

Biosynthesis of the Iron-Molybdenum Cofactor of Nitrogenase

Luis M. Rubio[1] and Paul W. Ludden[2]

[1]Department of Plant and Microbial Biology, University of California, Berkeley, California 94720; email: lrubio@nature.berkeley.edu

[2]Office of the Provost, Southern Methodist University, Dallas, Texas 75275; email: pludden@smu.edu

Annu. Rev. Microbiol. 2008. 62:93–111

First published online as a Review in Advance on April 22, 2008

The *Annual Review of Microbiology* is online at micro.annualreviews.org

This article's doi: 10.1146/annurev.micro.62.081307.162737

Key Words

nitrogen fixation, *nif*, FeMo-co, iron-sulfur, NifDK, MoFe protein

Abstract

The iron-molybdenum cofactor (FeMo-co), located at the active site of the molybdenum nitrogenase, is one of the most complex metal cofactors known to date. During the past several years, an intensive effort has been made to purify the proteins involved in FeMo-co synthesis and incorporation into nitrogenase. This effort is starting to provide insights into the structures of the FeMo-co biosynthetic intermediates and into the biochemical details of FeMo-co synthesis.

Contents

INTRODUCTION

NifDK: also designated MoFe protein, dinitrogenase, or component I

FeMo-co: iron-molybdenum cofactor

NifH: also designated Fe protein, dinitrogenase reductase, or component II

Biological nitrogen fixation accounts for roughly two-thirds of the nitrogen fixed globally, whereas the remaining portion is mostly contributed by the industrial Haber-Bosch process. Biological nitrogen fixation is performed, generally at mild temperatures, by diazotrophic microorganisms, which are widely distributed in nature (62). Most biological nitrogen fixation is carried out by the activity of the molybdenum nitrogenase, which is found in all diazotrophs. In addition to the molybdenum nitrogenase,

some diazotrophic microorganisms carry alternative vanadium and/or iron-only nitrogenases (7, 53).

The molybdenum nitrogenase enzyme complex has two component proteins (10) encoded by the *nifDK* and the *nifH* genes. The NifDK component is a heterotetrameric ($\alpha_2\beta_2$) protein formed by two $\alpha\beta$ dimers related by a twofold symmetry (**Figure 1**). NifDK carries one iron-molybdenum cofactor (FeMo-co) within the active site in each α-subunit (NifD) (47, 73) and one P-cluster at the interface of the α- and β-subunits in each $\alpha\beta$ pair. FeMo-co is located 10 Å beneath the protein surface. The Mo atom and the distal Fe atom at the other end of the cofactor are coordinated by one histidine and one cysteine residue from the NifD polypeptide, respectively. The P-cluster, also located beneath the protein surface, is coordinated by three cysteine residues from the NifD subunit and three cysteine residues from the NifK subunit.

The NifH component is a homodimer with twofold symmetry. NifH contains sites for $Mg \cdot ATP$ binding and hydrolysis at the dimer interface within each subunit and a single [Fe_4-S_4] cluster at the interface of both subunits (27). The [Fe_4-S_4] cluster of NifH is coordinated by two cysteine residues from each subunit. Binding and hydrolysis of $Mg \cdot ATP$ causes changes in NifH conformation and is coupled to electron transfer from the [Fe_4-S_4] cluster to the P-cluster of the NifDK component (49). The NifH and NifDK components associate and disassociate with each electron transfer cycle (31).

Electrons are subsequently transferred from the P-clusters to the FeMo-co embedded within the NifD subunits where reduction of substrates takes place (**Figure 1**). Dinitrogen (N_2) and protons are the physiological nitrogenase substrates. In the absence of N_2, proton reduction activity is maximal, as more electrons are allocated to this process. However, a minimal 2:1 ratio of proton to N_2 reduction is obligate even at high N_2 pressure (76), indicating that proton reduction is an intrinsic part of the mechanism of N_2 reduction. A number of additional triple-bonded molecules can serve

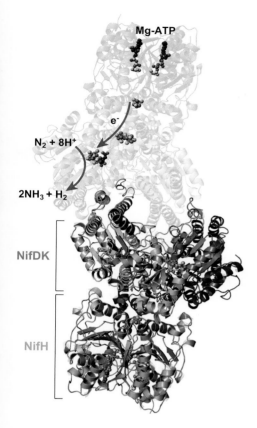

Mg-ATP

e⁻

N₂ + 8H⁺

2NH₃ + H₂

NifDK

NifH

Figure 1

Structure of the molybdenum nitrogenase enzyme complex. The nitrogenase reaction produces ammonia and hydrogen at the expense of ATP. The relative positions of Mg-ATP and the [Fe-S] clusters of the symmetric nitrogenase enzyme complex are shown in the upper half of the figure together with the electron transfer pathway to the active site. The NifH component is orange and the NifDK component is in blue and green.

as substrates for nitrogenase. Among these substrates, acetylene is the most relevant to this review because acetylene and its reduction product, ethylene, are easily detected by gas chromatography and are routinely used to assay nitrogenase activity.

No crystal structure of an alternative nitrogenase has been reported to date. Biochemical and genetic studies have indicated that, although genetically distinct from the molybdenum nitrogenase, the three enzymes have sim-

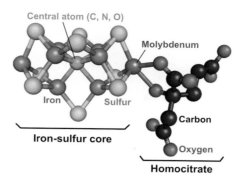

Central atom (C, N, O)

Molybdenum

Iron   Sulfur

Carbon

Oxygen

Iron-sulfur core

Homocitrate

Figure 2

Structure of iron-molybdenum cofactor (FeMo-co) of nitrogenase.

ilar subunit and cofactor composition. Attention has been directed to the use of the alternative vanadium and iron-only nitrogenases as catalysts for hydrogen production because they are less efficient in N_2 reduction and allocate more electrons to proton reduction than the molybdenum nitrogenase. However, the ATP dependence and the specific H_2 evolution activities are similar in molybdenum and alternative nitrogenases.

THE FeMo-co OF NITROGENASE

FeMo-co is required by the molybdenum nitrogenase to perform the chemically difficult lysis of the N_2 triple bond (reviewed in Reference 39). FeMo-co is unique and structurally different from all the other molybdenum cofactors known to date. FeMo-co is composed of an inorganic $Mo\text{-}Fe_7\text{-}S_9\text{-}X$ portion and the organic acid R-homocitrate, which is coordinated by its C-2 carboxyl and hydroxyl groups to the Mo atom (**Figure 2**) (11, 21, 35, 73). Although there is evidence for participation of the inorganic portion of FeMo-co in substrate binding and catalysis (5, 6), the exact role of homocitrate in catalysis is unknown. Replacement of homocitrate by citrate or other organic acid in FeMo-co drastically changes substrate specificity and results in an impairment in N_2 reduction (40).

The inorganic part of FeMo-co is regarded as one of the most complex iron-sulfur (plus

Molybdenum cofactors: with the exception of FeMo-co, biologically active molybdenum is always found in molybdopterin cofactors (Mo-co). Molybdoenzymes containing Mo-co are widely distributed in nature and participate in essential redox reactions of C, N, and S metabolism

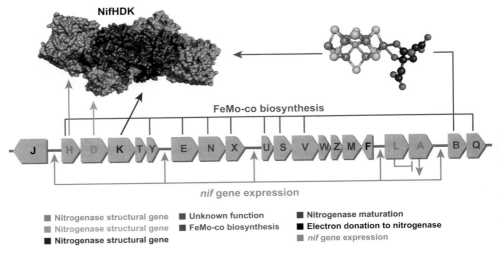

NifHDK

FeMo-co biosynthesis

J H D K T Y E N X U S V W Z M F L A B Q

nif gene expression

■ Nitrogenase structural gene   ■ Unknown function   ■ Nitrogenase maturation
■ Nitrogenase structural gene   ■ FeMo-co biosynthesis   ■ Electron donation to nitrogenase
■ Nitrogenase structural gene   ■ *nif* gene expression

Figure 3

Nitrogen fixation (*nif*) gene cluster of *Klebsiella pneumoniae*. Genes encoding the nitrogenase component proteins, *nifHDK*, are shown on the left. Genes whose products are involved in nitrogen fixation are color coded according to their functions.

a heterometal) clusters found in biology. The structural features of the inorganic part of FeMo-co can be analyzed from two different perspectives relevant to our understanding of FeMo-co biosynthesis and to the design of a synthetic chemistry approach to synthesize this cofactor. It can be described as one $MoFe_3$-S_3X cubane and one Fe_4-S_3X cubane sharing a single atom X at a corner common to both cubanes. These cubanes would be additionally bridged by three sulfur atoms. Alternatively, it can be described as a Fe_6-S_9 core (with the iron atoms symmetrically coordinating a central atom X) that is capped by one Mo and one Fe atom at the end sides of the core.

There are only two protein ligands to FeMo-co (α-Cys^{275} and α-His^{442} in the *Azotobacter vinelandii* nitrogenase enzyme), and the cofactor can be purified by extraction into the organic solvent *N*-methyl formamide after denaturing the NifDK protein (73). The purification of FeMo-co has been an indispensable tool to analyze the processes of FeMo-co biosynthesis and insertion into apo-dinitrogenase (apo-NifDK).

The other complex metallocluster present in the NifDK component of nitrogenase, the [Fe_8-S_7] P-cluster, is structurally similar to FeMo-co

in that it is formed by two [Fe_4-S_3] cubanes connected by a central sulfur atom. The structure of the P-cluster and its ligation to the polypeptide changes depending on the oxidation state of the NifDK protein.

THE *nif* GENE CLUSTER

Klebsiella pneumoniae was the first diazotroph whose nitrogen fixation (*nif*) genes were analyzed by a combination of genetic and biochemical techniques (4, 51, 64). **Figure 3** shows the *K. pneumoniae nif* gene cluster. Twenty *nif* genes, *nifJHDKTYENXUSVWZMFLABQ*, organized in several transcriptional units, are clustered in a single 23-kb region in the chromosome of *K. pneumoniae*. The *nif* gene cluster is simpler in *K. pneumoniae* than in other model organisms used to study the biochemistry and genetics of nitrogen fixation, such as *A. vinelandii*. This is probably due to the different physiological conditions faced by the oxygen-sensitive nitrogenase in these bacteria. Whereas *K. pneumoniae* fixes nitrogen only under strict anaerobic conditions, *A. vinelandii* combines in a single cell nitrogen fixation activity with strict aerobic metabolism.

nif: genes encoding proteins involved in molybdenum-dependent nitrogen fixation

The *A. vinelandii nif* genes are clustered into two different chromosomal linkage groups. The major *nif* cluster contains the *nifHDK-TYENX iscA^{nif} nifUSV cysE1^{nif} nifWZM nifF* genes and several open reading frames interspersed among these genes (42). The minor *nif* cluster contains the *rnfABCDGEH nafY* genes in one transcriptional direction (14, 69) and the *nifLAB fdxN nifOQ* genes in the opposite DNA orientation (44, 65).

The *nifHDK* genes encode the structural components of the molybdenum nitrogenase enzyme complex. The complexity and uniqueness of FeMo-co and the P-cluster demand complex biosynthetic pathways for cofactor biosynthesis and maturation of the NifDK protein. Thus, the products of at least 12 *nif* genes are involved in the biosynthesis of catalytically active molybdenum nitrogenase (68) (see **Figure 3**). Additional gene products are required to provide nitrogenase with electrons. The flavodoxin NifF and the pyruvate:flavodoxin oxidoreductase, NifJ, fulfill this role in *K. pneumoniae* (75). Finally, the products of the *nifA* and *nifL* genes form an activator/antiactivator regulatory system that controls *nif* gene expression to ensure it occurs only under appropriate physiological and environmental conditions (52).

OVERVIEW OF FeMo-co BIOSYNTHESIS

The products of the *nifD* and *nifK* genes do not seem to be required for FeMo-co biosynthesis. FeMo-co is first assembled by specialized biosynthetic machinery and then incorporated into FeMo-co-deficient apo-NifDK, generating the mature NifDK nitrogenase component that is competent for nitrogen fixation (see References 17, 50, and 68 for previous reviews). The proteins involved in FeMo-co biosynthesis can be functionally divided into three classes: molecular scaffolds (NifU, NifB, and NifEN) where FeMo-co is stepwise assembled, metallocluster carrier proteins (NifX and NafY) that carry FeMo-co precursors between assembly sites in the pathway,

and enzymes (NifS, NifQ, and NifV) that provide sulfur, molybdenum, and homocitrate as substrates for cofactor synthesis. The exact role of NifH remains controversial.

Figure 4 illustrates a schematic summary of our current model for FeMo-co biosynthesis. This scheme has been centered on NifEN, a scaffold protein that is proposed to function as a central node in the pathway to which [Fe-S]-containing FeMo-co precursors, molybdenum and homocitrate, might converge to complete the assembly of FeMo-co (13, 25, 29, 36, 38, 66, 78). Molybdenum is specifically donated by NifQ (J.A. Hernandez, L. Curatti, C.P. Aznar, Z. Perova, R.D. Britt & L.M. Rubio, unpublished results). The iron-sulfur core is provided by the sequential activities of NifS, NifU, and NifB. The cysteine desulfurase NifS directs the assembly on NifU of simple Fe-S clusters (probably $[Fe_2-S_2]$ or $[Fe_4-S_4]$) that will serve as metabolic substrates for NifB-cofactor (NifB-co) synthesis (84). The SAM radical protein NifB synthesizes NifB-co in a reaction that requires *S*-adenosylmethionine (SAM) (16). NifB-co comprises the Fe_6-S_9 core of FeMo-co (26) but lacks molybdenum and homocitrate (72). NifX would mobilize NifB-co from NifB to NifEN (32). Homocitrate is generated by the homocitrate synthase NifV. Although we hypothesize that the molybdenum, iron-sulfur, and homocitrate precursors converge on NifEN, it is not clear whether NifEN alone provides a homocitrate binding site. Current evidence indicates that both NifEN and NifH must be present to achieve homocitrate incorporation into the FeMo-co precursor (15, 36, 57). Finally, assembled FeMo-co would be transferred to apo-NifDK via NafY, the product of a non-*nif* gene that also stabilizes the target apo-NifDK protein (67).

Originally, the in vitro FeMo-co synthesis assay was carried out as a biochemical complementation of cell extracts from strains with lesions in different *nif* genes (74). Over the past several years, different laboratories have made extensive efforts to purify all Nif/Naf proteins involved in FeMo-co biosynthesis. **Figure 5** compiles much of this effort by

nafY: gene encoding the nitrogenase accessory factor Y (γ subunit)

NifB-cofactor (NifB-co): the metabolic product of NifB activity. It is an isolatable [Fe-S] cluster of unknown structure that serves as precursor to FeMo-co

SAM: *S*-adenosylmethionine

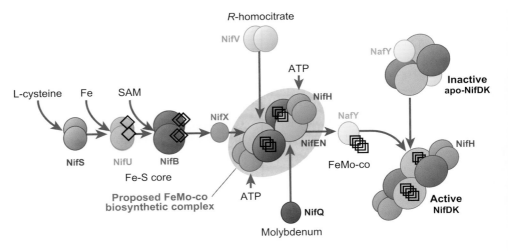

Figure 4

Schematic model of iron-molybdenum cofactor (FeMo-co) biosynthesis illustrating the convergence of FeMo-co precursors into the NifEN/NifH enzyme complex, a central node where FeMo-co synthesis is completed. De novo synthesized FeMo-co is inserted into apo-NifDK to generate catalytically active nitrogenase. The FeMo-co biosynthetic pathway involves enzymes, proteins that act as molecular scaffolds, and carriers of complex metalloclusters. The number of black squares for each FeMo-co precursor represents its level of structural complexity.

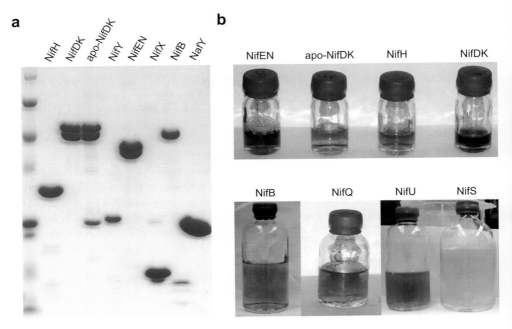

Figure 5

The colorful world of nitrogenase. Many Nif proteins carry [Fe-S] clusters or other cofactors, such as pyridoxal phosphate (PLP), that are essential to their functions. (*a*) SDS-PAGE analysis of purified Nif proteins. (*b*) Anaerobic solutions of purified Nif proteins.

showing purified preparations of most of these proteins, many of which carry [Fe-S] clusters or other colorful cofactors. Purification of all these proteins has allowed in vitro reconstitution of the complete FeMo-co biosynthetic pathway and has demonstrated that the reactions catalyzed by NifB, NifEN, and NifH are necessary and sufficient for FeMo-co synthesis from iron, sulfur, molybdenum, and homocitrate (15). Although not essential, the NifX and NafY proteins increase the synthetic efficiency of the pathway, possibly by providing protection to the oxygen-labile FeMo-co and its intermediates. The in vivo functions of the other Nif proteins involved in FeMo-co synthesis (e.g., NifU, NifS, NifV, and NifQ) can be replaced in vitro by adding the appropriate chemical substrates.

FORMATION OF THE IRON-SULFUR CORE OF FeMo-co

NifU and NifS Proteins

The first suggestion that NifU and NifS were involved in the assembly of [Fe-S] clusters for the nitrogenase component proteins, NifH and NifDK, came from the analysis of *A. vinelandii* *nifU* and *nifS* mutants. Mutations in either *nifS* or *nifU* affected NifH and NifDK activities, decreasing them by 15- and 4-fold, respectively (43). Because a common feature of NifH and NifDK was the requirement of their [Fe-S] clusters for activity, it was suggested that *nifU* and *nifS* mutants were impaired in the assembly of [Fe-S] clusters for the nitrogenase components. Similar phenotype had previously been observed in *K. pneumoniae* *nifS* mutants that exhibited negligible NifH activity and a 25-fold reduction in NifDK activity (64), but NifS was proposed to be involved exclusively in NifH processing.

Subsequently, a series of elegant studies in the laboratories of Dean and Johnson demonstrated that NifU and NifS compose cellular machinery for the assembly of [Fe$_2$-S$_2$] and [Fe$_4$-S$_4$] clusters under nitrogen-fixing condi-

tions (reviewed in Reference 45). NifS is a cysteine desulfurase that provides sulfur for the assembly of transient [Fe-S] clusters onto the molecular scaffold NifU. These transient labile clusters are then transferred to target apo-proteins. It was also shown that NifU and NifS were nitrogenase-specific homologs of the IscU and IscS proteins (85), which are involved in general [Fe-S] cluster assembly in a wide range of organisms (45). The special boost in [Fe-S] cluster assembly provided by the activities of NifU and NifS is required to fulfill the high demand imposed by nitrogenase, a catalytically slow enzyme that may represent up to 10% of the total cellular protein under nitrogen-fixing conditions.

NifU is a homodimer of 33-kDa subunits. The most interesting property of NifU is its modular structure. NifU comprises three well-defined highly conserved domains, all of which have the ability to coordinate an [Fe-S] cluster. The second (middle) domain carries four conserved cysteines that coordinate one [Fe$_2$-S$_2$] cluster (23). The first (N-terminal) and third (C-terminal) domains contain three and two conserved cysteine residues, respectively, that serve as ligands for the assembly of the transient [Fe-S] clusters that will be further delivered to target apo-proteins. The [Fe$_2$-S$_2$] cluster that is permanently bound to NifU has been proposed to have a redox function necessary for NifU to release the labile clusters (1).

The homodimeric NifS protein is a pyridoxal phosphate (PLP)-containing enzyme that catalyzes the desulfurization of L-cysteine, yielding sulfur destined to [Fe-S] cluster formation and L-alanine (87). A reaction intermediate in the form of cysteinyl persulfide is formed in the Cys^{325} residue of NifS (86). Consistently, substitution of Ala by Cys in this position eliminates cysteine desulfurase activity of the enzyme. Formation of a NifUS heterotetramer during cluster assembly has been observed. Within this protein complex, NifS activity directs the assembly of [Fe-S] clusters on the molecular scaffold NifU (77, 83).

In vitro experiments in which transient [Fe$_4$-S$_4$] clusters, assembled by NifS on NifU, were transferred to apo-NifH to reconstitute the dinitrogenase reductase activity of NifH have been reported (19). Additional in vivo experiments showing that NifU and NifS were necessary for the assembly of the $[Fe_4\text{-}S_4]$ cluster of NifH (18, 19, 46) provided physiological evidence to further support the roles of NifU and NifS in NifH maturation.

NifU and NifS are also involved in the assembly of the P-cluster and the FeMo-co of the NifDK component of nitrogenase. The involvement of NifU and NifS in P-cluster assembly is supported by the fact that *nifU* and *nifS* mutations cause the accumulation of a form of apo-NifDK whose activity cannot be recovered in vitro by the simple addition of FeMo-co (43). To understand the *nifUS* mutant phenotype, it is important to note that, during NifDK maturation, P-cluster assembly precedes FeMo-co insertion and that apo-NifDK containing P-clusters but lacking FeMo-co is readily activatable by FeMo-co. The synthesis of each P-cluster has been proposed to occur in two steps: (*a*) formation of two $[Fe_4\text{-}S_4]$ cluster units at the interface of the NifD and NifK subunits, and (*b*) NifH-dependent condensation of these clusters to generate the mature $[Fe_8\text{-}S_7]$ P-cluster (9, 63). It is likely that *nifUS* mutants are impaired in the assembly of the $[Fe_4\text{-}S_4]$ cluster units that serve as P-cluster precursors. An alternative interpretation is that the impairment of *nifUS* mutants in the P-cluster assembly is in fact reflecting a deficient assembly of the $[Fe_4\text{-}S_4]$ cluster of NifH. We do not favor this alternative interpretation because there is evidence showing that $[Fe_4\text{-}S_4]$ cluster-deficient NifH functions in apo-NifDK maturation (14, 61).

Direct participation of NifU and NifS in FeMo-co biosynthesis was demonstrated by analyzing the capability of *nifU* and *nifS* mutants to synthesize active NifB protein and its metabolic product, NifB-co (84). Most of the proteins involved in FeMo-co biosynthesis are [Fe-S] proteins, and therefore mutations in *nifU* and *nifS* were expected to have a pleiotropic effect in this pathway. However, NifB-co synthesis is an early step in the pathway that re-

quires only the activity of the [Fe-S]-containing protein NifB (15, 16). Whereas *K. pneumoniae* *nifU* and *nifS* mutants synthesized active NifB protein, they were unable to synthesize and accumulate measurable amounts of NifB-co. The simplest explanation for the *nifUS* phenotype is that the NifUS machinery acts as a major provider of [Fe-S] substrates for NifB-co biosynthesis by NifB.

NifB and The Formation of NifB-co

NifB is a homodimeric protein whose activity is essential to the biosynthesis of FeMo-co and those of the cofactors for the alternative vanadium and iron-only nitrogenases. According to its primary sequence, the N-terminal domain of NifB belongs to the SAM radical protein family (79). SAM radical proteins contain an $[Fe_4\text{-}S_4]$ cluster coordinated by three cysteine residues and a molecule of SAM. Although SAM radical enzymes catalyze diverse reactions, the mechanism of catalysis starts by reductive cleavage of SAM and generation of a 5′-deoxyadenosyl radical (82). The C-terminal domain of NifB is similar to the NifX/NafY family of proteins (see below) (69) and is hypothesized to be a NifB-co binding site (16).

A functional *nifB* gene is required for the synthesis of NifB-co in vivo, and consistently, the requirement for NifB in the in vitro FeMo-co synthesis assay was satisfied by addition of NifB-co (72). NifB-co, the metabolic product of NifB, is a molybdenum-free and homocitrate-free, oxygen-labile [Fe-S] cluster that serves as a precursor to FeMo-co. Allen et al. (2) have shown the specific incorporation of Fe and S from ^{55}Fe- and ^{35}S-labeled NifB-co into FeMo-co. NifB-co-dependent in vitro FeMo-co synthesis is catalyzed by NifEN and NifH and requires the presence of molybdate, homocitrate, sodium dithionite, and Mg · ATP in the reaction. **Figure 6** illustrates a NifB-co-dependent FeMo-co synthesis assay carried out with purified components. In this assay, FeMo-co is synthesized in vitro and inserted into apo-NifDK to generate active NifDK protein. The amount of synthesized FeMo-co is

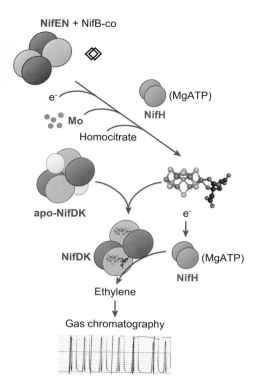

NifEN + NifB-co

e⁻

Mo

Homocitrate

(MgATP)

NifH

apo-NifDK

e⁻

NifDK

(MgATP)

NifH

Ethylene

Gas chromatography

Figure 6

In vitro iron-molybdenum cofactor (FeMo-co) synthesis assay. Purified Nif proteins transform FeMo-co precursors [such as NifB-cofactor (NifB-co)] into FeMo-co upon addition of molybdenum and homocitrate under reducing conditions. Formation of an active NifDK protein is then followed by its acetylene reduction activity in a gas chromatographer.

estimated by the acetylene reducing activity of de novo generated NifDK.

NifB seems to be a molecular scaffold for the assembly of NifB-co. As isolated from *A. vinelandii* cells, NifB did not support in vitro FeMo-co synthesis. However, incubation of purified NifB with ferrous iron, sulfide, and SAM under reducing conditions resulted in the incorporation of additional [Fe-S] clusters into NifB, which acquired the ability to substitute for NifB-co and support in vitro FeMo-co synthesis (16). Radical chemistry is thus required during NifB-co assembly. The reason for this requirement is unclear. Very low-potential rad-

ical chemistry could be needed to incorporate the central atom into NifB-co.

The properties of NifB, including its ability to readily support FeMo-co synthesis, were different when purified from *K. pneumoniae* or from *A. vinelandii* cells. The enzyme isolated from *K. pneumoniae* contained more iron and readily substituted for NifB-co in the in vitro FeMo-co synthesis assay without a requirement for iron, sulfide, or SAM (84). Additional evidence that the *K. pneumoniae* NifB carries NifB-co came from the ability of NifB to transfer an iron-containing moiety to NifX, which binds NifB-co but not free iron (32, 58). Comparison of spectroscopic signatures from NifB preparations obtained from *K. pneumoniae* or *A. vinelandii* cells are expected to provide insights into the nature of NifB-co precursors and into the mechanism of the reaction catalyzed by NifB.

MOLYBDENUM PROCESSING FOR FeMo-co BIOSYNTHESIS

Possible Role of NifQ as Physiological Molybdenum Donor

NifQ has been implicated in the processing of molybdenum specifically for the biosynthesis of FeMo-co (41). The *nifQ* mutants accumulate lower levels of molybdenum than the wild-type strain and exhibit a nitrogen-fixation-deficient phenotype, but they are not defective in the synthesis of molybdopterin cofactor (Mo-co).

The primary amino acid sequence of NifQ is unique and unrelated to other proteins involved in molybdenum trafficking, such as molbindins or molybdenum storage proteins (22, 30). NifQ proteins have a conserved $Cx_4Cx_2Cx_5C$ amino acid motif that was proposed to be a binding site for an [Fe-S] cluster, a molybdenum-containing metal cluster, or a Mo-S intermediate for FeMo-co synthesis (17, 67).

Because the phenotype of *nifQ* mutants is leaky, and because molybdate satisfies the molybdenum requirement for FeMo-co synthesis in vitro, the determination of a physiological provider of molybdenum for FeMo-co biosynthesis has been elusive. High concentrations of molybdate (41) or cysteine (81) (the sulfur source for [Fe-S] cluster biosynthesis) in the growth medium can suppress the *nifQ* phenotype. This was originally interpreted as an indication that a nonenzymatic reaction between molybdenum and sulfur could substitute for the reaction catalyzed by NifQ (81).

Recent results show that, as isolated from *A. vinelandii* cells, NifQ is an iron-sulfur protein with a redox-responsive [Fe-S] cluster. NifQ is also a molybdoprotein that serves as a direct molybdenum donor for FeMo-co synthesis, replacing molybdate in the in vitro FeMo-co synthesis assay (J.A. Hernandez, L. Curatti, C.P. Aznar, Z. Perova, R.D. Britt & L.M. Rubio, unpublished results). NifQ was unable to donate molybdenum for FeMo-co synthesis unless NifH and NifEN were simultaneously present in the reaction. One possible interpretation is that molybdenum delivery proceeds from NifQ to a NifEN/NifH complex. Electron paramagnetic resonance (EPR) spectroscopic studies indicated that NifQ carries a [Mo-Fe$_3$-S$_4$] cluster, and that the presence of this metal cluster in NifQ correlates with its ability to support in vitro FeMo-co synthesis. The chemical form of molybdenum donated by NifQ remains unknown.

Possible Role of NifH as Molybdenum Insertase

NifH has been proposed to serve as the entry point for molybdenum incorporation into the FeMo-co biosynthetic pathway based on ^{99}Mo radiolabeling experiments (57) or on extended X-ray absorption fine structure (EXAFS) analysis of FeMo-co precursors associated with NifH in vitro (38). In contrast, substoichiometric amounts of molybdenum were found in NifEN isolated from a Δ*nifH* strain, suggesting the existence of a NifH-independent molybdenum binding site within NifEN (78). The possible role of NifH in the incorporation of molybdenum into FeMo-co is discussed below.

THE HOMOCITRATE SYNTHASE NifV

The NifV protein is as a homocitrate synthase that catalyzes the condensation of acetyl coenzyme A and α-ketoglutarate to form *R*-homocitrate (88). *K. pneumoniae nifV* mutants incorporate citrate in place of homocitrate in FeMo-co (54) and exhibit altered substrate-reducing properties. The *nifV* nitrogenase variant reduces protons and acetylene effectively but is unable to reduce N$_2$ (55). The phenotype of a *K. pneumoniae nifV* strain could be reverted to wild type simply by the addition of homocitrate to the growth medium during nitrogenase derepression, further correlating the function of NifV in vivo to the synthesis of homocitrate (34).

^{99}Mo-radiolabeling experiments have indicated that addition of molybdenum to the [Fe-S] core of FeMo-co precedes homocitrate incorporation (57). It is not known by which mechanism homocitrate is coordinated to molybdenum and how the FeMo-co synthesis machinery discriminates against other organic acids present in the cell. Although homocitrate analogs can be incorporated into aberrant cofactors by increasing their concentration in the in vitro FeMo-co synthesis assay (40), there is a mechanism for homocitrate incorporation in vivo that appears to be specific enough to discriminate against similar organic acids that are not incorporated into FeMo-co.

The site of homocitrate incorporation into the cofactor is also a matter of debate. On the one hand, Rangaraj & Ludden (57) used ^{99}Mo-labeled molybdate in the purified in vitro FeMo-co synthesis assay to test the effect of various organic acids on ^{99}Mo incorporation into NifH or NifX and suggested that NifX had a role specifying the organic acid incorporated into the cofactor. On the other hand, Hu et al. (38) have recently used EXAFS analysis coupled with in vitro FeMo-co synthesis to propose NifH as responsible for homocitrate insertion into FeMo-co. In any case, the presence of NifEN in the reactions was required to achieve homocitrate incorporation in vitro, and we propose that the actual incorporation of homocitrate into the FeMo-co precursor would take place in the scaffold protein NifEN while associated with NifH. Given that NifX binds a variety of structurally related cofactors (NifB-co, VK-cluster, and FeMo-co), and exchanges some of them with NifEN, it would not be surprising that NifX accumulated an excess of molybdenum-containing FeMo-co precursor in reactions lacking homocitrate. The proposal of NifH as a homocitrate insertase is more puzzling and is discussed in the context of the FeMo-co biosynthetic factory model.

NifEN IS THE CENTRAL NODE IN FeMo-co BIOSYNTHESIS

Properties of the NifEN Protein

NifEN is a 200-kDa $\alpha_2\beta_2$ heterotetramer that contains [Fe-S] clusters (56). The function of NifEN is essential to FeMo-co biosynthesis. Because of the amino acid sequence similarity between NifEN and NifDK and the observation that NifDK was not required for FeMo-co biosynthesis (80), it was early proposed that NifEN could be a molecular scaffold to assemble FeMo-co (8).

An interesting characteristic of NifEN is the lability of some of its [Fe-S] clusters, which are lost during the purification of the protein, yielding NifEN preparations with different [Fe-S] cluster content. Over years of study, this property has led investigators to different conclusions about the role of NifEN in FeMo-co biosynthesis. From the first reported NifEN purification (4.6 iron atoms per NifEN tetramer) (56) to the last one (24 iron atoms per NifEN tetramer) (78), the iron content of NifEN preparations and the ability of NifEN to support FeMo-co biosynthesis in the absence of externally added iron, sulfur, and molybdenum substrates have increased significantly.

Originally, NifEN was purified from a *nifB* mutant strain, which yielded a NifEN protein able support in vitro FeMo-co synthesis when added to a *K. pneumoniae* cell extract lacking NifEN but containing NifB-co (56). NifEN appeared to carry one single [Fe$_4$-S$_4$] per tetramer that was suggested to be involved in electron transfer to a FeMo-co precursor (66). In vitro work demonstrated the incorporation in NifEN of ^{55}Fe or ^{35}S label from radioactive NifB-co (2), and the changes in electrophoretic mobility of NifEN upon NifB-co binding (66). Now we know that $\Delta nifB$ NifEN lacks the VK-cluster (37) and that its activity is dependent on processing the NifB-co present in the reaction (32).

Overexpression of His-tagged NifEN from a $\Delta nifHDK$ mutant strain, and application of a faster purification protocol, showed that NifEN contained two identical [Fe$_4$-S$_4$] clusters with $S = 1/2$ EPR signal in the reduced state (29). Because the His-tagged NifEN was purified from a strain having a functional *nifB* gene and contained more than eight iron atoms per tetramer, it was inspected for other EPR signals that could account for additional [Fe-S] clusters. Indeed, a different $S = 1/2$ EPR signal was observed in thionine-oxidized NifEN and suggested to arise from a FeMo-co precursor bound to the enzyme (28). This proposal was reinforced in a following study showing that, in contrast to the $\Delta nifHDK$ NifEN protein, indigo carmine (IDS)-oxidized preparations of the $\Delta nifB$ NifEN protein (which lacks FeMo-co precursor activity) lacked the $S = 1/2$ EPR signal (37).

The His-tagged NifEN tetramer purified from a $\Delta nifHDK$ mutant contains 24 iron atoms and, importantly, 0.3 molybdenum atoms (78). Although present in substoichiometric amounts, two lines of evidence suggest that the molybdenum present in NifEN is relevant to the FeMo-co biosynthesis pathway. First, the molybdenum was available for in vitro FeMo-co synthesis and supported apo-NifDK maturation (78). Second, EXAFS analysis showed that this molybdenum was not adventitiously bound molybdate, but that it was embedded in an [Fe-S] cluster ligand environment within the NifEN protein (25).

Vinod K Shah cluster (VK-cluster): represents an intermediate after NifB-co in FeMo-co synthesis and accumulates on NifEN in a $\Delta nifH$ mutant

IDS: indigo carmine

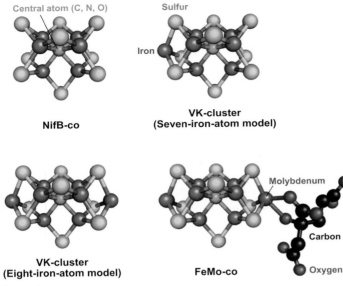

NifB-co

Central atom (C, N, O)

Sulfur

Iron

**VK-cluster
(Seven-iron-atom model)**

**VK-cluster
(Eight-iron-atom model)**

FeMo-co

Molybdenum

Carbon

Oxygen

Figure 7

Model of chemical transformations of the Fe-S core during FeMo-co biosynthesis. NifB-co would comprise a six-iron-atom core with a FeMo-co-like central atom signature. The VK-cluster would contain additional iron but no molybdenum or homocitrate.

The EPR signals from the NifB-co-derived FeMo-co precursor within NifEN were further analyzed by transferring it to purified NifX protein in vitro (32). Contrary to NifEN, NifX does not contain [Fe-S] clusters or any other EPR-active species that could interfere with the analysis. The [Fe-S] cluster transferred from NifEN to NifX served as a FeMo-co precursor in vitro and was designated VK-cluster to honor the pioneer Vinod K. Shah. The VK-cluster exhibited isotropic $S = 1/2$ and axial $S = 1/2$ EPR signals in the reduced and the oxidized states, respectively, with two-electron redox transitions that were fully reversible. The EPR signals from an isolated VK-cluster differed significantly from those attributed to the FeMo-co precursor bound to NifEN (37). The reason for this discrepancy is not clear, but it is our hypothesis that the $S = 1/2$ EPR signal detected in IDS-oxidized NifEN arises from an [Fe-S] cluster different from the VK-cluster. Given that this signal has properties resembling those of [Fe$_3$-S$_4$] clusters, perhaps the signal could

arise from the [MoFe$_3$S$_{3+X}$] cluster detected by EXAFS, from oxidative damage of NifEN permanent [Fe$_4$-S$_4$] clusters, or from additional clusters not yet identified.

Role of NifEN and Its Association with NifH

At least three reactions of the FeMo-co biosynthetic pathway have been proposed to occur within NifEN: incorporation of additional iron, insertion of molybdenum, and incorporation of homocitrate. The first reaction entails the conversion of NifB-co into the VK-cluster. Like NifB-co, the VK-cluster does not contain molybdenum or homocitrate. However, NifB-co and the VK-cluster are electronically different because NifB-co is EPR silent and the VK-cluster shows EPR signals in the reduced and oxidized states (32). Two lines of evidence suggest that additional iron atoms are added to NifB-co to convert it to the VK-cluster. (a) When NifDK protein was matured in a NifB-co-dependent FeMo-co synthesis and insertion assay in the presence of additional [55]FeCl$_3$, it incorporated some [55]Fe label, suggesting additional incorporation of iron after the stage of NifB-co (16). (b) A comparison of EXAFS data from isolated NifB-co and isolated VK-cluster shows differences that favor a six-iron-atom structure for NifB-co (26) and a seven- (or eight-) iron-atom structure for the VK-cluster (M. Demuez, Y. Guo, B. Soboh, S.J. George, R.Y. Igarashi, et al., unpublished results). Consistently, EXAFS analysis of purified NifEN protein loaded with a FeMo-co biosynthetic intermediate (the VK-cluster) fitted to a seven- or eight-iron-atom model (13). **Figure 7** shows plausible NifB-co and VK-cluster structures based on EXAFS analysis.

The second reaction that takes place within NifEN is the incorporation of molybdenum into the FeMo-co precursor, which appears to occur in two steps. Initially, molybdenum is incorporated into a transient site on NifEN. The presence of molybdenum in the $\Delta nifHDK$ NifEN protein implies that this step is not absolutely dependent on NifH, although a role for NifH enhancing molybdenum binding to NifEN cannot be ruled out (78). EXAFS analysis indicates that molybdenum is part of an [Fe-S] cluster at this stage (25). Subsequently, molybdenum is mobilized into the VK-cluster-derived FeMo-co precursor in a reaction that requires NifH and Mg·ATP (36, 38, 57). Reports on whether ATP hydrolysis is required at this step are contradictory. Although there is indirect evidence that NifEN might bind Mg·ATP, no ATP hydrolysis activity was detected in purified NifEN (56) and the attention was turned to the ATP-hydrolyzing protein NifH. In vitro experiments correlate ATP hydrolysis by NifH to molybdenum incorporation into the FeMo-co precursor (38). However, NifH mutant variants that bind but do not hydrolyze Mg·ATP were functional in FeMo-co synthesis in vitro (60) and in vivo (24). Whether the electron transfer ability of NifH is required for this role is also at disagreement. In two different reports, a NifH variant that had its [Fe$_4$-S$_4$] cluster removed by chelation was active (61) and inactive (38) in FeMo-co synthesis. If the role of NifH were to promote the transfer of molybdenum from the transient binding site on NifEN to the VK-cluster to generate the next FeMo-co biosynthetic intermediate, then the simplest interpretation fitting most of these data is that this role would be performed by docking with NifEN and exerting some sort of conformational change on it.

The third reaction would be the incorporation of homocitrate into the precursor to generate FeMo-co, which is thought to be NifH dependent. This proposal is speculative because the binding of homocitrate (e.g., ^{14}C-homocitrate) to NifEN has not been demonstrated, and the processing of homocitrate in vitro requires the presence of other substrates (the VK-cluster and molybdenum) and proteins (NifH). It is not clear whether incorporation of homocitrate directly requires NifH. Homocitrate incorporation follows molybdenum incorporation (57), which is dependent on NifH, and the incorporation of homocitrate into a molybdenum-containing precursor in the absence of NifH has not been reported.

The FeMo-co Biosynthetic Factory Model

Complete in vitro synthesis of FeMo-co from its basic constituents has been reported (15). NifB, NifEN, and NifH were sufficient for FeMo-co synthesis from iron, sulfur, molybdenum, and homocitrate under reducing conditions (the in vitro system also required SAM and Mg·ATP). As expected, NifX and NafY addition stimulated the efficiency of FeMo-co biosynthesis, and the addition of the appropriate chemical forms of homocitrate, iron, sulfur, and molybdenum could substitute for the roles played in vivo by NifV, NifS, NifU, and NifQ.

Because NifB, NifEN, and NifH constitute the essential catalytic components of the FeMo-co synthetic machinery, and because FeMo-co and its biosynthetic intermediates are labile (72, 73), the presence in a nitrogen-fixing bacterium of a multi-protein complex involved in FeMo-co biosynthesis seems likely. The structural core of this machinery would be NifEN in close association with NifH and NifB. Other proteins, such as NifX, NafY, NifQ, and NifV, would also interact with NifEN. There is some experimental evidence consistent with the existence of this complex, which we have designated the FeMo-co biosynthetic factory. First, NifEN and a L127Δ variant of NifH form a stable complex in the presence of NifB-co (59). The L127Δ NifH protein also shows strong interaction with NifDK (48), and it is reasonable to postulate that wild-type NifH would interact with NifEN as it does with NifDK. Second, in vitro FeMo-co synthesis experiments show that molybdenum donation by NifQ requires the concomitant presence of NifEN and NifH (J.A. Hernandez, L. Curatti, C.P. Aznar, Z. Perova, R.D. Britt & L.M. Rubio, unpublished results). Third, NifEN and NifH are simultaneously required for incorporation of molybdate and/or homocitrate into a FeMo-co precursor in vitro, and the products resulting from these reactions bind to both NifEN and NifH (57, 58). Fourth, the *nifB* and *nifN* genes are fused in a single open reading frame in several studied nitrogen-fixing *Clostridium* strains (12). Fifth, the transfer of FeMo-co biosynthetic precursors probably involves protein-protein interactions. Transfer of NifB-co from NifB to NifX (84), or from NifX to NifEN (32), as well as exchange of VK-cluster between NifEN and NifX (32), has been demonstrated in vitro.

The hypothesis of a FeMo-co biosynthetic factory has been formulated before (67) to explain the observed distribution of radioactively labeled FeMo-co precursors between the NifH, NifEN, NifX, and NafY proteins during the biosynthesis of FeMo-co in vitro (2, 3, 57, 58). This hypothesis is also consistent with the recently reported distribution of FeMo-co precursors between NifEN and NifH as analyzed by EXAFS (36, 38).

THE METALLOCLUSTER CARRIER PROTEINS NifX AND NafY

NifX and NafY are members of a family of nitrogenase cofactor binding proteins that would additionally include NifY, the C-terminal domain of NifB, and the VnfX and VnfY proteins [involved in assembly of iron-vanadium cofactor (FeV-co) for the vanadium nitrogenase] (69, 71). These proteins exhibit amino acid sequence conservation and common biochemical properties. NifX is the smallest member of the family and is formed by a single domain. This is also the case for VnfX and VnfY (which are not discussed further in this review). NifB, NifY, and NafY, however, are composed of a C-terminal NifX-like domain and an N-terminal domain that differ in origin and function.

NifX is a 17-kDa monomeric protein that binds structurally related FeMo-co precursors (NifB-co, VK-cluster) and transfers them to the scaffold protein NifEN (32, 58). NifX binds one metal cluster per monomer (32). Binding to NifX increased NifB-co stability in vitro (J.A. Hernandez & L.M. Rubio, unpublished results), and thus it is likely to provide protection to these labile metalloclusters during cofactor synthesis in vivo. Two compatible roles for

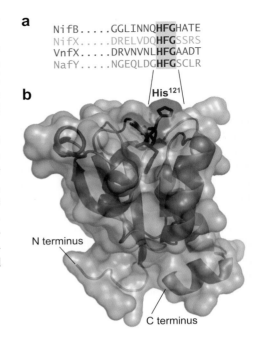

a

```
NifB.....GGLINNQHFGHATE
NifX.....DRELVDQHFGSSRS
VnfX.....DRVNVNLHFGAADT
NafY.....NGEQLDGHFGSCLR
```

His[121]

b

N terminus

C terminus

Figure 8

NafY represents a family of metallocluster carrier proteins. (*a*) The conserved sequence HFG proposed to be involved in metallocluster binding. (*b*) The structure of the metallocluster binding domain of NafY, in which residues 121-His-Phe-Gly-123 have been highlighted.

NifX have been proposed. First, NifX would carry FeMo-co precursors from one scaffold (NifB) to the next (NifEN) during cofactor assembly (32, 84). Second, NifX would serve as storage of FeMo-co precursors, buffering the flux of precursors and redirecting them toward NifEN to complete FeMo-co synthesis (32). NifX also binds FeMo-co, but this fact possibly reflects functional overlap between NifX and NafY, which is the major FeMo-co binding protein in the cell other than NifDK (33).

Although not essential, NafY has been proposed to assist during FeMo-co insertion into apo-NifDK. Two roles for NafY have been proposed: (*a*) stabilization of P-cluster-containing apo-NifDK protein in a conformation amenable to FeMo-co insertion, and (*b*) insertion of FeMo-co into apo-NifDK (33, 69). Consistently, purified NafY protein binds tightly either apo-NifDK or FeMo-co (70). NafY is a 26-kDa monomeric protein composed of two domains that can be separately expressed and purified. The 12-kDa N-terminal domain of NafY is required to bind apo-NifDK (J.A. Hernandez, K.W. Erbil, A. Phillips, D. Zhao, D.E. Wemmer & L.M. Rubio, unpublished results), whereas the 14-kDa C-terminal NifX-like domain is required to bind FeMo-co (20). The three-dimensional structure of the FeMo-co binding domain of NafY has been solved and folds differently than the FeMo-co binding site in NifDK (20) (**Figure 8**). Site-directed mutagenesis studies on NafY suggested the His[121] residue as a likely ligand to FeMo-co (70). This residue is part of a His-Phe-Gly sequence conserved in NifX, NifB, and VnfX, all of which bind either NifB-co or FeMo-co.

The analysis of the roles of NifX and NafY in vivo has been obscured for two reasons: First, there is functional overlap between members of this family and it is likely that they can partially substitute for each other. Second, neither NifX nor NafY plays an essential role in FeMo-co biosynthesis under standard laboratory growth conditions that provide excess iron and molybdenum (69).

SUMMARY POINTS

1. The biosynthesis of FeMo-co is a multi-step process that involves activities of molecular scaffolds, metallocluster carrier proteins, and enzymes.

2. The availability of all proteins involved in FeMo-co biosynthesis and insertion, in purified form, is providing the first insights into the structures of the FeMo-co biosynthetic intermediates.

3. The SAM radical protein NifB transforms common iron-sulfur clusters, generated by NifUS, into NifB-co, the first isolatable iron-sulfur FeMo-co biosynthetic intermediate with complex structure.

4. The central six-iron-atom cage of FeMo-co already exists at the stage of NifB-co, which also exhibits the spectroscopic signature that has been attributed to a central atom X in FeMo-co.

5. NifB, NifEN, and NifH are sufficient to perform all essential reactions to synthesize FeMo-co in vitro from its basic components: molybdenum, iron, sulfur, and homocitrate.

DISCLOSURE STATEMENT

The authors are not aware of any biases that might be perceived as affecting the objectivity of this review.

ACKNOWLEDGMENTS

This work was supported by grant 35332 from the National Institutes of Health. We thank Jose Hernandez and Leonardo Curatti for help preparing the figures.

LITERATURE CITED

1. Agar JN, Yuvaniyama P, Jack RF, Cash VL, Smith AD, et al. 2000. Modular organization and identification of a mononuclear iron-binding site within the NifU protein. *J. Biol. Inorg. Chem.* 5:167–77

2. Allen RM, Chatterjee R, Ludden PW, Shah VK. 1995. Incorporation of iron and sulfur from NifB cofactor into the iron-molybdenum cofactor of dinitrogenase. *J. Biol. Chem.* 270:26890–96

3. Allen RM, Roll JT, Rangaraj P, Shah VK, Roberts GP, Ludden PW. 1999. Incorporation of molybdenum into the iron-molybdenum cofactor of nitrogenase. *J. Biol. Chem.* 274:15869–74

4. Arnold W, Rump A, Klipp W, Priefer UB, Pühler A. 1988. Nucleotide sequence of a 24,206-base-pair DNA fragment carrying the entire nitrogen fixation gene cluster of *Klebsiella pneumoniae*. *J. Mol. Biol.* 203:715–38

5. Barney BM, Igarashi RY, Dos Santos PC, Dean DR, Seefeldt LC. 2004. Substrate interaction at an iron-sulfur face of the FeMo-cofactor during nitrogenase catalysis. *J. Biol. Chem.* 279:53621–24

6. Barney BM, Yang TC, Igarashi RY, Dos Santos PC, Laryukhin M, et al. 2005. Intermediates trapped during nitrogenase reduction of N triple bond N, CH_3-N = NH, and H_2N-NH_2. *J. Am. Chem. Soc.* 127:14960–61

7. Bishop PE, Joerger RD. 1990. Genetics and molecular biology of alternative nitrogen fixation systems. *Annu. Rev. Plant Physiol. Plant Mol. Biol.* 41:109–25

8. Brigle KE, Weiss MC, Newton WE, Dean DR. 1987. Products of the iron-molybdenum cofactor-specific biosynthetic genes, *nifE* and *nifN*, are structurally homologous to the products of the nitrogenase molybdenum-iron protein genes, *nifD* and *nifK*. *J. Bacteriol.* 169:1547–53

9. Broach RB, Rupnik K, Hu Y, Fay AW, Cotton M, et al. 2006. Variable-temperature, variable-field magnetic circular dichroism spectroscopic study of the metal clusters in the Δ*nifB* and Δ*nifH* MoFe proteins of nitrogenase from *Azotobacter vinelandii*. *Biochemistry* 45:15039–48

10. Bulen WA, LeComte JR. 1966. The nitrogenase system from *Azotobacter*: two enzyme requirements for N_2 reduction, ATP dependent H_2 evolution and ATP hydrolysis. *Proc. Natl. Acad. Sci. USA* 56:979–86

11. Chan MK, Kim J, Rees DC. 1993. The nitrogenase FeMo-cofactor and P-cluster pair: 2.2 A resolution structures. *Science* 260:792–94

12. Chen JS, Toth J, Kasap M. 2001. Nitrogen-fixation genes and nitrogenase activity in *Clostridium aceto-butylicum* and *Clostridium beijerinckii*. *J. Ind. Microbiol. Biotechnol.* 27:281–86

13. Corbett MC, Hu Y, Fay AW, Ribbe MW, Hedman B, Hodgson KO. 2006. Structural insights into a protein-bound iron-molybdenum cofactor precursor. *Proc. Natl. Acad. Sci. USA* 103:1238–43

14. Curatti L, Brown CS, Ludden PW, Rubio LM. 2005. Genes required for rapid expression of nitrogenase activity in *Azotobacter vinelandii*. *Proc. Natl. Acad. Sci. USA* 102:6291–96

15. Curatti L, Hernandez JA, Igarashi RY, Soboh B, Zhao D, Rubio LM. 2007. In vitro synthesis of the iron-molybdenum cofactor of nitrogenase from iron, sulfur, molybdenum and homocitrate using purified proteins. *Proc. Natl. Acad. Sci. USA* 104:17626–31

16. Curatti L, Ludden PW, Rubio LM. 2006. NifB-dependent in vitro synthesis of the iron-molybdenum cofactor of nitrogenase. *Proc. Natl. Acad. Sci. USA* 103:5297–301

17. Dos Santos PC, Dean DR, Hu Y, Ribbe MW. 2004. Formation and insertion of the nitrogenase iron-molybdenum cofactor. *Chem. Rev.* 104:1159–74

18. Dos Santos PC, Johnson DC, Ragle BE, Unciuleac MC, Dean DR. 2007. Controlled expression of *nif* and *isc* iron-sulfur protein maturation components reveals target specificity and limited functional replacement between the two systems. *J. Bacteriol.* 189:2854–62

19. Dos Santos PC, Smith AD, Frazzon J, Cash VL, Johnson MK, Dean DR. 2004. Iron-sulfur cluster assembly: NifU-directed activation of the nitrogenase Fe protein. *J. Biol. Chem.* 279:19705–11

20. Dyer DH, Rubio LM, Thoden JB, Holden HM, Ludden PW, Rayment I. 2003. The three-dimensional structure of the core domain of NafY from *Azotobacter vinelandii* determined at 1.8-A resolution. *J. Biol. Chem.* 278:32150–56

21. Einsle O, Tezcan FA, Andrade SL, Schmid B, Yoshida M, et al. 2002. Nitrogenase MoFe-protein at 1.16 A resolution: a central ligand in the FeMo-cofactor. *Science* 297:1696–700

22. Fenske D, Gnida M, Schneider K, Meyer-Klaucke W, Schemberg J, et al. 2005. A new type of metallo-protein: The Mo storage protein from *Azotobacter vinelandii* contains a polynuclear molybdenum-oxide cluster. *ChemBioChem* 6:405–13

23. Fu W, Jack RF, Morgan TV, Dean DR, Johnson MK. 1994. *nifU* gene product from *Azotobacter vinelandii* is a homodimer that contains two identical [2Fe-2S] clusters. *Biochemistry* 33:13455–63

24. Gavini N, Burgess BK. 1992. FeMo cofactor synthesis by a *nifH* mutant with altered Mg · ATP reactivity. *J. Biol. Chem.* 267:21179–86

25. George SJ, Igarashi RY, Piamonteze C, Soboh B, Cramer SP, Rubio LM. 2007. Identification of a Mo-Fe-S cluster on NifEN by Mo K-edge extended X-ray absorption fine structure. *J. Am. Chem. Soc.* 129:3060–61

26. George SJ, Igarashi RY, Xiao Y, Hernandez JA, Demuez M, et al. 2008. EXAFS and NRVS reveal that NifB-co, a FeMo-co precursor, comprises a 6 Fe core with an interstitial light atom. *J. Am. Chem Soc.* In press

27. Georgiadis MM, Komiya H, Chakrabarti P, Woo D, Kornuc JJ, Rees DC. 1992. Crystallographic structure of the nitrogenase iron protein from *Azotobacter vinelandii*. *Science* 257:1653–59

28. Goodwin PJ. 1999. *Biosynthesis of the nitrogenase FeMo-cofactor from* Azotobacter vinelandii: *involvement of the NifNE complex, NifX and the Fe protein*. PhD thesis. Virginia Tech. 210 pp.

29. Goodwin PJ, Agar JN, Roll JT, Roberts GP, Johnson MK, Dean DR. 1998. The *Azotobacter vinelandii* NifEN complex contains two identical [4Fe-4S] clusters. *Biochemistry* 37:10420–28

30. Grunden AM, Shanmugam KT. 1997. Molybdate transport and regulation in bacteria. *Arch. Microbiol.* 168:345–54

31. Hageman RV, Burris RH. 1978. Nitrogenase and nitrogenase reductase associate and dissociate with each catalytic cycle. *Proc. Natl. Acad. Sci. USA* 75:2699–702

32. Hernandez JA, Igarashi RY, Soboh B, Curatti L, Dean DR, et al. 2007. NifX and NifEN exchange NifB cofactor and the VK-cluster, a newly isolated intermediate of the iron-molybdenum cofactor biosynthetic pathway. *Mol. Microbiol.* 63:177–92

33. Homer MJ, Dean DR, Roberts GP. 1995. Characterization of the γ protein and its involvement in the metallocluster assembly and maturation of dinitrogenase from *Azotobacter vinelandii*. *J. Biol. Chem.* 270:24745–52

34. Hoover TR, Imperial J, Ludden PW, Shah VK. 1988. Homocitrate cures the NifV-phenotype in *Klebsiella pneumoniae*. *J. Bacteriol.* 170:1978–79

35. Hoover TR, Robertson AD, Cerny RL, Hayes RN, Imperial J, et al. 1987. Identification of the V factor needed for synthesis of the iron-molybdenum cofactor of nitrogenase as homocitrate. *Nature* 329:855–57

36. Hu Y, Corbett MC, Fay AW, Webber JA, Hodgson KO, et al. 2006. FeMo cofactor maturation on NifEN. *Proc. Natl. Acad. Sci. USA* 103:17119–24

37. Hu Y, Fay AW, Ribbe MW. 2005. Identification of a nitrogenase FeMo cofactor precursor on NifEN complex. *Proc. Natl. Acad. Sci. USA* 102:3236–41

38. Hu YL, Corbettt MC, Fay AW, Webber JA, Hodgson KO, et al. 2006. Nitrogenase Fe protein: a molybdate/homocitrate insertase. *Proc. Natl. Acad. Sci. USA* 103:17125–30

39. Igarashi RY, Seefeldt LC. 2003. Nitrogen fixation: the mechanism of the Mo-dependent nitrogenase. *Crit. Rev. Biochem. Mol. Biol.* 38:351–84

40. Imperial J, Hoover TR, Madden MS, Ludden PW, Shah VK. 1989. Substrate reduction properties of dinitrogenase activated in vitro are dependent upon the presence of homocitrate or its analogues during iron-molybdenum cofactor synthesis. *Biochemistry* 28:7796–99

41. Imperial J, Ugalde RA, Shah VK, Brill WJ. 1984. Role of the *nifQ* gene product in the incorporation of molybdenum into nitrogenase in *Klebsiella pneumoniae*. *J. Bacteriol.* 158:187–94

42. Jacobson MR, Brigle KE, Bennett LT, Setterquist RA, Wilson MS, et al. 1989. Physical and genetic map of the major *nif* gene cluster from *Azotobacter vinelandii*. *J. Bacteriol.* 171:1017–27

43. Jacobson MR, Cash VL, Weiss MC, Laird NF, Newton WE, Dean DR. 1989. Biochemical and genetic analysis of the *nifUSVWZM* cluster from *Azotobacter vinelandii*. *Mol. Gen. Genet.* 219:49–57

44. Joerger RD, Bishop PE. 1988. Nucleotide sequence and genetic analysis of the *nifB-nifQ* region from *Azotobacter vinelandii*. *J. Bacteriol.* 170:1475–87

45. Johnson DC, Dean DR, Smith AD, Johnson MK. 2005. Structure, function, and formation of biological iron-sulfur clusters. *Annu. Rev. Biochem.* 74:247–81

46. Johnson DC, Dos Santos PC, Dean DR. 2005. NifU and NifS are required for the maturation of nitrogenase and cannot replace the function of isc-gene products in *Azotobacter vinelandii*. *Biochem. Soc. Trans.* 33:90–93

47. Kim J, Rees DC. 1992. Crystallographic structure and functional implications of the nitrogenase molybdenum-iron protein from *Azotobacter vinelandii*. *Nature* 360:553–60

48. Lanzilotta WN, Fisher K, Seefeldt LC. 1996. Evidence for electron transfer from the nitrogenase iron protein to the molybdenum-iron protein without Mg·ATP hydrolysis: characterization of a tight protein-protein complex. *Biochemistry* 35:7188–96

49. Lanzilotta WN, Parker VD, Seefeldt LC. 1998. Electron transfer in nitrogenase analyzed by Marcus theory: evidence for gating by Mg·ATP. *Biochemistry* 37:399–407

50. Ludden PW, Rangaraj P, Rubio LM. 2004. Biosynthesis of the iron-molybdenum and iron-vanadium cofactors of the *nif-* and *vnf-*encoded nitrogenases. In *Catalysts for Nitrogen Fixation: Nitrogenases, Relevant Chemical Models, and Commercial Processes*, ed. BE Smith, RL Richards, WE Newton, pp. 219–53. Dordretch, The Neth.: Kluwer

51. MacNeil T, MacNeil D, Roberts GP, Supiano MA, Brill WJ. 1978. Fine-structure mapping and complementation analysis of *nif* (nitrogen fixation) genes in *Klebsiella pneumoniae*. *J. Bacteriol.* 136:253–66

52. Martinez-Argudo I, Little R, Shearer N, Johnson P, Dixon RA. 2004. The NifL-NifA system: a multidomain transcriptional regulatory complex that integrates environmental signals. *J. Bacteriol.* 186:601–10

53. Masepohl B, Schneider K, Drepper T, Muller A, Klipp W. 2002. Alternative nitrogenases. In *Nitrogen Fixation at the Millenium*, ed. GJ Leigh, pp. 191–222. Amsterdam: Elsevier

54. Mayer SM, Gormal CA, Smith BE, Lawson DM. 2002. Crystallographic analysis of the MoFe protein of nitrogenase from a *nifV* mutant of *Klebsiella pneumoniae* identifies citrate as a ligand to the molybdenum of iron molybdenum cofactor (FeMoco). *J. Biol. Chem.* 277:35263–66

55. McLean PA, Smith BE, Dixon RA. 1983. Nitrogenase of *Klebsiella pneumoniae nifV* mutants. *Biochem. J.* 211:589–97

56. Paustian TD, Shah VK, Roberts GP. 1989. Purification and characterization of the *nifN* and *nifE* gene products from *Azotobacter vinelandii* mutant UW45. *Proc. Natl. Acad. Sci. USA* 86:6082–86

57. Rangaraj P, Ludden PW. 2002. Accumulation of [99]Mo-containing iron-molybdenum cofactor precursors of nitrogenase on NifNE, NifH, and NifX of *Azotobacter vinelandii*. *J. Biol. Chem.* 277:40106–11

58. Rangaraj P, Ruttimann-Johnson C, Shah VK, Ludden PW. 2001. Accumulation of [55]Fe-labeled precursors of the iron-molybdenum cofactor of nitrogenase on NifH and NifX of *Azotobacter vinelandii*. *J. Biol. Chem.* 276:15968–74

59. Rangaraj P, Ryle MJ, Lanzilotta WN, Goodwin PJ, Dean DR, et al. 1999. Inhibition of iron-molybdenum cofactor biosynthesis by L127Δ NifH and evidence for a complex formation between L127Δ NifH and NifNE. *J. Biol. Chem.* 274:29413–19

60. Rangaraj P, Ryle MJ, Lanzilotta WN, Ludden PW, Shah VK. 1999. In vitro biosynthesis of iron-molybdenum cofactor and maturation of the *nif-*encoded apodinitrogenase. Effect of substitution for NifH with site-specifically altered forms of NifH. *J. Biol. Chem.* 274:19778–84

61. Rangaraj P, Shah VK, Ludden PW. 1997. ApoNifH functions in iron-molybdenum cofactor synthesis and apodinitrogenase maturation. *Proc. Natl. Acad. Sci. USA* 94:11250–55

62. Raymond J, Siefert JL, Staples CR, Blankenship RE. 2004. The natural history of nitrogen fixation. *Mol. Biol. Evol.* 21:541–54

63. Ribbe MW, Hu Y, Guo M, Schmid B, Burgess BK. 2002. The FeMoco-deficient MoFe protein produced by a *nifH* deletion strain of *Azotobacter vinelandii* shows unusual P-cluster features. *J. Biol. Chem.* 277:23469–76

64. Roberts GP, MacNeil T, MacNeil D, Brill WJ. 1978. Regulation and characterization of protein products coded by the *nif* (nitrogen fixation) genes of *Klebsiella pneumoniae*. *J. Bacteriol.* 136:267–79

65. Rodríguez-Quiñones F, Bosch R, Imperial J. 1993. Expression of the *nifBfdxNnifOQ* region of *Azotobacter vinelandii* and its role in nitrogenase activity. *J. Bacteriol.* 175:2926–35

66. Roll JT, Shah VK, Dean DR, Roberts GP. 1995. Characteristics of NIFNE in *Azotobacter vinelandii* strains. Implications for the synthesis of the iron-molybdenum cofactor of dinitrogenase. *J. Biol. Chem.* 270:4432–37

67. Rubio LM, Ludden PW. 2002. The gene products of the *nif* regulon. In *Nitrogen Fixation at the Millenium*, ed. GJ Leigh, pp. 101–36. Amsterdam: Elsevier

68. Rubio LM, Ludden PW. 2005. Maturation of nitrogenase: a biochemical puzzle. *J. Bacteriol.* 187:405–14

69. Rubio LM, Rangaraj P, Homer MJ, Roberts GP, Ludden PW. 2002. Cloning and mutational analysis of the γ gene from *Azotobacter vinelandii* defines a new family of proteins capable of metallocluster-binding and protein stabilization. *J. Biol. Chem.* 277:14299–305

70. Rubio LM, Singer SW, Ludden PW. 2004. Purification and characterization of NafY (apodinitrogenase γ subunit) from *Azotobacter vinelandii*. *J. Biol. Chem.* 279:19739–46

71. Ruttimann-Johnson C, Rubio LM, Dean DR, Ludden PW. 2003. VnfY is required for full activity of the vanadium-containing dinitrogenase in *Azotobacter vinelandii*. *J. Bacteriol.* 185:2383–86

72. Shah VK, Allen JR, Spangler NJ, Ludden PW. 1994. In vitro synthesis of the iron-molybdenum cofactor of nitrogenase. Purification and characterization of NifB cofactor, the product of NIFB protein. *J. Biol. Chem.* 269:1154–58

73. Shah VK, Brill WJ. 1977. Isolation of an iron-molybdenum cofactor from nitrogenase. *Proc. Natl. Acad. Sci. USA* 74:3249–53

74. Shah VK, Imperial J, Ugalde RA, Ludden PW, Brill WJ. 1986. In vitro synthesis of the iron-molybdenum cofactor of nitrogenase. *Proc. Natl. Acad. Sci. USA* 83:1636–40

75. Shah VK, Stacey G, Brill WJ. 1983. Electron transport to nitrogenase. Purification and characterization of pyruvate:flavodoxin oxidoreductase. The *nifJ* gene product. *J. Biol. Chem.* 258:12064–68

76. Simpson FB, Burris RH. 1984. A nitrogen pressure of 50 atmospheres does not prevent evolution of hydrogen by nitrogenase. *Science* 224:1095–97

77. Smith AD, Jameson GN, Dos Santos PC, Agar JN, Naik S, et al. 2005. NifS-mediated assembly of [4Fe-4S] clusters in the N- and C-terminal domains of the NifU scaffold protein. *Biochemistry* 44:12955–69

78. Soboh B, Igarashi RY, Hernandez JA, Rubio LM. 2006. Purification of a NifEN protein complex that contains bound Mo and a FeMo-co precursor from an *Azotobacter vinelandii* Δ*nifHDK* strain. *J. Biol. Chem.* 281:36701–9

79. Sofia HJ, Chen G, Hetzler BG, Reyes-Spindola JF, Miller NE. 2001. Radical SAM, a novel protein superfamily linking unresolved steps in familiar biosynthetic pathways with radical mechanisms: functional characterization using new analysis and information visualization methods. *Nucleic Acids Res.* 29:1097–106

80. Ugalde RA, Imperial J, Shah VK, Brill WJ. 1984. Biosynthesis of iron-molybdenum cofactor in the absence of nitrogenase. *J. Bacteriol.* 159:888–93

81. Ugalde RA, Imperial J, Shah VK, Brill WJ. 1985. Biosynthesis of the iron-molybdenum cofactor and the molybdenum cofactor in *Klebsiella pneumoniae*: effect of sulfur source. *J. Bacteriol.* 164:1081–87

82. Wang SC, Frey PA. 2007. S-adenosylmethionine as an oxidant: the radical SAM superfamily. *Trends Biochem. Sci.* 32:101–10

83. Yuvaniyama P, Agar JN, Cash VL, Johnson MK, Dean DR. 2000. NifS-directed assembly of a transient [2Fe-2S] cluster within the NifU protein. *Proc. Natl. Acad. Sci. USA* 97:599–604

84. Zhao D, Curatti L, Rubio LM. 2007. Evidence for *nifU* and *nifS* participation in the biosynthesis of the iron-molybdenum cofactor of nitrogenase. *J. Biol. Chem.* 282:37016–25

85. Zheng L, Cash VL, Flint DH, Dean DR. 1998. Assembly of iron-sulfur clusters. Identification of an *iscSUA-hscBA-fdx* gene cluster from *Azotobacter vinelandii*. *J. Biol. Chem.* 273:13264–72

86. Zheng L, White RH, Cash VL, Dean DR. 1994. Mechanism for the desulfurization of L-cysteine catalyzed by the *nifS* gene product. *Biochemistry* 33:4714–20

87. Zheng L, White RH, Cash VL, Jack RF, Dean DR. 1993. Cysteine desulfurase activity indicates a role for NIFS in metallocluster biosynthesis. *Proc. Natl. Acad. Sci. USA* 90:2754–58

88. Zheng L, White RH, Dean DR. 1997. Purification of the *Azotobacter vinelandii nifV*-encoded homocitrate synthase. *J. Bacteriol.* 179:5963–66

Chlamydiae as Symbionts in Eukaryotes

Matthias Horn

Department of Microbial Ecology, University of Vienna, A-1090 Vienna, Austria;
email: horn@microbial-ecology.net

Annu. Rev. Microbiol. 2008. 62:113–31

First published online as a Review in Advance on
May 12, 2008

The *Annual Review of Microbiology* is online at
micro.annualreviews.org

This article's doi:
10.1146/annurev.micro.62.081307.162818

Key Words

chlamydia, symbiosis, amoeba, evolution

Abstract

Members of the phylum *Chlamydiae* are obligate intracellular bacteria
that were discovered about a century ago. Although *Chlamydiae* are major pathogens of humans and animals, they were long recognized only
as a phylogenetically well-separated, small group of closely related microorganisms. The diversity of chlamydiae, their host range, and their
occurrence in the environment had been largely underestimated. Today, several chlamydia-like bacteria have been described as symbionts
of free-living amoebae and other eukaryotic hosts. Some of these environmental chlamydiae might also be of medical relevance for humans.
Their analysis has contributed to a broader understanding of chlamydial biology and to novel insights into the evolution of these unique
microorganisms.

Contents

INTRODUCTION

It has to be tried to collect by the term 'Chlamy-
dozoa' a group of peculiar microorganisms that do
belong neither to the protozoa nor to the bacteria.
They pass filters, cause inclusion bodies and will
multiply as contagium only in egg culture.

           Stanislaus von Prowazek, 1912 (58)

In Java, Indonesia, in 1907, the German radi-
ologist Ludwig Halberstädter and the Austrian
zoologist Stanislaus von Prowazek went on a
research expedition to find the causative agent
of syphilis. Among the discoveries they brought
back from this trip was a conspicuous agent they
considered responsible for trachoma, which
was at that time a global disease (47). Within
Giemsa-stained conjunctival epithelial cells of
trachoma patients they had found irregularly
blue-stained inclusions with small, dense particles, which they called "Chlamydozoa" (from
the Greek word χλαμυσ, meaning mantle or
cloak) (47). Originally considered neither protozoa nor bacteria and then regarded as viruses,
in the 1960s they were recognized as bacteria
(84). Later, these unique microorganisms were
found to be among the most important bacterial
pathogens of humankind.

    Halberstädter's and Prowazek's Chlamydo-
zoa are now called *Chlamydia trachomatis*, and it
is the most prominent representative of a small
group of closely related bacteria, the chlamydiae. Trachoma affects about 84 million people,
of whom about 8 million are visually impaired
as a consequence (103). *C. trachomatis* is also
the most common cause of sexually transmitted
diseases, with over 90 million new cases each
year (102). The second prime human pathogen
among the chlamydiae is *Chlamydophila pneumoniae*, a causative agent of pneumonia, which has
also been associated with a number of chronic
diseases such as atherosclerosis, asthma, and
Alzheimer's disease (67). In addition, several
other chlamydial species are primarily considered pathogens of animals, but some of them
also show zoonotic potential (67).

    Chlamydiae were long considered to com-
prise exclusively obligate intracellular bacterial
pathogens that show a characteristic developmental cycle, including metabolically inert elementary bodies (EBs) and actively dividing
reticulate bodies (RBs), which thrive within a
host-derived vacuole termed inclusion (1). This
phylogenetically well-isolated group of closely
related bacteria constituted the single family
Chlamydiaceae of the order *Chlamydiales*, which
form a separate phylum in the domain *Bacteria*, the *Chlamydiae* (**Figure 1**). Our perception
of chlamydial diversity changed substantially
when in the 1990s novel chlamydia-like bacteria
were discovered. This review summarizes our
current knowledge about these novel chlamydiae (also called environmental chlamydiae).

EB: elementary body

RB: reticulate body

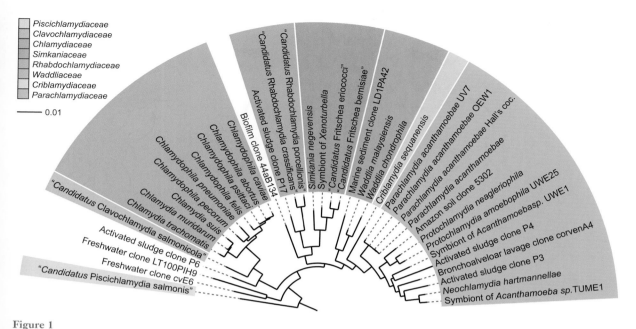

Figure 1

Phylogenetic 16S rRNA tree showing relationships among members of the phylum *Chlamydiae*. Representatives of all recognized families and a few clone sequences are shown. The tree was calculated with the ARB software, using a manually curated version of the SILVA database (16S/18S reference database, Version 92) and the PhyML treeing algorithm implemented in ARB, and visualized using the iTOL tool (43, 74, 77, 88). Bar, 10% estimated evolutionary distance.

DIVERSITY AND ECOLOGY OF CHLAMYDIA-LIKE BACTERIA

Ultrastructural evidence of the presence of unidentified bacteria resembling known chlamydiae was available in the 1980s (92). Yet it took another decade until the first chlamydia-like bacteria were identified on the molecular level, initially in laboratory cell culture (66) but mainly as symbionts of free-living amoebae (2, 5, 31, 54) [in this review the terms symbiosis and symbiont are used sensu deBary, i.e., to include mutualism, parasitism, and commensalism (19)]. These findings were the beginning of a series of reports describing a continuing increase of recognized diversity within the *Chlamydiae* (**Figure 2**) (16). Remarkably, these bacteria are not mere phylogenetic relatives of the *Chlamydiaceae*; they also show an obligate intracellular lifestyle and the unique chlamydial developmental cycle (**Figure 3**).

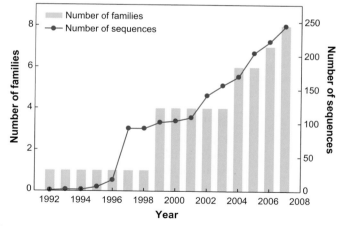

Figure 2

Increase in described diversity of the phylum *Chlamydiae* from 1992 to 2007. Numbers of 16S rRNA sequences deposited in public databases (>1000 nucleotides) and numbers of chlamydial families based on described chlamydiae [using the 90% 16S rRNA sequence similarity threshold suggested by Everett et al. (25) and Kuo et al. (73)] are shown.

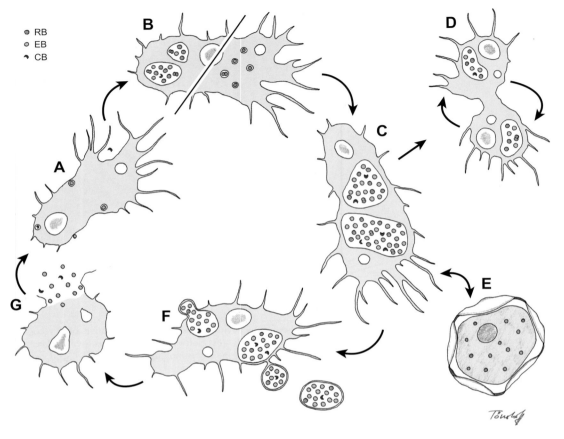

Figure 3

The developmental cycle of chlamydia-like bacteria in free-living amoebae. Elementary bodies (EBs) are shown in blue; reticulate bodies (RBs) are in red, the proposed crescent bodies (CBs) of *Parachlamydia acanthamoebae* are shown in dark purple. The occurrence of *P. acanthamoebae*, *Simkania negevensis*, and *Waddlia chondrophila* in larger inclusions compared with *Protochlamydia amoebophila* generally found in single-cell inclusion is indicated in step B; see text for further details. Figure by E. Toenshoff.

Symbionts of Free-Living Amoebae

Free-living amoebae are ubiquitous protozoa. They are the main predators controlling bacterial populations in soil, but they are also present in marine and freshwater habitats and the air (91). Frequently, they are found in human-made environments such as public water supplies, air-conditioning units, and sewage treatment plants. As a result, they can also be isolated from the human body, e.g., nasal cavities (69). Consistent with our common exposure to these amoebae, close to 90% of healthy individuals show specific immunoglobulin G re-

activity with free-living amoebae of the genus *Acanthamoeba* (10). Ubiquitous *Acanthamoeba* species frequently contain bacterial endosymbionts, and in about 5% of all *Acanthamoeba* isolates studied by Fritsche et al. (30, 31) these symbionts were novel chlamydia-like bacteria. To date, only a few of these chlamydial symbionts of amoebae have been characterized to some extent.

Parachlamydia acanthamoebae Bn$_9$ was discovered in an *Acanthamoeba* isolate recovered from a human nasal swab of a healthy volunteer (2, 25, 81). *Neochlamydia hartmannellae* A$_1$Hsp was detected in a *Hartmannella vermiformis* isolate from a water conduit system of a dental care unit (54), *Protochlamydia amoebophila* UWE25 was originally found in an *Acanthamoeba* isolate from a soil sample (13, 31), and *Protochlamydia naegleriophila* was identified in a *Naegleria* sp. (9). All these bacteria thrive as symbionts in amoebae, and thus it might be tempting to simply lump them together as chlamydia-like symbionts. It is, however, important to recognize that they are not closely related but that they are separated by roughly 300 million years of evolution [if a divergence rate of 1%–2% per 50 million years for the 16S ribosomal RNA gene is assumed (85)]. In fact, these chlamydia-like symbionts represent three distinct genera within the family *Parachlamydiaceae* (**Figure 1**) (25) and show remarkable differences with respect to their biology and the interaction with their host cells (see below).

Cocultivation with amoebae was used successfully to recover novel chlamydiae from complex environmental samples, although in this case the natural host remains unknown (12, 99). For example, the *P. acanthamoebae* strains UV7 and Seine and *Criblamydia sequanensis* (**Figure 1**) were identified by this method from activated sludge and water samples of the Seine river, respectively (12, 99).

Simkania negevensis was discovered as contaminant in a human cell culture (66). Similar to *C. sequanensis*, its original host is thus unknown, but it grows well in *Acanthamoeba* spp. and other amoebae (59, 82). Given the ubiquity of free-living amoebae, it is tempting to speculate that they serve as natural hosts for *S. negevensis* in the environment.

Other Nonhuman Hosts

In addition to chlamydial symbionts of amoebae, a number of other chlamydia-like bacteria were identified in phylogenetically diverse animal hosts. *Waddlia chondrophila* was recovered from an aborted bovine fetus (94). *Waddlia malaysiensis* was detected in urine of fruit bats (11). Representatives of two novel *Chlamydiales* families, the *Piscichlamydiaceae* and the *Clavochlamydiaceae*, were identified in gill tissues of fish suffering from epitheliocystis (22, 68). *Rhabdochlamydia porcellionis* and *R. crassificans* infect terrestrial isopods and cockroaches, respectively (15, 71). *Fritschea bemisiae* and *F. eriococci* were described in insects (98), and a yet unnamed chlamydial symbiont was detected in the enigmatic worms *Xenoturbella bocki* and *X. westbladi* (57). Taken together, the chlamydiae identified so far show a broad host range across the animal kingdom (**Table 1**).

Free-living amoebae seem to play a special role in the ecology of chlamydiae, as many of them, including *Chlamydophila pneumoniae* (23), survive and/or multiply in acanthamoebae (2, 31, 54, 59, 83). Amoebae, particularly *Acanthamoeba* species, might thus serve as major reservoirs and vehicles of dispersal for chlamydiae. Compared with other protozoa and most other amoebae, *Acanthamoeba* species, however, are easy to isolate, to adapt to, and to maintain in axenic culture in the laboratory. Our current view of *Acanthamoeba* species as major hosts for chlamydia-like bacteria thus might be biased toward these amoebae. In fact, as amoebae are a polyphyletic, largely unrelated assemblage of protists, it would not be surprising if other protozoa are hosts for chlamydial symbionts.

The Tip of the Iceberg

As illustrated above, known diversity of *Chlamydiae* has increased from one to currently eight families during the past decade [if the 90% 16S rRNA sequence similarity threshold suggested by Everett et al. (25) and Kuo et al. (73) is applied] (**Figures 1** and **2**). Eleven of 13 genera in these families are, however, represented currently by only one (7 genera) or two (4 genera) species or isolates, indicating that the actual diversity within the recognized chlamydial families is even larger and yet undiscovered. Indeed, molecular evidence suggests that the recognized diversity of the chlamydiae is just the tip of the iceberg (16–18, 53, 86). A large number of phylogenetically diverse

Table 1 Host range of the phylum *Chlamydiae*[a]

Group	Host	Chlamydia trachomatis	Chlamydia muridarum	Chlamydia suis	Chlamydophila abortus	Chlamydophila caviae	Chlamydophila felis	Chlamydophila pecorum	Chlamydophila pneumoniae	Chlamydophila psittaci	"Candidatus Clavochlamydia salmonicola"	Neochlamydia hartmannellae	Neochlamydia sp.	Parachlamydia acanthamoebae	Parachlamydia sp.	Protochlamydia amoebophila	Protochlamydia naegleriophila	"Candidatus Fritschea eriococci"	"Candidatus Fritschea bemisiae"	"Candidatus Piscichlamydia salmonis"	Simkania negevensis	Waddlia chondrophila	Waddlia malayensis	"Candidatus Rhabdochlamydia crassificans"	"Candidatus Rhabdochlamydia porcellionis"	Rhabdochlamydia sp.	Unidentified Chlamydiae
Vertebrates																											
Mammals	Human (*Homo sapiens*)	■							■				▨								■	▨					
	Cat (*Felis silvestris catus*)						■			■																	
	Pig (*Sus* sp.)			■	■			■							▨												■
	Cattle (*Bos taurus*)							■		■	▨																■
	African buffalo (*Syncerus caffer*)									■																	
	Water buffalo (*Bubauis* sp.)									■																	
	Chamois (*Rupicapra rupicapra*)				■																						
	Sheep (*Ovis aeries*)				■																						
	Arabian oryx (*Oryx leucoryx*)									▨																	
	Blackbuck (*Antilope cervicapra*)									▨																	
	Fallow deer (*Cervus dama*)									▨																	
	Fallow deer (*Dama dama*)																										▨
	Red deer (*Cervus elaphus*)																										▨
	Reindeer (*Rangifer tarandus*)																										▨
	Mule deer (*Odocoileus hemionus*)																										■
	Mouflon (*Ovis mussimon*)																										▨
	Spanish ibex (*Capra pyrenaica*)																										▨
	Goat (*Capra aegagrus hircus*)				■			■																			
	Mouse (*Mus musculus*)		■																								
	Hamster (*Mesocricetus auratus*)		■																								
	Horse (*Equus caballus*)								■	■																	
	Guinea pig (*Cavia porcellus*)					■																					
	Fruit bat (*Eonycteris spelaea*)									■													■				
Marsupials	Koala (*Phascolarctos cinereus*)							■		▨																	▨
	Great glider (*Petauroides volans*)							▨																			▨
	Mountain brushtail possum (*Trichosurus caninus*)							▨																			▨
	Greater gilby (*Macrotis lagotis*)															▨											
	Western barred bandicoot (*Perameles bouganville*)							▨																			
	Gilbert's potoroo (*Potorous gilbertii*)																							▨			
Amphibians	Great barred frog (*Mixophyes iteratus*)								■																		
	African clawed frog (*Xenopus tropicalis*)								■																		
	Blue Mountains tree frog (*Litoria citropa*)								■																		
	Common frog (*Rana temporaria*)			▨					■																		
	African clawed frog (*Xenopus laevis*)																										▨
Birds	>100 species									■																	
	Chicken (*Gallus gallus*)									■																	
Reptiles	Green sea turtle (*Chelonia mydas*)			■					■			■															
	Burmese python (*Python mulurus bivittatus*)			▨					▨																		
	Puff adder (*Bitis arietans*)			■					■																		
	Snake (unspecified)								■																		▨
	Chelonian (unspecified)								■																		
	Lizard (unspecified)																										▨
	Chameleon (*Chameleo dilepis*)						■																				
	Iguana (*Iguana iguana*)						■																				
	Nile crocodile (*Crocodylus niloticus*)									■																	
Fish	Atlantic salmon (*Salmo salar*)										■									■							
	Wild trout (*Salmo truttar*)										■																
	Leafy seadragon (*Phycodurus eques*)																									■	
	Silver perch (*Bidyanus bidyanus*)										▨																
	Barramundi (*Lates calcarifer*)										■																
	Arctic charr (*Salvelinus alpinus*)										▨																

(Continued)

Table 1 (*Continued*)

		C. trachomatis	C. muridarum	C. suis	Cp. abortus	Cp. caviae	Cp. felis	Cp. pecorum	Cp. pneumoniae	Cp. psittaci	"Ca. Clavochlamydia salmonicola"	Neochlamydia hartmannellae	Neochlamydia sp.	Parachlamydia acanthamoebae	Parachlamydia sp.	Protochlamydia amoebophila	Protochlamydia naegleriophila	"Ca. Fritschea eriococci"	"Ca. Fritschea bemisiae"	"Ca. Piscichlamydia salmonis"	Simkania negevensis	Waddlia chondrophila	Waddlia malayensis	"Ca. Rhabdochlamydia crassificans"	"Ca. Rhabdochlamydia porcellionis"	Rhabdochlamydia sp.	Unidentified Chlamydiae
Invertebrates																											
Insects	Sweetpotato whitefly (*Bemisia tabaci*)																		■								
	Imported elm bark louse (*Eriococcus spurius*)																	■									
	Oriental cockroach (*Blatta orientalis*)																							■			
Crustaceans	Rough woodlouse (*Porcellio scaber*)																								■		
Molluscs	Pacific oyster (*Crassostrea gigas*)				▨																						
Protozoa	*Acanthamoeba* sp.													■		■											
	Hartmannella vermiformis											■															
	Naegleria lovaniensis																■										

[a]Evidence for the presence of chlamydiae by 16S rRNA analysis in combination with microscopic analyses (immunofluorescence, electron microscopy, histology) or for the recovery of the respective organism is indicated by dark blue boxes. Evidence for the presence of chlamydiae by only 16S rRNA analysis or serology without microscopic data is indicated by light blue boxes. Due to the revision of chlamydial taxonomy in 1999 (25), evidence for *Chlamydophila psittaci* (formerly *Chlamydia psittaci*) could also refer to *Chlamydophila abortus*, *Chlamydophila felis*, or *Chlamydophila caviae*. Follow the Supplemental Material link from the Annual Reviews home page at http://www.annualreviews.org for a fully annotated table, including references.

rRNA sequences most similar to recognized chlamydiae have been detected in an impressive variety of habitats, including soils, sediments, aquatic environments, hydrothermal vent fluid, and engineered environments such as activated sludge and anaerobic bioreactors (**Supplemental Table 1** and **Supplemental Table 2**; follow the **Supplemental Material link** from the Annual Reviews home page at **http://www.annualreviews.org**). Owing to the varying length of the sequences obtained in the different studies (from <300 to ~1500 bp), it is impossible to estimate the true diversity of *Chlamydiae* from these data. However, if only sequences greater than 1000 nucleotides are considered, at least 11 operational taxonomic units (OTUs) at the family level and at least 27 OTUs at the genus level exist (corresponding to more than 90% and 95% 16S rRNA sequence similarity, respectively).

Taken together, there is molecular evidence of a previously unseen level of diversity of chlamydiae in the environment. *Chlamydiae* have a great and previously underestimated host range and show an almost ubiquitous distribution in the environment. It will be a major challenge to find, isolate and characterize those elusive microorganisms for which to date merely rRNA sequences exist.

CHLAMYDIA-LIKE BACTERIA: IMPLICATIONS FOR PUBLIC HEALTH?

Chlamydiae have been detected in diverse habitats; they are present in various animals and their major hosts, free-living amoebae, are ubiquitous. Therefore, humans are frequently exposed to chlamydia-like bacteria. Is this a concern for public health? Can chlamydia-like bacteria infect humans and cause disease, and thus should they be considered new emerging pathogens?

A number of reports suggest an association between chlamydia-like bacteria and human disease (14, 17, 18, 46, 86). Most recently, *Protochlamydia naegleriophila* was detected in a bronchoalveolar lavage sample of an immunocompromised pneumonia patient (9), and *Waddlia chondrophila* was implicated in a controlled, serology-based study with human fetal death (4). Accumulating data are available

rRNA: ribosomal RNA

for two chlamydia-like bacteria, *Parachlamydia acanthamoebae* and *Simkania negevensis*.

The Case of *Parachlamydia acanthamoebae*

In the laboratory, *P. acanthamoebae* infects and to a limited degree multiplies in simian and human cell lines (8, 12, 40). Proliferation of *P. acanthamoebae* in monocyte-derived macrophages, pneumocytes, and lung fibroblasts was, however, about three orders of magnitude slower than observed for the *Chlamydiaceae* or other pathogens such as *Legionella pneumophila* (8, 40). In addition, *P. acanthamoebae* showed major differences with respect to intracellular trafficking and host cell response compared with the *Chlamydiaceae* (36, 39).

P. acanthamoebae has been implicated primarily in respiratory disease (14), including community-acquired pneumonia, bronchitis, and aspiration pneumonia, but also in atherosclerosis (5, 18, 33, 34, 37, 79). The number of available studies is limited ($n = 5$), and the reported associations between *P. acanthamoebae* and human disease are based exclusively on serological (immunofluorescence) and molecular (PCR, nested PCR, real-time PCR) evidence. To date, *P. acanthamoebae* has not been isolated from a patient nor was its presence demonstrated at the site of infection.

The Case of *Simkania negevensis*

Simkania negevensis can thrive in a variety of human and simian cells, including Vero, BGM (buffalo green monkey), HeLa, human fibroblasts, McCoy, and HL cells (62), as well as in U973 macrophages, in epithelial cells originating from the respiratory or the genital tract, and in cells of vascular endothelial origin (61). However, *S. negevensis* takes longer to complete the developmental cycle than the *Chlamydiaceae* (7–12 versus 2–3 days) (60, 65). On the basis of serological and molecular evidence, *S. negevensis* has been associated with bronchiolitis and community-acquired pneumonia in children and with exacerbation of chronic obstructive pulmonary disease in adults (27, 32,

49, 63, 75, 76). A few studies also reported the isolation of *S. negevensis* from specimens of bronchiolitis and pneumonia patients (29, 64, 72). An association between *S. negevensis* and acute rejection in lung transplant recipients was supported not only by PCR but also by culturing the organism from bronchoalveolar lavage (56).

At first sight, these data appear to be convincing evidence of an association of *S. negevensis* with respiratory disease. However, the high seroprevalence of *S. negevensis* in healthy individuals, ranging from 39% to 68% and obviously increasing with age (29), shows that controlled studies are required for a reliable estimation of the contribution of *S. negevensis* to human disease. In one such study, no statistically significant association was found between respiratory disease and seropositivity or positive PCR for *S. negevensis* (72). Similarly, an association of *S. negevensis* with the onset of asthma in children was proposed but could not be confirmed in a controlled serology-based study (70). *S. negevensis* was cultured from 21 of 35 (60%) nasopharyngeal washing samples from healthy individuals and from 19 of 34 (56%) water samples analyzed in a recent study (64). Thus it is unclear whether an *S. negevensis*-positive culture from specimens such as bronchoalveolar lavage or nasopharyngeal washings indeed supports an association with disease or rather reflects the ubiquity of these bacteria in our surroundings.

Causal Relationships?

As illustrated above, the ubiquity of chlamydia-like bacteria and their amoeba hosts poses a serious problem for investigating the role of chlamydia-like bacteria as emerging pathogens. In addition, the lack of standardized diagnostic tests further aggravates the assessment of the actual prevalence of chlamydia-like bacteria in clinical specimens. Therefore, despite much work that has been done in this field, the key question remains, Is there a causal relationship between chlamydia-like bacteria and human disease? To answer this, the guidelines summarized by Fredericks & Relman (28) in their reconsideration of Koch's Postulates might be helpful. In brief, the putative pathogen should be present in most cases of an infectious disease and should be found preferentially in those anatomic sites known to be diseased. The pathogen should be visualized at the cellular level in these specimens, and it should be absent or less abundant in hosts or tissues without disease. With resolution of disease, the evidence of the presence of the pathogen should decrease. A causal relationship is more likely if the abundance of the pathogen correlates with severity of disease. The evidence of a causal relationship should be reproducible (28), i.e., concordant evidence should come from different approaches applied by different groups, at different time points, and in different places (3). To date, virtually none of these conditions has been fulfilled for chlamydia-like bacteria and human disease. Demonstrating a causal relationship thus remains a great challenge. Perhaps, chlamydia-like bacteria are opportunistic pathogens rather than a serious threat for healthy individuals.

Infections of Animals

Chlamydia-like bacteria infect a wide range of animals, in addition to humans (**Table 1**). In this context, the association of "*Candidatus* Piscichlamydia salmonis," "*Candidatus* Clavochlamydia salmonicola," and some unnamed chlamydia-like bacteria with epitheliocystis in fish and sea dragons seems to be supported particularly well (22, 68, 80). Although these organisms have not yet been obtained in cell culture, they were identified by sequencing their 16S rRNA genes, and they could be localized in epitheliocystis cysts in the gills using in situ hybridization and electron microscopy (22, 68). In addition, an association of *Waddlia chondrophila*, repeatedly isolated from aborted bovine fetuses (50, 94), with abortion in cattle was suggested (21), but more data are required to substantiate this. Future studies investigating the prevalence of chlamydia-like bacteria in animals and their role as potential veterinary pathogens are urgently needed.

BIOLOGY OF *CHLAMYDIA*-LIKE BACTERIA

Irrespective of their pathogenic potential, chlamydia-like bacteria deserve attention because they are the closest relatives of the *Chlamydiaceae*. Our knowledge about this diverse group is still scarce, primarily because of the obligate intracellular lifestyle, which hampers the ability to study these elusive microorganisms. Most available data come from the model organisms *Protochlamydia amoebophila*, *Parachlamydia acanthamoebae*, and *Simkania negevensis*.

Developmental Cycle

All chlamydia-like bacteria characterized so far show a developmental cycle similar to the *Chlamydiaceae*, with morphologically and physiologically distinct stages. Considerable diversity, however, exists with respect to the ultrastructure of the developmental forms. The coccoid EBs and RBs of *Parachlamydia acanthamoebae*, *Protochlamydia amoebophila*, *Neochlamydia hartmannellae*, *Simkania negevensis*, and *Waddlia chondrophila* resemble those of the *Chlamydiaceae*, whereas others like "*Candidatus* Clavochlamydia salmonicola," *Rhabdochlamydia* spp., and *Criblamydia sequanensis* show morphologically different developmental stages. In "*Candidatus* Clavochlamydia salmonicola" the proposed EBs were elongated and consisted of a pronounced head and tail region (68). *Rhabdochlamydia* spp. EBs were rod-shaped, possessed a five-layered cell wall, and showed elongate electron-translucent structures in the cytoplasm (15, 71). *C. sequanensis* EBs also possessed a five-layer cell wall but showed a star-like morphology (99).

The developmental cycle of *Parachlamydia acanthamoebae* and *Protochlamydia amoebophila* in their natural hosts, *Acanthamoeba* spp., starts with the uptake of EBs by phagocytosis (**Figure 3**, step A). Subsequently, EBs differentiate to RBs, which divide by binary fission to form multiple larger inclusions in the case of *P. acanthamoebae* (41, 81) or multiple single-cell inclusions in the case of *P. amoebophila* (13, 31)

(**Figure 3**, step B). Division of RBs can proceed coordinated with host cell division, leading to a long-term coexistence of the bacteria with their *Acanthamoeba* host (**Figure 3**, step D). At some point RBs redifferentiate back to EBs (**Figure 3**, step C) and are then released from the host cell within vesicles or by lysis of the amoebae to begin a new infection cycle (**Figure 3**, steps F and G) (41, 81). Lysis of the amoeba host is temperature dependent. *P. acanthamoebae* was lytic at temperatures above 32°C but showed a less cytopathic effect at temperatures between 25°C and 30°C, which are closer to the natural conditions in the environment (38). A third developmental form, the crescent body, has been observed for *P. acanthamoebae* and was mainly associated with prolonged incubation time (41). Like EBs, crescent bodies were also considered an infectious stage of *P. acanthamoebae*.

Although the inclusion is the preferred location of chlamydia-like bacteria in amoebae, *P. acanthamoebae* and *P. amoebophila* have been observed outside the inclusion too, residing directly in the cytoplasm (31, 41, 81). This was particularly pronounced for *N. hartmannellae*, in which no inclusion membrane was visible within the *Hartmannella* host (54). *P. acanthamoebae* largely inhibits cyst formation of its amoeba host (41, 81), whereas *P. amoebophila* and *S. negevensis* seem to survive the encystment and were also found within *Acanthamoeba* cysts (31, 59) (**Figure 3**, step E). *Acanthamoeba* cysts are stable and resistant against heat, desiccation, and disinfectants (78), thereby protecting intracellular bacteria from adverse environmental conditions and facilitating long-term survival in the environment (59). Developmental stages corresponding to persistent forms of the *Chlamydiaceae*, which have been implicated in chronic disease, have not been observed in acanthamoebae so far, but they were reported for *S. negevensis* in mammalian cells (61).

In summary, the existence of a developmental cycle is a feature conserved in all known chlamydiae, although differences with respect to developmental stages, their subcellular locations, and outcome are apparent. Time course analysis of gene expression using DNA microarrays has recently contributed to our understanding of the *Chlamydiaceae* developmental cycle (1). Future research should reveal the extent to which molecular mechanisms underlying the unique chlamydial developmental cycle are conserved in chlamydia-like bacteria.

Metabolism

Our current knowledge about the metabolism of chlamydia-like bacteria is based almost exclusively on comparative genome analysis of *Protochlamydia amoebophila*, to date the only chlamydia-like organism for which a genome sequence is available (51). Major differences may exist in other chlamydia-like bacteria.

The *Chlamydiaceae* have small (1 to 1.2 Mb), streamlined genomes, which are highly conserved in both gene content (sharing roughly 800 genes) and gene order (20, 51, 90, 96). The *Chlamydiaceae* are highly adapted to their intracellular lifestyle and they rely on their host cells with respect to a number of metabolic key intermediates. Despite the greater size of the *P. amoebophila* genome (2.4 Mb), a similar host dependence has been predicted (51, 52). Amino acid and cofactor biosynthesis pathways are largely absent in *P. amoebophila*, and most genes required for de novo synthesis of nucleotides are missing. Nevertheless, glycolysis and pentose phosphate pathways are encoded on the *P. amoebophila* genome, and in contrast to the *Chlamydiaceae* a complete gene set for the tricarboxylic acid cycle is present, suggesting that *P. amoebophila* is somewhat less dependent on its host cell compared with the *Chlamydiaceae* (51, 52). The respiratory chain predicted for *P. amoebophila* should be more versatile and efficient owing to the presence of a number of additional components (51, 52).

Although *P. amoebophila* is fully equipped to regenerate its own ATP, like the *Chlamydiaceae*, it encodes an ATP/ADP translocase. This transport protein catalyzes the import of host ATP into the bacterial cell in exchange for ADP and thus enables a life as energy parasite

(95). The ATP/ADP translocase of *P. amoebophila* has recently been characterized in detail and revealed the functional basis of bacterial energy parasitism by nucleotide transport proteins (100). The *P. amoebophila* ATP/ADP translocase is independent from the membrane potential and is stimulated by a high internal ADP/ATP ratio. However, it is functional only if the N terminus is directed toward the bacterial cytoplasm, thereby ensuring that it does not work in a mode detrimental to *P. amoebophila* (100). Four paralogs of this nucleotide transport protein are encoded in the *P. amoebophila* genome. Their concerted action facilitates the import of all RNA nucleotides and NAD^+ into *P. amoebophila* in a highly sophisticated manner (44, 45).

Host Cell Interaction

Attachment and entry are the earliest interactions between chlamydiae and their host cells during the developmental cycle, processes mediated primarily by the chlamydial cell envelope. In particular, the protein composition of the outer membrane of *P. amoebophila* seems fundamentally different from that of the *Chlamydiaceae* (48, 51, 52). Comparative genomics predicted only two conserved proteins, the cysteine-rich proteins OmcA and OmcB. Other important proteins present in the outer membrane of the *Chlamydiaceae*, including the major outer membrane protein (OmpA) and polymorphic membrane proteins, have no recognized homologs in *P. amoebophila*. These differences might reflect the different host range of the amoeba symbiont and its pathogenic counterparts, but further research is required to better characterize the cell envelope of *P. amoebophila* and other chlamydia-like bacteria.

The type three secretion system (T3SS) is another key mechanism of the *Chlamydiaceae* for host cell interaction (87). Analysis of the *P. amoebophila* genome and preliminary data from the *S. negevensis* genome sequence showed that it is well conserved among all chlamydiae (51, 52, 87). This complex, consisting of more than 20 different proteins, forms a molecular syringe that can be used to inject effector proteins into the host cytosol. Consistently, some effector proteins known from the *Chlamydiaceae* were also predicted in the *P. amoebophila* genome, for example, the serine/threonine protein kinase Pkn5, the inclusion protein IncA, and other proteins targeted to the host-derived inclusion membrane (51, 52). A few additional proteins that have been associated with pathogenicity of chlamydiae are also present in *P. amoebophila*, including the chlamydia protease-like activity factor (CPAF), whereas others such as the actin-recruiting protein Tarp and CADD (chlamydia protein associating with death domains) are notably absent.

Among proteins that were found in the genome of *P. amoebophila* but not in the *Chlamydiaceae* were a large number of leucine-rich repeat proteins, generally considered to be involved in protein-protein interactions and/or in recognition of bacterial motifs (24, 51, 52). *P. amoebophila* encodes a type four secretion system (T4SS) in a region with a G+C content significantly different from the genomic G+C content, suggesting that this genomic island has been acquired by lateral gene transfer (35, 51, 52). This is further supported by the finding of a similar transport system in the genomes of several *Rickettsia* species (6). The T4SS of *P. amoebophila* is most similar to that of the F plasmid of *Escherichia coli* (35) and could thus be used either for protein secretion or for conjugation. Experimental data for either hypothesis are not yet available.

Apart from the *Protochlamydia amoebophila* genome sequence, some in vitro data on host cell interaction is available for *Parachlamydia acanthamoebae* with human cells. Trafficking of *P. acanthamoebae* in macrophages (through the endocytic pathway) was fundamentally different from that of the *Chlamydiaceae* (39). Further differences between *P. acanthamoebae* and the *Chlamydiaceae* regarding cytokine production were observed. *P. acanthamoebae* did not induce an oxidative burst or the proinflammatory cytokines IL-6 and TNF-α in macrophages (36),

T3SS: type three secretion system

T4SS: type four secretion system

perhaps because of the differences in the cell envelope discussed above. However, in essence, despite large evolutionary differences between chlamydia-like bacteria and the *Chlamydiaceae* several basic mechanisms for host-cell interaction are conserved among all chlamydiae and are used by amoeba symbionts and human pathogens alike.

EVOLUTIONARY HISTORY OF THE *CHLAMYDIAE*

The evolution of the chlamydiae is intriguing. Although little is known about the early time points in their evolutionary history, the few hypotheses proposed have implications far beyond the chlamydiae.

A Last Common Intracellular Ancestor

Chlamydial genomes show hardly any evidence of recent lateral gene transfer (with the T4SS of *P. amoebophila* as a rare exception) (51, 90, 96). The genome sequence of *P. amoebophila*, twice as large as the genomes of the *Chlamydiaceae*, could thus be used to partially reconstruct the genome of the last common ancestor of this amoeba symbiont and its pathogenic counterparts (51, 52). On the basis of this analysis, the chlamydial ancestor already employed transport proteins, still present in all extant chlamydiae, such as the nucleotide transport proteins or a glucose-phosphate transporter. Development and conservation of such transport proteins can be explained most likely by an intracellular environment. It is thus most parsimonious that the last common ancestor of *P. amoebophila* and the *Chlamydiaceae* has already lived inside a eukaryotic host cell (51, 52). Consistent with this hypothesis, phylogenetic analysis of key mechanisms for host cell interaction, such as the T3SS or CPAF, shows that these mechanisms were already present in the last common chlamydial ancestor (51, 52). In the case of the T3SS, this is also supported by the highly conserved order of genes encoding this protein complex. In contrast to all other bacteria, chlamydial T3SS genes are located on multiple regions on the chromosome and are largely syntenic between the *Chlamydiaceae*, *P. amoebophila*, and *S. negevensis* (51, 87).

Roughly estimated, *Protochlamydia* and the *Chlamydiaceae* split at least 700 mya—at some time point in the Precambrian, when unicellular eukaryotes were abundant. The last common chlamydial ancestor has therefore most likely lived in some kind of primordial protist (51). If this is true, then the major mechanisms still used today by symbiotic and pathogenic chlamydiae were developed during this early interaction. Ancient protozoa could thus have served as evolutionary training grounds for the development of the intracellular lifestyle of chlamydiae.

Chlamydiae and the Origin of Plants

All chlamydial genomes share an intriguing feature: They possess more plant- and cyanobacteria-like proteins than most other bacterial genomes. Even more puzzling, most plant-like proteins of chlamydiae are targeted to and function in plastids (7, 51, 89, 93, 96, 104). Among the chlamydiae, this is most obvious in *P. amoebophila*, which showed more than 150 proteins with highest sequence similarity to proteins of plants or cyanobacteria (51). Phylogenetic analysis of these proteins did not reveal a consistent picture of their evolutionary history but suggested a primordial association of chlamydiae with cyanobacteria, plastids, and/or plants, including complex ancestral gene transfers between these groups (51). Such an association is further supported by the presence of a group I intron in the 23S rRNA of *Simkania negevensis* and "*Candidatus* Fritschea bemisiae" and by evidence of a past presence of a group I intron in the *Chlamydiaceae* (26). Group I introns are widespread in algal chloroplasts, in lower eukaryotes, and archaea, but they are notably absent in most bacteria.

Different scenarios have been put forward to explain these observations. It has been suggested that the presence of plant-like genes in chlamydiae reflects an evolutionary relationship between chlamydiae and the cyanobacterial ancestor of the chloroplast (7). Such a relationship is, however, not well supported by phylogenetic trees based on rRNA genes or ribosomal proteins (101). Alternatively, horizontal gene transfer events either from plants to chlamydiae (104) or from chlamydiae to plants (26, 93, 95) have been proposed, and it was speculated that chlamydiae participated in the ancient chimeric events that led to the formation of the plant and animal lineages (26). Indeed, recent phylogenomic analysis of the red alga *Cyanidioschyzon merolae* shed some light on this discussion and supported the latter hypothesis (55).

In this study, Huang & Gogarten detected 21 genes that were transferred between chlamydiae and *C. merolae* and green plants, with the donor most similar to *P. amoebophila*. They argued that the gene transfer events between chlamydiae and primary photosynthetic eukaryotes can be best explained by a stable association between these groups in the past, and they proposed a scenario in which a chlamydial symbiont similar to *P. amoebophila* facilitated the establishment of primary plastids (55). According to their hypothesis, the chlamydial symbiont entered a mitochondrion-containing eukaryote at about the same time the cyanobacterial ancestor of chloroplasts was captured. The eukaryotic host subsequently acquired transport proteins via lateral gene transfer from the chlamydial symbiont, which facilitated interactions between the cyanobacterial symbiont and its heterotrophic host cell. While the cyanobacterial symbiont was transformed to a photosynthetic organelle, the chlamydial symbiont might have gradually degenerated or transformed into an organelle not yet recognized in photosynthetic eukaryotes (55). Although it might not fully explain the observed high number of cyanobacteria-like proteins in chlamydial genomes, this is the most parsimonious scenario put forth. This hypothesis is intriguing because it implies that an ancient chlamydial symbiosis might have given rise to the first photosynthetic eukaryote, the ancestor of all extant plants on earth.

The Closest Free-Living Relatives

Even though chlamydiae have been detected in a wide variety of habitats by molecular methods, all known chlamydiae are obligate intracellular bacteria, and currently it seems questionable whether extant chlamydiae can live independent of a eukaryotic host cell. In view of the lack of such evidence, it might be worthwhile to search for the closest free-living relatives of chlamydiae. The phylum *Chlamydiae* is deeply branching in the domain Bacteria, and evolutionary relationships among most bacterial phyla are presently not resolved. Recent evidence, however, suggested a monophyletic grouping of the *Chlamydiae* with the phyla *Verrucomicrobia* and *Planctomycetes* (97, 101), as these groups share a number of unique features (101). A common ancestral origin of *Chlamydiae* and *Verrucomicrobia* is supported particularly well (42), indicating that the *Verrucomicrobia* are at present the closest free-living relatives of the *Chlamydiae*. The analysis of *Verrucomicrobia*, particularly of those members that live in association with eukaryotes, might reveal clues about evolutionary times at which chlamydiae were still free-living or facultative symbionts of eukaryotes.

PERSPECTIVES

From an evolutionary perspective, two chlamydia-like bacteria deserve special attention, the uncultured "*Candidatus* Piscichlamydia salmonis" and "*Candidatus* Clavochlamydia salmonicola." "*Ca.* Piscichlamydia salmonis" is particularly interesting because it currently represents the deepest branch in the *Chlamydiae* (**Figure 1**), i.e., it might still share features of the last common ancestor of all chlamydiae, which are absent in all other chlamydial lineages. "*Ca.* Clavochlamydia salmonicola" is the closest relative of the *Chlamydiaceae* (**Figure 1**) and might thus represent a transitional stage between the highly adapted human and animal pathogens of the *Chlamydiaceae* and all other chlamydia-like bacteria. Novel approaches and technologies such as whole-genome amplification and pyrosequencing might facilitate genome analysis of these organisms in the near future.

Global transcriptional and proteomic analysis of chlamydia-like bacteria will help us to better understand their biology, as well as differences and similarities to the *Chlamydiaceae*. The use of the well-characterized amoeba *Dictyostelium discoideum* as a surrogate host for chlamydial symbionts might help to shed new light on host-bacteria interaction. Further investigation of the role of some chlamydia-like bacteria, particularly *P. acanthamoebae*, *S. negevensis*, and *W. chondrophila*, with respect to their pathogenic potential toward humans is warranted; an animal model might be helpful in this regard as well. Without doubt, recent studies have greatly increased our knowledge of diversity, evolution, and biology of the chlamydiae and their distribution in nature, but we are only beginning to understand one of the most enigmatic groups in the domain *Bacteria*.

SUMMARY POINTS

1. Members of the phylum *Chlamydiae* are a phylogenetically diverse group of obligate intracellular bacteria. *Chlamydiae* consist of at least eight recognized families. Molecular evidence suggests an additional yet unexplored diversity of chlamydiae in the environment.

2. All known chlamydiae show the characteristic chlamydial developmental cycle, even though morphology, number, and presumably also the physiology of their developmental stages vary between different chlamydiae.

3. Free-living amoebae, particularly of the genus *Acanthamoeba*, are frequent hosts of chlamydiae in the environment and might play an important role for the survival and dispersal of chlamydiae in nature. The host range of chlamydiae includes diverse representatives across the animal kingdom.

4. *Parachlamydia acanthamoebae* and *Simkania negevensis* have been associated with respiratory disease in humans, yet a causal link has not been established.

5. The amoeba symbiont *Protochlamydia amoebophila* is auxotrophic for a number of essential cell building blocks. It has retained several proteins associated with virulence of the *Chlamydiaceae*, including a T3SS.

6. Nucleotide transport proteins, particularly ATP/ADP translocases, are a hallmark of chlamydiae and catalyze a sophisticated interaction with their host cells.

7. *Chlamydiae* have acquired basic mechanisms for host cell interaction in ancient unicellular eukaryotes and have lived in association with eukaryotic hosts for several hundreds of millions of years. *Chlamydiae* might have been involved in the ancient symbiotic events that led to the emergence of the first photosynthetic eukaryote.

DISCLOSURE STATEMENT

The authors are not aware of any biases that might be perceived as affecting the objectivity of this review.

ACKNOWLEDGMENTS

The author would like to sincerely apologize to authors of publications that were omitted for space considerations or accidentally. Susanne Haider, Eva Heinz, Jacqueline Montanaro, Stephan Schmitz-Esser, and Elena Toenshoff are greatly acknowledged for help with preparing figures and tables and for comments on the manuscript. The author is grateful to Tom Fritsche, who has initiated our work on chlamydia-like bacteria, and to Michael Wagner, an inspiring mentor and colleague. This work was supported by Austrian Science Fund (FWF) grant Y277-B03.

LITERATURE CITED

1. Abdelrahman YM, Belland RJ. 2005. The chlamydial developmental cycle. *FEMS Microbiol. Rev.* 29:949–59

2. Amann R, Springer N, Schonhuber W, Ludwig W, Schmid EN, et al. 1997. Obligate intracellular bacterial parasites of acanthamoebae related to *Chlamydia* spp. *Appl. Environ. Microbiol.* 63:115–21

3. Apfalter P. 2006. *Chlamydia pneumoniae*, stroke and serological associations. *Stroke* 37:756–58

4. Baud D, Thomas V, Arafa, Regan L, Greub G. 2007. *Waddlia chondrophila*, a potential agent of human fetal death. *Emerg. Infect. Dis.* 13:1239–43

5. Birtles RJ, Rowbotham TJ, Storey C, Marrie TJ, Raoult D. 1997. *Chlamydia*-like obligate parasite of free-living amoebae. *Lancet* 349:925–26

6. Blanc G, Ogata H, Robert C, Audic S, Claverie JM, Raoult D. 2007. Lateral gene transfer between obligate intracellular bacteria: evidence from the *Rickettsia massiliae* genome. *Genome Res.* 17:1657–64

7. Brinkman FS, Blanchard JL, Cherkasov A, Av-Gay Y, Brunham RC, et al. 2002. Evidence that plant-like genes in *Chlamydia* species reflect an ancestral relationship between *Chlamydiaceae*, cyanobacteria, and the chloroplast. *Genome Res.* 12:1159–67

8. Casson N, Medico N, Bille J, Greub G. 2006. *Parachlamydia acanthamoebae* enters and multiplies within pneumocytes and lung fibroblasts. *Microbes Infect.* 8:1294–300

9. Casson N, Michel R, Müller K-D, Aubert JD, Greub G. 2008. *Protochlamydia naegleriophila* as etiologic agent of pneumonia. *Emerg. Infect. Dis.* 14:168–72

10. Chappell CL, Wright JA, Coletta M, Newsome AL. 2001. Standardized method of measuring *Acanthamoeba* antibodies in sera from healthy human subjects. *Clin. Diagn. Lab. Immun.* 8:724–30

11. Chua PK, Corkill JE, Hooi PS, Cheng SC, Winstanley C, Hart CA. 2005. Isolation of *Waddlia malaysiensis*, a novel intracellular bacterium, from fruit bat (*Eonycteris spelaea*). *Emerg. Infect. Dis.* 11:271–77

12. Collingro A, Poppert S, Heinz E, Schmitz-Esser S, Essig A, et al. 2005. Recovery of an environmental chlamydia strain from activated sludge by cocultivation with *Acanthamoeba* sp. *Microbiology* 151:301–9

13. Collingro A, Toenshoff ER, Taylor MW, Fritsche TR, Wagner M, Horn M. 2005. 'Candidatus Protochlamydia amoebophila', an endosymbiont of *Acanthamoeba* spp. *Int. J. Syst. Evol. Microbiol.* 55:1863–66

14. Corsaro D, Greub G. 2006. Pathogenic potential of novel *Chlamydiae* and diagnostic approaches to infections due to these obligate intracellular bacteria. *Clin. Microbiol. Rev.* 19:283–97

15. Corsaro D, Thomas V, Goy G, Venditti D, Radek R, Greub G. 2007. 'Candidatus Rhabdochlamydia crassificans', an intracellular bacterial pathogen of the cockroach *Blatta orientalis* (Insecta: Blattodea). *Syst. Appl. Microbiol.* 30:221–28

16. Corsaro D, Valassina M, Venditti D. 2003. Increasing diversity within *Chlamydiae*. *Crit. Rev. Microbiol.* 29:37–78

17. Corsaro D, Venditti D, Le Faou A, Guglielmetti P, Valassina M. 2001. A new chlamydia-like 16S rDNA sequence from a clinical sample. *Microbiology* 147:515–16

18. Corsaro D, Venditti D, Valassina M. 2002. New parachlamydial 16S rDNA phylotypes detected in human clinical samples. *Res. Microbiol.* 153:563–67

19. De Bary A. 1879. *Die Erscheinung der Symbiose*. Strassburg: Verlag von Karl J. Trubner

20. Dean D, Myers GS, Read TD. 2006. Lessons and challenges arising from the 'first wave' of *Chlamydia* genome sequencing. In *Chlamydia: Genomics and Pathogenesis*, ed. PM Bavoil, PB Wyrick, pp. 1–24. Norfolk: Horizon Bioscience

21. Dilbeck-Robertson P, McAllister MM, Bradway D, Evermann JF. 2003. Results of a new serologic test suggest an association of *Waddlia chondrophila* with bovine abortion. *J. Vet. Diagn. Invest.* 15:568–69

22. **Draghi A 2nd, Popov VL, Kahl MM, Stanton JB, Brown CC, et al. 2004. Characterization of "*Candidatus* piscichlamydia salmonis" (order *Chlamydiales*), a chlamydia-like bacterium associated with epitheliocystis in farmed Atlantic salmon (*Salmo salar*). *J. Clin. Microbiol.* 42:5286–97**

23. Essig A, Heinemann M, Simnacher U, Marre R. 1997. Infection of *Acanthamoeba castellanii* by *Chlamydia pneumoniae*. *Appl. Environ. Microbiol.* 63:1396–99

24. Eugster M, Roten CA, Greub G. 2007. Analyses of six homologous proteins of *Protochlamydia amoebophila* UWE25 encoded by large GC-rich genes (lgr): a model of evolution and concatenation of leucine-rich repeats. *BMC Evol. Biol.* 7:231

25. Everett KD, Bush RM, Andersen AA. 1999. Emended description of the order *Chlamydiales*, proposal of *Parachlamydiaceae* fam. nov. and *Simkaniaceae* fam. nov., each containing one monotypic genus, revised taxonomy of the family *Chlamydiaceae*, including a new genus and five new species, and standards for the identification of organisms. *Int. J. Syst. Bacteriol.* 49:415–40

26. Everett KD, Kahane S, Bush RM, Friedman MG. 1999. An unspliced group I intron in 23S rRNA links *Chlamydiales*, chloroplasts, and mitochondria. *J. Bacteriol.* 181:4734–40

27. Fasoli L, Paldanius M, Don M, Valent F, Vetrugno L, et al. 2007. *Simkania negevensis* in community-acquired pneumonia in Italian children. *Scand. J. Infect. Dis.* 99999:1–4

28. **Fredericks DN, Relman DA. 1996. Sequence-based identification of microbial pathogens: a reconsideration of Koch's Postulates. *Clin. Microbiol. Rev.* 9:18–33**

29. Friedman MG, Kahane S, Dvoskin B, Hartley JW. 2006. Detection of *Simkania negevensis* by culture, PCR, and serology in respiratory tract infection in Cornwall, UK. *J. Clin. Pathol.* 59:331–33

30. **Fritsche TR, Gautom RK, Seyedirashti S, Bergeron DL, Lindquist TD. 1993. Occurrence of bacteria endosymbionts in *Acanthamoeba* spp. isolated from corneal and environmental specimens and contact lenses. *J. Clin. Microbiol.* 31:1122–26**

31. Fritsche TR, Horn M, Wagner M, Herwig RP, Schleifer KH, Gautom RK. 2000. Phylogenetic diversity among geographically dispersed *Chlamydiales* endosymbionts recovered from clinical and environmental isolates of *Acanthamoeba* spp. *Appl. Environ. Microbiol.* 66:2613–19

32. Greenberg D, Banerji A, Friedman MG, Chiu CH, Kahane S. 2003. High rate of *Simkania negevensis* among Canadian Inuit infants hospitalized with lower respiratory tract infections. *Scand. J. Infect. Dis.* 35:506–8

33. Greub G, Berger P, Papazian L, Raoult D. 2003. *Parachlamydiaceae* as rare agents of pneumonia. *Emerg. Infect. Dis.* 9:755–56

34. Greub G, Boyadjiev I, La Scola B, Raoult D, Martin C. 2003. Serological hint suggesting that *Parachlamydiaceae* are agents of pneumonia in polytraumatized intensive care patients. *Ann. N. Y. Acad. Sci.* 990:311–19

35. Greub G, Collyn F, Guy L, Roten CA. 2004. A genomic island present along the bacterial chromosome of the *Parachlamydiaceae* UWE25, an obligate amoebal endosymbiont, encodes a potentially functional F-like conjugative DNA transfer system. *BMC Microbiol.* 4:48

36. Greub G, Desnues B, Raoult D, Mege JL. 2005. Lack of microbicidal response in human macrophages infected with *Parachlamydia acanthamoebae*. *Microbes Infect.* 7:714–19

37. Greub G, Hartung O, Adekambi T, Alimi YS, Raoult D. 2006. *Chlamydia*-like organisms and atherosclerosis. *Emerg. Infect. Dis.* 12:705–6

38. Greub G, La Scola B, Raoult D. 2003. *Parachlamydia acanthamoebae* is endosymbiotic or lytic for *Acanthamoeba polyphaga* depending on the incubation temperature. *Ann. N. Y. Acad. Sci.* 990:628–34

39. Greub G, Mege JL, Gorvel JP, Raoult D, Meresse S. 2005. Intracellular trafficking of *Parachlamydia acanthamoebae*. *Cell Microbiol.* 7:581–89

40. **Greub G, Mege JL, Raoult D. 2003. *Parachlamydia acanthamoebae* enters and multiplies within human macrophages and induces their apoptosis. *Infect. Immun.* 71:5979–85**

41. Greub G, Raoult D. 2002. Crescent bodies of *Parachlamydia acanthamoebae* and its life cycle within *Acanthamoeba polyphaga*: an electron micrograph study. *Appl. Environ. Microbiol.* 68:3076–84

42. Griffiths E, Gupta RS. 2007. Phylogeny and shared conserved inserts in proteins provide evidence that *Verrucomicrobia* are the closest known free-living relatives of *Chlamydiae*. *Microbiology* 153:2648–54

22. Description of '*Candidatus* Piscichlamydia salmonis,' currently representing the deepest branch in the phylum *Chlamydiae*.

28. Revision of Koch's Postulates to meet current methods for the identification of pathogens.

30. First morphological description of chlamydia-like bacteria as symbionts in free-living amoebae.

40. First report showing survival and multiplication of *P. acanthamoebae* in human cells.

43. Guindon S, Lethiec F, Duroux P, Gascuel O. 2005. PHYML online—a web server for fast maximum likelihood-based phylogenetic inference. *Nucleic Acids Res.* 33:W557–59

44. Haferkamp I, Schmitz-Esser S, Linka N, Urbany C, Collingro A, et al. 2004. A candidate NAD⁺ transporter in an intracellular bacterial symbiont related to *Chlamydiae*. *Nature* 432:622–25

45. **Haferkamp I, Schmitz-Esser S, Wagner M, Neigel N, Horn M, Neuhaus HE. 2006. Tapping the nucleotide pool of the host: novel nucleotide carrier proteins of *Protochlamydia amoebophila*. *Mol. Microbiol.* 60:1534–45**

46. Haider S, Collingro A, Walochnik J, Wagner M, Horn M. 2008. Chlamydia-like bacteria in respiratory samples of community-acquired pneumonia patients. *FEMS Microbiol. Lett.* 281:198–202

47. **Halberstädter L, Prowazek SV. 1907. Über Zelleinschlüsse parasitärer Natur beim Trachom. *Arbeiten aus dem Kaiserlichen Gesundheitsamte, Berlin* 26:44–47**

48. Hatch TP. 1999. Developmental biology. In *Chlamydia*, ed. RS Stephens, pp. 29–67. Washington, DC: ASM

49. Heiskanen-Kosma T, Paldanius M, Korppi M. 2008. *Simkania negevensis* may be a true cause of community acquired pneumonia in children. *Scand. J. Infect. Dis.* 40:127–30

50. Henning K, Schares G, Granzow H, Polster U, Hartmann M, et al. 2002. *Neospora caninum* and *Waddlia chondrophila* strain 2032/99 in a septic stillborn calf. *Vet. Microbiol.* 85:285–92

51. **Horn M, Collingro A, Schmitz-Esser S, Beier CL, Purkhold U, et al. 2004. Illuminating the evolutionary history of chlamydiae. *Science* 304:728–30**

52. Horn M, Collingro A, Schmitz-Esser S, Wagner M. 2006. Environmental chlamydia genomics. In *Chlamydia: Genomics and Pathogenesis*, ed. PM Bavoil, PB Wyrick, pp. 25–44. Norfolk, UK: Horizon Bioscience

53. Horn M, Wagner M. 2001. Evidence for additional genus-level diversity of *Chlamydiales* in the environment. *FEMS Microbiol. Lett.* 204:71–74

54. Horn M, Wagner M, Muller KD, Schmid EN, Fritsche TR, et al. 2000. *Neochlamydia hartmannellae* gen. nov., sp. nov. (*Parachlamydiaceae*), an endoparasite of the amoeba *Hartmannella vermiformis*. *Microbiology* 146:1231–39

55. **Huang J, Gogarten J. 2007. Did an ancient chlamydial endosymbiosis facilitate the establishment of primary plastids? *Genome Biol.* 8:R99**

56. Husain S, Kahane S, Friedman MG, Paterson DL, Studer S, et al. 2007. *Simkania negevensis* in bronchoalveolar lavage of lung transplant recipients: a possible association with acute rejection. *Transplantation* 83:138–43

57. Israelsson O. 2007. Chlamydial symbionts in the enigmatic *Xenoturbella* (Deuterostomia). *J. Invertebr. Pathol.* 96:213–20

58. Jaenicke L. 2001. Stanislaus von Prowazek (1875–1915)—prodigy between working bench and coffee house. *Protist* 152:157–66

59. Kahane S, Dvoskin B, Mathias M, Friedman MG. 2001. Infection of *Acanthamoeba polyphaga* with *Simkania negevensis* and *S. negevensis* survival within amoebal cysts. *Appl. Environ. Microbiol.* 67:4789–95

60. Kahane S, Everett KD, Kimmel N, Friedman MG. 1999. *Simkania negevensis* strain ZT: growth, antigenic and genome characteristics. *Int. J. Syst. Bacteriol.* 49(Pt. 2):815–20

61. Kahane S, Fruchter D, Dvoskin B, Friedman MG. 2007. Versatility of *Simkania negevensis* infection in vitro and induction of host cell inflammatory cytokine response. *J. Infect.* 55:e13–21

62. Kahane S, Gonen R, Sayada C, Elion J, Friedman MG. 1993. Description and partial characterization of a new chlamydia-like microorganism. *FEMS Microbiol. Lett.* 109:329–34

63. Kahane S, Greenberg D, Friedman MG, Haikin H, Dagan R. 1998. High prevalence of "Simkania Z," a novel *Chlamydia*-like bacterium, in infants with acute bronchiolitis. *J. Infect. Dis.* 177:1425–29

64. Kahane S, Greenberg D, Newman N, Dvoskin B, Friedman MG. 2007. Domestic water supplies as a possible source of infection with *Simkania*. *J. Infect.* 54:75–81

65. Kahane S, Kimmel N, Friedman MG. 2002. The growth cycle of *Simkania negevensis*. *Microbiology* 148:735–42

66. **Kahane SEM, Friedman MG. 1995. Evidence that the novel microorganism 'Z' may belong to a new genus in the family *Chlamydiaceae*. *FEMS Microbiol. Lett.* 126:203–8**

45. Comprehensive biochemical analysis of nucleotide transport proteins of *P. amoebophila*.

47. The original report on the discovery of *Chlamydia trachomatis* as an agent of trachoma.

51. Analysis of the first genome sequence of a chlamydia-like organism, *Protochlamydia amoebophila* UWE25.

55. Elaborate analysis of chlamydia-like proteins in red algae and green plants, with implications for the origin of the first photosynthetic eukaryote.

66. First molecular identification of a chlamydia-like organism, *Simkania negevensis*.

67. Kalayoglu MV, Byrne GI. 2006. The genus *Chlamydia*—medical. In *The Prokaryotes*, ed. M Dworkin, S Falkow, E Rosenberg, K-H Schleifer, E Stackebrandt, pp. 741–54. New York: Springer

68. Karlsen M, Nylund A, Watanabe K, Helvik JV, Nylund S, Plarre H. 2007. Characterization of '*Candidatus* Clavochlamydia salmonicola': an intracellular bacterium infecting salmonid fish. *Environ. Microbiol.* 10:208–18

69. Khan NA. 2006. *Acanthamoeba*: biology and increasing importance in human health. *FEMS Microbiol. Rev.* 30:564–95

70. Korppi M, Paldanius M, Hyvarinen A, Nevalainen A. 2006. *Simkania negevensis* and newly diagnosed asthma: a case-control study in 1- to 6-year-old children. *Respirology* 11:80–83

71. Kostanjsek R, Strus J, Drobne D, Avgustin G. 2004. '*Candidatus* Rhabdochlamydia porcellionis', an intracellular bacterium from the hepatopancreas of the terrestrial isopod *Porcellio scaber* (Crustacea: Isopoda). *Int. J. Syst. Evol. Microbiol.* 54:543–49

72. Kumar S, Kohlhoff SA, Gelling M, Roblin PM, Kutlin A, et al. 2005. Infection with *Simkania negevensis* in Brooklyn, New York. *Pediatr. Infect. Dis. J.* 24:989–92

73. Kuo C-C, Horn M, Stephens RS. 2008. The order *Chlamydiales*. In *Bergey's Manual of Systematic Bacteriology: The Planctomycetes, Spriochaetes, Fibrobacteres, Bacteriodetes and Fusobacteria*, ed. B Hedlund, NR Krieg, W Ludwig, BJ Paster, JT Staley, et al. New York: Springer. In press

74. Letunic I, Bork P. 2007. Interactive Tree Of Life (iTOL): an online tool for phylogenetic tree display and annotation. *Bioinformatics* 23:127–28

75. Lieberman D, Dvoskin B, Lieberman DV, Kahane S, Friedman MG. 2002. Serological evidence of acute infection with the *Chlamydia*-like microorganism *Simkania negevensis* (Z) in acute exacerbation of chronic obstructive pulmonary disease. *Eur. J. Clin. Microbiol. Infect. Dis.* 21:307–9

76. Lieberman D, Kahane S, Lieberman D, Friedman MG. 1997. Pneumonia with serological evidence of acute infection with the *Chlamydia*-like microorganism 'Z'. *Am. J. Respir. Crit. Care Med.* 156:578–82

77. Ludwig W, Strunk O, Westram R, Richter L, Meier H, et al. 2004. ARB: a software environment for sequence data. *Nucleic Acids Res.* 32:1363–71

78. Marciano-Cabral F, Cabral G. 2003. *Acanthamoeba* spp. as agents of disease in humans. *Clin. Microbiol. Rev.* 16:273–307

79. Marrie TJ, Raoult D, La Scola B, Birtles RJ, de Carolis E. 2001. *Legionella*-like and other amoebal pathogens as agents of community-acquired pneumonia. *Emerg. Infect. Dis.* 7:1026–9

80. Meijer A, Roholl PJM, Ossewaarde JM, Jones B, Nowak BF. 2006. Molecular evidence for association of *Chlamydiales* bacteria with epitheliocystis in leafy seadragon (*Phycodurus eques*), silver perch (*Bidyanus bidyanus*), and barramundi (*Lates calcarifer*). *Appl. Environ. Microbiol.* 72:284–90

81. Michel R, Hauröder-Philippczyk B, Müller K-D, Weishaar I. 1994. *Acanthamoeba* from human nasal mucosa infected with an obligate intracellular parasite. *Eur. J. Protistol.* 30:104–10

82. Michel R, Müller KD, Zöller L, Walochnik J, Hartmann M, Schmid EN. 2005. Free-living amoebae serve as a host for the *Chlamydia*-like bacterium *Simkania negevensis*. *Acta Protozool.* 44:113–21

83. Michel R, Steinert M, Zöller L, Hauröder B, Henning K. 2004. Free-living amoebae may serve as hosts for the chlamydia-like bacterium *Waddlia chondrophila* isolated from a aborted bovine foetus. *Acta Protozool.* 43:37–42

84. Moulder JW. 1964. *The Psittacosis Group as Bacteria*. New York: Wiley. 95 pp.

85. Ochman H, Elwyn S, Moran NA. 1999. Calibrating bacterial evolution. *Proc. Natl. Acad. Sci. USA* 96:12638–43

86. Ossewaarde JM, Meijer A. 1999. Molecular evidence for the existence of additional members of the order *Chlamydiales*. *Microbiology* 145:411–17

87. Peters J, Wilson DP, Myers G, Timms P, Bavoil PM. 2007. Type III secretion a la *Chlamydia*. *Trends Microbiol.* 15:241–51

88. Pruesse E, Quast C, Knittel K, Fuchs BM, Ludwig W, et al. 2007. SILVA: a comprehensive online resource for quality checked and aligned ribosomal RNA sequence data compatible with ARB. *Nucleic Acids Res.* 35:7188–96

89. Read TD, Brunham RC, Shen C, Gill SR, Heidelberg JF, et al. 2000. Genome sequences of *Chlamydia trachomatis* MoPn and *Chlamydia pneumoniae* AR39. *Nucleic Acids Res.* 28:1397–406

90. Read TD, Myers GS, Brunham RC, Nelson WC, Paulsen IT, et al. 2003. Genome sequence of *Chlamydophila caviae* (*Chlamydia psittaci* GPIC): examining the role of niche-specific genes in the evolution of the *Chlamydiaceae*. *Nucleic Acids Res.* 31:2134–47

91. Rodriguez-Zaragoza S. 1994. Ecology of free living amoebae. *Crit. Rev. Microbiol.* 20:225–41

92. Rourke AW, Davis RW, Bradley TM. 1984. A light and electron microscope study of epitheliocystis in juvenile steelhead trout, *Salmo gairdneri* Richardson. *J. Fish Dis.* 7:301–9

93. Royo J, Gimez E, Hueros G. 2000. CMP-KDO synthetase: a plant gene borrowed from gram-negative eubacteria. *Trends Genet.* 16:432–33

94. Rurangirwa FR, Dilbeck PM, Crawford TB, McGuire TC, McElwain TF. 1999. Analysis of the 16S rRNA gene of microorganism WSU 86-1044 from an aborted bovine foetus reveals that it is a member of the order *Chlamydiales*: proposal of *Waddliaceae* fam. nov., *Waddlia chondrophila* gen. nov., sp. nov. *Int. J. Syst. Bacteriol.* 49:577–81

95. Schmitz-Esser S, Linka N, Collingro A, Beier CL, Neuhaus HE, et al. 2004. ATP/ADP translocases: a common feature of obligate intracellular amoebal symbionts related to *Chlamydiae* and *Rickettsiae*. *J. Bacteriol.* 186:683–91

96. Stephens RS. 1999. Genomic autobiographies of chlamydiae. In *Chlamydia*, ed. RS Stephens, pp. 9–27. Washington, DC: ASM

97. Strous M, Pelletier E, Mangenot S, Rattei T, Lehner A, et al. 2006. Deciphering the evolution and metabolism of an anammox bacterium from a community genome. *Nature* 440:790–94

98. Thao ML, Baumann L, Hess JM, Falk BW, Ng JC, et al. 2003. Phylogenetic evidence for two new insect-associated *Chlamydia* of the family *Simkaniaceae*. *Curr. Microbiol.* 47:46–50

99. Thomas V, Casson N, Greub G. 2006. *Criblamydia sequanensis*, a new intracellular *Chlamydiales* isolated from Seine River water using amoebal coculture. *Environ. Microbiol.* 8:2125–35

100. Trentmann O, Horn M, Van Scheltinga ACT, Neuhaus HE, Haferkamp I. 2007. Enlightening energy parasitism by analysis of an ATP/ADP transporter from *Chlamydiae*. *PLoS Biol.* 5:1938–51

101. Wagner M, Horn M. 2006. The *Planctomycetes*, *Verrucomicrobia*, *Chlamydiae* and sister phyla comprise a superphylum with biotechnological and medical relevance. *Curr. Opin. Biotechnol.* 17:241–49

102. WHO. 2001. *Global Prevalence and Incidence of Curable STIs.* Geneva: WHO

103. WHO. 2008. *Priority eye diseases.* **http://www.who.int/blindness/causes/priority/en/index2.html**

104. Wolf YI, Aravind L, Koonin EV. 1999. Rickettsiae and chlamydiae: evidence of horizontal gene transfer and gene exchange. *Trends Genet.* 15:173–75

100. Detailed biochemical analysis of the ATP/ADP translocase of *Protochlamydia amoebophila* revealing the molecular mechanism of energy exploitation.

Biology of *trans*-Translation

Kenneth C. Keiler

Department of Biochemistry and Molecular Biology, The Pennsylvania State University, University Park, Pennsylvania 16802; email: kkeiler@psu.edu

Annu. Rev. Microbiol. 2008. 62:133–51

First published online as a Review in Advance on June 16, 2008

The *Annual Review of Microbiology* is online at micro.annualreviews.org

This article's doi:
10.1146/annurev.micro.62.081307.162948

Key Words

tmRNA, SmpB, translation quality control, targeted proteolysis, mRNA degradation

Abstract

The *trans*-translation mechanism is a key component of multiple quality control pathways in bacteria that ensure proteins are synthesized with high fidelity in spite of challenges such as transcription errors, mRNA damage, and translational frameshifting. *trans*-Translation is performed by a ribonucleoprotein complex composed of tmRNA, a specialized RNA with properties of both a tRNA and an mRNA, and the small protein SmpB. tmRNA-SmpB interacts with translational complexes stalled at the 3′ end of an mRNA to release the stalled ribosomes and target the nascent polypeptides and mRNAs for degradation. In addition to quality control pathways, some genetic regulatory circuits use *trans*-translation to control gene expression. Diverse bacteria require *trans*-translation when they execute large changes in their genetic programs, including responding to stress, pathogenesis, and differentiation.

Contents

INTRODUCTION

trans-**Translation:**
synthesis of a protein
from two physically
distinct RNA
molecules

**Transfer messenger
RNA (tmRNA):**
RNA having tRNA-
and mRNA-like
properties

In bacteria, quality control pathways that eliminate aberrant proteins and mRNAs converge on a remarkable reaction, known as *trans*-translation, in which a translational complex is diverted to a specialized RNA that ultimately promotes the degradation of the nascent polypeptide and the mRNA and releases the ribosomal subunits. Bacteria also intentionally target some translational complexes for *trans*-translation as part of regulatory circuits. Re-

flecting the multiple roles of *trans*-translation in physiology and gene regulation, mutants with defects in *trans*-translation have a variety of phenotypes in different species.

The key molecule in *trans*-translation is transfer-messenger RNA (tmRNA), an RNA with properties of both tRNA and mRNA. tmRNA enters translational complexes stalled at the end of an mRNA and accepts the nascent polypeptide in the same manner as does tRNA (**Figure 1**). tmRNA then promotes the translation of a peptide tag encoded within tmRNA onto the C terminus of the nascent polypeptide (53). The peptide tag contains recognition determinants for many intracellular proteases, targeting the tagged protein for rapid degradation (11, 26, 34, 53). tmRNA is one of the most abundant RNAs in bacteria, and the *trans*-translation reaction occurs over 13,000 times per cell division in exponentially growing *Escherichia coli*, so on average, every ribosome in the cell translates tmRNA once every cell cycle (71). Moreover, tmRNA and other molecules required for *trans*-translation are found throughout the bacterial kingdom, and *trans*-translation is required for viability, virulence, development, and response to stresses in many bacterial systems. The ubiquity and abundance of tmRNA, as well as the phenotypes of mutants deficient in *trans*-translation, suggest that *trans*-translation confers a significant selective advantage to bacteria. Other recent reviews have covered structural and biochemical properties of tmRNA and *trans*-translation (17, 72). This review focuses on the biology of *trans*-translation and the consequences of mutations in components of this ubiquitous pathway.

HISTORY

tmRNA was first observed in 1978 in a survey of small, stable RNAs from *E. coli* (61). The molecule was named 10Sa RNA for its mobility in polyacrylamide gels, and the gene encoding 10Sa RNA was named *ssrA* (small, stable RNA A). Growth and motility defects in a Δ*ssrA* strain indicated that 10Sa RNA had some activity (78). Secondary structure predictions

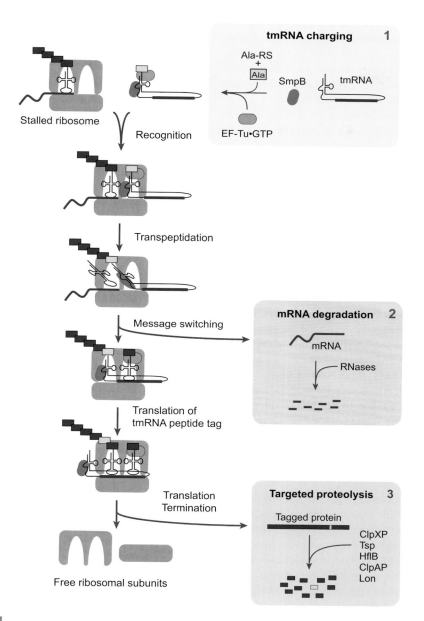

Figure 1

trans-Translation removes all components of stalled translation complexes. tmRNA binds to SmpB and is aminoacylated by alanyl-tRNA synthetase (AlaRS). EF-Tu in the GTP state binds to alanyl-tmRNA, activating the complex for ribosome interaction (*box 1*). The alanyl-tmRNA/SmpB/EF-Tu complex recognizes ribosomes at the 3′ end of an mRNA and enters the A-site as though it were a tRNA. The nascent polypeptide is transferred to tmRNA, and the tmRNA tag reading frame replaces the mRNA in the decoding center. The mRNA is rapidly degraded (*box 2*). Translation resumes, using tmRNA as a message, resulting in addition of the tmRNA-encoded peptide tag to the C terminus of the nascent polypeptide. Translation terminates at a stop codon in tmRNA, releasing the ribosomal subunits and the tagged protein. Multiple proteases recognize the tmRNA tag sequence and rapidly degrade the protein (*box 3*).

Nonstop mRNA:
mRNA with no
in-frame stop codon

suggested that the 5′ and 3′ ends of 10Sa RNA could fold together into a structure resembling alanyl-tRNA (55, 104, 106). The tRNA-like structure and alanine charging were experimentally confirmed (55, 106), but the function of this abundant RNA was still unclear.

The *trans*-translation reaction escaped detection until 1995, in large part because it is so efficient at removing all components of stalled translational complexes. In particular, the half-life of tmRNA-tagged proteins is less than 2 min in *E. coli* (53), and tagged proteins can be degraded by several different proteases (11, 26, 34, 53, 85); therefore, there was little chance of observing tagged proteins even in protease-deficient strains.

The first clue to the mechanism of tm-RNA activity came from studies of interleukin-6 (IL-6) that was overproduced in *E. coli* (103). IL-6 recovered from inclusion bodies included variants that were truncated at various distances from the C terminus and had the peptide AANDENYALAA attached to the end of the protein. The IL-6 mRNA was truncated from the 3′ end, accounting for the shortened proteins, and the last 10 residues of the attached peptide were found encoded in the *ssrA* gene, although not in a canonical open reading frame (103).

The second crucial piece of information for identifying the mechanism of tmRNA action came from studies of protein stability in *E. coli*. Proteins with sequence tags such as AAA or LAA at the C terminus were rapidly degraded in vivo, even though the protein structure was not disrupted (49, 81). The presence of a proteolysis signal within the tmRNA-encoded peptide tag led to the insight that the tag was added to target incomplete proteins for degradation, and the *trans*-translation model was proposed to explain the tag addition (49). The *trans*-translation hypothesis was tested by inserting a transcriptional terminator upstream of the stop codon into a reporter gene to generate a nonstop mRNA. The protein produced from this construct was tagged with the AANDENYALAA peptide by tmRNA, verifying that translation of nonstop mRNAs leads to tmRNA tagging (49).

The other key molecule in *trans*-translation is the protein SmpB, which binds tightly to tmRNA and is required for tmRNA structure, stability, and activity (47, 51, 109). SmpB was discovered because mutations in the *smpB* gene of *Salmonella enterica* produce a phenotype similar to that of *ssrA* mutations in *E. coli*, and the *smpB* gene is immediately upstream of *ssrA* in both species (47). In all cases that have been studied, SmpB is coregulated with tmRNA, and phenotypes caused by *smpB* deletions are the same as for *ssrA* deletions.

Subsequent biochemical and structural studies have validated and expanded the *trans*-translation model (17, 72). The current *trans*-translation model is shown in **Figure 1**.

PHYLOGENY

Genes encoding tmRNA and SmpB are present throughout the bacterial kingdom, and in some eukaryotic organelles descended from bacteria, suggesting that *trans*-translation arose early in bacterial evolution. To date, tmRNA has been identified in over 500 species of bacteria from all branches of the kingdom, including all species with a completed genome sequence (27). Plastid genomes of algae and diatoms contain tmRNA (24, 27, 44), and partial tmRNA genes lacking tag reading frames have been found in the mitochondrial genome of some protists (43, 52). tmRNA is also carried in some bacteriophage genomes (27, 83). Genes encoding SmpB have been found in all species that have tmRNA, although in eukaryotic plastids the *smpB* gene is encoded in the nuclear genome and the SmpB protein is imported into the organelle (24, 27, 44). tmRNA has not been identified in archaeal genomes or in the nuclear genomes of eukaryotes, suggesting that these species have evolved other mechanisms in place of *trans*-translation.

tmRNA STRUCTURE AND FUNCTION

Like tRNAs, tmRNA is synthesized as a precursor transcript and processed by ribonucleases to produce the mature RNA (9, 55, 67, 68, 97, 106). The 5' and 3' ends fold together to form the tRNA-like domain, which contains an acceptor stem and a TψC stem but lacks an anticodon stem. Instead, there are two to four pseudoknots and a specialized open reading frame encoding the peptide tag (20, 55, 110). The 3' end of the acceptor stem terminates with a CCA sequence that is found in all tRNAs. In *Bacillus subtilis* and other species that use tRNA nucleotidyl transferase to add CCA to tRNAs after transcription, CCA is added to tmRNA in the same fashion (106). The acceptor stem includes a G:U wobble base pair that is the primary discrimination determinant for alanyl-tRNA synthetase (38). tmRNA, when bound to SmpB, is efficiently charged with alanine by alanyl-tRNA synthetase (6, 55, 76, 106). Like a tRNA, alanyl-tmRNA is bound by EF-Tu, which protects the aminoacyl bond and facilitates productive interactions with the ribosome (6, 90, 107).

Although most *ssrA* genes are similar, a circular permutation has occurred in *ssrA* in three bacterial lineages. In these *ssrA* genes, the sequence encoding the mature 3' end is upstream of the sequence encoding the mature 5' end (23, 52, 92). Transcription and processing of the permuted tmRNA sequences produce a mature tmRNA composed of two distinct RNA molecules (52). Despite the two-piece composition, these tmRNA molecules have structures and activities similar to those of single-chain tmRNAs. Phylogenetic studies have indicated that the permuted *ssrA* genes from alpha-proteobacteria, cyanobacteria, and beta-proteobacteria are the result of independent circular permutation events and not lateral gene transfer (92). Therefore, either the permutation event is common, or there is some selective advantage to having a two-piece tmRNA in some organisms, perhaps to enable regulation of tmRNA activity.

SmpB specifically recognizes the tRNA-like domain of tmRNA, forming a 1:1 complex with a K_d in the low nanomolar range (4, 5, 8, 28, 29, 37, 47, 75, 113). In *E. coli* and *Caulobacter crescentus* during exponential growth, tmRNA is present at ~5%–10% of the concentration of rRNA, and SmpB is found at approximately equimolar amounts (52, 71). At these concentrations, it is likely that there is little unbound tmRNA or SmpB in vivo. The tmRNA-SmpB complex is required to carry out the *trans*-translation reaction because neither molecule interacts with the ribosome on its own in vivo (47, 51, 99, 109).

THE *trans*-TRANSLATION MODEL

The tmRNA/SmpB/EF-Tu complex enters the A-site of ribosomes that are stalled near the 3' end of an mRNA with the peptidyl-tRNA still in the P-site (**Figure 1**) (53, 99, 107). Structural, biochemical, and genetic experiments have suggested that SmpB provides contacts in place of the tRNA anticodon stem, interacting with the ribosome near the decoding center (8, 28, 57, 93, 98). The tRNA-like domain of tmRNA is located at the transpeptidation active site, and the nascent polypeptide is transferred to alanyl-tmRNA (53, 107). After transpeptidation, the tRNA-like domain is moved to the P-site and the message in the decoding center is switched from the mRNA to the tmRNA-encoded tag reading frame. After removal from the ribosome, the mRNA is degraded (41, 69, 86, 114). Translation of the tmRNA-encoded peptide tag begins at the resume codon (56, 60, 111). Unlike a canonical translation initiation event, no initiation factors are required to specify the resume codon, and the tag peptide does not begin with methionine. Instead, local sequence and secondary structure of tmRNA appear to be important for determining the first codon of the tag sequence (56, 60, 111). Translation continues, adding the tag peptide to the C terminus of the nascent polypeptide, and terminates at the encoded stop codon (53). The overall result of tmRNA activity is efficient removal of all components of the original stalled translation complex—the ribosome, the nascent polypeptide, and the mRNA.

RELEASE OF
STALLED RIBOSOMES

Ribosomes are released after translating the tmRNA tag reading frame even in the absence of ribosome recycling factor (RRF), a protein that dissociates ribosomal subunits after translation termination, suggesting that tmRNA may include an intrinsic ribosome recycling activity (35). Stalled translational complexes are stable in vitro (41), and releasing stalled ribosomes may be one of the important functions of *trans*-translation. However, ribosomes are still released in the absence of *trans*-translation activity in *E. coli*. Given the observed frequency of *trans*-translation during exponential growth in wild-type *E. coli*, all the ribosomes would be stuck at the end of an mRNA after one generation in a Δ*ssrA* strain if there was not an alternative mechanism for release (71). In fact, cells lacking *trans*-translation activity are viable and grow rapidly, indicating that ribosomes are not limiting (71, 78). This alternative mechanism is not known, but genetic experiments suggest that it may involve release of the peptidyl-tRNA from the ribosome followed by peptidyl-tRNA hydrolase activity (95).

PROTEIN TAGGING
AND PROTEOLYSIS

There appear to be few biochemical constraints on the sequence of the peptide tag added to proteins by tmRNA. Tags from different species vary in size from 8 to 17 residues (27). In addition, many different sequences have been engineered into the open reading frame of tmRNA and are successfully added to substrate proteins. The only residue of the tag that is conserved in all species is the first alanine, which is charged to the tRNA-like domain. tmRNA tags in most species end with a C-terminal alanine preceded by a nonpolar residue such as alanine, leucine, valine, or phenylalanine. These nonpolar residues at the C terminus are required for recognition by several intracellular proteases (11, 26, 34, 53).

The tmRNA-encoded peptide tag contains overlapping recognition determinants for proteases and proteolytic adaptors, and the tag is sufficient to target exogenous proteins for rapid degradation in vivo (11, 21, 26, 34, 53). The *E. coli* cytoplasmic proteases ClpXP, ClpAP, and Lon, the membrane-associated protease FtsH, and the periplasmic protease Tsp degrade tmRNA-tagged proteins in vitro and in vivo (11, 26, 34, 53). These proteases are widely conserved in bacteria and degrade tmRNA-tagged proteins in *B. subtilis* and *C. crescentus* (10, 109). The tmRNA-encoded peptide is also recognized by two proteolytic adaptors, SspB and ClpS (16, 21, 63). SspB binds to residues in the N-terminal region of the peptide tag, and it binds to the ClpX subunits of ClpXP, tethering the tagged protein to the protease (10, 62, 63). Although SspB enhances the rate of degradation of tagged proteins in both *E. coli* and *C. crescentus*, it is not required for proteolysis by ClpXP (10, 62, 63). In cells deleted for the *sspB* gene, tagged proteins are still degraded with a half-life of <10 min (10, 62, 63). ClpS, an adaptor for ClpAP, also binds the tmRNA tag peptide, but in this case binding prevents ClpAP degradation of tagged proteins (16). Although ClpAP and Lon degrade tmRNA-tagged proteins in vitro and in strains deleted for *clpX*, in exponentially growing wild-type *E. coli*, >90% of the degradation of tagged substrates in the cytoplasm is performed by ClpXP (19). Likewise, ClpXP accounts for most of the degradation of tmRNA-tagged proteins in *B. subtilis* (109).

mRNA DEGRADATION

The mRNA that is engaged in the stalled ribosome is rapidly degraded after *trans*-translation. In the absence of tmRNA, nonstop mRNAs are stabilized, but otherwise identical mRNAs that contain a stop codon are not affected, indicating that *trans*-translation facilitates the degradation of nonstop mRNAs (114). Moreover, some mutations in tmRNA decrease substrate mRNA turnover without altering tagging activity, and it has been suggested that ribonucleases are delivered to stalled translational complexes with tmRNA (69). RNase R, a 3′ to 5′ exonuclease, is required for some but not all tmRNA-stimulated degradation of mRNAs, indicating that multiple ribonucleases are involved (86, 114). *trans*-Translation may also facilitate the degradation of nonstop mRNAs by removing ribosomes from the 3′ end, thereby exposing them to the general nuclease activities in the cell. Removal of the mRNA presumably enhances the efficiency of *trans*-translation by preventing continued rounds of translation initiation followed by tagging.

TARGETING TRANSLATION COMPLEXES FOR *trans*-TRANSLATION

trans-Translation Substrates

The best substrates for *trans*-translation in vitro are translational complexes stalled at the 3′ end of the mRNA. Biochemical experiments showed that *trans*-translation was most efficient if the mRNA in stalled translation complexes contained no more than six nucleotides 3′ of the P-site codon (42). No tagging was detected with more than 15 nucleotides 3′ of the P-site codon, suggesting that tmRNA cannot enter translation complexes with mRNA extending past the leading edge of the ribosome (42).

Likewise, substrates for *trans*-translation in vivo are ribosomes at the 3′ end of an mRNA. *trans*-Translation occurs both on nonstop translational complexes, when the ribosome reads to the end of an mRNA without terminating, and on no-go translational complexes, when translation elongation or termination is stalled (**Figure 2**). In all examples of no-go substrates for *trans*-translation that have been examined, the mRNA is cleaved before *trans*-translation. Therefore, it appears that tmRNA-SmpB is excluded from translation elongation complexes. Consistent with this model, physiological measurements of *trans*-translation and tmRNA and SmpB levels in *E. coli* indicate that

trans-translation does not compete with translation elongation. Under normal culture conditions, there is a substantial excess of tmRNA-SmpB complex compared with the number of *trans*-translation substrates, but tmRNA-SmpB does not interfere with normal translation (71). Furthermore, overproduction of tmRNA and SmpB does not significantly increase the frequency of tagging, indicating that translating ribosomes are protected from tmRNA (71).

Nonstop Quality Control

Because translation initiation in bacteria does not require any sequences at the 3′ end of the mRNA (58), there is no way to ensure that an mRNA is intact before translation begins. Instead of recognizing mRNA problems before initiation, ribosomes detect aberrant translation when they reach the 3′ end of the mRNA with peptidyl-tRNA still in the P-site. Truncated mRNAs can be generated in the cell by premature termination of transcription or by physical damage, chemical damage, or nucleolytic degradation of a complete mRNA (**Figure 2**). For example, the first observed tmRNA tagging was on mRNAs truncated from the 3′ end, probably because of ribonuclease activity (53, 103). *trans*-Translation also occurs on mRNAs that have a transcriptional terminator before the stop codon (52, 53, 109). Translation of mRNAs after premature termination of transcription would also be expected to produce similar nonstop complexes that are targeted for *trans*-translation.

Nonstop translational complexes are also generated on full-length mRNAs by mistakes during translation that cause read-through of the normal stop codon, or by frameshifting that results in no in-frame stop codon (**Figure 2**). Similarly, suppressor tRNAs cause read-through of stop codons and promote *trans*-translation (105). In these cases, the mRNA is correct and full-length, but the nascent polypeptide is aberrant. Insertion or deletion errors during transcription could also put the stop codon out of frame and lead to nonstop translation complexes.

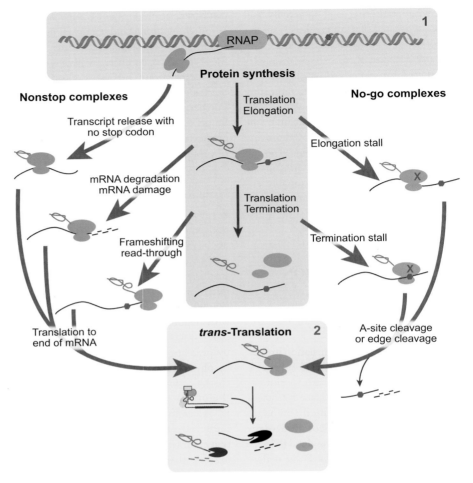

Figure 2

Nonstop and no-go translation complexes are targeted for *trans*-translation. Errors or programmed events during the normal process of protein synthesis (*box 1*) produce a nonstop translational complex when the mRNA has no in-frame stop codon. Translation of the ribosome to the 3′ end of the mRNA generates a substrate for *trans*-translation (*box 2*). Stalling during translation elongation or termination results in a no-go complex. The mRNA is cleaved in the A-site or at the leading edge of the ribosome, targeting the complex for *trans*-translation.

No-Go Quality Control

Translation reactions blocked during elongation or termination by consecutive rare codons, weak termination sequences, or artificial depletion of a required tRNA or release factor are recognized as no-go translation complexes and targeted for *trans*-translation (14, 30–32, 64, 66, 87, 100–102). In principle, no-go complexes could also be caused by damage to mRNA bases or by interactions between the nascent polypeptide and the ribosome exit channel that block translation.

The mRNA in no-go complexes can be cleaved in at least two different places to generate *trans*-translation substrates: in the A-site of the ribosome or 3′ of the leading edge of the ribosome (22, 32, 42, 64, 101, 102). Some A-site mRNA cleavage is caused by small

No-go translation complex: ribosome stalled in the middle of an mRNA during translation

protein toxins such as RelE, which can enter the A-site and promote cleavage of the mRNA at specific codons (82). However A-site mRNA cleavage has been observed even in strains with all known toxins deleted, leading to speculation that the nuclease activity is within the ribosome itself (32). Edge cleavage is the result of multiple ribonucleases degrading mRNA that is not protected by the ribosome. Deletion of individual RNases does not prevent edge cleavage, but alters the final 3′ end of the mRNA (22, 64, 101). Neither A-site cleavage nor edge cleavage requires tmRNA or SmpB, suggesting that mRNA cleavage in no-go complexes occurs independently of *trans*-translation (32).

Not all stalled translational complexes are targeted for mRNA cleavage and *trans*-translation. In *E. coli*, ribosomes paused by the leader peptide of *tnaC* are resistant to mRNA cleavage and tmRNA tagging (25). Likewise, programmed pausing of ribosomes by the SecM peptide does not lead to mRNA cleavage unless the pause is not released (22). In these examples, ribosome stalling has a regulatory purpose, and mRNA cleavage followed by *trans*-translation would defeat the control mechanism. In general, ribosomes stalled with an empty A-site are subject to A-site mRNA cleavage and no-go quality control, but if the A-site is occupied by tRNA or release factors, the mRNA is protected.

Regulation by *trans*-Translation

Translational complexes can be targeted to the nonstop or no-go pathways for regulation as well as quality control. *trans*-Translation activity on some substrates is important for control of individual regulatory circuits. In other cases, a regulatory role for *trans*-translation might explain the conserved use of rare codons, weak termination sequences, or toxin cleavage sites in some genes.

The best-understood use of *trans*-translation for genetic regulation is during the expression of the LacI transcriptional repressor in *E. coli* (1). LacI binds to *lac* operator sites in the *lacZYA* promoter to repress transcription of the operon. At high concentrations, a LacI tetramer binds to two *lac* operator sites, O$_1$ and O$_3$, looping the DNA between these sequences to exclude RNA polymerase. Because the *lacI* gene is immediately upstream of the *lacZ* promoter and the O$_3$ site lies within the *lacI* coding sequence, binding of the LacI tetramer to O$_1$ and O$_3$ prevents elongation of transcription through the 3′ end of the *lacI* gene. The *lacI* mRNA synthesized when LacI is bound to O$_1$ and O$_3$ has no stop codon, and LacI protein made from the nonstop mRNA is tagged by tmRNA and degraded. This autoregulatory circuit keeps LacI concentrations in an optimal range to respond to changes in nutrient conditions. In Δ*ssrA* cells, truncated LacI is produced from the nonstop mRNA but not degraded, and these truncated LacI variants are active repressors of the *lacZYA* promoter (1). *E. coli* mutants lacking tmRNA show delayed induction of the *lac* operon during diauxic growth, consistent with accumulation of excess LacI (1). Many DNA binding proteins contain a cognate binding site within their own coding sequence (89), so it is possible that other transcription factors require tmRNA for correct regulation in the same manner as does LacI.

A second example of regulation by *trans*-translation is control of the *kinA* gene in *B. subtilis*. KinA is a kinase that initiates sporulation under specific nutrient conditions (59). In some undomesticated strains of *B. subtilis* the *kinA* gene lacks the stop codon found in the laboratory strain *B. subtilis* 168, and the *kinA* open reading frame extends through a transcriptional terminator (54). mRNA transcribed from the variant *kinA* genes is nonstop, and it is likely that KinA protein synthesized from this mRNA is tagged by *trans*-translation and degraded. Strains that have the nonstop *kinA* do not sporulate under conditions that activate KinA kinase activity, presumably because there is not enough KinA protein. However, if *ssrA* is deleted in these strains, the cells sporulate normally (54). Wild-type cells that make nonstop *kinA* would be expected to sporulate under conditions that inactivate or overwhelm the *trans*-translation system and activate KinA.

Therefore, *trans*-translation may provide an additional level of regulation for sporulation in these strains.

Bacteria also produce no-go translational complexes for regulatory reasons. For example, under stress conditions RelE promotes widespread mRNA cleavage that halts cell growth, allowing cells to divert scarce resources to essential processes (82). After the stress is removed and RelE is inactivated, tmRNA is required for rapid return to normal growth rates (12). tmRNA also facilitates recovery from stasis induced by other protein toxins (13). These toxins are found in most bacteria and respond to a wide array of environmental signals (80); therefore cycles of toxin-induced stasis followed by *trans*-translation-dependent recovery may play an important physiological role in the wild.

Another potential mechanism for generating *trans*-translation substrates has been suggested from studies in *C. crescentus*. Forty-six *trans*-translation substrates identified in *C. crescentus* share a nucleotide sequence motif upstream of the tagging site (36). Mutation of this 16-mer sequence in endogenous substrates decreases the amount of tagging, indicating that it plays a role in tmRNA substrate selectivity. Because the motif is found upstream of the tagging site, it is not likely an RNase cleavage site or ribosome stalling site. The motif might promote transcription termination prior to the stop codon to make a nonstop complex, or stall translation after the ribosome has passed to make a no-go complex (36).

PHYSIOLOGY OF *trans*-TRANSLATION

What happens in cells with no *trans*-translation activity? tmRNA is essential in *Neisseria gonorrhoeae* (39) and *Shigella flexneri* (K.C. Keiler, unpublished results), and neither *ssrA* nor *smpB* was deleted in saturating mutagenesis screens of *Mycoplasma genitalium* or *Haemophilus influenza* (3, 40). The reasons why tmRNA activity is essential in these species, but not in most bacteria, are not known. In other species, lack of

trans-translation frequently causes growth defects when the cellular gene expression profile changes, such as during stress responses, virulence, and development (48).

Stress and Starvation Phenotypes

trans-Translation allows bacteria to efficiently respond to and recover from a variety of stresses. In addition to the diauxic lag caused by misregulation of LacI and delays in recovery from toxin-induced stasis, *E. coli* strains deficient in *trans*-translation have several phenotypes related to stress responses. Cells deleted for *ssrA* or *smpB* have a delayed recovery from stationary phase (78), are more sensitive to amino acid starvation (65) and heat shock (78), and are more sensitive to antibiotics that promote translational frameshifting, read-through, and stalling (2). The molecular bases for these phenotypes are not known, but some stress response phenotypes may be caused by misregulation of RpoS. RpoS is an alternative sigma factor that is expressed during stationary phase and other stress conditions (33). In Δ*ssrA* cells, truncated forms of RpoS accumulate and RpoS expression and activity are much lower during stationary phase (85). Other stress phenotypes may be the result of cumulative effects on many *trans*-translation substrates. Low concentrations of serine hydroxamate, which starves cells for serine, inhibit growth of Δ*ssrA* cells but do not alter the growth of cells that have tmRNA. Fragments of multiple mRNAs accumulate in Δ*ssrA* cells starved for serine or other amino acids, but these mRNA fragments are removed if *trans*-translation is active (65). Mild amino acid starvation may increase no-go translational complexes owing to depletion of charged tRNAs, creating an increased demand for *trans*-translation activity. Increased antibiotic sensitivity in cells lacking *trans*-translation activity is also likely due to a higher level of nonstop quality-control substrates caused by miscoding and read-through. *trans*-Translation may be important under most conditions that increase nonstop and no-go complexes. In fact, *E. coli* deleted for *ssrA* exhibits a constitutive heat-shock stress response during exponential growth, suggesting that it has difficulty even under optimal growth conditions (73).

Other species also display phenotypes related to stress survival when they lack *trans*-translation activity. *Synechocystis* sp. (15) and *Yersinia pseudotuberculosis* (79) strains deleted for *ssrA* or *smpB* have increased sensitivity to antibiotics. The *Y. pseudotuberculosis* mutants are also temperature sensitive and have increased sensitivity to oxidative and nitrosative stresses (79). In *B. subtilis*, mutants lacking tmRNA activity are sensitive to both high and low temperature, and tmRNA and SmpB levels increase during temperature stress (74, 94). tmRNA levels also increase eightfold during antibiotic challenge and biofilm formation in *Thermotoga maritima* (70). Like stationary phase, biofilm formation is associated with large changes in the gene expression profile and decreased availability of nutrients.

Role of *trans*-Translation in Pathogenesis

Large changes in the cellular gene expression profile are also important during pathogenesis, when bacteria face both starvation and attack by host defenses, and *trans*-translation is required for virulence in some pathogens. *Y. pseudotuberculosis* strains deleted for *ssrA* or *smpB* do not cause lethality in a mouse infection model and are unable to proliferate in macrophages (79). At least part of this virulence defect is due to a delay in the production of secreted effector proteins known as Yops. After invasion of macrophages, Yops are expressed and secreted through a dedicated Type III system and cause cytotoxicity in the host. The VirF transcription factor regulates production and secretion of Yops, but in the absence of tmRNA or SmpB VirF is misregulated and Yop secretion is delayed (79). It is not yet clear if VirF is regulated by a tmRNA-dependent genetic circuit, or if the requirement for *trans*-translation is less direct. In addition to virulence defects, *Y. pseudotuberculosis* cells lacking tmRNA activity are nonmotile and are sensitive to stress conditions as described above.

In *S. enterica*, tmRNA is required both for proliferation in macrophages and for virulence in mice (7, 46). The expression of several virulence genes is disrupted in the Δ*ssrA* strain, but the molecular basis for the virulence phenotype is not known.

Effects on Cell Cycle and Differentiation

Studies of *C. crescentus* indicate that *trans*-translation plays a role in the bacterial cell cycle. *C. crescentus* deleted for *ssrA* or *smpB* has a defect in the timing of DNA replication (51). *C. crescentus* coordinates initiation of DNA replication with morphological differentiation from a swarmer cell to a stalked cell (91). DNA replication normally initiates after proteolysis of CtrA, a response regulator that binds to the origin of replication. In cells deleted for *ssrA* or *smpB*, CtrA is degraded at the normal time, but replication does not initiate for 40 min (~30% of the time needed to complete a normal cell cycle) (51). After the initiation of replication, the cell cycle resumes with no other significant delays.

The maximum levels of tmRNA and SmpB in *C. crescentus* occur during the swarmer to stalked cell differentiation, when DNA replication initiates. Both tmRNA and SmpB increase approximately fivefold during this time and are removed from the cell early in S phase (50). tmRNA levels are controlled by both transcription and regulated ribonuclease activity. The *ssrA* promoter is active before the initiation of DNA replication and is turned off after replication initiates (50). In a complementary manner, tmRNA is stable during G1 phase but is degraded with a half-life of 5 min by the nuclease RNase R after replication initiates (37). RNase R processively degrades tmRNA in vitro, starting from the non-tRNA-like 3' end of the two-piece tmRNA molecule (37). In vivo, RNase R is constitutively present but will not degrade tmRNA when SmpB is present, presumably because SmpB blocks access to the 3' end of the coding RNA (37). Thus, the timing of degradation of tmRNA is controlled by proteolysis of SmpB. It is not yet clear if RNase R also degrades mRNAs during *trans*-translation in *C. crescentus* as it does in *E. coli*. The two-piece construction of tmRNA is crucial for its turnover in *C. crescentus*, and the ability to control degradation of tmRNA may account for the conservation of the circularly permuted form of *ssrA* in some species.

The cell cycle delay in Δ*ssrA* and Δ*smpB* strains may be due to misregulation of a single factor or to cumulative effects on many substrates. Seventeen DNA replication factors have been identified as substrates for *trans*-translation in *C. crescentus*, so it is possible that misregulation of one or more of these proteins in the absence of *trans*-translation causes the delay in initiation of DNA replication (36). *C. crescentus* cells lacking *trans*-translation activity also will not maintain a family of broad-host-range plasmids (51). The inability to maintain plasmids may be due to misregulation of DNA replication or to an unrelated defect.

Cellular differentiation is also blocked during symbiotic growth of *Bradyrhizobium japonicum* in the root nodules of soybean. *B. japonicum* mutants lacking tmRNA promote the formation of root nodules in the host but do not differentiate into the nitrogen-fixing bacteroid form (18). It remains unknown why tmRNA is required for this differentiation.

Phage Development in Mutant Hosts

Several bacteriophages of *E. coli* and *S. enterica* require the host *trans*-translation machinery for development. *E. coli* Δ*ssrA* strains do not support lytic growth of the hybrid phage λ*imm*P22 (96). Likewise, phage P22 cannot make plaques in *S. enterica* deleted for *ssrA* or *smpB* (46, 47). For both P22 and λ*imm*P22, phage lacking the C1 transcriptional activator develops correctly, even in the absence of *trans*-translation (46, 47, 112). Thus, C1 is required for the tmRNA phenotype, but it is not known if C1 is directly regulated by *trans*-translation.

Temperature-sensitive variants of phage Mu cannot be induced from lysates in an *E. coli* host deleted for *ssrA* or *smpB* (47, 84). The lysogenic state is maintained by binding of the Mu *c* repressor to the operator of the pE and pCM promoters. Mu *c* is tagged by tmRNA, and variants of Mu *c* lacking the last 18 amino acids have a noninducible phenotype, so production of truncated species in the absence of *trans*-translation might account for the observed lack of induction (77).

MOLECULAR BASES OF *trans*-TRANSLATION PHENOTYPES

What do the diverse phenotypes of *trans*-translation mutants in various species have in common? There are few shared properties, and this diversity may reflect the large differences in physiology among bacteria. Nevertheless, because tmRNA and SmpB appear to be conserved throughout the bacterial kingdom, it is tempting to speculate on universal roles for *trans*-translation. One possibility is that *trans*-translation is important during large changes in the cellular gene expression profile. In most species, *trans*-translation activity is dispensable or produces a weak phenotype during optimal growth conditions, for example, during exponential growth of *E. coli* and *B. subtilis*, or vegetative growth of *S. enterica*, *Y. pseudotuberculosis*, and *B. japonicum*. However, when the gene expression program changes as the cells respond to stress, switch carbon sources, differentiate, or initiate pathogenesis, *trans*-translation is frequently important. In addition, levels of tmRNA and SmpB typically increase during changes in gene expression patterns. tmRNA levels increase during entry into stationary phase in *T. maritima* and *Synechococcus* sp. (70, 108), during antibiotic challenge in *T. maritima* (70), in response to several stresses in *B. subtilis* (74), and during differentiation in *C. crescentus* (50). These stressful conditions may simply exacerbate mild defects in translation capacity caused by *trans*-translation mutations. Alternatively, steady-state gene expression may produce relatively few errors that require quality control pathways or genetic circuit regulation compared with conditions that have increased mRNA degradation and synthesis.

The ultimate result of *trans*-translation activity is release of stalled ribosomes, tagging and degradation of the nascent polypeptides, and degradation of the engaged mRNAs. Are any of these outcomes individually responsible for the observed phenotypes? It is not possible to separate tagging of the nascent polypeptide from ribosome release because message switching is required for translation termination. It is also not possible to examine the role of substrate proteolysis or mRNA degradation by eliminating particular proteins or nucleases because each of the enzymes implicated in these processes has additional functions independent of *trans*-translation. Therefore, variants of tmRNA that add a peptide lacking some proteolytic determinants have been used to try to deconvolute the effects of tagging from proteolysis. For example changing the last two alanine codons to aspartate codons (tmRNA-DD), or changing the last six codons to histidine (tmRNA-His6), results in tagged proteins that are not rapidly degraded (1, 52, 87, 88, 109). Phenotypes that are not complemented by these variants, including cell cycle control in *C. crescentus* (51) and motility in *Y. pseudotuberculosis* (79), are likely to require proteolysis of tagged substrates. However, the tmRNA-DD and tmRNA-His6 variants have decreased RNase R–dependent degradation of some mRNAs in *E. coli* (45). Therefore, effects of mRNA stability on the phenotypes need to be considered.

Many other phenotypes are complemented by the variant tmRNAs and are likely to require tagging but not proteolysis: viability for *N. gonorrhoeae* (39), the plating phenotypes in λ*imm*P22 and Mu (84, 112), stress phenotypes in *E. coli* and *B. subtilis* (2, 73, 74, 85), plasmid maintenance in *C. crescentus* (51), and virulence

in *Y. pseudotuberculosis* (79). Phenotypes complemented by the tmRNA-DD and tmRNA-His6 variants are frequently ascribed to the effects of inefficient ribosome release in the absence of *trans*-translation. Ribosome release may indeed be important for some phenotypes because sequestering ribosomes to stalled complexes might impair the translational capacity of the cell. However, stalled ribosomes have not been observed in the absence of *trans*-translation, and no loss of translation capacity has been demonstrated in any *trans*-translation mutant. It is also possible that stalled ribosomes exert a physiological effect independent of observable changes in translation capacity. For example, a small increase in the number of stalled ribosomes might initiate a stress response pathway, or toxic quantities of peptidyl-tRNA might be released from stalled ribosomes if *trans*-translation is not available. Further studies of the mechanistic details and physiological consequences of *trans*-translation on individual substrates will be required to understand how most phenotypes are produced.

FUTURE DIRECTIONS

Key issues that remain to be understood include how mutations in *trans*-translation produce the diverse phenotypes that have been observed, how *trans*-translation and its various components are controlled, and how this pathway fits into the global regulatory network of the cell. In addition, recent evidence from *C. crescentus* and *E. coli* indicates that tmRNA and SmpB are localized within the cell (K.C. Keiler, unpublished data), raising the possibility that *trans*-translation is spatially regulated. Finally, the *trans*-translation pathway is a potential target for development of novel antibiotics because it is essential in some human pathogens and required for virulence in others.

SUMMARY POINTS

1. *trans*-Translation releases ribosomes from the 3' end of mRNAs and targets the nascent polypeptide and mRNA for degradation.

2. Mistakes during protein synthesis can lead to nonstop complexes or no-go complexes. Both nonstop and no-go complexes are targeted for *trans*-translation.

3. *trans*-Translation is also used for genetic regulation. The best example of regulation is control of LacI repressor levels by autoregulatory production of a nonstop *lacI* mRNA, followed by tmRNA tagging of the LacI protein.

4. In the absence of *trans*-translation activity, *E. coli* and other species are more sensitive to many stress conditions, including stationary phase, carbon starvation, antibiotic exposure, and heat.

5. *trans*-Translation is important during changes to a genetic program, such as during pathogenesis and differentiation.

DISCLOSURE STATEMENT

The author is not aware of any biases that might be perceived as affecting the objectivity of this review.

ACKNOWLEDGEMENTS

I thank Sarah Ades, Wali Karzai, Bob Sauer, and Hiroji Aiba for helpful discussions and communicating unpublished results. This work is supported by NIH grant GM-68720.

LITERATURE CITED

1. **Abo T, Inada T, Ogawa K, Aiba H. 2000. SsrA-mediated tagging and proteolysis of LacI and its role in the regulation of *lac* operon. *EMBO J.* 19:3762–69**

2. Abo T, Ueda K, Sunohara T, Ogawa K, Aiba H. 2002. SsrA-mediated protein tagging in the presence of miscoding drugs and its physiological role in *Escherichia coli*. *Genes Cells* 7:629–38

3. Akerley BJ, Rubin EJ, Novick VL, Amaya K, Judson N, Mekalanos JJ. 2002. A genome-scale analysis for identification of genes required for growth or survival of *Haemophilus influenzae*. *Proc. Natl. Acad. Sci. USA* 99:966–71

4. Barends S, Bjork K, Gultyaev AP, de Smit MH, Pleij CW, Kraal B. 2002. Functional evidence for D- and T-loop interactions in tmRNA. *FEBS Lett.* 514:78–83

5. Barends S, Karzai AW, Sauer RT, Wower J, Kraal B. 2001. Simultaneous and functional binding of SmpB and EF-Tu-TP to the alanyl acceptor arm of tmRNA. *J. Mol. Biol.* 314:9–21

6. Barends S, Wower J, Kraal B. 2000. Kinetic parameters for tmRNA binding to alanyl-tRNA synthetase and elongation factor Tu from *Escherichia coli*. *Biochemistry* 39:2652–58

7. Baumler AJ, Kusters JG, Stojiljkovic I, Heffron F. 1994. *Salmonella typhimurium* loci involved in survival within macrophages. *Infect. Immun.* 62:1623–30

8. Bessho Y, Shibata R, Sekine S, Murayama K, Higashijima K, et al. 2007. Structural basis for functional mimicry of long-variable-arm tRNA by transfer-messenger RNA. *Proc. Natl. Acad. Sci. USA* 104:8293–98

9. Chauhan AK, Apirion D. 1989. The gene for a small stable RNA (10Sa RNA) of *Escherichia coli*. *Mol. Microbiol.* 3:1481–85

10. Chien P, Perchuk BS, Laub MT, Sauer RT, Baker TA. 2007. Direct and adaptor-mediated substrate recognition by an essential AAA+ protease. *Proc. Natl. Acad. Sci. USA* 104:6590–95

11. Choy JS, Aung LL, Karzai AW. 2007. Lon protease degrades transfer-messenger RNA-tagged proteins. *J. Bacteriol.* 189:6564–71

12. **Christensen SK, Gerdes K. 2003. RelE toxins from bacteria and archaea cleave mRNAs on translating ribosomes, which are rescued by tmRNA. *Mol. Microbiol.* 48:1389–400**

1. Shows that *trans*-translation regulates LacI and is required for *lac* operon control.

12. Shows that *trans*-translation counteracts the effects of mRNA cleavage by RelE and is required for recovery from the stringent response.

13. Christensen SK, Pedersen K, Hansen FG, Gerdes K. 2003. Toxin-antitoxin loci as stress-response-elements: ChpAK/MazF and ChpBK cleave translated RNAs and are counteracted by tmRNA. *J. Mol. Biol.* 332:809–19

14. Collier J, Binet E, Bouloc P. 2002. Competition between SsrA tagging and translational termination at weak stop codons in *Escherichia coli*. *Mol. Microbiol.* 45:745–54

15. de la Cruz J, Vioque A. 2001. Increased sensitivity to protein synthesis inhibitors in cells lacking tmRNA. *RNA* 7:1708–16

16. Dougan DA, Reid BG, Horwich AL, Bukau B. 2002. ClpS, a substrate modulator of the ClpAP machine. *Mol. Cell* 9:673–83

17. Dulebohn D, Choy J, Sundermeier T, Okan N, Karzai AW. 2007. *trans*-Translation: the tmRNA-mediated surveillance mechanism for ribosome rescue, directed protein degradation, and nonstop mRNA decay. *Biochemistry* 46:4681–93

18. Ebeling S, Kundig C, Hennecke H. 1991. Discovery of a rhizobial RNA that is essential for symbiotic root nodule development. *J. Bacteriol.* 173:6373–82

19. Farrell CM, Grossman AD, Sauer RT. 2005. Cytoplasmic degradation of ssrA-tagged proteins. *Mol. Microbiol.* 57:1750–61

20. Felden B, Himeno H, Muto A, McCutcheon JP, Atkins JF, Gesteland RF. 1997. Probing the structure of the *Escherichia coli* 10Sa RNA (tmRNA). *RNA* 3:89–103

21. Flynn JM, Levchenko I, Seidel M, Wickner SH, Sauer RT, Baker TA. 2001. Overlapping recognition determinants within the ssrA degradation tag allow modulation of proteolysis. *Proc. Natl. Acad. Sci. USA* 98:10584–89

22. Garza-Sanchez F, Janssen BD, Hayes CS. 2006. Prolyl-tRNAPro in the A-site of SecM-arrested ribosomes inhibits the recruitment of transfer-messenger RNA. *J. Biol. Chem.* 281:34258–68

23. Gaudin C, Zhou X, Williams KP, Felden B. 2002. Two-piece tmRNA in cyanobacteria and its structural analysis. *Nucleic Acids Res.* 30:2018–24

24. Gimple O, Schon A. 2001. In vitro and in vivo processing of cyanelle tmRNA by RNase P. *Biol. Chem.* 382:1421–29

25. Gong M, Cruz-Vera LR, Yanofsky C. 2007. Ribosome recycling factor and release factor 3 action promotes TnaC-peptidyl-tRNA dropoff and relieves ribosome stalling during tryptophan induction of *tna* operon expression in *Escherichia coli*. *J. Bacteriol.* 189:3147–55

26. Gottesman S, Roche E, Zhou Y, Sauer RT. 1998. The ClpXP and ClpAP proteases degrade proteins with carboxy-terminal peptide tails added by the SsrA-tagging system. *Genes Dev.* 12:1338–47

27. Gueneau de Novoa P, Williams KP. 2004. The tmRNA website: reductive evolution of tmRNA in plastids and other endosymbionts. *Nucleic Acids Res.* 32:D104–8

28. Gutmann S, Haebel PW, Metzinger L, Sutter M, Felden B, Ban N. 2003. Crystal structure of the transfer-RNA domain of transfer-messenger RNA in complex with SmpB. *Nature* 424:699–703

29. Hanawa-Suetsugu K, Takagi M, Inokuchi H, Himeno H, Muto A. 2002. SmpB functions in various steps of *trans*-translation. *Nucleic Acids Res.* 30:1620–29

30. Hayes CS, Bose B, Sauer RT. 2002. Proline residues at the C terminus of nascent chains induce SsrA tagging during translation termination. *J. Biol. Chem.* 277:33825–32

31. Hayes CS, Bose B, Sauer RT. 2002. Stop codons preceded by rare arginine codons are efficient determinants of SsrA tagging in *Escherichia coli*. *Proc. Natl. Acad. Sci. USA* 99:3440–45

32. **Hayes CS, Sauer RT. 2003. Cleavage of the A site mRNA codon during ribosome pausing provides a mechanism for translational quality control. *Mol. Cell* 12:903–11**

33. Hengge-Aronis R. 2002. Signal transduction and regulatory mechanisms involved in control of the sigma(S) (RpoS) subunit of RNA polymerase. *Microbiol. Mol. Biol. Rev.* 66:373–95

34. Herman C, Thevenet D, Bouloc P, Walker GC, D'Ari R. 1998. Degradation of carboxy-terminal-tagged cytoplasmic proteins by the *Escherichia coli* protease HflB (FtsH). *Genes Dev.* 12:1348–55

35. Hirokawa G, Inokuchi H, Kaji H, Igarashi K, Kaji A. 2004. In vivo effect of inactivation of ribosome recycling factor: fate of ribosomes after unscheduled translation downstream of open reading frame. *Mol. Microbiol.* 54:1011–21

32. Demonstrates that A-site mRNA cleavage targets no-go complexes for *trans*-translation, and that cleavage does not require tmRNA or SmpB.

36. Hong SJ, Lessner FH, Mahen EM, Keiler KC. 2007. Proteomic identification of tmRNA substrates. *Proc. Natl. Acad. Sci. USA* 104:17128–33

37. Hong SJ, Tran QA, Keiler KC. 2005. Cell-cycle dependent degradation of a regulatory RNA by RNase R. *Mol. Microbiol.* 57:565–75

38. Hou YM, Schimmel P. 1988. A simple structural feature is a major determinant of the identity of a transfer RNA. *Nature* 333:140–45

39. Huang C, Wolfgang MC, Withey J, Koomey M, Friedman DI. 2000. Charged tmRNA but not tmRNA-mediated proteolysis is essential for *Neisseria gonorrhoeae* viability. *EMBO J.* 19:1098–107

40. Hutchison CA, Peterson SN, Gill SR, Cline RT, White O, et al. 1999. Global transposon mutagenesis and a minimal *Mycoplasma* genome. *Science* 286:2165–69

41. Ivanova N, Pavlov MY, Ehrenberg M. 2005. tmRNA-induced release of messenger RNA from stalled ribosomes. *J. Mol. Biol.* 350:897–905

42. **Ivanova N, Pavlov MY, Felden B, Ehrenberg M. 2004. Ribosome rescue by tmRNA requires truncated mRNAs. *J. Mol. Biol.* 338:33–41**

43. Jacob Y, Seif E, Paquet PO, Lang BF. 2004. Loss of the mRNA-like region in mitochondrial tmRNAs of jakobids. *RNA* 10:605–14

44. Jacob Y, Sharkady SM, Bhardwaj K, Sanda A, Williams KP. 2005. Function of the SmpB tail in transfer-messenger RNA translation revealed by a nucleus-encoded form. *J. Biol. Chem.* 280:5503–9

45. Jentsch S. 1996. When proteins receive deadly messages at birth. *Science* 271:955–56

46. Julio SM, Heithoff DM, Mahan MJ. 2000. ssrA (tmRNA) plays a role in *Salmonella enterica* serovar Typhimurium pathogenesis. *J. Bacteriol.* 182:1558–63

47. Karzai AW, Susskind MM, Sauer RT. 1999. SmpB, a unique RNA-binding protein essential for the peptide-tagging activity of SsrA (tmRNA). *EMBO J.* 18:3793–99

48. Keiler KC. 2007. Physiology of tmRNA: What gets tagged and why? *Curr. Opin. Microbiol.* 10:169–75

49. Keiler KC, Sauer RT. 1996. Sequence determinants of C-terminal substrate recognition by the Tsp protease. *J. Biol. Chem.* 271:2589–93

50. Keiler KC, Shapiro L. 2003. tmRNA in *Caulobacter crescentus* is cell cycle regulated by temporally controlled transcription and RNA degradation. *J. Bacteriol.* 185:1825–30

51. **Keiler KC, Shapiro L. 2003. tmRNA is required for correct timing of DNA replication in *Caulobacter crescentus*. *J. Bacteriol.* 185:573–80**

52. Keiler KC, Shapiro L, Williams KP. 2000. tmRNAs that encode proteolysis-inducing tags are found in all known bacterial genomes: A two-piece tmRNA functions in *Caulobacter*. *Proc. Natl. Acad. Sci. USA* 97:7778–83

53. Keiler KC, Waller PR, Sauer RT. 1996. Role of a peptide tagging system in degradation of proteins synthesized from damaged messenger RNA. *Science* 271:990–93

54. Kobayashi K, Kuwana R, Takamatsu H. 2008. kinA mRNA is missing a stop codon in the undomesticated *Bacillus subtilis* strain ATCC 6051. *Microbiology* 154:54–63

55. Komine Y, Kitabatake M, Yokogawa T, Nishikawa K, Inokuchi H. 1994. A tRNA-like structure is present in 10Sa RNA, a small stable RNA from *Escherichia coli*. *Proc. Natl. Acad. Sci. USA* 91:9223–27

56. Konno T, Kurita D, Takada K, Muto A, Himeno H. 2007. A functional interaction of SmpB with tmRNA for determination of the resuming point of *trans*-translation. *RNA* 13:1723–31

57. Kurita D, Sasaki R, Muto A, Himeno H. 2007. Interaction of SmpB with ribosome from directed hydroxyl radical probing. *Nucleic Acids Res.* 35:7248–55

58. Laursen BS, Sorensen HP, Mortensen KK, Sperling-Petersen HU. 2005. Initiation of protein synthesis in bacteria. *Microbiol. Mol. Biol. Rev.* 69:101–23

59. LeDeaux JR, Yu N, Grossman AD. 1995. Different roles for KinA, KinB, and KinC in the initiation of sporulation in *Bacillus subtilis*. *J. Bacteriol.* 177:861–63

60. Lee S, Ishii M, Tadaki T, Muto A, Himeno H. 2001. Determinants on tmRNA for initiating efficient and precise *trans*-translation: Some mutations upstream of the tag-encoding sequence of *Escherichia coli* tmRNA shift the initiation point of *trans*-translation in vitro. *RNA* 7:999–1012

61. Lee SY, Bailey SC, Apirion D. 1978. Small stable RNAs from *Escherichia coli*: evidence for the existence of new molecules and for a new ribonucleoprotein particle containing 6S RNA. *J. Bacteriol.* 133:1015–23

42. Demonstrates that ribosomes must be at the 3′ end of an mRNA for *trans*-translation in vitro.

51. Shows that *trans*-translation and proteolysis of tagged substrates are required for cell cycle control in *C. crescentus*.

62. Lessner FH, Venters BJ, Keiler KC. 2007. Proteolytic adaptor for transfer-messenger RNA-tagged proteins from alpha-proteobacteria. *J. Bacteriol.* 189:272–75

63. Levchenko I, Seidel M, Sauer RT, Baker TA. 2000. A specificity-enhancing factor for the ClpXP degradation machine. *Science* 289:2354–56

64. Li X, Hirano R, Tagami H, Aiba H. 2006. Protein tagging at rare codons is caused by tmRNA action at the 3′ end of nonstop mRNA generated in response to ribosome stalling. *RNA* 12:248–55

65. Li X, Yagi M, Morita T, Aiba H. 2008. Cleavage of mRNAs and role of tmRNA system under amino acid starvation in Escherichia coli. *Mol. Microbiol.* 68:462–73

66. Li X, Yokota T, Ito K, Nakamura Y, Aiba H. 2007. Reduced action of polypeptide release factors induces mRNA cleavage and tmRNA tagging at stop codons in *Escherichia coli*. *Mol. Microbiol.* 63:116–26

67. Li Z, Pandit S, Deutscher MP. 1998. 3′ exoribonucleolytic trimming is a common feature of the maturation of small, stable RNAs in *Escherichia coli*. *Proc. Natl. Acad. Sci. USA* 95:2856–61

68. Lin-Chao S, Wei CL, Lin YT. 1999. RNase E is required for the maturation of ssrA RNA and normal ssrA RNA peptide-tagging activity. *Proc. Natl. Acad. Sci. USA* 96:12406–11

69. Mehta P, Richards J, Karzai AW. 2006. tmRNA determinants required for facilitating nonstop mRNA decay. *RNA* 12:2187–98

70. Montero CI, Lewis DL, Johnson MR, Conners SB, Nance EA, et al. 2006. Colocation of genes encoding a tRNA-mRNA hybrid and a putative signaling peptide on complementary strands in the genome of the hyperthermophilic bacterium *Thermotoga maritima*. *J. Bacteriol.* 188:6802–7

71. **Moore SD, Sauer RT. 2005. Ribosome rescue: tmRNA tagging activity and capacity in *Escherichia coli*. *Mol. Microbiol.* 58:456–66**

72. Moore SD, Sauer RT. 2007. The tmRNA system for translational surveillance and ribosome rescue. *Annu. Rev. Biochem.* 76:101–24

73. Munavar H, Zhou Y, Gottesman S. 2005. Analysis of the *Escherichia coli* Alp phenotype: heat shock induction in ssrA mutants. *J. Bacteriol.* 187:4739–51

74. Muto A, Fujihara A, Ito KI, Matsuno J, Ushida C, Himeno H. 2000. Requirement of transfer-messenger RNA for the growth of *Bacillus subtilis* under stresses. *Genes Cells* 5:627–35

75. Nameki N, Someya T, Okano S, Suemasa R, Kimoto M, et al. 2005. Interaction analysis between tmRNA and SmpB from *Thermus thermophilus*. *J. Biochem.* 138:729–39

76. Nameki N, Tadaki T, Muto A, Himeno H. 1999. Amino acid acceptor identity switch of *Escherichia coli* tmRNA from alanine to histidine in vitro. *J. Mol. Biol.* 289:1–7

77. O'Handley D, Nakai H. 2002. Derepression of bacteriophage mu transposition functions by truncated forms of the immunity repressor. *J. Mol. Biol.* 322:311–24

78. Oh BK, Apirion D. 1991. 10Sa RNA, a small stable RNA of *Escherichia coli*, is functional. *Mol. Gen. Genet.* 229:52–56

79. **Okan NA, Bliska JB, Karzai AW. 2006. A role for the SmpB-SsrA system in *Yersinia pseudotuberculosis* pathogenesis. *PLoS Pathog.* 2:e6**

80. Pandey DP, Gerdes K. 2005. Toxin-antitoxin loci are highly abundant in free-living but lost from host-associated prokaryotes. *Nucleic Acids Res.* 33:966–76

81. Parsell DA, Silber KR, Sauer RT. 1990. Carboxy-terminal determinants of intracellular protein degradation. *Genes. Dev.* 4:277–86

82. Pedersen K, Zavialov AV, Pavlov MY, Elf J, Gerdes K, Ehrenberg M. 2003. The bacterial toxin RelE displays codon-specific cleavage of mRNAs in the ribosomal A site. *Cell* 112:131–40

83. Pedulla ML, Ford ME, Houtz JM, Karthikeyan T, Wadsworth C, et al. 2003. Origins of highly mosaic mycobacteriophage genomes. *Cell* 113:171–82

84. Ranquet C, Geiselmann J, Toussaint A. 2001. The tRNA function of SsrA contributes to controlling repression of bacteriophage Mu prophage. *Proc. Natl. Acad. Sci. USA* 98:10220–25

85. Ranquet C, Gottesman S. 2007. Translational regulation of the *Escherichia coli* stress factor RpoS: a role for SsrA and Lon. *J. Bacteriol.* 189:4872–79

86. Richards J, Mehta P, Karzai AW. 2006. RNase R degrades nonstop mRNAs selectively in an SmpB-tmRNA-dependent manner. *Mol. Microbiol.* 62:1700–12

71. Finds *trans*-translation does not compete with translation elongation and proposes the existence of an alternative ribosome release mechanism.

79. Uncovers the importance of *trans*-translation in pathogenesis.

87. Roche ED, Sauer RT. 1999. SsrA-mediated peptide tagging caused by rare codons and tRNA scarcity. *EMBO J.* 18:4579–89

88. Roche ED, Sauer RT. 2001. Identification of endogenous SsrA-tagged proteins reveals tagging at positions corresponding to stop codons. *J. Biol. Chem.* 276:28509–15

89. Roy S, Sahu A, Adhya S. 2002. Evolution of DNA binding motifs and operators. *Gene* 285:169–73

90. Rudinger-Thirion J, Giege R, Felden B. 1999. Aminoacylated tmRNA from *Escherichia coli* interacts with prokaryotic elongation factor Tu. *RNA* 5:989–92

91. Ryan KR, Shapiro L. 2003. Temporal and spatial regulation in prokaryotic cell cycle progression and development. *Annu. Rev. Biochem.* 72:367–94

92. Sharkady SM, Williams KP. 2004. A third lineage with two-piece tmRNA. *Nucleic Acids. Res.* 32:4531–38

93. Shimizu Y, Ueda T. 2006. SmpB triggers GTP hydrolysis of elongation factor Tu on ribosomes by compensating for the lack of codon-anticodon interaction during *trans*-translation initiation. *J. Biol. Chem.* 281:15987–96

94. Shin JH, Price CW. 2007. The SsrA-SmpB ribosome rescue system is important for growth of *Bacillus subtilis* at low and high temperatures. *J. Bacteriol.* 189:3729–37

95. Singh NS, Varshney U. 2004. A physiological connection between tmRNA and peptidyl-tRNA hydrolase functions in *Escherichia coli*. *Nucleic Acids Res.* 32:6028–37

96. Strauch MA, Baumann M, Friedman DI, Baron LS. 1986. Identification and characterization of mutations in *Escherichia coli* that selectively influence the growth of hybrid lambda bacteriophages carrying the immunity region of bacteriophage P22. *J. Bacteriol.* 167:191–200

97. Subbarao MN, Apirion D. 1989. A precursor for a small stable RNA (10Sa RNA) of *Escherichia coli*. *Mol. Gen. Genet.* 217:499–504

98. Sundermeier TR, Dulebohn DP, Cho HJ, Karzai AW. 2005. A previously uncharacterized role for small protein B (SmpB) in transfer messenger RNA-mediated *trans*-translation. *Proc. Natl. Acad. Sci. USA* 102:2316–21

99. Sundermeier TR, Karzai AW. 2007. Functional SmpB-ribosome interactions require tmRNA. *J. Biol. Chem.* 282:34779–86

100. Sunohara T, Abo T, Inada T, Aiba H. 2002. The C-terminal amino acid sequence of nascent peptide is a major determinant of SsrA tagging at all three stop codons. *RNA* 8:1416–27

101. Sunohara T, Jojima K, Tagami H, Inada T, Aiba H. 2004. Ribosome stalling during translation elongation induces cleavage of mRNA being translated in *Escherichia coli*. *J. Biol. Chem.* 279:15368–75

102. Sunohara T, Jojima K, Yamamoto Y, Inada T, Aiba H. 2004. Nascent-peptide-mediated ribosome stalling at a stop codon induces mRNA cleavage resulting in nonstop mRNA that is recognized by tmRNA. *RNA* 10:378–86

103. Tu GF, Reid GE, Zhang JG, Moritz RL, Simpson RJ. 1995. C-terminal extension of truncated recombinant proteins in *Escherichia coli* with a 10Sa RNA decapeptide. *J. Biol. Chem.* 270:9322–26

104. Tyagi JS, Kinger AK. 1992. Identification of the 10Sa RNA structural gene of *Mycobacterium tuberculosis*. *Nucleic Acids Res.* 20:138

105. Ueda K, Yamamoto Y, Ogawa K, Abo T, Inokuchi H, Aiba H. 2002. Bacterial SsrA system plays a role in coping with unwanted translational readthrough caused by suppressor tRNAs. *Genes Cells* 7:509–19

106. Ushida C, Himeno H, Watanabe T, Muto A. 1994. tRNA-like structures in 10Sa RNAs of *Mycoplasma capricolum* and *Bacillus subtilis*. *Nucleic Acids Res.* 22:3392–96

107. Valle M, Gillet R, Kaur S, Henne A, Ramakrishnan V, Frank J. 2003. Visualizing tmRNA entry into a stalled ribosome. *Science* 300:127–30

108. Watanabe T, Sugita M, Sugiura M. 1998. Identification of 10Sa RNA (tmRNA) homologues from the cyanobacterium *Synechococcus* sp. strain PCC6301 and related organisms. *Biochim. Biophys. Acta* 1396:97–104

109. Wiegert T, Schumann W. 2001. SsrA-mediated tagging in *Bacillus subtilis*. *J. Bacteriol.* 183:3885–89

110. Williams KP, Bartel DP. 1996. Phylogenetic analysis of tmRNA secondary structure. *RNA* 2:1306–10

111. Williams KP, Martindale KA, Bartel DP. 1999. Resuming translation on tmRNA: a unique mode of determining a reading frame. *EMBO J.* 18:5423–33

112. Withey J, Friedman D. 1999. Analysis of the role of *trans*-translation in the requirement of tmRNA for lambdaimmP22 growth in *Escherichia coli*. *J. Bacteriol.* 181:2148–57

113. Wower J, Zwieb CW, Hoffman DW, Wower IK. 2002. SmpB: a protein that binds to double-stranded segments in tmRNA and tRNA. *Biochemistry* 41:8826–36

114. Yamamoto Y, Sunohara T, Jojima K, Inada T, Aiba H. 2003. SsrA-mediated *trans*-translation plays a role in mRNA quality control by facilitating degradation of truncated mRNAs. *RNA* 9:408–18

Regulation and Function of Ag43 (Flu)

Marjan W. van der Woude[1] and Ian R. Henderson[2]

[1]Department of Biology and the Hull York Medical School, University of York, York YO10 5YW, United Kingdom; email: mvdw1@york.ac.uk

[2]Division of Immunity and Infection, University of Birmingham, B15 2TT United Kingdom; email: i.r.henderson@bham.ac.uk

Annu. Rev. Microbiol. 2008. 62:153–69

The *Annual Review of Microbiology* is online at micro.annualreviews.org

This article's doi: 10.1146/annurev.micro.62.081307.162938

Key Words

Escherichia coli, autotransporter, outer membrane protein, phase variation, pathogenesis, biofilm

Abstract

Antigen 43 (Ag43) is an abundant outer membrane protein in *Escherichia coli* belonging to the autotransporter family. Structure-function relationships of Ag43 proposed on the basis of experimental work and in silico analysis are discussed in context of insights derived from molecular modeling. New sequence analysis sheds light on the phylogeny of the allelic variants of the Ag43-encoding gene and identifies two distinct families that appear to be distributed between specific pathogenic and commensal isolates. The molecular mechanism that controls expression by phase variation to create population heterogeneity is discussed. Proposed roles of Ag43 expression for *E. coli* are summarized and the studies are put into perspective regarding the role of allelic variants, genetic background of the bacterial strain, and control of expression by phase variation. We conclude that future studies need to take into account these variables to obtain a complete understanding of the contribution of Ag43 expression to *E. coli* biology.

Contents

INTRODUCTION

Studies on outer membrane proteins in *Escherichia coli* identified a protein described as "the most abundant phase varying outer membrane protein" (21). This protein, antigen 43 (Ag43), is encoded by a single gene originally identified in 1980 and designated *flu* (for fluffing) (11) in relation to the aggregative property Ag43 expression confers upon the cells. The original designation *flu+*, as nonfluffing (11), actually corresponds to a phenotype associated with a lack of expression of the gene. The gene is now frequently referred to as *agn43* (Antigen43), and because this adheres more closely to the convention of having a similar gene name and protein name, this designation is preferred. In the 1980 study, it was noted as a "metastable" gene, something now known as phase variation. Phase variation describes a gene expression pattern that results in a mixed phenotype in a clonal population of cells. The regulatory mechanism controlling this for *agn43* has been one area of research. Additionally, Ag43 is an autotransporter protein, and this class of proteins in general has generated interest for potential use in autodisplay and to uncover the mechanism of secretion and translocation. Further interest in Ag43 has been generated because the protein is abundant when it is expressed, and multiple studies have addressed how this affects bacterial populations in the context of biofilms and infection. In this chapter, insights into the distribution and diversity of the *agn43* alleles based on genome analysis are presented and results and conclusions on structure, function, and regulation are discussed in this context. Together, these findings may help focus future studies in directions that can inform the biology of *E. coli*.

Ag43 PROTEIN

Ag43 is an outer membrane protein belonging to the autotransporter family, which is defined by shared structural features and a common pathway of secretion. Thus, many of the general features and questions regarding structure, secretion, and processing of autotransporters are also relevant for Ag43. These issues have been discussed in detail in an excellent review (10). Indeed, much of what is known about the Ag43 protein is inferred from studies on other autotransporters.

Ag43 possesses the typical autotransporter protein domains: an N-terminal signal peptide; an N-proximal passenger domain, or α^{43}, that is secreted; an autochaperone domain that facilitates folding of the passenger domain; and a C-terminal β-barrel domain that forms an integral outer membrane protein (β^{43}) (**Figure 1a**). The 52-amino-acid signal peptide of the Ag43 protein consists of a C-terminal domain that resembles a classic signal peptide and an N-terminal extension that is conserved among many members of the autotransporter family (10). The significance of extended signal peptides in general has not been resolved. Initial studies on the hemoglobin protease autotransporter Hbp indicated the extended signal sequence mediated cotranslational targeting of the protein to the inner membrane translocation machinery (46). However, subsequent analyses of the autotransporters Pet and EspP suggested that the signal peptide mediates targeting in a posttranslational fashion and that the N-terminal extension indirectly affects folding of the protein in the periplasm and thus subsequent translocation across the outer membrane (38).

The passenger domain (α domain) of Ag43 confers the autoaggregation phenotype associated with many of the Ag43 variants. Domain swapping between the aggregating Ag43^{K12}

Antigen 43 (Ag43): an outer membrane protein of *E. coli*

***agn43*:** gene encoding the Ag43 protein; also called *flu*

Phase variation: when a gene is regulated by a mechanism that allows switching between On and Off expression states

Autotransporter: a protein defined by its secretion mechanism, correlating with structural features consisting of a signal peptide and passenger (α^{43}), autochaperone, and β-barrel translocator domains (β^{43})

Passenger domain: the domain of an autotransporter that is secreted to the outer surface

α^{43}: alpha domain of Ag43 protein

β^{43}: C-terminal β domain of Ag43 protein

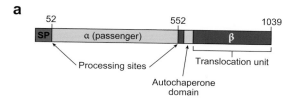

a

52   552   1039

SP | α (passenger) | | β

Processing sites

Autochaperone domain

Translocation unit

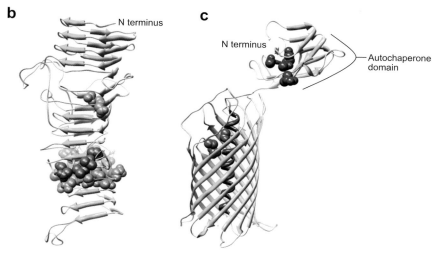

b

N terminus

c

N terminus

Autochaperone domain

Figure 1

Domain organization of Antigen 43 (Ag43). (*a*) Schematic showing the organization of the protein domains of Ag43. Shown are the signal peptide (*red*), the α domain (*green*), the β domain (*blue*), and the autochaperone domain (*light blue*). Numbers designate the numbers of amino acids at the domain boundary. Drawing is not to scale. Modeled structures of (*b*) α^{43} and (*c*) β^{43}, shown in ribbon representation. For clarity the domain models are depicted as separate molecules. β-strands in each case are colored yellow, loops are gray, and the central α-helix of the translocator domain is blue. Sequence motifs previously identified by Henderson et al. (22) are highlighted in space-filling representation: R^{208}GD (*orange*), aspartyl protease (processing) site G^{384}GVLLADSGA (*green*), leucine zipper L^{500}TVSNTTLTQKAVNLNEGTLTNDS (*pink*), P-loop G^{631}GRATGKT (*blue*), and R^{970}GD (*purple*). The position of the autochaperone domain responsible for folding of the β-helix passenger domain is indicated.

variant and the nonaggregating Ag43b^{CFT073} variant mapped this property to the N-terminal one-third of the passenger domain (30). Like the passenger domains of all members of the autotransporter family, α^{43} is predicted to form a β-helix, which in Ag43 consists of 18 rungs of 16 to 19 amino acids (21, 27) (**Figure 1*b***). The β-helical rod of the passenger domain of Ag43 is predicted to extend at least 85 Å (27).

Several amino acid sequence motifs were previously identified in the passenger domain of Ag43$^{ML308-225}$, e.g., an aspartyl protease site,

a leucine zipper motif implicated in protein-protein interactions, and an RGD motif that can facilitate binding to host cells by interactions with integrin, as is the case for the autotransporter AIDA-I (5, 21) (**Figure 1*b***). Molecular modeling of α^{43} revealed a parallel β-helix structure for α^{43} that encompasses the predicted RGD, aspartyl protease, and leucine zipper motifs. In contrast to the RGD motif of pertactin, which is present in a large discursive surface loop, the RGD motif of α^{43} maps to a position within a β-strand, indicating that it

α domain: the secreted passenger domain of Ag43

Ag43^{K12} or *agn43*^{K12}: Ag43 variant or *agn43* allele in which the superscript denotes isolate origin

Translocator
domain: the
membrane-bound
domain of an
autotransporter that
facilitates translocation
of the passenger
domain across the
outer membrane

β domain: the
translocation domain
of Ag43

could not partake in receptor-mediated interactions within host cells. Similarly, the predicted aspartyl protease and leucine zipper motifs are modeled at the C terminus of α^{43}, composing the rungs of the parallel β-helical stem. Further examination of the amino acid sequences revealed that the aspartyl protease and leucine zipper motifs are reminiscent of the amphipathic nature of β-strands within the β-helix, whereby alternating hydrophobic side chains are oriented into the hydrophobic core of the structure and separated by defined turn residues such as asparagines and threonines (26).

Comparison of the amino acid sequences of all available *agn43* alleles revealed that none of these motifs was absolutely conserved. Furthermore, the RGD motif of Ag43^{K12} at amino acid 208 is dispensable for Ag43-mediated uptake of this strain into polymorphonuclear neutrophils when Ag43 is expressed from a high-copy plasmid in a lab strain (13). Mutagenesis of the predicted motifs and subsequent functional analyses of the mutant proteins have indicated that these motifs do not correlate with their predicted functional role but rather appear to play a role in the structural integrity of the molecule (I.R. Henderson, unpublished data). To summarize, none of the predicted motifs are conserved, and none are likely to function as one might predict from in silico analyses.

The α^{43} passenger domain can be released from the cell surface by heat shock (37), indicating cleavage by an as yet unidentified protease. The relevance of the cleavage is not clear because the passenger domain remains associated with the cell surface under experimental conditions. Whether this is true for all growth conditions, including in an animal host, has not been examined. In Ag43^{K12} and proteins from similar alleles the processing site between α^{43} and β^{43} (37) is 46 residues upstream of the autochaperone region described by Oliver et al. (35) (**Figure 1a**); however, the site of processing is not conserved among all Ag43 alleles. In contrast, recent investigations of the autotransporter EspP revealed that cleavage occurs further C-terminal within the pore of the β-barrel and that this cleavage modifies

the structure of the translocator domain (2, 9). Yet other autotransporters undergo processing through cleavage by outer membrane proteases or by autocatalytic cleavage owing to the presence of an integral proteolytic domain within the passenger domain (10). However, in the absence of an apparent functional proteolytic domain, and with the clear demonstration that Ag43 continues to be processed in strains lacking a variety of outer membrane proteases (21), it appears plausible that Ag43 undergoes an intrabarrel cleavage event similar to that of EspP. Further investigation of the cleavage mechanism, specifically of Ag43, could provide additional insight into structure-function relations in autotransporters.

Modeling of β^{43} from Ag43^{K12} indicates that it consists of the integral outer membrane translocator domain and a surface-exposed β-helical structure reminiscent of autotransporter domains described by other groups and that it encompasses the autochaperone domain required for folding of the β-helix found in the passenger domains (10) (**Figure 1a,c**). Like α^{43}, β^{43} also possesses putative functional motifs: a P-loop motif, which is implicated in ATP-binding and thus energy provision, and a second RGD motif. The second predicted RGD motif lies within a large extracellular loop modeled onto the autotransporter translocator domain component, whereas the predicted P-loop motif resides centrally within the autochaperone domain (**Figure 1c**). Like the α^{43} motifs, mutagenesis of the β^{43} motifs does not appear to impact significantly the known functions of Ag43. However, protein modeling explained why the membrane-embedded β domain of Ag43 is larger than that of most autotransporters. For Ag43, processing occurs between the passenger domain and the autochaperone domain; thus the autochaperone domain remains covalently associated with the β-barrel after processing. A similar processing event and domain organization was also recently predicted for AIDA-I, an autotransporter homologous to Ag43 that is involved in diffuse adherence of a subset of diarrheagenic *E. coli* strains (32). This is in direct contrast to the serine protease autotransporters, including EspP, of the *Enterobacteriaceae*, in which processing occurs between the autochaperone domain and the β-barrel, giving a smaller β domain than β[43].

On the basis of conserved features, including the number of amino acids per rung, the occurrence of particular amino acids, and the predicted three-dimensional structure of the passenger domain, Ag43 can be placed in a group of similar autotransporters that includes AIDA-I and TibA, an adhesin of enterotoxigenic *E. coli* (28). The AIDA-I and TibA autotransporters are glycosylated by AaH and TibC, respectively, by the addition of a heptose at several amino acids of the passenger domain (4). The modification genes are adjacent to the corresponding autotransporter gene. In contrast, no modification genes are found adjacent to any of the *agn43* alleles, and electrospray mass spectroscopy of α[43] from Ag43[ML308-225] did not reveal the presence of any posttranslational modification (21). However, Ag43 can be glycosylated by the Aah and TibC enzymes when they are supplied in *trans* (45), and there is limited evidence that one of the Ag43[UTI536] variants is glycosylated, presumably because of a non-Ag43-specific glycosyltransferase. A comparison of purified glycosylated (AaH-mediated) and nonglycosylated α[K12] identified 16 glycosylated threonine and serine residues (31). This modification of Ag43 may affect interaction with eukaryotic cells (45); however, recent investigations have demonstrated that glycosylation of AIDA-I is not required for its ability to adhere to eukaryotic cells but that it is required for stable localization of the AIDA-I passenger domain on the cell surface (4). In contrast, Ag43[K12], Ag43[ML308-225], and presumably other variants are stably localized on the cell surface in the absence of glycosylation (21). A range of in vitro analyses indicated that α[K12] glycosylation protected against denaturation by heat and chemicals and enhanced refolding properties after denaturation (31). Together, this suggests glycosylation may affect assembly, structure, and possibly function. Thus, it will be important to determine the extent of natural occurrence and the role of this modification in Ag43, especially in the context of the genetic background of pathogenic isolates and the specific allele of *agn43*.

A FAMILY OF *agn43* ALLELES

E. coli ML308-225 is a derivative of *E. coli* ML originally isolated by Monods laboratory and was the workhorse for those studying cellular architecture and biochemical processes in *E. coli* during the 1960s and 1970s (58). It was from this strain that Ag43 was first isolated (36). Early on it was noted that multiple immunoreactive bands could be detected in Western blotting experiments when cell envelope fractions derived from a variety of *E. coli* strains were probed with anti-Ag43 antisera (37). This suggested the existence of multiple copies of *agn43* within single strains of *E. coli* and multiple alleles of *agn43* within single strains and across the breadth of the species. This finding was later confirmed when a number of different groups detected multiple copies of *agn43* in *E. coli* ML308-225 and in a variety of pathogenic *E. coli* strains (30, 39, 40, 48). The expansion in the number of *E. coli* genome-sequencing projects has revealed the distribution and number of *agn43* alleles associated with *E. coli* and the existence of *agn43* within the closely related *Citrobacter rodentium*. Previously, it was suggested that the diversity of autotransporter proteins had arisen through speciation rather than horizontal gene transfer (59). However, Ag43 is located within a CP4 phage-like element that undergoes precise deletion from the *Shigella flexneri* chromosome (34, 42, 50), and comparison of gene synteny within sequenced versions of *E. coli* reveals that *agn43* is located in different strains at different points in the chromosome, strongly suggesting that *agn43* is acquired horizontally and that differences in gene sequences or copy number are not due to clonal expansion.

Phylogenetic analyses used to investigate the genetic relationship between the different alleles of *agn43* and their host strains revealed a dendrogram with a bifurcating distribution of *agn43* alleles, indicating the existence of two subfamilies of Ag43 (**Figure 2a**). Further split decomposition analyses of *agn43* alleles revealed evidence of multiple recombination events among the alleles or their progenitors. This dichotomous grouping can be explained by a region of significant diversity encompassing the C-terminal and N-terminal ends of α^{43} and β^{43}, respectively (**Figure 2c**). This region of diversity maps to the point of cleavage between

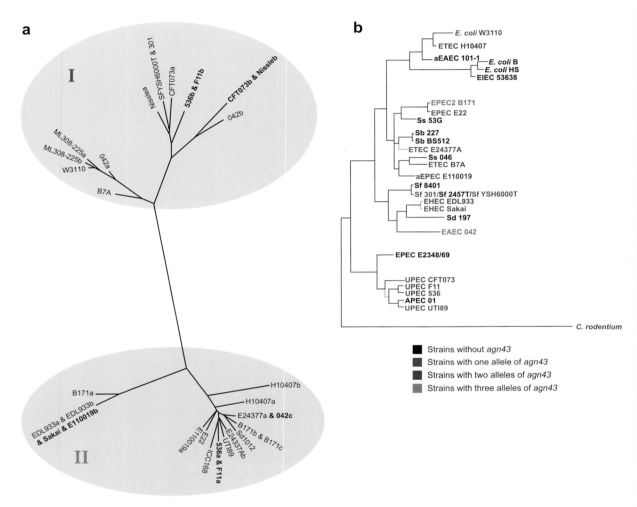

Figure 2

Phylogenetic analyses of Antigen 43 (Ag43). (*a*) The amino acid sequences of full-length Ag43 proteins were aligned using ClustalX and trees were generated using the neighbor-joining method. The hybridization network demonstrates a bifurcating pattern with two distinct subfamilies of Ag43 proteins. Subfamily I includes the protein encoded by the *agn43* allele from *Escherichia coli* K12. Subfamily II includes the *characterized cah* alleles from *E. coli* O157:H7 EDL933. Labeling of each protein is represented by the strain designation for the *E. coli*, *Shigella flexneri*, or *Citrobacter rodentium* strain from which each protein is derived. Where multiple proteins are present within a single strain, these are further designated by the presence of "a," "b," or "c." (*b*) Phylogram of the genome-sequenced strains of *E. coli*, *S. flexneri*, and *C. rodentium* based on multilocus sequence typing. Strains that do not possess *agn43* are depicted in black, and those harboring one, two, or three alleles are highlighted in blue, red, or green, respectively. (*c*) Amino acid sequence alignment of the Ag43 proteins. Representatives from each subfamily are shown with the corresponding, partial amino acid sequences, which determine the subfamily to which each protein belongs. (*d*) DNA sequence alignment of the promoter regions of *agn43*. Representatives from each subfamily are shown. An arrow designates the transcriptional start site as shown for *agn43*[K12].

c

E. coli W3110	EKGSSFTLNA	GDTATDTTVN	--GGLFTARG	GTLAGTTTLN	NGAILTLSGK	TVNNDTLTIR
EAEC 042a	EKGSSFTLNA	GDTATDTTVN	--GGLFTARG	GTLAGTTTLN	NGAILTLSGK	TVNNDTLTIR
UPEC CFT073b	EKGSSFTLNA	GDTATDTTVN	--GGLFTARG	GSLAGTTTLN	NGATFFTLAGK	TVNNDTLTIR
SF 301	EKGSSFTLNA	GDTATDTTVN	--GGLFTARG	GTLAGTTTLN	NGAILTLSGK	TVNNDTLTIR
EAEC 042c	ENGGSFTVNA	GGLASNTTVG	HRGTLTLAAG	GSLSGRTQLS	KGASMVLNGD	VVS-------
EHEC EDL933b	ENGGSFTVNA	GGQAGNTTVG	HRGTLTLAAG	GSLSGRTQLS	KGASMVLNGD	VVS-------
C. rodentium	ENGGSFTVNA	GGQAGNTTVG	HRGTLTLAAG	GSLSGRTQLS	KGASMVLNGD	VVS-------
UPEC UTI89	ENGGSFTVNA	GGLASNTTVG	HRGTLTLAAG	GSLSGRTQLS	KGASMVLNGD	VVS-------

E. coli W3110	EGDALLQGGS	LTGNGSVEKS	GSGTLTVSNT	TLTQKAVNLN	EGTLTLNDST	VTTDVIAQRG
EAEC 042a	EGDALLQGGS	LTGNGSVEKS	GSGTLTVSNT	TLTQKAVNLN	EGTLTLNDST	VTTDVIAQRG
UPEC CFT073b	EGDALLQGGA	LTGNGRVEKS	GSGTLTVSNT	TLTQKAVNLN	EGTLTLNDST	VTTDIIAHRG
SF 301	EGDALLQGGA	LTGNGSVEKS	GSGTLTVSNT	TLTQKAVNLN	EGTLTLNDST	VTTDVIAQRG
EAEC 042c	----------	----------	----------	----------	----------	----------
EHEC EDL933b	----------	----------	----------	----------	----------	----------
C. rodentium	----------	----------	----------	----------	----------	----------
UPEC UTI89	----------	----------	----------	----------	----------	----------

                                     ★★

E. coli W3110	TALKLTGSTV	LNGAIDPTNV	TLASGATWNI	PDNATVQSVV	DDLSHAGQIH	FTSTRTGKFV
EAEC 042a	TALKLTGSTV	LNGAIDPTNV	TLASGATWNI	PDNATVQSVV	DDLSHAGQIH	FTSTRTGKFV
UPEC CFT073b	TALKLTGSTV	LNGAIDPTNV	TLTSGATWNI	PDNATVQSVV	DDLSHAGQIH	FTSARTGKFV
SF 301	TALKLTGSTV	LNGAIDPTNV	TLASGATWNI	PDNATVQSVV	DDLSHAGQIH	FTSTRTGKFV
EAEC 042c	-----TGDIV	NAGEIRFDNQ	TT--------	PDAALSRAVA	KGDSPVTFHK	LTTS------
EHEC EDL933b	-----TGDIV	NAGEIRFDNQ	TT--------	PNAALSRAVA	KSNSPVTFHK	LTTT------
C. rodentium	-----TGDIV	NAGEIHFDNQ	TT--------	PDAALSRAVA	KGDSPVTFHK	LTTS------
UPEC UTI89	-----TGDIV	NAGEIRFDNQ	TT--------	PDAALSRAVA	KGDSPVTFHK	LTTS------

E. coli W3110	PATLKVKNLN	GQNGTISLRV	RPDMAQNNAD	RLVIDGGRAT	GKTILNLVNA	GNSASGLATS
EAEC 042a	PATLKVKNLN	GQNGTISLRV	RPDMAQNNAD	RLVIDGGRAT	GKTILNLVNA	GNSASGLATS
UPEC CFT073b	PTTLQVKNLN	GQNGTISLRV	RPDMAQNNAD	RLVIDGGRAT	GKTILNLVNA	GNSGTGLATT
SF 301	PATLKVKNLN	GQNGTISLRV	RPDMAQNNAD	RLVIDGGRAT	GKTILNLVNA	GNSASGLATS
EAEC 042c	-------NLT	GQGGTINMRV	RLD-GSNTSD	QLVINGGQAT	GKTWLAFTNV	GNSNLGVATS
EHEC EDL933b	-------NLT	GQGGTINMRV	RLD-GSNASD	QLVINGGQAT	GKTWLAFTNV	GNSNLGVATT
C. rodentium	-------NLT	GQGGTINMRV	RLD-GSNASD	QLVINGGQAT	GKTWLAFTNV	GNSNLGVATS
UPEC UTI89	-------NLT	GQGGTINMRV	RLD-GSNASD	QLVINGGQAT	GKTWLAFTNV	GNSNLGVATS

Key to parts c and d:

■ Conserved in both subfamilies
■ Conserved only in Subfamily I
■ Conserved only in Subfamily II
■ Nonconserved

★ Empirically determined cleavage site of the α[43] and β[43] subunits of Subfamily I

\# GATC sites responsible for epigenetic phase variation

^ ATG translational start

d

                  ↓

                                ####   ####   ####

```
E. coli W3110   TAGCTCAATAATAGAATAAAACGATCAATATCTATTTTATCGATCGTTTATATCGATCGAT---------------AAGCTAATAA-----TAACCTTT
EAEC 042a       TAGCTCAATAATAGAATAAAACGATCGATATCTATTTTATCGATCGTTTATATCGATCGAT---------------AAGCTAATAA-----TAACCTTT
UPEC CFT073b    TAGCTTGATAATAGAATAAAACGATCAATACCTATTTTACCGATCGTTAATATCGATCGTT---------------ATGCTAATAA-----TAACTCCT
SF 301          TAGACTGATAATAGAATAAAACGATCGATACCTATTTTACCGATCGTTAATATCGATCGTT---AATATCGATCGTTATGCTAATAA-----TAACTCCT
EAEC 042c       CTTGCAATCAATAGAATAAAACGATCGATAAAACAGGTATCGATCGTTACTATCGATCGTTTATAGTGGTTAATGGTCTGCGAAAAGGCAGCTGAATCTC
EHEC EDL933b    TCTACTGTTAATAGAATAAAACGATCGATAAAACAGGTATCGATCGTTTATATCGATCGTTTATAGTGGTTAATGGCCTGCGAAAAGGCAGCTGAATCTC
C. rodentium    CCTGCCATCAATAGAATAAAACGATCGATAAAACAGGTATCGATCGTTACTATCGATCGTTTATAGTGGTTAATGGTCTGCGAAAAGGCAGCTGAATCTC
UTI89           CTTGTTATTAATAGAATAAAACGATCGATAAAACAGGTATCGATCGTTACCATCGATCGTTTATAGTGGTTAATG-CCTGCGAAAAGGCAGCTGAATCTC
```

```
E. coli W3110   GTCAGTAACATGCACAGA--TACGTACA--GA-AAGACA-TTC---AGGGAACAACAGAACCACA-ATTCAGAAACTCCCA-CAGCCGGACCTCCGGCAC
EAEC 042a       GTCAGTAACATGCACAGA--TACGTACA--GA-AAGACA-TTC---AGGGAACAACAGAACCACA-ATTCAGAAACTCCCA-CAGCCGGACCTCCGGCAC
UPEC CFT073b    GTTAGCAACGTGCGCAGA--TACACACA--GACATGAGA-TTC---AGGGAACAACAGAGCCACA-CGTCAGAAACTTCCGTCAGCCGGACCTCCGGCAC
SF 301          GTTAGCAACGTGCGCAGA--TACACACA--GACATGAGA-TTC---AGGGAACAACAGAGCCACA-CGTCAGAAACTTCCGTCAGCCGGACCTCCGGCAC
EAEC 042c       TTCATCATGCAGAACGGAATTGCACACAACGGACTGATACTTCTCTGGCGGATGAGAGAGGGGAAGCATTTATGCCCGGGAAAAACCACTTTATCTGACC
EHEC EDL933b    TTCATCATGCAGAACGGAATTGCACACAACAGACTGATACTTTCTCTGTCTGATGA-AGGGGGAGAGCGTTTACGCCCGGGACGTATCGCTGTGCCCGATA
C. rodentium    TTCATCATGCAGAACGGAATTGCACACAACGGACTGATACTTCTCTGGCGGATGAGAGAGGGGAAGCATTTATGTCCGGGAAAAACCACTTTATCTGACC
UTI89           TTCATCATGCAGAACGGAATTGCACACAACGGACTGATACTTCTCTGGCGGATGAGAGAGGGGAAGCATTTATGTCCGGGAAAAACCACTTTATCTGACC
```

```
E. coli W3110   -TGTAACCCTTTACCTGCCGGTATCCACGTTTGTGGGTACCGGCTTTTTTATTCACCCTCAATCT----AAGGAAAAGCTGATGAAACGACATCTGAATA
EAEC 042a       -TGTAACCCTTTACCTGCCGGTATCCACGTTTGTGGGTACCGGCTTTTTTATTCACCCTCAATCT----AAGGAAAAGCTGATGAAACGACATCTGAATA
UPEC CFT073b    -TGTAACCCTTTACCTGCCGGTATCCACATCTGTGGATACCGGCTTTTTTATTCACCCTCACTCTGATTAAGGAAATGCTGATGAAACGACATCTGAATA
SF 301          -TGTAACCCTTTACCTGCCGGTATCCACATCTGTGGATACCGGCTTTTTTATTCACCCTCACTCTGATTAAGGAAATGCTGATGAAACGACATCTGAATA
EAEC 042c       ATGCTGTTTTGTACCTGCCGGTATCCACATTTGTGGGTACCGGCTTTTTTATTCACCCTCTGT------AAGGAAAAGCTGATGAAACGACATCTGAACA
EHEC EDL933b    ACTCTGTTTTGTACCTGCCGGTATCCACTTTTGTGGGTACCGGCTTTTTTATTCACCCTCTGT------AAGGAAAAGCTGATGAAACGACATCTGAACA
C. rodentium    ATGCTGTTTTGTACCTGCCGGTATCCACATTTGTGGGTACCGGCTTTTTTATTCACCCTCTGT------AAGGAAAAGCTGATGAAACGACATCTGAACA
UTI89           ATGCTGTTTTGTACCTGCCGGTATCCACATTTGTGGGTACCGGCTTTTTTATTCACCCTCTGT------AAGGAAAAGCTGATGAAACGACATCTGAACA
```

                                                               ^^^

the two domains, and although one might intuitively reason that such proteins are uncleaved and thus represent the 94-kDa immunoreactive bands previously described by Owen et al. (37), no defect in the processing of these proteins has been detected (48). Comparison of the promoter regions of the two allelic families also reveals that the alleles possess variation in the regulatory region responsible for phase variation (**Figure 2d**). Despite these differences both allelic families possess three GATC sites with similar spacing, suggesting that members of both families undergo reversible phase variation in a deoxyadenosine methyltransferase (Dam)- and OxyR-dependent fashion (also see below).

Comparison of the phylogenetic tree of Ag43 with the multilocus sequence typing (MLST) profiles for genome-sequenced versions of *E. coli* reveals no correlation between *E. coli* evolution and possession of members of either allelic subfamily of *agn43*, further demonstrating that *agn43* is acquired by horizontal acquisition (**Figure 2b**). Several of these strains either do not possess *agn43* alleles or possess remnants of *agn43*. Because Ag43 exists in the close relative *C. rodentium*, it thus remains possible that Ag43 predates the differentiation of *E. coli* and *C. rodentium* and that those strains which do not possess *agn43* have lost the gene through mutational attrition.

No absolute correlation exists between the clinical disease manifested by a particular strain of *E. coli* and the presence or absence of a particular allele of *agn43*. However, the limited number of sequenced *E. coli* genomes indicates *agn43* is found more frequently among uropathogens (**Figure 2b**). Using a multiplex PCR-based screening protocol, Restieri et al. (39) noted that although *agn43* sequences were present in approximately 56% of commensal isolates, they were present in over 90% of uropathogens and diarrheagenic *E. coli*, suggesting Ag43 may be more prevalent in pathogenic *E. coli* strains than in commensal *E. coli* strains and by implication associated with disease. In contrast to the data revealed by Restieri et al. (39), analyses of the genome-sequenced *E. coli* strains suggest that the *agn43* Subfamily II is more prevalent than Subfamily I. All members of Subfamily II are associated with pathogenic *E. coli*; however, this may simply reflect the bias toward sequencing pathogenic versions of *E. coli* rather than nonpathogens. Thus, phylogenetic analyses do not reveal conserved features indicative of a functional role for Ag43.

ROLES ATTRIBUTED TO Ag43

Where Ag43 is localized on the cell surface may determine in part how it affects the characteristics of the bacteria. In immunofluorescence studies of Ag43-producing *E. coli* K12 and *E. coli* ML308-225, the protein is seen evenly distributed over the surface of the entire cell (20). However, various autotransporters are secreted at the pole of the cell and remain localized there in strains with a full-length lipopolysaccharide (LPS) but not in strains with a short, rough LPS lacking the O-antigen (25). If this is also true for Ag43, it is possible that Ag43 localization varies between strains, which could affect the traits Ag43 expression confers on the bacteria.

The Ag43-encoding gene was identified as a result of the autoaggregative property it confers on the bacterial population (11), yet the benefits of this property for an individual bacterial cell or the population are still not well defined, nor is it clear how some of the roles ascribed to Ag43 expression relate to this trait. This phenotype is visible in laboratory cultures as a rapid settling of the cells [the original fluffing phenotype (11)], but it is apparent in strains that overproduce Ag43 or contain mutations locking the phase variation into the On state and much less so in strains where Ag43 expression is controlled by the native promoter and thus is under phase variation control (see below) (**Figure 3**). This raises the question of how relevant the findings are regarding the role of Ag43 and Ag43-mediated aggregation because many studies have been performed with isolates constitutively producing Ag43 from a plasmid, which abrogates phase variation. Furthermore, the degree of autoaggregation may depend on

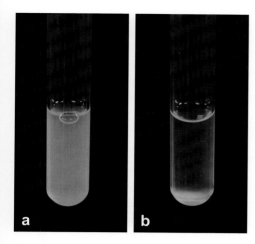

Figure 3

Antigen 43 (Ag43) expression mediates cellular autoaggregation. (*a*) Culture of an *E. coli* K12 strain with Ag43 expression under phase variation control by its own promoter. (*b*) Same strain with plasmid pTP166 resulting in increased Dam production, which relieves transcriptional repression by OxyR. Thus, in this culture 100% of cells are in the On phase, or Ag43-expressing phase. Note bacterial aggregation at the bottom of the tube.

the *agn43* allele. Using strains that expressed Ag43 from a multicopy plasmid, Klemm et al. (30) determined that alleles with the closest homology to Ag43^{K12} cause the most rapid aggregation, whereas other alleles did not confer autoaggregation, including the two alleles in uropathogenic *E. coli* (UPEC) isolate CFT073. However, in a different study using a similar approach the two *agn43* CFT073 alleles did confer aggregation (51). The incomplete understanding of the significance of the sequence diversity among *agn43* alleles and of the genetic background of the strain necessitates a discussion on the role of Ag43 in this context.

Biofilms are communities of cells attached to a surface and enveloped in an extracellular matrix, and they have been implicated in many diseases. Ag43 expression increases the quantity of biofilm formation by *E. coli* in vitro, although not as significantly as other factors (7, 29, 41). Presumably enhanced biofilm formation is as a result of Ag43-mediated autoaggregation, even though this has not been shown. The relatively

moderate effect of Ag43 may be in part due to opposite, indirect effects on biofilm formation of Ag43 production. Type 1 fimbriae are required for initiating biofilm formation but they also block Ag43-mediated autoaggregation (7, 18). Furthermore, cells producing Ag43 show decreased motility (52), and motility also enhances biofilm formation (7, 29, 41).

The role of Ag43 expression for *E. coli* virulence has been addressed by both in vitro studies with eukaryotic cells (13) and model systems for *E. coli* pathogenesis, mostly for urinary tract infections (UTIs). However, different UPEC strains are also used, with CFT073 and UTI89 the most prevalent. These two strains differ in the *agn* alleles, among others. Specifically, CFT073 has two alleles that cluster in Subfamily I, whereas UTI89 has just one that allele that belongs to Subfamily II (**Figure 2a**). The amino acid identity between *agn43*^{UTI89} and the two CFT073 alleles c3655 (*agn43*^{CFT073a}) and c1273 (*agn43*^{CFT073b}) is 67% and 66%, respectively. These differences make it difficult to compare results and generalize conclusions.

There is evidence that Ag43 is expressed during infection, at least during an experimental model for UTI. First, antibodies against Ag43^{CFT073} are found in the serum in a mouse model of infection (17). This provides one possible raison d'etre for phase variation of Ag43 in facilitating immune evasion, but other putative roles should be addressed experimentally (53). Second, although considered noninvasive, UTI89 can be found as intercellular pods that consist of a tightly packed group of bacteria in urothelial cells and Ag43^{UTI89} is expressed on bacterial cells within these pods (1). Whether Ag43-mediated autoaggregation contributes to this process remains to be addressed.

Ulett et al. (51) addressed the role of the two *agn43* alleles in UPEC isolate CFT073. When overexpressed, one allele promoted biofilm formation better than the other in a K12 background, but in CFT073 neither allele appeared to contribute to in vitro biofilm formation on an abiotic surface as determined by deletion analysis, nor did constitutive expression promote biofilm formation. This illustrates that the

UPEC: uropathogenic *E. coli*

UTI: urinary tract infection

allele, level of expression, and the genetic background of the strain need to be taken into account when examining the role of this protein. In this same study, the role of the *agn43* alleles in pathogenesis was examined using a mouse model for UTI (51). The *agn43*[CFT073a] allele had a measurable effect by promoting long-term colonization when expressed constitutively, which is consistent with the finding that deletion of this allele decreased successful colonization of the bladder at day 5 postinfection. Ag43[CFT073a] is 81.7% identical to the Ag43[K12] protein and 85% identical to the Ag43[CFT073b] protein (51). Several lines of evidence suggest Ag43 might function as an adhesin, particularly because of its high level of homology to the AIDA-I adhesin. However, recent in vitro and in vivo investigations of Ag43 variants from *E. coli* ML308-225 and from the enteroaggregative *E. coli*–type strain (*E. coli* 042) do not support a role for Ag43 in mediating adherence to a variety of cell lines or in promoting colonization of the intestine (8). However, the few percent of amino acid variation between the Ag43 variants may have significant biological impact that has not been resolved in these assays.

An additional proposed role for Ag43 is that expression promotes uptake and survival in polymorphonuclear neutrophils, as was determined by an in vitro assay (13) based on the effect of *agn43* expressed from a high-copy plasmid. To evaluate the significance of these results, it will be essential to determine if a similar effect is observed with isolates in which Ag43 is expressed from the chromosomal, native promoter, if this effect is allele specific, and to confirm this effect in an animal model for disease with relevant isolates.

Whether any of these proposed in vivo roles relate to Ag43-mediated autoaggregation or even biofilm formation is not known. Furthermore, the significance of phase variation of Ag43 on the proposed roles has yet to be addressed, possibly with a similar approach that has been used to examine the role of fimbrial phase variation (47). However, the UTI model may not be well-suited because it can require a high dosage of infection of a virulent strain into the colonization site and thus bypass potential bottlenecks where phase variation could play a significant role. Thus, although some *agn43* alleles promote bacterial autoaggregation, the relevancy and the actual contribution of the Ag43 protein for *E. coli* biology are far from clear.

REGULATION OF EXPRESSION OF THE *agn43* GENE

The main regulatory feature of Ag43 expression is phase variation (11, 20, 21). Phase variation is a common feature mainly of surface proteins of commensal and pathogenic bacteria. As a result of phase variation, cells in a clonal population either express the gene (On phase) or they do not (Off phase). This expression state is heritable yet reversible (53, 54). The switch frequency is the number of cells that switch expression state per generation, and thus provides an indication of the stability of the gene expression state over time. For Ag43 this lies around 10^{-3} cells for the switch from the On to Off direction as well as from Off to On. Because the rates are of a similar order of magnitude, the composition of the population regarding Ag43 expression can change, but only slowly. More rapid changes can occur with selection against or enrichment for cells in either expression phase.

Phase variation of Ag43 requires specific sequence elements in the regulatory region and two proteins: the global regulatory protein OxyR and Dam (19), which methylates the adenine at GATC sequences. The results of extensive analyses of OxyR-DNA interactions in vitro, of the effect of *cis* and *trans* mutations and of overexpression of the regulatory factors on *agn43* expression, and of analysis of the DNA methylation state at *agn43* in cells are all consistent with the following model for the On and Off states (16, 55, 57) (**Figure 4**). In an Off phase, OxyR is bound to its recognition site in the regulatory region of *agn43*, which overlaps the −10 of the promoter sequence, and represses *agn43* transcription. However, within this OxyR binding site

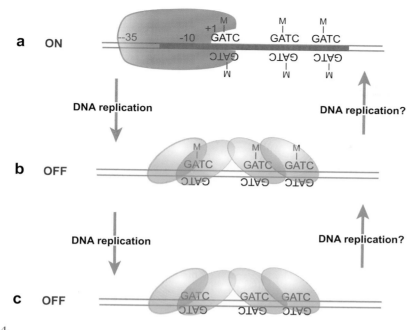

a ON −35 −10 +1 M | GATC   M | GATC   M | GATC

b OFF   M | GATC   M | GATC   M | GATC

c OFF   GATC   GATC   GATC

DNA replication   DNA replication?

Figure 4

Schematic depicting regulatory control of transcription of *agn43* by OxyR and Dam leading to phase variation. Shown are both strands of DNA with the Dam target sequences GATC. Methylation is depicted with an M, RNA polymerase is shown in green, OxyR as a dimer of dimers is in blue, the promoter is labeled −10 and −35, and the transcription start site is +1. (*a*) The On phase with fully methylated sequences that block OxyR binding. (*b*) An intermediate Off phase with hemimethylated DNA and bound OxyR. (*c*) The Off phase with unmethylated DNA and OxyR bound with higher affinity than in panel *b*. DNA replication must occur for new, unmethylated sequences and thus is required for the On to Off switch. It may also be required for the Off to On switch to displace OxyR and allow Dam access to sites.

there are three Dam target sequences (GATC), and OxyR can only occupy this binding site when these are not methylated. Furthermore, when OxyR is bound, it prevents Dam from methylating these three GATC sequences. Because methylation of these specific GATC sequences blocks OxyR from binding, once these sites are methylated transcription occurs, giving rise to an On phase. Thus, methylation of the *agn43* GATC sequences and OxyR binding are mutually exclusive, and the expression state is determined by the outcome of the competition between these two proteins for the *agn43* regulatory region. OxyR and Dam are present in the cell at all times. DNA synthesis is required to generate the unmethylated DNA in *E. coli* that occurs in the Off phase, and thus the hemimethylated state plays a significant role in

this phase variation. Both OxyR and the sequestration protein SeqA can bind the *agn43* regulatory region when it is hemimethylated (6). The latter may not be a defining feature, however, since Dam as well as OxyR and excess DNA can readily disrupt the SeqA-*agn43* DNA complex (R. Kaminska & M. van der Woude, unpublished data).

The involvement of OxyR as a repressor initially suggested that oxidative stress, which alters the oxidation state of OxyR and its DNA binding properties, could affect its role as a repressor of *agn43* transcription. Indeed, it would be an attractive model if oxidative stress initiates Ag43 expression and thus autoaggregation, which in turn could provide the population a measure of protection. However, analyses of interactions of oxidized OxyR with *agn43*

DNA, and the effect of oxidative stress on phase variation of *agn43*, do not support this model (56). In vitro analysis with the *agn43* regulatory region from *E. coli* K12 showed there is no difference in affinity for the nonmethylated regulatory region between oxidized, wild type OxyR and OxyR(C199S), which mimics the reduced form, and that this affinity is biologically relevant since it is similar to the affinity of genes induced upon oxidative stress by oxidized OxyR. Furthermore, GATC methylation abrogates binding and oxidized OxyR represses in vitro transcription (56). However, the possibility that OxyR oxidation may have different effects on the various *agn43* alleles owing to sequence variations in the OxyR binding region could be explored (**Figure 2*d***).

Because of its apparent simplicity in creating stochastic behavior in cells, *agn43* phase variation has generated interest as a system to model. In a modeling study by Lim & van Oudenaarden (33), a signal amplification step was developed to enable GFP to be used as a reporter for *agn43* expression, and thus to allow easy analysis of expression at the single-cell level using flow cytometry. With this approach, a third expression state was identified with a low level of GFP expression, which did not require the *agn43* promoter or regulatory region. Based on the modeling of phase variation in this study, this state helps define the distribution between On- and Off-phase cells in populations and the transition rates between these (33). It will be important to define and confirm this expression state experimentally in molecular biology terms, including identifying the origin of transcription initiation, DNA methylation state, and OxyR binding status. The modeling study also suggests sequences upstream of the *agn43* gene in K12 can modulate the level of expression and are essential for the Off phase. However, this sequence is not conserved owing to the cryptic nature of the prophage insertion carrying *agn43*, and a requirement for specific upstream sequences was not observed using a less-sensitive reporter system (16, 55). With the queries that remain regarding the molecular mechanisms, testing predictions based on models of the epigenetic regulation of *agn43* reiteratively with modeling studies opens an interesting and new way forward for regulatory studies on this gene. Complementary approaches should address whether any process or cellular event is required to initiate a switch in gene expression state.

The overall level of expression of the Ag43 protein in a cell population can theoretically change by altering phase variation to bias the On or Off phase or by influencing the level of protein expression in a cell with the switch in the On position. As discussed above, the oxidation state of OxyR does not affect *agn43*[K12] regulation, and there is no evidence that OxyR or Dam cellular concentrations vary and so there is no obvious mechanism to alter the switch frequency. Thus, regulation will most likely be by posttranscriptional initiation. The most plausible point of additional regulation is transcriptional elongation, especially because all *agn43* alleles have long leader regions preceding the coding sequence with variable length and sequence. However, there is a highly conserved 32-nucleotide stretch immediately upstream of the start codon of *agn43* that is an effective terminator of transcription in in vitro assays. However, deletion of this *rho*-independent terminator in *agn43*[K12] does not affect expression of a reporter gene placed downstream of the terminator, indicating antitermination occurs (A. Wallecha & M. van der Woude, unpublished data) that does not involve RfaH (3). Identifying the antitermination mechanism will be important to determine if modulation of Ag43 expression can take place in response to external signals.

Differential expression of *agn43* has been noted in studies using microarrays, often when examining gene expression in biofilms or in the context of quorum sensing (3, 12, 14, 23, 43). However, differential expression in a microarray of genes that phase-vary can be due to a difference in the random distribution of the percentage of On cells between two independent bacterial populations. This would be apparent as inconsistent signals between truly independent experimental duplicates, as was described for *agn43* expression in a study on the small RNAs OmrA and OmrB (15). A second aspect to consider is that enrichment for or selection against either the On or Off phase will also bias the distribution of On and Off cells, resulting in an apparent regulation of expression. For example, enrichment of On cells was proposed to underlie a 13-fold increase in *agn43* expression in a *ceuO* mutant, which encodes for a multicopper oxidase, where it was shown that the percentage of On cells had changed and not the level of expression per On cell (49). This proposed enrichment could be due to protection mediated by Ag43-dependent autoaggregation. Thus, microarray-dependent analysis of gene expression is not well-suited to study regulation of Ag43 expression or indeed any phase-variable protein.

In *E. coli*, phase variation and the level of expression of many fimbriae are subject to environmental signals as well as cross-regulation by regulatory proteins from other fimbrial operons. This has led to a model in which a regulatory network controls phase variation of different fimbrial types, leading to sequential expression in a cell population (24). On the basis of complementary function in adhesion, it would not be illogical if Ag43 expression were part of this regulatory network, but there are conflicting data regarding coordinated expression of Ag43 and Type 1 fimbriae (18, 44). No molecular mechanism has been identified that could control this, and thus regulated expression of Ag43 as part of a network of genes controlling cellular behavior remains questionable.

CONCLUSION

Studies on the autotransporter Ag43 have revealed that this abundant protein shares many features with other members of this group of secreted proteins, which will help clarify details on its synthesis, secretion, and translocation. Analysis of its regulation has elucidated a relatively simple epigenetic mechanism that is attractive for modeling stochastic behavior and that will assist researchers to further probe aspects of epigenetic regulation. Perhaps somewhat surprisingly, it has been difficult to assign significant in vivo biological importance to Ag43 production, even though the aggregation that results appears to encourage biofilm formation. Recent insights into the allelic variations and possible effects of the genetic background of the strain on modification and localization may indicate how future studies can be more defined. It seems as though Ag43 is not a major virulence factor, which may lead to a loss of interest in this protein. However, this protein likely plays a role in persistence, something that is not easily determined. Persistence is increasingly considered an important aspect of *E. coli* biology, both in infection and in commensalism, and thus studies on this protein should continue to be valued.

SUMMARY POINTS

1. Ag43 is an autotransporter that uses the Type V secretion pathway.

2. The *agn43* gene, which encodes the outer membrane protein Ag43, exists as a family of alleles consisting of two major subgroups.

3. Isolates may differ in *agn43* content, ranging from 0 to 3 copies, and a correlation may exist that links the copy number and perhaps variant to pathogenesis, specifically to isolates associated with UTIs.

4. The genetic background of the strain that expresses Ag43 may affect protein localization and modification, and thus function.

5. Expression is controlled by an epigenetic mechanism that results in phase variation, which involves the DNA binding protein OxyR and the maintenance DNA methyltransferase

Dam. Phase variation results in heterogeneous expression of Ag43 in a clonal bacterial population. The biological significance of this regulation remains to be addressed.

6. According to in silico structural predictions and mutational analyses, the RGD domains do not appear to be significant for determining Ag43-mediated interactions between the bacterium and eukaryotic cells.

7. Ag43 expression mediates bacterial autoaggregation. Ag43 expression has been implicated in biofilm formation and long-term colonization of the bladder by *E. coli* in a mouse model, presumably as a result of mediating autoaggregation. However, these roles of Ag43 need to be studied in the context of host strain and allelic variant.

DISCLOSURE STATEMENT

The authors are not aware of any biases that might be perceived as affecting the objectivity of this review.

ACKNOWLEDGMENTS

Work in the van der Woude lab is supported by the Biotechnology and Biological Sciences Research Council, U.K., and the Wellcome Trust. Work in the Henderson lab is supported by the Biotechnology and Biological Sciences Research Council, U.K., and Medical Research Council, U.K.

LITERATURE CITED

1. First study to demonstrate expression of Ag43 can occur in vivo during a UTI with *E. coli*.

7. First time it is shown that biofilm formation can be modulated by Ag43 expression.

1. **Anderson G, Palermo J, Schilling J, Roth R, Heuser J, Hultgren S. 2003. Intracellular bacterial biofilm-like pods in urinary tract infections.** *Science* **301:105–7**
2. Barnard TJ, Dautin N, Lukacik P, Bernstein HD, Buchanan SK. 2007. Autotransporter structure reveals intra-barrel cleavage followed by conformational changes. *Nat. Struct. Mol. Biol.* 14:1214–20
3. Beloin C, Michaelis K, Lindner K, Landini P, Hacker J, et al. 2006. The transcriptional antiterminator RfaH represses biofilm formation in *Escherichia coli*. *J. Bacteriol.* 188:1316–31
4. Charbonneau ME, Girard V, Nikolakakis A, Campos M, Berthiaume F, et al. 2007. O-linked glycosylation ensures the normal conformation of the autotransporter adhesin involved in diffuse adherence. *J. Bacteriol.* 189:8880–89
5. Charbonneau ME, Mourez M. 2007. Functional organization of the autotransporter adhesin involved in diffuse adherence. *J. Bacteriol.* 189:9020–29
6. Correnti J, Munster V, Chan T, Woude M. 2002. Dam-dependent phase variation of Ag43 in *Escherichia coli* is altered in a *seqA* mutant. *Mol. Microbiol.* 44:521–32
7. **Danese P, Pratt L, Dove S, Kolter R. 2000. The outer membrane protein, antigen 43, mediates cell-to-cell interactions within *Escherichia coli* biofilms.** *Mol. Microbiol.* **37:424–32**
8. das Graças de Luna M, Scott-Tucker A, Desvaux M, Ferguson P, Morin N, et al. 2008. The *Escherichia coli* biofilm-promoting protein Antigen 43 does not contribute to intestinal colonization. *FEMS Microbiol. Lett.* 284:237–46
9. Dautin N, Barnard TJ, Anderson DE, Bernstein HD. 2007. Cleavage of a bacterial autotransporter by an evolutionarily convergent autocatalytic mechanism. *EMBO J.* 26:1942–52
10. Dautin N, Bernstein HD. 2007. Protein secretion in gram-negative bacteria via the autotransporter pathway. *Annu. Rev. Microbiol.* 61:89–112
11. Diderichsen B. 1980. *flu*, a metastable gene controlling surface properties of *Escherichia coli*. *J. Bacteriol.* 141:858–67

12. Ferrieres L, Clarke DJ. 2003. The RcsC sensor kinase is required for normal biofilm formation in *Escherichia coli* K-12 and controls the expression of a regulon in response to growth on a solid surface. *Mol. Microbiol.* 50:1665–82

13. Fexby S, Bjarnsholt T, Jensen P, Roos V, Høiby N, et al. 2007. Biological Trojan horse: Antigen 43 provides specific bacterial uptake and survival in human neutrophils. *Infect. Immun.* 75:30–34

14. Gonzalez Barrios AF, Zuo R, Hashimoto Y, Yang L, Bentley WE, Wood TK. 2006. Autoinducer 2 controls biofilm formation in *Escherichia coli* through a novel motility quorum-sensing regulator (MqsR, B3022). *J. Bacteriol.* 188:305–16

15. Guillier M, Gottesman S. 2006. Remodelling of the *Escherichia coli* outer membrane by two small regulatory RNAs. *Mol. Microbiol.* 59:231–47

16. Haagmans W, van der Woude M. 2000. Phase variation of Ag43 in *Escherichia coli*: Dam-dependent methylation abrogates OxyR binding and OxyR-mediated repression of transcription. *Mol. Microbiol.* 35:877–87

17. Hagan E, Mobley H. 2007. Uropathogenic *Escherichia coli* outer membrane antigens expressed during urinary tract infection. *Infect. Immun.* 75:3941–49

18. Hasman H, Chakraborty T, Klemm P. 1999. Antigen-43-mediated autoaggregation of *Escherichia coli* is blocked by fimbriation. *J. Bacteriol.* 181:4834–41

19. Henderson I, Meehan M, Owen P. 1997. A novel regulatory mechanism for a novel phase-variable outer membrane protein of *Escherichia coli*. *Adv. Exp. Med. Biol.* 412:349–55

20. Henderson I, Meehan M, Owen P. 1997. Antigen 43, a phase-variable bipartite outer membrane protein, determines colony morphology and autoaggregation in *Escherichia coli* K-12. *FEMS Microbiol. Lett.* 149:115–20

21. Henderson I, Owen P. 1999. The major phase-variable outer membrane protein of *Escherichia coli* structurally resembles the immunoglobulin A1 protease class of exported protein and is regulated by a novel mechanism involving Dam and oxyR. *J. Bacteriol.* 181:2132–41

22. Henderson I, Owen P, Nataro J. 1999. Molecular switches—the ON and OFF of bacterial phase variation. *Mol. Microbiol.* 33:919–32

23. Herzberg M, Kaye I, Peti W, Wood T. 2006. YdgG (TqsA) controls biofilm formation in *Escherichia coli* K-12 through autoinducer 2 transport. *J. Bacteriol.* 188:587–98

24. Holden NJ, Gally DL. 2004. Switches, cross-talk and memory in *Escherichia coli* adherence. *J. Med. Microbiol.* 53:585–93

25. Jain S, van Ulsen P, Benz I, Schmidt MA, Fernandez R, et al. 2006. Polar localization of the autotransporter family of large bacterial virulence proteins. *J. Bacteriol.* 188:4841–50

26. Junker M, Schuster CC, McDonnell AV, Sorg KA, Finn MC, et al. 2006. Pertactin beta-helix folding mechanism suggests common themes for the secretion and folding of autotransporter proteins. *Proc. Natl. Acad. Sci. USA* 103:4918–23

27. Kajava AV, Cheng N, Cleaver R, Kessel M, Simon MN, et al. 2001. Beta-helix model for the filamentous haemagglutinin adhesin of *Bordetella pertussis* and related bacterial secretory proteins. *Mol. Microbiol.* 42:279–92

28. Kajava AV, Steven AC. 2006. The turn of the screw: variations of the abundant beta-solenoid motif in passenger domains of Type V secretory proteins. *J. Struct. Biol.* 155:306–15

29. Kjaergaard K, Schembri M, Hasman H, Klemm P. 2000. Antigen 43 from *Escherichia coli* induces inter- and intraspecies cell aggregation and changes in colony morphology of *Pseudomonas fluorescens*. *J. Bacteriol.* 182:4789–96

30. Klemm P, Hjerrild L, Gjermansen M, Schembri M. 2004. Structure-function analysis of the self-recognizing Antigen 43 autotransporter protein from *Escherichia coli*. *Mol. Microbiol.* 51:283–96

31. Knudsen SK, Stensballe A, Franzmann M, Westergaard UB, Otzen DE. 2008. Effect of glycosylation on the extracellular domain of the Ag43 bacterial autotransporter: enhanced stability and reduced cellular aggregation. *Biochem. J.* 412:563–77

32. Konieczny MPJ, Benz I, Hollinderbaumer B, Beinke C, Niederweis M, Schmidt MA. 2001. Modular organization of the AIDA autotransporter translocator: the N-terminal beta1-domain is surface-exposed and stabilizes the transmembrane beta2-domain. *Antonie Van Leeuwenhoek* 80:19–34

19. OxyR and Dam are identified as regulators of phase variation of *agn43*.

33. As a result of building a model for transition rates in phase variation of Ag43, new and testable features of the regulation are proposed.

33. **Lim H, van Oudenaarden A. 2007. A multistep epigenetic switch enables the stable inheritance of DNA methylation states.** *Nat. Genet.* **39:269–75**

34. Luck SN, Turner SA, Rajakumar K, Sakellaris H, Adler B. 2001. Ferric dicitrate transport system (Fec) of *Shigella flexneri* 2a YSH6000 is encoded on a novel pathogenicity island carrying multiple antibiotic resistance genes. *Infect. Immun.* 69:6012–21

35. Oliver DC, Huang G, Nodel E, Pleasance S, Fernandez RC. 2003. A conserved region within the *Bordetella pertussis* autotransporter BrkA is necessary for folding of its passenger domain. *Mol. Microbiol.* 47:1367–83

36. Owen P, Kaback HR. 1978. Molecular structure of membrane vesicles from *Escherichia coli. Proc. Natl. Acad. Sci. USA* 75:3148–52

37. Owen P, Meehan M, de Loughry-Doherty H, Henderson I. 1996. Phase-variable outer membrane proteins in *Escherichia coli. FEMS Immunol. Med. Microbiol.* 16:63–76

38. Peterson JH, Szabady RL, Bernstein HD. 2006. An unusual signal peptide extension inhibits the binding of bacterial presecretory proteins to the signal recognition particle, trigger factor, and the SecYEG complex. *J. Biol. Chem.* 281:9038–48

39. Isolates were analyzed to determine the distribution of *agn43* alleles and address a correlation of the presence of the gene with virulence.

39. **Restieri C, Garriss G, Locas M, Dozois C. 2007. Autotransporter-encoding sequences are phylogenetically distributed among *Escherichia coli* clinical isolates and reference strains.** *Appl. Environ. Microbiol.* **73:1553–62**

40. Roche A, McFadden J, Owen P. 2001. Antigen 43, the major phase-variable protein of the *Escherichia coli* outer membrane, can exist as a family of proteins encoded by multiple alleles. *Microbiology* 147:161–69

41. Roux A, Beloin C, Ghigo J. 2005. Combined inactivation and expression strategy to study gene function under physiological conditions: application to identification of new *Escherichia coli* adhesins. *J. Bacteriol.* 187:1001–13

42. Sakellaris H, Luck SN, Al-Hasani K, Rajakumar K, Turner SA, Adler B. 2004. Regulated site-specific recombination of the she pathogenicity island of *Shigella flexneri. Mol. Microbiol.* 52:1329–36

43. Schembri M, Kjaergaard K, Klemm P. 2003. Global gene expression in *Escherichia coli* biofilms. *Mol. Microbiol.* 48:253–67

44. Schembri M, Ussery D, Workman C, Hasman H, Klemm P. 2002. DNA microarray analysis of *fim* mutations in *Escherichia coli. Mol. Genet. Genomics* 267:721–29

45. Sherlock O, Dobrindt U, Jensen J, Munk Vejborg R, Klemm P. 2006. Glycosylation of the self-recognizing *Escherichia coli* Ag43 autotransporter protein. *J. Bacteriol.* 188:1798–807

46. Sijbrandi R, Urbanus ML, ten Hagen-Jongman CM, Bernstein HD, Oudega B, et al. 2003. Signal recognition particle (SRP)-mediated targeting and Sec-dependent translocation of an extracellular *Escherichia coli* protein. *J. Biol. Chem.* 278:4654–59

47. Snyder JA, Lloyd AL, Lockatell CV, Johnson DE, Mobley HL. 2006. Role of phase variation of type 1 fimbriae in a uropathogenic *Escherichia coli* cystitis isolate during urinary tract infection. *Infect. Immun.* 74:1387–93

48. Torres AG, Perna NT, Burland V, Ruknudin A, Blattner FR, Kaper JB. 2002. Characterization of Cah, a calcium-binding and heat-extractable autotransporter protein of enterohaemorrhagic *Escherichia coli. Mol. Microbiol.* 45:951–66

49. Tree J, Ulett G, Hobman J, Constantinidou C, Brown N, et al. 2007. The multicopper oxidase (CueO) and cell aggregation in *Escherichia coli. Environ. Microbiol.* 9:2110–16

50. Turner SA, Luck SN, Sakellaris H, Rajakumar K, Adler B. 2004. Role of *attP* in integrase-mediated integration of the *Shigella* resistance locus pathogenicity island of *Shigella flexneri. Antimicrob. Agents Chemother.* 48:1028–31

51. Addresses the role of Ag43 in an infection model and shows Ag43 may facilitate bacterial persistence.

51. **Ulett G, Valle J, Beloin C, Sherlock O, Ghigo J, Schembri M. 2007. Functional analysis of antigen 43 in uropathogenic *Escherichia coli* reveals a role in long-term persistence in the urinary tract.** *Infect. Immun.* **75:3233–44**

52. Ulett G, Webb R, Schembri M. 2006. Antigen-43-mediated autoaggregation impairs motility in *Escherichia coli. Microbiology* 152:2101–10

53. van der Woude MW. 2006. Re-examining the role and random nature of phase variation. *FEMS Microbiol. Lett.* 254:190–97

54. van der Woude MW, Baumler AJ. 2004. Phase and antigenic variation in bacteria. *Clin. Microbiol. Rev.* 17:581–611

55. Waldron D, Owen P, Dorman C. 2002. Competitive interaction of the OxyR DNA-binding protein and the Dam methylase at the antigen 43 gene regulatory region in *Escherichia coli*. *Mol. Microbiol.* 44:509–20

56. Wallecha A, Correnti J, Munster V, van der Woude M. 2003. Phase variation of Ag43 is independent of the oxidation state of OxyR. *J. Bacteriol.* 185:2203–9

57. Wallecha A, Munster V, Correnti J, Chan T, van der Woude M. 2002. Dam- and OxyR-dependent phase variation of agn43: essential elements and evidence for a new role of DNA methylation. *J. Bacteriol.* 184:3338–47

58. Winkler HH, Wilson TH. 1966. The role of energy coupling in the transport of beta-galactosides by *Escherichia coli*. *J. Biol. Chem.* 241:2200–11

59. Yen MR, Peabody CR, Partovi SM, Zhai Y, Tseng YH, Saier MH. 2002. Protein-translocating outer membrane porins of gram-negative bacteria. *Biochim. Biophys. Acta* 1562:6–31

57. Sequence elements and DNA-protein interactions required for *agn43* phase variation are identified and analyzed in detail.

RELATED RESOURCES

Casadesús J, Low D. 2006. Epigenetic gene regulation in the bacterial world. *Microbiol. Mol. Biol. Rev.* 70:830–56

Henderson IR, Nataro JP. 2005. Chapter 8.7.3. Autotransporter proteins. In *EcoSal—Escherichia coli and* Salmonella: *Cellular and Molecular Biology*, ed. A Böck, R Curtiss III, JB Kaper, FC Neidhardt, T Nyström, KE Rudd, CL Squires. Washington, DC: ASM Press. **http://www.ecosal.org/ecosal/index.jsp**

Henderson IR, Navarro-Garcia F, Desvaux M, Fernandez RC, Ala'Aldeen. 2004. Type V protein secretion pathway: the autotransporter story. *Microbiol. Mol. Biol. Rev.* 68:692–744

Klemm P, Schembri M. 2004. Chapter 8.3.2.6. Type 1 fimbriae, curli, and antigen 43: adhesion, colonization, and biofilm formation. In *EcoSal—Escherichia coli and* Salmonella: *Cellular and Molecular Biology*, ed. A Böck, R Curtiss III, JB Kaper, FC Neidhardt, T Nyström, KE Rudd, CL Squires. Washington, DC: ASM Press. **http://www.ecosal.org/ecosal/index.jsp**

van der Woude M. 2008. Some types of bacterial phase variation are epigenetic. *Microbe* 3:21–26

Wells T, Tree J, Ulett G, Schembri M. 2007. Autotransporter proteins: novel targets at the bacterial cell surface. *FEMS Microbiol. Lett.* 274:163–72

Viral Subversion of Apoptotic Enzymes: Escape from Death Row*

Sonja M. Best

Laboratory of Persistent Viral Diseases, Rocky Mountain Laboratories, National Institute of Allergy and Infectious Diseases, National Institutes of Health, Hamilton, Montana 59840; email: sbest@niaid.nih.gov

Annu. Rev. Microbiol. 2008. 62:171–92

First published online as a Review in Advance on June 6, 2008

The *Annual Review of Microbiology* is online at micro.annualreviews.org

This article's doi: 10.1146/annurev.micro.62.081307.163009

Key Words

caspase inhibition, serine protease, CrmA, p35, IAP, vFLIP

Abstract

To prolong cell viability and facilitate replication, viruses have evolved multiple mechanisms to inhibit the host apoptotic response. Cellular proteases such as caspases and serine proteases are instrumental in promoting apoptosis. Thus, these enzymes are logical targets for virus-mediated modulation to suppress cell death. Four major classes of viral inhibitors antagonize caspase function: serpins, p35 family members, inhibitor of apoptosis proteins, and viral FLICE-inhibitory proteins. Viruses also subvert activity of the serine proteases, granzyme B and HtrA2/Omi, to avoid cell death. The combined efforts of viruses to suppress apoptosis suggest that this response should be avoided at all costs. However, some viruses utilize caspases during replication to aid virus protein maturation, progeny release, or both. Hence, a multi-faceted relationship exists between viruses and the apoptotic response they induce. Examination of these interactions contributes to our understanding of both virus pathogenesis and the regulation of apoptotic enzymes in normal cellular functions.

Contents

INTRODUCTION

The successful replication of a virus within an individual cell requires a remarkable cascade of interactions between virus and host, beginning at the first engagement of the cell receptor to the final release of progeny virions. Viruses utilize everything from cellular enzymes and transcription factors to membranes and organelles to facilitate their replication (11). However, hijacking of the cell by a virus is not without a potent antiviral response. One such response is apoptosis, the genetically and biochemically controlled process of cell death that functions in development and homeostasis of multicellular organisms through selective removal of unwanted or damaged cells (120). Apoptosis can also serve as an innate cellular response to infection that limits both the time and cellular machinery available for virus replication (120). Many of the key biochemical events that occur during apoptosis are mediated by proteases (62), the most important of which are caspases (cysteine-dependent aspartate-specific proteases). Not surprisingly, viruses from diverse families have evolved mechanisms to evade or delay cell death by suppressing the activity of caspases and other enzymes with central roles in the implementation of apoptosis. Examination of interactions between viruses and the host apoptotic machinery has contributed extensively to our understanding of virus replication and pathogenesis. Furthermore, the elucidation of how viruses modulate these responses has furthered our knowledge of apoptotic pathways and how they contribute to both normal and disease states. This review enumerates the specific mechanisms underlying virus-induced suppression of enzyme activity during apoptosis.

In its simplest form, apoptosis can be considered a two-step proteolytic pathway (92). The first is an initiating phase resulting in activation of initiator caspases. These caspases are responsible for the second execution phase by cleaving and activating executioner or effector caspases. During this second phase, effector caspases cleave target host proteins, culminating in the step-wise demise of the cell (66). Apoptosis is initiated through two general mechanisms: from outside the cell (extrinsic) or from within (intrinsic) (**Figure 1**). Extrinsic triggering of apoptosis occurs following ligation of death receptors by the tumor necrosis factor (TNF) superfamily including TNF, Fas ligand (FasL), and TNF-related apoptosis-inducing ligand (TRAIL) (**Figure 1**). Ligation of the TNF receptor family results in the formation of the death-inducing signaling complex (DISC) (63) required for activation of initiator caspases, particularly caspase-8 and caspase-10.

Intrinsic activation of apoptosis is triggered following translocation of a proapoptotic Bcl-2 family member, Bid or Bax, to the mitochondria (133, 135) (**Figure 1**). In particular, cytosolic Bid is cleaved by caspases or other proapoptotic proteases to form truncated Bid (tBid), which then localizes to the mitochondria. Here, tBid interacts with other proapoptotic Bcl-2 member proteins such as Bax and/or Bak, resulting in pore formation and permeabilization of the outer mitochondrial membrane (45, 133). This initiates the release of proteins from the

---

Apoptosis: refers to morphological changes occurring in a controlled process of cell death, resulting in membrane-bound cell fragments usually eliminated by phagocytosis

Caspase: cysteine-dependent aspartate-specific protease

Death induced signaling complex (DISC): multimeric assembly platform required for caspase activation induced after death receptor ligation

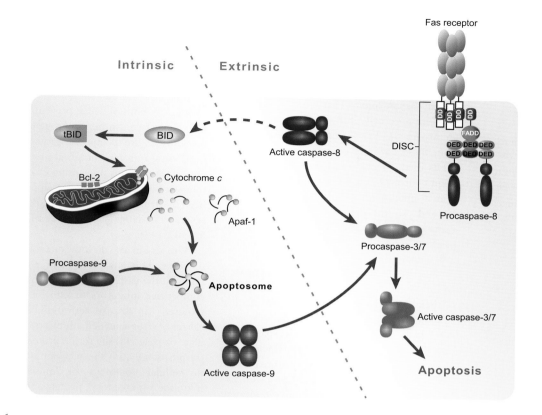

Intrinsic   Extrinsic

Fas receptor

tBID ← BID

DD DD DD DD
FADD
DED DED DED
DED DED DED

DISC

Active caspase-8

Procaspase-8

Bcl-2   Cytochrome c

Apaf-1

Procaspase-9

Procaspase-3/7

Apoptosome

Active caspase-3/7

Active caspase-9

Apoptosis

Figure 1

Apoptosis signaling pathways. Apoptosis is initiated extrinsically following ligation of death receptors including the Fas receptor. This results in the recruitment of the adaptor protein FADD and the formation of the death-inducing signaling complex (DISC). Procaspase-8 is also recruited to the DISC, where it is activated, enabling it to cleave and activate caspase-3 and caspase-7. Intrinsic initiation of apoptosis is triggered following cleavage and translocation of a proapoptotic Bcl-2 family member such as Bid to the mitochondria, where it is involved in altering membrane permeabilization and pore formation, enabling the release of cytochrome c. Cytosolic cytochrome c binds Apaf-1 to induce formation of the apoptosome. This complex recruits and activates caspase-9, enabling it to cleave and activate caspase-3 and caspase-7. Caspase-8 can also activate intrinsic signaling through the cleavage of Bid. Abbreviations: Apaf-1, apoptosis protease-activating factor; DD, death domain; DED, death effector domain; FADD, Fas-associated death domain; tBID, truncated Bid.

inner membrane into the cytosol. One such protein, cytochrome c, is central to this apoptosis program (45) because it binds to apoptosis protease-activating factor (Apaf-1) to induce the formation of an oligomeric assembly platform, termed the apoptosome (7, 92). The initiator caspase-9 is subsequently recruited to the apoptosome, resulting in its activation (92). A wide variety of cellular insults result in loss of mitochondrial membrane integrity, including aberrant calcium signaling, growth factor

withdrawal, treatment with various cytotoxic agents, and endoplasmic reticulum stress. However, receptor-mediated cell death can also result in caspase-8-mediated cleavage of Bid, mitochondrial membrane permeabilization, and activation of caspase-9, thus amplifying apoptotic signaling pathways (133) (**Figure 1**). Active caspase-8, -10, and -9 then target effector caspases for cleavage. The individual pathways involved in initiating and regulating apoptosis are expanded upon in the following sections.

Apoptosome: multimeric assembly platform required for caspase-9 activation induced following disruption of mitochondrial membrane potential

CASPASES

As highlighted above, the key effector proteins activated during apoptosis and targeted by viruses for inhibition are caspases (1). There are currently 13 members of the mammalian caspase family, characterized by a near-absolute specificity for substrates containing an Asp in the P_1 cleavage position. In addition, these enzymes contain both a Cys and a His in the active site that assist in peptide bond hydrolysis (24, 115). Caspases that function in apoptosis include caspase-2, -8, -9, -10, and -12 (the initiator caspases) as well as caspase-3, -6, -7, and -14 (the effector caspases). Caspase-1, -4, -5, and -11 function in inflammation.

Caspases exist in the cell as inactive procaspases, or zymogens, and based on the length of their prodomains, they can be classified in two main categories (66). The first category is the large-prodomain caspases (caspase-1, -2, -4, -5, -8, -9, -10, -11, and -12), which have a long prodomain characterized by death domain (DD) motifs including death effector domains (DEDs) and caspase activation and recruitment domains (CARDs) (90). These caspases are present in the cell as monomers and, following recruitment into multimeric protein platforms such as the DISC (63) or apoptosome (7, 92), dimerize and undergo a conformational change resulting in activation. The second category is the short-prodomain caspases (the initiator caspase-3, -6, -7, and -14). These caspases exist as preformed dimers and, in contrast to the large-prodomain enzymes, require proteolysis to be activated (99). Active caspases are obligate homodimers with each monomer composed of a large (17- to 20-kDa) subunit and a small (10- to 12-kDa) subunit (42).

VIRUS-MEDIATED INHIBITION OF CASPASES

Because of their central role in controlling cell fate decisions, caspases are subject to regulation at multiple levels, through prevention of activation as well as inhibition of activity following maturation (28). This property renders caspases a logical target for inhibition by viruses. Inhibition can be achieved through indirect mechanisms such as the virus-induced downregulation of death receptor expression (116) or the expression of secreted viral TNF receptor homologs (9), both of which prevent signaling at the cell surface. An important general mechanism viruses use to suppress caspase activation is inhibition of mitochondrial membrane permeabilization, thus preventing or delaying the release of cytochrome c. These approaches utilized by viruses to prevent apoptosis have been reviewed in detail elsewhere (2, 14, 17, 39, 47, 61, 107, 113).

Viruses can also directly antagonize enzyme function. This is achieved either by interacting with the caspase active site or by acting as competitive inhibitors of signaling molecules required for caspase activation. Four major classes of virus-encoded inhibitors directly regulate caspases: (*a*) the serine protease inhibitor (serpin) family, (*b*) the p35 family (for which no cellular homologs are known), (*c*) viral inhibitor of apoptosis proteins (vIAPs), and (*d*) viral FLICE (Fas-associated death domain-like interleukin-1β-converting enzyme) inhibitory proteins, or vFLIPs (**Table 1**). Each class of inhibitor is discussed below.

CrmA: Building a Better Mousetrap

The first caspase inhibitor discovered was cytokine response modifier A (CrmA) encoded by cowpox virus, an orthopoxvirus. CrmA is a member of the serpin superfamily. It was originally identified as an inhibitor of interleukin-1β (IL-1β)-converting enzyme (ICE), now known as caspase-1, by preventing its ability to process the precursor of IL-1β into its active, secreted form (89). This discovery was made before it was understood that caspases were the central proteases of cell death (16). However, when the critical nature of caspases in apoptosis was determined following the discovery of CED-3 (a caspase that functions as both an initiator and an executioner) from *Caenorhabditis elegans*

Table 1 Virus-encoded caspase inhibitors

	Protein family	Specific protein	Virus	Caspases targeted	Reference(s)
Direct caspase inhibitor	Serpin	CrmA/SPI-1	Cowpox virus/ orthopoxvirus	Caspase-1, -8, and -10	(44, 64, 140)
		SERP2	Leporipoxvirus	Caspase-1, -8, and -10	(77)
	p35	p35	Baculoviruses	Broad spectrum including caspase-1, -3, -6, -7, -8, and -10	(130, 139)
		p49	Baculoviruses	Similar to p35 plus caspase-9 and DRONC	(56, 142)
		p33	*Amsacta moorei* entomopoxvirus	Similar to p35 (as far as tested)	(78)
Competitive caspase inhibitor	IAP	Op-IAP AMV-IAP A224L	Baculoviruses Entomopoxvirus African swine fever virus	Indirect inhibition of caspase-3, -7, -9, DRONC, and DrICE	(79, 131)
	FLIP	vFLIP MC159	γ-herpesviruses Molluscum contagiosum virus	Indirect inhibition of caspase-8 and caspase-10 activation	(10, 55)

Abbreviations: CrmA, cytokine response modifier A; DrICE, *Drosophila* interleukin-1β converting enzyme; DRONC, *Drosophila* Nedd-2-like caspase; FLIP, FLICE (Fas-associated death domain-like interleukin-1β converting enzyme) inhibitory protein; IAP, inhibitor of apoptosis protein; SERP-2, serine protease inhibitor-2; SPI-1, serine protease inhibitor-1; vFLIP, viral FLIP.

(84, 136), the ability of CrmA to inhibit apoptosis was tested. CrmA can inhibit apoptosis triggered by overexpression of caspase-1 (84) as well as by treatment of cells with TNF or ligation of Fas receptors. In addition to direct inhibition of caspase-1 in the low picomolar range, CrmA can also suppress caspase-8 and caspase-10 activity (44, 44, 139, 140), thus explaining its protective properties when apoptosis is initiated through death receptors (**Table 1**). CrmA can inhibit a wide range of caspases including caspase-3, -6, -7, and -9 (140). However, the efficiency of inhibition is generally lower than that demonstrated for caspase-1 and caspase-8, and caspase-9 is not thought to be an important target in vivo (27).

CrmA is unique among the serpins, as this family normally functions to suppress the activity of serine proteases rather than that of cysteine proteases. Serpins have a conserved core globular domain consisting of three β-sheets and eight to nine α-helices. CrmA maintains the serpin architecture and fold even though it lacks conserved features of the serpin superfamily including the D-helix, half of the A-

helix, and part of the E-helix (91, 101). In addition to these differences, CrmA contains a pseudosubstrate of the caspases (Leu-Val-Ala-Asp^{303}) within the serpin flexible reactive site loop (RSL), with Asp^{303} in the P$_1$ position, an unusual residue for a serpin.

CrmA inhibits caspase-1 by a mechanism of conformational trapping similar to that determined for serpin inhibition of serine proteases (106, 110). The exposed RSL of CrmA is inserted into the caspase active site where the caspase catalytic cysteine is covalently linked to the serpin P$_1$ residue. The RSL is then cleaved at the P$_1$-P$_1$' bond, leading to a radical conformational change whereby the N terminus of the newly cleaved loop inserts into the center of the major β-sheet of CrmA (sheet A) (**Figure 2**). This causes the caspase tethered to the P$_1$ residue to be translocated to the opposite end of the serpin. The caspase is inactivated as its active site becomes conformationally distorted and trapped (106). In addition, the small subunit of caspase-1 and caspase-8 can disassociate from the large subunit during the trapping process, providing an additional

RSL: reactive site loop

a Native state

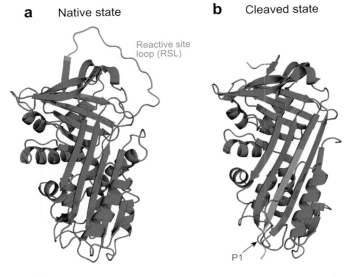

Reactive site
loop (RSL)

b Cleaved state

P1

Figure 2

Model of CrmA-mediated caspase inhibition. (*a*) Ribbon diagram of a serpin in its native state demonstrating the position of the reactive site loop (RSL). The crystal structure of native CrmA is not known so the structure of the related α_1-antitrypsin (PDB# 1psi) is depicted to illustrate the mechanism of inhibition by CrmA (32, 37). (*b*) Diagram of CrmA following cleavage and insertion of the RSL into the center of the molecule (PDB# 1c8o) (91, 101). A caspase covalently linked to the P_1 residue (*labeled*) within the RSL would be translocated to the opposite end of CrmA during this conformational rearrangement. Protein structures were generated using PyMOL.

tively. This property is dependent on the P_1 Asp residue present in both proteins (77). Despite these similarities, SERP-2 and CrmA are not functionally interchangeable. For example, replacement of CrmA by SERP-2 enabled the resulting recombinant cowpox virus to inhibit apoptosis and reach wild-type virus yields in infected CAMs. However, SERP-2 did not block inflammation in this context (86), which is the original hallmark of CrmA function (89). Similarly, the reciprocal recombinant myxoma virus containing CrmA in place of SERP-2 was not fully virulent in its European rabbit host and did not cause the fulminant skin lesions characteristic of infection with wild-type virus (86). Both recombinant viruses could inhibit apoptosis of infected cells, suggesting that poxvirus serpins may have host-specific roles in control of inflammation.

p35 Inhibition of Caspases: A Ménage à Trois

p35 is a broad-spectrum caspase inhibitor identified from the baculovirus, *Autographa californica* multiple *Nucleopolyhedrovirus* (AcMNPV) (23). This protein inhibits caspases from *C. elegans* (CED-3), *Drosophila melanogaster* (the effector caspase DrICE), *Spodoptera frugiperda* (the effector Sf-caspase-1), and mammals (caspase-1, -3, -6, -8, -7, and -10) (67, 130, 139) (**Table 1**). p35 is least effective against caspase-2 and caspase-9. The related p49 protein from *Spodoptera littoralis Nucleopolyhedrovirus* (SlNPV) (34) inhibits a wider range of caspases compared with p35, including caspase-2 and caspase-9 as well as *Drosophila* Nedd2-like caspase (DRONC), a caspase-9-like initiator caspase (56, 142) (**Table 1**). However, unlike p35, p49 cannot suppress DrICE activity, suggesting host-specific roles for this class of inhibitors (70). Recently, a homolog of the p35 protein, called p33, was discovered in *Amsacta moorei* entomopoxvirus (AmEPV), and it is the first homolog found outside of the baculoviruses (78). p33 is only 25% identical to p35 at the amino

irreversible inhibition of the protease (32). The structure of the serpin in complex with its target is less energy rich than its native structure (101), thus providing the energetics for this virus-encoded mousetrap.

Other members of the poxviruses encode serpins homologous to CrmA, including serine protease inhibitor (SPI)-1 and SPI-2 [in orthopoxviruses including vaccinia (31), ectromelia (118), and rabbitpox (76) viruses] or serpin-2 (SERP-2) from myxoma virus, a *Leporipoxvirus* (87). SERP-2 shares 34.9% identity at the amino acid level, and 57.3% similarity (87), with CrmA and can inhibit caspase-1, -8, and -10 in vitro (77) (**Table 1**). Both CrmA and SERP-2 are required for full virus virulence in their models of infection, chicken chorioallantoic membranes (CAMs) and rabbits, respec-

DRONC: *Drosophila* Nedd2-like caspase

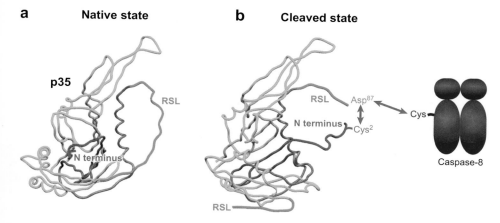

a   **Native state**   **b**   **Cleaved state**

p35

RSL

N terminus

RSL

RSL Asp^{87}

N terminus Cys^2

Cys

Caspase-8

RSL

Figure 3

Mechanism of p35-mediated inhibition of caspase-8. (*a*) Worm diagram of p35 in its native state. The extreme N terminus (*red*) is buried in the core of the protein. (*b*) Diagram of p35 after complex formation with caspase-8. Cleavage of the reactive site loop (RSL) enables Asp^{87} to form a thioester intermediate with the caspase active site Cys residue. The liberated Cys^2 of the p35 N terminus inserts into the caspase active site, trapping Asp^{87} at the thioester intermediate step and establishing equilibrium of reversible *trans*-thioesterification between Asp^{87}-Cys^{casp8} and Asp^{87}-Cys^2. In panels *a* and *b*, the C terminus of p35 is depicted in light blue and the remaining N terminus is dark blue. This figure was reproduced with modifications from Reference 74 with permission from the publisher.

acid level but has a strikingly similar predicted secondary structure and can inhibit similar effector caspases in vitro (78).

Similar to CrmA, p35 is a suicide inhibitor of caspases, acting as bait for the active caspase to attack. In addition, the exposed RSL of p35 contains P_4-P_1 residues, Asp-Gln-Met-Asp^{87}, with the critical Asp residue in the P_1 position. However, the molecular mechanism of inhibition differs between p35 and CrmA, as demonstrated for the p35-mediated inhibition of caspase-8 (74). Following insertion of the RSL into the caspase active site and cleavage of p35, the cleavage residue Asp^{87} forms a thioester intermediate with the caspase active site Cys (Cys^{casp8}). This enables the portion of the RSL distal to Asp^{87} to dissociate from the caspase, leading to dramatic conformational changes in p35 (**Figure 3**). This conformational change liberates the N terminus of p35 from where it is buried in the core of the molecule and enables it to insert into the caspase active site (74). During normal substrate cleavage by

caspase-8, a water molecule is activated by the active site His to quickly hydrolyze the thioester intermediate. However, the p35 N-terminal cysteine residue (Cys^2) attacks the thioester bond of Asp^{87} and excludes solvent, thereby trapping Asp^{87} at the thioester intermediate step. In addition, once attacked by the thiol of Cys^2, the carbonyl carbon of Asp^{87} is no longer accessible to the free amino group of Cys^2, much in the same way that water is excluded from attacking the Asp^{87}-Cys^{casp8} bond (74). In the absence of this second step, chemical ligation of the two ends is not completed and p35 does not become circularized. Instead, an equilibrium of reversible *trans*-thioesterification between Asp^{87}-Cys^{casp8} and Asp^{87}-Cys^2 is established (74), trapping the caspase active site in a futile ménage à trois (**Figure 3**). It is currently unknown if this mechanism of inhibition is limited to p35/caspase-8 interactions or if this is a general mechanism that contributes to the promiscuity of the p35 class of caspase inhibitors (103).

Suicide inhibitor: a protein that inhibits by irreversibly binding to its target

Viral Inhibitor of Apoptosis Proteins: The Ol' Bait and Switch

BIR: baculovirus IAP repeat

RING domain: specialized type of zinc-finger domain involved in protein-protein interactions that possesses E3-ubiquitin ligase activity

The first IAP was discovered in a genetic screen of baculoviruses searching for genes that could complement the deletion of *p35* from AcMNPV (26). In addition to the baculoviruses, homologous IAPs are encoded by entomopoxviruses, iridoviruses, and African swine fever virus (21, 22, 29). Cellular homologs of IAPs also exist in diverse organisms, from yeast to humans, with eight identified in the human genome. However, all IAPs do not function in modulating apoptotic responses, as they have roles in diverse cellular functions including innate immunity to bacterial infection and the regulation of chromosome segregation during mitosis (reviewed in References 95 and 121). The IAP family is characterized by the presence of baculovirus IAP repeat domains (BIRs), a novel zinc binding fold of approximately 70 amino acids (52, 83). Cellular IAPs (cIAPs) contain up to three BIRs while viral IAPs (vIAPs) contain only one or two. In addition, the IAPs contain a really interesting new gene (RING) domain that possesses E3 ubiquitin ligase activity and is required for IAP-mediated suppression of apoptosis (121, 122).

To understand how vIAPs function to suppress caspase activity, it is important to understand the function and regulation of cIAPs during apoptosis. Of the human IAPs, X-linked IAP (XIAP), cellular IAP1 (c-IAP1), and c-IAP2 can directly bind specific caspases including caspase-3, -7, and -9 in vitro (35, 98, 109). However, only XIAP can directly inhibit caspase activity, and therefore it is thought to be the primary IAP responsible for caspase inhibition in vivo (36). XIAP has two distinct domains that bind either to caspase-9 (via its BIR3) or to caspase-3 and caspase-7 (via the region immediately N-terminal to BIR2) (30), resulting in caspase inhibition (**Figure 4a**). Similar to XIAP, *Drosophila* IAP1 (DIAP1) is essential for negative regulation of the initiator caspase DRONC and the effector caspases DrICE and *Drosophila* caspase-1 (DCP-1). However, in this case, binding of caspases by the BIR1 domain of DIAP1 alone is not sufficient for inhibition. Following binding, DIAP1 is cleaved by *Drosophila* effector caspases at its N terminus, resulting in the targeting of DIAP1 for ubiquitin-dependent proteasome-mediated degradation. Presumably, the active caspase tethered to DIAP1 is also degraded (108, 109, 131).

Activity of the cellular antiapoptotic IAPs is modulated by IAP-interacting proteins often termed IAP antagonists. In humans and mice, the major IAP antagonist is termed SMAC (second mitochondrial activator of caspases) or DIABLO (direct IAP binding protein with low pI), respectively, and resides in the intermembrane space of the mitochondria (33, 123). The release of SMAC/DIABLO occurs in response to apoptotic stimuli along with cytochrome *c*, at which time it can bind to XIAP (72, 129) and displace the active caspase(s) to promote apoptosis. *Drosophila* has multiple IAP antagonists, including Reaper, HID (head-involution defective), Grim, and Sickle, that can bind to DIAP1 and antagonize its ability to bind to DRONC. These IAP antagonists share a common IAP binding motif [IBM (98), also known as the Reaper-HID-Grim (RHG) motif] that enables them to bind to and suppress the function of DIAP1 (128) and XIAP (**Figure 4b**), suggesting a mechanism of binding conserved between flies and humans. One function of Reaper, HID, and Grim, as well as the *Drosophila* ubiquitin conjugase-related protein termed Morgue, is to promote degradation of *Drosophila* IAPs (54, 95, 134). Degradation is dependent on a functional RING domain in the IAP itself (93, 127), suggesting that binding by the antagonist induces IAP autoubiquitination (95).

Given the complex roles of cIAPs in apoptosis suppression, how do vIAPs with antiapoptotic activity function? vIAPs such as the baculovirus Op-IAP (from *Orgyia pseudotsugata* NPV) do not bind to insect or mammalian caspases (127), nor do they undergo caspase-mediated cleavage (108). As these two events are essential for XIAP- and DIAP1-mediated inhibition of caspases, respectively, it is

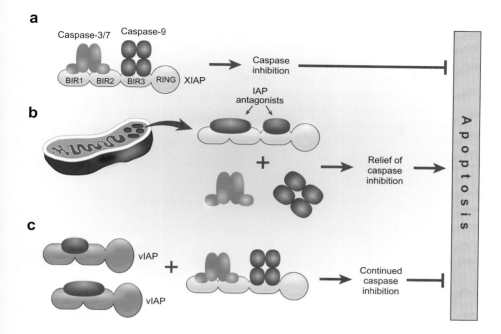

Figure 4

Mechanism of vIAP inhibition of caspase activity. (*a*) Direct inhibition of caspases by cellular XIAP. The BIR3 domain of XIAP binds to caspase-9, and the region proximal to the XIAP BIR2 binds to caspase-3 and caspase-7 and suppresses their activity. (*b*) IAP antagonists (such as SMAC/DIABLO) are released from the mitochondria during apoptosis and competitively bind the IAP, resulting in displacement of active caspases. (*c*) vIAPs function by binding to IAP antagonists, permitting endogenous IAPs such as XIAP to continue to inhibit caspases. Abbreviations: BIR, baculovirus IAP repeat domain; IAP, inhibitor of apoptosis protein; vIAP, viral IAP; XIAP, X-linked IAP.

unlikely that vIAPs function as direct caspase inhibitors. Instead, it is thought that vIAPs act as decoys for the IAP antagonists. Op-IAP contains two BIR domains and one RING domain. The BIR2 domain of Op-IAP can bind to Reaper, HID, and Grim (124–126). Op-IAP can also efficiently bind SMAC/DIABLO via its BIR2 domain, a property required for its ability to prevent apoptosis when expressed in human cells (127). Furthermore, the RING domain of Op-IAP also contributes to IAP function, facilitating ubiquitination of itself and of HID (22, 49) or SMAC/DIABLO (127) and presumably degradation. Thus, Op-IAP and probably vIAPs in general function in an old-fashioned bait-and-switch tactic. They deceive the IAP antagonists and then disable them, per-mitting endogenous IAPs such as DIAP1 to in-hibit caspases (**Figure 4c**).

The observations that cIAPs are conserved in diverse genomes, including flies and humans, and function in apoptosis by conserved mech-anisms suggest that vIAPs might be encoded by a wide variety of viruses. However, such ho-mologs have been found only in viruses that infect arthropods (22). This restricted presence of viral homologs is consistent with the rela-tive role of IAPs in apoptotic programs of in-sects versus humans. The prevailing view is that induction of apoptosis in humans is regulated by the activation of inactive zymogens, whereas apoptosis in *Drosophila* is initiated by the libera-tion of constitutively active caspases from their complex with IAPs (94). Hence, vIAPs may

FADD: Fas-associated
via death domain

impart a greater degree of caspase control if the virus's host is an arthropod rather than a higher vertebrate, thus creating the selective pressure required to retain IAPs as principal modulators of apoptosis. This suggestion is supported by the observation that known genomes of two entomopoxviruses do not encode serpins like their mammalian counterparts, although at least one of these, *Amsacta moorei* (AmEPV), has a functional IAP (71).

Viruses FLIP Out Over the DISC

Signaling through death receptors, including TNF and Fas, generates oligomeric signaling assemblies such as the DISC (reviewed in References 7, 15, and 92) (**Figure 5**). The intracellular region of death receptors contains a DD, which in the case of Fas recruits the Fas-associated via death domain (FADD) adaptor protein via a homotypic interaction with

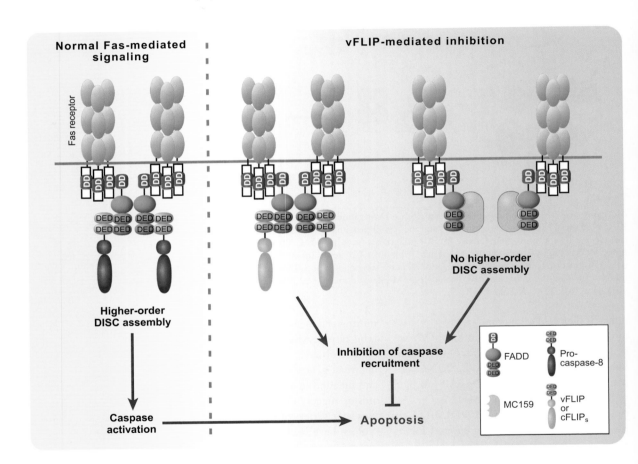

Figure 5

Mechanism of vFLIP-mediated inhibition of caspase activation. Death-receptor induced apoptosis occurs following ligation of the receptor, recruitment of the adaptor protein FADD via its DD and higher-order DISC assembly. Procaspase-8 or procaspase-10 are subsequently recruited and activated. DISC assembly and caspase activation are negatively regulated by cFLIP$_S$ or by γ-herpesvirus vFLIP that also contains DEDs and directly compete with caspase-8 for recruitment to FADD. The molluscum contagiosum virus MC159 protein binds at least two FADD molecules to disrupt FADD oligomerization and higher-order DISC assembly. Abbreviations: vFLIP, viral FLICE (Fas-associated death domain-like interleukin-1β-converting enzyme) inhibitory protein; cFLIP$_S$, short isoform of cellular FLIP; DD, death domain; DED, death effector domain; DISC, death induced signaling complex; FADD, Fas-associated via death domain.

the DD of FADD. FADD contains an additional DED that interacts with DEDs in the prodomains of caspase-8 and caspase-10. The recruitment of these initiator caspases into the DISC results in their dimerization and maturation. During this activation process, caspase N-terminal DEDs are removed by intermolecular processing, resulting in the release of caspases into the cytosol where they activate effector caspases (24).

DISC assembly and caspase activation are negatively regulated by cellular FLIP (cFLIP) that directly competes with caspase-8 for recruitment to FADD (65) (**Figure 5**). Two major isoforms of cFLIP exist, the 26-kDa short form (cFLIP$_S$) and the 55-kDa long form (cFLIP$_L$) (112). cFLIP$_S$ contains two DEDs and completely inhibits proteolytic processing of caspase-8 (65). In contrast, in addition to two DEDs, cFLIP$_L$ contains a longer C terminus than does cFLIP$_S$, which closely resembles the overall structure of caspase-8 and caspase-10 (65). cFLIP$_L$ can form a heterodimer with caspase-8 through interactions between the DEDs and caspase-like domains. This association causes a conformational change in caspase-8, exposing the caspase active site (19, 81). The result is a partial autoprocessing of caspase-8 that in turn cleaves cFLIP$_L$ (19, 82). Cleavage of cFLIP$_L$ augments its ability to recruit TNF-receptor-associated factor 2 (TRAF2) and receptor-interacting protein 1 (RIP1) (60). These events facilitate NF-κB activation, which is the quintessential role of TNF signaling. A similar partial activation of caspase-8 may also be important in T-cell activation (reviewed in Reference 15).

vFLIPs (111) are encoded by several γ-herpesviruses and by the molluscum contagiosum virus (MCV) (10, 10, 55, 112). vFLIPs contain two DED domains and, similar to their cellular counterparts, interfere with recruitment, processing, and release of caspase-8 and caspase-10 following ligation of death receptors. However, the precise mechanism utilized by various viruses differs, as illustrated by their interactions with Fas receptor-induced DISC formation. vFLIP encoded by γ-herpesviruses

is analogous to cFLIP$_S$ outcompeting caspases for recruitment to FADD (132) (**Figure 5**). In contrast, the MCV genome contains two vFLIPs, termed MC159 and MC160. MC159 interacts with both FADD and Fas (100) but does not compete with caspase-8 for recruitment. Instead, it binds at least two FADD molecules and disrupts FADD oligomerization (132). Hence, caspase inhibition results from the absence of FADD self-association and DISC formation (132) (**Figure 5**). In addition, MCV FLIPs are unique in that they have C-terminal extensions responsible for binding TRAF2 and TRAF3 and recruiting them to the Fas DISC in a FasL-dependent manner. The binding of TRAF3 in particular contributes to inhibition of Fas receptor internalization (114), a step that occurs after DISC formation and caspase-8 activation and is required for optimal induction of apoptosis. MCV-encoded vFLIPs also interact with proteins that regulate NF-κB activation including TRAFs, RIP1, NF-κB-inducing kinase (NIK), and inhibitor of κB-kinase 2 (IKK2) (20). Thus, MCV vFLIPs may impart multiple blocks following death receptor ligation, resulting in the inhibition of initiator caspase activation as well as the promotion of cellular survival signals such as NF-κB activation.

Although not a vFLIP, the E6 protein of human papillomavirus type 16 (HPV-16) also targets DISC assembly at multiple levels to suppress death receptor-induced apoptosis. The long isoform of E6 can bind to the DD of TNF receptor 1 and interfere with the recruitment of TNF receptor-associated via death domain (TRADD) and therefore DISC formation (41). In addition, E6 can bind to the DEDs of both FADD and procaspase-8. The consequences of this binding are twofold. First, E6 prevents the normal association of caspase-8 with FADD, and second, it accelerates the degradation of both molecules (40, 41) and prevents DISC formation. Furthermore, the large isoform of E6 accelerates degradation of the tumor suppressor p53 (40, 97), thus rendering cells less sensitive to apoptosis initiated by genotoxic stress. E6 is therefore a multifunctional protein (with functions additional to those described here) that blocks numerous initiating signals of apoptosis. Finally, the human cytomegalovirus UL36 gene product termed vICA also associates with procaspase-8 and blocks its activation, although it possesses no sequence homology with vFLIPs (102).

EVADING SERINE PROTEASES: THRUST AND PARRY

Effective antiviral immune responses involve the generation of virus-specific cytotoxic T lymphocytes (CTLs), whose function is to kill virus-infected cells. The two main weapons utilized by CTLs are perforin-/granzyme- and Fas-mediated apoptosis. Following recognition of virus-infected cells, perforin and granzymes are released by CTLs and are endocytosed by the target cell (5, 73). Perforin then polymerizes and inserts into the membrane of the endocytic vesicle, forming a pore through which granzymes including granzyme B are transported into the cytosol (73). Granzyme B is a unique serine protease owing to its strict requirement for Asp in the substrate P_1 position. Apoptosis is initiated following direct cleavage and activation of caspase-3 by granzyme B, or following cleavage and activation of Bid (8, 51, 73). Truncated Bid translocates to the mitochondria and recruits Bax to facilitate alterations in membrane permeabilization, release of cytochrome c, and activation of downstream caspases (8, 51).

In addition to potently inhibiting caspases, CrmA can antagonize the proteolytic activity of granzyme B (88). Granzyme B is structurally and mechanistically distinct from the caspases, demonstrating a remarkable cross-class ability of CrmA to inhibit proteolysis (64). The 100,000-kDa (100 K) assembly protein of human adenovirus type 5 (Ad5-100K) is also a strong inhibitor of granzyme B (3, 5). The Ad5-100K protein has several essential functions in the adenovirus life cycle, including virus assembly, activation of late viral protein synthesis, and inhibition of cellular protein synthesis (53). Inhibition of granzyme B is absolutely depen-

dent on Asp[48] within a granzyme B consensus recognition sequence in the Ad5-100K protein (3). Although granzyme B from multiple species (human, rat, and mouse) can cleave Ad5-100K, these proteins cleave at different sites within the viral protein and only activity of human granzyme B is inhibited (4). The precise molecular mechanisms by which Ad5-100K prevents granzyme B activity are not known, although at least two intermolecular interactions are required. These are between the viral protein RSL and the protease active site, resulting in the cleavage of the former, as well as an additional interaction involving the extreme C-terminal residues of Ad5-100K (residues 688–781) that may stabilize interactions between protease and inhibitor (4).

The fencing match between host and virus does not end following the initial thrust with granzyme B and the parry of the adenovirus inhibitor. The human host has a riposte in the form of granzyme H that is also delivered to cells by activated CTLs. Granzyme H cleaves and inactivates Ad5-100K, as well as the adenoviral DNA binding protein (6). Cleavage of these proteins not only significantly restricts virus replication, but also relieves virus-mediated inhibition of granzyme B and restores its ability to process caspase-3 (6). Granzyme H does not have orthologs in species other than humans, and the inhibitory activity of Ad5-100K is specifically directed toward human granzyme B. Thus, despite generally well-conserved processes of apoptosis, these findings suggest a remarkably complex evolutionary relationship between the induction and suppression of host cell apoptosis and virus replication.

Viruses target proapoptotic serine proteases other than granzymes to suppress the cell death response. The high-temperature requirement protein A2 (HtrA2/Omi) is a serine protease released from the mitochondrial inner membrane space along with cytochrome c and SMAC/DIABLO following membrane permeabilization (50). Once released, HtrA2/Omi can degrade XIAP and thereby relieve its inhibition of caspases (105). When used in combination as clinical therapy for some tumors, interferon (IFN) and retinoic acid promote apoptosis in part through the upregulation of genes associated with retinoid-interferon-induced mortality (GRIM) protein expression (59). In particular, GRIM-19 associates with HtrA2/Omi and augments XIAP degradation (75). The Kaposi's sarcoma–associated herpesvirus oncoprotein termed viral IFN-regulatory factor 1 (vIRF1) binds to GRIM-19 and prevents its interaction with HtrA2/Omi (75). This preserves cellular XIAP levels in response to IFN and retinoic acid and ablates caspase-9 activation.

The original function identified for vIRF1 was a suppressor of IFN signaling (141), a potent antiviral response that viruses devote large amounts of their genomes to evade. It is well known that IFN stimulation can eventually lead to apoptosis through the increased expression of proapoptotic genes. However, the targeting of GRIM-19 and HtrA2/Omi by a viral homolog of the IFN signal transduction pathway suggests that there may be direct regulation, or cross-talk, between the two cellular pathways. The recent identification of HtrA2/Omi as a mitochondria-resident protein with roles in apoptosis in *Drosophila* (18) suggests a level of conservation that facilitates virus coevolution. Therefore, it is of interest to determine if additional oncogenic viruses, which may persist in the presence of chronic expression of proapoptotic cytokines such as IFN, also target HtrA2/Omi to suppress apoptosis.

WHEN RESISTANCE IS FUTILE: UTILIZATION OF CASPASES FOR VIRUS REPLICATION

The combined efforts of viruses to suppress apoptosis in general and caspase activity in particular suggest that this cellular response is to be avoided at all costs. However, caspases also have roles in cell proliferation, differentiation, and NF-κB activation (69). These roles for caspase activation raise the question of whether viruses can positively modulate or utilize caspase activity to facilitate replication. The role of the HPV-16 E6 long isoform in caspase-8

degradation has already been highlighted. The E6 short isoform, which is only produced by so-called high-risk HPV types, can also bind to caspase-8. However, this molecular interaction results in caspase-8 stabilization (40). The implications from these findings are that, like inhibition of caspase activity, stabilization of limited caspase-8 activity may be important for cell survival or maintaining a cell at a particular stage of differentiation. Thus, caspases may be targeted by certain viruses to prolong cell viability or aid in cellular transformation.

Examples of viruses that directly utilize caspase activity to facilitate replication also exist. Permissive replication of Aleutian mink disease parvovirus (ADV) in cell culture is associated with an apoptotic response that can be blocked by treatment with caspase inhibitors. There are numerous examples of virus infection in which blockage of apoptosis increases the virus titer recovered from treated cells, consistent with the antiviral role of apoptosis (23, 38, 137). However, in the case of ADV, inhibition of specific caspases results in decreased virus yield, suggesting that caspase activity facilitates virus replication (13). The major nonstructural protein of ADV, NS1, is required for many replication functions including control of viral and cellular gene transcription, viral DNA replication, and capsid assembly. Caspases mediate cleavage of NS1 early in virus replication, particularly by effector caspase-3, caspase-7, or both (12; S.M. Best, unpublished data). In the absence of cleavage, translocation of NS1 to the site of virus replication in the nucleus is impaired. It appears that the role of caspases is to generate a C-terminal cleavage product that contains the NS1 nuclear localization sequence. This product forms oligomers with full-length NS1 and facilitates transport of the latter to the nucleus (M.E. Bloom & S.M. Best, unpublished data).

Although a clear requirement exists for caspases in ADV replication, it is currently unknown how this contributes to virus pathogenesis. In infected mink kits, permissive replication of ADV occurs in pulmonary type II pneumocytes, resulting in high levels of cytopathology and mortality. In contrast, infection of adult mink results in restricted replication in lymph node macrophages and persistent infection. It is possible that cell-type-specific regulation of caspase activity modulates the degree of NS1 nuclear translocation and hence its function in DNA replication and viral gene expression. In adult mink, for example, tight regulation of caspase activity in macrophages may restrict nuclear translocation of NS1 and virus replication, contributing to persistent infection. However, apoptosis may be triggered during virus infection of pneumocytes, resulting in elevated caspase activation and permissive replication in mink kits.

Adenoviral proteins are also cleaved during virus replication. The early transcription units E1A (which encodes two proteins, 12S and 13S) and E1B encode proteins involved in transactivation of both cellular and viral transcription as well as cellular transformation. Following transient expression of the E1A proteins from Ad2 or Ad12, caspases cleave both 12S and 13S at multiple sites, resulting in progressive truncation from the N termini (48). This cleavage disrupts interactions of E1A with cellular transcription-regulating proteins that bind to the N terminus but not to the C terminus (48). Thus, caspases have a potential regulatory role in E1A-mediated gene expression. E1A is also proteolytically cleaved by caspases during Ad5 replication. However, cleavage of E1A proteins, as well as cellular proteins normally cleaved during apoptosis, is limited and dependent on cell type (48). The degree of cleavage is likely influenced by temporal expression of adenovirus-encoded inhibitors of apoptosis such as E3, the protein responsible for the downregulation of cell surface death receptors (116). Thus, an intricate interplay between caspase activation and inhibition in a particular cell type may be required for optimal virus replication.

Additional examples of viruses that utilize caspase activity include human astroviruses that cause viral gasteroenteritis. In this case, caspase-mediated cleavage of the capsid precursor protein facilitates the release of viral particles from the cell (80). Other examples of virus-encoded proteins cleaved by caspases include NS5A of hepatitis C virus (46, 58, 96), the nucleocapsid protein of influenza A virus (138), and ICP22 of herpes simplex virus (85), although the role of protein cleavage in replication of these viruses is unclear (11). Cleavage may facilitate protein degradation and negatively impact virus replication as described for the granzyme H-mediated cleavage of the Ad5-100K protein. However, because mutation of the crucial Asp residue would eliminate the enzyme recognition site, conservation of these sequences suggests a selection for caspase-mediated cleavage in virus replication (11). Hence, in addition to facilitating virus release, caspases may have roles in regulation of virus protein maturation and function as suggested by studies using ADV and adenoviruses.

CLOSING REMARKS

To prolong cell viability after infection, the function of enzymes essential for the promotion of host cell death, such as caspases and serine proteases, are obvious targets for suppression by viruses. Although not discussed here, additional proteases including cathepsins, calpains (104, 117, 119), phosphatidylinositol 3-kinase (PI3K) (25), mitogen-activated protein kinases (MAPK) (57), and protein kinase R (PKR) (43) have important roles in regulating apoptosis. PI3K, MAPK and PKR are all targeted by various viruses to modulate cell death. It is possible that other proteases like the cathepsins or calpains are also targeted by viruses for this purpose.

The discovery of virus-encoded inhibitors of apoptotic proteases has been invaluable, both in our understanding of virus requirements for replication and for tools used to study apoptosis. In this arena, CrmA, p35, vIAPs, and vFLIPs have yielded enormous insight into apoptotic signaling pathways stimulated by a plethora of insults. These proteins also have potential use as therapeutics to treat conditions in which apoptosis is a major contributor to disease. Owing to the observation that many viruses encode several inhibitors of apoptosis and that individual proteins have multiple functions, the full spectrum of viral protein activity in modulating cell death is probably not fully realized. Furthermore, caspases have roles in nonapoptotic cellular processes (68, 69) as well as in the facilitation of virus replication (11). Thus, it is likely that further knowledge of both cell death and cell survival programs will be gained from understanding specific mechanisms by which virus proteins modulate these enzymes.

Taken together, the studies highlighted in this review suggest that complex relationships exist between enzyme activity, virus replication, and cell survival and transformation. Furthermore, although general mechanisms of apoptosis are conserved from *Drosophila* to humans, many of the viral proteins that modulate enzyme activity during apoptosis do so in a host-specific context, including CrmA, p35 and p49, vIAPs, and Ad5-100K. Combined with the fact that apoptosis is a strong antiviral response, these studies suggest that virus-encoded proteins are important contributors to virus cell tropism and host range specificity. Thus, challenges remain in understanding how these multifaceted virus-host interactions within the individual cell contribute to the outcome of virus infection at the level of the organism.

SUMMARY POINTS

1. Viruses suppress the function of proteases with central roles in promoting apoptosis to prolong cell viability and facilitate replication. The functions of caspases and serine proteases are most commonly modulated by viruses to suppress apoptosis.

2. Four classes of viral proteins directly suppress caspase function. Two of these, serpins and p35 family members, have in common a RSL that is inserted into the caspase active site. However, serpins function by conformational distortion of the caspase active site, whereas p35 engages the active site in a process of chemical ligation.

3. Two other classes of virus inhibitors suppress caspase activity by competing with signaling molecules involved in caspase activation. vIAPs act as decoys for cellular IAP antagonists, thus enabling cellular IAPs to inhibit caspase activity. vFLIPs prevent DISC assembly in response to ligation of death receptors, resulting in suppression of caspase-8 and caspase-10 activation.

4. Viral gene products also antagonize the proapoptotic function of the serine proteases, granzyme B and Htr2A/Omi.

5. Despite the antiviral role of apoptosis, some viruses appear to exploit this response by utilizing caspase activity to cleave viral proteins and facilitate replication. Thus, complex relationships exist between caspase activity, virus replication, and cell survival.

FUTURE ISSUES

1. Elucidation of protein structures in complex with their targets has provided enormous insight into distinct mechanisms utilized by p35 to suppress apoptosis. Additional structural studies, for example, of Ad5-100K in complex with granzyme B, are required to fully elucidate the mechanisms underlying virus protein function in apoptosis suppression.

2. The diverse functions of cIAPs in the regulation of apoptosis are beginning to be revealed. Hence, functions of vIAPs in addition to their ability to subvert cellular IAP antagonists may exist and remain to be identified.

3. In examples in which viruses utilize caspases for replication, little is known regarding how caspase activity is positively regulated while limited to complete replication. Thus, the temporal relationships between viral pro- and antiapoptotic signals are poorly understood. Furthermore, it is unknown how caspase-dependent replication contributes to virus pathogenesis.

4. A major challenge continues to be converting knowledge of viral protease inhibitors into viable therapeutics to treat conditions involving aberrant apoptosis or inflammation.

DISCLOSURE STATEMENT

The author is not aware of any biases that might be perceived as affecting the objectivity of this review.

ACKNOWLEDGMENTS

Thank you to Marshall Bloom, Dana Mitzel, Shelly Robertson, Travis Taylor, and Dan Voth for critique of the manuscript and to Anita Mora for graphical expertise. This work was supported by the Intramural Research Program of the National Institute of Allergy and Infectious Diseases, National Institutes of Health.

LITERATURE CITED

1. Alnemri ES, Livingston DJ, Nicholson DW, Salvesen G, Thornberry NA, et al. 1996. Human ICE/CED-3 protease nomenclature. *Cell* 87:171
2. Andoniou CE, Degli-Esposti MA. 2006. Insights into the mechanisms of CMV-mediated interference with cellular apoptosis. *Immunol. Cell Biol.* 84:99–106
3. **Andrade F, Bull HG, Thornberry NA, Ketner GW, Casciola-Rosen LA, Rosen A. 2001. Adenovirus L4-100K assembly protein is a granzyme B substrate that potently inhibits granzyme B-mediated cell death. *Immunity.* 14:751–61**

3. Together with Reference 6 reports the interplay between Ad5-100K and granzymes B and H.

4. Andrade F, Casciola-Rosen LA, Rosen A. 2003. A novel domain in adenovirus L4-100K is required for stable binding and efficient inhibition of human granzyme B: possible interaction with a species-specific exosite. *Mol. Cell Biol.* 23:6315–26
5. Andrade F, Casciola-Rosen LA, Rosen A. 2004. Granzyme B-induced cell death. *Acta Haematol.* 111:28–41
6. Andrade F, Fellows E, Jenne DE, Rosen A, Young CS. 2007. Granzyme H destroys the function of critical adenoviral proteins required for viral DNA replication and granzyme B inhibition. *EMBO J.* 26:2148–57
7. Bao Q, Shi Y. 2007. Apoptosome: a platform for the activation of initiator caspases. *Cell Death Differ.* 14:56–65
8. Barry M, Heibein JA, Pinkoski MJ, Lee SF, Moyer RW, et al. 2000. Granzyme B short-circuits the need for caspase 8 activity during granule-mediated cytotoxic T-lymphocyte killing by directly cleaving Bid. *Mol. Cell Biol.* 20:3781–94
9. Benedict CA, Norris PS, Ware CF. 2002. To kill or be killed: viral evasion of apoptosis. *Nat. Immunol.* 3:1013–18

10. Together with Reference 111 provided the first description of vFLIPs before the discovery of cFLIP.

10. **Bertin J, Armstrong RC, Ottilie S, Martin DA, Wang Y, et al. 1997. Death effector domain-containing herpesvirus and poxvirus proteins inhibit both Fas- and TNFR1-induced apoptosis. *Proc. Natl. Acad. Sci. USA* 94:1172–76**
11. Best SM, Bloom ME. 2004. Caspase activation during virus infection: more than just the kiss of death? *Virology* 320:191–94

12. Best SM, Shelton JF, Pompey JM, Wolfinbarger JB, Bloom ME. 2003. Caspase cleavage of the non-structural protein NS1 mediates replication of Aleutian mink disease parvovirus. *J. Virol.* 77:5305–12

13. Best SM, Wolfinbarger JB, Bloom ME. 2002. Caspase activation is required for permissive replication of Aleutian mink disease parvovirus in vitro. *Virology* 292:224–34

14. Boya P, Pauleau AL, Poncet D, Gonzalez-Polo RA, Zamzami N, Kroemer G. 2004. Viral proteins targeting mitochondria: controlling cell death. *Biochim. Biophys. Acta* 1659:178–89

15. Budd RC, Yeh WC, Tschopp J. 2006. cFLIP regulation of lymphocyte activation and development. *Nat. Rev. Immunol.* 6:196–204

16. Callus BA, Vaux DL. 2007. Caspase inhibitors: viral, cellular and chemical. *Cell Death Differ.* 14:73–78

17. Cassens U, Lewinski G, Samraj AK, von BH, Baust H, et al. 2003. Viral modulation of cell death by inhibition of caspases. *Arch. Immunol. Ther. Exp.* 51:19–27

18. Challa M, Malladi S, Pellock BJ, Dresnek D, Varadarajan S, et al. 2007. *Drosophila* Omi, a mitochondrial-localized IAP antagonist and proapoptotic serine protease. *EMBO J.* 26:3144–56

19. Chang DW, Xing Z, Pan Y, Geciras-Schimnich A, Barnhart BC, et al. 2002. c-FLIP(L) is a dual function regulator for caspase-8 activation and CD95-mediated apoptosis. *EMBO J.* 21:3704–14

20. Chaudhary PM, Jasmin A, Eby MT, Hood L. 1999. Modulation of the NF-kappa B pathway by virally encoded death effector domains-containing proteins. *Oncogene* 18:5738–46

21. Clem RJ. 2005. The role of apoptosis in defense against baculovirus infection in insects. *Curr. Top. Microbiol. Immunol.* 289:113–29

22. Clem RJ. 2007. Baculoviruses and apoptosis: a diversity of genes and responses. *Curr. Drug Targets* 8:1069–74

23. Clem RJ, Fechheimer M, Miller LK. 1991. Prevention of apoptosis by a baculovirus gene during infection of insect cells. *Science* 254:1388–90

24. Cohen GM. 1997. Caspases: the executioners of apoptosis. *Biochem. J.* 326(Pt. 1):1–16

25. Cooray S. 2004. The pivotal role of phosphatidylinositol 3-kinase-Akt signal transduction in virus survival. *J. Gen. Virol.* 85:1065–76

26. Crook NE, Clem RJ, Miller LK. 1993. An apoptosis-inhibiting baculovirus gene with a zinc finger-like motif. *J. Virol.* 67:2168–74

27. Datta R, Kojima H, Banach D, Bump NJ, Talanian RV, et al. 1997. Activation of a CrmA-insensitive, p35-sensitive pathway in ionizing radiation-induced apoptosis. *J. Biol. Chem.* 272:1965–69

28. Degterev A, Boyce M, Yuan J. 2003. A decade of caspases. *Oncogene* 22:8543–67

29. Delhon G, Tulman ER, Afonso CL, Lu Z, Becnel JJ, et al. 2006. Genome of invertebrate iridescent virus type 3 (mosquito iridescent virus). *J. Virol.* 80:8439–49

30. Deveraux QL, Leo E, Stennicke HR, Welsh K, Salvesen GS, Reed JC. 1999. Cleavage of human inhibitor of apoptosis protein XIAP results in fragments with distinct specificities for caspases. *EMBO J.* 18:5242–51

31. Dobbelstein M, Shenk T. 1996. Protection against apoptosis by the vaccinia virus SPI-2 (B13R) gene product. *J. Virol.* 70:6479–85

32. Dobo J, Swanson R, Salvesen GS, Olson ST, Gettins PG. 2006. Cytokine response modifier a inhibition of initiator caspases results in covalent complex formation and dissociation of the caspase tetramer. *J. Biol. Chem.* 281:38781–90

33. Du C, Fang M, Li Y, Li L, Wang X. 2000. Smac, a mitochondrial protein that promotes cytochrome c-dependent caspase activation by eliminating IAP inhibition. *Cell* 102:33–42

34. Du Q, Lehavi D, Faktor O, Qi Y, Chejanovsky N. 1999. Isolation of an apoptosis suppressor gene of the *Spodoptera littoralis* nucleopolyhedrovirus. *J. Virol.* 73:1278–85

35. Eckelman BP, Salvesen GS. 2006. The human antiapoptotic proteins cIAP1 and cIAP2 bind but do not inhibit caspases. *J. Biol. Chem.* 281:3254–60

36. Eckelman BP, Salvesen GS, Scott FL. 2006. Human inhibitor of apoptosis proteins: why XIAP is the black sheep of the family. *EMBO Rep.* 7:988–94

37. Elliott PR, Lomas DA, Carrell RW, Abrahams JP. 1996. Inhibitory conformation of the reactive loop of alpha 1-antitrypsin. *Nat. Struct. Biol.* 3:676–81

38. Everett H, McFadden G. 1999. Apoptosis: an innate immune response to virus infection. *Trends Microbiol.* 7:160–65

23. The first description of p35.

26. The first description of an IAP as a viral product before the discovery of cIAPs.

39. Everett H, McFadden G. 2001. Viruses and apoptosis: meddling with mitochondria. *Virology* 288:1–7

40. Filippova M, Johnson MM, Bautista M, Filippov V, Fodor N, et al. 2007. The large and small isoforms of human papillomavirus type 16 E6 bind to and differentially affect procaspase 8 stability and activity. *J. Virol.* 81:4116–29

41. Filippova M, Parkhurst L, Duerksen-Hughes PJ. 2004. The human papillomavirus 16 E6 protein binds to Fas-associated death domain and protects cells from Fas-triggered apoptosis. *J. Biol. Chem.* 279:25729–44

42. Fuentes-Prior P, Salvesen GS. 2004. The protein structures that shape caspase activity, specificity, activation and inhibition. *Biochem. J.* 384:201–32

43. Garcia MA, Meurs EF, Esteban M. 2007. The dsRNA protein kinase PKR: virus and cell control. *Biochimie* 89:799–811

44. Garcia-Calvo M, Peterson EP, Leiting B, Ruel R, Nicholson DW, Thornberry NA. 1998. Inhibition of human caspases by peptide-based and macromolecular inhibitors. *J. Biol. Chem.* 273:32608–13

45. Garrido C, Galluzzi L, Brunet M, Puig PE, Didelot C, Kroemer G. 2006. Mechanisms of cytochrome c release from mitochondria. *Cell Death Differ.* 13:1423–33

46. Goh PY, Tan YJ, Lim SP, Lim SG, Tan YH, Hong WJ. 2001. The hepatitis C virus core protein interacts with NS5A and activates its caspase-mediated proteolytic cleavage. *Virology* 290:224–36

47. Goldmacher VS. 2005. Cell death suppression by cytomegaloviruses. *Apoptosis* 10:251–65

48. Grand RJ, Schmeiser K, Gordon EM, Zhang X, Gallimore PH, Turnell AS. 2002. Caspase-mediated cleavage of adenovirus early region 1A proteins. *Virology* 301:255–71

49. Green MC, Monser KP, Clem RJ. 2004. Ubiquitin protein ligase activity of the antiapoptotic baculovirus protein Op-IAP3. *Virus Res.* 105:89–96

50. Hegde R, Srinivasula SM, Zhang Z, Wassell R, Mukattash R, et al. 2002. Identification of Omi/HtrA2 as a mitochondrial apoptotic serine protease that disrupts inhibitor of apoptosis protein-caspase interaction. *J. Biol. Chem.* 277:432–38

51. Heibein JA, Goping IS, Barry M, Pinkoski MJ, Shore GC, et al. 2000. Granzyme B-mediated cytochrome c release is regulated by the Bcl-2 family members Bid and Bax. *J. Exp. Med.* 192:1391–402

52. Hinds MG, Norton RS, Vaux DL, Day CL. 1999. Solution structure of a baculoviral inhibitor of apoptosis (IAP) repeat. *Nat. Struct. Biol.* 6:648–51

53. Hodges BL, Evans HK, Everett RS, Ding EY, Serra D, Amalfitano A. 2001. Adenovirus vectors with the 100K gene deleted and their potential for multiple gene therapy applications. *J. Virol.* 75:5913–20

54. Holley CL, Olson MR, Colon-Ramos DA, Kornbluth S. 2002. Reaper eliminates IAP proteins through stimulated IAP degradation and generalized translational inhibition. *Nat. Cell Biol.* 4:439–44

55. Hu S, Vincenz C, Buller M, Dixit VM. 1997. A novel family of viral death effector domain-containing molecules that inhibit both CD-95- and tumor necrosis factor receptor-1-induced apoptosis. *J. Biol. Chem.* 272:9621–24

56. Jabbour AM, Ekert PG, Coulson EJ, Knight MJ, Ashley DM, Hawkins CJ. 2002. The p35 relative, p49, inhibits mammalian and *Drosophila* caspases including DRONC and protects against apoptosis. *Cell Death Differ.* 9:1311–20

57. Junttila MR, Li SP, Westermarck J. 2008. Phosphatase-mediated crosstalk between MAPK signaling pathways in the regulation of cell survival. *FASEB J.* 22:1–12

58. Kalamvoki M, Georgopoulou U, Mavromara P. 2006. The NS5A protein of the hepatitis C virus genotype 1a is cleaved by caspases to produce C-terminal-truncated forms of the protein that reside mainly in the cytosol. *J. Biol. Chem.* 281:13449–62

59. Kalvakolanu DV. 2004. The GRIMs: a new interface between cell death regulation and interferon/retinoid induced growth suppression. *Cytokine Growth Factor Rev.* 15:169–94

60. Kataoka T, Tschopp J. 2004. N-terminal fragment of c-FLIP(L) processed by caspase 8 specifically interacts with TRAF2 and induces activation of the NF-kappaB signaling pathway. *Mol. Cell Biol.* 24:2627–36

61. Keckler MS. 2007. Dodging the CTL response: viral evasion of Fas and granzyme induced apoptosis. *Front. Biosci.* 12:725–32

62. Kidd VJ, Lahti JM, Teitz T. 2000. Proteolytic regulation of apoptosis. *Semin. Cell Dev. Biol.* 11:191–201

63. Kischkel FC, Hellbardt S, Behrmann I, Germer M, Pawlita M, et al. 1995. Cytotoxicity-dependent APO-1 (Fas/CD95)-associated proteins form a death-inducing signaling complex (DISC) with the receptor. *EMBO J.* 14:5579–88

40. Provides insight into the role of HPV E6 in modulation of cell survival and transformation.

64. Komiyama T, Ray CA, Pickup DJ, Howard AD, Thornberry NA, et al. 1994. Inhibition of interleukin-1 beta converting enzyme by the cowpox virus serpin CrmA. An example of cross-class inhibition. *J. Biol. Chem.* 269:19331–37

65. Krueger A, Schmitz I, Baumann S, Krammer PH, Kirchhoff S. 2001. Cellular FLICE-inhibitory protein splice variants inhibit different steps of caspase-8 activation at the CD95 death-inducing signaling complex. *J. Biol. Chem.* 276:20633–40

66. Kumar S. 2007. Caspase function in programmed cell death. *Cell Death Differ.* 14:32–43

67. LaCount DJ, Hanson SF, Schneider CL, Friesen PD. 2000. Caspase inhibitor P35 and inhibitor of apoptosis Op-IAP block in vivo proteolytic activation of an effector caspase at different steps. *J. Biol. Chem.* 275:15657–64

68. Lamkanfi M, Declercq W, Vanden BT, Vandenabeele P. 2006. Caspases leave the beaten track: caspase-mediated activation of NF-kappaB. *J. Cell Biol.* 173:165–71

69. Lamkanfi M, Festjens N, Declercq W, Vanden BT, Vandenabeele P. 2007. Caspases in cell survival, proliferation and differentiation. *Cell Death Differ.* 14:44–55

70. Lannan E, Vandergaast R, Friesen PD. 2007. Baculovirus caspase inhibitors P49 and P35 block virus-induced apoptosis downstream of effector caspase DrICE activation in *Drosophila melanogaster* cells. *J. Virol.* 81:9319–30

71. Li Q, Liston P, Moyer RW. 2005. Functional analysis of the inhibitor of apoptosis (iap) gene carried by the entomopoxvirus of *Amsacta moorei*. *J. Virol.* 79:2335–45

72. Liu Z, Sun C, Olejniczak ET, Meadows RP, Betz SF, et al. 2000. Structural basis for binding of Smac/DIABLO to the XIAP BIR3 domain. *Nature* 408:1004–8

73. Lord SJ, Rajotte RV, Korbutt GS, Bleackley RC. 2003. Granzyme B: a natural born killer. *Immunol. Rev.* 193:31–38

74. Lu M, Min T, Eliezer D, Wu H. 2006. Native chemical ligation in covalent caspase inhibition by p35. *Chem. Biol.* 13:117–22

75. Ma X, Kalakonda S, Srinivasula SM, Reddy SP, Platanias LC, Kalvakolanu DV. 2007. GRIM-19 associates with the serine protease HtrA2 for promoting cell death. *Oncogene* 26:4842–49

76. Macen JL, Garner RS, Musy PY, Brooks MA, Turner PC, et al. 1996. Differential inhibition of the Fas- and granule-mediated cytolysis pathways by the orthopoxvirus cytokine response modifier A/SPI-2 and SPI-1 protein. *Proc. Natl. Acad. Sci. USA* 93:9108–13

77. MacNeill AL, Turner PC, Moyer RW. 2006. Mutation of the *Myxoma* virus SERP2 P1-site to prevent proteinase inhibition causes apoptosis in cultured RK-13 cells and attenuates disease in rabbits, but mutation to alter specificity causes apoptosis without reducing virulence. *Virology* 356:12–22

78. Means JC, Penabaz T, Clem RJ. 2007. Identification and functional characterization of AMVp33, a novel homolog of the baculovirus caspase inhibitor p35 found in *Amsacta moorei* entomopoxvirus. *Virology* 358:436–47

79. Meier P, Silke J, Leevers SJ, Evan GI. 2000. The *Drosophila* caspase DRONC is regulated by DIAP1. *EMBO J.* 19:598–611

80. Mendez E, Salas-Ocampo E, Arias CF. 2004. Caspases mediate processing of the capsid precursor and cell release of human astroviruses. *J. Virol.* 78:8601–8

81. Micheau O, Thome M, Schneider P, Holler N, Tschopp J, et al. 2002. The long form of FLIP is an activator of caspase-8 at the Fas death-inducing signaling complex. *J. Biol. Chem.* 277:45162–71

82. Micheau O, Tschopp J. 2003. Induction of TNF receptor I-mediated apoptosis via two sequential signaling complexes. *Cell* 114:181–90

83. Miller LK. 1999. An exegesis of IAPs: salvation and surprises from BIR motifs. *Trends Cell Biol.* 9:323–28

84. Miura M, Zhu H, Rotello R, Hartwieg EA, Yuan J. 1993. Induction of apoptosis in fibroblasts by IL-1 beta-converting enzyme, a mammalian homolog of the *C. elegans* cell death gene *ced-3*. *Cell* 75:653–60

85. Munger J, Hagglund R, Roizman B. 2003. Infected cell protein no. 22 is subject to proteolytic cleavage by caspases activated by a mutant that induces apoptosis. *Virology* 305:364–70

86. Nathaniel R, MacNeill AL, Wang YX, Turner PC, Moyer RW. 2004. Cowpox virus CrmA, myxoma virus SERP2 and baculovirus P35 are not functionally interchangeable caspase inhibitors in poxvirus infections. *J. Gen. Virol.* 85:1267–78

74. Reports the structure of p35 complexed with caspase-8 and demonstrates inhibition by chemical ligation.

84. The original demonstration that caspase-1 could induce apoptosis and that CrmA could inhibit it.

87. Petit F, Bertagnoli S, Gelfi J, Fassy F, Boucraut-Baralon C, Milon A. 1996. Characterization of a myxoma virus-encoded serpin-like protein with activity against interleukin-1 beta-converting enzyme. *J. Virol.* 70:5860–66

88. Quan LT, Caputo A, Bleackley RC, Pickup DJ, Salvesen GS. 1995. Granzyme B is inhibited by the cowpox virus serpin cytokine response modifier A. *J. Biol. Chem.* 270:10377–79

89. Ray CA, Black RA, Kronheim SR, Greenstreet TA, Sleath PR, et al. 1992. Viral inhibition of inflammation: cowpox virus encodes an inhibitor of the interleukin-1 beta converting enzyme. *Cell* 69:597–604

90. The original description of CrmA.

90. Reed JC, Doctor KS, Godzik A. 2004. The domains of apoptosis: a genomics perspective. *Sci. STKE* 2004:re9

91. Renatus M, Zhou Q, Stennicke HR, Snipas SJ, Turk D, et al. 2000. Crystal structure of the apoptotic suppressor CrmA in its cleaved form. *Structure* 8:789–97

91. Together with Reference 101 reports the crystal structure of cleaved CrmA.

92. Riedl SJ, Salvesen GS. 2007. The apoptosome: signalling platform of cell death. *Nat. Rev. Mol. Cell Biol.* 8:405–13

93. Ryoo HD, Bergmann A, Gonen H, Ciechanover A, Steller H. 2002. Regulation of *Drosophila* IAP1 degradation and apoptosis by reaper and ubcD1. *Nat. Cell Biol.* 4:432–38

94. Salvesen GS, Abrams JM. 2004. Caspase activation: stepping on the gas or releasing the brakes? Lessons from humans and flies. *Oncogene* 23:2774–84

95. Salvesen GS, Duckett CS. 2002. IAP proteins: blocking the road to death's door. *Nat. Rev. Mol. Cell Biol.* 3:401–10

96. Satoh S, Hirota M, Noguchi T, Hijikata M, Handa H, Shimotohno K. 2000. Cleavage of hepatitis C virus nonstructural protein 5A by a caspase-like protease(s) in mammalian cells. *Virology* 270:476–87

97. Scheffner M, Werness BA, Huibregtse JM, Levine AJ, Howley PM. 1990. The E6 oncoprotein encoded by human papillomavirus types 16 and 18 promotes the degradation of p53. *Cell* 63:1129–36

98. Scott FL, Denault JB, Riedl SJ, Shin H, Renatus M, Salvesen GS. 2005. XIAP inhibits caspase-3 and -7 using two binding sites: evolutionarily conserved mechanism of IAPs. *EMBO J.* 24:645–55

99. Shi Y. 2002. Mechanisms of caspase activation and inhibition during apoptosis. *Mol. Cell* 9:459–70

100. Shisler JL, Moss B. 2001. Molluscum contagiosum virus inhibitors of apoptosis: The MC159 v-FLIP protein blocks Fas-induced activation of procaspases and degradation of the related MC160 protein. *Virology* 282:14–25

101. Simonovic M, Gettins PGW, Volz K. 2000. Crystal structure of viral serpin crmA provides insights into its mechanism of cysteine proteinase inhibition. *Protein Sci.* 9:1423–27

102. Skaletskaya A, Bartle LM, Chittenden T, McCormick AL, Mocarski ES, Goldmacher VS. 2001. A cytomegalovirus-encoded inhibitor of apoptosis that suppresses caspase-8 activation. *Proc. Natl. Acad. Sci. USA* 98:7829–34

103. Stennicke HR, Salvesen GS. 2006. Chemical ligation—an unusual paradigm in protease inhibition. *Mol. Cell* 21:727–28

104. Stoka V, Turk V, Turk B. 2007. Lysosomal cysteine cathepsins: signaling pathways in apoptosis. *Biol. Chem.* 388:555–60

105. Suzuki Y, Imai Y, Nakayama H, Takahashi K, Takio K, Takahashi R. 2001. A serine protease, HtrA2, is released from the mitochondria and interacts with XIAP, inducing cell death. *Mol. Cell* 8:613–21

106. Swanson R, Raghavendra MP, Zhang W, Froelich C, Gettins PG, Olson ST. 2007. Serine and cysteine proteases are translocated to similar extents upon formation of covalent complexes with serpins. Fluorescence perturbation and fluorescence resonance energy transfer mapping of the protease binding site in CrmA complexes with granzyme B and caspase-1. *J. Biol. Chem.* 282:2305–13

107. Taylor JM, Barry M. 2006. Near death experiences: poxvirus regulation of apoptotic death. *Virology* 344:139–50

108. Tenev T, Ditzel M, Zachariou A, Meier P. 2007. The antiapoptotic activity of insect IAPs requires activation by an evolutionarily conserved mechanism. *Cell Death Differ.* 14:1191–201

109. Tenev T, Zachariou A, Wilson R, Ditzel M, Meier P. 2005. IAPs are functionally nonequivalent and regulate effector caspases through distinct mechanisms. *Nat. Cell Biol.* 7:70–77

110. Tesch LD, Raghavendra MP, Bedsted-Faarvang T, Gettins PG, Olson ST. 2005. Specificity and reactive loop length requirements for crmA inhibition of serine proteases. *Protein Sci.* 14:533–42

111. Thome M, Schneider P, Hofmann K, Fickenscher H, Meinl E, et al. 1997. Viral FLICE-inhibitory proteins (FLIPs) prevent apoptosis induced by death receptors. *Nature* 386:517–21

112. Thome M, Tschopp J. 2001. Regulation of lymphocyte proliferation and death by FLIP. *Nat. Rev. Immunol.* 1:50–58

113. Thomson BJ. 2001. Viruses and apoptosis. *Int. J. Exp. Pathol.* 82:65–76

114. Thurau M, Everett H, Tapernoux M, Tschopp J, Thome M. 2006. The TRAF3-binding site of human molluscipox virus FLIP molecule MC159 is critical for its capacity to inhibit Fas-induced apoptosis. *Cell Death Differ.* 13:1577–85

115. Timmer JC, Salvesen GS. 2007. Caspase substrates. *Cell Death Differ.* 14:66–72

116. Tollefson AE, Hermiston TW, Lichtenstein DL, Colle CF, Tripp RA, et al. 1998. Forced degradation of Fas inhibits apoptosis in adenovirus-infected cells. *Nature* 392:726–30

117. Turk B, Stoka V. 2007. Protease signalling in cell death: caspases versus cysteine cathepsins. *FEBS Lett.* 581:2761–67

118. Turner SJ, Silke J, Kenshole B, Ruby J. 2000. Characterization of the ectromelia virus serpin, SPI-2. *J. Gen. Virol.* 81:2425–30

119. Vandenabeele P, Orrenius S, Zhivotovsky B. 2005. Serine proteases and calpains fulfill important supporting roles in the apoptotic tragedy of the cellular opera. *Cell Death Differ.* 12:1219–24

120. Vaux DL, Haecker G, Strasser A. 1994. An evolutionary perspective on apoptosis. *Cell* 76:777–79

121. Vaux DL, Silke J. 2005. IAPs, RINGs and ubiquitylation. *Nat. Rev. Mol. Cell Biol.* 6:287–97

122. Vaux DL, Silke J. 2005. IAPs—the ubiquitin connection. *Cell Death Differ.* 12:1205–7

123. Verhagen AM, Ekert PG, Pakusch M, Silke J, Connolly LM, et al. 2000. Identification of DIABLO, a mammalian protein that promotes apoptosis by binding to and antagonizing IAP proteins. *Cell* 102:43–53

124. Vucic D, Kaiser WJ, Harvey AJ, Miller LK. 1997. Inhibition of reaper-induced apoptosis by interaction with inhibitor of apoptosis proteins (IAPs). *Proc. Natl. Acad. Sci. USA* 94:10183–88

125. Vucic D, Kaiser WJ, Miller LK. 1998. A mutational analysis of the baculovirus inhibitor of apoptosis Op-IAP. *J. Biol. Chem.* 273:33915–21

126. Vucic D, Kaiser WJ, Miller LK. 1998. Inhibitor of apoptosis proteins physically interact with and block apoptosis induced by *Drosophila* proteins HID and GRIM. *Mol. Cell Biol.* 18:3300–9

127. Wilkinson JC, Wilkinson AS, Scott FL, Csomos RA, Salvesen GS, Duckett CS. 2004. Neutralization of Smac/Diablo by inhibitors of apoptosis (IAPs). A caspase-independent mechanism for apoptotic inhibition. *J. Biol. Chem.* 279:51082–90

128. Wing JP, Schwartz LM, Nambu JR. 2001. The RHG motifs of *Drosophila* Reaper and Grim are important for their distinct cell death-inducing abilities. *Mech. Dev.* 102:193–203

129. Wu G, Chai J, Suber TL, Wu JW, Du C, et al. 2000. Structural basis of IAP recognition by Smac/DIABLO. *Nature* 408:1008–12

130. Xu G, Rich RL, Steegborn C, Min T, Huang Y, et al. 2003. Mutational analyses of the p35-caspase interaction. A bowstring kinetic model of caspase inhibition by p35. *J. Biol. Chem.* 278:5455–61

131. Yan N, Wu JW, Chai J, Li W, Shi Y. 2004. Molecular mechanisms of DrICE inhibition by DIAP1 and removal of inhibition by Reaper, Hid and Grim. *Nat. Struct. Mol. Biol.* 11:420–28

132. Yang JK, Wang L, Zheng L, Wan F, Ahmed M, et al. 2005. Crystal structure of MC159 reveals molecular mechanism of DISC assembly and FLIP inhibition. *Mol. Cell* 20:939–49

133. Yin XM. 2006. Bid, a BH3-only multi-functional molecule, is at the cross road of life and death. *Gene* 369:7–19

134. Yoo SJ, Huh JR, Muro I, Yu H, Wang L, et al. 2002. Hid, Rpr and Grim negatively regulate DIAP1 levels through distinct mechanisms. *Nat. Cell Biol.* 4:416–24

135. Youle RJ, Strasser A. 2008. The BCL-2 protein family: opposing activities that mediate cell death. *Nat. Rev. Mol. Cell Biol.* 9:47–59

136. Yuan J, Shaham S, Ledoux S, Ellis HM, Horvitz HR. 1993. The *C. elegans* cell death gene *ced-3* encodes a protein similar to mammalian interleukin-1 beta-converting enzyme. *Cell* 75:641–52

137. Zaragoza C, Saura M, Padalko EY, Lopez-Rivera E, Lizarbe TR, et al. 2006. Viral protease cleavage of inhibitor of kappaBalpha triggers host cell apoptosis. *Proc. Natl. Acad. Sci. USA* 103:19051–56

123. The first study to suggest that vIAPs interact with IAP antagonists.

138. Zhirnov OP, Konakova TE, Garten W, Klenk H. 1999. Caspase-dependent N-terminal cleavage of influenza virus nucleocapsid protein in infected cells. *J. Virol.* 73:10158–63

139. Zhou Q, Krebs JF, Snipas SJ, Price A, Alnemri ES, et al. 1998. Interaction of the baculovirus antiapoptotic protein p35 with caspases. Specificity, kinetics, and characterization of the caspase/p35 complex. *Biochemistry* 37:10757–65

140. Zhou Q, Snipas S, Orth K, Muzio M, Dixit VM, Salvesen GS. 1997. Target protease specificity of the viral serpin CrmA. Analysis of five caspases. *J. Biol. Chem.* 272:7797–800

141. Zimring JC, Goodbourn S, Offermann MK. 1998. Human herpesvirus 8 encodes an interferon regulatory factor (IRF) homolog that represses IRF-1-mediated transcription. *J. Virol.* 72:701–7

142. Zoog SJ, Schiller JJ, Wetter JA, Chejanovsky N, Friesen PD. 2002. Baculovirus apoptotic suppressor P49 is a substrate inhibitor of initiator caspases resistant to P35 in vivo. *EMBO J.* 21:5130–40

Bistability, Epigenetics, and Bet-Hedging in Bacteria

Jan-Willem Veening,[1,3] Wiep Klaas Smits,[2,3]
and Oscar P. Kuipers[3]

[1]Institute for Cell and Molecular Biosciences, Newcastle University, Newcastle upon Tyne NE2 4HH, United Kingdom; email: j.w.veening@ncl.ac.uk

[2]Department of Biology, Massachusetts Institute of Technology, Cambridge, Massachusetts 02139; email: smitswk@mit.edu

[3]Molecular Genetics Group, Groningen Biomolecular Sciences and Biotechnology Institute, University of Groningen, 9751 NN Haren, The Netherlands; email: o.p.kuipers@rug.nl

Annu. Rev. Microbiol. 2008. 62:193–210

First published online as a Review in Advance on June 6, 2008

The *Annual Review of Microbiology* is online at micro.annualreviews.org

This article's doi:
10.1146/annurev.micro.62.081307.163002

Copyright © 2008 by Annual Reviews.
All rights reserved

0066-4227/08/1013-0193$20.00

Key Words

Bacillus subtilis, competence, sporulation, AND gate, phenotypic variation, synthetic biology

Abstract

Clonal populations of microbial cells often show a high degree of phenotypic variability under homogeneous conditions. Stochastic fluctuations in the cellular components that determine cellular states can cause two distinct subpopulations, a property called bistability. Phenotypic heterogeneity can be readily obtained by interlinking multiple gene regulatory pathways, effectively resulting in a genetic logic-AND gate. Although switching between states can occur within the cells' lifetime, cells can also pass their cellular state over to the next generation by a mechanism known as epigenetic inheritance and thus perpetuate the phenotypic state. Importantly, heterogeneous populations can demonstrate increased fitness compared with homogeneous populations. This suggests that microbial cells employ bet-hedging strategies to maximize survival. Here, we discuss the possible roles of interlinked bistable networks, epigenetic inheritance, and bet-hedging in bacteria.

Contents

PHENOTYPIC VARIATION AND ITS ORIGINS

lacZ: β-galactosidase, traditional reporter that cleaves colorless X-gal, resulting in bright blue products

Bet-hedging: a risk spreading strategy to diversify phenotypes with the aim to increase fitness in temporally variable conditions

Bacterial growth is traditionally viewed as the result of (symmetrical) cell division yielding siblings that are genetically identical. Consequently, the results from reporter studies such as those employing *lacZ* have traditionally been interpreted using the assumption that all cells in a culture behave in an identical manner. However, it has long been recognized that within isogenic populations, bacterial cells can display various phenotypes. This microbial cell individuality or phenotypic variation is receiving increased attention because of its rele-

vance for cellular differentiation and implications for the treatment of bacterial infections (92). Phenotypic variation is a widespread phenomenon in the bacterial realm. Some of the well-characterized examples include the lysis-lysogeny switch of phage lambda, lactose utilization and chemotaxis in *Escherichia coli*, phase variation in a number of pathogens, and cellular differentiation in *Bacillus subtilis* (for recent reviews see References 11, 25, and 92). Strikingly, many documented cases of phenotypic variability relate to responses to environmental stresses, suggesting that phenotypic variation aids in the survival of cells under adverse conditions and therefore may be an evolvable trait. The potential function of phenotypic variation as a bet-hedging strategy is further elaborated upon in other parts of this review.

Various different mechanisms are involved in phenotypic variation. Phenotypic differences can be due to mutation, variations in the microenvironment, mutation, phase variation, cell cycle, and the wiring of the network that governs a specific stress response (11, 92). The focus of this review is on the role of phenotypic variability that results from amplified noise in gene expression.

NETWORK TOPOLOGY

As early as 1961, Monod and Jacob postulated that the differences in the response of individual cells to a stimulus could in theory be explained by the architecture of the underlying gene regulatory network (66). However, their hypotheses could not be experimentally addressed until the development of single-cell techniques and were not computationally tractable until recently. Considering the importance of this type of mechanism in generating phenotypic variation (92), it is discussed in more detail below.

Noise, Hysteresis, and Bistability

In biological systems, signals are never discrete because of random fluctuations in the biochemical reactions in the cell. This stochastic variation is called noise and is a key determinant of phenotypic variation (49, 81, 85). Noise is predicted to be most dominant when the number of molecules involved is small (finite number effect). Experimental verification of this notion came from fluorescent reporter studies (28, 78, 98). This effect is notable for two reasons. First, transcription and translation are thought to generally involve relatively small numbers of molecules compared with, for instance, the numbers of molecules participating in protein-protein interactions. Second, when not activated, transcription factors are usually in low abundance. Moreover, many stress responses are accompanied by a reduction in general transcriptional and/or translational efficiency (38). This potentially leads to an induction of phenotypic variation under these conditions. Generating variable phenotypes may be beneficial for the survival of populations under adverse conditions, and stimulating noisy expression might be an elegant way of achieving this (72).

Noise can be exploited under certain conditions to generate phenotypic heterogeneity. For example, noise in the regulatory cascade that governs the chemotactic response of *E. coli* results in behavioral individuality with respect to the rotational direction of the flagella (54). When a noisy signal is amplified by net positive feedback, gene expression levels can be further bifurcated and this situation deserves special attention. In the presence of positive feedback, a graded response (i.e., with intermediate levels of expression) can be converted to a binary response, in which cells express a certain gene at high or low levels (13). At the population level, this switch-like behavior can result in a bimodal distribution in gene expression because some cells switch, whereas others do not. This type of gene expression pattern is commonly referred to as bistability (25, 92).

In physics, multistationarity describes a network that has more than one stable state. Extending this to biology, it means that a gene regulatory network potentially exhibits two (or more) discrete levels of gene expression (a high state and a low state). Bistability describes a parameter regime in which a dynamic system can rest in either of two stable states. Anal-

ogous to the previous definition, it refers to conditions under which cells can be in a high-expressing or low-expressing state for biological systems. Multistationarity at the cellular level is an intuitive explanation for population bistability; hence, the terms are frequently used interchangeably. Although most biological systems that demonstrate population bistability involve noise amplified by some form of net positive feedback, they are not necessarily bistable in a deterministic sense (95).

The requirements for a gene network to exhibit multistationarity have been explored in detail (29, 92). In summary, the system needs to display nonlinear kinetics in addition to positive feedback. For transcriptional regulators, nonlinearity can be the result of multimerization, cooperative binding to target sequences on the DNA, or phosphorylation of certain amino acid residues. In many cases nonlinearity is evident as a sharp increase in the expression of a downstream target gene above a certain threshold level of the regulator. Only networks that include an even number of negative-feedback loops and/or any number of positive-feedback loops are capable of causing multistationarity (8). Experimentally, some bistable gene expression patterns rely on positive feedback as well as double-negative feedback (toggle switch) (92 and references therein). However, positive feedback in itself is no guarantee for bistability (29), and bistability is also possible when based on other types of network architecture (8) or mechanisms such as multisite phosphorylation (55, 77).

A common feature of bistability is hysteresis (74). Hysteresis refers to the situation in which the transition from one state to the other requires an induction (or relief of induction) greater than that for the reverse transition. This imposes memory-like characteristics onto the network (see also Epigenetic Inheritance of Phenotypic Variation, below), making the response of cells dependent on their recent history. Hysteresis in biological systems can reside, for instance, in the stability of one of the proteins involved. When Novick & Weiner (76) described the all-or-none enzyme induction

Multistationarity: multiple stationary stable states within a (genetic) network between which switching is possible

Bistable: a network with two steady states, or two distinguishable phenotypes within a clonal population

in lactose utilization they noted that at near-threshold concentrations of inducer the population of *E. coli* cells segregated into two subpopulations, which is now regarded as one of the earliest examples of bistability. Subsequent experiments revealed that the history of the inoculum influenced the fraction of cells in each subpopulation (23). The hysteretic behavior of the multistable lactose utilization network is a result of the stability and abundance of the lactose permease (79, 107). Hysteresis can act as a buffer, reducing accidental switching between states due to minor perturbations (1, 16).

Although bistable systems are in principle reversible, the time required for a cell to revert to the initial state (escape time) may exceed the duration of the experiment or even the lifetime of the organism. Moreover, phenotypic switches can be rendered unidirectional by downstream signaling events. For instance, the bistable switch governing sporulation in *B. subtilis* becomes irreversible after its earliest stages owing to an orchestrated sequence of events (26).

COMPETENCE FOR GENETIC TRANSFORMATION IN *BACILLUS SUBTILIS*

To further explore general mechanisms by which phenotypic variation can arise, we discuss one of the best-understood naturally occurring bistable systems in bacteria: competence development in *B. subtilis*. The first evidence for the existence of subpopulations in a competent culture of *B. subtilis* came from elegant experiments that demonstrate biosynthetic latency of competent cells (17, 41, 73). Subsequently, the expression of the key regulator of competence development, ComK, was limited to the competent fraction of the culture (42).

ComK is a multimeric transcription factor that is necessary and sufficient to activate the expression of all genes that encode the DNA uptake and integration machinery by binding to a consensus motif in the target promoters (44). Key features of the complex regulatory network that controls ComK levels are transcriptional regulation at the *comK* promoter and proteolytic degradation of ComK protein (44). ComK stimulates its own expression by reversing the effects of at least two repressors, one of them named Rok (for repressor of *comK*), establishing a positive-feedback loop (91). Additionally, ComK is believed to repress transcription of *rok*. This interaction forms a putative toggle switch. Proteolytic degradation of ComK is antagonized by the anti-adaptor protein ComS, which is required for the initiation of competence. Evidence suggests an indirect negative-feedback loop, as overproduction of ComK inhibits ComS expression (95). The features described above are summarized in **Figure 1**, and they all form modules that are potentially involved in phenotypic variation.

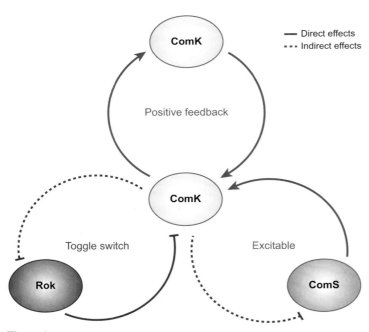

Figure 1

Regulatory elements featured in current models on competence development in *Bacillus subtilis*. A highly simplified representation of the three core elements of the competence regulatory network: (*a*) ComK autostimulation is responsible for a positive-feedback loop required for a bimodal expression pattern, (*b*) a putative toggle switch is dispensable for a bimodal expression pattern, and (*c*) interlinked positive- and negative-feedback loops that result in excitable behavior are implicated in the temporal nature of competence. Solid lines represent direct or well-characterized interactions; dotted lines represent putative or indirect effects.

ComK autostimulation is necessary and can be sufficient to establish a bimodal expression pattern (61, 90), but it is independent of Rok, excluding a toggle switch–like mechanism. The transition between low- and high-expressing states was attributed to stochastic fluctuations in conjunction with the positive-feedback loop that would amplify the signal as the concentrations of ComK exceed a certain threshold (103). The role of noise was experimentally addressed in two studies. Süel and coworkers averaged out the noise of multiple cells by depleting cells for *ftsW*, which is required for septation. An analysis of competence development under those conditions revealed that the chance of initiation of competence was greatly reduced (96). In a more direct approach, Maamar and coworkers (62) adopted a method derived from Elowitz et al. (28) to show that intrinsic noise in *comK* expression selects cells for competence. Reducing intrinsic noise, by increasing transcriptional efficiency and reducing translational efficiency, caused significantly less cells to enter competence. Their findings are consistent with another report that demonstrated significant variation in basal promoter activity of *comK* (57). Because ComK is responsible for the activation of the late competence genes (such as *comG*), intrinsic noise in *comK* expression results in pathway-specific extrinsic noise in *comG* expression.

Competence is a transient process; under laboratory conditions it is limited to several hours in stationary growth phase or until cells are resuspended in fresh growth medium. Although the molecular mechanisms responsible for escape from the competent state remain elusive, mathematical modeling has recently shed some light on potential mechanisms and has led to the development of two predominant models. Both models share the notion that noise is amplified by the ComK autostimulatory loop. In the bistable model, intrinsic noise of *comK* expression (57, 62) is critical for the switching of cells from the noncompetent to the competent state. In the excitable model, the source of the noise that triggers the excursion from the vegetative state remains undefined (95, 96). Although both models can result in a bimodal distribution at the population level (as both involve stochastic switching), only the first model is bistable in a deterministic sense. The excitable model generates a bimodal gene expression pattern because the transition to the high-expressing state is fast compared with the slowly acting negative-feedback loop, but the high expression level does not represent a stable state (95).

Both models offer a different explanation for the temporal nature of competence. In the bistable model, two mechanisms are at play. First, cells can revert from the high-expressing to the vegetative state by stochastic transitions. Second, the basal promoter activity of *comK*, as measured by the number of mRNA molecules per cell, is greatly reduced in stationary growth phase (57, 62). This causes a window of opportunity for cells to switch to the competent state and generates conditions under which the saturated proteolytic complex reduces ComK levels enough to escape the competent state. The validity of this hypothesis was confirmed through mathematical modeling (62).

The excitable model offers an attractive hypothesis for the limited time span during which cells are competent for DNA uptake. In contrast to the bistable model, the competent state is not stable owing to the action of a slowly acting negative-feedback loop. As a result cells will always return to the vegetative and stable state. The model makes some predictions about the dynamics of the competence network that are experimentally addressed using time-lapse fluorescent microscopy (96).

The elegance of the excitable model has attracted a lot of attention, as it resembles the dynamics of oscillatory systems such as cell cycle and circadian rhythms. However, it fails to couple back to the observations made in single-cell analyses of competent cultures that demonstrate a limited time frame during which competence occurs in a culture and does not take the observed decrease in basal *comK* transcription into account. Although certain features of

Excitable: a transient excursion from a stable state leading to the expression of a phenotype in a limited period of time

the two models are not reconcilable, it is possible that both mechanisms occur in nature under different conditions, for example, the timescale on which they occur could vary. Moreover, it has been suggested that stochastic activation of *comK* in combination with positive feedback could result in a bimodal expression pattern, even in the absence of bistability in the deterministic sense (9, 50). It is a challenge for future investigators to address these unanswered questions.

PROSPECTS OF USING BISTABLE SWITCHES FOR BIOTECHNOLOGY AND SYNTHETIC BIOLOGY

Construction of synthetic genetic circuits using naturally occurring cellular components in living cells allows them to be tested separately from the context of other physiological processes. Synthetic switches are operational in prokaryotic and mammalian cells and valuable for gaining insight into naturally occurring genetic circuitries (45, 46). Synthetic biology also allows the creation of entirely new, or rerouted, networks, such as toggle switches, oscillatory networks, and even synthetic multicellular clocks based on quorum sensing (10, 27, 34, 36). Some of these findings made it to patents (37), showing the realistic prospect of industrial utilization of engineered circuitries leading to phenotypic variation.

Combinatorial promoter design also is effective for engineering noisy gene expression (71), and various successful examples of combinatorial promoter design have been published (24, 43). Global transcription machinery engineering (gTME) is a compatible strategy for improving metabolic engineering efforts. Instead of direct enzyme or metabolic pathway engineering, gTME reprograms the transcription machinery, resulting for example in increased ethanol tolerance and production in yeast (7). This method could be well combined with the strategies outlined above to engineer novel regulatory circuits.

CELL AGE AND ITS ROLE IN PHENOTYPIC VARIATION

Although aging has already been described to cause phenotypic variability in yeast (4), *Caulobacter crescentus* was the first bacterium for which aging was demonstrated (3). It was found that the reproductive output of cells decreased with age. Asymmetric division is a hallmark of the life cycle of this bacterium, and these observations are therefore consistent with the hypothesis that mortality requires asymmetry (80).

In many other prokaryotes, however, cell division leads to two visibly identical daughter cells, and as a result, they have been regarded as nonsenescent. Yet, the subcellular localization of a set of proteins may distinguish old and new poles in morphologically symmetrical bacteria. By following single *E. coli* cells through several rounds of cell division, Stewart and coworkers showed that growth rate inversely correlates to cell pole age, demonstrating that aging is not limited to organisms with asymmetric division (94). It was recently found that aggregated proteins and chaperones preferentially accumulate at the old cell pole (59), reminiscent of the situation in yeast in which oxidatively damaged proteins accumulate in the mother cell (4).

Recently, time-lapse microscopy has been used to follow the growth, division, and cellular differentiation of individual cells of *B. subtilis* (104), an organism that is well known for asymmetric division prior to the formation of an endospore. The study revealed that *B. subtilis*, like *E. coli* and *C. crescentus*, suffers from aging but that spore formation is not biased toward either the old or the new cell pole (104). Interestingly, the magnitude of this aging effect is nearly identical to that seen in *E. coli* and *C. crescentus*.

EPIGENETIC INHERITANCE OF PHENOTYPIC VARIATION

Epigenetic inheritance (EI) (or non-Mendelian inheritance) is the passage of cellular states from one generation to the next, without alterations of the genome (48). The classic example of EI is the stable transfer of a phenotype by modifications to the DNA such as methylation (19). This modification can be stable over multiple rounds of cell division but it does not involve actual changes in the DNA sequence of the organism. Other epigenetic phenomena include prions, genomic imprinting, and histone modification (19 and references therein).

It has been proposed that autophosphorylating kinases have the potential to store memory. In this scenario, a specific stimulus activates the kinase, and because of its autocatalytic properties the kinase stays in its active state, regardless of the presence or absence of the stimulus (60). As a result, the progeny of cells in the ON-state will also be in the ON-state because the activated kinase is passed on to the offspring. Using artificial bistable gene regulatory circuits in both *E. coli* and *Saccharomyces cerevisiae*, autostimulatory regulation systems can function as memory devices in microorganisms (13, 36).

EI of phenotypic variation can also be based on the transfer of active transcriptional regulators during cell division via positive feedback (18, 60, 84). When cells divide, not only DNA but also cellular factors such as proteins and RNA are partitioned, and importantly, this can dictate future life-history decisions of the new offspring. Valuable knowledge on the molecular mechanism responsible for EI and the minimal requirements to generate stable inheritance of phenotypic variation is arising from studies using well-defined artificial gene networks (6, 51). The simplest network that demonstrates EI is one in which a positive regulator autostimulates its own promoter upon stimulation by an exogenous signal. Once activated, the positive feedback of the system will ensure high intracellular levels of the positive regulator, regardless of the absence or presence of the signal. In such a system, the degradation rate of the regulator and the growth rate of the cell are determining factors of the stability of the memory response (6).

An example of a simple (but general) network motif that putatively generates EI is depicted in **Figure 2**. A number of requirements need to be met before EI can occur. The network should show two stable steady states (activator OFF and activator ON). This depends on activator production/decay rate and growth rate, and activator production should be cooperative (6). In addition to this, the basal activator levels should be at a level lower than required to autoactivate its own synthesis; otherwise cells will always be in the ON state. Furthermore, once the system is activated, activator levels should be high enough to drive its own expression; if not, cells will quickly switch back from the ON to the OFF state and EI cannot be established. Even cell fates driven by a semistable stochastic switch with reduced positive feedback inherit epigenetically. This is likely caused by initial bursts of activator protein in the mother cell, which maintains at high levels through multiple rounds of division (51). Two examples of the significance of EI of phenotypic variation in bacteria are discussed below. Other instances, primarily from eukaryotes, fall outside the scope of this review.

Memory Within the *lac* Operon

As discussed above, bistable systems depend on some form of positive feedback within the gene network. The first epigenetic system described in bacteria is the *lac* operon of *E. coli* (76). The genes that encode the proteins required for the uptake and utilization of lactose are induced in the presence of the gratuitous (nonmetabolizable) lactose analogue, isopropyl-D-thio-β-galactopyranoside (IPTG). At high IPTG concentrations the *lac* operon is fully derepressed and cells highly express the IPTG permease protein and thus remain highly activated. At low concentrations, however, cells that were previously uninduced and do not have any permease in their membranes do not respond to the low level of IPTG and remain in the OFF state. Cells that were previously induced and still have some permease are activated by the low level of IPTG and remain in the ON state. Reculturing of single cells results in populations that either give high or low *lac* expression (70 and references therein). This phenomenon

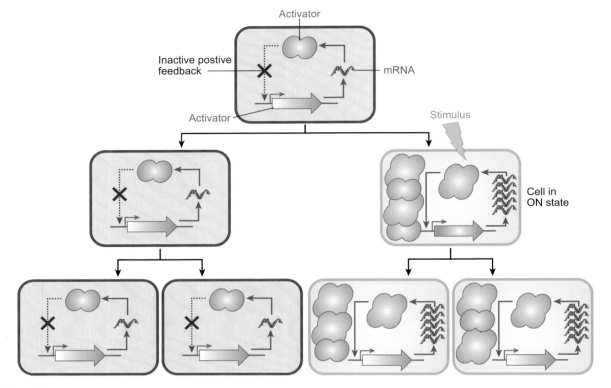

Figure 2

Epigenetic inheritance by positive feedback. A basal level of activator protein and mRNA (*single helix*) is always present regardless of the absence of stimulus (*lightning symbol*). However, this basal level is insufficient to activate the positive-feedback loop (*red X*) and activator protein levels remain low. When the signal is present, however (which might be caused by noise), activator protein multimerizes and stimulates its own expression, resulting in high concentrations of activator, and in this example, high activator concentrations induce multimerization. Because of the positive-feedback loop, intracellular activator concentrations remain above the threshold required to stimulate transcription and cells remain in the ON state (*green cells*) for multiple generations even in the absence of stimulus. Cell growth and division can dilute activator, but as long as the concentrations remain high enough to drive promoter firing, cells will remain in the ON state.

is called all-or-none enzyme induction (76) and is indicative of the presence of two coexisting subpopulations. The permease plays a pivotal role and constitutes the positive-feedback loop in this system: High permease levels keep the levels of intracellular IPTG high, thus inducing permease gene expression. Importantly, under low inducer conditions, either the ON or OFF state can be epigenetically inherited by the offspring through multiple rounds of growth and division. In this situation, the physiological state of the offspring is a reflection of the past state of its ancestor. A possible explanation for such a positive-feedback loop in the *lac* operon is that in the presence of (metabolizable) lactose, the

IPTG: isopropyl-
ᴅ-thio-β-
galactopyranoside

E. coli population can quickly drain the sugar pool even when the sugar concentration starts to decrease (18).

Sporulation in *B. subtilis*

Sporulation of *B. subtilis* has been described as a bistable process because two distinct subpopulations can be distinguished within an isogenic population of stationary-phase cells: sporulating and nonsporulating cells (reviewed in References 25 and 92). Initiation of sporulation is driven by the master sporulation regulator Spo0A. A basal level of Spo0A is always present, and upon specific environmental signals such as high cell density and nutrient deprivation, Spo0A is phosphorylated and directly activates expression of more than 100 genes, including its own gene (31, 65). Sporulation bistability is not a simple ON/OFF switch, because the levels of Spo0A~P increase gradually after activation (32). Recent research has shown that although the positive feedback of Spo0A~P on *spo0A* transcription plays an important role in the distribution of cellular states (31, 101), it is not critical in establishing sporulation bistability (104). Rather it seems that the activity of the phosphorelay dictates sporulation bistability because cells constructed to express a mutant form of Spo0A (Sad67) (47) that does not require activation no longer show bistability (104).

A recent study using time-lapse microscopy found a strong correlation between cell lineage and the decision to sporulate or not sporulate (104). Close relatives often demonstrate a similar phenotype (to either sporulate or not sporulate). Phylogenetic reconstruction of sporulating microcolonies using parsimony analyses showed that the decision to sporulate could often be traced back more than two generations before the actual appearance of the phenotype (**Figure 3**). This finding indicates that the signal to sporulate already occurs during the logarithmic growth phase and is epigenetically passed on. Again, an important role for the sporulation phosphorelay was identified for this epigenetic effect (104), indicating that bistability is a prerequisite for EI of the sporulation signal.

The putative benefits of EI within a sporulating population are complex. For cold-shock adaptation in bacteria, cells pretreated by a mild cold shock memorize this stress and are better prepared for a harsher cold shock, which would otherwise be lethal (40). In analogy to this, it can be envisaged that propagation of the sporulation signal from the mother cell to its descendants helps the progeny to be prepared for potential nutritional limitations in the future in such a way that they can rapidly respond

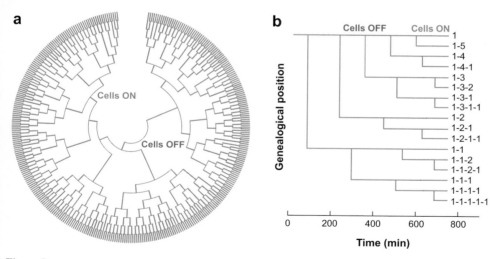

Figure 3

Lineage reconstruction to plot cell fate distributions within isogenic populations. (*a*) Parsimony reconstruction of the sporulation signal within a *Bacillus subtilis* microcolony. Every node in the radial tree represents one cell division event. Every endpoint in the tree represents one offspring cell. Orange tips are cells that have activated Spo0A. Parsimony reconstruction shows the first appearance of a mother cell that creates offspring of mostly cells in the ON state (*orange lines*). Figure from Reference 104. (*b*) Family tree of *Saccharomyces cerevisiae* harboring an artificial bistable switch. Gray lines indicate cells in the OFF state, whereas orange lines represent cells after they have switched to the ON state. In this genealogy graph, in contrast to panel *a*, line length is a direct measure of time. Figure from Reference 51.

CRIF: *cis*-regulatory input function

Altruism: a behavior that decreases the fitness of the altruistic individual while benefiting others

and commit to spore formation when required. Alternatively, EI may serve to coordinate multicellular behavior (104), a process which in *B. subtilis* is also dictated by Spo0A (5).

GENETIC LOGIC-AND GATES

Often, transcription of a gene is regulated by more than one regulator (input). The way these inputs control the transcription rate (output) is described by the *cis*-regulatory input function (CRIF) (64). CRIFs can often be described by Boolean-type functions such as logic-AND gates and logic-OR gates (64 and references therein). Synthetic logic-AND gates can be exploited to program specific responses of cells (75). If one of the inputs of a CRIF is heterogeneous and the target gene is under control of a logic-AND gate, then by definition the output is also heterogeneous.

A number of studies recognize that certain genes of one pathway are heterogeneously expressed because their regulation is interlinked with another (bistable) network through a logic-AND gate. The use of an AND gate system is a simple strategy to generate phenotypic variability without the necessity to create complex switches with multiple steady states (**Figure 4a**). Here we consider a few examples in which heterogeneity in gene expression can be ascribed to the logic of the underlying circuitry. We discuss the putative physiological relevance of the observed heterogeneity as a result of the AND circuit.

Heterogeneity in Exoprotease and Biofilm Matrix Production

Recently, it was found that high expression of *aprE* (subtilisin) and *bpr* (bacillopeptidase), two important extracellular proteases (exoproteases) of *B. subtilis*, is limited to only a small part of the population (**Figure 4b**) (102). Exoprotease production has been described as a survival strategy under nutrient-limiting conditions, and these enzymes act as scavenging proteins that degrade (large) proteins into smaller fragments that can be subsequently taken up as

a new nutrient source (68). Studies using wild *B. subtilis* strains also indicate a role for exoproteases during biofilm formation (53, 105).

Expression of both *aprE* and *bpr* is under the control of the DegS-DegU two-component system (68). To activate *aprE* gene expression, DegU needs to be phosphorylated by the DegS sensor protein (69). In addition, *aprE* is under direct negative control of at least three other transcriptional regulators (AbrB, SinR, and ScoC), all of which are under direct or indirect negative control by the key sporulation regulator, Spo0A~P (31). The result of this intertwinement with the sporulation pathways is that *aprE* will only be derepressed in a subpopulation when nutrients become limiting. Together, the *aprE* gene regulatory network acts as a logic-AND circuit in which a threshold level of dimerized DegU~P and Spo0A~P is integrated to activate gene expression (102) (**Figure 4c**).

It has been hypothesized that cells that produce and secrete these proteases help not only themselves, but all clonal cells within the local growth medium. This might be regarded as a simple form of altruism. One explanation for altruism is when the cooperation is directed toward individual cells that are genetically similar (kin selection) (63). Heterogeneity in gene expression ensures that not all cells commence into the costly production of Bpr and AprE, but all cells within the clonal population benefit from the activity of these extracellular proteases.

Similarly, the extracellular matrix within biofilms of *B. subtilis* is produced by a small fraction of cells within the population (20, 106). Expression of the products that form the extracellular matrix (EpsA-O, YqxM, and TasA) is under direct negative control of SinR, the master biofilm regulator in *B. subtilis* (52). This regulator is antagonized by SinI, a protein under control of Spo0A. *sinI* seems to be activated by low levels of Spo0A~P but repressed at high levels of Spo0A~P (20), although this still awaits experimental validation. Thus, expression of *sinI* and, as a result, the genes responsible for the extracellular matrix

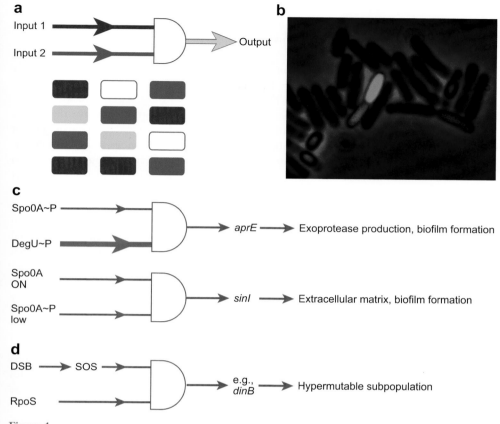

a

Input 1 ➜
Input 2 ➜

Output

b

c

Spo0A~P ➜
DegU~P ➜

aprE ➜ Exoprotease production, biofilm formation

Spo0A
ON ➜
Spo0A~P
low ➜

sinI ➜ Extracellular matrix, biofilm formation

d

DSB ➜ SOS ➜
RpoS ➜

e.g.,
dinB ➜ Hypermutable subpopulation

Figure 4

Naturally occurring genetic logic-AND gates. Arrows indicate positive actions. (*a*) Input 1 (*red arrow*) and input 2 (*blue arrow*) are active only in ~50% of the isogenic population (*red and blue cells*, respectively). Cells not active for input 1 or 2 are depicted in white. The output of the system (*yellow arrow*) is expressed only when both input 1 and input 2 are active within the same cell (*yellow*). As a result of this AND gate, the isogenic population of four distinguishable phenotypes can exist: white, red, blue, or yellow cells. (*b*) Heterogeneous expression of *aprE*. Expression of *aprE* is monitored by a fusion of the *aprE* promoter to *gfp*. Within stationary-phase cultures, three distinct phenotypes can readily be observed: sporulating cells, vegetative cells, and *aprE-gfp*-expressing cells. (*c*) The genetic circuit responsible for *aprE* and biofilm heterogeneity. Thick arrows indicate that the system can be overridden by (artificial) induction of the activator protein. For simplicity, we depict the effect of Spo0A~P on the multiple repressors that act on *aprE* as a single positive arrow. (*d*) SOS response and RpoS requirement for the formation of a hypermutable subpopulation in *Escherichia coli*. See text for details.

within biofilms are activated only when two conditions are met: (*a*) Spo0A needs to be activated and (*b*) Spo0A~P levels cannot be too high. Although this is not a true logic-AND circuit, the result of the network wiring is that only a small subpopulation of cells expresses *sinI*. Because the production of the extracellular matrix within biofilms is energetically costly, the division of labor might enhance the total fitness of

the entire bacterial community. It will be interesting to see how this labor is divided in multispecies biofilms.

Hypermutable Subpopulations in *E. coli*

Clonal populations of cells may diverge owing to changes in their genetic makeup. The

Adaptive mutagenesis: describes a set of conditions under which mutations appear to occur more often when selective pressure is present than when not

HMS: hypermutable subpopulation

DSB: double-strand break

Persistence: the phenomenon of the existence of a small subpopulation of cells that do not grow compared with the rest of the isogenic culture, and as a result are antibiotic resistant

occurrence of mutations may give certain cells a selective advantage over others, and this may cause a subpopulation to form or even take over the culture (108).

Under conditions of stress, adaptive mutagenesis (and/or stationary-phase mutagenesis) can occur (for recent review see Reference 30). In *E. coli*, adaptive mutations were associated with other, unselected mutations, indicating the existence of a hypermutable subpopulation (HMS) (100). The observed hypermutation is not caused by a stable mutator phenotype that could result from genetic differences, but reflects a transient differentiated state (39, 87).

The mechanisms involved in hypermutation include double-strand break (DSB) repair, SOS response, and a general stress response, of which the first two have a causal relationship (33). The critical factor in HMS, though not the only one, is that cells are continuously facing DNA double-strand breaks, even in the absence of external DNA-damaging agents. The induction of DSB repair is evidenced by the formation of foci of RecA protein, a key protein in the repair pathway, in a subset of cells (86, 88). DSBs can lead to induction of the SOS response, and ~1% of growing *E. coli* cells is SOS induced under steady-state conditions (82).

The switch to HMS requires an additional requirement to be satisfied, as artificially induced DSBs do not lead to HMS until cells enter stationary phase (83). At that time, the levels of the general stress sigma factor RpoS rise, and it was found that artificially inducing RpoS can lead to HMS in exponential growth phase (83). Thus, the preexisting heterogeneous input of (at least) the SOS response, together with RpoS, forms a logic-AND gate that leads to the formation of the HMS (**Figure 4d**).

PHENOTYPIC VARIATION AS A BET-HEDGING STRATEGY

A major question that arises from the finding of population heterogeneity is, Why do bacteria display phenotypic variation? The most apparent hypothesis is that this strategy is a form of bet-hedging. Under challenging conditions, the production of offspring with variable phenotypes ensures that at least one offspring will be appropriate (fit) under a given situation (22). This is a risk-spreading or bet-hedging strategy, because not every offspring will be optimally suited for the future environment. However, the overall fitness of the genotype will increase because some offspring will have the proper adaptation. Although heterogeneity might not be ideal under homogenous, steady-state conditions, mathematical studies support the notion that in a variable environment a heterogeneous population outcompetes (or is fitter than) a homogeneous population (56, 99). Importantly, it was suggested that phenotypic variation is an evolvable trait. This was recently underscored in an elegant study on *S. cerevisiae*, in which interphenotype switching rates, like those between the two stable states of gene expression in a bistable system, are tuned to the frequency of changes in the environment (2).

Experimental evidence for the benefits of phenotypic variation is limited. In yeast, clonal populations with increased variability in stress resistance are more successful than strains with limited variability under conditions of stress (15). Moreover, heterogeneous populations of yeast outcompete homogenous populations under cadmium stress conditions (89).

Bacterial Persistence

Originally identified in 1944 (14) persistence is one of the best-documented examples of a bacterial bet-hedging strategy (for a recent review see 58). Persister cells are not simply antibiotic resistant but rather reflect a transient growth arrested state. Persister cells can be grown to form a population that once again consists of antibiotic-sensitive cells and a small subpopulation of persisters (67). The switch from normal growth to persistence and vice versa is stochastic and epigenetic in nature (12). At least in *Mycobacterium tuberculosis*, the regulation of persistence appears to involve noise in gene expression amplified by positive feedback (97). Persistence is a form of bet-hedging as it ensures survival during catastrophes (56). In addition, persistence of a subpopulation of cells might indirectly benefit other cells in a population as the growth-arrested cells do not compete for limited resources (35). Recent mathematical modeling suggests that bacterial persistence can be regarded as a social trait and can be influenced by kin selection (35).

Sporulation Bistability as a Bet-Hedging Strategy

Recently, quantitative time-lapse microscopy was used to generate lineage and cell fate maps of single *B. subtilis* cells growing out to a sporulating microcolony (**Figure 3**). The study demonstrated that under these conditions *B. subtilis* employs a bet-hedging strategy whereby some cells sporulate and others utilize alternative metabolites to continue growth (and can putatively pursue other survival tactics) (104).

For individual cells the benefit of sporulation is clear; spores are resistant to various environmental conditions and can ensure the preservation of the clonal lineage, whereas vegetative cells could not. In the laboratory strain, however, a significant fraction of cells do not use the remaining energy sources for sporulation but rather delay spore formation or avoid it. The potential advantage for these cells is twofold. First, these cells increase in number and may sporulate later using nutrients re-

leased by cells that have lysed. This resource use, termed cannibalism or fratricide, has been demonstrated in a number of studies (21 and references therein). Second, these cells are capable of rapidly resuming growth in the event of a new flux of nutrients. In contrast, cells that have sporulated are committed to a long-term process of spore formation and subsequent germination. Each of these paths is a form of specialization that increases efficiency in one area at the expense of the other.

OUTLOOK

The strategies and mechanisms discussed in this review are not limited to the microorganisms mentioned here. Many other bacterial and fungal species display phenotypic variation that may reflect a form of bet-hedging (see 11, 25, 92, 93 and references therein). These include processes that affect intra- or interspecies competition, as well as host-pathogen interactions, such as mucoidy and cytotoxicity of *Pseudomonas aeruginosa* and bacteriocin production in *E. coli*. A major challenge for future research will be to assess the effects of variable phenotypes on the interactions between organisms under steady-state and fluctuating conditions. This finding(s) may shed light on the pressures responsible for the evolution of genetic networks that directly or indirectly result in population multistability.

SUMMARY POINTS

1. Research increasingly acknowledges the presence and importance of cell-to-cell variability for the perpetuation of clonal populations.

2. Multistability is a ubiquitous feature of bacteria involving many different processes.

3. Phenotypic variable populations show increased fitness compared with homogeneous populations under fluctuating environments.

4. Genetic logic-AND gates are common network motifs in bacteria to generate heterogeneity.

5. Cell states can be passed on from one generation to the next via EI and this process might be important in bacterial development.

6. Synthetic biology and qualitative analyses of network motifs are promising for biotechnological and medical applications.

DISCLOSURE STATEMENT

The authors are not aware of any biases that might be perceived as affecting the objectivity of this review.

ACKNOWLEDGMENTS

We apologize to those whose research could not be cited due to space limitations. JWV was supported by an Intra-European Marie-Curie Fellowship from the European Commission, and by a grant from the Biotechnology and Biological Sciences Research Council awarded to J. Errington. WKS was supported by a Rubicon fellowship from the Netherlands Organization of Scientific Research (NWO). JWV and WKS contributed equally to this manuscript.

LITERATURE CITED

1. Acar M, Becskei A, van Oudenaarden A. 2005. Enhancement of cellular memory by reducing stochastic transitions. *Nature* 435:228–32
2. **Acar M, Mettetal JT, van Oudenaarden A. 2008. Stochastic switching as a survival strategy in fluctuating environments. *Nat. Genet.* 40:471–75**

   2. Elegant experimental study that shows how switching affects population growth.

3. Ackermann M, Stearns SC, Jenal U. 2003. Senescence in a bacterium with asymmetric division. *Science* 300:1920
4. Aguilaniu H, Gustafsson L, Rigoulet M, Nystrom T. 2003. Asymmetric inheritance of oxidatively damaged proteins during cytokinesis. *Science* 299:1751–53
5. Aguilar C, Vlamakis H, Losick R, Kolter R. 2007. Thinking about *Bacillus subtilis* as a multicellular organism. *Curr. Opin. Microbiol.* 10:638–43
6. Ajo-Franklin CM, Drubin DA, Eskin JA, Gee EP, Landgraf D, et al. 2007. Rational design of memory in eukaryotic cells. *Genes Dev.* 21:2271–76
7. Alper H, Moxley J, Nevoigt E, Fink GR, Stephanopoulos G. 2006. Engineering yeast transcription machinery for improved ethanol tolerance and production. *Science* 314:1565–68
8. Angeli D, Ferrell JE Jr, Sontag ED. 2004. Detection of multistability, bifurcations, and hysteresis in a large class of biological positive-feedback systems. *Proc. Natl. Acad. Sci. USA* 101:1822–27
9. Artyomov MN, Das J, Kardar M, Chakraborty AK. 2007. Purely stochastic binary decisions in cell signaling models without underlying deterministic bistabilities. *Proc. Natl. Acad. Sci. USA* 104:18958–63
10. Atkinson MR, Savageau MA, Myers JT, Ninfa AJ. 2003. Development of genetic circuitry exhibiting toggle switch or oscillatory behavior in *Escherichia coli*. *Cell* 113:597–607
11. Avery SV. 2006. Microbial cell individuality and the underlying sources of heterogeneity. *Nat. Rev. Microbiol.* 4:577–87
12. Balaban NQ, Merrin J, Chait R, Kowalik L, Leibler S. 2004. Bacterial persistence as a phenotypic switch. *Science* 305:1622–25
13. Becskei A, Seraphin B, Serrano L. 2001. Positive feedback in eukaryotic gene networks: cell differentiation by graded to binary response conversion. *EMBO J.* 20:2528–35
14. Bigger JW. 1944. Treatment of *Staphylococcal* infections with penicillin by intermittent sterilisation. *Lancet* 244:497–500
15. Bishop AL, Rab FA, Sumner ER, Avery SV. 2007. Phenotypic heterogeneity can enhance rare-cell survival in 'stress-sensitive' yeast populations. *Mol. Microbiol.* 63:507–20
16. Bren A, Eisenbach M. 2001. Changing the direction of flagellar rotation in bacteria by modulating the ratio between the rotational states of the switch protein FliM. *J. Mol. Biol.* 312:699–709
17. Cahn FH, Fox MS. 1968. Fractionation of transformable bacteria from competent cultures of *Bacillus subtilis* on renografin gradients. *J. Bacteriol.* 95:867–75
18. Casadesus J, D'Ari R. 2002. Memory in bacteria and phage. *Bioessays* 24:512–18
19. Casadesus J, Low D. 2006. Epigenetic gene regulation in the bacterial world. *Microbiol. Mol. Biol. Rev.* 70:830–56

20. Chai Y, Chu F, Kolter R, Losick R. 2007. Bistability and biofilm formation in *Bacillus subtilis*. *Mol. Microbiol.* 67:254–63

21. Claverys JP, Havarstein LS. 2007. Cannibalism and fratricide: mechanisms and raisons d'etre. *Nat. Rev. Microbiol.* 5:219–29

22. Cohen D. 1966. Optimizing reproduction in a randomly varying environment. *J. Theor. Biol.* 12:119–29

23. Cohn M, Horibata K. 1959. Analysis of the differentiation and of the heterogeneity within a population of *Escherichia coli* undergoing induced beta-galactosidase synthesis. *J. Bacteriol.* 78:613–23

24. Cox RS 3rd, Surette MG, Elowitz MB. 2007. Programming gene expression with combinatorial promoters. *Mol. Syst. Biol.* 3:145

25. Dubnau D, Losick R. 2006. Bistability in bacteria. *Mol. Microbiol.* 61:564–72

26. Dworkin J, Losick R. 2005. Developmental commitment in a bacterium. *Cell* 121:401–9

27. Elowitz MB, Leibler S. 2000. A synthetic oscillatory network of transcriptional regulators. *Nature* 403:335–38

28. Elowitz MB, Levine AJ, Siggia ED, Swain PS. 2002. Stochastic gene expression in a single cell. Science 297:1183–86

29. Ferrell JE Jr. 2002. Self-perpetuating states in signal transduction: positive feedback, double-negative feedback and bistability. *Curr. Opin. Cell Biol.* 14:140–48

30. Foster PL. 2007. Stress-induced mutagenesis in bacteria. *Crit. Rev. Biochem. Mol. Biol.* 42:373–97

31. Fujita M, Gonzalez-Pastor JE, Losick R. 2005. High- and low-threshold genes in the Spo0A regulon of *Bacillus subtilis*. *J. Bacteriol.* 187:1357–68

32. Fujita M, Losick R. 2005. Evidence that entry into sporulation in *Bacillus subtilis* is governed by a gradual increase in the level and activity of the master regulator Spo0A. *Genes Dev.* 19:2236–44

33. Galhardo RS, Hastings PJ, Rosenberg SM. 2007. Mutation as a stress response and the regulation of evolvability. *Crit. Rev. Biochem. Mol. Biol.* 42:399–435

34. Garcia-Ojalvo J, Elowitz MB, Strogatz SH. 2004. Modeling a synthetic multicellular clock: repressilators coupled by quorum sensing. *Proc. Natl. Acad. Sci. USA* 101:10955–60

35. Gardner A, West SA, Griffin AS. 2007. Is bacterial persistence a social trait? *PLoS ONE* 2:e752

36. Gardner TS, Cantor CR, Collins JJ. 2000. Construction of a genetic toggle switch in *Escherichia coli*. *Nature* 403:339–42

37. Gardner TS, Collins JJ. 2005. Bistable genetic toggle switch. *U.S. Patent No. 6841376*

38. Gerdes K, Christensen SK, Lobner-Olesen A. 2005. Prokaryotic toxin-antitoxin stress response loci. *Nat. Rev. Microbiol.* 3:371–82

39. Godoy VG, Gizatullin FS, Fox MS. 2000. Some features of the mutability of bacteria during nonlethal selection. *Genetics* 154:49–59

40. Goldstein J, Pollitt NS, Inouye M. 1990. Major cold shock protein of *Escherichia coli*. *Proc. Natl. Acad. Sci. USA* 87:283–87

41. Hadden C, Nester EW. 1968. Purification of competent cells in the *Bacillus subtilis* transformation system. *J. Bacteriol.* 95:876–85

42. Haijema BJ, Hahn J, Haynes J, Dubnau D. 2001. A ComGA-dependent checkpoint limits growth during the escape from competence. *Mol. Microbiol.* 40:52–64

43. Hammer K, Mijakovic I, Jensen PR. 2006. Synthetic promoter libraries—tuning of gene expression. *Trends Biotechnol.* 24:53–55

44. Hamoen LW, Venema G, Kuipers OP. 2003. Controlling competence in *Bacillus subtilis*: shared use of regulators. *Microbiology* 149:9–17

45. Hasty J, McMillen D, Collins JJ. 2002. Engineered gene circuits. *Nature* 420:224–30

46. Heinemann M, Panke S. 2006. Synthetic biology—putting engineering into biology. *Bioinformatics* 22:2790–99

47. Ireton K, Rudner DZ, Siranosian KJ, Grossman AD. 1993. Integration of multiple developmental signals in *Bacillus subtilis* through the Spo0A transcription factor. *Genes Dev.* 7:283–94

48. Jablonka E, Lamb MJ. 2006. *Evolution in Four Dimensions: Genetic, Epigenetic, Behavioral, and Symbolic Variation in the History of Life*. Cambridge, MA: MIT Press

49. Kærn M, Elston TC, Blake WJ, Collins JJ. 2005. Stochasticity in gene expression: from theories to phenotypes. *Nat. Rev. Genet.* 6:451–64

28. Together with Reference 78, this study pioneered the use of an in vivo method to analyze and quantify noise.

48. Thought-provoking book in which the authors argue that there is more to heredity than genes.

50. Karmakar R, Bose I. 2007. Positive feedback, stochasticity and genetic competence. *Phys. Biol.* 4:29–37
51. Kaufmann BB, Yang Q, Mettetal JT, van Oudenaarden A. 2007. Heritable stochastic switching revealed by single-cell genealogy. *PLoS Biol.* 5:e239
52. Kearns DB, Chu F, Branda SS, Kolter R, Losick R. 2005. A master regulator for biofilm formation by *Bacillus subtilis. Mol. Microbiol.* 55:739–49
53. Kobayashi K. 2007. Gradual activation of the response regulator DegU controls serial expression of genes for flagellum formation and biofilm formation in *Bacillus subtilis. Mol. Microbiol.* 66:395–409
54. Korobkova E, Emonet T, Vilar JM, Shimizu TS, Cluzel P. 2004. From molecular noise to behavioural variability in a single bacterium. *Nature* 428:574–78
55. Krishnamurthy S, Smith E, Krakauer D, Fontana W. 2007. The stochastic behavior of a molecular switching circuit with feedback. *Biol. Direct.* 2:13
56. Kussell E, Leibler S. 2005. Phenotypic diversity, population growth, and information in fluctuating environments. *Science* 309:2075–78
57. Leisner M, Stingl K, Radler JO, Maier B. 2007. Basal expression rate of *comK* sets a 'switching-window' into the K-state of *Bacillus subtilis. Mol. Microbiol.* 63:1806–16
58. Lewis K. 2007. Persister cells, dormancy and infectious disease. *Nat. Rev. Microbiol.* 5:48–56
59. Lindner AB, Madden R, Demarez A, Stewart EJ, Taddei F. 2008. Asymmetric segregation of protein aggregates is associated with cellular aging and rejuvenation. *Proc. Natl. Acad. Sci. USA.* 105:3076–81
60. Lisman JE. 1985. A mechanism for memory storage insensitive to molecular turnover: a bistable autophosphorylating kinase. *Proc. Natl. Acad. Sci. USA* 82:3055–57
61. Maamar H, Dubnau D. 2005. Bistability in the *Bacillus subtilis* K-state (competence) system requires a positive feedback loop. *Mol. Microbiol.* 56:615–24

62. Experimental study showing directly for the first time that noise in *comK* transcription determines entry into the competent state in *B. subtilis.*

62. **Maamar H, Raj A, Dubnau D. 2007. Noise in gene expression determines cell fate in *Bacillus subtilis. Science* 317:526–29**
63. Maynard Smith J. 1964. Group selection and kin selection. *Nature* 201:1145–47
64. Mayo AE, Setty Y, Shavit S, Zaslaver A, Alon U. 2006. Plasticity of the *cis*-regulatory input function of a gene. *PLoS. Biol.* 4:e45
65. Molle V, Fujita M, Jensen ST, Eichenberger P, Gonzalez-Pastor JE, et al. 2003. The Spo0A regulon of *Bacillus subtilis. Mol. Microbiol.* 50:1683–701
66. Monod J, Jacob F. 1961. Teleonomic mechanisms in cellular metabolism, growth, and differentiation. *Cold Spring Harb. Symp. Quant. Biol.* 26:389–401
67. Moyed HS, Broderick SH. 1986. Molecular cloning and expression of *hipA*, a gene of *Escherichia coli* K-12 that affects frequency of persistence after inhibition of murein synthesis. *J. Bacteriol.* 166:399–403
68. Msadek T. 1999. When the going gets tough: survival strategies and environmental signaling networks in *Bacillus subtilis. Trends Microbiol.* 7:201–7
69. Mukai K, Kawata M, Tanaka T. 1990. Isolation and phosphorylation of the *Bacillus subtilis degS* and *degU* gene products. *J. Biol. Chem.* 265:20000–6
70. Muller-Hill B. 1996. *The* lac *Operon: A Short History of a Genetic Paradigm.* Berlin: Walter de Gruyter
71. Murphy KF, Balazsi G, Collins JJ. 2007. Combinatorial promoter design for engineering noisy gene expression. *Proc. Natl. Acad. Sci. USA* 104:12726–31
72. Neildez-Nguyen TM, Parisot A, Vignal C, Rameau P, Stockholm D, et al. 2008. Epigenetic gene expression noise and phenotypic diversification of clonal cell populations. *Differentiation* 76:33–40
73. Nester EW, Stocker BA. 1963. Biosynthetic latency in early stages of deoxyribonucleic acid transformation in *Bacillus subtilis. J. Bacteriol.* 86:785–96
74. Ninfa AJ, Mayo AE. 2004. Hysteresis vs graded responses: the connections make all the difference. *Sci. STKE* 2004:e20
75. Ninfa AJ, Selinsky S, Perry N, Atkins S, Xiu SQ, et al. 2007. Using two-component systems and other bacterial regulatory factors for the fabrication of synthetic genetic devices. *Methods Enzymol.* 422:488–512
76. Novick A, Weiner M. 1957. Enzyme induction as an all-or-none phenomenon. *Proc. Natl. Acad. Sci. USA* 43:553–66
77. Ortega F, Garces JL, Mas F, Kholodenko BN, Cascante M. 2006. Bistability from double phosphorylation in signal transduction. Kinetic and structural requirements. *FEBS J.* 273:3915–26

78. Ozbudak EM, Thattai M, Kurtser I, Grossman AD, van Oudenaarden A. 2002. Regulation of noise in the expression of a single gene. *Nat. Genet.* 31:69–73

79. Ozbudak EM, Thattai M, Lim HN, Shraiman BI, van Oudenaarden A. 2004. Multistability in the lactose utilization network of *Escherichia coli*. *Nature* 427:737–40

80. Partridge L, Barton NH. 1993. Optimality, mutation and the evolution of ageing. *Nature* 362:305–11

81. Paulsson J. 2004. Summing up the noise in gene networks. *Nature* 427:415–18

82. Pennington JM, Rosenberg SM. 2007. Spontaneous DNA breakage in single living *Escherichia coli* cells. *Nat. Genet.* 39:797–802

83. Ponder RG, Fonville NC, Rosenberg SM. 2005. A switch from high-fidelity to error-prone DNA double-strand break repair underlies stress-induced mutation. *Mol. Cell* 19:791–804

84. Rando OJ, Verstrepen KJ. 2007. Timescales of genetic and epigenetic inheritance. *Cell* 128:655–68

85. Raser JM, O'Shea EK. 2005. Noise in gene expression: origins, consequences, and control. *Science* 309:2010–13

86. Renzette N, Gumlaw N, Nordman JT, Krieger M, Yeh SP, et al. 2005. Localization of RecA in *Escherichia coli* K-12 using RecA-GFP. *Mol. Microbiol.* 57:1074–85

87. Rosche WA, Foster PL. 1999. The role of transient hypermutators in adaptive mutation in *Escherichia coli*. *Proc. Natl. Acad. Sci. USA* 96:6862–67

88. Simmons LA, Grossman AD, Walker GC. 2007. Replication is required for the RecA localization response to DNA damage in *Bacillus subtilis*. *Proc. Natl. Acad. Sci. USA* 104:1360–65

89. Smith MC, Sumner ER, Avery SV. 2007. Glutathione and Gts1p drive beneficial variability in the cadmium resistances of individual yeast cells. *Mol. Microbiol.* 66:699–712

90. Smits WK, Eschevins CC, Susanna KA, Bron S, Kuipers OP, Hamoen LW. 2005. Stripping *Bacillus*: ComK auto-stimulation is responsible for the bistable response in competence development. *Mol. Microbiol.* 56:604–14

91. Smits WK, Hoa TT, Hamoen LW, Kuipers OP, Dubnau D. 2007. Antirepression as a second mechanism of transcriptional activation by a minor groove binding protein. *Mol. Microbiol.* 64:368–81

92. Smits WK, Kuipers OP, Veening JW. 2006. Phenotypic variation in bacteria: the role of feedback regulation. *Nat. Rev. Microbiol.* 4:259–71

93. Smits WK, Veening JW, Kuipers OP. 2008. Phenotypic variation and bistable switching in bacteria. In *Bacterial Physiology: A Molecular Approach*, ed. W El-Sharoud, pp. 339–65. Berlin/Heidelberg: Springer-Verlag

94. Stewart EJ, Madden R, Paul G, Taddei F. 2005. Aging and death in an organism that reproduces by morphologically symmetric division. *PLoS Biol.* 3:e45

95. Süel GM, Garcia-Ojalvo J, Liberman LM, Elowitz MB. 2006. An excitable gene regulatory circuit induces transient cellular differentiation. *Nature* 440:545–50

96. Süel GM, Kulkarni RP, Dworkin J, Garcia-Ojalvo J, Elowitz MB. 2007. Tunability and noise dependence in differentiation dynamics. *Science* 315:1716–19

97. Sureka K, Ghosh B, Dasgupta A, Basu J, Kundu M, Bose I. 2008. Positive feedback and noise activate the stringent response regulator rel in mycobacteria. *PLoS ONE* 3:e1771

98. Swain PS, Elowitz MB, Siggia ED. 2002. Intrinsic and extrinsic contributions to stochasticity in gene expression. *Proc. Natl. Acad. Sci. USA* 99:12795–800

99. Thattai M, van Oudenaarden A. 2004. Stochastic gene expression in fluctuating environments. *Genetics* 167:523–30

100. Torkelson J, Harris RS, Lombardo MJ, Nagendran J, Thulin C, Rosenberg SM. 1997. Genome-wide hypermutation in a subpopulation of stationary-phase cells underlies recombination-dependent adaptive mutation. *EMBO J.* 16:3303–11

101. Veening JW, Hamoen LW, Kuipers OP. 2005. Phosphatases modulate the bistable sporulation gene expression pattern in *Bacillus subtilis*. *Mol. Microbiol.* 56:1481–94

102. Veening JW, Igoshin OA, Eijlander RT, Nijland R, Hamoen LW, Kuipers OP. 2008. Transient heterogeneity in extracellular protease production by *Bacillus subtilis*. *Mol. Syst. Biol.* 4:184

103. Veening JW, Smits WK, Hamoen LW, Kuipers OP. 2006. Single cell analysis of gene expression patterns of competence development and initiation of sporulation in *Bacillus subtilis* grown on chemically defined media. *J. Appl. Microbiol.* 101:531–41

78. Together with Reference 28, this study pioneered the use of an in vivo method to analyze and quantify noise.

94. Shows that even symmetrical dividing bacteria suffer from aging.

95. Provides evidence for the excitable model that dictates competence development.

104. Time-lapse study that shows that sporulation in *B. subtilis* is a bet-hedging strategy and that the signal to sporulate can be epigenetically inherited.

104. Veening JW, Stewart EJ, Berngruber TW, Taddei F, Kuipers OP, Hamoen LW. 2008. Bethedging and epigenetic inheritance in bacterial cell development. *Proc. Natl. Acad. Sci. USA* **105:4393–98**

105. Verhamme DT, Kiley TB, Stanley-Wall NR. 2007. DegU co-ordinates multicellular behaviour exhibited by *Bacillus subtilis*. *Mol. Microbiol.* 65:554–68

106. Vlamakis H, Aguilar C, Losick R, Kolter R. 2008. Control of cell fate by the formation of an architecturally complex bacterial community. *Genes Dev.* 22:945–53

107. Yildirim N, Santillan M, Horike D, Mackey MC. 2004. Dynamics and bistability in a reduced model of the *lac* operon. *Chaos* 14:279–92

108. Zambrano MM, Kolter R. 1996. GASPing for life in stationary phase. *Cell* 86:181–84

RNA Polymerase
Elongation Factors

Jeffrey W. Roberts, Smita Shankar, and Joshua J. Filter

Department of Molecular Biology and Genetics, Cornell University, Ithaca, New York 14853; email: jwr7@cornell.edu

Annu. Rev. Microbiol. 2008. 62:211–33

First published online as a Review in Advance on June 6, 2008

The *Annual Review of Microbiology* is online at micro.annualreviews.org

This article's doi:
10.1146/annurev.micro.61.080706.093422

Key Words

antiterminator, termination, pausing

Abstract

The elongation phase of transcription by RNA polymerase is highly regulated and modulated. Both general and operon-specific elongation factors determine the local rate and extent of transcription to coordinate the appearance of transcript with its use as a messenger or functional ribonucleoprotein or regulatory element, as well as to provide operon-specific gene regulation.

Contents

INTRODUCTION

The movement of RNA polymerase (RNAP) as it transcribes DNA is not uniform and inevitable, but instead it is modulated by regulatory influences that accelerate it or slow it or determine if it stops altogether. For bacterial RNAP the best-characterized examples are antiterminators, operon-specific genetic regulatory elements that allow genes downstream of a terminator to be expressed. Other processes coordinate transcription with the use of the transcript—usually of course as messenger or "structural" RNA, although RNA can be a regulatory element itself. Thus, in regulation by an attenuator RNA or certain riboswitches, the rate of RNA synthesis is critical to the function. A global view is that utilization of the emerging transcript is a precisely evolved pathway of interactions with the translation apparatus, processing factors, and the RNA degradation machinery. Disruption of this pathway can interfere with RNA function and cause potentially deleterious free RNA to accumulate; in response, the cell has a mechanism to stop futile transcription. Furthermore, transcription can be blocked by accident, for example, when a noncoding lesion like a thymine dimer is encountered in the DNA template strand or during clashes with processes of replication and recombination. In these cases cellular processes remove the transcription complex.

We consider first the establishment and structure of the transcription elongation complex, and then elongation factors that act upon it. Structural and functional studies have revealed how nucleic acids move through RNAP and what interactions between nucleic acids and enzyme stabilize the complex. This provides a basis to understand the activity of both intrinsic modulatory signals in the DNA—i.e., pause and termination signals—and regulatory proteins. Transcription termination and release of the transcript can be described in some detail, and several specific classes of pause-inducing sequences that provide signals and substrates for regulatory interactions are known. The most complete description of a regulatory function in transcription elongation is antitermination in *Escherichia coli*, mediated by the bacteriophage λ gene *N* and *Q* proteins, the cellular RfaH protein, and the cellular ribosomal RNA gene antitermination system that is analogous to that of λN. Several proteins discovered through studies of antitermination—primarily NusA and NusG—determine the nature of pausing. These are essential in *E. coli* and highly conserved in bacteria, and although some biochemical activities and even atomic structures are known, it is difficult to describe their precise cellular function. Specialized proteins deal with accidental disruption of the elongation reaction: Gre and Mfd factors catalyze, respectively, the removal of an aberrant RNA 3' end so that RNA synthesis can be restarted from an active primer, and the release of enzyme irretrievably blocked by a lesion in the DNA template strand. Other recent reviews have considered in greater detail some of the topics discussed below, as well as topics not discussed (e.g., attenuation) (11, 94, 109, 150).

BASIC PROCESSES OF TRANSCRIPTION

Initiation of Transcription

Once an open promoter complex is formed, the initial stage of RNA synthesis occurs with melting of downstream DNA and scrunching of DNA of the growing bubble into the enzyme, during which process the contacts of the initiation factor sigma with promoter elements are maintained. Energy stored as scrunched DNA may be used to break sigma-promoter contacts, thus effecting the transition to elongation (49, 108, 135). When this transition fails, the RNA aborts, a feature common to in vitro transcription by most RNAPs. Importantly, Gre proteins (see below) suppress abortive initiation and promote elongation.

Elongation of the RNA Chain

The elongation cycle comprises three basic steps (142): (*a*) binding of a template-complementary nucleoside triphosphate (NTP) into the active site; (*b*) chemical reaction of the RNA chain 3'-OH with the NTP α-PO$_4$, catalyzed by a pair of bound Mg^{2+} ions, resulting in one NMP addition to the RNA and liberation of pyrophosphate; and (*c*) translocation of the nucleic acid assemblage to place the next template base in the active center. The elongation complex of RNAP (**Figure 1**) is stabilized by several sets of interactions among protein and nucleic acids (58, 93, 141): (*a*) downstream duplex DNA is bound within the enzyme; (*b*) about nine nucleotides of RNA at the growing end are annealed to the template DNA strand, forming a 9-bp RNA/DNA hybrid enclosed by protein; and (*c*) an additional ~5 nucleotides of RNA upstream of the hybrid are bound in a protein channel until the RNA emerges 14 nucleotides from the growing end. Despite the high stability of the complex to dissociation, the various interactions allow lateral mobility of DNA and RNA through the complex during translocation. In fact, the RNA alone has some

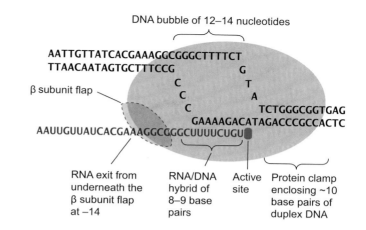

DNA bubble of 12–14 nucleotides

AATTGTTATCACGAAAGGCGGGCTTTTCT
TTAACAATAGTGCTTTCCG

β subunit flap

AAUUGUUAUCACGAAAGGCGGGCUUUUCUGU

RNA exit from underneath the β subunit flap at −14

RNA/DNA hybrid of 8–9 base pairs

Active site

Protein clamp enclosing ~10 base pairs of duplex DNA

Escherichia coli RNAP elongation complex

Figure 1

Structure of the transcription elongation complex.

mobility, showing an ability to slip relative to the template; this is observed at positions where a template homopolymer sequence is longer than 9 bp, so that the slipped position is not destabilized by DNA/RNA mismatches in the hybrid (145).

A remarkable activity of transcribing RNAP is to move backward along the template, or backtrack, reversing the translocation steps that assembled the RNA chain but not depolymerizing the chain itself; the RNA 3' end protrudes as single-stranded RNA from the secondary channel, an aperture that connects the active center to the outside and is believed also to be the entry site of the NTP substrate (57, 95). Backtracking is likely a response to some failure of the RNA 3' end to be elongated, as might occur, for example, through chemical damage or misincorporation. Backtracking is strongly stimulated by weak or disrupted pairing in the RNA/DNA hybrid where templating of the RNA occurs (95), consistent with its being a response to an aberrant hybrid.

Transcription Pausing

The average transcription rate of *E. coli* RNAP is in the range of 50–100 nucleotides per second, equaling a translocation step time of

10–20 ms. But the enzyme pauses frequently, remaining at the same template site for up to tens of seconds or more. Some of these pause events are significant and are detected both in vivo and in vitro, as described below. In addition to known authentic pause sites, however, single-molecule assays of transcription in vitro display continuous, low-efficiency pausing, apparently from common sequence features that modulate the elongation rate (2, 44, 88). Possibly this ubiquitous pausing reflects a finely detailed evolution of transcription rate to match the particular fate of the transcript. However, it cannot be excluded that much of this is an unnatural property of the purified transcription system.

A major reason for interest in transcription pausing is that antiterminators antipause (6, 53, 106, 153), meaning they reduce the half-life and/or efficiency of pausing at some sites. Furthermore, mutations in RNAP core subunits can either enhance or reduce pausing (29). It has long been thought that antipausing could explain antitermination (153), because the uridine-rich segment of intrinsic terminators induces pausing at the release site even in the absence of the hairpin, and certain Rho terminators are strong pausing sites in the absence of Rho (64, 85). Furthermore, it is plausible that the pause is kinetically necessary to provide time for the termination event. It is impossible by standard methods to measure a pause associated with a termination event of an active intrinsic terminator, so that no computational resolution of this proposal by considering pausing kinetics is possible, but other evidence that antipausing can underlie antitermination is described below. For neither core subunit mutations nor most antiterminators is it known if the effect is only on specific classes of pause sites, where their effects are well documented, or if there is a general effect on elongation in every translocation step.

In addition to pausing at terminators, there exist several well-defined types of sequence-based pausing with established regulatory consequences. Before it dissociates from the transcribing core enzyme, the σ^{70} initiation factor can recognize and bind a near repeat of the –10 component of the promoter sequence early in the transcription unit, inducing a transcription pause of the order of seconds in vivo; such a paused RNAP is recognized by the phage λ gene Q antiterminator as it engages RNAP (111, 114). High concentrations of σ^{70} can induce pausing in vitro at such sequences far from the promoter (82), but this may not occur in vivo. There is a significant occurrence of promoter-proximal sigma-dependent pauses in bacterial operons, but their regulatory significance is unknown (13, 42, 89).

Early in the 5'-leader RNA portions of the his and trp operons of E. coli and Salmonella (60) there is a transcription pause dependent mostly upon an RNA hairpin that forms in the emerging RNA and binds the β flap of core RNAP (138). Its function is to synchronize translation with transcription, which is required for operon regulation mediated by uncharged tRNA (59). The efficiency of pausing is determined partially by sequence elements surrounding the hairpin-coding segment (66). This pause is not backtracked; instead, there is evidence that pausing occurs because structural rearrangement of the active site prevents nucleotide addition to the RNA 3' end (139).

A pause at the operon polarity sequence (OPS) site of E. coli, where the RfaH antiterminator binds and engages the elongating complex, is backtracked, so that the RNA 3' end is unavailable until the complex isomerizes forward (5). Although there may be other important sequence elements, the major one would appear to be a GC-rich region preceding a U-rich region. This structure, similar to that at the release site of intrinsic terminators, promotes backtracking because the RNA/DNA hybrid containing GC-rich RNA is much more stable than the U-rich hybrid.

Termination. Three processes destabilize the elongation complex and release the transcript.

Intrinsic (or hairpin) termination. Intrinsic (or hairpin) termination is encoded in nucleic acid. Intrinsic terminators include most defined termination sites in bacterial transcription units, although this may be in part because they are fairly well recognized by sequence analysis. The intrinsic terminator has two essential components: a GC-rich hairpin that forms in the emerging transcript and is closed about nine nucleotides upstream of the RNA release site, and an adjacent U-rich segment that extends through most of the hybrid region. Release accompanies formation of the hairpin, an event that is believed to shorten the RNA/DNA hybrid, thus destabilizing this critical region of the complex, followed by dissociation (41, 56, 119, 156). The upstream portion of the hairpin can be replaced experimentally by a DNA oligonucleotide added in *trans* (156), implying that the function of the hairpin formation is probably mechanical, i.e., the hairpin as a structure is not required. The weakness of the rU/dA hybrid in the terminal region is thought to facilitate its unwinding and dissociation (78). Rewinding of the upstream hybrid region is essential to efficient termination, demonstrated by the failure of artificial heteroduplex templates to support termination (116). Hybrid shortening can occur by forward translocation of the enzyme, i.e., rewinding of upstream DNA and unwinding downstream without elongation of the RNA (62, 119), which also occurs in dissociation of stopped elongation complexes of bacteriophage T7 RNAP (160). Other pathways of RNA release exist; in particular, the homopolymeric terminal sequence can support slippage of the transcript past the template in RNA release by the T7 terminator (62, 74). Single-molecule analysis of termination suggests an important distinction among intrinsic terminators: If the uridine-rich terminal segment is interrupted by other bases, forward translocation occurs, whereas a homopolymeric sequence allows RNA release through slippage (62). A terminator that otherwise acts by forward translocation supports (slower) RNA release, presumably by slippage, if forward translocation is prevented by an interstrand DNA crosslink (119), suggesting that both mechanisms might apply to a terminator in different conditions.

Rho-dependent termination. Rho-dependent termination depends upon the Rho termination factor, an essential (in *E. coli*) hexameric ATP-dependent RNA translocase that binds the emerging transcript and is believed to pull it out of the transcription complex; in effect, ATP provides the energy that, in the case of the intrinsic terminator, is provided by hairpin formation (21, 109, 128). Rho provides the natural termination mechanism for many transcription units, as detected by extensive transcription into intergenic regions when Rho is inhibited by the antibiotic bicyclomycin (18). More broadly, Rho prevents accumulation of RNA that is not being utilized, either by the translation apparatus or by incorporation into an RNA-protein complex, as occurs normally when the noncoding ribosomal RNA is bound by ribosomal protein, tRNA is bound by synthetases and translation elongation factors, and small regulatory RNAs are bound by Hfq (37). Thus, Rho is the agent of operon polarity, the interruption of downstream gene expression by a nonsense codon; polarity results from Rho binding the ribosome-free RNA and stopping transcription (3, 30, 110). There is evidence that the problem with free RNA is its incorporation through strand displacement into DNA to form R-loops, the deleterious consequence of which is not understood (38). R-loop formation would be facilitated by the unconstrained negative supercoiling of the *E. coli* chromosome, but in fact R-loops are deleterious in eukaryotic cells as well (47). Rho-dependent terminators are the targets of the antitermination system of ribosomal RNA transcription (23), the bacteriophage λ gene *N* antiterminator (113), and the *E. coli* tryptophanase operon antitermination control (154).

Mfd-dependent termination. The transcription repair coupling factor Mfd is an ATP-dependent DNA translocase that dissociates a transcription complex by binding both RNAP and DNA emerging upstream from the complex (25, 100, 122). Mfd recognizes stalled RNAP rather than a particular genetic site, as can be demonstrated with any stalled complex in vitro. Mfd acts in DNA repair by recognizing RNAP stalled by a noncoding lesion in DNA, removing it from DNA, and mediating recruitment of the UvrABC excision repair enzymes to the site, using a domain believed to have affinity to UvrA (122).

Antitermination. Regulation of gene expression through antitermination was discovered in bacteriophage λ (113). Detailed genetic analysis of N-mediated antitermination uncovered the essential elongation proteins NusA and NusG, in addition to the involvement of other cellular proteins such as NusB and the ribosomal protein S10 (150). A common feature of antitermination by λN, λQ, *E. coli* RfaH, and the RNA-based *put* system of the λ relative HK022 is that the antiterminator binds at a genetically specified site and forms a persistent complex with RNAP that causes it to bypass terminators. Furthermore, this modification inhibits RNAP pausing (6, 53, 106, 153), implying some basic effect on the elongation reaction. This discovery, along with the fact that both intrinsic and Rho-dependent terminators have associated pausing, suggested that antipausing can underlie antitermination, which is at least partly true (126). All four systems are studied with purified components in vitro, leading to mechanistic insights at the molecular level; these are considered in detail below.

CELLULAR MACHINERY OF ELONGATION CONTROL

Gre Proteins

Like their eukaryotic counterpart transcription factor IIS (TFIIS), the universal bacterial Gre proteins mediate activities of the active center of RNAP by binding in and near the secondary channel and projecting a finger into the region of the active center (61, 98). Gre proteins (GreA and GreB of *E. coli*) were discovered through their activity to stimulate hydrolysis of a backtracked elongation complex by the RNAP active center of a backtracked elongation complex (12, 99). The reaction yields a 3'-OH primer end in the active center and a 3'-terminal oligonucleotide that diffuses away through the secondary channel. Gre enables catalysis of the reaction by providing two carboxylate residues at the inserted end that chelate an Mg^{2+} (61, 98) and complete the two Mg^{2+} set required for all reactions catalyzed by the active center, including hydrolysis (132). During chain elongation, this function is believed to be provided by the incoming NTP (132). Gre-stimulated cleavage of RNA would rescue complexes trapped in a backtracked state, although only a few specific instances of natural backtracking are known. One type of obstruction that gives rise to backtracking is the σ^{70}-dependent pause of bacteriophage λ (see below) (111). This paused complex, which presumably achieves a scrunched state as downstream DNA moves through the active center while the core framework is restrained by σ^{70}, can collapse into a Gre-sensitive structure that impedes release of enzyme from the pause both in vivo and in vitro (77). Other sorts of binding events that restrain RNAP from moving along DNA could induce backtracking and require rescue. In addition, the elongation complex might respond to collisions with barriers downstream by eventually diffusing backward into an arrested state that requires rescue by cleavage.

A specific proposal relates backtracking to proofreading of misincorporated NMP at the growing point (158). Because the mispaired base at the end provides little stabilizing energy, misincorporation favors not hydrolysis of the single mismatched terminal nucleotide, but instead backtracking, so that the active center is positioned at the previous phosphodiester bond; hydrolysis then removes a terminal dinucleotide (132). It is proposed that the second Mg^{2+} required to catalyze the hydrolysis is chelated by these two terminal nucleotides, an activity that promotes proofreading. This chelation of Mg^{2+} might be a primitive function of RNA that was usurped later by protein when the Gre factors evolved (158).

Gre proteins also promote the initial step of elongation by inhibiting abortive initiation (46), a presumably aberrant pathway in which the emerging RNA is lost and the incipient elongation complex collapses. The failure of a Gre protein altered to lack the terminal Mg^{2+}-binding carboxylates to inhibit abortive initiation (134) suggests that cleavage is required—but it does not prove it, because some other change in the environment of the active site could stabilize the scrunched intermediate in the initial elongation step and promote elongation. A possible precedent is the activity of the initiation factor DksA, a Gre-like protein that also projects into the active center through the secondary channel (105) and acts not to mediate cleavage, but instead to destabilize the open complex through interactions around the active site (104).

NusA

The highly conserved bacterial transcription elongation factor NusA, discovered over three decades ago because of its role in antitermination (31), is an intriguing protein with diverse and apparently contradictory effects on transcription elongation and termination. Its most apparent biochemical activities are to enhance pausing during transcription with purified RNAP (120), particularly at RNA hairpin-induced pauses (6), and to increase the activity of intrinsic terminators (120). Contrarily, NusA also is an essential component of the λ gene N antitermination complex and important for the λ gene Q antiterminator, both of which inhibit pausing; thus the activity of NusA is subverted to a distinct function by the antitermination proteins.

E. coli NusA is a 55-kDa monomeric RNA binding protein. Structural studies of the closely related *Mycobacterium tuberculosis* NusA either alone or in combination with RNA (10, 36) reveal an elongated protein with three distinct domains: (*a*) an N-terminal RNAP binding domain; (*b*) a middle portion comprising three RNA binding globular domains, S1, KH1, and KH2; and (*c*) a C-terminal autoinhibitory domain. RNA binding is cryptic in full-length NusA (71) but revealed by deletion of its C-terminal domain (CTD) (75). A con-

tact point of NusA to RNAP is the α-CTD, which binds the NusA CTD and potentiates RNA binding by the S1 and KH domains; cross-linking experiments suggest that RNA previously bound to the α-CTD now binds NusA (71). Deletion of the α-CTD leads to a loss of NusA functions such as enhanced pausing, Q-mediated antitermination, and Q-mediated occlusion of the RNA transcript (72, 126). NusA does not require the α-CTD for its function as a cofactor of antiterminator λN (72). Binding assays and affinity chromatography show that λN binds the NusA CTD and thus presumably provides a function for which the α-CTD is needed in other NusA activities (75).

In addition to the NusA CTD binding site on the RNAP subunit α-CTD, the NusA N-terminal domain (NTD) has been suggested to have homology to σ^{70} region 2 (11) and thus to bind the same N-terminal coiled-coil of the β′ subunit of RNAP to which the σ^{70} region 2 binds; some cross-linking evidence supports this (140). Consistently, NusA binds to core RNAP in vitro but not to holoenzyme-containing σ^{70} (40). Other cross-linking experiments and RNA protection studies with functional elongation complexes suggest that NusA likely makes contacts near the RNA exit channel, around the β-flap domain (41, 71, 126, 138); conceivably, an extended conformation of the NusA protein contacts both regions, not unlike the wide reach of σ^{70}. NusA no longer enhances pausing when the flap-tip helix of the β subunit is deleted, again suggesting an interaction with the β-flap domain (138).

The biochemical basis of pause enhancement by NusA is not understood. One view is that pausing frequently derives from RNA structure in the emerging transcript, like that of the *trp* and *his* hairpins but perhaps less extensive; NusA stimulation of the formation of such structures would increase pausing (138). The hairpin, in turn, is proposed to act through long-range allosteric changes that distort the active center and inhibit polymerization. Alternatively, NusA binding might allosterically reshape protein structures adjacent to the nucleic acid binding pocket, including the active center, to affect pausing; a precedent is mutations in the rifampicin binding pocket that incidentally affect pausing (29). Whatever the mechanism, this activity of NusA is fundamentally altered when an antiterminator (λQ or λN) is present. For example, NusA binds reversibly to RNAP in affecting pausing and termination (120) but is believed to bind irreversibly in an antitermination complex with N (8, 45) or Q of bacteriophage 82 (82Q) (126). Presumably, interactions with the antiterminator and other factors provide the extra stability in the complex. 82Q actively inhibits pausing by itself, and incorporation of NusA into a complex with 82Q does not increase pausing; 82Q is dominant over NusA in the complex (153). One interpretation is that emerging RNA in a complex containing both 82Q and NusA is structured differently from RNA in a complex containing NusA alone; a further aspect of this structure is that emerging RNA is protected for about 10 nucleotides from nuclease attack in a completely NusA-dependent way, an activity that 82Q alone does not have (see below).

How does NusA enhance intrinsic termination (120, 127)? The effect could be secondary to stimulation of pausing, through simple kinetic effects that provide more time for the intrinsic termination mechanisms to act; there is a competition between termination and elongation at the site of release of intrinsic terminators. However, NusA probably acts directly to stabilize the terminator hairpin: In a model system that uses a small oligonucleotide to mimic a terminator hairpin, NusA enhances RNA release from static transcription elongation complexes. Presumably, NusA stimulates formation of the DNA/RNA hybrid that substitutes for the terminator hairpin (126), consistent with previous views of NusA function (41, 138).

Despite the presumption that NusA is universally involved in cellular transcription, its major known role is in antitermination of bacterial ribosomal RNA (rRNA) operons, an activity discerned both in vitro and in vivo and probably similar to the N antitermination system (23, 144). Furthermore, there is direct evidence for NusA influence on mRNA elonga-

tion in vivo from a study of response to starvation. In the stringent response, the nucleotide guanosine $3'5'$-bisphosphate (ppGpp) accumulates to high concentration through synthesis from GTP, and one effect of ppGpp is to slow mRNA elongation. In a NusA mutant, this effect disappears, suggesting that NusA mediates a ppGpp-dependent slowing of transcription (144). Despite an obvious similarity, it is not clear how this ppGpp-dependent slowing is related to the pausing stimulated by NusA in vitro; perhaps other cellular factors are involved.

NusA is essential in wild-type *E. coli*, although in combination with other mutations a *nusA* deletion can survive (159). A recent discovery clarifies this: Deletion from *E. coli* of horizontally acquired DNA, including cryptic prophages and transposons, allows *nusA* and *nusG* deletions to survive, although they grow slowly. The single genetic locus making *nusA* essential is the cryptic prophage *Rac* (18). Furthermore, the reduced genome strain is less sensitive to the Rho inhibitor bicyclomycin, suggesting that high activity of Rho is required for cell survival because it terminates transcription of these prophages. Microarray expression analysis of the reduced genome cell reveals strikingly similar patterns of gene expression in three conditions: inhibition of Rho and deletion of either *nusA* or *nusG* (18). Because NusG stimulates Rho-dependent termination in vitro (14), the result is consistent with NusG acting as a cofactor of Rho in vivo. This result also clarifies the effect of NusA in Rho-dependent termination: Although NusA somewhat inhibits Rho function in vitro, this activity is probably irrelevant. Instead, NusA acts with Rho to stimulate termination in vivo, and an attractive model is that stimulation of pausing by NusA synchronizes Rho function with the emerging transcript by slowing RNAP and allowing Rho time to act.

NusG

NusG is a bacterial transcription elongation factor that was identified biochemically as an essential component of the phage λN-mediated antitermination complex (68) and genetically as a suppressor of *nusA* or *nusE* mutants that impair N function (137). It is highly conserved among bacteria and archaea and is homologous to elongation factor Spt5 of eukaryotes. NusG has two known activities detectable both in vivo and in vitro: It inhibits pausing and increases the rate of elongation (17), and it enhances Rho-dependent termination (14), particularly in suboptimal conditions. NusG binds both Rho and RNAP, providing a possible link between them; however, Rho is active in vivo in the absence of NusG (18), so that this link cannot be essential. Stable NusG association with a transcription elongation complex has been detected only in the presence of either NusA or Rho. That these are RNA binding proteins suggests that NusG recognizes RNA-bound protein in the complex (68). However, NusG has no reported affinity for RNA by itself. NusG inhibits a backtracked pause but not an RNA hairpin-stabilized pause, suggesting that it acts by stabilizing the forward translocated state of the transcription complex (5, 102).

The ability of NusG to stimulate Rho activity but inhibit pausing appears paradoxical, because pausing is believed to be an essential prelude to Rho activity. Most likely, NusG stimulates Rho function in some way independent of its effect on pausing. Kinetic studies revealed that NusG is required for Rho-dependent termination when RNAP is elongating rapidly but not when elongation is slowed by low NTP concentrations (14), suggesting that NusG stimulates interaction of Rho with the elongation complex. In fact, NusG enhances the rate of Rho-mediated RNA release from stalled elongation complexes (15, 86). NusG also causes a promoter-proximal shift in the end points of Rho-terminated RNA and allows Rho to use shorter segments of upstream RNA than NusG can use alone, both effects indicating that NusG enhances Rho interaction with RNA (87). As discussed below, studies of the operon-specific transcription factor RfaH, whose NTD is a close paralog of NusG, suggest strongly that NusG binds the same coiled-coil structure of the RNAP β' subunit as does σ^{70} region 2.

Rho

Several early studies uncovered the biochemical basis of Rho action. First, Rho is an RNA-dependent ATPase (73), reflecting energy input in termination that depends upon binding the transcript. Second, the amount of transcript required for termination is a substantial 60–80 nucleotides of relatively unstructured RNA (63, 84). There is a distinct preference for cytidine and, to a smaller extent, uridine in the RNA, originally discovered as the high activity of polyribocytidylic acid in the ATPase assay (73). Assays with different polynucleotide activators revealed two types of RNA interaction sites, termed primary and secondary; the primary sites reflect the preference for cytidine (35, 81). Despite the presence of some recurring oligonucleotide motifs in a few instances, there is no sequence-specified Rho binding site: sites of Rho action are determined mostly by the functional context of the RNA, e.g., the absence or presence of translation. Nonetheless, these sites may be well determined, as in the phage λ early operons, in the *E. coli* tryptophanase leader region (154), or presumably in the ends of a large fraction of *E. coli* operons (18). Or, the sites may be contingent upon the failure of translation, for example, after a nonsense mutation in a structural gene or accident or stress that reduces the availability of charged tRNA.

Structural analysis of Rho in complex with RNA gives a detailed picture of its function (128). The protein is a hexamer of 48 kDa subunits, each of which comprises an N-terminal RNA binding domain and a CTD with both RNA binding and ATPase activities. The primary and secondary RNA binding sites can be associated with the two domains: Primary binding to the NTDs occupies the major 5'-terminal segment of the RNA, and the secondary sites in the CTDs bind the 3' portion of the RNA and direct it through the cavity in the center of the hexamer (128). The otherwise closed hexamer ring can exist as a lock washer variant in which the ring is separated at one monomer-monomer interface, allowing RNA to enter the cavity without a free end (128).

ATP binding and hydrolysis drive a sequential binding of RNA segments that propels RNA through the hexamer in a 5'-3' direction. In effect, the protein is an ATP-dependent RNA translocase (1, 128). A widespread view is that Rho tracks along the RNA in this direction, presumably faster than RNAP emits RNA, and then effects termination when it catches up to RNAP. Although relative movement of RNA and protein is central to RNA function, it is not known that Rho moves a substantial distance from its initial binding site to release RNA; instead, the movements may reflect the act of RNA release after Rho binds initially to its site of action in the elongation complex. It is also unknown how and where Rho contacts RNAP. The only demonstrated protein interactions are between Rho and NusG (69) and between NusG and RNAP (68), so that NusG could be a natural bridge. However, Rho works both in vitro and in vivo in the absence of NusG (18). Rho may act in an untargeted mechanical fashion against RNAP (121), a notion supported by the ability of Rho to release transcripts from distantly related RNAP II of eukaryotes in vitro (26).

A unifying view of the activities of intrinsic (hairpin) terminators and Rho is that both hairpin formation and the ATP-dependent Rho translocase exert force to extract RNA from the complex (101). As for the intrinsic terminator, collapse of the upstream segment of the open transcription bubble at the site of release is necessary for efficient RNA release by Rho, i.e., the energy of DNA rewinding helps drive release (101, 116). This evidence for branch migration at the back end of the transcription bubble, along with the absence of homopolymeric sequences at the release sites of Rho-dependent terminators, supports a forward translocation model for Rho function.

Mfd

Mfd stands for mutation frequency decline, reflecting its function in transcription-coupled DNA repair, a universal cellular process that removes transcribing complexes obstructed and trapped by nontemplating DNA lesions (122). In fact, any stopped transcription complex is a target of Mfd in vitro, and a nonrepair function is revealed by *mfd* mutants that facilitate transcription through transcription roadblocks in vivo (157), presumably by allowing RNAP to remain undisturbed until the block diffuses away. Mfd not only removes stopped transcription complexes, but it also recruits the DNA excision repair machinery to the site, an activity that is little understood.

Mfd is a 130-kDa multidomain protein with several well-characterized functions: an ATP-dependent DNA translocase, an RNAP binding domain, a putative UvrA binding domain, and a C-terminal regulatory domain (25). A high-resolution atomic structure of Mfd provides a basis for modeling Mfd function and its interaction with RNAP (25) and suggests how the regulatory domain may control interaction with excision repair enzymes. Mutational studies reveal the details of its interaction with the β subunit of RNAP and the molecular motions that underlie the DNA translocase activity, and show that the regulatory domain inhibits translocase activity until Mfd is bound to an elongation complex (19, 130, 131).

The mechanism of Mfd-mediated RNA release was revealed by the finding that Mfd induces forward translocation of RNAP in stopped complexes in vitro, as demonstrated with persistently backtracked complexes (100). Mfd simultaneously binds about 20 bp of DNA emerging upstream from the enzyme and a site on the β subunit, using the translocase activity and the energy of ATP to force the enzyme downstream. Because backtracked complexes are released if no NTP substrates are available, but are rescued into productive elongation in the presence of NTPs, release must be preceded by translocation to the fully forward position and must also occur from this position; this result indicates that a continuation of the action that induced forward movement is responsible for release (100, 101). Further, as for the intrinsic terminator and Rho, RNA release is facilitated by rewinding of DNA in the upstream transcription bubble region (101). These results suggest two likely complementary models of release: If there is a blockage that prevents further RNAP movement, collapse of the bubble within the complex from the torque imposed by the DNA translocase activity induces release, whereas failure of elongation (e.g., from a nontemplating lesion) where there is no blockage induces release by hybrid shortening through continued forward translocation with unwinding of downstream DNA.

ANTITERMINATION SYSTEMS

N Protein of Phage λ

Bacteriophage λN and λQ proteins modify *E. coli* RNAP into a termination-resistant form that allows readthrough of downstream termination signals. Antitermination was discovered through studies of the λ early promoters *pL* and *pR*, whose service to distal regions depends upon N allowing readthrough of nearby Rho-dependent terminators (113); the persistence (sometimes called processivity) of the N modification was revealed by showing that the sites of engagement and action of N are different (34). The N binding/modification sites *nutL* and *nutR* (N-utilization) (117), which are downstream of *pL* and *pR* but before the terminators, function as RNA. A complex including *nut*, N, and the *E. coli* accessory factors NusA, NusB, NusE, and NusG remains associated with RNAP after *nut* is transcribed, endowing persistence (**Figure 2**) (8, 45). The *nut* sites consist of the *boxA* sequence and *boxB* stem loop, separated by a short spacer region (20, 97) (**Figure 2**).

Numerous interactions construct the stable N antitermination complex: the ARM (arginine-rich motif) of the N-terminal portion of the 107-amino-acid λN protein binds to *boxB* (20, 65, 67); a complex of NusB and NusE binds to *boxA* (69, 79, 92); the core subunit α-CTD binds to NusE (103); N binds to NusA, as well as to some unknown core contact (40); and NusA binds to RNAP. Furthermore, NusG binds RNAP. There is no information about N contacts to RNAP core subunits. Whereas

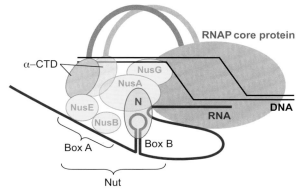

λN-modified elongation complex

Figure 2

Structure of the λN-modified elongation complex. Bacterial accessory proteins are shown in orange (NusA, NusB, NusE, NusG), RNAP core protein in gray, RNA in red, and DNA in black. Contacts between most accessory proteins and the RNAP core protein are based on known interactions, but the overall configuration is arbitrary. No contact of NusG with other accessory proteins is known, although NusG binds to the RNAP core.

N alone can modify elongation complexes in vitro (107), and a complex of N and NusA alone works better (151), stable antitermination at distances from the promoter requires the complex (24, 80, 107).

Several of the host factor requirements, including NusA, NusB, and NusE (ribosomal protein S10), were identified genetically in screens for mutations that abolished N function (31–33). The absolute requirement of NusA for N function in vivo is shown by the failure of λ, but the ability of a λN-independent variant, to grow on *nusA* deletion strains (18, 159).

N antiterminates at both Rho-dependent and intrinsic termination sites. For a Rho-dependent site near *nutR*, the mechanism may be simply obstruction of the Rho binding site by proteins of the *nut* complex (143). However, for the persistent antitermination activity downstream there are two potential mechanisms: antipausing that speeds RNAP through the critical release sites faster than some rate-limiting step in either termination mechanism, and stabilization of the complex, for example, by blocking some step of termination. Measurements of pausing at terminators and antipausing induced by N purport to rule out antipausing as a mechanism (41, 106); however, these experiments do not determine the actual kinetics at normal terminator release sites in real time. There is evidence that N-modification suppresses termination by directly preventing the formation of the hairpin at an intrinsic terminator (41). We argue below that Q protein antiterminates by both mechanisms.

Q Protein of Phage λ

The phage λ family Q proteins are antiterminators that regulate phage late gene expression by becoming subunits of RNAP (27, 156) through a pathway distinct from that of N protein: They bind to a transcription elongation complex held at a σ^{70}-dependent pause site near the promoter, specifically recognizing a site in DNA (**Figure 3**) (111, 114, 155). The paused complexes, which contain 16–25 nucleotides of RNA among the related phage, form independently of Q and are induced by a reiteration of the −10 promoter element that binds σ^{70} region 2 in the open promoter complex. In the paused complex σ^{70} regions 2 and 4 occupy positions similar to those in the open promoter complex, even though there is no sequence like the promoter −35 element present (76). The linker between σ^{70} regions 3 and 4 is displaced in the

paused complex, relative to the open complex, which is expected because emerging RNA occupies the same channel. Complexes stopped artificially at this site in conditions in which σ^{70} is not present are not modified by Q, so that σ^{70} provides an important structural or functional role in addition to stopping RNAP (155).

In the engagement complex, Q contacts a DNA site overlapping that occupied by σ^{70}-region 4, necessitating its displacement. In particular, λQ stabilizes the binding of region 4 to a DNA site adjacent to that bound by σ^{70} region 2 through a specific protein-protein contact (91). λQ binding to the DNA in the context of the paused early elongation complex (which contains a 16-nucleotide nascent RNA) is strengthened or weakened by σ^{70} region 4 mutations that, respectively, weaken or strengthen binding of region 4 to its natural site on the β subunit flap. Thus, before λQ displaces it, σ^{70} region 4 is bound to the β flap of core RNAP in the paused complex (90). Because Q-modified complexes that proceed downstream in vitro lack σ^{70} (156), Q engagement probably is accompanied by the release of σ^{70} from the complex. The binding site of Q in the core subunits has been elusive; a search for RNAP mutations that impair Q function revealed numerous sites believed to underlie conformation changes that enable the Q modification but no plausible binding site (118). Recently a binding site has been found in the β flap (P. Deighan, C. Diez, M. Leibman, A. Hochschild, and B.E. Nickels, unpublished data), which also is the locus of σ^{70} region 4 displacement by Q and of the RNA barrier described below. Contact with the α-CTD also can be detected (B.E. Nickels and A. Hochschild, unpublished data), although Q has activity in the absence of the α-CTD.

The σ^{70}-dependent paused complex is homologous in structure to the open promoter complex, in which scrunching of DNA allows initial stages of synthesis to occur without sigma release (49, 108). Thus, the λ +16 σ^{70}-dependent paused complex has a natural extension to only +12 according to the position where σ^{70} binds, implying a scrunch of

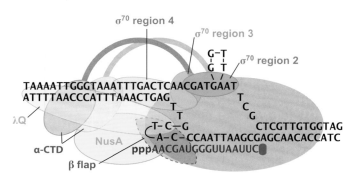

Paused complex with λQ

Figure 3

The promoter-proximal σ^{70}-dependent paused complex of the λ late gene promoter, with the antiterminator λQ bound. As described in the text, the DNA is shown scrunched by four nucleotides (49), allowing synthesis to +16.

four nucleotides of each DNA strand (77). Just as scrunching energy is believed to be used for the open promoter complex to break its sigma-DNA bonds, we suggest that the scrunching energy of the σ[70]-dependent paused complex is used for escape from the pause to produce the Q-modified complex. In this case, scrunching energy may enable breaking bonds between DNA and both σ[70] region 2 and Q in order to allow escape. (A variant model is that only one monomer of a Q dimer binds DNA, and only the second monomer binds and travels with RNAP; in this case, scrunching energy might be used to separate the Q subunits.) Q binding initiates the escape process, because addition of Q to a preformed paused complex accelerates escape. An interesting speculation is that in the initial stage of escape, scrunching continues beyond the usual stable extension (e.g., +16), storing more energy. A possible by-product of such extended scrunching is a backtracked λQ-dependent pause at about +25 (39). Collapse at the +25 site into a backtracked state could be analogous to collapse of the +16 complex when it extends (in the absence of Q) to +17, and to collapse of the open promoter complex in the process of abortive initiation.

The final element of the Q-bound paused complex is NusA. NusA stimulates the activity of Q (even though Q is distinctly active in its absence) (39, 152) and stabilizes the Q-bound paused complex against exonuclease digestion (155). For 82Q from the λ relative phage 82, but likely also for λQ, the modified complex that eventually escapes the pause site has distinct properties if NusA is present during modification (see below). Thus, NusA also can be present at the σ[70]-dependent paused complex. Because the NusA-dependent effect also depends on the RNAP core subunit α-CTD—a binding site of NusA—both NusA and the α-CTD can be functional elements of the Q-modified paused complex (**Figure 3**).

The properties of the 82Q-modified elongation complex provide a new view of how an antiterminator can modify RNAP, as well as a potential resolution of the roles of NusA and antipausing in antitermination (**Figure 4**).

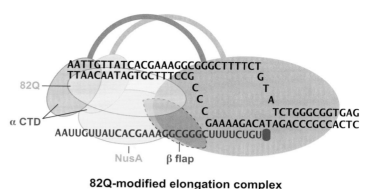

82Q-modified elongation complex

Figure 4

The 82Q-modified elongation complex. NusA and 82Q are shown obscuring about 10 nucleotides of emerging RNA to represent the protection of this segment from nuclease digestion, but the actual relative configuration of RNA and proteins is unknown.

Although NusA stimulates antitermination, Q antiterminates without it, and in fact Q is active in vivo in a *nusA* deletion strain. What NusA contributes was revealed by nuclease probing of RNA emerging from the 82Q-modified complex: If NusA is present, the usual RNA emergence at about 14 nucleotides from the growing end is extended to about 25 nucleotides, implying that NusA and 82Q construct a barrier where the hairpin of an intrinsic terminator forms (126). Static complexes modified by 82Q in the presence of NusA but not in its absence resist both an oligonucleotide-mediated release reaction that models the intrinsic terminator (126, 156) and the activity of Rho (126); the latter suggests either that Rho has an obligatory target site in the region where RNA emerges, or that the Q-modified complex is strengthened enough to resist the force of the Rho translocase.

If NusA is not present, antitermination still occurs, but in this case there is evidence that antipausing is the mechanism: Restricting elongation rate at the critical release site of an intrinsic terminator by reducing the substrate NTP concentration impairs antitermination in the absence of NusA, but not in its presence (126). The role of these two different modes of antitermination in vivo remains to be understood.

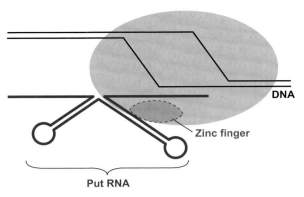

HK022 Put-modified elongation complex

Figure 5

The phage HK022 Put-modified elongation complex. The RNA secondary structure is indicated by two hairpin stems and loops, and the length of RNA is shown as it might be when Put first binds as it emerges from RNAP.

Put RNA of HK022

The phage λ relative HK022 also regulates expression of its early and late genes by antitermination. Like λN and its Nut sites, the Put system requires two RNA sites, PutL and PutR (polymerase utilization, left and right), encoded downstream of the early promoters *pL* and *pR*. The novelty of the Put system is that these nascent RNAs alone are the antiterminators (22, 53, 96, 124): Mutational analysis, structural probing by RNase cleavage, and prediction methods show that PutL and PutR each folds into a pair of stem loops that binds to and modifies RNAP to a termination-resistant form (**Figure 5**) (7, 53). The downstream or proximal 3′ loop of the Put RNA interacts directly with RNAP, whereas the distal 5′ loop facilitates this interaction, without making detectable contacts with the elongation complex itself (124). Put RNA interacts with RNAP through the zinc finger motif of the β′ subunit, and mutations in this motif eliminate *put* function (53, 55, 123). Put-modified elongation complexes are similar to λN-modified or Q-modified complexes: They read through both Rho-dependent and intrinsic terminators, and they display an increased elongation rate (53, 124). Presumably this effect on elongation rate is equivalent to antipausing, meaning that the

Put modification could act by affecting the elongation/release competition at a terminator, similarly to λQ or N. Put also of course could directly block emerging RNA, similarly to λQ and possibly λN.

The PutL modification of RNAP strongly prevents pausing at a uridine-rich, backtracked pause site just downstream of PutL itself. This mechanism, however, is unique: Formation and binding of the PutL structure to RNAP constrain the transcript such that its retraction into RNAP to produce a backtracked complex close to PutL is prevented (T. Velikodvorskaya, N. Komissarova, R. Sen, R. King, S. Banik-Maiti, and R. Weisberg, unpublished data). Moving the pause sequence a few base pairs downstream prevents this PutL effect on pausing. Because antitermination works at any distance, and because the elongation effect persists for long distances downstream, the processive effects of Put must have a different basis, presumably like that of other antiterminators.

Nun Protein of Phage HK022

The λ-related phage HK022 mounts a striking assault in the evolutionary wars: It expresses a protein that binds RNA as does λN, but with the opposite activity. Nun binding to Nut sites causes termination, not antitermination. Furthermore, Nun binding blocks translation of N protein mRNA (52). The HK022 lysogen constitutively expresses Nun, which arrests transcribing RNAP and destroys the transcription program of a sensitive phage (54, 112), no doubt acting to prevent superinfection by competing phages (20, 96, 112, 115). In a purified system with RNAP and λ DNA, Nun alone can cause transcription arrest (48), but NusA, NusB, NusE, and NusG enhance this arrest activity in vitro (115). Correspondingly, several of the *nus* mutations that interfere with N-mediated anititermination also disrupt Nun function in vivo. The requirements at *nutL* and *nutR* differ: NusA and NusG are not required for Nun-mediated arrest at *nutL* (16, 51, 147). Despite this difference, the set of Nus factors required by N also is recruited by Nun to construct a particle around the *nut* site.

The N-terminal segments (ARMs) of N and Nun are similar and contact *nut* identically (20, 28, 136). The N-terminal portions of N and Nun can be interchanged without loss of function in either case, implying that the termination function of Nun resides in the C terminus (43). In fact, the C terminus of Nun contains residues responsible for interaction with both RNAP and the DNA template, contacting RNAP in a Zn^{2+}-dependent manner through a cluster of histidine residues (148, 149). The C terminus of Nun also contains residues responsible for its interaction with the DNA template just downstream of the transcription bubble: A pair of charged residues (K106 and K107) binds the phosphate backbone of the DNA, and the penultimate residue (W108) is proposed to intercalate into the DNA strands (50, 149), causing the elongation complex to arrest just downstream of the *nut* site (149). In vivo, arrested transcripts are terminated and released, whereas in a minimal in vitro system the transcripts remain associated with the elongation complex (48, 112, 129). The minimal system lacks the *E. coli* Mfd protein, which releases stalled transcription elongation complexes (100, 122) and facilitates the release of Nun-arrested complexes both in vivo and in vitro (146).

E. coli RfaH Protein

E. coli RfaH, a regulator of virulence genes, is an antiterminator that inhibits transcription pausing (6), an activity that probably underlies its antitermination mechanism. Like λQ, RfaH engages RNAP at a natural transcription pause site (OPS), although this pause is induced by sequence elements that affect RNAP core subunits rather than σ^{70}; the paused complex at OPS is stabilized by backtracking (6). One underlying similarity is that both σ^{70} and RfaH recognize the nontemplate DNA strand in the paused complex. The mechanism of engagement is well understood: RfaH has an NTD similar to the close paralog NusG that binds RNAP and DNA, and a distinct CTD that sequesters the RNAP binding site (9). Binding of RfaH to OPS site DNA in the paused complex induces a rearrangement that exposes its RNAP binding surface, which then associates with RNAP and allows RfaH to travel along with the elongation complex.

As this pathway would predict, the isolated RNAP binding domain of RfaH associates with RNAP independently of the OPS pause site. The binding site of RfaH on RNAP is the β' subunit coiled coil that also is the binding site of σ^{70} region 2 (9). Mutations that block RfaH binding and function without interfering with σ^{70} function can be found. An important implication of this binding site is that the close paralog NusG likely binds to the β' coiled coil. A further implication is that because RfaH, σ^{70} region 2, and NusG compete for the same site, the ability of σ^{70} region 2 to rebind during elongation after its initial dissociation and to induce downstream pauses by the same mechanism as the promoter proximal pause where λQ protein engages would depend upon its outcompeting NusG or RfaH. In vitro, RfaH wins this competition, suggesting that σ^{70}-dependent pausing in downstream elongation might not occur through σ^{70} rebinding (125).

E. coli Ribosomal RNA Antitermination

The ribosomal RNA operons of *E. coli* include an antitermination system similar to that of λN protein, in that the same accessory factors, NusA, NusB, NusE, and NusG, appear to be involved (23, 133). Thus, mutations in *nusB* and *nusG* inhibit antitermination in ribosomal operons. Furthermore, there is a defined site similar to λ *nut*, including an essential *boxA* sequence (70). However, there is no known analogue of N itself, so either the analog has not been found, or, more likely, the system operates differently.

The existence of the ribosomal RNA antitermination system was implied by the discovery that insertion sequences in the ribosomal operons are not polar, i.e., do not support the expected Rho-dependent termination (83). The system is active in antiterminating at Rho-dependent terminators both in vitro and in vivo with a reporter system (4, 133), although it has not been reconstituted with entirely purified components, unlike the other antitermination systems. Clear evidence for the function of the ribosomal system in vivo derives from measurements of elongation rates of either ribosomal RNA synthesis or mRNA synthesis in constructs to which the ribosomal antitermination site is fused: These rates are about twice as great in the presence of antitermination (144). Presumably this greater speed reflects the absence of pausing, which also underlies antitermination; however, pausing is generally measured in vitro, and there is no precise correlation between the in vivo and in vitro phenomena. A biological rationale for antitermination in transcription of noncoding RNA is clear: Ribosomes are not present to inhibit the activity of Rho. Nonetheless, the system is not essential, as shown, for example, by the viability of *nusA* deletion strains (18). Ribosomal proteins assembling on the emerging RNA might also inhibit Rho activity, at least sufficiently for viability.

SUMMARY POINTS

1. Transcription elongation is highly regulated through encoded signals recognized by regulatory factors and by RNAP itself.

2. Transcription is terminated by intrinsic terminators, by Rho, and by the release factor Mfd, which removes RNAP stalled by DNA damage.

3. Regulation of transcription elongation frequently is mediated through transcription pausing, the property of RNAP to stay at a single template site for much longer than the average elongation time. Pausing is a first step of transcription termination.

4. The cellular proteins NusA and NusG modulate pausing and are components of antitermination systems.

5. Antiterminators are operon-specific regulators consisting of either protein or RNA.

6. Antitermination occurs through both antipausing and direct shielding of emerging RNA of the elongation complex.

FUTURE ISSUES

1. How do transcription elongation regulatory proteins contact RNAP core subunits?

2. What are the relative roles of antipausing and complex stabilization in antitermination in vivo?

3. What is the molecular mechanism by which antiterminators inhibit pausing?

4. What is the atomic structure of an elongation complex modified by an antiterminator?

DISCLOSURE STATEMENT

The authors are not aware of any biases that might be perceived as affecting the objectivity of this review.

ACKNOWLEDGMENTS

We thank members of the laboratory and A. Hochschild for comments. Research in this laboratory was supported by grant 21941 from the National Institutes of Health. Present address for Smita Shankar is Department of Biochemistry and Biophysics, University of California, San Francisco, California 94158.

LITERATURE CITED

1. Adelman JL, Jeong YJ, Liao JC, Patel G, Kim DE, et al. 2006. Mechanochemistry of transcription termination factor Rho. *Mol. Cell* 22:611–21
2. Adelman K, La Porta A, Santangelo TJ, Lis JT, Roberts JW, Wang MD. 2002. Single molecule analysis of RNA polymerase elongation reveals uniform kinetic behavior. *Proc. Natl. Acad. Sci. USA* 99:13538–43
3. Adhya S, Gottesman M, De Crombrugghe B. 1974. Release of polarity in *Escherichia coli* by gene N of phage lambda: termination and antitermination of transcription. *Proc. Natl. Acad. Sci. USA* 71:2534–38
4. Albrechtsen B, Squires CL, Li S, Squires C. 1990. Antitermination of characterized transcriptional terminators by the *Escherichia coli* rrnG leader region. *J. Mol. Biol.* 213:123–34
5. Artsimovitch I, Landick R. 2000. Pausing by bacterial RNA polymerase is mediated by mechanistically distinct classes of signals. *Proc. Natl. Acad. Sci. USA* 97:7090–95
6. Artsimovitch I, Landick R. 2002. The transcriptional regulator RfaH stimulates RNA chain synthesis after recruitment to elongation complexes by the exposed nontemplate DNA strand. *Cell* 109:193–203
7. Banik-Maiti S, King RA, Weisberg RA. 1997. The antiterminator RNA of phage HK022. *J. Mol. Biol.* 272:677–87
8. Barik S, Ghosh B, Whalen W, Lazinski D, Das A. 1987. An antitermination protein engages the elongating transcription apparatus at a promoter-proximal recognition site. *Cell* 50:885–99
9. Belogurov GA, Vassylyeva MN, Svetlov V, Klyuyev S, Grishin NV, et al. 2007. Structural basis for converting a general transcription factor into an operon-specific virulence regulator. *Mol. Cell* 26:117–29
10. Beuth B, Pennell S, Arnvig KB, Martin SR, Taylor IA. 2005. Structure of a *Mycobacterium tuberculosis* NusA-RNA complex. *EMBO J.* 24:3576–87
11. Borukhov S, Lee J, Laptenko O. 2005. Bacterial transcription elongation factors: new insights into molecular mechanism of action. *Mol. Microbiol.* 55:1315–24
12. Borukhov S, Polyakov A, Nikiforov V, Goldfarb A. 1992. GreA protein: a transcription elongation factor from *Escherichia coli*. *Proc. Natl. Acad. Sci. USA* 89:8899–902
13. Brodolin K, Zenkin N, Mustaev A, Mamaeva D, Heumann H. 2004. The sigma 70 subunit of RNA polymerase induces lacUV5 promoter-proximal pausing of transcription. *Nat. Struct. Mol. Biol.* 11:551–57
14. Burns CM, Nowatzke WL, Richardson JP. 1999. Activation of Rho-dependent transcription termination by NusG. Dependence on terminator location and acceleration of RNA release. *J. Biol. Chem.* 274:5245–51
15. Burns CM, Richardson LV, Richardson JP. 1998. Combinatorial effects of NusA and NusG on transcription elongation and Rho-dependent termination in *Escherichia coli*. *J. Mol. Biol.* 278:307–16
16. Burova E, Hung SC, Chen J, Court DL, Zhou JG, et al. 1999. *Escherichia coli* nusG mutations that block transcription termination by coliphage HK022 Nun protein. *Mol. Microbiol.* 31:1783–93
17. Burova E, Hung SC, Sagitov V, Stitt BL, Gottesman ME. 1995. *Escherichia coli* NusG protein stimulates transcription elongation rates in vivo and in vitro. *J. Bacteriol.* 177:1388–92
18. Cardinale CJ, Washburn RS, Tadigotla VR, Brown LM, Gottesman ME, Nudler E. 2008. Termination factor Rho and its cofactors NusA and NusG silence foreign DNA in *E. coli*. *Science* 320:935–38
19. Chambers AL, Smith AJ, Savery NJ. 2003. A DNA translocation motif in the bacterial transcription–repair coupling factor, Mfd. *Nucleic Acids Res.* 31:6409–18
20. Chattopadhyay S, Garcia-Mena J, DeVito J, Wolska K, Das A. 1995. Bipartite function of a small RNA hairpin in transcription antitermination in bacteriophage lambda. *Proc. Natl. Acad. Sci. USA* 92:4061–65

21. Ciampi MS. 2006. Rho-dependent terminators and transcription termination. *Microbiology* 152:2515–28
22. Clerget M, Jin DJ, Weisberg RA. 1995. A zinc-binding region in the beta subunit of RNA polymerase is involved in antitermination of early transcription of phage HK022. *J. Mol. Biol.* 248:768–80
23. Condon C, Squires C, Squires CL. 1995. Control of rRNA transcription in *Escherichia coli*. *Microbiol. Rev.* 59:623–45
24. Das A, Wolska K. 1984. Transcription antitermination in vitro by lambda N gene product: requirement for a phage nut site and the products of host nusA, nusB, and nusE genes. *Cell* 38:165–73
25. Deaconescu AM, Chambers AL, Smith AJ, Nickels BE, Hochschild A, et al. 2006. Structural basis for bacterial transcription-coupled DNA repair. *Cell* 124:507–20
26. Dedrick RL, Kane CM, Chamberlin MJ. 1987. Purified RNA polymerase II recognizes specific termination sites during transcription in vitro. *J. Biol. Chem.* 262:9098–108
27. Deighan P, Hochschild A. 2007. The bacteriophage lambdaQ antiterminator protein regulates late gene expression as a stable component of the transcription elongation complex. *Mol. Microbiol.* 63:911–20
28. Faber C, Scharpf M, Becker T, Sticht H, Rosch P. 2001. The structure of the coliphage HK022 Nun protein-lambda-phage boxB RNA complex. Implications for the mechanism of transcription termination. *J. Biol. Chem.* 276:32064–70
29. Fisher RF, Yanofsky C. 1983. Mutations of the beta subunit of RNA polymerase alter both transcription pausing and transcription termination in the *trp* operon leader region in vitro. *J. Biol. Chem.* 258:8146–50
30. Franklin NC. 1974. Altered reading of genetic signals fused to the *N* operon of bacteriophage lambda: genetic evidence for modification of polymerase by the protein product of the N gene. *J. Mol. Biol.* 89:33–48
31. Friedman DI, Baron LS. 1974. Genetic characterization of a bacterial locus involved in the activity of the N function of phage lambda. *Virology* 58:141–48
32. Friedman DI, Baumann M, Baron LS. 1976. Cooperative effects of bacterial mutations affecting lambda N gene expression. I. Isolation and characterization of a *nusB* mutant. *Virology* 73:119–27
33. Friedman DI, Schauer AT, Baumann MR, Baron LS, Adhya SL. 1981. Evidence that ribosomal protein S10 participates in control of transcription termination. *Proc. Natl. Acad. Sci. USA* 78:1115–18
34. Friedman DI, Wilgus GS, Mural RJ. 1973. Gene N regulator function of phage lambda immun21: evidence that a site of N action differs from a site of N recognition. *J. Mol. Biol.* 81:505–16
35. Galluppi GR, Richardson JP. 1980. ATP-induced changes in the binding of RNA synthesis termination protein Rho to RNA. *J. Mol. Biol.* 138:513–39
36. Gopal B, Haire LF, Gamblin SJ, Dodson EJ, Lane AN, et al. 2001. Crystal structure of the transcription elongation/antitermination factor NusA from *Mycobacterium tuberculosis* at 1.7 A resolution. *J. Mol. Biol.* 314:1087–95
37. Gottesman S. 2004. The small RNA regulators of *Escherichia coli*: roles and mechanisms. *Annu. Rev. Microbiol.* 58:303–28
38. Gowrishankar J, Harinarayanan R. 2004. Why is transcription coupled to translation in bacteria? *Mol. Microbiol.* 54:598–603
39. Grayhack EJ, Yang XJ, Lau LF, Roberts JW. 1985. Phage lambda gene *Q* antiterminator recognizes RNA polymerase near the promoter and accelerates it through a pause site. *Cell* 42:259–69
40. Greenblatt J, McLimont M, Hanly S. 1981. Termination of transcription by nusA gene protein of *Escherichia coli*. *Nature* 292:215–20
41. Gusarov I, Nudler E. 2001. Control of intrinsic transcription termination by N and NusA: the basic mechanisms. *Cell* 107:437–49
42. Hatoum A, Roberts J. 2008. Prevalence of RNA polymerase stalling at *Escherichia coli* promoters after open complex formation. *Mol. Microbiol.* 68:17–28
43. Henthorn KS, Friedman DI. 1996. Identification of functional regions of the Nun transcription termination protein of phage HK022 and the N antitermination protein of phage lambda using hybrid nun-N genes. *J. Mol. Biol.* 257:9–20
44. Herbert KM, La Porta A, Wong BJ, Mooney RA, Neuman KC, et al. 2006. Sequence-resolved detection of pausing by single RNA polymerase molecules. *Cell* 125:1083–94
45. Horwitz RJ, Li J, Greenblatt J. 1987. An elongation control particle containing the N gene transcriptional antitermination protein of bacteriophage lambda. *Cell* 51:631–41

46. Hsu LM, Vo NV, Chamberlin MJ. 1995. *Escherichia coli* transcript cleavage factors GreA and GreB stimulate promoter escape and gene expression in vivo and in vitro. *Proc. Natl. Acad. Sci. USA* 92:11588–92

47. Huertas P, Aguilera A. 2003. Cotranscriptionally formed DNA:RNA hybrids mediate transcription elongation impairment and transcription-associated recombination. *Mol. Cell* 12:711–21

48. Hung SC, Gottesman ME. 1995. Phage HK022 Nun protein arrests transcription on phage lambda DNA in vitro and competes with the phage lambda N antitermination protein. *J. Mol. Biol.* 247:428–42

49. Kapanidis AN, Margeat E, Ho SO, Kortkhonjia E, Weiss S, Ebright RH. 2006. Initial transcription by RNA polymerase proceeds through a DNA-scrunching mechanism. *Science* 314:1144–47

50. Kim HC, Gottesman ME. 2004. Transcription termination by phage HK022 Nun is facilitated by COOH-terminal lysine residues. *J. Biol. Chem.* 279:13412–17

51. Kim HC, Washburn RS, Gottesman ME. 2006. Role of E. coli NusA in phage HK022 Nun-mediated transcription termination. *J. Mol. Biol.* 359:10–21

52. Kim HC, Zhou JG, Wilson HR, Mogilnitskiy G, Court DL, Gottesman ME. 2003. Phage HK022 Nun protein represses translation of phage lambda N (transcription termination/translation repression). *Proc. Natl. Acad. Sci. USA* 100:5308–12

53. King RA, Banik-Maiti S, Jin DJ, Weisberg RA. 1996. Transcripts that increase the processivity and elongation rate of RNA polymerase. *Cell* 87:893–903

54. King RA, Madsen PL, Weisberg RA. 2000. Constitutive expression of a transcription termination factor by a repressed prophage: promoters for transcribing the phage HK022 nun gene. *J. Bacteriol.* 182:456–62

55. King RA, Markov D, Sen R, Severinov K, Weisberg RA. 2004. A conserved zinc binding domain in the largest subunit of DNA-dependent RNA polymerase modulates intrinsic transcription termination and antitermination but does not stabilize the elongation complex. *J. Mol. Biol.* 342:1143–54

56. Komissarova N, Becker J, Solter S, Kireeva M, Kashlev M. 2002. Shortening of RNA:DNA hybrid in the elongation complex of RNA polymerase is a prerequisite for transcription termination. *Mol. Cell* 10:1151–62

57. Komissarova N, Kashlev M. 1997. RNA polymerase switches between inactivated and activated states by translocating back and forth along the DNA and the RNA. *J. Biol. Chem.* 272:15329–38

58. Korzheva N, Mustaev A, Kozlov M, Malhotra A, Nikiforov V, et al. 2000. A structural model of transcription elongation. *Science* 289:619–25

59. Landick R, Carey J, Yanofsky C. 1985. Translation activates the paused transcription complex and restores transcription of the *trp* operon leader region. *Proc. Natl. Acad. Sci. USA* 82:4663–67

60. Landick R, Turnbough J, Yanofsky C. 1996. Transcription attenuation. In Escherichia coli *and* Salmonella: *Cellular and Molecular Biology*, ed. F Neidhardt, R Curtiss, JL Ingraham, ECC Lin, KB Low, et al., pp. 1263–86. Washington, DC: ASM Press

61. Laptenko O, Lee J, Lomakin I, Borukhov S. 2003. Transcript cleavage factors GreA and GreB act as transient catalytic components of RNA polymerase. *EMBO J.* 22:6322–34

62. Larson MH, Greenleaf WJ, Landick R, Block SM. 2008. Applied force reveals mechanistic and energetic details of transcription termination. *Cell* 132:971–82

63. Lau LF, Roberts JW. 1985. Rho-dependent transcription termination at lambda R1 requires upstream sequences. *J. Biol. Chem.* 260:574–84

64. Lau LF, Roberts JW, Wu R. 1983. RNA polymerase pausing and transcript release at the lambda tR1 terminator in vitro. *J. Biol. Chem.* 258:9391–97

65. Lazinski D, Grzadzielska E, Das A. 1989. Sequence-specific recognition of RNA hairpins by bacteriophage antiterminators requires a conserved arginine-rich motif. *Cell* 59:207–18

66. Lee DN, Phung L, Stewart J, Landick R. 1990. Transcription pausing by *Escherichia coli* RNA polymerase is modulated by downstream DNA sequences. *J. Biol. Chem.* 265:15145–53

67. Legault P, Li J, Mogridge J, Kay LE, Greenblatt J. 1998. NMR structure of the bacteriophage lambda N peptide/boxB RNA complex: recognition of a GNRA fold by an arginine-rich motif. *Cell* 93:289–99

68. Li J, Horwitz R, McCracken S, Greenblatt J. 1992. NusG, a new *Escherichia coli* elongation factor involved in transcriptional antitermination by the N protein of phage lambda. *J. Biol. Chem.* 267:6012–19

69. Li J, Mason SW, Greenblatt J. 1993. Elongation factor NusG interacts with termination factor Rho to regulate termination and antitermination of transcription. *Genes Dev.* 7:161–72

70. Li SC, Squires CL, Squires C. 1984. Antitermination of *E. coli* rRNA transcription is caused by a control region segment containing lambda nut-like sequences. *Cell* 38:851–60

71. Liu K, Hanna MM. 1995. NusA interferes with interactions between the nascent RNA and the C-terminal domain of the alpha subunit of RNA polymerase in *Escherichia coli* transcription complexes. *Proc. Natl. Acad. Sci. USA* 92:5012–16

72. Liu K, Zhang Y, Severinov K, Das A, Hanna MM. 1996. Role of *Escherichia coli* RNA polymerase alpha subunit in modulation of pausing, termination and antitermination by the transcription elongation factor NusA. *EMBO J.* 15:150–61

73. Lowery-Goldhammer C, Richardson JP. 1974. An RNA-dependent nucleoside triphosphate phosphohydrolase (ATPase) associated with Rho termination factor. *Proc. Natl. Acad. Sci. USA* 71:2003–7

74. Macdonald LE, Zhou Y, McAllister WT. 1993. Termination and slippage by bacteriophage T7 RNA polymerase. *J. Mol. Biol.* 232:1030–47

75. Mah TF, Li J, Davidson AR, Greenblatt J. 1999. Functional importance of regions in *Escherichia coli* elongation factor NusA that interact with RNA polymerase, the bacteriophage lambda N protein and RNA. *Mol. Microbiol.* 34:523–37

76. Marr MT, Datwyler SA, Meares CF, Roberts JW. 2001. Restructuring of an RNA polymerase holoenzyme elongation complex by lambdoid phage Q proteins. *Proc. Natl. Acad. Sci. USA* 98:8972–78

77. Marr MT, Roberts JW. 2000. Function of transcription cleavage factors GreA and GreB at a regulatory pause site. *Mol. Cell* 6:1275–85

78. Martin FH, Tinoco I Jr. 1980. DNA-RNA hybrid duplexes containing oligo(dA:rU) sequences are exceptionally unstable and may facilitate termination of transcription. *Nucleic Acids Res.* 8:2295–99

79. Mason SW, Li J, Greenblatt J. 1992. Direct interaction between two *Escherichia coli* transcription antitermination factors, NusB and ribosomal protein S10. *J. Mol. Biol.* 223:55–66

80. Mason SW, Li J, Greenblatt J. 1992. Host factor requirements for processive antitermination of transcription and suppression of pausing by the N protein of bacteriophage lambda. *J. Biol. Chem.* 267:19418–26

81. McSwiggen JA, Bear DG, von Hippel PH. 1988. Interactions of *Escherichia coli* transcription termination factor Rho with RNA. I. Binding stoichiometries and free energies. *J. Mol. Biol.* 199:609–22

82. Mooney RA, Landick R. 2003. Tethering sigma70 to RNA polymerase reveals high in vivo activity of sigma factors and sigma70-dependent pausing at promoter-distal locations. *Genes Dev.* 17:2839–51

83. Morgan EA. 1980. Insertions of Tn 10 into an *E. coli* ribosomal RNA operon are incompletely polar. *Cell* 21:257–65

84. Morgan WD, Bear DG, Litchman BL, von Hippel PH. 1985. RNA sequence and secondary structure requirements for Rho-dependent transcription termination. *Nucleic Acids Res.* 13:3739–54

85. Morgan WD, Bear DG, von Hippel PH. 1983. Rho-dependent termination of transcription. II. Kinetics of mRNA elongation during transcription from the bacteriophage lambda PR promoter. *J. Biol. Chem.* 258:9565–74

86. Nehrke KW, Platt T. 1994. A quaternary transcription termination complex. Reciprocal stabilization by Rho factor and NusG protein. *J. Mol. Biol.* 243:830–39

87. Nehrke KW, Zalatan F, Platt T. 1993. NusG alters Rho-dependent termination of transcription in vitro independent of kinetic coupling. *Gene Expr.* 3:119–33

88. Neuman KC, Abbondanzieri EA, Landick R, Gelles J, Block SM. 2003. Ubiquitous transcriptional pausing is independent of RNA polymerase backtracking. *Cell* 115:437–47

89. Nickels BE, Mukhopadhyay J, Garrity SJ, Ebright RH, Hochschild A. 2004. The sigma 70 subunit of RNA polymerase mediates a promoter-proximal pause at the lac promoter. *Nat. Struct. Mol. Biol.* 11:544–50

90. Nickels BE, Roberts CW, Roberts JW, Hochschild A. 2006. RNA-mediated destabilization of the sigma(70) region 4/beta flap interaction facilitates engagement of RNA polymerase by the Q antiterminator. *Mol. Cell* 24:457–68

91. Nickels BE, Roberts CW, Sun H, Roberts JW, Hochschild A. 2002. The sigma(70) subunit of RNA polymerase is contacted by the (lambda)Q antiterminator during early elongation. *Mol. Cell* 10:611–22

92. Nodwell JR, Greenblatt J. 1993. Recognition of boxA antiterminator RNA by the *E. coli* antitermination factors NusB and ribosomal protein S10. *Cell* 72:261–68

93. Nudler E, Avetissova E, Markovtsov V, Goldfarb A. 1996. Transcription processivity: protein-DNA interactions holding together the elongation complex. *Science* 273:211–17

94. Nudler E, Gottesman ME. 2002. Transcription termination and antitermination in *E. coli. Genes Cells* 7:755–68

95. Nudler E, Mustaev A, Lukhtanov E, Goldfarb A. 1997. The RNA-DNA hybrid maintains the register of transcription by preventing backtracking of RNA polymerase. *Cell* 89:33–41

96. Oberto J, Clerget M, Ditto M, Cam K, Weisberg RA. 1993. Antitermination of early transcription in phage HK022. Absence of a phage-encoded antitermination factor. *J. Mol. Biol.* 229:368–81

97. Olson ER, Flamm EL, Friedman DI. 1982. Analysis of nutR: a region of phage lambda required for antitermination of transcription. *Cell* 31:61–70

98. Opalka N, Chlenov M, Chacon P, Rice WJ, Wriggers W, Darst SA. 2003. Structure and function of the transcription elongation factor GreB bound to bacterial RNA polymerase. *Cell* 114:335–45

99. Orlova M, Newlands J, Das A, Goldfarb A, Borukhov S. 1995. Intrinsic transcript cleavage activity of RNA polymerase. *Proc. Natl. Acad. Sci. USA* 92:4596–600

100. Park JS, Marr MT, Roberts JW. 2002. *E. coli* transcription repair coupling factor (Mfd protein) rescues arrested complexes by promoting forward translocation. *Cell* 109:757–67

101. Park JS, Roberts JW. 2006. Role of DNA bubble rewinding in enzymatic transcription termination. *Proc. Natl. Acad. Sci. USA* 103:4870–75

102. Pasman Z, von Hippel PH. 2000. Regulation of Rho-dependent transcription termination by NusG is specific to the *Escherichia coli* elongation complex. *Biochemistry* 39:5573–85

103. Patterson TA, Zhang Z, Baker T, Johnson LL, Friedman DI, Court DL. 1994. Bacteriophage lambda N-dependent transcription antitermination. Competition for an RNA site may regulate antitermination. *J. Mol. Biol.* 236:217–28

104. Paul BJ, Barker MM, Ross W, Schneider DA, Webb C, et al. 2004. DksA: a critical component of the transcription initiation machinery that potentiates the regulation of rRNA promoters by ppGpp and the initiating NTP. *Cell* 118:311–22

105. Perederina A, Svetlov V, Vassylyeva MN, Tahirov TH, Yokoyama S, et al. 2004. Regulation through the secondary channel–structural framework for ppGpp-DksA synergism during transcription. *Cell* 118:297–309

106. Rees WA, Weitzel SE, Das A, von Hippel PH. 1997. Regulation of the elongation-termination decision at intrinsic terminators by antitermination protein N of phage lambda. *J. Mol. Biol.* 273:797–813

107. Rees WA, Weitzel SE, Yager TD, Das A, von Hippel PH. 1996. Bacteriophage lambda N protein alone can induce transcription antitermination in vitro. *Proc. Natl. Acad. Sci. USA* 93:342–46

108. Revyakin A, Liu C, Ebright RH, Strick TR. 2006. Abortive initiation and productive initiation by RNA polymerase involve DNA scrunching. *Science* 314:1139–43

109. Richardson JP, Greenblatt J. 1996. Control of RNA chain elongation and termination. In Escherichia coli *and* Salmonella: *Cellular and Molecular Biology*, ed. FC Neidhardt, R Curtiss, JL Ingraham, ECC Lin, KB Low, et al., pp. 822–48. Washington, DC: ASM Press

110. Richardson JP, Grimley C, Lowery C. 1975. Transcription termination factor Rho activity is altered in *Escherichia coli* strains with suA gene mutations. *Proc. Natl. Acad. Sci. USA* 72:1725–28

111. Ring BZ, Yarnell WS, Roberts JW. 1996. Function of *E. coli* RNA polymerase sigma factor sigma 70 in promoter-proximal pausing. *Cell* 86:485–93

112. Robert J, Sloan SB, Weisberg RA, Gottesman ME, Robledo R, Harbrecht D. 1987. The remarkable specificity of a new transcription termination factor suggests that the mechanisms of termination and antitermination are similar. *Cell* 51:483–92

113. Roberts JW. 1969. Termination factor for RNA synthesis. *Nature* 224:1168–74

114. Roberts JW, Yarnell W, Bartlett E, Guo J, Marr M, et al. 1998. Antitermination by bacteriophage lambda Q protein. *Cold Spring Harb. Symp. Quant. Biol.* 63:319–25

115. Robledo R, Atkinson BL, Gottesman ME. 1991. *Escherichia coli* mutations that block transcription termination by phage HK022 Nun protein. *J. Mol. Biol.* 220:613–19

116. Ryder AM, Roberts JW. 2003. Role of the nontemplate strand of the elongation bubble in intrinsic transcription termination. *J. Mol. Biol.* 334:205–13

117. Salstrom JS, Szybalski W. 1978. Coliphage lambdanutL-: a unique class of mutants defective in the site of gene N product utilization for antitermination of leftward transcription. *J. Mol. Biol.* 124:195–221

118. Santangelo TJ, Mooney RA, Landick R, Roberts JW. 2003. RNA polymerase mutations that impair conversion to a termination-resistant complex by Q antiterminator proteins. *Genes Dev.* 17:1281–92

119. Santangelo TJ, Roberts JW. 2004. Forward translocation is the natural pathway of RNA release at an intrinsic terminator. *Mol. Cell* 14:117–26

120. Schmidt MC, Chamberlin MJ. 1987. nusA protein of *Escherichia coli* is an efficient transcription termination factor for certain terminator sites. *J. Mol. Biol.* 195:809–18

121. Schwartz A, Margeat E, Rahmouni AR, Boudvillain M. 2007. Transcription termination factor Rho can displace streptavidin from biotinylated RNA. *J. Biol. Chem.* 282:31469–76

122. Selby CP, Sancar A. 1993. Molecular mechanism of transcription-repair coupling. *Science* 260:53–58

123. Sen R, King RA, Mzhavia N, Madsen PL, Weisberg RA. 2002. Sequence-specific interaction of nascent antiterminator RNA with the zinc-finger motif of *Escherichia coli* RNA polymerase. *Mol. Microbiol.* 46:215–22

124. Sen R, King RA, Weisberg RA. 2001. Modification of the properties of elongating RNA polymerase by persistent association with nascent antiterminator RNA. *Mol. Cell* 7:993–1001

125. Sevostyanova A, Svetlov V, Vassylyev DG, Artsimovitch I. 2008. The elongation factor RfaH and the initiation factor sigma bind to the same site on the transcription elongation complex. *Proc. Natl. Acad. Sci. USA* 105:865–70

126. Shankar S, Hatoum A, Roberts JW. 2007. A transcription antiterminator constructs a NusA-dependent shield to the emerging transcript. *Mol. Cell* 27:914–27

127. Sigmund CD, Morgan EA. 1988. NusA protein affects transcriptional pausing and termination in vitro by binding to different sites on the transcription complex. *Biochemistry* 27:5622–27

128. Skordalakes E, Berger JM. 2006. Structural insights into RNA-dependent ring closure and ATPase activation by the Rho termination factor. *Cell* 127:553–64

129. Sloan SB, Weisberg RA. 1993. Use of a gene encoding a suppressor tRNA as a reporter of transcription: analyzing the action of the Nun protein of bacteriophage HK022. *Proc. Natl. Acad. Sci. USA* 90:9842–46

130. Smith AJ, Savery NJ. 2005. RNA polymerase mutants defective in the initiation of transcription-coupled DNA repair. *Nucleic Acids Res.* 33:755–64

131. Smith AJ, Szczelkun MD, Savery NJ. 2007. Controlling the motor activity of a transcription-repair coupling factor: autoinhibition and the role of RNA polymerase. *Nucleic Acids Res.* 35:1802–11

132. Sosunov V, Sosunova E, Mustaev A, Bass I, Nikiforov V, Goldfarb A. 2003. Unified two-metal mechanism of RNA synthesis and degradation by RNA polymerase. *EMBO J.* 22:2234–44

133. Squires CL, Greenblatt J, Li J, Condon C, Squires CL. 1993. Ribosomal RNA antitermination in vitro: requirement for Nus factors and one or more unidentified cellular components. *Proc. Natl. Acad. Sci. USA* 90:970–74

134. Stepanova E, Lee J, Ozerova M, Semenova E, Datsenko K, et al. 2007. Analysis of promoter targets for *Escherichia coli* transcription elongation factor GreA in vivo and in vitro. *J. Bacteriol.* 189:8772–85

135. Straney DC, Crothers DM. 1987. A stressed intermediate in the formation of stably initiated RNA chains at the *Escherichia coli* lac UV5 promoter. *J. Mol. Biol.* 193:267–78

136. Stuart AC, Gottesman ME, Palmer AG 3rd. 2003. The N-terminus is unstructured, but not dynamically disordered, in the complex between HK022 Nun protein and lambda-phage BoxB RNA hairpin. *FEBS Lett.* 553:95–98

137. Sullivan SL, Ward DF, Gottesman ME. 1992. Effect of *Escherichia coli* nusG function on lambda N-mediated transcription antitermination. *J. Bacteriol.* 174:1339–44

138. Toulokhonov I, Artsimovitch I, Landick R. 2001. Allosteric control of RNA polymerase by a site that contacts nascent RNA hairpins. *Science* 292:730–33

139. Toulokhonov I, Zhang J, Palangat M, Landick R. 2007. A central role of the RNA polymerase trigger loop in active-site rearrangement during transcriptional pausing. *Mol. Cell* 27:406–19

140. Traviglia SL, Datwyler SA, Yan D, Ishihama A, Meares CF. 1999. Targeted protein footprinting: where different transcription factors bind to RNA polymerase. *Biochemistry* 38:15774–78

141. Vassylyev DG, Vassylyeva MN, Perederina A, Tahirov TH, Artsimovitch I. 2007. Structural basis for transcription elongation by bacterial RNA polymerase. *Nature* 448:157–62

142. Vassylyev DG, Vassylyeva MN, Zhang J, Palangat M, Artsimovitch I, Landick R. 2007. Structural basis for substrate loading in bacterial RNA polymerase. *Nature* 448:163–68

143. Vieu E, Rahmouni AR. 2004. Dual role of boxB RNA motif in the mechanisms of termination/antitermination at the lambda tR1 terminator revealed in vivo. *J. Mol. Biol.* 339:1077–87

144. Vogel U, Jensen KF. 1997. NusA is required for ribosomal antitermination and for modulation of the transcription elongation rate of both antiterminated RNA and mRNA. *J. Biol. Chem.* 272:12265–71

145. Wagner LA, Weiss RB, Driscoll R, Dunn DS, Gesteland RF. 1990. Transcriptional slippage occurs during elongation at runs of adenine or thymine in *Escherichia coli*. *Nucleic Acids Res.* 18:3529–35

146. Washburn RS, Wang Y, Gottesman ME. 2003. Role of *E. coli* transcription-repair coupling factor Mfd in Nun-mediated transcription termination. *J. Mol. Biol.* 329:655–62

147. Watnick RS, Gottesman ME. 1998. *Escherichia coli* NusA is required for efficient RNA binding by phage HK022 nun protein. *Proc. Natl. Acad. Sci. USA* 95:1546–51

148. Watnick RS, Gottesman ME. 1999. Binding of transcription termination protein nun to nascent RNA and template DNA. *Science* 286:2337–39

149. Watnick RS, Herring SC, Palmer AG 3rd, Gottesman ME. 2000. The carboxyl terminus of phage HK022 Nun includes a novel zinc-binding motif and a tryptophan required for transcription termination. *Genes Dev.* 14:731–39

150. Weisberg RA, Gottesman ME. 1999. Processive antitermination. *J. Bacteriol.* 181:359–67

151. Whalen W, Ghosh B, Das A. 1988. NusA protein is necessary and sufficient in vitro for phage lambda N gene product to suppress a Rho-independent terminator placed downstream of nutL. *Proc. Natl. Acad. Sci. USA* 85:2494–98

152. Yang XJ, Hart CM, Grayhack EJ, Roberts JW. 1987. Transcription antitermination by phage lambda gene Q protein requires a DNA segment spanning the RNA start site. *Genes Dev.* 1:217–26

153. Yang XJ, Roberts JW. 1989. Gene Q antiterminator proteins of *Escherichia coli* phages 82 and lambda suppress pausing by RNA polymerase at a Rho-dependent terminator and at other sites. *Proc. Natl. Acad. Sci. USA* 86:5301–5

154. Yanofsky C. 2007. RNA-based regulation of genes of tryptophan synthesis and degradation, in bacteria. *Rna* 13:1141–54

155. Yarnell WS, Roberts JW. 1992. The phage lambda gene Q transcription antiterminator binds DNA in the late gene promoter as it modifies RNA polymerase. *Cell* 69:1181–89

156. Yarnell WS, Roberts JW. 1999. Mechanism of intrinsic transcription termination and antitermination. *Science* 284:611–15

157. Zalieckas JM, Wray LV Jr, Ferson AE, Fisher SH. 1998. Transcription-repair coupling factor is involved in carbon catabolite repression of the *Bacillus subtilis hut* and *gnt* operons. *Mol. Microbiol.* 27:1031–38

158. Zenkin N, Yuzenkova Y, Severinov K. 2006. Transcript-assisted transcriptional proofreading. *Science* 313:518–20

159. Zheng C, Friedman DI. 1994. Reduced Rho-dependent transcription termination permits NusA-independent growth of *Escherichia coli*. *Proc. Natl. Acad. Sci. USA* 91:7543–47

160. Zhou Y, Navaroli DM, Enuameh MS, Martin CT. 2007. Dissociation of halted T7 RNA polymerase elongation complexes proceeds via a forward-translocation mechanism. *Proc. Natl. Acad. Sci. USA* 104:10352–57

Base J: Discovery, Biosynthesis, and Possible Functions

Piet Borst[1] and Robert Sabatini[2]

[1]Center of Biomedical Genetics, Division of Molecular Biology, The Netherlands Cancer Institute, 1066 CX Amsterdam, The Netherlands; email: p.borst@nki.nl

[2]Department of Biochemistry and Molecular Biology, University of Georgia, Athens, Georgia 30602-7229; email: rsabatini@bmb.uga.edu

Annu. Rev. Microbiol. 2008. 62:235–51

First published online as a Review in Advance on June 6, 2008

The *Annual Review of Microbiology* is online at micro.annualreviews.org

This article's doi: 10.1146/annurev.micro.62.081307.162750

Key Words

epigenetics, DNA hypermodification, antigenic variation, kinetoplastids

Abstract

In 1993, a new base, β-D-glucopyranosyloxymethyluracil (base J), was identified in the nuclear DNA of *Trypanosoma brucei*. Base J is the first hypermodified base found in eukaryotic DNA. It is present in all kinetoplastid flagellates analyzed and some unicellular flagellates closely related to trypanosomatids, but it has not been found in other protozoa or in metazoa. J is invariably present in the telomeric repeats of all organisms analyzed. Whereas in *Leishmania* nearly all J is telomeric, there are other repetitive DNA sequences containing J in *T. brucei* and *T. cruzi*, and most J is outside telomeres in *Euglena*. The biosynthesis of J occurs in two steps: First, a specific thymidine in DNA is converted into hydroxymethyldeoxyuridine (HOMedU), and then this HOMedU is glycosylated to form J. This review discusses the identification and localization of base J in the genome of kinetoplastids, the enzymes involved in J biosynthesis, possible biological functions of J, and J as a potential target for chemotherapy of diseases caused by kinetoplastids.

Contents

DISCOVERY OF BASE J

J: β-D-glucosyl-hydroxymethyluracil

VSG: variant surface glycoprotein

ES: expression site

The first indications for the existence of a modified base in trypanosome nuclear DNA came from an analysis of variant surface glycoprotein (VSG) expression sites (ESs) in the bloodstream form of *Trypanosoma brucei*. Trypanosomes can switch off a site and simultaneously switch on another one. In this way they can alter the VSG gene expressed as well as the cotranscribed expression site associated genes (ESAGs) during infection of the mammalian host, as indicated schematically in **Figure 1**. Independent analyses by Bernards et al. (3) and Pays et al. (23) showed that silent telomeric VSG genes were partially digested by some restriction endonucleases. When the ES was activated, the VSG gene became fully digestible. The very same gene in a chromosome-internal position, where it is never transcribed, was fully digestible, and this was also found for the VSG genes in procyclic (insect-form) trypanosomes (23), in which all ESs are turned off. The modification of silent ESs, which is due to the presence of base J, has a number of peculiarities that are worth recalling, as they remain unexplained today:

- The modification involved restriction enzymes that do not share a common sequence: Hind III (AAGCTT), Pst I (CTGCAG), Pvu II (CAGCTG), and SphI (GCATGC). Initially, Pays et al. (23) suggested that the C in GC doublets would be modified, but later work by Van Leeuwen et al. (39) showed that Nco I (CCATGG) and Dra I (TTTAAA) sites could also be affected, as could the T's in

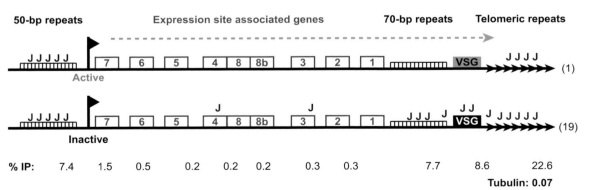

Figure 1

Location of base J in and around active and inactive telomeric variant surface glycoprotein (VSG) gene expression sites in *Trypanosoma brucei*. The presence of base J was determined by immunoprecipitation of J-containing DNA fragments with anti-J antibodies followed by dot-blot hybridization. The % IP shows a quantitation of the immunoprecipitation (IP) efficiency, i.e., the immunoprecipitated fraction of input DNA. Modified from Reference 39.

the telomeric sequences (CCCAAT) and (GGGTTA). Remarkably, only the second T in the latter telomeric sequence was modified. As other restriction sites were never found to be uncleavable, the sequences modified are highly nonrandom, but we have been unable to determine what the modified sequences have in common.

- Modification is invariably partial. For a given site, modification varies from less than 5% to over 20%. If only 20% is modified, that means that only 1 in 5 trypanosomes in a trypanosome population contains J at that site. Cloning of the trypanosome population does not change the frequency (3). Therefore, there is a statistical chance of 20% for this site to be modified.
- In silent bloodstream-form ESs there is a gradient of modification; the sites near the telomere are the most highly modified (3).
- Telomere length affects the degree of modification of the adjacent silent VSG genes. The trypanosome telomeres are unusual in that they grow in time by addition of hexamer units (2). In this way one can obtain trypanosome clones containing the same silent ES with different lengths of telomeric repeats attached to it. Using the silent telomere containing the 221 VSG gene, Bernards et al. (3) showed that modification of a given PstI site var-

ied from undetectable when the telomere was 2 kb to greater than 20% when it was 20 kb.

These results led Bernards et al. (3) to propose a model in which a modifying enzyme(s) recognizes telomeric sequences and then modifies the adjacent subtelomeric sequences in a stochastic fashion, resulting in a partial modification of modifiable sites. Given the link of modification to inactive ESs, it seemed probable that modification is somehow responsible for transcriptional inactivation of the ES. We shall return to these ideas at the end of this review.

Initial attempts to determine the nature of the modified base went nowhere. It did not seem to be MeC (5-methylcytosine), and standard DNA base analysis did not yield any unusual base (9). Only when we turned to ^{32}P-postlabeling of enzymatic DNA hydrolysates and analyzed the labeled nucleotides by 2-D thin-layer chromatography did we find an additional minor nucleotide. The fact that this spot did not comigrate with any known nucleotide (17) unleashed a major hunt for the elusive modified base culminating in its identification as β-D-glucosyl-hydroxymethyluracil, or J for short (**Figure 2**) (19). J was a new base, as it had never been found before as a normal constituent of DNA and it was also the first hypermodified DNA base ever found in eukaryotes (18).

In addition to base J, trypanosome DNA contains small amounts of hydroxymethyluracil

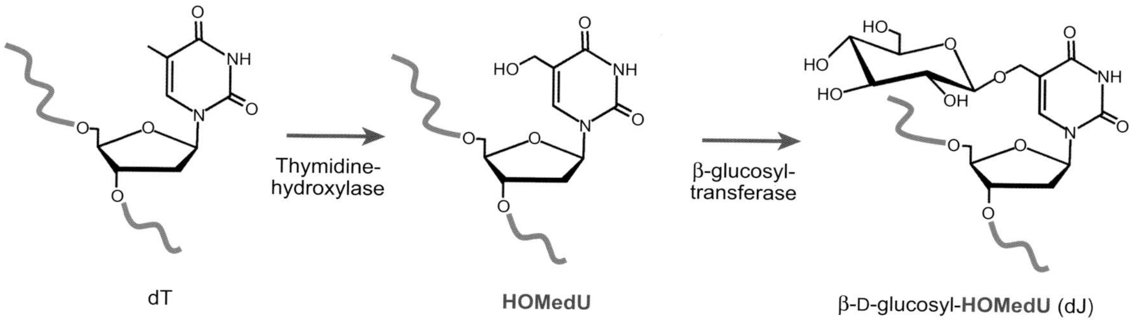

MeC: 5-methylcytosine

Figure 2

Biosynthesis of base J by modification of a specific thymine base in DNA.

HOMeU:
hydroxymethyluracil

HOMedU: hydro-
xymethyldeoxyuridine

SMUG: single-
strand selective
monofunctional uracil
DNA glycosylase

(HOMeU) (17), which is an intermediate in J biosynthesis (see below). We never found any indication for the presence of other minor nucleotides in trypanosome DNA, and the incomplete digestion of bloodstream-form *T. brucei* DNA by restriction endonucleases can be explained by the presence of J. Raibaud et al. (24) have reported the presence of a modified base replacing C in *T. equiperdum*, but this report has not been confirmed by later work (9).

The early work on base J has been reviewed by Borst & Van Leeuwen (6) and by Simpson (28). An overview of hypermodified bases is presented by Gommers-Ampt & Borst (18).

Which Organisms Contain J?

Analyzing J content by postlabeling of enzymic DNA hydrolysates is laborious and technically difficult because J-containing oligonucleotides are poorly digested by some of the nucleases used in the postlabeling analysis (34). On average, only 50% of the total J is recovered. After the nature of base J had been solved, Van Boom and coworkers (30, 40) and De Kort et al. (10) developed a method for the chemical synthesis of dJMP. This allowed Van Leeuwen et al. (39) to raise rabbit polyclonal antisera against dJMP coupled to keyhole limpet hemocyanin or to bovine serum albumin. The specificity and sensitivity of these antisera were analyzed with the help of defined J-containing oligonucleotides synthesized by De Kort et al. (10). In dot-blot experiments the antisera detected less than 1 J moiety in 10^6 bases, raising the sensitivity of J detection 100-fold relative to postlabeling. Moreover, the antisera immunoprecipitated J-containing duplex DNA fragments, allowing a rough determination of the distribution of J in DNA.

With the anti-J dot-blot test, J was found in all kinetoplastids analyzed, in the related marine flagellate *Diplonema* (38), and in *Euglena gracilis* (12), a unicellular algae closely related to the *Kinetoplastida*. No J was found in a variety of other protozoa, fungi, or vertebrates (38). We are often asked how sure we are that we have not missed minute amounts of J in human cells, yeast, or *Escherichia coli*. One should never say never in biology, but we like to stress that most organisms that lack J contain DNA glycosylases attacking hydroxymethyldeoxyuridine (HOMedU) (33), a freely accessible intermediate in J synthesis (32). Mammals even contain a highly active dedicated HOMedU glycosylase called single-strand selective monofunctional uracil DNA glycosylase (SMUG). The name is confusing, because it is now known that this enzyme also efficiently tackles HOMedU in duplex DNA. Introduction of human SMUG into bloodstream-form *T. brucei* kills the parasite through excessive DNA repair (32). From these experiments we conclude that J synthesis would not be compatible with chromosome integrity in mammals, given the presence of interfering DNA glycosylases.

The complete absence of base J in the insect form of *T. brucei* (23) was confirmed by Van Leeuwen et al. (39). This absence appears to be unique for *T. brucei*. Other trypanosomatids that shuttle between mammal and insect, such as *T. cruzi* and *Leishmania donovani*, contain J in both life cycle stages (38). The loss of J when *T. brucei* enters the insect (as monitored by in vitro differentiation into insect-form trypanosomes) is not due to active removal of J, but to downregulation of the modification machinery, resulting in the gradual decrease of J levels due to the synthesis of unmodified DNA during DNA replication (4). J cannot be detected in trypanosomes in the fly midgut or in the fly salivary gland (35). Modification is therefore only resumed after the trypanosomes are ejected from the salivary gland into the mammalian host.

The first two rabbits recruited for production of antisera against protein-coupled J-deoxyribosemonophosphate (dJMP) duly responded and generated the antisera still in use in our labs (39). Later attempts were less successful. We have tried to generate monoclonal antibodies against protein-coupled JMP in mice and failed. Even our attempts to generate more rabbit polyclonal antisera were unsuccessful. This is why we are less than generous to other investigators with the precious antibodies in our freezer.

Distribution of J in *Trypanosoma* and *Leishmania*

With antisera against J that specifically precipitate J-containing DNA, it became possible to roughly map the presence of J in DNA. Initial experiments were done on the VSG gene ESs of *T. brucei* (39). As shown in **Figure 1**, the semiquantitative immunoprecipitation analysis with anti-J antiserum confirmed that VSG genes in inactive ESs were modified, whereas those in transcriptionally active sites were not. Also the gradient of modification, noted by Bernards et al. (3), was confirmed: DNA fragments close to the telomere were modified the most, whereas fragments located more upstream were only weakly modified. Obviously, the semiquantitative immunoprecipitation of DNA fragments does not prove that the partially cleavable restriction enzyme recognition sites in inactive ESs are actually modified by J insertion. However, this is most likely the case. J is a hypermodified base that can be expected to interfere with cleavage by all restriction endonucleases, and this was experimentally verified for Pvu II (39). Moreover, no modified bases other than J (and traces of HOMedU) were found in *T. brucei* DNA. It would be nice to be able to map precisely the presence of J along a DNA strand, but nobody has yet come up with a method to do so.

The loss of J from a VSG gene ES when it is activated only affects the ES itself. As shown in **Figure 1**, the 50-bp repeat upstream of the promoter remain modified, just like the telomeric repeats downstream of the VSG gene (39). This restricted loss of J upon transcriptional activation may simply reflect the limits of the polymerase I transcription unit within the ES (22, 26), as discussed below.

A survey of sequences containing J in bloodstream-form *T. brucei* has identified a number of repetitive sequences in addition to the telomeric and 50-bp repeats: the minichromosomal 177-bp repeats, the 70-bp repeats upstream of VSG genes present in silent ESs, the 5S RNA repeats, and the mini-exon (spliced leader) repeats (37). From this semiquantitative

analysis the degree of modification appears to be highest in the telomeric repeats, followed by the 50-bp, 177-bp, and 70-bp repeats, with the lowest degree of modification found in the mini-exon and 5S RNA repeats.

A more quantitative analysis of J content was done on the telomeric repeats using postlabeling. Telomeric fragments were isolated by digesting nontelomeric DNA with multiple restriction enzymes, and the strands of the purified telomeres were separated by equilibrium centrifugation in alkaline CsCl. In telomeres 13% of T was replaced by J, compared with 0.8% in total DNA. In the (TAACCC) strand 14% of T was replaced by J; in the (GGGTTA) strand 36% of the second T was replaced and none of the first T. These values are minimal estimates, given the underestimation of J in the postlabeling analysis (34).

Because of the large number of minichromosomes, *T. brucei* has about 130 chromosomes with telomeric repeats varying in size between 2 and 20 kb. Therefore, telomeric repeats encompass about 3% of the *T. brucei* genome. With the approximate 16-fold enrichment of J in telomeres versus bulk DNA, it follows that ~50% of the total J in *T. brucei* is in the telomeric repeats.

The prominence of J in telomeric repeats is a feature of all kinetoplastid flagellates analyzed (38). However, in *Euglena* the bulk of J does not colocalize with telomeric repeats. This was first found with anti-J antibodies in permeabilized nuclei (12), and the same result was obtained by Southern blotting of restriction digests of DNA after separating telomeric repeats from the nontelomeric DNA (15). The nontelomeric sequences in which J resides in *Euglena* have not been characterized.

Leishmania is the other extreme: More than 98% of J appears to be telomeric and a similar result was obtained with *Crithidia fasciculata* (15). None of the subtelomeric repeats of *Leishmania major* Friedlin contains detectable amounts of J. Although we do not want to lightly dismiss the 1% nontelomeric J in *Leishmania* as insignificant (as long as the sequences in which this 1% resides have not been

TH: thymidine
hydroxylase

identified), in *Leishmania* nearly all J is in telomeric repeats, firmly linking the function of J to telomeres in this organism.

In contrast, substantial amounts of nontelomeric J were found in *Trypanosoma equiperdum*, *Trypanosoma cruzi*, and *Trypanoplasma borreli* (13, 15). Although *T. cruzi* lacks the VSG ESs and the repetitive DNA sequences that contain the majority of nontelomeric J in *T. brucei*, J localization analysis has indicated that 25% of the J is nontelomeric (13). A significant fraction of this nontelomeric J is localized in the subtelomeric repeat sequences (DIRE and VIPER) and in subtelomeric members of the stage-specific surface glycoprotein gene families of *T. cruzi* that have been implicated in pathogenesis (13). The potential role of J in regulating telomeric gene expression/diversity in *T. cruzi* remains to be explored. The same repeat sequences (i.e., DIRE and VIPER) in a chromosome-internal position do not contain the modified base. This finding shows that it is not the sequence or repetitive nature per se that determines whether a particular DNA region contains J, but only its localization near the end of a chromosome in *T. cruzi*. The same situation is found in *T. brucei*, in which a silent VSG gene located near a telomere contains J, whereas the same silent VSG gene in a chromosome-internal position is not modified (3, 39).

BIOSYNTHESIS OF BASE J

All available evidence now supports the hypothesis that J is synthesized in two separate steps, as depicted in **Figure 2**. In the first step a thymidine hydroxylase (TH) recognizes a specific thymidine residue in DNA and oxidizes its exocyclic methyl group, resulting in HOMedU. In the second step the HOMedU is converted into β-D-glucosyl-5-(hydroxymethyl)uracil (dJ) by the addition of a glucose group by a glucosyl transferase. Three lines of evidence support this model.

At an early stage it was obvious that J is made by modification of T in DNA and not by incorporation of J-deoxyribosetriphosphate (dJTP) into DNA, as J is found at specific positions in the genome. In contrast, precursors of unusual bases made at the nucleotide level are incorporated (semi-) randomly into DNA and incorporation cannot be restricted to unique regions/sequences. A case in point is the replacement of C by glucosylated hydroxy-5-methylcytosine (HOMeC) in T-even bacteriophages, in which the modified base is made at the nucleotide level and the incorporated HOMedCMP replaces dCMP throughout the DNA (18).

When trypanosomes are grown in the presence of HOMedU, HOMedUMP is randomly incorporated into DNA (36). In bloodstream-form trypanosomes nearly all the HOMedU is converted into J, showing that the two steps in the formation of J are not tightly coupled (36).

If the gene for a mammalian DNA repair enzyme, SMUG, which specifically excises HOMeU from DNA, is introduced into bloodstream-form *T. brucei*, expression of this gene kills the trypanosomes by excessive DNA repair (32). As SMUG does not excise J (33), the enzyme must excise the HOMeU that is continuously made in the trypanosome. This HOMeU is therefore a freely accessible intermediate in J biosynthesis. Insect-form trypanosomes are impervious to SMUG unless they are grown in HOMedU (32). This confirms that procyclic form trypanosomes are unable to convert T into HOMedU, the first step in J biosynthesis.

J-Binding Protein 1 Enters the Scene

After the structure of J was solved, a major effort started in the Borst lab to identify the enzymes involved in J biosynthesis, as disruption of J biosynthesis should help to reveal J function. When this effort remained fruitless, we started to look for other ways to understand J function. In mammalian DNA, MeC exerts its function by binding MeC-binding proteins that translate the presence of the minor base in DNA into chromatin alteration. It seemed reasonable to expect that kinetoplastids would contain an analogous protein binding to J. Using J-containing oligonucleotides (10), Cross et al. (8) identified J-binding protein 1 (JBP1) in nuclear extracts of *Crithidia fasciculata*. The 90-kDa protein was purified, the corresponding gene was cloned, and genes for homologous proteins were identified in *T. brucei* and *Leishmania tarentolae*. The initial analysis of JBP1 showed it to be unlike any known protein.

A knock out of the single diploid JBP1 gene in bloodstream-form *T. brucei* was produced to obtain information on the possible function of J (7). The disruption had no effect on the degree of repression of silent VSG genes or on the rate of switching from one active VSG gene ES to another. The disruption also had no effect on the stability of J-containing repetitive sequences, notably telomeric sequences and the 50-bp sequences upstream of all VSG gene ESs.

Unexpectedly, however, JBP1 disruption resulted in a profound decrease in J level, down to 5% of wild type. All sequences normally containing base J seemed to be affected to a similar degree by this decrease (7). This result suggested that JBP1 is involved in the maintenance of J in DNA rather than translating the presence of J into chromatin action. J maintenance might be accomplished either by promoting increased synthesis or by preventing turnover. The latter alternative was eliminated by experiments in which the DNA of bloodstream-form *T. brucei* was loaded up with J by culturing the cells in medium with HOMedU (5, 36). This results in J levels 10-fold higher than those of wild-type *T. brucei* (5, 36). When this experiment was done with JBP1 knock out trypanosomes, the incorporated J was lost by simple dilution as the trypanosomes multiplied (7). This shows that JBP1 does not protect J against turnover but uses existing J to promote J biosynthesis. Indeed, in trypanosomes containing JBP1, the excess J made during growth in HOMedU was lost only sluggishly, even from tubulin gene sequences, which normally do not contain J (7, 11).

The function of JBP1 was further analyzed in *Leishmania*. In contrast to the ability of *T. brucei* to live happily without JBP1, a knock out of JBP1 is lethal both in *L. tarentolae* and

L. major (14, 16). Heterozygote JBP1+/− are fine, however, and even conditional knock out mutants of *L. tarentolae* producing only 10%–15% of wild-type JBP1 have no growth defect, although their J level is reduced twofold (14). Although the data support the role of JBP1 in stimulating J synthesis, the ability of *L. tarentolae* to survive with these minimal amounts of JBP1 has thwarted all attempts to produce conditional knock outs to test the effects of the controlled complete removal of JBP1 from the cell. All tetracycline-inducible constructs made thus far have proven too leaky to reduce JBP1 levels to less than 10% of wild-type levels (14).

The conclusion that JBP1 plays a catalytical role, promoting local J synthesis in regions already containing a basal level of J, is supported by the low ratio of JBP1 to J in *Trypanosoma*, *Leishmania*, and *Crithidia*. There are only 1.0–2.6×10^3 molecules of JBP1 per cell and this is 30- to 60-fold lower than the number of JBP1-binding sites, J residues, in the genome (29). A significant fraction (>95%) of the total J is not bound to JBP1 in vivo.

How JBP1 Binds J DNA

JBP1 was discovered by its ability to bind to J DNA. To examine how it recognizes and binds J DNA, gel-shift experiments were performed using recombinant JBP1 protein expressed in *E. coli* and J-DNA oligonucleotides (10). The gel-shift analysis showed that JBP1 binds specifically to J-containing duplex DNA with high affinity (40–140 nM) (25). The protein does not bind to single-stranded J DNA or to the free base J and requires J in the context of one helical turn of duplex DNA (B-form DNA helix) for optimal binding (25). Presumably, the interaction is sensitive to the nature of the helix because JBP1 does not bind base J when present in the J DNA/RNA duplex (A-form DNA helix). Initial studies using the gel-shift-binding assay did not indicate any significant binding of JBP1 to unmodified DNA. However, recent measurements of JBP1 binding using a more sensitive fluorescence anisotropy approach indicate that JBP1 can bind to unmodified DNA

JBP: J-binding protein

albeit with 100-fold-less affinity than to J DNA (20). The fluorescence anisotropy assay also allowed an analysis of the binding of JBP1 to J DNA under equilibrium conditions. The K_d values obtained are about threefold lower than in the band-shift assay (20). This is not surprising as some of the JBP1 dissociates from the DNA during electrophoresis, resulting in an overestimation of the K_d value.

Methylation interference and DNA footprinting analysis indicate that JBP1 does not make any sequence-specific contacts with the bases surrounding the modified base J. JBP1 should therefore recognize J in any sequence context. Experimental analysis of the JBP1/J DNA interaction utilizing various J-DNA substrates supports this conclusion (25). However, JBP1 binds J with higher affinity when present in the context of telomeric repetitive DNA. This apparent preference of JBP1 for repetitive DNA elements correlates with the higher levels of J in the telomeres and subtelomeric repeats in vivo. Preferential binding of JBP1 to a particular sequence class could in principle lead to increased J levels in these regions of the genome.

DNA footprinting techniques further indicate that the only critical interactions between JBP1 and J DNA occur via major and minor groove contacts at base J and a sequence-independent major groove contact at the nucleotide immediately 5′ of base J (called the J-1 position) (26). This requirement for major and minor groove interactions has suggested that DNA structure is an essential component of the recognition of J DNA by JBP1 (25, 26). This is supported both by the inability of JBP1 to bind to an A-form DNA helix and by its increased affinity for repetitive DNA. The essential J-1 interaction may be explained by the recent in-depth analysis of the molecular recognition between the glucose component of base J and JBP1 (20). Molecular dynamic simulations and sensitive K_d measurements of various J analogs indicate that optimal JBP1 binding occurs when the sugar moiety of base J adopts a conformation in the major groove that allows the 2′ and 3′ hydroxyl groups to make hydrogen bonds

with the nonbridging phosphoryl oxygen of the nucleotide base in the J-1 position (20). Any slight modification of this orientation, by utilizing epimeric modifications of single hydroxyl groups around the glucose moiety of J, led to a major (~100-fold) decrease in JBP1-binding affinity. It is proposed that the J-1 phosphoryl oxygen interaction serves to lock the glucose moiety into an edge-on conformation necessary for optimal JBP1 binding (20).

The domain of JBP1 that would recognize an edge-on presentation of base J has not been fully characterized. Initial truncation analyses of JBP1 were inconclusive. Minimal deletions from the N terminus (40 amino acids) or C terminus (10 amino acids) of JBP1 abolished J DNA binding in gel-shift assays (R. Sabatini, unpublished data). However, limited proteolysis of JBP1 bound to J DNA has recently been used to identify a 20-kDa minimal J-DNA-binding domain within the middle of JBP1 (**Figure 3**). This mini-JBP1 is located approximately at residues 380–460 of *Crithidia* JBP1 and it binds to J DNA with about the same affinity as full-length JBP1 (E. Christodoulou & A. Perrakis, personal communication). Mutation of residues within this domain inhibits the JBP1/J DNA binding in vitro (R. Sabatini, unpublished data). However, it is unclear whether these mutations affect the structure of the polypeptide or inhibit specific contacts with J DNA. In contrast, several

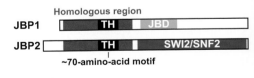

Figure 3

Functional domains of JBP1 and JBP2. The homologous region shared between JBP1 and JBP2 at the N terminus is indicated in red. Indicated in black is the ~70-amino-acid motif within this region, which is related to the functional domain of members of the oxygenase/hydroxylase family. The light blue rectangle labeled JBD represents the 20-kDa minimal J-binding domain. The C terminus of JBP2 is homologous to the SWI2/SNF2 family of ATPase/DNA helicases (depicted by the *dark blue box*). TH; thymidine hydroxylase.

mutations within the N-terminal region of JBP1 (the region that has homology with JBP2 and that contains the proposed TH domain) do not affect the ability to bind J DNA (41).

Identification of JBP2: The De Novo J Synthesis Factor

Analysis of nuclear extracts of the JBP1-null cell by gel shift indicated that JBP1 is the only detectable JBP in trypanosomes (7). However, during the search for additional JBPs, a homolog of JBP1 was identified in the *T. brucei* genome database (11). This 120-kDa protein, called JBP2, contains an N terminus with homology to JBP1 (34% identity, 47% similarity) (**Figure 3**). The remaining C-terminal half of JBP2 is homologous (24% identity, 45% similarity) to the SWI2/SNF2 family of ATPase DNA helicase proteins involved in chromatin remodeling. JBP2 (like JBP1) is developmentally regulated. It is present in bloodstream-form trypanosomes, but undetectable in insect-form trypanosomes. JBP2 does not bind to J DNA and it is therefore not a real JBP. It is present in the nucleus, however, and it binds to chromatin whether base J is present or not.

A remarkable result was obtained when JBP2 was ectopically expressed in insect-form trypanosomes, which normally lack J. This resulted in de novo site-specific synthesis of a basal level of J (11). This basal level of J could be amplified by concomitant expression of JBP1. The J produced in this way in insect-form trypanosomes, whether containing JBP2 alone or JBP2 and JBP1, was located in the same repetitive sequences as in wild-type bloodstream-form trypanosomes.

While these initial studies allowed the direct analysis of JBP2 function in a J minus background, they relied on overexpression of JBP2 in a life stage that normally lacks the modified base. Follow-up studies were therefore done in bloodstream-form trypanosomes. The JBP2-null bloodstream-form trypanosome contains fivefold less J and is unable to insert J

de novo into newly generated telomeric DNA (22). These results suggest that JBP2 is the key regulator of J biosynthesis: regulating the developmental and site-specific de novo synthesis of J in the genome.

Results with *L. tarentolae* support this conclusion. JBP2-null parasites are initially fine but have difficulty maintaining J levels, which gradually decrease over many generations, down to about fourfold lower than in wild-type parasites (14). The *L. tarentolae* JBP2-null cells containing fourfold-less J are hypersensitive to growth in BrdU (S. Vainio, B. ter Riet, H. van Luenen, and P. Borst., unpublished results), a treatment previously shown to result in a decrease in J levels in *T. brucei* (36). Indeed, growth in BrdU also results in a decrease in J levels in *L. tarentolae* DNA. Although we think that the BrdU-treated cells die owing to the apparent essential nature of J in this organism, we are unsure why they die. They do not stop at a specific point in the cell cycle, nor do they undergo massive DNA rearrangements. Both light and electron microscope analyses failed to reveal any clues (S. Vainio, B. ter Riet, H. van Luenen, and P. Borst, unpublished data). We do not know why incorporation of relatively low levels of BrdU into DNA results in a decrease in J levels. The simple explanation that JBP1 tightly binds to BrdU-DNA, preventing JBP1 to carry out its normal role in J biosynthesis, is incorrect (P.A. Genest, B. ter Riet, H. van Luenen, and P. Borst, unpublished results).

Although the studies discussed above implicate JBP2 in the J synthesis pathway, it is unclear how JBP2 stimulates the site-specific localization of the modified base in the chromosome. Mutation of residues within the SWI2/SNF2 motif, which is involved in ATPase function, inhibits JBP2 stimulation of J synthesis (11). On the basis of these results it was initially proposed that chromatin remodeling by JBP2 is required for de novo J biosynthesis. Additionally, the SWI2/SNF2 domain might recognize distinct chromatin elements unique to the telomeric and subtelomeric repeats in the trypanosome genome.

Role of JBP1 and JBP2 in J Biosynthesis

Both JBP1 and JBP2 influence the level of J in DNA. This could be explained by indirect effects, e.g., the recruitment of a TH to DNA (11). This explanation did not account, however, for the substantial homology between JBP1 and JBP2 in their N-terminal halves (34% identity) but none in the remainder of the protein. Yu et al. (41) therefore raised the possibility that JBP1 and JBP2 are directly involved in the synthesis of HOMedU and that their homologous N-terminal halves contain the TH activity. Sequence comparison with known hydroxylases initially yielded no match, but when bioinformatician Daniel Rigden repeated the exercise by hand, an interesting match turned up, with enzymes belonging to the family of Fe^{2+}- and 2-oxoglutarate-dependent dioxygenases (hydroxylases) (41). These enzymes catalyze oxidation reactions, with molecular oxygen as the oxidizing agent and ferrous iron and 2-oxoglutarate as a cofactor and cosubstrate, respectively (27). One member of this superfamily, AlkB, oxidizes the harmful methyl group in 1-MeA or 3-MeC, inserted by DNA-methylating agents. The hydroxymethyl moiety formed is spontaneously released as formaldehyde, regenerating the normal base. This oxidation of a methyl group by AlkB resembles the reaction catalyzed by the putative TH involved in J synthesis, which oxidizes the methyl moiety in 5-MedU (thymidine), forming HOMedU as a stable endproduct (**Figure 2**).

Enzymes in this superfamily have a β-strand fold that contains a highly conserved motif consisting of four amino acids that bind Fe^{2+} and 2-oxoglutarate (1, 21). These four amino acids are conserved in appropriate positions in the N-terminal halves of all known versions of JBP1 and JBP2 and are essential for JBP1 function (41). Replacement of any of the four conserved amino acids by alanine or serine abolishes the ability of the resulting altered JBP1 to stimulate J synthesis. This is not due to the inability of the altered (mutant) JBP1 to enter the nucleus or to bind to J DNA (41). Whereas the individual JBP1- and JBP2-null mutant trypanosomes have reduced J levels (20- and 4-fold, respectively), the JBP1/JBP2-double-null mutant completely lacks base J (R. Kieft, L. Cliffe, and R. Sabatini, unpublished data). These cells still contain glucosyltransferase activity because HOMedU feeding leads to high levels of J. These results fully support the hypothesis that JBP1 and JBP2 are Fe^{2+}- and 2-oxoglutarate-dependent oxygenases (hydroxylases) that catalyze the hydroxylation of T in DNA to yield HOMedU. However, all attempts to demonstrate hydroxylase activity of JBP1 in vitro have failed thus far (14, 41). These negative results do not mean much, as nothing is known about the nature of the DNA substrate used by JBP1 for its hydroxylase activity.

But why do these organisms require two distinct THs involved in the regulation of J synthesis? On the basis of the analysis of JBP1 and JBP2 function discussed above, we think that the two THs work together to allow the proper localization, levels, and maintenance of J in the genome. According to this model (**Figure 4**), JBP2 recognizes specific regions of the chromatin (primarily telomeric) via its SWI2/SNF2 domain, allowing the TH domain to oxidize a particular thymidine residue to form HOMedU. It is not clear whether the hydrolysis of ATP by the SWI2/SNF2 domain is required for chromatin binding or modification of chromatin structure to allow recognition of the thymidine residue by the TH domain. The nonspecific glucosyl transferase (GT) then adds the glucose that converts HOMedU into dJ. Once the specific basal level of dJ is laid down by JBP2, JBP1 will bind and oxidize adjacent thymidine residues, which are subsequently converted into dJ by the GT. Upon cell division, JBP1 prevents the twofold dilution of J by recognizing hemi-J-lated regions of the chromosome and oxidizes a thymidine on the opposing DNA strand.

In the model of **Figure 4** the specificity of J insertion is provided by JBP2. It is JBP2 that determines the region and the sequence into which J is inserted. JBP1 only provides

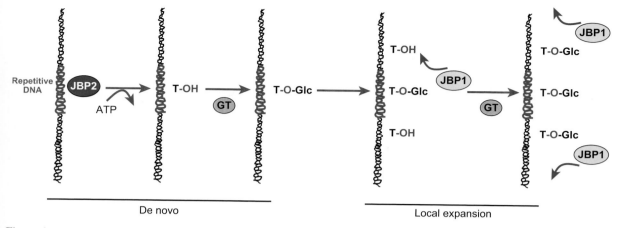

Figure 4

Speculative scheme delineating the roles of JBP1 and JBP2 in the regulation of J biosynthesis. JBP2 is proposed to recognize and bind specific chromatin domains (like telomeric repeats), hydrolyze ATP, and stimulate de novo hydroxylation of a thymidine residue in DNA. The glucosyl transferase (GT) converts the hydroxymethyl deoxyuridine (T-OH) into base J. JBP1 then binds base J and hydroxylates adjacent thymidines followed by the conversion into J by the GT enzyme. It is currently unknown what factors limit the local expansion of J synthesis by JBP1 along the chromatin.

maintenance and local amplification. There are now several indications that this model is an oversimplification.

Our recent observations show that JBP1 is also capable of de novo J synthesis (P. Borst and R. Sabatini, unpublished data). The initial analysis of JBP1 function in procyclic cells utilized a GFP fusion molecule, which apparently interfered with the nonspecific binding of JBP1. Overexpression of an untagged version of JBP1 leads to low levels of J synthesis in procyclic cells. Preliminary analysis indicates that the localization of J made de novo by JBP1 is nonspecific. It is not a surprise that JBP1 catalyzes some thymidine hydroxylation, even in DNA that does not contain J. After all, JBP1 does bind to DNA without J, albeit with a K_d value 100-fold higher than when binding to J DNA (20). Obviously, simply bringing the TH domain of JBP1 in close proximity to DNA is sufficient for modification of a thymidine residue. No J-specific interaction/conformational change is required. This conclusion is supported by the ability of a LacI-JBP1 fusion to stimulate de novo J synthesis in LacO DNA repeats introduced at specific sites within the genome (S. Birkeland and R. Sabatini, unpublished data).

Some of the DNA sequence specificity of J insertion must be contributed by JBP1. Without JBP1, the J level in *T. brucei* bloodstream-form trypanosomes drops to 5% of the wild-type level, and the regional distribution of J remains approximately the same (7). Although this shows that JBP2 by itself can determine the regional specificity of J insertion, most of the J in wild-type cells is inserted by JBP1 and this enzyme must therefore also have sequence specificity. This is most obvious in the case of telomeric repeats in which only the second T is modified in the (GGGTTA) strand.

The model predicts that it should be possible to locally convert all suitable T into J, if the concentration of JBP1 is high enough. This was not observed in experiments with forced overexpression of JBP1 in *T. brucei* (7) or in *Leishmania* (14). Obviously there are unknown limitations to the replacement of T by J that result in replacement of a given T only in a small part of the parasite population.

Although the present version of the model for the regulation of J synthesis by two TH enzymes presented in **Figure 4** is an oversimplification, it may nevertheless be used to explain why a silent telomeric ES loses J once it becomes activated. As previously mentioned, J is present in repeats that immediately flank the active ES (telomeric and 50-bp repeats), but it is absent from the ~20-kb polymerase I transcription unit. However, when new telomeric repeats are placed in an active ES, in the *JBP2*-null cell, they also lack J (22). In the absence of JBP2 the cell is unable to stimulate de novo J in the new repeats, and apparently JBP1 is unable to propagate J from the 50-bp repeat down through the actively transcribed ES. These data, along with the finding that JBP1 fails to bind single-stranded J DNA or a J DNA/RNA duplex (25), suggest that highly processive polymerase I transcription in the active ES excludes the J synthesis machinery (or minimally the propagation/maintenance of J). If JBP1 is not present or unable to bind DNA, J will be diluted out (7, 36). What effect transcription has on the ability of JBP2 to stimulate de novo J synthesis remains to be tested.

The Elusive Glucosyl Transferase

When bloodstream-form *T. brucei* is cultured in medium containing HOMedU, HOMedUMP is inserted randomly into the DNA instead of dTMP and converted into JMP (5, 36). This suggests that the putative GT that puts a glucose moiety on HOMeU anywhere in DNA does not have stringent requirements for the packaging or context of its DNA substrate. Hence, we expected that this enzyme would be readily detectable in nuclear extracts using HOMedU-containing oligonucleotides as substrate and UDPG as glucose donor. However, notwithstanding the availability of increasingly sophisticated assays, no enzyme activity was found. Some of these negative experiments have been summarized by Ulbert (31). M.A.J. Ferguson (personal communication) has pointed out that the putative GT is unique in that it operates in the nucleus, whereas all known GTs are cytoplasmic. This unique location could entail an unusual glucose donor or unusual cofactor requirements. Unfortunately, GTs are poorly conserved in evolution and there is only a weak protein signature to work with. In the Borst lab

the expression of some candidate genes has been knocked down by RNAi thus far without an effect on J levels.

FUNCTIONS OF BASE J

The functions of base J remain unknown and we can only list here the speculations generated over the years. The first idea proposed was that J is involved in gene silencing (3). This idea came from the association of J with silent telomeric VSG genes in *T. brucei*, and it received support from the finding that a decrease in J, induced by BrdU incorporation, led to some derepression of the silent VSG genes (36). However, more recent experiments make the association between J and VSG gene repression tenuous at best. The first blow came when Cross et al. (7) showed that the JBP1 knock out of *T. brucei* has 20-fold-less J than does wild-type *T. brucei*, but that this decrease does not result in derepression of silent telomeric VSG genes. Obviously, the derepression of VSG genes induced by incorporation of BrdU is not due to the 10-fold reduction in J, but to other effects of the presence of BrdU in the DNA. The second blow to the idea that J causes VSG repression came recently when the Sabatini lab constructed *T. brucei* JBP1/JBP2-null strains devoid of J. In these strains the silent telomeric VSG genes remain repressed in bloodstream-form *T. brucei* (R. Sabatini, unpublished results). The idea that J is involved in gene repression, analogous to the role of MeC in vertebrates and plants, lacks support and is moribund.

A second idea is that J helps to repress homologous recombination between repetitive sequences, which is analogous to the role postulated for MeC in other eukaryotes. This has been checked in the JBP1 knock out, in which J levels are 20-fold decreased. No alterations in the size of representative repeats, the telomeric repeats, and the 50-bp repeats upstream of VSG gene ESs were found (7). This finding is suggestive evidence against a role for J in regulating the stability of repetitive elements, and a recent analysis of repeats that completely lack J supports this conclusion (R. Kieft, L. Cliffe, and R. Sabatini, unpublished results). In contrast, the analysis of these J-null bloodstream-form trypanosomes implicates base J in the regulation of telomeric VSG homologous recombination events and thus in regulating antigenic variation. Whether this upregulation of DNA rearrangements is due to the loss of J in the 70-bp repeats immediately upstream of the silent VSG genes needs to be addressed.

The recent finding that in *Leishmania* J is nearly completely restricted to telomeres obviously suggests that J must have a telomeric function in this organism (15). This function is presumably conserved, as all organisms in which J has been found contain J in their telomeres (12, 38). In the case of *Leishmania*, the telomeric function of J is apparently essential, but how J functions at telomeric repeats remains unknown and poses a challenge for future research.

The evidence is also not generalizable, as *T. brucei* is unique regarding J function. It is the only organism that lacks all J in one phase of its life cycle, the insect phase (17, 19, 23). Moreover, recent work in the Sabatini lab has shown that even bloodstream-form *T. brucei* can survive in the lab without any J, in stark contrast to *Leishmania*. The possibility that *T. brucei* has managed to become less dependent on J than other *Kinetoplastida* should therefore be kept in mind.

Although it is plausible that J will have a telomeric function in all organisms in which J is found, it is likely that J has been pressed into other services in the course of evolution. It is probably not accidental that so many non-telomeric repetitive sequences in *T. brucei* and in *T. cruzi* contain J. There is also the intriguing case of *Euglena*, in which the bulk of J is in non-telomeric sequences of unknown nature. Even if the telomeric function of J is identified soon, there are puzzles that remain to be solved.

IS J BIOSYNTHESIS EXPLOITABLE FOR CHEMOTHERAPY?

J is unique to *Kinetoplastida* and *Euglena*. It is present in all major kinetoplastid pathogens of humans and their live stock thus far analyzed, but not in their hosts. Although *T. brucei* lab strains seem to do fairly well without J or JBP1/2, the absence of JBP1 is lethal in *Leishmania* and we think this is due to loss of J. This makes JBP1, and possibly J biosynthesis in general, a target for parasite-specific chemotherapy.

The Wentworth group (20) has initiated a search for compounds that inhibit the binding of JBP1 to J DNA. Usually DNA-binding proteins are not druggable, but JBP1 may be an exception, given the need to bind to a highly hydrophilic molecule, glucose, in a stereospecific fashion. Stollar has developed a fluorescence-polarization assay for determining JBP1 binding (20) that has been adapted to high-throughput screening (D. Stollar, personal communication). To support the drug-finding project, Perrakis and colleagues have attempted to determine the 3-D structure of JBP1 of *Crithidia* or *Leishmania* and to get more information on how JBP1 binds to J DNA (A. Perrakis, personal communication). If our current hypothesis that JBP1 is a hydroxylase required for J biosynthesis is correct (and we are convinced that it is) and if an assay can be set up for this hydroxylase activity that is suitable for high-throughput screening, this assay may provide another way to exploit J for drug finding.

SUMMARY POINTS

1. A new minor modified T base, base J, is present in the nuclear DNA of kinetoplastid flagellates and in the green algae *Euglena*, but not in the other protozoa or metazoa investigated.

2. Base J replaces about 1% of thymidine; it is mainly found in repetitive DNA sequences and most prominently in telomeric DNA repeats. Although J is found in unique sequences, such as the second T in the telomeric repeat $(GGGTTA)_n$, no consensus sequence recognized by the modifying enzymes is apparent.

3. Biosynthesis of J in DNA occurs in two steps. In the first step a specific T residue is oxidized to yield HOMeU, which is converted in a second step into J by the addition of glucose. Strong indirect evidence suggests that the first step is catalyzed by two THs, JBP1 and JBP2, that are members of the Fe^{2+}- and 2-oxoglutarate-dependent oxygenase family. JBP2 appears to be mainly responsible for de novo and site-specific J synthesis. JBP1, which is a J-DNA-binding protein, mediates the loco-regional amplification/maintenance of the J present. The putative GT catalyzing the addition of the glucose moiety of J has not been identified.

4. In *Leishmania* JBP1 is essential, and the available evidence indicates that this is due to the inability of this organism to maintain J in DNA. As nearly all J is located in the telomeric repeats of *Leishmania* DNA, J appears to be an essential modification of telomeres in *Leishmania*.

5. *T. brucei* lacking both JBP1 and JBP2 is unable to synthesize base J but remains viable. The defects in J-null *T. brucei* remain to be delineated.

6. In the trypanosomatids *T. brucei* and *T. cruzi*, base J is associated with subtelomeric genes for surface antigens. Whether J is directly involved in gene silencing remains to be determined, but the available evidence suggests it is not.

FUTURE ISSUES

1. The most pressing issue is to unambiguously determine the function of base J. The recent generation of *T. brucei* mutants devoid of J provides a handle on this. So does the fact that JBP1 is essential in *Leishmania* and that JBP2-null *Leishmania* mutants become hypersensitive to the toxic effects of growth in BrdU, which leads to loss of J from DNA.

2. JBP1 and JBP2 should be further characterized. Unambiguous proof should be obtained that these proteins are indeed THs by demonstrating hydroxylase activity in vitro. The cofactor requirements of both enzymes should be determined. The relative roles of JBP1 and JBP2 in specifying the sequence and regional distribution of base J in DNA should be elucidated.

3. The putative GT required for J biosynthesis should be found and characterized.

4. The structure of JBP1 bound to J DNA should be determined, as well as the determinants of binding, both in the DNA and in the protein.

5. How JBP2 modifies specific sequences and regions in DNA should be determined. What is the exact role of SWI2/SNF2 motifs in this protein in its function?

6. The basis for the different regional distributions of base J in *T. brucei*, *T. cruzi*, *Leishmania* spp., and *Euglena* spp. should be determined. The possibility that base J has been recruited for different functions over the course of evolution should be explored.

7. The possible role of J in antigenic variation, in both *T. brucei* and *T. cruzi*, needs further investigation.

8. The J biosynthesis machinery should be further tested as a potentially druggable target against diseases caused by kinetoplastid flagellates, such as sleeping sickness (African trypanosomes), Chagas' disease (*T. cruzi*), and the various forms of leishmaniases.

DISCLOSURE STATEMENT

The authors are not aware of any biases that might be perceived as affecting the objectivity of this review.

ACKNOWLEDGMENTS

We are grateful to the current members of our labs and to Dr. Anastassis Perrakis (NKI-AVL) for allowing us to quote unpublished results and for their helpful comments on the manuscript. We are also indebted to Dr. P.A. Genest, Dr. L. Cliffe, Dr. H. van Luenen and Dr. F. van Leeuwen for constructive critique. Recent work in our labs was supported by a grant from The Netherlands Organization for Scientific Research—Chemical Sciences (NWO—CW) to PB and from NIH (A1063523) to RS.

LITERATURE CITED

1. Aravind L, Koonin EV. 2001. The DNA-repair protein AlkB, EGL-9, and leprecan define new families of 2-oxoglutarate- and iron-dependent dioxygeneases. *Genome Biol.* 2:research 7.1–research 7.8
2. Bernards A, Michels PA, Lincke CR, Borst P. 1983. Growth of chromosome ends in multiplying trypanosomes. *Nature* 303:592–97
3. **Bernards A, Van Harten-Loosbroek N, Borst P. 1984. Modification of telomeric DNA in *Trypanosoma brucei*: a role in antigenic variation? *Nucleic Acids Res.* 12:4153–70**
4. Blundell PA, Van Leeuwen F, Brun R, Borst P. 1998. Changes in expression site control and DNA modification in *Trypanosoma brucei* during differentiation of the bloodstream form to the procyclic form. *Mol. Biochem. Parasitol.* 93:115–30
5. Borst P, Gommers-Ampt JH, Ligtenberg MJL, Rudenko G, Kieft R, et al. 1993. Control of antigenic variation in African trypanosomes. *Cold Spring Harb. Symp. Quant. Biol.* 58:105–15
6. Borst P, Van Leeuwen F. 1997. Beta-D-glucosyl-hydroxymethyluracil, a novel base in African trypanosomes and other *Kinetoplastida. Mol. Biochem. Parasitol.* 90:1–8
7. Cross M, Kieft R, Sabatini R, Dirks-Mulder A, Chaves I, Borst P. 2002. J-binding protein increases the level and retention of the unusual base J in trypanosome DNA. *Mol. Microbiol.* 46:37–47
8. **Cross M, Kieft R, Sabatini R, Wilm M, De Kort M, van der Marel GA, et al. 1999. The modified base J is the target for a novel DNA-binding protein in kinetoplastid protozoans. *EMBO J.* 18:6573–81**
9. Crozatier M, De Brij RJ, Den Engelse L, Johnson PJ, Borst P. 1988. Nucleoside analysis of DNA from *Trypanosoma brucei* and *Trypanosoma equiperdum*. *Mol. Biochem. Parasitol.* 31:127–32
10. De Kort M, Ebrahimi E, Wijsman ER, Van der Marel GA, Van Boom JH. 1999. Synthesis of oligodeoxynucleotides containing 5-(beta-D-glucopyranosyloxymethyl)-2'deoxyuridine, a modified nucleoside in the DNA of *Trypanosoma brucei*. *Eur. J. Org. Chem.* 9:2337–44
11. **DiPaolo C, Kieft R, Cross M, Sabatini R. 2005. Regulation of trypanosome DNA glycosylation by a SWI2/SNF2-like protein. *Mol. Cell* 17:441–51**
12. Dooijes D, Chaves I, Kieft R, Dirks-Mulder A, Martin W, Borst P. 2000. Base J originally found in *Kinetoplastida* is also a minor constituent of nuclear DNA of *Euglena gracilis*. *Nucleic Acids Res.* 28:3017–21

3. Proposed the existence of an unusual form of DNA modification in bloodstream-form *T. brucei* inferred from partial resistance to cleavage with PstI and PvuII.

8. Discovery of JBP1, which binds specifically to J-containing duplex DNA.

11. Presents a model that postulates that JBP2 determines the sites at which J is inserted, whereas JBP1 mediates the maintenance and local amplification of J in DNA.

13. Ekanayake DK, Cipriano M, Sabatini R. 2007. Telomeric localization of the modified base J and contingency genes in the genome of the protozoan parasite *Trypanosoma cruzi*. *Nucleic Acids Res.* 35:6367–77

14. Genest PA. 2007. *Analysis of the modified DNA base J and the J-binding proteins in* Leishmania. PhD thesis. Univ. Amsterdam. 112 pp.

15. Genest PA, ter Riet B, Cysouw T, van Luenen H, Borst P. 2007. Telomeric localization of the modified DNA base J in the genome of the protozoan parasite *Leishmania*. *Nucleic Acids Res.* 35:2116–24

16. Genest PA, ter Riet B, Dumas C, Papadopoulou B, Van Luenen HG, Borst P. 2005. Formation of linear inverted repeat amplicons following targeting of an essential gene in *Leishmania*. *Nucleic Acids Res.* 33:1699–709

17. Gommers-Ampt J, Lutgerink J, Borst P. 1991. A novel DNA nucleotide in *Trypanosoma brucei* only present in the mammalian phase of the life cycle. *Nucleic Acids Res.* 19:1745–51

18. Gommers-Ampt JH, Borst P. 1995. Hypermodified bases in DNA. *FASEB J.* 9:1034–42

19. Gommers-Ampt JH, Van Leeuwen F, de Beer AL, Vliegenthart JF, Dizdaroglu M, et al. 1993. Beta-d-glucosyl-hydroxymethyluracil: a novel modified base present in the DNA of the parasitic protozoan *T. brucei*. Cell 75:1129–36

20. Grover RK, Pond SJ, Cui Q, Subramaniam P, Case DA, et al. 2007. O-glycoside orientation is an essential aspect of base J recognition by the kinetoplastid DNA-binding protein JBP1. *Angew. Chem. Int. Ed. Engl.* 46:2839–43

21. Hausinger RP. 2004. FeII/alpha-ketoglutarate-dependent hydroxylases and related enzymes. *Crit. Rev. Biochem. Mol. Biol.* 39:21–68

22. Kieft R, Brand V, Ekanayake DK, Sweeney K, DiPaolo C, et al. 2007. JBP2, a SWI2/SNF2-like protein, regulates de novo telomeric DNA glycosylation in bloodstream form *Trypanosoma brucei*. *Mol. Biochem. Parasitol.* 156:24–31

23. Pays E, Delauw MF, Laurent M, Steinert M. 1984. Possible DNA modification in GC dinucleotides of *Trypanosoma brucei* telomeric sequences: relationship with antigen transcription. *Nucleic Acids Res.* 12:5235–47

24. Raibaud A, Gaillard C, Longacre S, Hibner U, Buck G, et al. 1983. Genomic environment of variant surface antigen genes of *Trypanosoma equiperdum*. *Proc. Natl. Acad. Sci. USA* 80:4306–10

25. Sabatini R, Meeuwenoord N, Van Boom JH, Borst P. 2002. Recognition of base J in duplex DNA by J-binding protein. *J. Biol. Chem.* 277:958–66

26. Sabatini R, Meeuwenoord N, Van Boom JH, Borst P. 2002. Site-specific interactions of JBP with base and sugar moieties in duplex J-DNA. Evidence for both major and minor groove contacts. *J. Biol. Chem.* 277:28150–56

27. Schofield CJ, Zhang Z. 1999. Structural and mechanistic studies on 2-oxoglutarate-dependent oxygenases and related enzymes. *Curr. Opin. Struct. Biol.* 9:722–31

28. Simpson L. 1998. A base called J. *Proc. Natl. Acad. Sci. USA* 95:2037–38

29. Toaldo CB, Kieft R, Dirks-Mulder A, Sabatini R, Van Luenen HG, Borst P. 2005. A minor fraction of base J in kinetoplastid nuclear DNA is bound by the J-binding protein 1. *Mol. Biochem. Parasitol.* 143:111–15

30. Turner JJ, Meeuwenoord NJ, Rood A, Borst P, van der Marel GA, Van Boom JH. 2003. Reinvestigation into the synthesis of oligonucleotides containing 5-(β-d-glucopyranosyloxymethyl)-2′deoxyuridine. *Eur. J. Org. Chem.* 3832–39

31. Ulbert S. 2003. *DNA repair and antigenic variation in* Trypanosoma brucei. PhD thesis. Univ. Amsterdam. 120 pp.

32. Ulbert S, Cross M, Boorstein RJ, Teebor GW, Borst P. 2002. Expression of the human DNA glycosylase hSMUG1 in *Trypanosoma brucei* causes DNA damage and interferes with J biosynthesis. *Nucleic Acids Res.* 30:3919–26

33. Ulbert S, Eide L, Seeberg E, Borst P. 2004. Base J, found in nuclear DNA of *Trypanosoma brucei*, is not a target for DNA glycosylases. *DNA Repair* 3:145–54

34. Van Leeuwen F, De Kort M, van der Marel GA, Van Boom JH, Borst P. 1998. The modified DNA base beta-D-glucosyl-hydroxymethyluracil confers resistance to micrococcal nuclease and is incompletely recovered by 32P-postlabeling. *Anal. Biochem.* 258:223–29

19. Identification of base J.

20. Oligonucleotides containing analogs of base J were used to infer that interactions of the glucose moiety of J with the nucleotide 5′ of J allows the optimal edge-on presentation of the sugar to JBP1.

23. Independently identified a potential modification of telomeric DNA in African trypanosomes and its correlation with VSG silencing (see Reference 3). Also showed absence of modification in insect-form *T. brucei*.

35. Van Leeuwen F, Dirks-Mulder A, Dirks RW, Borst P, Gibson W. 1998. The modified DNA base beta-D-glucosyl-hydroxymethyluracil is not found in the tsetse fly stage of *Trypanosoma brucei*. *Mol. Biochem. Parasitol.* 94:127–30

36. **Van Leeuwen F, Kieft R, Cross M, Borst P. 1998. Biosynthesis and function of the modified DNA base beta-D-glucosyl-hydroxymethyluracil in *Trypanosoma brucei*. *Mol. Cell. Biol.* 18:5643–51**

37. Van Leeuwen F, Kieft R, Cross M, Borst P. 2000. Tandemly repeated DNA is a target for the partial replacement of thymine by β-D-glucosyl-hydroxymethyluracil in *Trypanosoma brucei*. *Mol. Biochem. Parasitol.* 109:133–45

38. **Van Leeuwen F, Taylor MC, Mondragon A, Moreau H, Gibson W, et al. 1998. Beta-D-glucosyl-hydroxymethyluracil is a conserved DNA modification in kinetoplastid protozoans and is abundant in their telomeres. *Proc. Natl. Acad. Sci. USA* 95:2366–71**

39. **Van Leeuwen F, Wijsman ER, Kieft R, Van Der Marel GA, Van Boom JH, Borst P. 1997. Localisation of the modified base J in telomeric VSG gene expression sites of *Trypanosoma brucei*. *Genes Dev.* 11:3232–41**

40. Wijsman ER, Van Den Berg O, Kuyl-Yeheskiely E, Van Der Marel GA, Van Boom JH. 1994. Synthesis of 5-(beta-D-glucopyranosyloxymethyl)-2′-deoxyuridine and derivatives thereoff. A modified D-nucleoside from the DNA of *Trypanosoma brucei*. *Recl. Trav. Chim. Pays-Bas* 113:337–38

41. **Yu Z, Genest PA, ter Riet B, Sweeney K, DiPaolo C, et al. 2007. The protein that binds to DNA base J in trypanosomatids has features of a thymidine hydroxylase. *Nucleic Acids Res.* 35:2107–15**

36. Established the two-step synthesis of J and provided the first evidence that base J is involved in telomeric VSG gene expression.

38. Demonstration that base J is present in the telomeres of all kinetoplastids analyzed, including the human pathogens *T. cruzi* and *L. donovani*.

39. Reported the isolation of antibodies specific for J DNA and their use to demonstrate that J can account for the uncleavable Pvu II and Pst I sites in silent VSG gene expression sites.

41. Mutagenesis analyses of JBP1 led to the proposal that JBP1 and JBP2 are THs catalyzing the first step in J biosynthesis.

A Case Study for Microbial Biodegradation: Anaerobic Bacterial Reductive Dechlorination of Polychlorinated Biphenyls— From Sediment to Defined Medium

Donna L. Bedard

Department of Biology, Rensselaer Polytechnic Institute, Troy, New York 12180;
email: Bedard@rpi.edu

Annu. Rev. Microbiol. 2008. 62:253–70

First published online as a Review in Advance on
June 6, 2008

The *Annual Review of Microbiology* is online at
micro.annualreviews.org

This article's doi:
10.1146/annurev.micro.62.081307.162733

Key Words

Dehalococcoides, PCBs, dehalogenation, halorespiration, Aroclor, PCB
bioremediation

Abstract

The history of anaerobic microbial polychlorinated biphenyl (PCB)
dechlorination is traced over 20 years using a case study of PCB dechlorination in the Housatonic River (Massachusetts) as an example. The
history progresses from the characterization of the PCBs in the sediment, to cultivation in sediment microcosms, to the identification of
four distinct types of PCB dechlorination, to a successful field test, to the
cultivation in defined medium of the organisms responsible for extensive dechlorination of Aroclor 1260, and finally to the identification of a
Dehalococcoides population that links its growth to the dechlorination of
Aroclor 1260. Other PCB dechlorinators have also been identified. Two
bacterial strains, *o*-17 and DF-1, that link their growth to the dechlorination of several PCB congeners belong to a novel clade of putative
dechlorinating bacteria within the phylum *Chloroflexi*. *Dehalococcoides
ethenogenes* strain 195 also dechlorinates several PCB congeners when
grown on chlorinated ethenes. Evidence is mounting that *Dehalococcoides*
and other dechlorinating *Chloroflexi* may play a significant role in the
dechlorination of commercial PCBs in situ.

Contents

INTRODUCTION

Three decades after the manufacture of polychlorinated biphenyls (PCBs) was banned in the United States and many other countries, PCBs remain a concern. They are included among the 12 worldwide priority persistent organic pollutants (or POPs) (41) and are ranked fifth on the U.S. Environmental Protection Agency Superfund Priority List of Hazardous Compounds (4). PCBs contaminate the sediments of many lakes, rivers, and harbors where they bioaccumulate and biomagnify in the food chain. They are suspected of contributing to many health effects (3). PCBs are a family of compounds with a biphenyl backbone chlorinated at 1 to 10 positions (**Figure 1**). There are 209 different PCB molecules, known as congeners, that differ in the number and position of chlorines. Commercial PCB mixtures were made by catalytic chlorination of biphenyl to a specified weight percent of chlorine. These were

manufactured and used worldwide for a variety of applications for nearly 50 years beginning in 1929. Each commercial PCB mixture contains 60 to 90 different PCB congeners and typically has the consistency or an oil or resin (25, 29). In the United States and Great Britain, commercial PCB mixtures are known as Aroclors.

Anaerobic reductive dechlorination of PCBs is a natural process (16, 17) that provides a means of detoxification and, when coupled with aerobic degradation, completely destroys PCBs (5). The process occurs naturally in many aquatic sediments (5, 10) but typically does not achieve the potential demonstrated in laboratory microcosms (15, 39). The third decade of research in this area continues with the conviction that learning more about the organisms that carry out PCB dechlorination will facilitate the development of efficient and cost-effective methods of in situ PCB bioremediation.

Polychlorinated biphenyl (PCB) dechlorination process: a set of microbially mediated dechlorination reactions exhibiting a particular pattern of congener selectivity and chlorophenyl reactivity

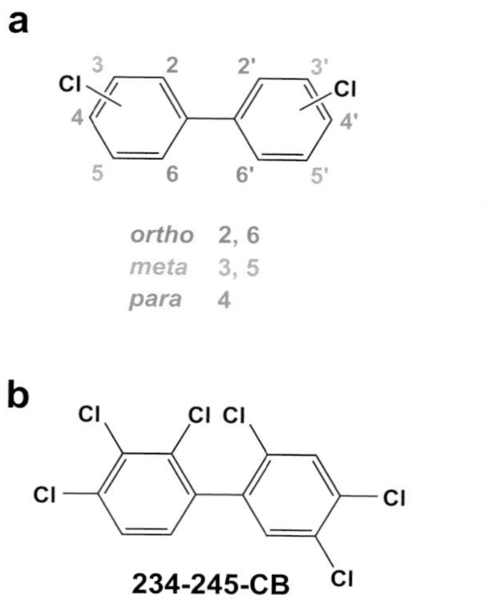

a

ortho   2, 6
meta   3, 5
para   4

b

234-245-CB

Figure 1

Structure of and numbering system for polychlorinated biphenyls (PCBs). (*a*) PCB structure showing the *ortho*, *meta*, and *para* chlorines. (*b*) Structure of 2,2′,3,4,4′,5′-tetrachlorobiphenyl. To facilitate visualization of the chlorination pattern of PCB congeners, we designate the chlorination substitution pattern on each ring separated by a hyphen and followed by -CB. Thus the PCB congener shown is called 234-245-CB.

SCOPE OF THIS REVIEW

There have been several reviews of microbial PCB dechlorination and it is not the intent of this review to repeat what has already been published. For a comprehensive description of dechlorination in situ and a detailed description of seven of the eight known microbial PCB dechlorination processes, the reader is referred to Reference 10. That review also discusses mass balance and quantitation issues, as well as concerns about drawing conclusions regarding the dechlorination of complex PCB mixtures in sediment microcosms on the basis of experiments with single congeners. The eighth dechlorination process, LP, is described in Reference 5 and updated in Reference 9. For a review of PCB dechlorination and degradation from a remediation perspective, see Reference 5. Wiegel & Wu (48) have also provided an excellent review of microbial PCB dechlorination.

This review concentrates on providing an assessment of the data underlying recent successes in developing sediment-free cultures of PCB-dechlorinating bacteria, and in identifying several bacteria responsible for PCB dechlorination. The review also follows the progression of our understanding of PCB dechlorination in the Woods Pond/Housatonic River system from the initial characterization of dechlorination in situ to the recent cultivation in defined medium and identification of the organisms responsible for the major dechlorination process in that system.

CB: chlorinated biphenyl

PCB CONTAMINATION IN WOODS POND AND THE HOUSATONIC RIVER

The Housatonic River is contaminated with highly chlorinated PCBs and weathered unidentified hydrocarbon oil, possibly fuel oil. Woods Pond (Lenox, Massachusetts) is a shallow impoundment on the Housatonic River 13 miles downstream from the source of the PCB contamination. The sediments of Woods Pond are rich in organic content and are highly methanogenic. PCB concentrations in Woods Pond range from 15 to 180 μg gram^{-1} (sediment dry weight), and the oil concentrations range from 5.3 to 32.4 mg gram^{-1} (sediment dry weight) (8). The PCBs are a mixture of partially dechlorinated Aroclors 1260 and 1254 (95%:5%) (8). Aroclor 1260 is primarily composed of hexa- and heptachlorobiphenyls (hepta-CBs), and 85% to 90% of the congeners in this PCB mixture have six or more chlorines (25). Congener-specific analysis of the PCBs in Woods Pond sediment revealed that the relative proportions of major hexa- and hepta-CBs were decreased by as much as 45% relative to Aroclor 1260, whereas the proportions of tri-, tetra-, and penta-CBs were increased (8). In addition, many of the tetra-, penta-, and hexa-CBs contained 24- and 246-chlorophenyl rings (8), which are uncommon in Aroclors (29) but frequently result from *meta* dechlorination of PCBs. However, overall dechlorination was modest. Even the most extensively dechlorinated samples had lost only 11% to 19% of the *meta* chlorines and only 2% to 7% of the *para* chlorines (8).

MULTIPLE PCB DECHLORINATION ACTIVITIES PRESENT IN WOODS POND

We sought to understand why PCB dechlorination in Woods Pond was so modest. Experiments with sediment microcosms showed that PCB dechlorinators were present and could be activated to dechlorinate the PCBs associated with the sediment by the addition of high concentrations of individual PCB congeners. We discovered that the dechlorinating community within Woods Pond is complex.

Dechlorination Processes N and P

The dominant dechlorination activity in Woods Pond is known as Process N. This activity targets all flanked *meta* chlorines except those on 23-chlorophenyl rings (10, 39, 45). This activity could best be activated in sediment microcosms by the addition of 236-CB or 2346-CB (350 μM) (15). A second major activity that could be stimulated in sediment microcosms is Process P (7). This activity exclusively removes flanked *para* chlorines (7, 10). The two dechlorination processes have distinct products; the key dechlorination products for Process N are 24-24-CB, 24-26-CB, and 246-24-CB (10, 45), and the key dechlorination products for Process P are 25-25-CB, 235-25-CB, and 23-25-CB (7, 10). The relative proportions of these dechlorination products in the Woods Pond sediments indicate that both dechlorination processes have occurred in situ but that Process N dominates (8).

Ortho Dechlorination and Process LP

Experiments with single congeners in sediment microcosms also led to the discovery of two other PCB dechlorination activities in Woods Pond: *ortho* dechlorination and unflanked *para* dechlorination (**Figure 2**). *Ortho* dechlorina-

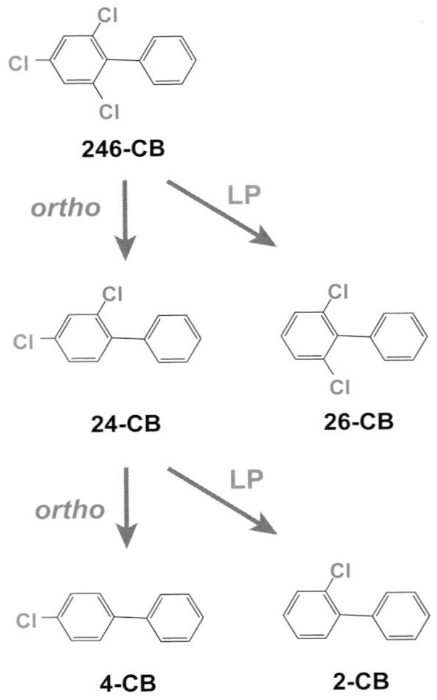

Figure 2

Ortho and LP dechlorination of 246-CB observed in microcosms of Woods Pond sediment.

tion was observed for only a few congeners; 2356-CB was dechlorinated to 235-CB (44), and 246-CB was dechlorinated to 24-CB and then to 4-CB (49, 50). Because this activity has never been observed in situ in Woods Pond or in any of our studies of Aroclor 1260 dechlorination with sediment microcosms from that site, it is unlikely that it plays a significant role in the dechlorination of Aroclor 1260. However, because it has the ability to remove *ortho* chlorines from 24- and 246-chlorophenyl rings, it could potentially further dechlorinate many of the terminal products of Process N dechlorination. For example, 246-24-CB → 24-24-CB → 24-4-CB → 4-4-CB.

The second activity, Process LP, removes unflanked *para* chlorines from 24- and 246-chlorophenyl groups (5, 50, 51). We recently discovered that Process LP can also remove the isolated *para* chlorine on the 4-chlorophenyl group of some congeners and the *meta* chlorine in position 3 from 23-, 234-, and 235-chlorophenyl groups (9). Although Process LP does not efficiently dechlorinate Aroclor

1260, it does further dechlorinate the terminal products of Process N dechlorination of Aroclor 1260 to primarily *ortho*-substituted di- and tri-CBs: 2-2-CB, 25-2-CB, and 26-2-CB (L.A. Smullen & D.L. Bedard, unpublished data). These products are, in turn, degradable by aerobic bacteria (14). Process LP activity has been observed in situ in the Housatonic River upstream of Woods Pond (13), but not in Woods Pond itself, even though sediment microcosm studies showed that the organisms responsible are present in Woods Pond (9).

PRIMING PCB DECHLORINATION IN SITU: A SUCCESSFUL FIELD TEST IN WOODS POND

In seeking alternative substrates that would stimulate Process N activity in sediment microcosms, we discovered that certain di- and tri-bromobiphenyls were highly effective (15). The brominated biphenyls were completely dehalogenated to biphenyl. Most-probable-number serial dilution experiments indicated that 2,6-dibromobiphenyl (26-BB) stimulated a 2000-fold growth in both the organisms that dehalogenated the 26-BB and the organisms that dechlorinated Aroclor 1260, suggesting that the same population was responsible for both dehalogenations (52). Certain other halogenated aromatics also stimulated Process N activity in sediment microcosms, but not as effectively or reproducibly as 26-BB; these included several brominated benzoates, brominated benzonitriles, and brominated nitrobenzenes, as well as other brominated aromatics (24).

Beginning in June 1992 we carried out a field test in Woods Pond to determine if it was feasible to activate PCB dechlorination under field conditions. The field test was carried out in steel caissons (6 ft in diameter) driven 7 ft into the clay subsoil. The caissons contained approximately 400 kg of sediment (dry weight) and a total volume of 3500 liters of water and sediment—a 10,000-fold scale-up from laboratory microcosms (12). PCB concentration was 26 μg per gram (dry sediment basis).

The field experiment was conducted using the same amendments as laboratory experiments (15). To ensure a homogeneous distribution, the sediment was briefly suspended with a 17-in impeller in the water column during amendments. Disodium malate was added to a final concentration of 10 mM and 26-BB was added to a final concentration of 350 μM. The control caisson received only the malate. Additions and sampling were done under positive nitrogen flow to prevent introduction of oxygen.

The experiment was conducted, with no further stirring, at ambient temperatures that ranged from 15°C to 17°C at the bottom of the sediment, and from 16°C to 19°C at the top, during the first three months (12). Dehalogenation of the 26-BB to 2-BB and then to biphenyl began within two weeks and was 98% to 99% complete at four months (12). Within 93 days the PCBs with six or more chlorines had decreased from an initial value of 68 mol% of the total PCBs to 26 mol%, a 62% decrease (12). By the following summer the hexa- through nona-CBs had decreased by 69% to 74%, respectively, at the bottom and top of the sediment (12). The dechlorination was Process N (**Figure 3**), the same pattern observed in the laboratory; thus the laboratory results were successfully transferred to the field despite a large excess of oil in the sediment (8) and a cold New England climate. This field test unequivocally demonstrated that the dechlorinating microorganisms indigenous to Woods Pond could be activated by a single addition of a suitable halogenated substrate. It also showed that in situ bioremediation of highly chlorinated PCBs is feasible if environmentally friendly substrates that activate indigenous dechlorinators can be found.

SEDIMENT-FREE PCB-DECHLORINATING CULTURES

The next major obstacle to overcome was developing the means to sustain and transfer PCB dechlorinators in defined medium, a necessary prerequisite to isolation of the

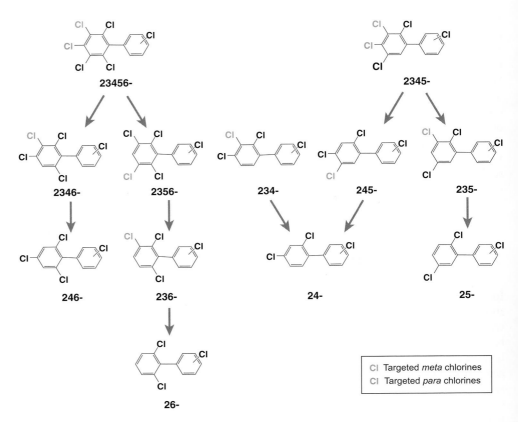

Figure 3

Dechlorination of target chlorophenyl rings of polychlorinated biphenyls (PCBs) by dechlorination Process N as observed in situ in Woods Pond, and in sediment microcosms and the sediment-free JN cultures derived from Woods Pond.

dechlorinators. For more than 10 years, laboratory investigations of microbial PCB dechlorination were severely hampered by the apparent requirement of sediment to sustain and transfer PCB-dechlorinating activity. For example, in 1997 Wu & Wiegel (55) reported the development of two distinct PCB-dechlorinating enrichments. The first, derived from PCB-contaminated sediment and enriched on 2346-CB, could remove only flanked or doubly flanked *para* chlorines from 3 of 25 congeners tested: 2346-CB, 234-CB, and 2345-CB, (underlines indicate the chlorine removed) (55). The second enrichment, derived from a PCB-free pond and enriched on 246-CB, could remove unflanked, flanked, and doubly flanked *para* chlorines from 22 of 25

congeners tested, including many substituted on both rings (55). Both of these enrichments were sustained for more than 20 transfers to autoclaved soil slurries, but neither activity could be transferred in the absence of soil or sediment (55).

The *Ortho*-Dechlorinating and DF-1 Cultures

However, in 1998 Cutter et al. (21) announced the development of a culture that dechlorinated PCBs in the complete absence of soil or sediment. This culture, subsequently designated the *ortho*-dechlorinating culture, was derived from Baltimore Harbor sediment and enriched on 2356-CB in an estuarine medium with gradual removal of sediment over four transfers. The enrichment removed both *ortho* chlorines from 2356-CB, but more than a 100-day lag time preceded dechlorination in sediment-free medium versus a much shorter lag time in the presence of 1% sterile sediment. Nineteen months later the same laboratory announced the development of a second sediment-free PCB-dechlorinating culture (53). This one, subsequently designated the DF culture, was also enriched on estuarine medium but was derived from Charleston Harbor (South Carolina), enriched on 2345-CB, and specifically removed doubly flanked *meta* or *para* chlorines (indicated by underline) from 2345-CB, 234-CB, 2346-CB, and 345-CB (53).

The development of sediment-free PCB-dechlorinating cultures was a major breakthrough that paved the way for microscopic studies, studies to determine if the observed PCB dechlorination was linked to growth, and studies to identify and isolate the organisms responsible for dechlorination.

The JN Cultures

Our laboratory specifically set out to develop stable sediment-free cultures from Woods Pond that retained the ability to dechlorinate the same broad spectrum of PCB congeners dechlorinated in situ, i.e., Process N dechlorination of Aroclor 1260. Aroclor 1260 has an average of 6.3 chlorines per biphenyl and consequently has a solubility of only 2.7 µg per liter in water and an octanol/water partition coefficient (log K_{ow}) of 6.8 (4). Because of these characteristics, we faced two major obstacles: (*a*) ensuring the bioavailability of Aroclor 1260 and (*b*) ensuring that the PCBs could be uniformly and reproducibly sampled. We solved these problems by adsorbing Aroclor 1260 onto very fine silica powder (about 240 mesh) as a carrier (6). We reasoned that the inert silica would mimic sediment in its ability to provide a large amount of surface area to which PCBs could adsorb and subsequently desorb. Furthermore, when the culture is shaken immediately before sampling, the silica stays in suspension long enough to ensure uniform sampling. We gradually removed

the sediment in a series of transfers, each with 10% inoculum, and used silica (10 mg per ml of culture) as a carrier for Aroclor 1260 (6).

Our sediment-free cultures, known as the JN cultures, grow with acetate and H_2 and carry out extensive Process N dechlorination (6), the same activity previously observed in sediment microcosms and in the Woods Pond field test (10, 12). When incubated with Aroclor 1260 at 5 to 500 µg per ml (13.5 µM to 1.35 mM), the JN cultures decrease the proportion of PCBs with six or more chlorines from 84.7% of the total PCBs to 20%, a 76% decrease (6). At least 64 PCB congeners chlorinated on both rings are dechlorinated, and 47 of these have six or more chlorines. All congeners containing 234-, 235-, 236-, 245-, 2345-, 2346-, and 2356-chlorophenyl rings are dechlorinated. Thus we achieved our objective and another major milestone in the investigation of anaerobic PCB dechlorination.

The JN cultures have now been maintained completely sediment free and transferred repeatedly for more than four years without loss of activity. As of this writing, they are the only sediment-free cultures that dechlorinate commercial PCB mixtures. However, this same strategy of substituting inert silica for sediment as a carrier for PCBs should enable successful development of sediment-free cultures exhibiting other PCB dechlorination processes from other sites. This in turn will enable investigations into the growth requirements and identity of the organisms responsible for other microbial PCB dechlorination processes that occur in nature.

IDENTIFICATION OF PCB-DECHLORINATING BACTERIA

Early studies of PCB-dechlorinating enrichments suggested that the organisms responsible for dechlorination processes H, M, and Q might be methanogens, spore-forming sulfate reducers, or nonspore-forming sulfate reducers, respectively (57–60). No such PCB dechlorinators have yet been identified. It was also suggested that *Dehalococcoides* might dechlorinate PCBs (5). The development of sediment-free cultures has led to optimization of growth conditions and the identification of several PCB dechlorinators by molecular techniques.

Bacterium *o*-17

An organism responsible for *ortho* dechlorination of 2356-CB grows optimally on 20 mM acetate in the presence of 2356-CB (22). It was identified without isolation by analysis of the denaturing gradient gel electrophoresis (DGGE) profiles of 16S rRNA genes amplified from the genomic DNA of the highly enriched

ortho-dechlorinating culture (21) incubated under various conditions that promoted or inhibited PCB dechlorination. The DGGE band corresponding to bacterium *o*-17 was always present when PCB dechlorination occurred but disappeared when PCB dechlorination was inhibited or when the culture was transferred without PCBs. On the basis of these data, the investigators concluded that bacterium *o*-17 was the dechlorinator and that its growth was linked to the dechlorination of 2356-CB (22). Phylogenetic analysis of the 16S rRNA gene of *o*-17 placed it on a deep branch within the *Chloroflexi* (**Figure 4**) (22). Its closest cultured relatives were *D. ethenogenes* strain 195 and

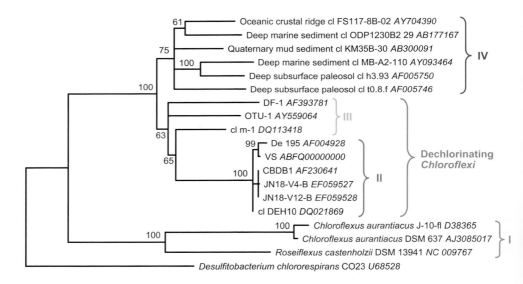

Figure 4

Phylogenetic tree showing the relationship of various PCB-dechlorinating organisms. Taxa were aligned over 1146 bp of the 16S rRNA gene sequences using Greengenes software (23). A full heuristic search to find the best tree was conducted using the distance (minimum evolution) optimality criterion of PAUP* 4.0b (43) with 1000 bootstrap replicates. *Desulfitobacterium chlororespirans* was used to root the tree. The distance measure used was Jukes-Cantor. The starting tree was obtained via neighbor-joining. Branch lengths were constrained to non-negative values and the branch-swapping algorithm used was tree-bisection-reconnection (TBR). Four clusters fall within the phylum *Chloroflexi*. Group I consists of the anoxygenic photosynthetic *Chloroflexus* and *Roseiflexus* spp. Group II consists entirely of *Dehalococcoides* strains. Group III is the *o*-17/DF-1 clade. Together Groups II and III form the putative dechlorinating *Chloroflexi*. Group IV is a group of uncultivated *Chloroflexi* about which nothing is known. The *Dehalococcoides* sequences designated JN18 are from the JN cultures. SF1 and *o*-17 are not shown because their sequences in GenBank are too short. However, their positions on the tree are effectively represented by clone m-1 (identical to SF1 over 470 bp) and clone OTU-1 (only 4 bp differences from *o*-17 over 714 bp).

Dehalococcoides sp. strain CBDB1, which grow by using chlorinated ethenes and chlorinated benzenes, respectively, as terminal electron acceptors (2, 38). However, with less than 89% sequence identity over 879 bp, bacterium *o-17* most likely belongs to a different genus than *Dehalococcoides*. Because most halorespiring organisms are Proteobacteria or Clostridia, the discovery that *o-17* belongs to a previously unknown genus within the *Chloroflexi* was unexpected and intriguing.

Bacterium DF-1

The second PCB dechlorinator to be identified was bacterium DF-1. Both microscopic and PCR amplification/DGGE analysis of the 16S rRNA genes in genomic DNA from the sediment-free DF culture, which exclusively removes doubly flanked chlorines, revealed that the DF culture is a co-culture of a vibrio and a small (<1 μm) coccus (54). The vibrio was isolated and identified as a sulfate-reducing *Desulfovibrio* that could not dechlorinate PCBs. Attempts to isolate the coccus failed, but PCR amplification/DGGE analysis of 16S rRNA genes from genomic DNA of DF culture incubated with and without its PCB substrates indicated that the coccus, subsequently designated bacterium DF-1, dechlorinates PCBs with doubly flanked chlorines when incubated with formate or H_2-CO_2 (80:20) (54). Phylogenetic analysis of the 16S rRNA gene of bacterium DF-1 also placed this organism deep within the *Chloroflexi*, with bacterium *o-17* as its closest relative (91% identity over 733 bp) and 87% to 88% similarity to *Dehalococcoides* strains 195 and CBDB1 (54). DGGE analyses also indicated that DF-1 links its growth to reductive dechlorination, but DGGE analysis is not quantitative and could not determine how much cell growth is supported per μmole of chlorine removed.

The identification and cultivation of two completely novel *Chloroflexi* PCB dechlorinators were highly significant. However, without in any way diminishing this accomplishment, it seems unlikely that either bacterium *o-17* or DF-1 plays a significant role in the in situ dechlorination of Aroclors or will play a role in PCB bioremediation efforts. This is because the reported activity for both organisms appears to be limited to congeners chlorinated on a single ring that are not prominent in any of the Aroclors (21, 36, 53).

Dehalococcoides ethenogenes Strain 195

D. ethenogenes strain 195 is best known for its ability to dehalogenate chlorinated ethenes and ethanes (37). However, when preliminary analyses of its genome sequence revealed that this organism had 18 different putative reductive dehalogenase genes (40, 46), experiments were conducted to determine if it could also dechlorinate chlorinated aromatic compounds (28). The amazing result was that it dechlorinated hexachlorobenzene, 1,2,3,4-tetrachloronapthalene, and 1,2,3,4-tetradibenzo-*p*-dioxin (28). In addition, when grown on tetrachloroethene it dechlorinated 23456-CB to 2346-CB/2356-CB and 246-CB (28). Unfortunately, no experiments were conducted to determine whether this versatile dechlorinator could dechlorinate PCBs chlorinated on both rings, such as those found in commercial PCB mixtures, and no experiments were conducted to determine if *D. ethenogenes* strain 195 could obtain energy for growth from the dechlorination of PCBs (see sidebar, *Dehalococcoides*).

Denaturing gradient gel electrophoresis (DGGE) analysis: 100- to 500-bp fragments of PCR-amplified 16S rRNA genes are analyzed by gel electrophoresis under increasingly denaturing conditions to distinguish phylotypes

PCR: polymerase chain reaction

DEHALOCOCCOIDES

Dehalococcoides are fascinating organisms that can grow only by halorespiration of chlorinated organic compounds. They are strictly anaerobic tiny disk-shaped cells that do not form cysts or spores and lack cell walls. Cell doubling time is about 2 days, and the cultures grow to cell densities of 10^7 to 10^8 with acetate, H_2, and a suitable chlorinated organic. Based on 16S rRNA phylogeny, the *Dehalococcoides* fall within the phylum *Chloroflexi*.

The *Dehalococcoides* Population in the JN Cultures

The dechlorinating organisms in the JN mixed cultures were identified by molecular techniques. The 16S rRNA genes were amplified from genomic DNA of a sediment-free JN culture grown in defined medium with acetate, H$_2$, and Aroclor 1260 in the presence of 5 µg per ml of vancomycin. On the basis of restriction fragment length polymorphism (RFLP) analysis of more than 100 clones and sequencing of each of the RFLP groups, it was estimated that the *Dehalococcoides* population comprised approximately 25% of the cells in the JN culture (11). Several slightly different *Dehalococcoides* 16S rRNA gene sequences were identified, all of which were >99% identical to the sequence of *Dehalococcoides* sp. strain CBDB1 over 1350 bp (**Figure 4**). [Unfortunately *Dehalococcoides* strains cannot be distinguished on the basis of 16S rRNA. Four physiologically distinct *Dehalococcoides* strains have 16S rRNA genes that are identical or differ by only one nucleotide (2, 30, 31, 42).] To determine whether other known dechlorinators were present, nested PCR with universal 16S rRNA primers and primers specifically targeting six other known halorespirers was carried out. These analyses ruled out any involvement of *Dehalobacter*, *Desulfitobacterium*, *Desulfuromonas*, *Sulfurospirillum*, *Anaeromyxobacter*, and *Geobacter* in the dechlorination of Aroclor 1260. In addition, PCR analysis with primers specifically targeting the *o*-17/DF-1 clade of *Chloroflexi* (47) ruled out any involvement of that group (11).

Quantitative real-time PCR revealed that the *Dehalococcoides* cell titer in the JN cultures increased 23.6-fold, from $7.07 \times 10^6 \pm 0.42 \times 10^6$ cells per ml to $1.67 \times 10^8 \pm 0.04 \times 10^8$ cells per ml, concomitant with the dechlorination of Aroclor 1260, whereas no growth of *Dehalococcoides* occurred in controls incubated without PCBs (11). Furthermore, the *Dehalococcoides* cell titer increased 200-fold (from 2.44×10^5 to 4.94×10^7) when Aroclor 1260 was restored to one culture that had

been transferred three times without PCBs (11). These studies unequivocally demonstrated that Process N dechlorination of Aroclor 1260 in the JN culture supported the growth of $9.25 \times 10^8 \pm 0.04 \times 10^8$ *Dehalococcoides* cells per µmole of chlorine removed (11). Finally, extensive Process N activity was recovered from 10^8 dilution tubes established from active JN cultures in dilution-to-extinction experiments (11). Microscopic examination of these cultures shows a high proportion of tiny disk-shaped cells typical of *Dehalococcoides* (D.L. Bedard & A.D. Fricker, unpublished results). Collectively, these results finally and definitively prove the hypothesis first put forward by John Brown Jr. and colleagues 20 years earlier (17): Commercial PCBs can act as terminal electron acceptors that support the growth of the organisms that dechlorinate them, a process now known as halorespiration.

EVIDENCE FROM SEDIMENT MICROCOSMS SUPPORTING THE ROLE OF *CHLOROFLEXI* IN PCB DECHLORINATION

In addition to the results reported above with PCB dechlorinators grown in the absence of sediment, several laboratories have reported a correlation of *Dehalococcoides* and *o*-17/DF-1-type *Chloroflexi* with PCB dechlorination in sediment microcosms (26, 27, 47, 56). For example, the Novak laboratory developed 2345-CB-dechlorinating anaerobic sediment microcosms from Baltimore Harbor (estuarine), Palos Verdes (California) (marine), and the Hudson River (New York) (freshwater) (56). Two electron donors, fatty acids and elemental iron, were used separately for each sediment. The investigators used 16S rRNA PCR/DGGE analysis and cloning/RFLP analysis/sequencing to examine the microbial populations of the sediment microcosms at 24- to 59-day intervals for 294 days in both PCB-amended microcosms and control microcosms with no PCBs. In all cases 2345-CB was dechlorinated to both 235-CB and 245-CB. Both of these products result from the removal of doubly flanked chlorines. As expected, the microbial populations in the enrichments from each site and on each electron donor differed substantially. In contrast, *Dehalococcoides*-like organisms (99.7% identity to the 16S rRNA gene of *Dehalococcoides* sp. strain CBDB1) were found in all enrichments (56). These organisms were not detected prior to PCB dechlorination or in PCB-free controls, but only when PCB dechlorination occurred. Thus the data are consistent with the hypothesis that the *Dehalococcoides*-like organisms dechlorinated the 2345-CB and that their numbers were increased by that dechlorination.

A second set of recently reported experiments conducted with sediment microcosms derived from Baltimore Harbor (BH microcosms) present strong evidence that both *Dehalococcoides* and other *Chloroflexi* organisms may contribute to the dechlorination of Aroclor 1260. In these investigations Fagervold et al. (26) used universal 16S rRNA DGGE primers to amplify genomic DNA collected from sediment microcosms incubated with and without Aroclor 1260 (50 µg per ml) over time. Four phylotypes were detected only when PCBs were actively dechlorinating and not in PCB-free controls. These phylotypes were DEH10 (*Dehalococcoides*), SF1 (*o*-17/DF-1 clade of *Chloroflexi*), spirochetes, and *Bacteroidetes* (26). The investigators focused their experiments on DEH10 and SF1 because these organisms group with known dechlorinators, but a role in dechlorination for the spirochetes and *Bacteroidetes*, which were also detected, cannot be ruled out.

The observed dechlorination of Aroclor 1260 in the BH microcosms was predominantly flanked *meta* dechlorination similar to Process N. Fagervold et al. (26) designed DGGE primers to specifically target all *Dehalococcoides* and *o*-17/DF-1 type *Chloroflexi*, i.e., all putative dechlorinating *Chloroflexi*, and used these and competitive PCR (cPCR) to demonstrate a 25-fold increase in these phylotypes during dechlorination of Aroclor 1260. The relative proportions of the two phylotypes were not reported, but the DGGE profiles suggest that they were present in approximately equal amounts. Unfortunately, no data were provided for dechlorinator cell concentrations in the cultures or for the amount of cell growth supported per µmole of chlorine removed. In addition, although the data presented (26) are important and convincing, they are necessarily weakened because full documentation of the cPCR assay used for quantitation has not been published as of this writing.

To further corroborate their findings and to elucidate the dechlorination pathways, Fagervold et al. (26) also set up triplicate enrichments with the 12 most abundant congeners in Aroclor 1260 and, subsequently, with dechlorination intermediates produced from these congeners. These enrichments were also analyzed by DGGE and cPCR as above. Eleven of the 12 most abundant congeners in Aroclor 1260 were dechlorinated, as were 12 of 13 intermediate products of these congeners. The data, compiled in **Table 1**, show which of the two phylotypes, DEH10 or SF1, were preferentially enriched on the various congeners and the preferred site of initial dechlorination. It appears from these data that congeners with 2345- and 236-chlorophenyl groups were equally likely to select for either phylotype, whereas congeners with 234- and 245-chlorophenyl groups were more likely to enrich DEH10 and congeners with 235- and 2356-chlorophenyl groups were more likely to enrich SF1. However, it appears that both phylotypes are capable of dechlorinating all of these chlorophenyl groups, with the possible exception of the 235-chlorophenyl group. These studies also provided further insights into the dechlorination potential of DEH10 and SF1 and the ability of the various congeners to support their growth. The cPCR amplification/DGGE analysis showed a 5- to 9-fold increase of dechlorinating *Chloroflexi* grown on five individual congeners and a $\geq$10-fold increase for 13 more congeners; thus it appears that 18 of 25 tested PCB congeners may be able to support growth.

Taken together, these results from the Fagervold et al. (26) studies provide convincing data that, at least in BH microcosms, both

Table 1 Initial dechlorination of individual PCB congeners by phylotype[a]

DEH10 and SF1		Predominantly DEH10		Predominantly SF1	
PCB IUPAC no.	**PCB congener[b]**	**PCB IUPAC no.**	**PCB congener[b]**	**PCB IUPAC no.**	**PCB congener[b]**
180	2345-245[c]	153	245-245	187	2356-245
174	2345-236	151	2356-25	170	2345-234[d]
149	236-245	138	234-245	154	245-246
146	235-245	132	234-236	147	2356-24
135	235-236	102	245-26	137	2345-24[e]
		101	245-25	130	234-235
		99	245-24	92	235-25
		95	236-25	90	235-24
				91	236-24

[a]Data are compiled from figures 2 and 5 and table 1 from Reference 26.
[b]PCB congener used for enrichment.
[c]Position of chlorine removed is indicated by underline. The initial chlorine loss in this congener was from either doubly flanked position. The *meta* chlorine was removed 67% of the time.
[d]The initial chlorine loss in this congener was observed from each of the three doubly flanked positions.
[e]The initial chlorine loss in this congener occurred with equal frequency from each of the doubly flanked positions.

Dehalococcoides and dechlorinating *Chloroflexi* such as SF1 may play a role in the dechlorination of Aroclor 1260. Other than this it is impossible to draw conclusions about whether these organisms actually act in a synergistic manner in situ. It is also impossible to infer the route of dechlorination of individual congeners in Aroclor 1260 from single-congener experiments in microcosms, especially when more than one PCB dechlorinator is known to be present. The likely scenario is that the actual biological, physical, and chemical conditions of each sediment microniche will determine whether either of these bacteria (or possibly some other dechlorinator) dominates or whether they act synergistically. Even if they do act synergistically, the actual dechlorination pathways used will likely be influenced by all the individual PCB congeners present and by their concentrations. There is also ample evidence that the composition of a PCB mixture affects the dechlorination of the individual components (9, 26, 36).

One other phylotype, designated SF2, was enriched on 2 of the 25 congeners tested in the studies described above. The 16S rRNA of this organism was identical to that of bacterium *o*-17 over 470 bp. SF2 was enriched with 2346-245-CB and, in at least one of the triplicate incubations, with 2356-25-CB (chlorines removed are underlined). Consistent with previous observations for *o*-17, this bacterium removed the flanked *ortho* chlorine from the 2356-chlorophenyl ring and the doubly flanked *meta* chlorine from the 2346-chlorophenyl ring (36). Small amounts of *ortho* dechlorination of the 235-chlorophenyl group of 235-24-CB and 235-25-CB were also observed and attributed to SF2. However, SF2 was not detected in the Aroclor 1260 enrichments and likely plays no significant role in the dechlorination of Aroclor 1260 (26).

PROCESS N DECHLORINATION: A SINGLE DECHLORINATION PROCESS?

It seems appropriate to compare the dechlorination of Aroclor 1260 in the JN cultures and the BH microcosms because such a comparison may lead to a better understanding of Process N dechlorination. Both cultures remove predominantly flanked and doubly flanked *meta* chlorines as well as some doubly flanked *para* chlorines from 2345-chlorophenyl groups. Both cultures dechlorinate congeners with 234-, 235-, 236-, 2345-, 2346-, and 2356-chlorophenyl rings. And the three most prominent terminal dechlorination products for both cultures are 24-24-CB, 24-26-CB, and 24-25-CB. In the JN cultures a typical end result is that these three products constitute 23.78 mol%, 16.70 mol%, and 8.24 mol% of the total PCBs after 54 to 154 days, depending on the culture (11). In the BH microcosms these dechlorination products constituted 12.4 mol%, 8.74 mol%, and 7.5 mol% after 400 days incubation (26). Obviously, the dechlorination in the JN cultures has progressed further, which is likely an indication of higher enrichment of the dechlorinators in the JN cultures. But more importantly, although the ratio of 24-24-CB to 24-26-CB in both cultures is identical at 1.42, the ratio of 24-24-CB to 24-25-CB is different: 1.65 for the BH microcosm but 2.88 for the JN cultures. Even when the JN cultures are at the same stage of dechlorination as the BH microcosm (as determined from the mol% values for 24-24-CB and 24-26-CB) the ratio of 24-24-CB to 24-25-CB is at least 2.15 (D.L. Bedard, unpublished data).

The difference in the relative proportions of 24-24-CB and 24-25-CB in these cultures suggests that the activity previously described as Process N dechlorination may actually represent the activity of more than one organism,

more than one reductive dehalogenase, or both. The data suggest that in the BH microcosms two different organisms, DEH10 and SF1, play a role in the dechlorination of Aroclor 1260 (26). In the JN cultures all of the dechlorination is due to the *Dehalococcoides* population, but we cannot rule out the possibility that two or more distinct strains of *Dehalococcoides* may play a role in the dechlorination (11). Ultimately it will be necessary to isolate these organisms to determine whether one or more organisms and one or more reductive dehalogenases are responsible for the activity that we call Process N dechlorination.

Table 2 gives the range of rates of dechlorination for penta-, hexa-, and hepta-CBs in the BH microcosms and for Aroclor 1260 in the JN cultures. There are important differences to consider in evaluating this comparison: (*a*) BH microcosms are fourth-transfer sediment enrichments, whereas JN cultures are sediment-free and have been enriched for much longer. (*b*) The dechlorination activity in the BH cultures is due to two different genera of dechlorinating *Chloroflexi*, whereas all dechlorination in the JN cultures is due to the *Dehalococcoides* population. (*c*) The rates in the BH microcosms were calculated for PCB congeners incubated individually, whereas those in the JN cultures were calculated for Aroclor 1260 as a whole. (*d*) The rates for the BH culture were calculated over the linear proportion of a curve

Table 2 Estimated rate of dechlorination of PCBs

Culture	Dechlorination substrate	PCB conc. (μMol)	Estimated rate of dechlorination (nmol liter^{-1} day^{-1})			
			PCB congener[a]	Lowest observed	PCB congener	Highest observed
BH microcosm	Hepta-CBs	126[b]	(187) 2356-245	164[b]	(174) 2345-236	403[b]
BH microcosm	Hexa-CBs	139[b]	(154) 245-246	97[b]	(137) 2345-24	1640[b]
BH microcosm	Penta-CBs	154[b]	(90) 235-24	245[b]	(91) 236-24	2600[b]
JN culture	Aroclor 1260[c]	13.5	NA	118[d]	NA	163[d]
JN culture	Aroclor 1260	135	NA	3300[d]	NA	5633[d]
JN culture	Aroclor 1260	675	NA	9667[d]	NA	18133[d]

[a]Parentheses indicate PCB IUPAC number.
[b]Calculated from data presented in table 1 from Reference 26 based on molecular weight and concentration of 50 μg ml^{-1} used in incubations.
[c]Aroclor 1260 has an average of 6.3 chlorines per biphenyl.
[c]Recalculated from data presented in table 1 from Reference 11.

generated from sampling at 50-day intervals, whereas those for the JN cultures are the maximum observed rates. (*e*) The BH microcosms were incubated at 30°C and the JN cultures at 22°C to 24°C. Despite all these differences, the data obtained from the two systems are reasonably consistent. The closest comparison is that of the hexa-CBs to 135 μM Aroclor 1260. The dechlorination in the JN cultures was two to three times the highest rate observed for hexa-CBs in the BH microcosms, but the fact that the rates for both sets of data were this close strengthens the data. Both sets of experiments (see also table 2 in Reference 6) indicate lower rates of dechlorination for hepta-CBs than for hexa-CBs. This is likely due to lower solubility of the hepta-CBs. The data from the JN cultures (6) indicate that the rate of dechlorination increased at higher concentrations of PCBs. This is consistent with what would be expected; more bacteria should be enriched at the higher PCB concentration because the PCBs are used as terminal electron acceptors (11).

THE QUEST FOR PCB REDUCTIVE DEHALOGENASES

No PCB dechlorinase has yet been identified. However, evidence from three different sources that some *Dehalococcoides* strains dechlorinate PCBs provides us with a wealth of potential candidates. So far six *Dehalococcoides* strains have been isolated (2, 30, 31, 38, 42), and genomic studies have revealed that each strain carries a suite of at least 10 and as many as 35 related, but distinct, putative reductive dehalogenase (RDase) genes (19, 20, 32, 33, 40). More than 100 different RDase genes have been identified in *Dehalococcoides*, but so far the function is known for only five of these. A chlorobenzene RDase gene was recently identified in *Dehalococcoides* sp. strain CBDB1 (1). Might the RDase encoded by this gene also dechlorinate PCBs? So far the complete genome has been published for four *Dehalococcoides* isolates (19, 20, 33, 40), and another will soon be available. These will likely yield even more potential candidates for PCB dechlorinases. It will be

interesting to learn if the *Chloroflexi* strains *o*-17 and DF-1 also harbor multiple putative RDase genes.

The identification of functions of RDase genes is severely hampered by the inability to express active RDases in a heterologous system. A breakthrough in this area would make it possible to rapidly assay the products of all the cloned RDase genes for PCB-dechlorinating activity. Perhaps then methods could be found to stabilize these PCB-dechlorinating enzymes for remediation or even to engineer them to increase their PCB substrate spectrum.

CONCLUSIONS

The identification and cultivation of bacteria responsible for the dechlorination of commercial PCB mixtures has taken 20 years. But we now have strong leads and techniques that should greatly facilitate the discovery of additional PCB dechlorinators.

The methods applied for cultivation in defined medium, and identification of the organisms responsible for Process N dechlorination of Aroclor 1260 in the Housatonic River (6, 11), should facilitate the cultivation and identification of the organisms responsible for other PCB dechlorination activities such as Processes M and Q dechlorination of Aroclor 1242 in the Hudson River (New York) (10) or Lake Hartwell (South Carolina) (18, 35). Identifying and cultivating the organisms responsible for dechlorination processes M and Q are especially important because these organisms can remove unflanked *meta* or *para* chlorines, respectively, which means that they can dechlorinate PCBs to the point where they can be aerobically degraded.

The development of several primer sets (27, 47) has revealed a monophyletic clade of putative dechlorinating *Chloroflexi* that includes *Dehalococcoides*, as well as other organisms that have not yet been classified. The intriguing bacterium SF1 falls within this group. It is important that we learn more about what role these organisms play in PCB dechlorination. For example, are they found in freshwater sites as well as estuarine sites? The isolation or cultivation of bacterium SF1 and other members of this group in defined medium is a necessary prerequisite for the study of their PCB dechlorination capabilities.

The discovery that some *Dehalococcoides* populations reductively dechlorinate Aroclor 1260 (11, 26) holds promise for the development of effective in situ bioremediation technologies for PCBs because biostimulation of and bioaugmentation with *Dehalococcoides* at sites contaminated with chlorinated ethenes have proved effective means of treating these contaminants (34).

SUMMARY POINTS

1. The successful stimulation of Process N dechlorination in a field test in Woods Pond proved that it is feasible to bioremediate PCBs in situ if environmentally friendly substrates that biostimulate the dechlorinators can be found.

2. Highly enriched cultures of PCB-dechlorinating bacteria in defined media offer opportunities to identify these bacteria and study their nutritional requirements.

3. Molecular techniques have been used to identify several PCB-dechlorinating organisms belonging to a new and diverse clade of putative dechlorinating bacteria within the *Chloroflexi*. These bacteria, not yet isolated, appear to link their growth to PCB dechlorination.

4. Highly enriched cultures (the JN cultures) that extensively dechlorinate Aroclor 1260 have been cultivated in defined medium. The *Dehalococcoides* population in these cultures carries out the same PCB dechlorination activity, Process N, observed in situ.

5. The dechlorination of Aroclor 1260 in the JN cultures supports the growth of 9.25×10^8 *Dehalococcoides* cells per μmole of chlorine removed.

6. Evidence is growing that *Dehalococcoides* and other dechlorinating *Chloroflexi* may play a role in PCB dechlorination at various sites.

FUTURE ISSUES

1. Can organisms other than *Chloroflexi* dechlorinate PCBs? Known dechlorinators should be systematically tested for their ability to dechlorinate PCBs.

2. The six pure strains of *Dehalococcoides* provide an ideal opportunity to determine whether any of these strains can dechlorinate the broad spectrum of PCBs that constitute one of the recognized PCB dechlorination processes or even a previously undescribed PCB dechlorination process.

3. Finding more soluble halogenated substrates that PCB dechlorinators can use would make it easier to grow and isolate these bacteria.

4. Many of our harbors, coasts, lakes, and rivers remain contaminated with PCBs. Cost-effective engineering methods to grow large quantities of PCB dechlorinators for bioaugmentation or to stimulate their growth in situ (biostimulation) are needed.

DISCLOSURE STATEMENT

The author was a Research Scientist at GE Corporate Research and Development, Niskayuna, NY, prior to her move to RPI in 1999. Since that time she has had no obligations to GE of any kind.

ACKNOWLEDGMENTS

This work was supported by NSF grant 0641743.

LITERATURE CITED

1. Adrian L, Rahnenfuhrer J, Gobom J, Hölscher T. 2007. Identification of a chlorobenzene reductive dehalogenase in *Dehalococcoides* sp. strain CBDB1. *Appl. Environ. Microbiol.* 73:7717–24

2. Adrian L, Szewzyk U, Wecke J, Görisch H. 2000. Bacterial dehalorespiration with chlorinated benzenes. *Nature* 408:580–83

3. Agency for Toxic Substances and Disease Registry. 2000. *Toxicological Profile for Polychlorinated Biphenyls (Update)*. Atlanta: U.S. Dep. Health Human Serv. Agency Toxic Subst. Dis. Regist.

4. Agency for Toxic Substances and Disease Registry. 2006. *CERCLA priority list of hazardous compounds*. Washington, DC. **http://www.atsdr.cdc.gov/cercla/05list.html.**

5. Bedard DL. 2003. Polychlorinated biphenyls in aquatic sediments: environmental fate and outlook for biological treatment. In *Dehalogenation: Microbial Processes and Environmental Applications*, ed. MM Häggblom, I Bossert, pp. 443–65. Boston: Kluwer

6. Bedard DL, Bailey JJ, Reiss BL, Jerzak GVS. 2006. Development and characterization of stable sediment-free anaerobic bacterial enrichment cultures that dechlorinate Aroclor 1260. *Appl. Environ. Microbiol.* 72:2460–70

7. Bedard DL, Bunnell SC, Smullen LA. 1996. Stimulation of microbial *para*-dechlorination of polychlorinated biphenyls that have persisted in Housatonic River sediment for decades. *Environ. Sci. Technol.* 30:687–94

8. Bedard DL, May RJ. 1996. Characterization of the polychlorinated biphenyls in the sediments of Woods Pond: evidence for microbial dechlorination of Aroclor 1260 in situ. *Environ. Sci. Technol.* 30:237–45

9. Bedard DL, Pohl EA, Bailey JJ, Murphy A. 2005. Characterization of the PCB substrate range of microbial dechlorination process LP. *Environ. Sci. Technol.* 39:6831–39

10. Bedard DL, Quensen JF III. 1995. Microbial reductive dechlorination of polychlorinated biphenyls. In *Microbial Transformation and Degradation of Toxic Organic Chemicals*, ed. LY Young, CE Cerniglia, pp. 127–216. New York: Wiley-Liss

11. Bedard DL, Ritalahti KM, Löffler FE. 2007. The *Dehalococcoides* population in sediment-free mixed cultures metabolically dechlorinates the commercial polychlorinated biphenyl mixture Aroclor 1260. *Appl. Environ. Microbiol.* 73:2513–21

12. Bedard DL, Smullen LA, DeWeerd KA, Dietrich DK, Frame GM, et al. 1995. Chemical activation of microbially mediated PCB dechlorination. A field study. *Organohalogen Compd.* 24:23–28

13. Bedard DL, Smullen LA, May RJ. 1996. Microbial dechlorination of highly chlorinated PCBs in the Housatonic River. In *Proceedings of the 1996 International Symposium on Subsurface Microbiology*, pp. 117. Davos, Switz.: Swiss Soc. Microbiol., Inst. Plant Biol., Univ. Zurich

14. Bedard DL, Unterman R, Bopp LH, Brennan MJ, Haberl ML, Johnson C. 1986. Rapid assay for screening and characterizing microorganisms for the ability to degrade polychlorinated biphenyls. *Appl. Environ. Microbiol.* 51:761–68

15. Bedard DL, Van Dort H, DeWeerd KA. 1998. Brominated biphenyls prime extensive microbial reductive dehalogenation of Aroclor 1260 in Housatonic River sediment. *Appl. Environ. Microbiol.* 64:1786–95

16. Brown JF Jr, Bedard DL, Brennan MJ, Carnahan JC, Feng H, Wagner RE. 1987. Polychlorinated biphenyl dechlorination in aquatic sediments. *Science* 236:709–11

17. Brown JF Jr, Wagner RE, Feng H, Bedard DL, Brennan MJ, et al. 1987. Environmental dechlorination of PCBs. *Environ. Toxicol. Chem.* 6:579–93

18. Bzdusek PA, Christensen ER, Lee CM, Pakdeesusuk U, Freedman DL. 2006. PCB congeners and dechlorination in sediments of Lake Hartwell, South Carolina, determined from cores collected in 1987 and 1998. *Environ. Sci. Technol.* 40:109–19

19. Copeland A, Lucas S, Lapidus A, Barry K, Detter JC, et al. 2007. Complete sequence of Dehalococcoides sp. BAV1. U.S. Dep. Energy Joint Genome Inst. **http://www.jgi.doe.gov/**

20. Copeland A, Lucas S, Lapidus A, Barry K, Glavina del Rio T, et al. 2007. Sequencing of the draft genome and assembly of Dehalococcoides sp. VS. U.S. Dep. Energy Joint Genome Inst. **http://www.jgi.doe.gov/**

21. Cutter L, Sowers KR, May HD. 1998. Microbial dechlorination of 2,3,5,6-tetrachlorobiphenyl under anaerobic conditions in the absence of soil or sediment. *Appl. Environ. Microbiol.* 64:2966–69

22. Cutter LA, Watts JEM, Sowers KR, May HD. 2001. Identification of a microorganism that links its growth to the reductive dechlorination of 2,3,5,6-chlorobiphenyl. *Environ. Microbiol.* 3:699–709

23. DeSantis TZ, Hugenholtz P, Larsen N, Rojas M, Brodie EL, et al. 2006. Greengenes, a chimera-checked 16S rRNA gene database and workbench compatible with ARB. *Appl. Environ. Microbiol.* 72:5069–72

24. DeWeerd KA, Bedard DL. 1999. Use of halogenated benzoates and other halogenated aromatic compounds to stimulate the microbial dechlorination of PCBs. *Environ. Sci. Technol.* 33:2057–63

25. Erickson MD. 1997. *Analytical Chemistry of PCBs*. New York: Lewis. 2nd ed.

26. Fagervold SK, May HD, Sowers KR. 2007. Microbial reductive dechlorination of Aroclor 1260 in Baltimore Harbor sediment microcosms is catalyzed by three phylotypes within the phylum *Chloroflexi*. *Appl. Environ. Microbiol.* 73:3009–18

27. Fagervold SK, Watts JEM, May HD, Sowers KR. 2005. Sequential reductive dechlorination of *meta*-chlorinated polychlorinated biphenyl congeners in sediment microcosms by two different *Chloroflexi* phylotypes. *Appl. Environ. Microbiol.* 71:8085–90

28. Fennell DE, Nijenhuis I, Wilson SF, Zinder SH, Häggblom MM. 2004. *Dehalococcoides ethenogenes* strain 195 reductively dechlorinates diverse chlorinated aromatic pollutants. *Environ. Sci. Technol.* 38:2075–81

29. Frame GM, Cochran JW, Bøwadt SS. 1996. Complete PCB congener distributions for 17 Aroclor mixtures determined by 3 HRGC systems optimized for comprehensive, quantitative, congener-specific analysis. *J. High Resolut. Chromatogr.* 19:657–68

30. He J, Sung Y, Krajmalnik-Brown R, Ritalahti KM, Löffler FE. 2005. Isolation and characterization of *Dehalococcoides* sp. strain FL2, a trichloroethene (TCE)- and 1,2-dichloroethene-respiring anaerobe. *Environ. Microbiol.* 7:1442–50

31. He JZ, Ritalahti KM, Yang KL, Koenigsberg SS, Löffler FE. 2003. Detoxification of vinyl chloride to ethene coupled to growth of an anaerobic bacterium. *Nature* 424:62–65

32. Hölscher T, Krajmalnik-Brown R, Ritalahti KM, von Wintzingerode F, Görisch H, et al. 2004. Multiple nonidentical reductive-dehalogenase-homologous genes are common in *Dehalococcoides*. *Appl. Environ. Microbiol.* 70:5290–97

33. Kube M, Beck A, Zinder SH, Kuhl H, Reinhardt R, Adrian L. 2005. Genome sequence of the chlorinated compound respiring bacterium *Dehalococcoides* species strain CBDB1. *Nat. Biotechnol.* 23:1269–73

34. Löffler FE, Edwards EA. 2006. Harnessing microbial activities for environmental cleanup. *Curr. Opin. Biotechnol.* 17:274–84

35. Magar VS, Johnson GW, Brenner RC, Quensen JF III, Foote EA, et al. 2005. Long-term recovery of PCB-contaminated sediments at the Lake Hartwell superfund site: PCB dechlorination. 1. End-member characterization. *Environ. Sci. Technol.* 39:3538–47

36. May HD, Cutter LA, Miller GS, Milliken CE, Watts JE, Sowers KR. 2006. Stimulatory and inhibitory effects of organohalides on the dehalogenating activities of PCB-dechlorinating bacterium *o*-17. *Environ. Sci. Technol.* 40:5704–9

37. Maymó-Gatell X, Anguish T, Zinder SH. 1999. Reductive dechlorination of chlorinated ethenes and 1,2-dichloroethane by *Dehalococcoides ethenogenes* 195. *Appl. Environ. Microbiol.* 65:3108–13

38. Maymó-Gatell X, Chien Y-T, Gossett JM, Zinder SH. 1997. Isolation of a bacterium that reductively dechlorinates tetrachloroethene to ethene. *Science* 276:1568–71

39. Quensen JF III, Boyd SA, Tiedje JM. 1990. Dechlorination of four commercial polychlorinated biphenyl mixtures (Aroclors) by anaerobic microorganisms from sediments. *Appl. Environ. Microbiol.* 56:2360–69

40. Seshadri R, Adrian L, Fouts DE, Eisen JA, Phillippy AM, et al. 2005. Genome sequence of the PCE-dechlorinating bacterium *Dehalococcoides ethenogenes*. *Science* 307:105–8

41. Stockholm Convention on Persistent Organic Pollutants (POPs). 1971. *The 12 POPs under the Stockholm Convention*. **http://www.pops.int/documents/pops/default.htm**

42. Sung Y, Ritalahti KM, Apkarian RP, Löffler FE. 2006. Quantitative PCR confirms purity of strain GT, a novel trichloroethene-to-ethene-respiring *Dehalococcoides* isolate. *Appl. Environ. Microbiol.* 72:1980–87

43. Swofford DL. 1998. *PAUP* Phylogenetic Analysis Using Parsimony (* and Other Methods). Version 4.* Sunderland, MA: Sinauer

44. Van Dort HM, Bedard DL. 1991. Reductive *ortho* and *meta* dechlorination of a polychlorinated biphenyl congener by anaerobic microorganisms. *Appl. Environ. Microbiol.* 57:1576–78

45. Van Dort HM, Smullen LA, May RJ, Bedard DL. 1997. Priming microbial *meta*-dechlorination of polychlorinated biphenyls that have persisted in Housatonic River sediments for decades. *Environ. Sci. Technol.* 31:3300–7

46. Villemur R, Saucier M, Gauthier A, Beaudet R. 2002. Occurrence of several genes encoding putative reductive dehalogenases in *Desulfitobacterium hafniense/frappieri* and *Dehalococcoides ethenogenes*. *Can. J. Microbiol.* 48:697–706

47. Watts JEM, Fagervold SK, May HD, Sowers KR. 2005. A PCR-based specific assay reveals a population of bacteria within the *Chloroflexi* associated with the reductive dehalogenation of polychlorinated biphenyls. *Microbiology* 151:2039–46

48. Wiegel J, Wu Q. 2000. Microbial reductive dehalogenation of polychlorinated biphenyls. *FEMS Microbiol. Ecol.* 32:1–15

49. Williams WA. 1994. Microbial reductive dechlorination of trichlorobiphenyls in anaerobic sediment slurries. *Environ. Sci. Technol.* 28:630–35

50. Wu Q, Bedard DL, Wiegel J. 1997. Effect of incubation temperature on the route of microbial reductive dechlorination of 2,3,4,6-tetrachlorobiphenyl in polychlorinated biphenyl (PCB)-contaminated and PCB-free freshwater sediments. *Appl. Environ. Microbiol.* 63:2836–43

51. Wu Q, Bedard DL, Wiegel J. 1997. Temperature determines the pattern of anaerobic microbial dechlorination of Aroclor 1260 primed by 2,3,4,6-tetrachlorobiphenyl in Woods Pond sediment. *Appl. Environ. Microbiol.* 63:4818–25

52. Wu Q, Bedard DL, Wiegel J. 1999. 2,6-Dibromobiphenyl primes extensive dechlorination of Aroclor 1260 in contaminated sediment at 8–30°C by stimulating growth of PCB-dehalogenating microorganisms. *Environ. Sci. Technol.* 33:595–602

53. Wu Q, Sowers KR, May HD. 2000. Establishment of a polychlorinated biphenyl-dechlorinating microbial consortium, specific for doubly flanked chlorines, in a defined, sediment-free medium. *Appl. Environ. Microbiol.* 66:49–53

54. Wu Q, Watts JEM, Sowers KR, May HD. 2002. Identification of a bacterium that specifically catalyzes the reductive dechlorination of polychlorinated biphenyls with doubly flanked chlorines. *Appl. Environ. Microbiol.* 68:807–12

55. Wu Q, Wiegel J. 1997. Two anaerobic polychlorinated biphenyl-dehalogenating enrichments that exhibit different *para*-dechlorination specificities. *Appl. Environ. Microbiol.* 63:4826–32

56. Yan T, LaPara TM, Novak PJ. 2006. The reductive dechlorination of 2,3,4,5-tetrachlorobiphenyl in three different sediment cultures: evidence for the involvement of phylogenetically similar *Dehalococcoides*-like bacterial populations. *FEMS Microbiol. Ecol.* 55:248–61

57. Ye D, Quensen JF III, Tiedje JM, Boyd SA. 1992. Anaerobic dechlorination of polychlorobiphenyls (Aroclor 1242) by pasteurized and ethanol-treated microorganisms from sediments. *Appl. Environ. Microbiol.* 58:1110–14

58. Ye D, Quensen JF III, Tiedje JM, Boyd SA. 1995. Evidence for *para* dechlorination of polychlorobiphenyls by methanogenic bacteria. *Appl. Environ. Microbiol.* 61:2166–71

59. Ye D, Quensen JF III, Tiedje JM, Boyd SA. 1999. 2-Bromoethanesulfonate, sulfate, molybdate, and ethanesulfonate inhibit anaerobic dechlorination of polychlorobiphenyls by pasteurized microorganisms. *Appl. Environ. Microbiol.* 65:327–29

60. Zwiernik MJ, Quensen JF III, Boyd SA. 1998. FeSO$_4$ amendments stimulate extensive anaerobic PCB dechlorination. *Environ. Sci. Technol.* 32:3360–65

Molecular Mechanisms of the Cytotoxicity of ADP-Ribosylating Toxins

Qing Deng and Joseph T. Barbieri

Department of Microbiology and Molecular Genetics, Medical College of Wisconsin, Milwaukee, Wisconsin 53226; email: dengqing@mcw.edu; jtb01@mcw.edu

Annu. Rev. Microbiol. 2008. 62:271–88

The *Annual Review of Microbiology* is online at micro.annualreviews.org

This article's doi:
10.1146/annurev.micro.62.081307.162848

Key Words

Corynebacterium diphtheriae, diphtheria toxin, *Pseudomonas aeruginosa*, Exo S, ADP-ribosyltransferase, ADP-ribose hydrolase

Abstract

Bacterial pathogens utilize toxins to modify or kill host cells. The bacterial ADP-ribosyltransferases are a family of protein toxins that covalently transfer the ADP-ribose portion of NAD to host proteins. Each bacterial ADP-ribosyltransferase toxin modifies a specific host protein(s) that yields a unique pathology. These toxins possess the capacity to enter a host cell or to use a bacterial Type III apparatus for delivery into the host cell. Advances in our understanding of bacterial toxin action parallel the development of biophysical and structural biology as well as our understanding of the mammalian cell. Bacterial toxins have been utilized as vaccines, as tools to dissect host cell physiology, and more recently for the development of novel therapies to treat human disease.

Contents

ADP-ribosylation:
enzyme reaction
catalyzed by bacterial
toxins where the
ADP-ribose portion of
NAD^+ is covalently
transferred to a host
protein

A domain: catalytic
domain

T domain:
translocation domain

R domain: receptor
binding domain

INTRODUCTION

Bacterial pathogens invade and then damage their host through the action of toxins that act at the site of infection or at a location that is distanced from the site of infection. Invasive properties of bacteria involve systematic growth within the host and the ability to evade the host innate immune system, where toxins neutralize the phagocytic capacity of immune cells or disrupt the physiology of immune cells. Toxins act directly through the targeting of specific cells by covalently or noncovalently modifying the function of intracellular targets,

which are typically proteins but may also be other macromolecular molecules such as RNA. The covalent modifications to host targets that are catalyzed by bacterial toxins include ADP-ribosylation (22), glucosylation (90), deamidation (89), deadenylation (31, 86), proteolysis (87), and acetylation (72). These modifications often result in an inactivation of the target, which changes cell physiology or may lead to necrotic or apoptotic cell death. In contrast, some bacterial toxins upregulate cell physiology. Cytotoxic necrotizing factor deamidates a glutamine of the host Rho GTPases, which leads to a constitutively active Rho with the elevation of polymerized intracellular actin (37, 89).

Bacterial toxins are classified according to their structure-function organization: exotoxins (20); exoenzymes (1); toxins that are delivered directly into the host cell through a complex secretion system, such as the Type III, Type IV, and Type V cytotoxins (42); pore-forming toxins (41); and superantigens, which activate immune cells through the antigen-independent mechanism (6). Exotoxins are secreted proteins that comprise three structural domains. The **A** domain encodes the catalytic domain, and the **B** domain comprises a translocation (**T**) domain that facilitates translocation of the **A** domain into the cytoplasm of the host cells and a receptor binding (**R**) domain that binds the toxin to host cell surface receptors. Representative structure-function properties of bacterial exotoxins include single proteins (**AB**); complexes of noncovalently bound proteins (**A5B**); or proteins that encode the **A** and **B** domains, which are independently synthesized and secreted from the bacterium and associate only when the **B** domain binds to the surface receptor of sensitive cells (**A-B**). Exoenzymes are relatively small proteins with molecular masses of ~20–25 kDa that comprise an **A** domain but lack either a **T** or **R** domain. Bacterial secretion systems can effectively deliver **A**-domain-containing toxins from the cytoplasm of the bacterium to the intracellular environment of the host cell. Thus, the secretion apparatus supplies the **B** domain for toxin action.

Table 1 Representative bacterial ADP-ribosyltransferases

Exotoxin	Bacterium	AB	Target	Role in pathogenesis
Diphtheria toxin	*Corynebacterium diphtheriae*	AB	EF-2	Inhibition of protein synthesis
Exotoxin A	*Pseudomonas aeruginosa*	AB	EF-2	Inhibition of protein synthesis
Cholera toxin	*Vibrio cholerae*	A5B	Gs	Inhibition of GTPase activity of Gs
Heat-labile enterotoxin	*Escherichia coli*	A5B	Gs	Inhibition of GTPase activity of Gs
Pertussis toxin	*Bordetella pertussis*	A5B	Gi	Uncoupled Gi protein–mediated signal transduction
C2	*Clostridium botulinum*	A-B	Actin	Actin depolymerization

Abbreviation: EF-2, elongation factor-2.

One difference between exotoxins and secretion apparatus-delivered toxins, such as the Type III cytotoxins, is that whereas only a few molecules of exotoxins are thought to be delivered into the host cell, Type III secretion has the capacity to deliver thousands of molecules of preformed cytotoxins into the host cell (88).

ADP-ribosylation was the first covalent modification shown to be performed by a bacterial toxin, in which diphtheria toxin (DT) transferred the ADP-ribose moiety of NAD$^+$ to the R-group of a posttranslationally modified histidine on elongation factor-2 (EF-2) (20). Yates et al. (108) recently provided an overview of the molecular basis for ADP-ribosyltransferase reaction. Subsequently, several toxins were identified as catalyzing the ADP-ribosylation of host target proteins. **Table 1** provides a representative listing of members of the ADP-ribosylating toxins. Note that each ADP-ribosylating toxin has a unique property, including the host protein that is targeted for ADP-ribosylation, the amino acid that is ADP-ribosylated, and the outcome of ADP-ribosylation on the modified protein. While DT has an absolute host target specificity for EF-2, other toxins, such as pertussis toxin (68, 70, 101) and ExoS (17), have less precise substrate specificity, in which several or numerous host proteins are targeted for ADP-ribosylation, respectively.

Early studies on bacterial toxins focused on their structural organization at the protein level and on the development of chemical methods to detoxify the toxins (toxoid) for vaccine development (76). Subsequent studies on bacterial toxins addressed the mechanisms utilized for entry into the host cell and how toxins can be utilized in medicine (24) or for malicious activities (45). In addition, bacterial toxins have been used to decipher basic concepts of host cell physiology. For example, the determination that the clostridial neurotoxins cleaved host proteins involved in synaptic vesicle fusion to the plasma membrane in neuronal cells (87) provided a functional role for these proteins in the fusion process.

Recent reviews have addressed several aspects of the action of the ADP-ribosylating toxins, including the mode of action of DT (108) and cholera toxin intracellular trafficking (13). In this review, we use the bacterial ADP-ribosylating toxins DT and *Pseudomonas aeruginosa* ExoS and ExoT to present our current understanding of how toxins recognize their substrates, the intracellular trafficking of bacterial toxins, and the physiological outcomes of host protein modification. In addition, we discuss the relationship between mammalian mono-ADP-ribosyltransferase and ADP-ribose hydrolases with bacterial ADP-ribosylating toxins.

DIPHTHERIA TOXIN

Pathogenesis

Corynebacterium diphtheriae, a facultative anaerobic gram-positive bacterium, is the causative agent of respiratory or cutaneous diphtheria. DT, the only virulence factor secreted by *C. diphtheriae*, has a lethal dose for humans of ~0.1 μg of toxin per kilogram. Whereas DT is proposed to damage epithelial cells to assist in bacterial colonization, nontoxinogenic strains

Type III secretion: a complex of bacterial proteins evolved by gene duplication from the bacterial flagellum that forms a tunnel to facilitate the translocation of bacterial toxins from the cytoplasm to the exterior of the cell

DT: diphtheria toxin

EF-2: elongation factor-2

Toxoid: chemically inactivated toxin that is used as a vaccine

of *C. diphtheriae* can colonize the airway without systemic pathology. Elegant studies showed that delivery of a single molecule of the **A** domain into the cytosol is sufficient to kill that cell (107). The current diphtheria vaccine is chemically inactivated DT; the vaccine is effective and has been used for decades without significant adverse reactions in children or adults (9). Prior to vaccination, diphtheria was a major disease of the young, but the incidence of diphtheria is almost eradicated in countries with a routine vaccine program. Between 1998 and 2004, only seven cases of respiratory diphtheria were reported to the Centers for Disease Control and Prevention (CDC) (9). In contrast, in countries without a diphtheria vaccination program, the disease remains a major source of morbidity and mortality. With the breakup of the Soviet Union in 1991, routine childhood vaccination programs were interrupted owing to the disruption of vaccine distribution. This resulted in a diphtheria epidemic, leading to 150,000 cases and 4000 deaths. Through the efforts of the World Health Organization, which reestablished vaccination, the diphtheria epidemic is under control (8). The utility of the DT vaccine has been extended and is currently used as a conjugate vaccine to deliver additional immunogens, such as the *Haemophilus influenzae* type B polysaccharide (32) and the meningococcal C carbohydrate (30), for human immunization.

DT, the prototype for the family of ADP-ribosylating toxins, is secreted by *C. diphtheriae* as a single protein of 535 amino acids (**Figure 1**). The N-terminal domain (193 amino acids) represents the A domain in the crystal structure, and the C-terminal domain (342 amino acids) represents the B domain and is functionally divided into a T domain and a R domain (15). Activities of all three domains, A-T-R, are required for the observed toxicity of DT. DT is cleaved at the surface of sensitive eukaryotic cells by furin or a furin-like protease to produce a di-chain protein that is linked by a single disulfide bond. Receptor binding triggers the entry of DT into the lumen of a developing endosome by receptor-mediated endocytosis. Upon acidification, the "daggers of death" (3) within the **T** domain facilitate the translocation of the **A** domain across the endosomal membrane and into the host cell cytoplasm (**Figure 1**), where the **A** domain catalyzes the ADP-ribosylation of eukaryotic elongation factor-2 (eEF-2) to inhibit protein synthesis.

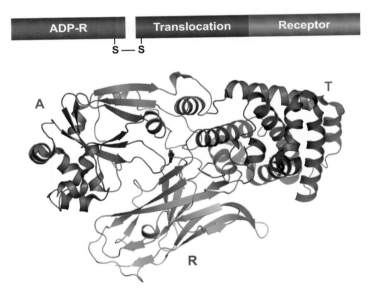

Figure 1

Structural properties of diphtheria toxin (DT). DT is organized as an AB toxin (PDB # 1DDT). The N-terminal domain (193 amino acids) represents the catalytic (A) domain (*red*), and the C-terminal domain (342 amino acids) represents the B domain and is functionally divided into a translocation (T) domain (*blue*) and a receptor binding (R) domain (*green*). Each domain, A-T-R, contributes to the toxicity of DT for sensitive cells.

Exotoxin A of *P. aeruginosa* ADP-Ribosylates Diphthamide on EF-2

Shortly after the identification of EF-2 as the substrate of DT, Iglewski et al. (51) showed that exotoxin A of *P. aeruginosa* also ADP-ribosylated EF-2. Exotoxin A is a 613-amino-acid protein with **AB** structure-function properties, but the functional domains are organized in the reverse order of DT (**Figure 2**). The N terminus encodes the **R** domain, the internal domain encodes the **T** domain, and the C terminus possesses the ADP-ribosyltransferase

A domain. Exotoxin A binds to the LDL-Receptor Related Protein 1 (36, 63), enters cells via receptor-mediated endocytosis (34), and retrograde traffics within the cell to the endoplasmic reticulum, where the **A** domain is translocated across the membrane and delivered into the host cytoplasm (94). Exotoxin A has been used in conjugate immunotoxin therapies to target and kill cancer cells (35). Comparative studies between DT and exotoxin A (15, 102) have provided insight into the general molecular and cellular properties of bacterial toxins. For example, a recent commentary (82) points out that, unlike DT, exotoxin A is not a useful carrier for conjugate vaccines.

The DT Receptor

Early studies showed the saturable binding of DT to sensitive eukaryotic cells (67), indicating that the initial interaction with the host cell was through a specific toxin-receptor interaction. The DT receptor was identified first by Eidels and coworkers (73), who used an expression cloning approach that was designed to identify a functional protein receptor. In this assay, the gene encoding the cell surface–expressed heparin-binding epidermal growth factor (HB-EGF)-like precursor was cloned from a DT-sensitive monkey cDNA library into naturally resistant mouse L-M cells (73), yielding mouse cell transformants that were sensitive to the action of DT. Subsequent studies identified a second membrane protein, CD-9, that did not directly bind to DT but increased the affinity of DT for the HB-EGF precursor, indicating that the two proteins functioned together as the DT-receptor complex (10, 68). The DT-receptor complex was internalized to an endosomal compartment (71).

DT Translocation Across the Endosome Membrane

The molecular and cellular basis for the translocation of bacterial toxins across the endosomal membrane is the least-understood aspect of the intoxication process. This is due in part to the

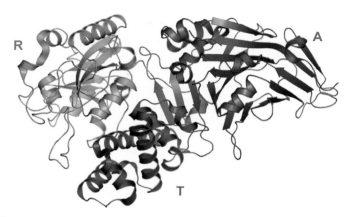

Receptor | Translocation | ADP-R

Figure 2

Structural properties of *Pseudomonas aeruginosa* exotoxin A. Exotoxin A is organized as an AB toxin (PBD# 1ikq). The N-terminal domain comprises the B domain, which is functionally divided into an N-terminal receptor binding (R) domain (*green*) and a translocation (T) domain (*blue*). The C-terminal domain comprises the catalytic (A) domain (*red*). While exotoxin A and DT ADP-ribosylate the identical substrate, EF-2, the toxins are organized in reverse structural order.

transient nature of the translocation event, in which the toxin or a component of the toxin exists in a folded form within the lumen of the endosome and, upon signaling, unravels and is transported across the lipid bilayer of the endosome and then refolds upon entry into the cytoplasm of the host cell. An early clue into the nature of the translocation process was obtained upon resolution of the crystal structure of the DT (15), in which the **T** domain was observed to consist of nine α-helices, of which two pairs were particularly apolar. This led the authors to propose a helices insertion mechanism for pH-triggered membrane insertion and translocation of the **A** domain (**Figure 1**) across the endosomal membrane. DT (93), anthrax toxin (21), and the botulinum neurotoxins (50) utilize an acidification trigger for **A** domain translocation. The vesicular (v)-ATPase has been implicated in contributing to the acidification of the lumen of the endosome, which facilitates the protonation of acidic amino acids at the tips of the apolar α-helices within the **T** domain to allow insertion and channel formation of the helices into the membrane. These conformational changes provide a model for the insertion and translocation of the **A** domain across the endosomal membrane via a mechanism that remains to be determined but appears to involve the reversible unfolding of the **A** domain during the translocation process. The direct translocation of the **A** domain through a protein channel has not been observed but has been inferred as the mechanism for translocation on the basis of a compilation of studies performed on several toxin translocation systems (83, 102, 104, 110).

Recently, two hypotheses have been proposed for the translocation of the **A** domain of DT across the endosomal membrane. The first hypothesis proposes that acidification of the endosome lumen stimulates **T** domain insertion into the vesicle membrane and that this constitutes a chaperone-like property to autonomously translocate the **A** domain through a formed channel into the cytosol, where the **A** domain refolds into the active conformation without the aid of host factors (75). The second hypothesis proposes that in addition to the **T** domain, host cytosolic factors, both ATP and a complex (cytosolic translocation factor) that includes thioredoxin reductase and heat shock protein 90, contribute to the chaperone function for the delivery of the **A** domain across the endosome membrane and refold within the cytosol (100). Although the **T** domain is sufficient for the translocation of the **A** domain across a membrane, it is conceivable that host proteins could contribute a chaperone function in the translocation process. Future studies that quantify the contribution of the host protein(s) to the translocation process, along with knockout and efficient knockdown of host factors, may clarify the role of host proteins in the translocation process.

ADP-Riboyslation of EF-2 by DT

EF-2 is a GTP-hydrolyzing motor protein essential for the elongation step in protein synthesis whereby tRNAs are translocated along the mRNA after peptidyl bond transfer on the 80S ribosome (57). EF-2 has a unique posttranslationally modified histidine (His715 in human and His699 in yeast), called diphthamide, as the 2-[3-carboxyamido-3-(trimethylammonio)propyl] modification provides the exclusive cellular substrate for ADP-ribosylation by DT and exotoxin A. The biological and biochemical properties of diphthamide have been reviewed (56) and are nonessential but required for an optimal growth phenotype. The diphthamide modification is found only in EF-2 and is conserved from archaea to humans, but it is not present in eubacteria, providing the molecular basis for DT action on EF-2 without inactivating the analogous bacterial elongation factor (EF-G).

Diphthamide is located close to the proposed interaction site (residue 694–698) of EF-2 with the codon-anticodon duplex. The determination that diphthamide was the substrate for DT was followed by genetic studies to test the role of EF-2 in protein synthesis and the specificity of DT toxicity through the ADP-ribosylation of EF-2. Initial screening identified yeast mutants that were resistant to the action of DT through the selection of spheroplasts challenged with DT (12) or by the intracellular expression of the **A** domain (79). These selections yielded DT-resistant mutants that had defects in the diphthamide biosynthesis pathway, which showed that diphthamide was required for the ADP-ribosylation of EF-2 by DT and that EF-2 was the sole substrate targeted by DT that yielded a toxic phenotype in yeast. Studies with mammalian Chinese hamster ovary cells also implicated a role for EF-2 in the unique targeting of host cells for the cytotoxicity elicited by DT (69). This finding also showed that diphthamide was necessary and sufficient to elicit the sensitivity of mammalian cells for DT through ADP-ribosylation. The crystal structure of ADP-ribosylated EF-2 with homologues in yeast in the presence of the fungi inhibitor sordarin and GDP (58) has been reported. In the cocrystal, the N3 atom of the diphthamide imidazole ring of EF-2 interacts with the β-phosphate of ADP-ribose.

Thus, although the inhibitory mechanism of ADP-ribosylation on EF-2 is not clear, these data suggest that ADP-ribosylation may elicit erroneous interaction of the translation factor with the codon-anticodon area in the P-site of the ribosome, leading to frameshift mutations.

A novel deoxyribonuclease activity is associated with DT (11). Whereas the initial model proposed that DT stimulated a host nuclease to facilitate DNA hydrolytic activity, subsequent studies implicated the presence of a second activity within the **A** domain of DT. The current model is that the nucleolytic activity stimulates cell lysis independently of the inhibition elicited by ADP-ribosylation of EF-2. Recent studies indicate that the DNA nuclease activity of DT is independent of the ADP-ribosyltransferase activity of the **A** domain and propose a second cation-dependent nuclease activity within the **A** domain (65). Although the crystal structure of the **A** domain of DT does not contain a typical cation binding pocket linked to a DNA binding pocket, the proposal that this nuclease activity is independent of ADP-ribosylation presents the possibility of generating a mutated form of the **A** domain that retains cation-dependent nuclease activity but lacks ADP-ribosyltransferase activity. This would provide a tool to establish the potency of the cation-dependent nuclease activity relative to that of the ADP-ribosylation of EF-2 elicited by the **A** domain, as studies in yeast and mammalian cells suggest that the ADP-ribosylation of EF-2 is necessary and sufficient for the toxicity elicited by DT.

Substrate Recognition by Exotoxin A of *P. aeruginosa*

Recent structural data of the cocrystal of exotoxin A with an NAD^+ analog (βTAD) and EF-2 provide insight into the mechanism of substrate recognition by ADP-ribosylating toxins (57). The toxin-EF-2 interface is involved in the interaction of the L4 loop of exotoxin A with a cleft between domains III and IV of EF-2. The L4 loop is directly connected to a tyrosine that stacks with the nicotinamide of NAD^+,

which may facilitate scission of the glycosidic bond of NAD^+. Electrostatic interactions direct the quaternary ammonium group of diphthamide toward the β-phosphate of the βTAD and therefore may be important for the recognition of EF-2. These interactions might introduce additional strain on NAD^+ by pulling the β-phosphate away from the substrate. Within the complex, the intermolecular interface between EF-2 and exotoxin A showed only a few contacts, which may explain why both exotoxin A and DT can ADP-ribosylate EF-2 from archaea to mammals.

PSEUDOMONAS AERUGINOSA ExoS AND ExoT

Pathogenesis

P. aeruginosa is an opportunistic human pathogen associated with multi-drug-resistant nosocomial infections, including pneumonia, urinary tract infections, surgical wounds, and sepsis (74) in immunocompromised individuals, such as patients with cystic fibrosis, cancer, burn wounds, and bone marrow transplantation (2, 25, 47, 54, 95). Unlike diphtheria, there are no approved vaccines to prevent *P. aeruginosa* infections (29, 81) and the increased prevalence of antibiotic-resistant clinical isolates (49) complicates strategies to control infections in the compromised host. The pathogenesis of *P. aeruginosa* is complex, with the production of cell-associated factors, including a single polar flagellum, pili, adhesins, alginate, and secreted factors, including a hemolysin, lipases, and proteases, to facilitate colonization. In addition, exotoxin A and the Type III cytotoxins ExoS, ExoT, ExoU, and ExoY are produced to damage the host. Expression of Type III cytotoxins has been correlated to clinical disease elicited by *P. aeruginosa* (48), compromising the ability of the host cell to phagocytose the bacterium. Segregation of Type III cytotoxins occurs where coexpression of ExoU and ExoS is excluded in clinical isolates (52, 92).

Type III cytotoxins are delivered across the bacterial cell membrane and cell wall into a

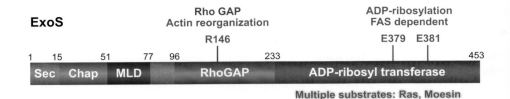

ExoS

Rho GAP
Actin reorganization
R146

ADP-ribosylation
FAS dependent
E379 E381

1	15	51	77	96		233			453
Sec	Chap	MLD			RhoGAP		ADP-ribosyl transferase		

Multiple substrates: Ras, Moesin

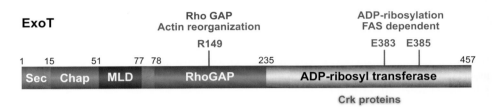

ExoT

Rho GAP
Actin reorganization
R149

ADP-ribosylation
FAS dependent
E383 E385

1	15	51	77	78		235			457
Sec	Chap	MLD			RhoGAP		ADP-ribosyl transferase		

Crk proteins

Figure 3

Pseudomonas aeruginosa ExoS and ExoT, bifunctional Type III–secreted cytotoxins. The N terminus (Sec) is used to secrete the cytotoxins from the bacterium. A chaperone binding region (Chap) maintains the cytotoxins in an extended conformation within the bacterium poised for translocation through the Type III secretion apparatus. The membrane localization domain (MLD) targets the cytotoxins with mammalian cells for efficient modification of host proteins. ExoS and ExoT possess identical RhoGAP activities, inactivating Rho, Rac, and Cdc42. The ADP-ribosyltransferase domains are unique, with ExoS possessing the ability to ADP-ribosylate numerous host proteins and ExoT ADP-ribosylating the Crk proteins. Expression of ADP-ribosyltransferase activity depends upon the binding of a 14-3-3 protein (FAS) to the C terminus of the ADP-ribosyltransferase domain.

mammalian cell through a secretion apparatus that is composed of 20–30 proteins. Next, the pseudomonas proteins PopB, PopD, and PcrV facilitate the translocation of the Type III cytotoxins across the host cell membrane into the cytosol of the host cell. Type III secretion is conserved among gram-negative bacteria (16). The genes encoding the pseudomonas Type III secretion system clustered on the *P. aeruginosa* PA01 chromosome have been designated *psc*, *pcr*, *exs*, and *pop* (38). Functions of the Type III cytotoxins of *P. aeruginosa* have been resolved. ExoU is a phospholipase (85) and correlates with acute cytotoxicity (92, 99). ExoY is an adenylate cyclase, which elevates intracellular cAMP levels in cultured mammalian cells and causes actin cytoskeleton reorganization (106). ExoS and ExoT are similar yet distinct (4). ExoS and ExoT share 76% amino acid identity (**Figure 3**) and are bifunctional Type III cytotoxins that comprise a Rho GTPase-

activating protein (RhoGAP) domain and a ADP-ribosyltransferase domain. The RhoGAP activities of ExoS and ExoT are similar, whereas ExoS and ExoT ADP-ribosylate unique host substrates.

Intracellular Trafficking of ExoS

Membrane localization of ExoS was first observed during the transfection of DNA encoding the N terminus of ExoS (78). Subsequent studies showed that the membrane localization was independent of the catalytic activity of ExoS and mapped to residues 51–72 (called the membrane localization domain, MLD) (78). The MLD was not required for Type III secretion of ExoS into cultured cells, but it was needed for the intracellular expression of Rho GAP (113) and ADP-ribosyltransferase activities (84) in which subcellular localization contributes to the targeting of ExoS to

Rho GTPase-activating protein (RhoGAP): a protein that aligns the catalytic residues within monomeric G proteins to facilitate the hydrolysis of the γ-phosphate of GTP within the G protein

Rho GTPases and Ras GTPases. The primary amino acid sequences between the MLD of ExoS and the analogous region of ExoT are highly homologous, implicating the presence of an ExoS-equivalent MLD in ExoT. This is supported by the determination that the Type III–delivered ExoT has a intracellular fractionation pattern similar to that of ExoS and that the RhoGAP domains of ExoS and ExoT have the same eukaryotic targets, Rho, Rac, and Cdc42 (61, 64).

Recent studies have addressed how ExoS utilizes the MLD to traffic within mammalian cells. Type III–delivered ExoS is initially delivered to the plasma membrane with subsequent movement to the perinuclear region of the cell (112). Type III–delivered ExoS coimmunoprecipitated 14-3-3 proteins and late endosome markers, Rab9 and Tip47, which indicated that Type III–delivered ExoS colocalized with late endosomes. Subsequent studies showed that the trafficking of Type III–delivered ExoS was inhibited by methyl-beta-cyclodextrin, a cholesterol-depleting reagent, and by nocodazole, an inhibitor of microtubules. This showed that the endocytosis of Type III–delivered ExoS is cholesterol dependent and utilizes host microtubules (**Figure 4**) (28). Future studies will address how ExoS associates with the endosome, which may identify new targets for therapeutic intervention toward ExoS and other gram-negative bacterial pathogens.

RhoGAP Domain of ExoS and ExoT

Heterologous expression of ExoS in *Yersinia pseudotuberculosis* provided the first indication that ExoS could act through two different mechanisms to reorganize the actin cytoskeleton in cultured cells (39). Subsequently, residues 96–234 possess a RhoGAP activity for Rho, Rac, and Cdc42 (46). A cocrystal of a Rac-ExoSRhoGAP complex showed that ExoS-RhoGAP and mammalian RhoGAPs did not share structural homology, but like the eukaryotic RhoGAPs, ExoSRhoGAP stabilized

the transition state of GTPase reaction, indicating ExoSRhoGAP was a functional mimic of eukaryotic GAPs (105). ExoSRhoGAP and mammalian RhoGAPs are examples of proteins evolving similar function through convergent evolution.

ADP-Ribosylation of Host Proteins by ExoS

ExoS was first described to have an ADP-ribosyltransferase activity that was distinct from that of exotoxin A, catalyzing the transfer of ADP-ribose from NAD^+ to several eukaryotic proteins (51), and where expression of ExoS ADP-ribosyltransferase activity was dependent upon the presence of a mammalian protein termed factor activating exoenzyme S (FAS) (18) that Fu et al. (40) identified as a mammalian 14-3-3 protein. Several small monomeric GTPases were identified as cellular targets of ExoS (17, 19). ExoS was determined to ADP-ribosylate Ras at Arg41, which was adjacent to the switch 1 region of the GTPase (43), and subsequent studies showed that ADP-ribosylation inhibited guanine nucleotide exchange factor (GEF)-catalyzed nucleotide exchange, which uncoupled Ras signal transduction (44). Early studies showed that expression of ADP-ribosyltransferase activity by Type III–delivered ExoS stimulated the early uptake of trypan blue, which indicated that ExoS caused a necrotic cell death (77). Recent studies by Jansson et al. (53) showed that the toxic phenotype elicited by ExoS ADP-ribosylation in cultured cells could be rescued by activated Ras, which indicated that the inhibition of Ras signaling is responsible for the cytotoxic features of ExoS. Expression of the ADP-ribosylation activity of ExoS also stimulates an apoptotic cell response (60). Because ExoS elicited limited caspase cleavage relative to agents that stimulate apoptosis (55), this effect may be indirect, which may be related to the observation that caspase cleavage did not follow a typical apoptotic signaling pathway. The limited amount of caspase cleavage elicited

Necrotic cell death: cell death caused by the release of intracellular enzymes that damage the cell membrane and other intracellular components that can lead to the death of the entire cell and surrounding cells

Apoptosis: a form of programmed cell death that leads to specific morphological changes, including cell shrinkage and nuclear fragmentation

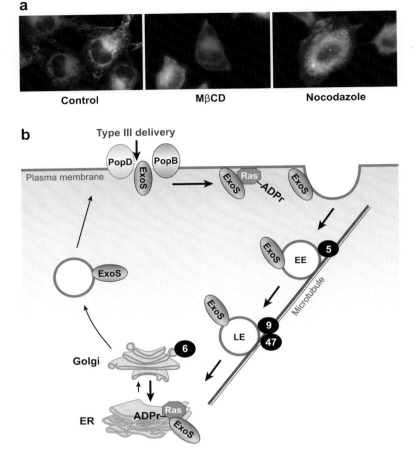

a

Control   MβCD   Nocodazole

b

Type III delivery

PopD   ExoS   PopB

Plasma membrane

ExoS

ExoS—ADPr   Ras   ExoS

ExoS   EE   5

ExoS   LE   9   47

Golgi   6

Microtubule

ER   ADPr—   Ras   ExoS

Figure 4

(*a*) Trafficking of Type III–delivered ExoS within cultured cells. HeLa cells were pretreated with nocodazole or methyl-beta-cyclodextrin (MβCD) as indicated or left untreated (control), and then infected with *Pseudomonas aeruginosa* PA103, Δ*exoU*, *exoT::Tc* carrying pUCP-ExoS (G⁻A⁻)-HA, a catalytic null form of ExoS with an influenza hemagglutinin (HA) epitope for immunofluorescence detection. ExoS traffics from plasma membrane to perinuclear region in control cells (112). MβCD traps a fraction of ExoS on plasma membrane, and nocodazole disrupts ExoS perinuclear localization. (*b*) Model for the trafficking of Type III–delivered ExoS in cultured cells. The initial entry of Type III–delivered ExoS is cholesterol dependent. Intracellular movement of ExoS is microtubule dependent but actin independent and is associated with Rab GTPases. Most intracellular ExoS is associated with late endosomal vesicles, and a fraction of ExoS appears to traffic from the late endosome to the plasma membrane. This figure is adapted from Deng et al. (28). PopB/PopD: channel forming proteins of the *P. aeruginosa* Type III secretion system; Ras, GTPase ADP-ribosylated by ExoS; ADPr, ADP-ribose; EE, early/sorting endosome; LE, late endosome. Rab GTPases are indicated as black ovals: 5, Rab5; 6, Rab6; 9, Rab; 47, tail-interacting protein 47.

by Type III–delivered ExoS may be due to the intracellular targeting of the toxin. Previous studies observed the differential ADP-ribosylation of host proteins by ExoS (111).

These studies show not only the complexity of characterizing the cytotoxic properties of ExoS, but also the potential utility of ExoS as a tool to study host cell physiology.

ADP-Ribosylation of Host Proteins by ExoT

The restricted substrate specificity of ExoT is a feature that differentiates the functional potential of ExoS and ExoT and allows a more direct analysis of the cellular basis for ExoT intoxication by ADP-ribosylation. ExoT was originally thought to be an inactive precursor of ExoS, but later studies implicated a role for the ADP-ribosyltransferase domain in eliciting cell rounding. Subsequently, ExoT was determined to ADP-ribosylate CT10-regulator of kinase (Crk) (96). Crk proteins are SH2-SH3 domain-containing adaptor proteins that function in integrin-mediated phagocytosis and focal adhesion. ADP-ribosylated Crk fails to interact with phosphorylated focal adhesion proteins, which uncouples signaling to the Rac GTPase proteins (27) to block Rap1- and Rac1-mediated focal adhesion and phagocytosis. This finding indicated that the RhoGAP and ADP-ribosyltransferase activities of ExoT had redundant properties in the inhibition of Rho GTPase activity. Engel and coworkers (91) have recently observed that prolonged expression of ExoT elicits an apoptotic response in cultured cells that is due primarily to the expression of ADP-ribosyltransferase activity. This is consistent with the observations that Crk signaling is linked to apoptosis (14, 33).

Yersinia YopH, a Type III–secreted phosphatase, dephosphorylates focal adhesion complex proteins and downregulates Crk-mediated phagocytosis (80, 103). Moreover, *Shigella flexneri*, by an undefined mechanism, activates the Abl family of tyrosine kinases (Abl and Arg) to facilitate bacterial invasion through the modulation of phosphorylation of Crk to regulate Rac-mediated phagocytosis (7).

Structural modeling and biochemical studies defined the molecular basis for the substrate specificity of ExoS and ExoT ADP-ribosylation (97). The loops that surround the NAD binding region allow ExoS to efficiently ADP-ribosylate Ras, and the molecular site for Crk recognition by ExoT was located on an α-helix that was distanced from the NAD binding region. Thus,

ExoS and ExoT recognize their substrates through different and independent mechanisms, in contrast to the redundant RhoGAP activities of ExoS and ExoT. This suggests that strains of *P. aeruginosa* that coexpress ExoS and ExoT modulate host cell physiology, because the spectrum of host proteins that ExoS and ExoT target for ADP-ribosylation is unique.

MAMMALIAN MONO-ADP-RIBOSYLTRANSFERASES AND ADP-RIBOSE HYDROLASES

Mammalian Mono-ADP-Ribosyltransferases

Similar to protein phosphorylation, the eukaryotic ADP-ribosylation cycle possesses characteristics that are crucial for cell regulation, such as temporal spatial control and reversibility, through the action of mono-ADP-ribosyltransferases and ADP-ribose hydrolases. **Table 2** shows the properties of known mammalian ADP-ribosyltransferases and ADP-ribose hydrolases. Host mono-ADP-ribosylation and hydrolysis may contribute to the observed physiological effects elicited by bacterial ADP-ribosylating toxins.

ADP-ribosyltransferases are an expanding family of proteins that have homologues across the prokaryotic and eukaryotic species (23). Among mammals, the human mono-ADP-ribosyltransferases (ART1–7) are the best-characterized mono-ADP-ribosyltransferases. ARTs are ectoenzymes that ADP-ribosylate extracellular proteins. ART1–4 are glycosylphosphatidylinositol (GPI)-anchored membrane proteins with an extracellular **A** domain, whereas ART5–7 are secreted proteins. ARTs appear to participate in the regulation of the immune response, because inhibitors of mono-ADP-ribosylation block the release of inflammatory cytokines from LPS-activated alveolar epithelia cells (26). Several proteins are substrates for ART1, including integrin α7, whose mono-ADP-ribosylation by ART1 is proposed to play a role in myogenesis (109), and defensin, an antimicrobial peptide

Crk: CT10-regulator of kinase

Table 2 Represenatative eukaryotic mono-ADP-ribosyltransferase and ADP-ribose hydrolase

Enzyme	Substrate	Activity
ADP-ribosyltransferase		
ART1	Integrin, defensin	Inhibits substrate activity
ART2	Unknown	T cell proliferation
ART6	p33/actin	Inhibits p33/actin activity
Sirtuin2	Albumin	Histone deacetylation
Arginine specific	Gβ	Inhibits Gβ activity
Unknown	GRP78/BiP	Inhibits GRP78/BiP activity
BFA sensitive	GAPDH,CtBP3/BARS	Inhibits GAPDH,CtBP3/BARS activity
ADP-ribose hydrolase		
ARH1	ADP-ribose-ʟ-arginine	Reverse ADP-ribosylation of proteins

Abbreviations: BFA, brefeldin A; CtBP3/BARS, C-terminal binding protein 3/brefeldin A-ribosylated substrate; GAPDH, glyceraldehyde-3-phosphate dehydrogenase.

secreted by immune cells (62). Host mono-ADP-ribosylation also affects other cellular processes. ADP-ribosylation inhibits the polymerization of nonmuscle actin, which can inhibit the release of azurophilic granules and can contribute to apoptosis. In vitro, ADP-ribosylation inhibits the assembly of muscle-specific desmin into intermediate filaments, and nutritional stress stimulates the ADP-ribosylation of GRP78/BiP, an endoplasmic reticulum lumen resident molecular chaperone that regulates protein synthesis and processing.

The fungal toxin brefeldin A (BFA) affects the structure and function of the Golgi complex by inactivating ADP-ribosylation factor 1, a low-molecular-weight GTPase required for membrane fission. Two proteins are ADP-ribosylated in BFA-treated cells, glyceraldehyde-3-phosphate dehydrogenase (GAPDH) and C-terminal binding protein 3/brefeldin A-ribosylated substrate (CtBP3/BARS). CtBP3/BARS, a lysophosphatidic acid-specific acyl transferase and potent activator of the fission of Golgi tubular-reticular domains, is a cytosolic factor that can counteract the effect of BFA on the Golgi complex, because it abolishes the organelle tubulation induced by BFA in permeabilized cells. ADP-ribosylated CtBP3/BARS inhibits the ability to promote the fission of Golgi tubules. CtBP1/BARS has a regulatory role in neutral lipid storage (5).

Mammalian ADP-Ribose Hydrolysis

The reversibility of mono-ADP-ribosylation in eukaryotes was first suggested by Moss et al. (70), who identified an ADP-ribose-ʟ-arginine cleavage enzyme from turkey erythrocytes. The hydrolase catalyzed the hydrolysis of ADP-ribose-arginine, but not ribosylarginine (98), showing the importance of ADP-ribose for substrate recognition. Identified ADP-ribose hydrolases are specific for ADP-ribose-arginine, but not for other ADP-ribose-modified amino acids (66).

ADP-ribose hydrolase activity may contribute to the regulation of intracellular signaling. It may also be a defense mechanism against the action of ADP-ribosylating bacterial toxins. Kato et al. (59) observed that the expression of ADP-ribose hydrolase influenced mouse intoxication by cholera toxin such that ADP-ribose hydrolase −/− cells were more sensitive to the effects of cholera toxin intoxication than were wild-type cells and that intestinal loops from ADP-ribose hydrolase −/− mice showed more fluid accumulation than did wild-type tissue. Thus, during cholera, cholera toxin-catalyzed ADP-ribosylation overwhelms the endogenous ADP-ribose hydrolase, resulting in an overall increase in activated ADP-ribosylated-Gs and intoxication of the host. In addition to Gs that is ADP-ribosylated by cholera toxin, nonmuscle actin ADP-ribosylated by botulinum C2 toxin also served as a substrate of the glycohydrolase. This demonstrates enzymatic cross-talk between bacterial toxin ADP-ribosyltransferases and host ADP-ribose hydrolases and presents a novel avenue for the development of therapeutics to control the pathology associated with bacterial ADP-ribosylating toxins.

CONCLUSION

Bacterial pathogens have evolved the use of protein toxins as potent modifiers of host cell physiology. Bacterial toxins continue to serve as targets for vaccine development and therapeutic intervention and as tools to dissect host cell physiology.

SUMMARY POINTS

1. Bacterial toxins modify specific host targets that provide an advantage for the pathogen in establishing an infection or damaging the host.

2. Bacterial toxins are organized into discrete structural domains that allow the toxin to enter and deliver an A domain into the host cell.

3. Each ADP-ribosylating toxin ADP-ribosylates a specific host protein(s), which leads to a unique cellular effect. DT shows an absolute specificity for the ADP-ribosylation for the diphthamide residue on EF-2.

4. ExoS and ExoT are bifunctional cytotoxins that possess RhoGAP and ADP-ribosyltransferase domains.

5. ExoS ADP-ribosylates numerous host proteins, whereas ExoT is more specific and ADP-ribosylates Crk proteins.

6. ExoS can usurp host intracellular trafficking pathways to travel to specific locations within the cell to efficiently ADP-ribosylate target proteins.

7. Advances in our understanding of the action of bacterial toxins parallel the development of structure-based studies and our understanding of the signaling pathways of the mammalian cell.

8. Bacterial toxins are utilized for vaccines and therapies against bacterial infections and for the development of drugs to control medical diseases.

FUTURE ISSUES

1. Toxin derivatives are to be developed for vaccine and therapeutic utilization.

2. The mechanism of translocation of the A domain across the endosomal membrane is to be determined.

3. The molecular basis for ExoS cytotoxicity is to be further characterized.

4. How ExoS efficiently ADP-ribosylates multiple host proteins is to be determined.

5. The cellular basis for the trafficking of ExoS within mammalian cells is to be determined.

6. How host ADP-ribosylation and ADP-ribose hydrolysis enzymes influence bacterial ADP-ribosylating toxins is to be explored.

DISCLOSURE STATEMENT

The authors are not aware of any biases that might be perceived as affecting the objectivity of this review.

ACKNOWLEDGMENTS

Research conducted by the Barbieri laboratory is supported by NIH-AI-30162. The authors thank Michael Baldwin for assistance in the preparation of illustrations.

LITERATURE CITED

1. Aktories K, Mohr C, Koch G. 1992. *Clostridium botulinum* C3 ADP-ribosyltransferase. *Curr. Top. Microbiol. Immunol.* 175:115–31
2. Armour AD, Shankowsky HA, Swanson T, Lee J, Tredget EE. 2007. The impact of nosocomially-acquired resistant *Pseudomonas aeruginosa* infection in a burn unit. *J. Trauma* 63:164–71
3. Baldwin MR, Kim J-JP, Barbieri JT. 2007. Botulinum neurotoxin B–host receptor recognition: It takes two receptors to tango. *Nat. Struct. Mol. Biol.* 14:9–10
4. Barbieri JT, Sun J. 2004. *Pseudomonas aeruginosa* ExoS and ExoT. *Rev. Physiol. Biochem. Pharmacol.* 152:79–92
5. Bartz R, Seemann J, Zehmer JK, Serrero G, Chapman KD, et al. 2007. Evidence that mono-ADP-ribosylation of CtBP1/BARS regulates lipid storage. *Mol. Biol. Cell* 18:3015–25
6. Bohach GA, Stauffacher CV, Ohlendorf DH, Chi YI, Vath GM, Schlievert PM. 1996. The staphylococcal and streptococcal pyrogenic toxin family. *Adv. Exp. Med. Biol.* 391:131–54
7. Burton EA, Plattner R, Pendergast AM. 2003. Abl tyrosine kinases are required for infection by *Shigella flexneri*. *EMBO J.* 22:5471–79
8. CDC. 1995. Diphtheria acquired by U.S. citizens in the Russian Federation and Ukraine-1994. *MMWR* 44:243–44
9. CDC. 2006. Preventing tetanus, diphtheria, and pertussis among adolescents: use of tetanus toxoid, reduced diphtheria toxoid and acellular pertussis vaccines. *MMWR* 55:1–37
10. Cha JH, Brooke JS, Ivey KN, Eidels L. 2000. Cell surface monkey CD9 antigen is a coreceptor that increases diphtheria toxin sensitivity and diphtheria toxin receptor affinity. *J. Biol. Chem.* 275:6901–7
11. Chang MP, Bramhall J, Graves S, Bonavida B, Wisnieski BJ. 1989. Internucleosomal DNA cleavage precedes diphtheria toxin-induced cytolysis. Evidence that cell lysis is not a simple consequence of translation inhibition. *J. Biol. Chem.* 264:15261–67
12. **Chen JY, Bodley JW, Livingston DM. 1985. Diphtheria toxin-resistant mutants of *Saccharomyces cerevisiae*. *Mol. Cell Biol.* 5:3357–60**
13. Chinnapen DJ, Chinnapen H, Saslowsky D, Lencer WI. 2007. Rafting with cholera toxin: endocytosis and trafficking from plasma membrane to ER. *FEMS Microbiol. Lett.* 266:129–37
14. Cho SY, Klemke RL. 2000. Extracellular-regulated kinase activation and CAS/Crk coupling regulate cell migration and suppress apoptosis during invasion of the extracellular matrix. *J. Cell Biol.* 149:223–36
15. **Choe S, Bennett MJ, Fujii G, Curmi PM, Kantardjieff KA, et al. 1992. The crystal structure of diphtheria toxin. *Nature* 357:216–22**
16. Coburn B, Sekirov I, Finlay BB. 2007. Type III secretion systems and disease. *Clin. Microbiol. Rev.* 20:535–49
17. Coburn J, Dillon ST, Iglewski BH, Gill DM. 1989. Exoenzyme S of *Pseudomonas aeruginosa* ADP-ribosylates the intermediate filament protein vimentin. *Infect. Immun.* 57:996–98
18. Coburn J, Kane AV, Feig L, Gill DM. 1991. *Pseudomonas aeruginosa* exoenzyme S requires a eukaryotic protein for ADP-ribosyltransferase activity. *J. Biol. Chem.* 266:6438–46
19. **Coburn J, Wyatt RT, Iglewski BH, Gill DM. 1989. Several GTP-binding proteins, including p21c-H-ras, are preferred substrates of *Pseudomonas aeruginosa* exoenzyme S. *J. Biol. Chem.* 264:9004–8**

12. Describes the selection of yeast mutants resistant to DT by mutations in EF-2.

15. Describes the crystal structure of DT and how DT can translocate the A domain across the endosome membrane.

19. Provides an early indication that ExoS has the capability of ADP-ribosylating numerous host proteins.

20. Collier RJ. 1975. Diphtheria toxin: mode of action and structure. *Bacteriol. Rev.* 39:54–85

21. Collier RJ. 1999. Mechanism of membrane translocation by anthrax toxin: insertion and pore formation by protective antigen. *J. Appl. Microbiol.* 87:283

22. Collier RJ, Cole HA. 1969. Diphtheria toxin subunit active in vitro. *Science* 164:1179–81

23. Corda D, Di Girolamo M. 2003. Functional aspects of protein mono-ADP-ribosylation. *EMBO J.* 22:1953–58

24. Cordivari C, Misra VP, Catania S, Lees AJ. 2004. New therapeutic indications for botulinum toxins. *Mov. Disord.* 19(Suppl. 8):S157–61

25. da Silva Filho LV, Tateno AF, Martins KM, Azzuz Chernishev AC, Garcia Dde O, et al. 2007. The combination of PCR and serology increases the diagnosis of *Pseudomonas aeruginosa* colonization/infection in cystic fibrosis. *Pediatr. Pulmonol.* 42:938–44

26. Del Vecchio M, Balducci E. 2008. Mono ADP-ribosylation inhibitors prevent inflammatory cytokine release in alveolar epithelial cells. *Mol. Cell Biochem.* 310:77–83

27. Deng Q, Sun J, Barbieri JT. 2005. Uncoupling Crk signal transduction by *Pseudomonas* exoenzyme T. *J. Biol. Chem.* 280:35953–60

28. Deng Q, Zhang Y, Barbieri JT. 2007. Intracellular trafficking of *Pseudomonas* ExoS, a type III cytotoxin. *Traffic* 8:1331–45

29. Doring G, Meisner C, Stern M. 2007. A double-blind randomized placebo-controlled phase III study of a *Pseudomonas aeruginosa* flagella vaccine in cystic fibrosis patients. *Proc. Natl. Acad. Sci. USA* 104:11020–25

30. El Bashir H, Heath PT, Papa T, Ruggeberg JU, Johnson N, et al. 2006. Antibody responses to meningococcal (groups A, C, Y and W135) polysaccharide diphtheria toxoid conjugate vaccine in children who previously received meningococcal C conjugate vaccine. *Vaccine* 24:2544–49

31. Endo Y, Tsurugi K. 1987. RNA N-glycosidase activity of ricin A-chain. Mechanism of action of the toxic lectin ricin on eukaryotic ribosomes. *J. Biol. Chem.* 262:8128–30

32. Eskola J, Kayhty H, Peltola H, Karanko V, Makela PH, et al. 1985. Antibody levels achieved in infants by course of *Haemophilus influenzae* type B polysaccharide/diphtheria toxoid vaccine. *Lancet* 1:1184–86

33. Evans EK, Lu W, Strum SL, Mayer BJ, Kornbluth S. 1997. Crk is required for apoptosis in *Xenopus* egg extracts. *EMBO J.* 16:230–41

34. FitzGerald D, Morris RE, Saelinger CB. 1980. Receptor-mediated internalization of *Pseudomonas* toxin by mouse fibroblasts. *Cell* 21:867–73

35. Fitzgerald D, Pastan I. 1993. *Pseudomonas* exotoxin and recombinant immunotoxins derived from it. *Ann. N. Y. Acad. Sci.* 685:740–45

36. Fitzgerald DJ, Fryling CM, Zdanovsky A, Saelinger CB, Kounnas M, et al. 1994. Selection of *Pseudomonas* exotoxin-resistant cells with altered expression of alpha 2MR/LRP. *Ann. N. Y. Acad. Sci.* 737:138–44

37. Flatau G, Lemichez E, Gauthier M, Chardin P, Paris S, et al. 1997. Toxin-induced activation of the G protein p21 Rho by deamidation of glutamine. *Nature* 387:729–33

38. Frank DW. 1997. The exoenzyme S regulon of *Pseudomonas aeruginosa*. *Mol. Microbiol.* 26:621–29

39. Frithz-Lindsten E, Du Y, Rosqvist R, Forsberg A. 1997. Intracellular targeting of exoenzyme S of *Pseudomonas aeruginosa* via type III-dependent translocation induces phagocytosis resistance, cytotoxicity and disruption of actin microfilaments. *Mol. Microbiol.* 25:1125–39

40. Fu H, Coburn J, Collier RJ. 1993. The eukaryotic host factor that activates exoenzyme S of *Pseudomonas aeruginosa* is a member of the 14-3-3 protein family. *Proc. Natl. Acad. Sci. USA* 90:2320–24

41. Fussle R, Bhakdi S, Sziegoleit A, Tranum-Jensen J, Kranz T, Wellensiek HJ. 1981. On the mechanism of membrane damage by *Staphylococcus aureus* alpha-toxin. *J. Cell Biol.* 91:83–94

42. Galan JE. 2001. *Salmonella* interactions with host cells: type III secretion at work. *Annu. Rev. Cell Dev. Biol.* 17:53–86

43. Ganesan AK, Frank DW, Misra RP, Schmidt G, Barbieri JT. 1998. *Pseudomonas aeruginosa* exoenzyme S ADP-ribosylates Ras at multiple sites. *J. Biol. Chem.* 273:7332–37

44. Ganesan AK, Vincent TS, Olson JC, Barbieri JT. 1999. *Pseudomonas aeruginosa* exoenzyme S disrupts Ras-mediated signal transduction by inhibiting guanine nucleotide exchange factor-catalyzed nucleotide exchange. *J. Biol. Chem.* 274:21823–29

22. Provides an early indication of the mechanism of action of DT.

27. Showed that ExoT stimulated cell rounding based on the inactivation of Crk proteins.

43. Shows how ADP-ribosylation by ExoS uncouples Ras signal transduction.

45. Garland T, Bailey EM. 2006. Toxins of concern to animals and people. *Rev. Sci. Tech.* 25:341–51

46. Goehring UM, Schmidt G, Pederson KJ, Aktories K, Barbieri JT. 1999. The N-terminal domain of *Pseudomonas aeruginosa* exoenzyme S is a GTPase-activating protein for Rho GTPases. *J. Biol. Chem.* 274:36369–72

47. Hachem RY, Chemaly RF, Ahmar CA, Jiang Y, Boktour MR, et al. 2007. Colistin is effective in treatment of infections caused by multidrug-resistant *Pseudomonas aeruginosa* in cancer patients. *Antimicrob. Agents Chemother.* 51:1905–11

48. Hauser AR, Cobb E, Bodi M, Mariscal D, Valles J, et al. 2002. Type III protein secretion is associated with poor clinical outcomes in patients with ventilator-associated pneumonia caused by *Pseudomonas aeruginosa*. *Crit. Care Med.* 30:521–28

49. Hauser AR, Sriram P. 2005. Severe *Pseudomonas aeruginosa* infections. Tackling the conundrum of drug resistance. *Postgrad. Med.* 117:41–48

50. Hoch DH, Romero-Mira M, Ehrlich BE, Finkelstein A, DasGupta BR, Simpson LL. 1985. Channels formed by botulinum, tetanus, and diphtheria toxins in planar lipid bilayers: relevance to translocation of proteins across membranes. *Proc. Natl. Acad. Sci. USA* 82:1692–96

51. Iglewski BH, Liu PV, Kabat D. 1977. Mechanism of action of *Pseudomonas aeruginosa* exotoxin adenosine diphosphate-ribosylation of mammalian elongation factor 2 in vitro and in vivo. *Infect. Immun.* 15:138–44

52. Jain M, Ramirez D, Seshadri R, Cullina JF, Powers CA, et al. 2004. Type III secretion phenotypes of *Pseudomonas aeruginosa* strains change during infection of individuals with cystic fibrosis. *J. Clin. Microbiol.* 42:5229–37

53. Jansson AL, Yasmin L, Warne P, Downward J, Palmer RH, Hallberg B. 2006. Exoenzyme S of *Pseudomonas aeruginosa* is not able to induce apoptosis when cells express activated proteins, such as Ras or protein kinase B/Akt. *Cell Microbiol.* 8:815–22

54. Jelsbak L, Johansen HK, Frost AL, Thogersen R, Thomsen LE, et al. 2007. Molecular epidemiology and dynamics of *Pseudomonas aeruginosa* populations in lungs of cystic fibrosis patients. *Infect. Immun.* 75:2214–24

55. Jia J, Wang Y, Zhou L, Jin S. 2006. Expression of *Pseudomonas aeruginosa* toxin ExoS effectively induces apoptosis in host cells. *Infect. Immun.* 74:6557–70

56. Jorgensen R, Merrill AR, Andersen GR. 2006. The life and death of translation elongation factor 2. *Biochem. Soc. Trans.* 34:1–6

57. Jorgensen R, Merrill AR, Yates SP, Marquez VE, Schwan AL, et al. 2005. Exotoxin A-eEF2 complex structure indicates ADP ribosylation by ribosome mimicry. *Nature* 436:979–84

58. Jorgensen R, Yates SP, Teal DJ, Nilsson J, Prentice GA, et al. 2004. Crystal structure of ADP-ribosylated ribosomal translocase from *Saccharomyces cerevisiae*. *J. Biol. Chem.* 279:45919–25

59. Kato J, Zhu J, Liu C, Moss J. 2007. Enhanced sensitivity to cholera toxin in ADP-ribosylarginine hydrolase-deficient mice. *Mol. Cell Biol.* 27:5534–43

60. Kaufman MR, Jia J, Zeng L, Ha U, Chow M, Jin S. 2000. *Pseudomonas aeruginosa* mediated apoptosis requires the ADP-ribosylating activity of exoS. *Microbiology* 146(Pt. 10):2531–41

61. Kazmierczak BI, Engel JN. 2002. *Pseudomonas aeruginosa* ExoT acts in vivo as a GTPase-activating protein for RhoA, Rac1, and Cdc42. *Infect. Immun.* 70:2198–205

62. Kim C, Slavinskaya Z, Merrill AR, Kaufmann SH. 2006. Human alpha-defensins neutralize toxins of the mono-ADP-ribosyltransferase family. *Biochem. J.* 399:225–29

63. Kounnas MZ, Morris RE, Thompson MR, FitzGerald DJ, Strickland DK, Saelinger CB. 1992. The alpha 2-macroglobulin receptor/low density lipoprotein receptor-related protein binds and internalizes *Pseudomonas* exotoxin A. *J. Biol. Chem.* 267:12420–23

64. Krall R, Schmidt G, Aktories K, Barbieri JT. 2000. *Pseudomonas aeruginosa* ExoT is a Rho GTPase-activating protein. *Infect. Immun.* 68:6066–68

65. Lee JW, Nakamura LT, Chang MP, Wisnieski BJ. 2005. Mechanistic aspects of the deoxyribonuclease activity of diphtheria toxin. *Biochim. Biophys. Acta* 1747:121–31

66. Maehama T, Nishina H, Katada T. 1994. ADP-ribosylarginine glycohydrolase catalyzing the release of ADP-ribose from the cholera toxin-modified alpha-subunits of GTP-binding proteins. *J. Biochem.* 116:1134–38

59. First article to show that host ADP-ribose hydrolases can influence the outcome of ADP-ribosylation by bacterial toxins.

67. Middlebrook JL, Dorland RB, Leppla SH. 1978. Association of diphtheria toxin with Vero cells. Demonstration of a receptor. *J. Biol. Chem.* 253:7325–30

68. Mitamura T, Iwamoto R, Umata T, Yomo T, Urabe I, et al. 1992. The 27-kDa diphtheria toxin receptor-associated protein (DRAP27) from Vero cells is the monkey homologue of human CD9 antigen: Expression of DRAP27 elevates the number of diphtheria toxin receptors on toxin-sensitive cells. *J. Cell Biol.* 118:1389–99

69. Moehring JM, Moehring TJ. 1979. Characterization of the diphtheria toxin-resistance system in Chinese hamster ovary cells. *Somatic Cell Genet.* 5:453–68

70. Moss J, Bruni P, Hsia JA, Tsai SC, Watkins PA, et al. 1984. Pertussis toxin-catalyzed ADP-ribosylation: effects on the coupling of inhibitory receptors to the adenylate cyclase system. *J. Recept. Res.* 4:459–74

71. Moya M, Dautry-Varsat A, Goud B, Louvard D, Boquet P. 1985. Inhibition of coated pit formation in Hep2 cells blocks the cytotoxicity of diphtheria toxin but not that of ricin toxin. *J. Cell Biol.* 101:548–59

72. Mukherjee S, Keitany G, Li Y, Wang Y, Ball HL, et al. 2006. *Yersinia* YopJ acetylates and inhibits kinase activation by blocking phosphorylation. *Science* 312:1211–14

73. **Naglich JG, Metherall JE, Russell DW, Eidels L. 1992. Expression cloning of a diphtheria toxin receptor: identity with a heparin-binding EGF-like growth factor precursor. *Cell* 69:1051–61**

74. Obritsch MD, Fish DN, MacLaren R, Jung R. 2005. Nosocomial infections due to multidrug-resistant *Pseudomonas aeruginosa*: epidemiology and treatment options. *Pharmacotherapy* 25:1353–64

75. Oh KJ, Senzel L, Collier RJ, Finkelstein A. 1999. Translocation of the catalytic domain of diphtheria toxin across planar phospholipid bilayers by its own T domain. *Proc. Natl. Acad. Sci. USA* 96:8467–70

76. Park WH, Banzhaf EJ, Zingher A, Schroder MC. 1924. Observations on diphtheria toxoid as an immunizing agent. *Am. J. Public Health* 14:1047–49

77. Pederson KJ, Krall R, Riese MJ, Barbieri JT. 2002. Intracellular localization modulates targeting of ExoS, a type III cytotoxin, to eukaryotic signalling proteins. *Mol. Microbiol.* 46:1381–90

78. **Pederson KJ, Pal S, Vallis AJ, Frank DW, Barbieri JT. 2000. Intracellular localization and processing of *Pseudomonas aeruginosa* ExoS in eukaryotic cells. *Mol. Microbiol.* 37:287–99**

79. Perentesis JP, Genbauffe FS, Veldman SA, Galeotti CL, Livingston DM, et al. 1988. Expression of diphtheria toxin fragment A and hormone-toxin fusion proteins in toxin-resistant yeast mutants. *Proc. Natl. Acad. Sci. USA* 85:8386–90

80. Persson C, Carballeira N, Wolf-Watz H, Fallman M. 1997. The PTPase YopH inhibits uptake of *Yersinia*, tyrosine phosphorylation of p130Cas and FAK, and the associated accumulation of these proteins in peripheral focal adhesions. *EMBO J.* 16:2307–18

81. Pier G. 2005. Application of vaccine technology to prevention of *Pseudomonas aeruginosa* infections. *Expert Rev. Vaccines* 4:645–56

82. Pier GB. 2007. Is *Pseudomonas aeruginosa* exotoxin A a good carrier protein for conjugate vaccines? *Hum. Vaccin.* 3:39–40

83. Ren J, Sharpe JC, Collier RJ, London E. 1999. Membrane translocation of charged residues at the tips of hydrophobic helices in the T domain of diphtheria toxin. *Biochemistry* 38:976–84

84. Riese MJ, Barbieri JT. 2002. Membrane localization contributes to the in vivo ADP-ribosylation of Ras by *Pseudomonas aeruginosa* ExoS. *Infect. Immun.* 70:2230–32

85. Sato H, Frank DW, Hillard CJ, Feix JB, Pankhaniya RR, et al. 2003. The mechanism of action of the *Pseudomonas aeruginosa*-encoded type III cytotoxin, ExoU. *EMBO J.* 22:2959–69

86. Saxena SK, O'Brien AD, Ackerman EJ. 1989. Shiga toxin, Shiga-like toxin II variant, and ricin are all single-site RNA N-glycosidases of 28 S RNA when microinjected into *Xenopus* oocytes. *J. Biol. Chem.* 264:596–601

87. Schiavo G, Benfenati F, Poulain B, Rossetto O, Polverino de Laureto P, et al. 1992. Tetanus and botulinum-B neurotoxins block neurotransmitter release by proteolytic cleavage of synaptobrevin. *Nature* 359:832–35

88. Schlumberger MC, Muller AJ, Ehrbar K, Winnen B, Duss I, et al. 2005. Real-time imaging of type III secretion: *Salmonella* SipA injection into host cells. *Proc. Natl. Acad. Sci. USA* 102:12548–53

89. Schmidt G, Sehr P, Wilm M, Selzer J, Mann M, Aktories K. 1997. Gln 63 of Rho is deamidated by *Escherichia coli* cytotoxic necrotizing factor-1. *Nature* 387:725–29

73. Identifies the host receptor for DT.

78. Shows the intracellular trafficking of Type III–delivered ExoS.

90. Sehr P, Joseph G, Genth H, Just I, Pick E, Aktories K. 1998. Glucosylation and ADP ribosylation of Rho proteins: effects on nucleotide binding, GTPase activity, and effector coupling. *Biochemistry* 37:5296–304

91. Shafikhani SH, Morales C, Engel J. 2007. The *Pseudomonas aeruginosa* type III secreted toxin ExoT is necessary and sufficient to induce apoptosis in epithelial cells. *Cell Microbiol.* In press

92. Shaver CM, Hauser AR. 2004. Relative contributions of *Pseudomonas aeruginosa* ExoU, ExoS, and ExoT to virulence in the lung. *Infect. Immun.* 72:6969–77

93. Silverman JA, Mindell JA, Zhan H, Finkelstein A, Collier RJ. 1994. Structure-function relationships in diphtheria toxin channels. I. Determining a minimal channel-forming domain. *J. Membr. Biol.* 137:17–28

94. Smith DC, Spooner RA, Watson PD, Murray JL, Hodge TW, et al. 2006. Internalized *Pseudomonas* exotoxin A can exploit multiple pathways to reach the endoplasmic reticulum. *Traffic* 7:379–93

95. Stanzani M, Tumietto F, Giannini MB, Bianchi G, Nanetti A, et al. 2007. Successful treatment of multi-resistant *Pseudomonas aeruginosa* osteomyelitis after allogeneic bone marrow transplantation with a combination of colistin and tigecycline. *J. Med. Microbiol.* 56:1692–95

96. Sun J, Barbieri JT. 2003. *Pseudomonas aeruginosa* ExoT ADP-ribosylates CT10 regulator of kinase (Crk) proteins. *J. Biol. Chem.* 278:32794–800

97. Sun J, Maresso AW, Kim JJ, Barbieri JT. 2004. How bacterial ADP-ribosylating toxins recognize substrates. *Nat. Struct. Mol. Biol.* 11:868–76

98. Takada T, Okazaki IJ, Moss J. 1994. ADP-ribosylarginine hydrolases. *Mol. Cell Biochem.* 138:119–22

99. Tam C, Lewis SE, Li WY, Lee E, Evans DJ, Fleiszig SM. 2007. Mutation of the phospholipase catalytic domain of the *Pseudomonas aeruginosa* cytotoxin ExoU abolishes colonization promoting activity and reduces corneal disease severity. *Exp. Eye Res.* 85:799–805

100. Trujillo C, Ratts R, Tamayo A, Harrison R, Murphy JR. 2006. Trojan horse or proton force: finding the right partner(s) for toxin translocation. *Neurotox. Res.* 9:63–71

101. Van Dop C, Yamanaka G, Steinberg F, Sekura RD, Manclark CR, et al. 1984. ADP-ribosylation of transducin by pertussis toxin blocks the light-stimulated hydrolysis of GTP and cGMP in retinal photoreceptors. *J. Biol. Chem.* 259:23–26

102. Wedekind JE, Trame CB, Dorywalska M, Koehl P, Raschke TM, et al. 2001. Refined crystallographic structure of *Pseudomonas aeruginosa* exotoxin A and its implications for the molecular mechanism of toxicity. *J. Mol. Biol.* 314:823–37

103. Weidow CL, Black DS, Bliska JB, Bouton AH. 2000. CAS/Crk signalling mediates uptake of *Yersinia* into human epithelial cells. *Cell Microbiol.* 2:549–60

104. Wu Z, Jakes KS, Samelson-Jones BS, Lai B, Zhao G, et al. 2006. Protein translocation by bacterial toxin channels: a comparison of diphtheria toxin and colicin Ia. *Biophys. J.* 91:3249–56

105. Wurtele M, Wolf E, Pederson KJ, Buchwald G, Ahmadian MR, et al. 2001. How the *Pseudomonas aeruginosa* ExoS toxin downregulates Rac. *Nat. Struct. Biol.* 8:23–26

106. Yahr TL, Vallis AJ, Hancock MK, Barbieri JT, Frank DW. 1998. ExoY, an adenylate cyclase secreted by the *Pseudomonas aeruginosa* type III system. *Proc. Natl. Acad. Sci. USA* 95:13899–904

107. Yamaizumi M, Mekada E, Uchida T, Okada Y. 1978. One molecule of diphtheria toxin fragment A introduced into a cell can kill the cell. *Cell* 15:245–50

108. Yates SP, Jorgensen R, Andersen GR, Merrill AR. 2006. Stealth and mimicry by deadly bacterial toxins. *Trends Biochem. Sci.* 31:123–33

109. Yau L, Litchie B, Zahradka P. 2004. MIBG, an inhibitor of arginine-dependent mono(ADP-ribosyl)ation, prevents differentiation of L6 skeletal myoblasts by inhibiting expression of myogenin and p21(cip1). *Exp. Cell Res.* 301:320–30

110. Young JA, Collier RJ. 2007. Anthrax toxin: receptor binding, internalization, pore formation, and translocation. *Annu. Rev. Biochem.* 76:243–65

111. Zhang Y, Barbieri JT. 2005. A leucine-rich motif targets *Pseudomonas aeruginosa* ExoS within mammalian cells. *Infect. Immun.* 73:7938–45

112. Zhang Y, Deng Q, Barbieri JT. 2007. Intracellular localization of type III-delivered *Pseudomonas* ExoS with endosome vesicles. *J. Biol. Chem.* 282:13022–32

113. Zhang Y, Deng Q, Porath JA, Williams CL, Pederson-Gulrud KJ, Barbieri JT. 2007. Plasma membrane localization affects the RhoGAP specificity of *Pseudomonas* ExoS. *Cell Microbiol.* 9:2192–201

107. Shows that one molecule of DT can kill a mammalian cell.

Ins and Outs of Major Facilitator Superfamily Antiporters

Christopher J. Law,[1] Peter C. Maloney,[2] and Da-Neng Wang[1]

[1] The Helen L. and Martin S. Kimmel Center for Biology and Medicine at the Skirball Institute of Biomolecular Medicine, and Department of Cell Biology, New York University School of Medicine, New York, NY 10016; email: law@saturn.med.nyu.edu; wang@saturn.med.nyu.edu

[2] Department of Physiology, Johns Hopkins University School of Medicine, Baltimore, Maryland 21205; email: pmaloney@jhmi.edu

Annu. Rev. Microbiol. 2008. 62:289–305

First published online as a Review in Advance on June 6, 2008

The *Annual Review of Microbiology* is online at micro.annualreviews.org

This article's doi: 10.1146/annurev.micro.61.080706.093329

Key Words

secondary membrane transporter proteins, rocker-switch mechanism, GlpT, UhpT, OxlT, EmrD

Abstract

The major facilitator superfamily (MFS) represents the largest group of secondary active membrane transporters, and its members transport a diverse range of substrates. Recent work shows that MFS antiporters, and perhaps all members of the MFS, share the same three-dimensional structure, consisting of two domains that surround a substrate translocation pore. The advent of crystal structures of three MFS antiporters sheds light on their fundamental mechanism; they operate via a single binding site, alternating-access mechanism that involves a rocker-switch type movement of the two halves of the protein. In the *sn*-glycerol-3-phosphate transporter (GlpT) from *Escherichia coli*, the substrate-binding site is formed by several charged residues and a histidine that can be protonated. Salt-bridge formation and breakage are involved in the conformational changes of the protein during transport. In this review, we attempt to give an account of a set of mechanistic principles that characterize all MFS antiporters.

Contents

INTRODUCTION

Control of substrate movement across cytoplasmic or internal membranes is vital to the long-term survival of a cell. To achieve this, nature has evolved a diverse system of channel and transporter proteins that effect the translocation of ions and small hydrophilic molecules across these membranes. Bacteria employ three main types of transporter to fulfill this function: (a) so-called primary transporters, such as P-type ATPases and ATP-binding cassette transporters, which use the energy released from ATP hydrolysis to drive ions or solutes across the membrane; (b) secondary transporters, which drive substrate translocation by exploiting the free energy stored in the ion or solute gradients generated by primary transporters; and (c) group translocation systems, which couple translocation of a substrate to its chemical modification, resulting in release of a modified substrate at the opposite side of the membrane.

About 25% of all known membrane transport proteins in prokaryotes belong to the major facilitator superfamily (MFS) (68), the largest and most diverse superfamily of secondary active transporters. The MFS contains 58 distinct families (http://www.tcdb.org/), with about 15,000 sequenced members identified to date (http://www.membranetransport. org/) (13, 53, 59, 64, 65, 68), a number that is expected to swell as more genomes are sequenced. This superfamily is ubiquitous in all kingdoms of life and in all biological cells, and it includes members of direct medical and pharmaceutical significance. Mechanistically, MFS transporters display three distinct kinetic mechanisms: (a) uniporters, which transport only one type of substrate and are energized solely by the substrate gradient; (b) symporters, which translocate two or more substrates in the same direction simultaneously, making use of the electrochemical gradient of one substrate as the driving force; and (c) antiporters, which transport two or more substrates but in opposite directions across the membrane. Individual members within the MFS show stringent specificity, yet as a group the superfamily accepts an enormous diversity of substrate types (ions, sugars, sugar phosphates, drugs, neurotransmitters, nucleosides, amino acids, and peptides, among others). As a result, mechanistic studies of the MFS allows one to focus on fundamental questions related to how proteins recognize substrates and transport them across a membrane. Our understanding of these issues is well advanced in some model systems. This is especially true of the *Escherichia coli* lactose/H^+ permease (LacY), an MFS symporter that has been the subject of intense experimental scrutiny for many decades by a number of laboratories (1, 21, 38, 39). By contrast, insight into the molecular details of mechanism is less advanced for antiporters, which are the main topic of this review. Here, we focus on studies of bacterial MFS antiporters, in particular the closely related *sn*-glycerol-3-phosphate:phosphate (GlpT) and hexosephosphate:phosphate (UhpT) antiporters from

---

MFS: major facilitator superfamily

Antiporter: a membrane transport protein that transports two or more substrates in opposite directions across the cell membrane using the gradient of one to drive movements of the other(s)

GlpT: *sn*-glycerol-3-phosphate transporter from *E. coli*

UhpT: hexose phosphate:phosphate antiporter from *E. coli*

E. coli and the oxalate:formate transporter from *Oxalobacter formigenes* (OxlT). We try to provide a snapshot of contemporary understanding of structural and mechanistic aspects of these important and fascinating integral membrane proteins.

AN ABBREVIATED HISTORY

From the time of Hippocrates and Galen, who believed substances were absorbed through orifices in blood vessels in the intestinal wall, hypotheses to account for the translocation of even large substrates across biological membranes invoked processes such as osmosis and simple diffusion (20). But as it stands today, that sector of membrane biology concerned with solute transport more properly owes its identity to theoretical and practical studies dating from the early 1950s. Until that time, those interested in transport of sugars, amino acids, or similar metabolites were constrained by a conceptual framework focused on the flow and distribution of ions (Na, K, Cl) across membranes, with findings evaluated in terms of coefficients of diffusion or permeability, flux ratio measurements, and whether or not a system could be considered at equilibrium (33). This all changed in the early 1950s, when W.F. Widdas offered a new perspective in his study of glucose transport by the sheep placenta (81) and human erythrocytes (82), and when others applied this same view to sugar transport by rat small intestine (19) and to phosphate transport across bacterial membranes (54, 55). Widdas' particular contribution was to posit an intermediate—a complex reflecting the direct and stoichiometric combination of substrate and a carrier, much like the intermediate complexes then being invoked in the context of enzymatic catalysis. Accordingly, to frame the process of transport, he proposed that descriptions of substrate (glucose) transport must recognize terms relevant to ligand binding (which legitimized use of biochemistry as a practical tool); he also imagined that the liganded carrier diffused across the membrane, so that a kinetic formulation would include terms describing mobility of the carrier/

substrate complex. At the opposite surface, the bound passenger would be released, and the unoccupied carrier would return to the original membrane surface to complete a cycle, in which one cycle (turnover) leads to net transfer of one molecule of substrate (81). Of course, we no longer believe that protein carriers literally diffuse across the membrane, although this is true of peptide carrier ionophores (e.g., valinomycin). Instead, in the spirit of Occam's razor, we accept the concept of a single binding site, alternating-access mechanism, and variations thereof, in which the transporter can possess two major alternating conformations, inward facing (C_i) and outward facing (C_o) (36, 73, 74, 77, 80). In this case, translocation of substrate across the membrane is catalyzed by the interconversion of these two conformations.

Indeed, much of the current emphasis in the field is dedicated to developing experimental approaches that describe this conformational change. Such efforts now appear to validate two general kinds of models discussed over the years. On the one hand, structural insights provided by the study of MFS family members are nicely summarized by the rocker-switch mechanism (35), a term meant to suggest that substrate might be fixed in space at its binding site. On the other hand, alternating access is provided as protein conformational changes alternately generate pathways of access to either surface (51). As well, one might imagine a gated pore (60), in which substrate moves as through an ion channel, but with gates that alternately open and close at either end. This is a view more suited to structural features revealed by certain non-MFS transporters. The now-accelerating pace of structural work should provide enough examples to populate a continuum between these two extremes and perhaps even define new axes to consider.

STRUCTURAL CONSIDERATIONS

General Structure of MFS Proteins

Until the recent arrival of high-resolution three-dimensional (3D) structures of

OxlT: oxalate:formate antiporter from *Oxalobacter formigenes*

Alternating-access: general mechanism by which the single binding site of MFS antiporters is sequentially exposed to the cytoplasmic and periplasmic sides of the membrane

C_i: inward- or cytoplasmic-facing conformation of bacterial antiporters

C_o: outward- or periplasmic-facing conformation of bacterial antiporters

Rocker-switch mechanism: the motion of the N- and C-terminal domains of MFS transporters that results in alternating-access of the substrate-binding site

TMs: transmembrane α-helices

OPA: organophosphate:phosphate antiporter

Pᵢ: inorganic phosphate

G3P: glycerol-3-phosphate

representatives of the MFS, the field relied on indirect methods, such as phylogeny and sequence analysis, as a guide to the structure and function of these proteins (13). Hydropathy analysis of protein sequences and topological studies with gene fusion constructs predicted that almost all MFS proteins possess a uniform topology of 12 transmembrane α-helices (TMs) connected by hydrophilic loops, with both their N and C termini located in the cytoplasm (59, 67, 68). Often, the N-terminal half of the protein (TM1–TM6) displays weak sequence homology to the C-terminal half (TM7–TM12), suggesting that the molecule may have arisen following a gene duplication/fusion event (49). This prediction has implications, now confirmed, regarding an underlying structural symmetry. Exceptions to the 12-TM rule do exist—a few MFS families have 14 TMs, one family has only 6 TMs, and yet another has 24 TMs (59). The extra two helices in the 14-TM members probably arose via insertion of the central cytoplasmic loop into the membrane, whereas the 24-TM member is likely a consequence of a gene fusion event (67); the lone example with six TMs likely functions as a homodimer.

MFS proteins typically consist of 400–600 amino acids, and analysis of their primary sequences revealed that within any single family, sequence similarity is highly significant (59). By contrast, at the level of the superfamily, individual MFS members share low sequence identity or similarity and are united only by a pair of conserved signature sequences, DRXXRR, at equivalent positions in the N- and C-terminal halves of the proteins in loops that join TM2 to TM3 and TM8 to TM9, respectively (49). Along with the conserved transmembrane topology and the internal sequence homology between the two halves of the proteins, this duplicated signature sequence suggests a common ancestral gene for the MFS.

The importance of these earlier studies to a structural perspective was underlined with the publication of high-resolution 3D crystal structures of *E. coli* GlpT (35), LacY (2), and EmrD (84), and a lower-resolution structure of *O. formigenes* OxlT (30, 32). This revealed that all these MFS proteins—despite their sequence divergence—do indeed share almost the same 3D structures. Essentially all existing biochemical and biophysical data for these and other MFS proteins are in agreement with these structures (45, 47, 57). This finding leads to the notion of a shared fold that acts as a scaffold for all MFS proteins, irrespective of their particular function as a symporter, uniporter, or antiporter.

Structure of MFS Antiporters

Detailed 3D structural information is currently available for three MFS antiporters, all from prokaryotes (30, 35, 84). The structure of GlpT (35), along with that of the related LacY symporter (2, 22), has been determined for the C_i conformation. Two additional structures—those of the oxalate exchanger, OxlT, and the multidrug transporter, EmrD—may represent transporters in an occluded state (30, 32, 84). There is as yet no structure of an MFS transporter in the C_o conformation. Nevertheless, the available structures emphasize the general conservation of fold and architecture, consisting of two domains with a pore between them.

Publication of these structures, in combination with previous mutagenesis, biochemical, and biophysical studies (6–8, 17, 18), has enabled a dramatic enrichment of our understanding of transporter mechanism. Furthermore, such structural information has been invaluable for the design of new studies to test mechanistic aspects of these proteins.

Structure of GlpT in the C_i Conformation

As a member of the organophosphate:phosphate antiporter (OPA) family of the MFS, GlpT functions to couple an outward flow of internal inorganic phosphate (P_i) to the uptake of glycerol-3-phosphate (G3P) into the cell (26). G3P is an important intermediate in both glycolysis and phospholipid biosynthesis, and it can act as the sole energy source for bacterial growth (48, 63). In GlpT and other antiporters that utilize P_i-linked exchange, the net reaction is enabled by the outwardly directed P_i gradient. In addition to bacterial OPA proteins, homologues of GlpT have been identified in a range of eukaryotes, including plants (12), fruit flies (3), mice (72), and humans (11). The 3.3 Å structure of GlpT showed the molecule to have a silhouette reminiscent of a Mayan temple (35). The periplasmic side of the protein is plateau-like and protrudes only slightly into the external surface. In contrast, several TMs extend beyond the membrane interface at the cytoplasmic side.

GlpT consists of N- and C-terminal domains related by a pseudo twofold symmetry (**Figure 1a**). Each domain is composed of

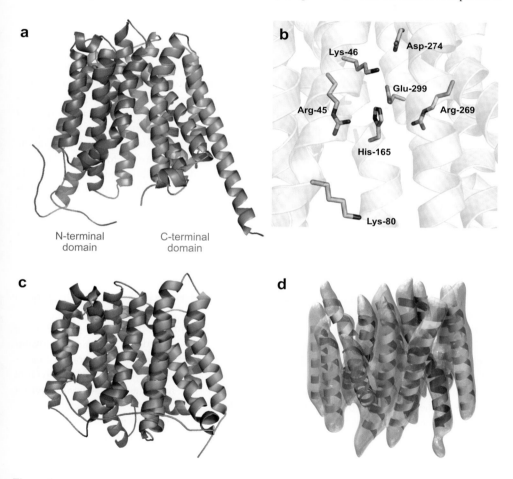

Figure 1

Three-dimensional structures of GlpT, EmrD, and OxlT viewed parallel to the membrane. (*a*) The 3.3 Å structure of GlpT in the C_i conformation. Transmembrane α-helices of the N-terminal domain are colored magenta, and those of the C-terminal domain are green. The substrate translocation pore is situated between the two domains (35). (*b*) The GlpT substrate-binding site, depicting the basic residues intimately involved in binding (Lys-80, Arg-45, His-165, and Arg-269) and those that participate in intra- and interhelical salt-bridge formation (Lys-46, Asp-274, and Glu-299). (*c*) The 3.5 Å structure of EmrD in a compact, occluded state (84). (*d*) The 6.5 Å density map of OxlT, also in an occluded state, with 12 TMs modeled into it (30). In all panels, the periplasmic side of each transporter is at the top.

six transmembrane α-helices, divided into two three-helix bundles inserted into the membrane in opposite orientations. These TMs are packed loosely in the membrane, presumably to facilitate helix and domain mobility that occurs during the substrate translocation process (28, 29). A long cytoplasmic loop (L6-7) of 45 residues, part of which is disordered in the crystal structure, links the N- and C-terminal domains (35). Although this loop acts as a constraint for movement of the N- and C-terminal domains, it likely still permits a relatively large interdomain movement, which has implications for the substrate translocation mechanism.

Although MFS proteins such as the lactose transporter LacS from *Streptococcus thermophilus* (76) and the *Bacillus subtilis* tetracycline transporter TetL (66) are dimeric, GlpT and its close homolog UhpT function as monomers (5, 10). For this reason, the permeation pathway is presumed to reside within the monomer itself, and the 3D GlpT structure immediately suggests that the substrate translocation pathway is the pore saddled by the GlpT N- and C-terminal domains (**Figure 1a**) (35). When viewed perpendicular to the membrane, the pore is corralled by a rectangular fence composed of 8 of 12 TMs (two on each of the two sides, two at the front, and two at the back), which act as a scaffolding for the remaining 4 TMs. The TMs that form the front and back of the fence (TM2 and TM11, and TM5 and TM8, respectively) are curved like a banana, so that each helix pair is shaped like an hourglass, in which the helices make contact with one another at the periplasmic side of the membrane but are separated from each other in the cytosol. The curvature of these helices holds significance for the rocker-switch type of movement that the GlpT N- and C-terminal domains undergo during the transport reaction cycle (35).

In the crystal structure of GlpT, the substrate translocation pore is open only to the cytoplasmic side of the membrane (**Figure 1a**) (35). The pore is prevented from forming a portal through the membrane by portions of TM1 and TM7, which fills gaps in the barrier with the side chains of nine aromatic residues. This arrangement ensures there is no communication between opposite sides of the membrane when the transporter is in the C_i conformation, an important structural feature consistent with the substrate translocation mechanism. From the middle of the membrane, the substrate translocation pore opens out to form a well about 30 Å deep, with a funnel-shaped outer section and a more cylinder-shaped inner section. The base of this inner section is lined by side chains that yield a positive surface electrostatic potential, and this likely enhances binding of GlpT to its negatively charged, oxyanionic substrates (10). The interior surface of the funnel section, however, is greasy and lined with mostly hydrophobic residues that would prevent adhesion of ions and water molecules and thereby direct hydrophilic substrates down to the electropositive binding site. The lower and outermost sides of the pore are formed by the cytoplasmic ends of the N-terminal domain TM4 and TM5 helices, their connecting loop L4-5, and the corresponding C-terminal helices and loop, TM10 and TM11 and L10-11 (35). The cytoplasmic ends of TM1 and TM7 make contacts with these loops (L4-5 and L10-11, respectively), and this arrangement appears optimal for transmitting substrate-induced conformational changes throughout the protein (35).

The information provided by the 3D structure of GlpT (**Figure 1a**) has been invaluable to efforts to unravel general principles of substrate binding and protein architecture in the MFS. But when considered together with the structures of two other MFS antiporters, the *O. formigenes* oxalate:formate transporter OxlT (30, 32) and the *E. coli* drug:H[+] antiporter EmrD (84), it is also possible to derive important clues regarding the conformational change that MFS antiporters undergo during the transport cycle.

Structures of OxlT and EmrD in an Occluded Conformation

In contrast to the overall Mayan temple-shape of GlpT in the C_i conformation, the OxlT and

EmrD structures show these proteins in a more compact, occluded conformation. A projection map at 6 Å (27) and a 3D map at 6.5 Å of OxlT (**Figure 1d**) (30) were determined using electron crystallography of two-dimensional crystals grown in the presence of saturating concentrations of substrate (oxalate), giving us a first glimpse of what MFS proteins look like at the molecular level in terms of helix organization. The overall topology of the transporter revealed 12 TMs arranged around a central pore that is closed to both sides of the membrane (30). Like GlpT, the OxlT protein consists of two six-helix bundles representing the N- and C-terminal domains. As might be expected if this structure reflected a conformation different from that of GlpT, there are subtle differences in the relative displacement of OxlT helices compared to those of GlpT. Although the low resolution of the OxlT structure did not permit visualization of bound substrate, the overall structure is much more compact and closed to both sides of the membrane, suggesting it was crystallized in an occluded state (30). This conformation likely corresponds to a key intermediate state in the reaction pathway somewhere between the C_i and C_o conformations (40). With the GlpT structure as a template, the 6.5 Å OxlT density map was used subsequently to model a structure of the latter in the substrate-bound occluded state (32). This too suggested that substrate binding initiates a global conformational change imperative for function (31). Similarly, biochemical and biophysical studies have also provided evidence of a more compact form of GlpT upon substrate binding (10, 35).

The other MFS protein whose 3D structure represents an occluded state is EmrD (84). With respect to substrate specificity, EmrD differs significantly from GlpT and OxlT in that it can export a broad range of hydrophobic substrates from the cell (59). Nevertheless, the 3.5 Å crystal structure of EmrD (**Figure 1c**) showed that, like other MFS proteins, it consists of 12 TMs organized as a pair of six-helix domains resembling those found in OxlT and GlpT. The pore surrounded by these helices is closed to both sides of the membrane. Although the rel-

ative orientation of some of the helices (TM3, TM6, TM9, and TM12) is similar to that of their GlpT counterparts, the other helices deviate substantially from the observed positions of their GlpT equivalents. The EmrD structure also shows two long helical regions (composed of TM4 and TM5, and TM10 and TM11, and the loops that connect them) on the cytoplasmic side of the membrane that are arranged much closer to the substrate translocation pore and that extend farther into the cytoplasm than their GlpT counterparts (84). It was suggested that these regions, and a run of positively charged residues located at the end of TM4, act as a substrate specificity filter (84). Akin to the OxlT structure, the EmrD helices form a compact structure of ~50 Å in the plane of the membrane by ~45 Å along the membrane normal. Consistent with its function of catalyzing the translocation of hydrophobic compounds across the bacterial inner membrane, the substrate translocation pore of EmrD is much more hydrophobic than that of GlpT, with stacked aromatic side chains lining the pore surface probably to aid in substrate binding (84). That this transporter has been crystallized with its substrate translocation pore closed to both sides of the membrane makes it likely that it has been captured in an occluded state (84). It is surprising, then, that no substrate is visible in this structure, because in the absence of substrate an occluded state is a high-energy, unstable state that would in principle be refractory to crystallization. The absence of substrate, therefore, is likely a consequence of insufficient resolution. In summary, we now have access to three 3D structures of MFS antiporters, one in the substrate-free C_i conformation (35) and two in a substrate-bound, occluded state (30, 84).

SUBSTRATE SPECIFICITY OF MFS ANTIPORTERS

One intriguing aspect of MFS transporters as a whole is their ability to differentiate between vast selections of often similar substrates. Indeed, there is probably an individual MFS transporter for each small- or medium-sized

MD: molecular dynamics

G6P: glucose-6-phosphate

hydrophilic molecule of biological relevance in the cell. The simplicity of the MFS fold design, that of two symmetrical domains saddling a substrate-specific pathway, has been used repeatedly by nature, so clearly it is the presence of only a few amino acid residues at the substrate-binding site that determines the specificity of each transporter for its cognate substrate.

Among MFS antiporters, substrate specificity determinants have been well studied in UhpT (23–25). An intrahelical ion pair in UhpT formed between an asparagine (D388) and lysine (K391) of TM11 is essential for normal UhpT function (23). Because TM11 lines the substrate translocation pore in UhpT, it was suggested that residues that compose sections of it could function as determinants of and regulate substrate specificity of the transporter (24). Furthermore, an uncompensated cationic charge at position 388 or 391 in UhpT resulted in gain-of-function mutants that preferred divalent phosphoenolpyruvate as a substrate to the monovalent hexosephosphates normally transported by UhpT (24). In contrast, an uncompensated anionic charge at position 388 increased the preference of UhpT for monovalent hexosephosphates rather than divalent sugar species (23). A related antiporter from *Salmonella typhimurium*, PgtP, which transports phosphoenolpyruvate, lacks the equivalent D388-K391 salt bridge but does possess an uncompensated R391, indicating that residues at positions 388 and 391 in these transporters act as determinants for substrate selectivity (23).

More recent work has used structural, computational, and biochemical analyses to describe a more detailed mechanism for substrate binding by GlpT (43). GlpT, like other MFS transporters, operates via a single binding site, alternating-access mechanism (77) involving a rocker-switch type movement of the protein (35, 44). Because the published structure of GlpT is of the molecule in the absence of substrate, in the C_i conformation, the nature of the substrate-binding site could only be inferred (**Figure 1b**) (35). Nevertheless, strong arguments attest to the validity of this inference. Two conserved, positively charged residues, arginine 45 (R45) from TM1 and arginine 269 (R269) from TM7, are located at the inner end of the substrate translocation pore visualized in the GlpT structure, and these basic residues are strategically placed to interact with substrate as central components of the binding site (35). In the crystal structure, the shortest distance between the R45 and R269 side chains is 9.9 Å. For both side chains to form a hydrogen bond length of 2.9 Å with the negatively charged oxygen atoms of the P_i or G3P substrate, they must move 1.4 Å closer to each other. This could occur if substrate binding pulls TM1 and TM7 closer together (35). Molecular dynamics (MD) simulations support the notion that R45 and R269 directly bind to substrates (43). As expected, mutation of either arginine into a lysine eliminates transport activity of GlpT reconstituted into proteoliposomes.

These presumptive ligand-binding residues are conserved between GlpT and its close homolog from *E. coli*, UhpT, and also in G6PT, the human microsomal glucose-6-phosphate transporter (4, 17). Importantly, the two equivalent arginine residues (R46 and R275) in UhpT have also been identified as substrate-binding elements, as only these two among all UhpT arginines are essential (17). This finding is entirely consistent with the binding chemistry suggested by the GlpT crystal structure. Similarly, in G6PT, mutation of R28 (equivalent to GlpT R45) to cysteine or histidine abolishes transport activity; indeed, such mutations cause glycogen storage disease type Ib (4, 14). Paired basic residues are also essential for function in several other MFS anion transporters such as the nitrate transporter, NrtA, from *Aspergillus nidulans* (75), the putative *E. coli* nitrate:nitrite antiporter, NarU (37), and OxlT (79).

Other residues are also involved directly in substrate binding in OPA family proteins. Studies on UhpT have implicated a lysine (K82) and a histidine (H168) residue as important to substrate binding (23, 25), and equivalent residues in GlpT (K80 and H165) line the substrate translocation pathway (**Figure 1b**).

In a recent analysis of the GlpT structure, it has been suggested that K80, in partnership with R45, R269, and H165, plays a vital role in substrate binding (43). This proposal was supported by MD simulations (43) suggesting that H165 is involved in substrate coordination only when it is protonated. The maintenance of a positive charge on the H165 residue during substrate binding presumably enables stronger binding to the GlpT anionic substrates.

While changes in the ionization state of histidine are common in many enzyme-catalyzed reactions (69), observations in LacY (58), UhpT (25), and now GlpT (43) highlight its importance in membrane transporter proteins in general. In GlpT, H165 is surrounded by several conserved aromatic residues (Y38, Y42, Y76, W138, W161, Y362, and Y393) (35, 46). Aromatic residues such as tyrosine stabilize positive charges within the membrane electric field (16); thus, it is possible that the primary function of the conserved tyrosines and other aromatic residues surrounding the binding site of GlpT is not in substrate binding but rather in stabilizing the basicity of the binding site. Preservation of the basicity of the substrate-binding site is vital for function of OxlT as well (79, 83), and a membrane-embedded positive charge is proposed to be essential for multidrug transport by the *E. coli* MFS protein MdfA (70). Moreover, several aromatic residues are required for substrate transport in the human organic anion transporter hOAT1 (62), and both histidine and tyrosine residues appear to participate in substrate binding by another MFS protein, the mammalian H^+:peptide transporter PepT1 (15).

Taken together, a reasonably detailed mechanism for substrate binding by the C_i conformation of GlpT can be summarized as follows: (*a*) An initial weak binding of substrate occurs at the inner end of the substrate-binding pore and involves three basic residues, K80, R45, and R269 (**Figure 1b**). At this stage H165 is unprotonated and does not participate strongly in binding. (*b*) H165 then undergoes protonation, perhaps facilitated by the proximity of P_i, and its side chain moves closer toward and inter-

acts more strongly with the substrate. At the same time, the R45 and R269 side chains also move 1.4 Å closer to each other, pulling TM1 and TM7 closer together on the cytoplasmic side (35). Coupled with a rotation of the N- and C-terminal domains (see below), this elicits stronger, more stable interactions between the substrate and K80, R45, R269, and protonated H165 and thus tighter binding to the transporter. The intrinsic binding energy released by this stronger interaction is utilized to overcome the energy barrier to conformational change of the unloaded transporter (**Figure 2d**). (*c*) Subsequent deprotonation of H165 weakens the interactions with the substrate, now bound in the C_o conformation, allowing it to be released into the periplasm.

Such a mechanism agrees with earlier energetics considerations of membrane transport. It has been suggested before for membrane transporter systems that reduction in the energy barrier to conformational change—paid for by intrinsic binding energy between substrate and its binding site—is dependent on formation of an initial loose complex followed by a tight complex that forms in the transition state (41, 42). This mechanism is probably conserved among all OPA family antiporters.

ROCKER-SWITCH MECHANISM

A fundamental problem encountered in the study of membrane transport concerns the mechanism of substrate translocation; following substrate binding, how can a membrane-embedded protein catalyze movement of the substrate from one side of the membrane to the other? In both MFS antiporters and symporters, the substrate-binding site has access to each side of the membrane in an alternating fashion. This can now be seen as a direct consequence of domain movement that allows the N- and C-terminal portions of the transporter to rock back and forth against each other along an axis that runs along the domain interface in the membrane. This rocker-switch mechanism, implied from the crystal structures of GlpT (35) and LacY (2), is the contemporary version of what Mitchell referred to as the mobile barrier model (56), a name intended to replace the idea (and implications) of the mobile carrier (51). Recent work on LacY has supported the no-

tion of a global conformational change during substrate translocation (50, 71).

Salt-Bridge Formation and Breakage during Substrate Translocation

The current model of GlpT-mediated transport invokes a single binding site, alternating-access mechanism accompanied by rocker-switch type movement of the N- and C-terminal domains

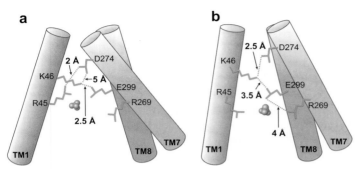

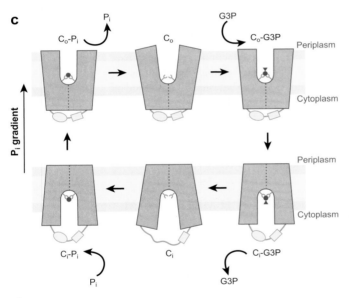

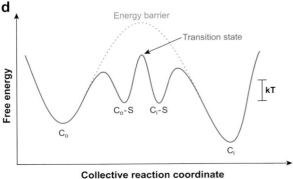

Figure 2

Substrate binding and transport cycle of GlpT. (*a*) Schematic diagram illustrating the salt bridges that are formed and broken upon initial loose and subsequent tight binding of substrate to GlpT. When substrate binds loosely to GlpT in the C_i conformation and H165 is unprotonated, interhelical salt bridges are formed between R45-D274 and K46-D274 (43). (*b*) Protonation of H165 elicits tighter substrate binding, and R45 and R269 move closer to each other, pulling TM1 and TM7 closer together. The R45-D274 interhelical salt bridge breaks and a new intrahelical salt bridge forms between R269 and E299. The existing interhelical salt bridge formed between K46 and D274 becomes stronger, and the transporter takes on a more compact conformation (43). TM1, TM7, and TM8 are depicted as cylinders, and the amino acid side chains that participate in salt-bridge formation are depicted as stick models, and salt bridges as dashed lines. The substrate molecule is represented as blue van der Waals spheres. (*c*) Schematic diagram of the single binding site, alternating-access mechanism with a rocker-switch type of movement for the GlpT-mediated G3P-P_i exchange reaction. The diagram describes the proposed conformational changes that the transporter undergoes during the reaction cycle. C_o, outward-facing conformation; C_i, inward-facing conformation. The G3P substrate is represented by a small disk and triangle, and P_i is represented by a small disk (35). (*d*) A schematic free-energy diagram illustrating the energy levels of the different conformations of GlpT that occur during the transport reaction cycle under physiological conditions. In the absence of substrate binding, the energy barrier prevents the conformational interconversion between the C_o and C_i states of the transporter. Substrate binding lowers the energy barrier sufficiently to allow Brownian motion (kT) to drive the conformational interconversion. The energy barrier is represented by a dotted line. S denotes substrate (44).

of the transporter; in kinetic terms, this reflects that interconversion between the C_i and C_o (and vice versa) conformations is favored upon substrate binding. Recent work using a combination of MD simulations and biochemical experiments suggests that formation and breaking of inter- and intradomain salt bridges are important in controlling the helical motions necessary for the large-scale conformational change between C_i and C_o (43). Examination of the crystal structure of GlpT implicated three conserved, charged residues, K46, D274, and E299, that line the periplasmic end of the substrate-binding pore to be of potential importance to function (**Figure 1b**). MD simulations and biochemical experiments then suggested that interdomain salt bridges formed by the negatively charged D274 with two separate basic residues (K46 and R45) and by E299 with K46 act in concert as springs to stabilize the different conformations during the transport reaction cycle (**Figure 2a,b**). Initially, when H165 is unprotonated and P_i binds loosely to the binding site, there is only a weak interaction between D274 and K46 (**Figure 2a**). At the same time, much stronger salt bridges exist between D274 and R45 and between E299 and K46. This would maintain the transporter in the C_i conformation, in which the binding site allows substrate approach from the cytoplasmic side of the substrate translocation pore. Subsequent tighter binding of substrate, accompanied by protonation of H165, pulls R45 of TM1 and R269 of TM7 closer together. Tight substrate binding causes complete disruption of the interdomain D274-R45 salt bridge, a weakening of the E299-K46 salt bridge, and the formation of stronger inter- and intradomain salt bridges between D274 and K46 and between E299 and R269, respectively, resulting in interdomain movement and relative rotation of the helices in each domain (**Figure 2b**).

In this model, substrate binding effectively weakens interactions that stabilize the C_i conformation. This enables new interactions involved in domain rotation, possibly involving yet other salt bridges. In this model, the salt-bridge dynamics are part of the spring that permits the delicately poised GlpT molecule to flip via the rocker-switch mechanism (35, 47) from the C_i-S conformation to the C_o-S conformation. The salt bridges may even act as the pivot upon which the teeter-totter motion of the two domains of the protein is centered. Interestingly, the *E. coli* outer membrane protein OmpA uses the formation and breaking of salt bridges to control the gating of its substrate translocation pore (34).

Evidence for conformational changes during substrate translocation by antiporters is available from the structure of GlpT (35), and from its comparisons with the OxlT (30, 31) and EmrD (84) structures. As noted above, a model of OxlT in substrate-bound form gave insight into the conformational changes that occur upon substrate binding to antiporters (32), suggesting that the main consequence of transition from the unloaded, C_i state to the closed, substrate-bound C_i-S state is a closing of the cytoplasmic ends of the N- and C-terminal domains, involving a swiveling movement of each domain relative to the other (**Figure 2c**). Further, it was suggested that the continuation of this movement brings the protein into the C_o-S state, in which the cytoplasmic ends of the domains are closer together than the periplasmic ones, thereby revealing the binding site to the periplasm and exposing the substrate to it. Because of the symmetry of the global architecture of OxlT, the C_o state is likely to mirror the C_i state (31, 32).

Kinetic Considerations

Recent kinetic studies on GlpT provide further evidence for the rocker-switch mechanism, and the findings have allowed the reaction cycle to be described in detailed physical chemistry terms (44). The rate of the complete reaction cycle is associated with and dictated by large conformational changes in the protein. An occluded state of the protein is also expected during substrate translocation (40), although it is unknown whether this state is as stable in GlpT as it is in OxlT and EmrD. As the entire antiport process must occur at a physiologically realistic rate and be responsive to changes in substrate concentrations on both sides of the membrane, reasonable values for the rate constants are needed for the individual stages of the reaction cycle (**Figure 2c**). Like other antiporters, GlpT obeys Michaelis-Menten kinetics and operates via a ping-pong mechanism (80). This allows one to conceptually distinguish six discrete steps, with three steps dedicated to the binding, translocation, and release of one of the transported substrates and the other three steps describing the same events for the transport of the countersubstrate. This represents the minimum number of steps required for antiport. The half-reaction for translocation of one substrate (P_i in the case of GlpT) can therefore be written as

$$C_i + P_i \underset{k_{-1}}{\overset{k_1}{\rightleftharpoons}} C_i - P_i \underset{k_{-2}}{\overset{k_2}{\rightleftharpoons}} C_o - P_i \overset{k_3}{\longrightarrow} C_o + P_i. \quad 1.$$

The three stages of the half-reaction are (*a*) binding of substrate to the protein in the C_i conformation; (*b*) switching of the substrate-bound form of the protein from the C_i-P_i state to the C_o-P_i state, which allows the pathway to face the opposing surface; and (*c*) release of the translocated substrate into the periplasm. A detailed description of the antiport reaction mechanism begins with GlpT in the C_i conformation, with its substrate-binding site accessible only to P_i on the cytoplasmic side of the membrane (**Figure 2c**). In this state, the antiporter is in a stable configuration of low energy. The conformational change from the C_i state to the C_o state (or vice versa) of unloaded, substrate-free transporter over the energy barrier is slow compared with the substrate-bound form (**Figure 2c**). Thus, one may derive from Equation 1,

$$K_m = \frac{(k_{-1}/k_1)(k_3 + k_{-2} + k_2(k_3/k_{-1}))}{(k_3 + k_{-2} + k_2)}, \quad 2.$$

where $K_d = (k_{-1}/k_1)$ and K_m is dependent on the ratios of (k_{-1}/k_1) and (k_3/k_{-1}).

Biochemical experiments that exploit the quenching of intrinsic tryptophan fluorescence of GlpT upon binding of G3P substrate in detergent solution give an apparent substrate-binding dissociation constant of ~0.8 μM (44). Significantly, the binding of substrate to GlpT is independent of temperature, as is the Michaelis constant for transport, K_m, calculated from assays performed using both *E. coli* whole cells and purified GlpT reconstituted into proteoliposomes (44). The next kinetic event (the k_2/k_{-2} step) represents the interconversion of the protein from the C_i-P_i to the C_o-P_i state, which is accompanied by exposure of the substrate-binding site to the opposite side of the membrane (**Figure 2c**). This step encompasses the transition state and is the rate-limiting step of the whole transport process (18). Crucially, transport assays showed that this step is highly temperature dependent, with the maximal velocity of transport, V_{max}, increasing with temperature (44). This led to the conclusion that release of binding energy alone is insufficient to drive the C_i-P_i to C_o-P_i transition, and that an energy input of roughly 35 kJ mol^{-1} (at 37°C) is derived as thermal energy from the environment (**Figure 2d**). The final step of the half-reaction described in Equation 1 involves release of bound substrate into the periplasmic compartment and presentation of the binding site to countersubstrate (G3P in the case of GlpT) to allow the antiport reaction to recur by reversal of the three steps described above (**Figure 2c**). In the periplasm, the relatively low affinity of the transporter for P_i (10, 26) allows its replacement by G3P, whereas in the cytoplasm P_i replaces G3P at the binding site owing to its much higher cellular concentration (4 mM under nongrowing conditions) (78), thus driving the entire transport reaction. This mechanism effectively allows substrate-bound GlpT to randomly sample the C_i and C_o states via simple Brownian motion; the net direction of transport is then determined by substrate association/dissociation reactions in accordance with the rules of mass action at either surface. We believe that all antiporters of the MFS likely utilize this same mechanistic strategy.

Substrate Exchange Stoichiometry

Even today, ambiguity exists as to the actual stoichiometry of exchange in MFS antiporters. For technical reasons discussed by Maloney et al. (53), investigations into the stoichiometry of G6P:phosphate exchange were performed on membrane vesicles of *Streptococcus lactis*. A striking finding from this work was the effect of pH on the substrate-exchange process (9, 52). First, a reduction in pH slowed the reaction owing to a velocity effect. Dropping the pH from 7 to 5.2 reduced the V_{max} 10-fold, whereas the K_m remained constant for both heterologous and homologous exchange reactions, implying that monobasic and dibasic sugars were equally effective as substrates. Second, the pH had a direct impact on stoichiometry, with a 2:1 P_i:sugar phosphate exchange measured at pH 7, and a 1:1 exchange measured at pH 5.2 (9). On the basis of these findings, along with the presumption that monovalent P_i but not divalent P_i is the preferred exchange substrate, a model was forwarded that incorporated a bifunctional substrate-binding site that accepted either two monovalent sugar phosphates ($2HG6P^{1-}$) or a single divalent anion ($1G6P^{2-}$), but not both (53). This seemed to reconcile the observed 2:1 ($2P_i$ versus $1G6P$) stoichiometry at alkaline pH and a 1:1 ($2P_i^{1-}$ versus $2G6P^{1-}$) ratio at acid pH. The arrival of the crystal structure of GlpT casts some doubt on these interpretations because there does not appear to be sufficient room in the translocation pore for two molecules of P_i, much less G3P. Nevertheless, because the GlpT structure has no bound substrate (35), the issue remains unsettled, and it will be of great interest to reexamine how these transporters select among the various ionic forms of their substrates. For example, the finding that H165 in GlpT is likely to accept a proton during the binding of P_i (43) immediately raises the question of whether the proton is abstracted from the medium or from P_i and how this might affect selectivity among mono- and dibasic forms of P_i.

FINAL CONSIDERATIONS

As a final point in this review, it is worth noting that redundancy with respect to membrane transporters is built into the cell. Often, the cell contains both primary and secondary active transporters for the same substrate. As the largest primary transporter superfamily in the cell (61), the ATP-binding cassette proteins, like their MFS counterparts, display diverse substrate specificity. Although secondary active transporters do not establish as steep a substrate gradient as primary transporters do, they may be viewed as energetically more frugal. This allows the cell to modulate expression of each type of transporter for a particular substrate according to cellular and environmental needs (46).

SUMMARY POINTS

1. All MFS proteins share essentially the same 3D structure, with two domains that saddle the substrate translocation pore.

2. Conservation of the blueprint for the fold of MFS antiporters implies that the diversity of substrate preference seen in the MFS is a result of changes in only a few residues in the substrate-binding site and translocation pathway.

3. Release of intrinsic binding energy upon binding of substrate lowers the activation energy barrier sufficiently to allow Brownian motion to drive substrate translocation across the membrane.

4. Substrate translocation, via the rocker-switch mechanism, is accompanied by large, global conformational change of the protein that involves formation and breaking of inter- and intrahelical salt bridges.

FUTURE ISSUES

1. How many conformational states of the transporter are there? There is an urgent need to define the transition states in conformational space. Characterization of these states and the ability to achieve an individual state in vitro are the main hurdles to realizing a full structural characterization of any secondary active transporter.

2. We eagerly await the arrival of high-resolution structures of an MFS antiporter in the C_o conformation and in the substrate-bound C_i conformation.

3. Are the same residues involved in binding substrate in both the C_i and C_o conformations, and what are the exact determinants of substrate specificity and ionic selectivity?

DISCLOSURE STATEMENT

The authors are not aware of any biases that might be perceived as affecting the objectivity of this review.

ACKNOWLEDGMENTS

CJL and DNW would like to thank past members of the lab; Jonas Almqvist, Manfred Auer, Yafei Huang, Myon Jin Kim, Joanne Lemieux, Xiaodan Li, Jinmei Song, Céline Soudant, and Xiaorong Zhang for their dedicated and enthusiastic work on GlpT. The authors' work was supported in part by NIH grants DK-053973 (to DNW) and GM-24195 (to PCM).

LITERATURE CITED

1. Abramson J, Kaback HR, Iwata S. 2004. Structural comparison of lactose permease and the glycerol-3-phosphate antiporter: members of the major facilitator superfamily. *Curr. Opin. Struct. Biol.* 14:413–19

2. Abramson J, Smirnova I, Kasho V, Verner G, Kaback HR, Iwata S. 2003. Structure and mechanism of the lactose permease of *Escherichia coli*. *Science* 301:610–15

3. Adams MD, Celniker SE, Holt RA, Evans CA, Gocayne JD, et al. 2000. The genome sequence of *Drosophila melanogaster*. *Science* 287:2185–95

4. Almqvist J, Huang Y, Hovmöller S, Wang DN. 2004. Homology modeling of the human microsomal glucose-6-phosphate transporter explains the mutations that cause the glycogen storage disease type Ib. *Biochemistry* 43:9289–97

5. Ambudkar SV, Anantharam V, Maloney PC. 1990. UhpT, the sugar phosphate antiporter of *Escherichia coli*, functions as a monomer. *J. Biol. Chem.* 265:12287–92

6. Ambudkar SV, Larson TJ, Maloney PC. 1986. Reconstitution of sugar phosphate transport systems of *Escherichia coli*. *J. Biol. Chem.* 261:9083–86

7. Ambudkar SV, Maloney PC. 1984. Characterization of phosphate:hexose 6-phosphate antiport in membrane vesicles of *Streptococcus lactis*. *J. Biol. Chem.* 259:12576–85

8. Ambudkar SV, Maloney PC. 1985. Reconstitution of phosphate-linked antiport from *Streptococcus lactis*. *Biochem. Biophys. Res. Commun.* 129:568–75

9. Ambudkar SV, Sonna LA, Maloney PC. 1986. Variable stoichiometry of phosphate-linked anion exchange in *Streptococcus lactis*: implications for the mechanism of sugar phosphate transport by bacteria. *Proc. Natl. Acad. Sci. USA* 83:280–84

10. Auer M, Kim MJ, Lemieux MJ, Villa A, Song J, et al. 2001. High-yield expression and functional analysis of *Escherichia coli* glycerol-3-phosphate transporter. *Biochemistry* 40:6628–35

11. Bartoloni L, Wattenhofer M, Kudoh J, Berry A, Shibuya K, et al. 2000. Cloning and characterization of a putative human glycerol 3-phosphate permease gene (SLC37A1 or G3PP) on 21q22.3: mutation analysis in two candidate phenotypes, DFNB10 and a glycerol kinase deficiency. *Genomics* 70:190–200

12. Bevan M, Bancroft I, Bent E, Love K, Goodman H, et al. 1998. Analysis of 1.9 Mb of contiguous sequence from chromosome 4 of *Arabidopsis thaliana*. *Nature* 391:485–88

13. Chang AB, Lin R, Studley WK, Tran CV, Saier MH. 2004. Phylogeny as a guide to structure and function of membrane transport proteins. *Mol. Membr. Biol.* 21:171–81

14. Chen LY, Pan CJ, Shieh JJ, Chou JY. 2002. Structure-function analysis of the glucose-6-phosphate transporter deficient in glycogen storage disease type Ib. *Hum. Mol. Genet.* 11:3199–207

15. Chen XZ, Steel A, Hediger MA. 2000. Functional roles of histidine and tyrosine residues in the H^+-peptide transporter PepT1. *Biochem. Biophys. Res. Comm.* 272:726–30

16. Dougherty DA. 1996. Cation-π interactions in chemistry and biology: a new view of benzene, Phe, Tyr, and Trp. *Science* 271:163–68

17. Fann M, Davies AH, Varadhachary A, Kuroda T, Sevier C, et al. 1998. Identification of two essential arginine residues in UhpT, the sugar phosphate antiporter of *Escherichia coli*. *J. Membr. Biol.* 164:187–95

18. Fann MC, Maloney PC. 1998. Functional symmetry of UhpT, the sugar phosphate transporter of *Escherichia coli*. *J. Biol. Chem.* 273:33735–40

19. Fisher RB, Parsons DS. 1953. Glucose movements across the wall of the rat small intestine. *J. Physiol.* 119:210–23

20. Goldschmidt S. 1921. On the mechanism of absorption from the intestine. *Physiol. Rev.* 1:421–53

21. Guan L, Kaback HR. 2006. Lessons from lactose permease. *Annu. Rev. Biophys. Biomol. Struct.* 35:67–91

22. Guan L, Mirza O, Verner G, Iwata S, Kaback HR. 2007. Structural determination of wild-type lactose permease. *Proc. Natl. Acad. Sci. USA* 104:15294–98

23. Hall JA, Fann MC, Maloney PC. 1999. Altered substrate selectivity in a mutant of an intrahelical salt bridge in UhpT, the sugar phosphate carrier of *Escherichia coli*. *J. Biol. Chem.* 274:6148–53

24. Hall JA, Maloney PC. 2001. Transmembrane segment 11 of UhpT, the sugar phosphate carrier of *Escherichia coli*, is an alpha-helix that carries determinants of substrate selectivity. *J. Biol. Chem.* 276:25107–13

25. Hall JA, Maloney PC. 2005. Altered oxyanion selectivity in mutants of UhpT, the Pi-linked sugar phosphate carrier of *Escherichia coli*. *J. Biol. Chem.* 280:3376–81

26. Hayashi S, Koch JP, Lin ECC. 1964. Active transport of L-α-glycerophosphate in *Escherichia coli*. *J. Biol. Chem.* 239:3098–105

27. Heymann JA, Sarker R, Hirai T, Shi D, Milne JL, et al. 2001. Projection structure and molecular architecture of OxlT, a bacterial membrane transporter. *EMBO J.* 20:4408–13

28. Hildebrand PW, Günther S, Goede A, Forrest L, Frömmel C, Preissner R. 2008. Hydrogen-bonding and packing features of membrane proteins: functional implications. *Biophys. J.* 94:1945–53

29. Hildebrand PW, Rother K, Goede A, Preissner R, Frommel C. 2005. Molecular packing and packing defects in helical membrane proteins. *Biophys. J.* 88:1970–77

30. Hirai T, Heymann JA, Shi D, Sarker R, Maloney PC, Subramaniam S. 2002. Three-dimensional structure of a bacterial oxalate transporter. *Nat. Struct. Biol.* 9:597–600

31. Hirai T, Heymann JAW, Shi D, Subramaniam S. 2004. Comparative structural analysis of the "symmetric", substrate-bound state of the oxalate transporter OxlT with the "cytoplasmically-open" state of the MFS transporters GlpT and LacY. *Biophys. J.* 86:611–11

32. Hirai T, Subramaniam S. 2004. Structure and transport mechanism of the bacterial oxalate transporter OxlT. *Biophys. J.* 87:3600–7

33. Hodgkin AL, Huxley AF. 1952. Currents carried by sodium and potassium ions through the membrane of the giant axon of Loligo. *J. Physiol.* 116:449–72

34. Hong H, Szabo G, Tamm LK. 2006. Electrostatic couplings in OmpA ion-channel gating suggest a mechanism for pore opening. *Nat. Chem. Biol.* 2:627–35

35. Huang Y, Lemieux MJ, Song J, Auer M, Wang DN. 2003. Structure and mechanism of the glycerol-3-phosphate transporter from *Escherichia coli*. *Science* 301:616–20

36. Jencks WP. 1980. The utilization of binding energy in coupled vectorial processes. *Adv. Enzymol. Relat. Areas Mol. Biol.* 51:75–106

37. Jia W, Cole JA. 2005. Nitrate and nitrite transport in *Escherichia coli*. *Biochem. Soc. Trans.* 33:159–61

38. Kaback HR. 2005. Structure and mechanism of the lactose permease. *C. R. Biol.* 328:557–67

39. Kaback HR, Wu J. 1997. From membrane to molecule to the third amino acid from the left with a membrane transport protein. *Q. Rev. Biophys.* 30:333–64

40. Klingenberg M. 2005. Ligand-protein interaction in biomembrane carriers. The induced transition fit of transport catalysis. *Biochemistry* 44:8563–70

41. Krupka RM. 1989. Role of substrate binding forces in exchange-only transport systems. I. Transition-state theory. *J. Membr. Biol.* 109:151–58

42. Krupka RM. 1989. Role of substrate binding forces in exchange-only transport systems. II. Implications for the mechanism of the anion exchanger of red cells. *J. Membr. Biol.* 109:159–71

43. Law CJ, Almqvist J, Bernstein A, Goetz RM, Huang Y, Soudant C, Laaksonen A, Hovmöller S, Wang DN. 2008. Salt-bridge dynamics control substrate-induced conformational change in the membrane transporter GlpT. *J. Mol. Biol.* 378:826–37

44. Law CJ, Yang Q, Soudant C, Maloney PC, Wang DN. 2007. Kinetic evidence is consistent with the rocker-switch mechanism of membrane transport by GlpT. *Biochemistry* 46:12190–97

45. Lemieux MJ. 2007. Eukaryotic major facilitator superfamily transporter modeling based on the prokaryotic GlpT crystal structure. *Mol. Membr. Biol.* 24:333–41

46. Lemieux MJ, Huang Y, Wang DN. 2004. Glycerol-3-phosphate transporter of *Escherichia coli*: structure, function and regulation. *Res. Microbiol.* 155:623–29

47. Lemieux MJ, Huang Y, Wang DN. 2004. Structural basis of substrate translocation by the *Escherichia coli* glycerol-3-phosphate transporter: a member of the major facilitator superfamily. *Curr. Opin. Struct. Biol.* 14:405–12

48. Lin ECC. 1976. Glycerol dissimilation and its regulation in bacteria. *Annu. Rev. Microbiol.* 30:535–78

49. Maiden MCJ, Davis EO, Baldwin SA, Moore DCM, Henderson PJF. 1987. Mammalian and bacterial sugar-transport proteins are homologous. *Nature* 325:641–43

50. Majumdar DS, Smirnova I, Kasho V, Nir E, Kong XX, et al. 2007. Single-molecule FRET reveals sugar-induced conformational dynamics in LacY. *Proc. Natl. Acad. Sci. USA* 104:12640–45

51. Maloney PC. 1994. Bacterial transporters. *Curr. Opin. Cell. Biol.* 6:571–82

52. Maloney PC, Ambudkar SV. 1985. Anion exchange in bacteria. Variable stoichiometry of phosphate: sugar 6-phosphate antiport. *Ann. N. Y. Acad. Sci.* 456:245–47

53. Maloney PC, Ambudkar SV, Anatharam V, Sonna LA, Varadhachary A. 1990. Anion-exchange mechanisms in bacteria. *Microbiol. Rev.* 54:1–17

54. Mitchell P. 1953. Transport of phosphate across the surface of *Micrococcus pyogenes*; nature of the cell inorganic phosphate. *J. Gen. Microbiol.* 9:273–87

55. Mitchell P. 1954. Transport of phosphate across the osmotic barrier of *Micrococcus* pyogenes; specificity and kinetics. *J. Gen. Microbiol.* 11:73–82

56. Mitchell P. 1991. Foundations of vectorial metabolism and osmochemistry. *Biosci. Rep.* 11:387–435

57. Mueckler M, Makepeace C. 2006. Transmembrane segment 12 of the Glut1 glucose transporter is an outer helix and is not directly involved in the transport mechanism. *J. Biol. Chem.* 281:36993–98

58. Padan E, Sarkar HK, Viitanen PV, Poonian MS, Kaback HR. 1985. Site-specific mutagenesis of histidine residues in the *lac* permease of *Escherichia coli*. *Proc. Natl. Acad. Sci. USA* 82:6765–68

59. Pao SS, Paulsen IT, Saier MH. 1998. Major facilitator superfamily. *Microbiol. Mol. Biol. Rev.* 62:1–34

60. Patlak C. 1957. Contributions to the theory of active transport. II. The gate type noncarrier mechanism and generalizations concerning tracer flow, efficiency, and measurement of energy expenditure. *Bull. Math. Biophys.* 19:209–35

61. Paulsen IT, Nguyen L, Sliwinski MK, Rabus R, Saier MH. 2000. Microbial genome analyses: comparative transport capabilities in eighteen prokaryotes. *J. Mol. Biol.* 301:75–100

62. Perry JL, Dembla-Rajpal N, Hall LA, Pritchard JB. 2006. A three-dimensional model of human organic anion transporter 1: aromatic amino acids required for substrate transport. *J. Biol. Chem.* 281:38071–79

63. Rao NN, Roberts MF, Torriani A, Yashphe J. 1993. Effect of glpT and glpD mutations on expression of the phoA gene in *Escherichia coli*. *J. Bacteriol.* 175:74–79

64. Ren Q, Paulsen IT. 2005. Comparative analyses of fundamental differences in membrane transport capabilities in prokaryotes and eukaryotes. *PLoS Comput. Biol.* 1:e27

65. Ren Q, Paulsen IT. 2007. Large-scale comparative genomic analyses of cytoplasmic membrane transport systems in prokaryotes. *J. Mol. Microbiol. Biotechnol.* 12:165–79

66. Safferling M, Griffith H, Jin J, Sharp J, De Jesus M, et al. 2003. The TetL tetracycline efflux protein from *Bacillus subtilis* is a dimer in the membrane and in detergent solution. *Biochemistry* 42:13969–76

67. Saier MH. 2003. Tracing pathways of transport protein evolution. *Mol. Microbiol.* 48:1145–56

68. Saier MH Jr, Beatty JT, Goffeau A, Harley KT, Heijne WH, et al. 1999. The major facilitator superfamily. *J. Mol. Microbiol. Biotechnol.* 1:257–79

69. Schneider F. 1978. Histidine in enzyme active centers. *Angew. Chem. Int. Ed. Engl.* 17:583–92

70. Sigal N, Vardy E, Molshanski-Mor S, Eitan A, Pilpel Y, et al. 2005. 3D model of the *Escherichia coli* multidrug transporter MdfA reveals an essential membrane-embedded positive charge. *Biochemistry* 44:14870–80

71. Smirnova I, Kasho V, Choe JY, Altenbach C, Hubbell WL, Kaback HR. 2007. Sugar binding induces an outward-facing conformation of LacY. *Proc. Natl. Acad. Sci. USA* 104:16504–9

72. Takahashi Y, Miyata M, Zheng P, Imazato T, Horwitz A, Smith JD. 2000. Identification of cAMP analogue inducible genes in RAW264 macrophages. *Biochim. Biophys. Acta* 1492:385–94

73. Tanford C. 1982. Simple model for the chemical potential change of a transported ion in active transport. *Proc. Natl. Acad. Sci. USA* 79:2882–84

74. Tanford C. 1983. Mechanism of free energy coupling in active transport. *Annu. Rev. Biochem.* 52:379–409

75. Unkles SE, Rouch DA, Wang Y, Siddiqi MY, Glass AD, Kinghorn JR. 2004. Two perfectly conserved arginine residues are required for substrate binding in a high-affinity nitrate transporter. *Proc. Natl. Acad. Sci. USA* 101:17549–54

76. Veenhoff LM, Heuberger EH, Poolman B. 2001. The lactose transport protein is a cooperative dimer with two sugar translocation pathways. *EMBO J.* 20:3056–62

77. Vidavar GA. 1966. Inhibition of parallel flux and augmentation of counter flux shown by transport models not involving a mobile carrier. *J. Theor. Biol.* 10:301–6

78. Vink R, Bendall MR, Simpson SJ, Rogers PJ. 1984. Estimation of H^+ to adenosine 5′-triphosphate stoichiometry of *Escherichia coli* ATP synthase using ^{31}P NMR. *Biochemistry* 23:3667–75

79. Wang X, Sarker RI, Maloney PC. 2006. Analysis of substrate-binding elements in OxlT, the oxalate:formate antiporter of *Oxalobacter formigenes*. *Biochemistry* 45:10344–50

80. West IC. 1997. Ligand conduction and the gated-pore mechanism of transmembrane transport. *Biochim. Biophys. Acta* 1331:213–34

81. Widdas WF. 1952. Inability of diffusion to account for placental glucose transfer in the sheep and consideration of the kinetics of a possible carrier transfer. *J. Physiol.* 118:23–39

82. Widdas WF. 1953. Kinetics of glucose transfer across the human erythrocyte membrane. *J. Physiol.* 120:23–24

83. Yang Q, Wang X, Ye L, Mentrikoski M, Mohammadi E, et al. 2005. Experimental tests of a homology model for OxlT, the oxalate transporter of *Oxalobacter formigenes*. *Proc. Natl. Acad. Sci. USA* 102:8513–18

84. Yin Y, He X, Szewczyk P, Nguyen T, Chang G. 2006. Structure of the multidrug transporter EmrD from *Escherichia coli*. *Science* 312:741–44

Evolutionary History and Phylogeography of Human Viruses

Edward C. Holmes

Center for Infectious Disease Dynamics, Department of Biology, The Pennsylvania State University, University Park, Pennsylvania 16802; email: ech15@psu.edu

Fogarty International Center, National Institutes of Health, Bethesda, Maryland 20892

Annu. Rev. Microbiol. 2008. 62:307–28

The *Annual Review of Microbiology* is online at micro.annualreviews.org

This article's doi:
10.1146/annurev.micro.62.081307.162912

Key Words

evolution, phylogeny, coalescent, emergence, epidemic

Abstract

Understanding the evolutionary history of human viruses, along with the factors that have shaped their spatial distributions, is one of the most active areas of study in the field of microbial evolution. I give an overview of our current knowledge of the genetic diversity of human viruses using comparative studies of viral populations, particularly those with RNA genomes, to highlight important generalities in the patterns and processes of viral evolution. Special emphasis is given to the major dichotomy between RNA and DNA viruses in their epidemiological dynamics and the different types of phylogeographic pattern exhibited by human viruses. I also consider a central paradox in studies of viral evolution: Although epidemiological theory predicts that RNA viruses have ancestries dating back millennia, with major ecological transitions facilitating their emergence, the genetic diversity in currently circulating viral populations has a far more recent ancestry, indicative of continual lineage turnover.

Contents

INTRODUCTION

The study of the origins, emergence, and spread of viral infections in human populations is one of the most active and productive areas of research in modern evolutionary biology (109). This success is due in large part to the rapid rate at which many viruses evolve, particularly those with RNA genomes, allowing genetic variation to accumulate within an epidemiological time frame and thus adding power to phylogenetic analyses (30, 82). In more formal terms, the rapid rate of evolutionary change in RNA viruses means that the epidemiological and ecological processes that shape their genetic diversity act on approximately the same timescale as mutations fixed in viral populations (55). Consequently, the genetic variation generated by RNA viruses can be used to infer the patterns, processes, and dynamics of viral evolution, providing a unique molecular per-

Phylogeography: the study of biogeography revealed by the comparison of phylogenetic trees of populations with the geographic distribution of each population

spective on their ancestry and mechanisms of change (82). Indeed, the rapidity of RNA virus evolution is such that phylogenetic relationships can often be resolved among isolates that have been sampled only days apart (24, 139), providing data on questions of forensic importance (26). Although rather less evolutionary work has been undertaken on the DNA viruses that infect humans, particularly as their lower rates of nucleotide substitution mean that evolutionary change cannot normally be observed in real time, gene and genome sequence data from DNA viruses are growing in abundance and provide a powerful means by which to reconstruct phylogenetic history and reveal the mechanisms of viral evolution (39).

I review our current understanding of the evolutionary history and phylogeography of human viral infections. Because RNA viruses are the most common class of viral pathogen in humans, and for which most genome sequence data are available, I necessarily devote most attention to detailing their evolution. Further, rather than considering case studies of individual viruses, which are relatively commonplace, I take a broader perspective, reviewing the general properties that underlie viral evolution.

The study of viral phylogeography and evolution is not only of historical significance. By revealing the rules of viral evolution, it might also be possible to shed light on one of the most important topics in disease epidemiology: predicting what new infections, from what reservoir species, and in what locations will emerge and spread in human populations in the future (66). As emergence is in reality only one aspect of the evolutionary process exhibited by viruses, understanding the rules of viral evolution must in turn give insight into the mechanics of emergence. I therefore highlight a number of important and recurrent themes in the evolutionary biology of human viruses. (*a*) There is a fundamental division between viruses that cause acute infections, frequently jump species boundaries, and evolve rapidly, and viruses that cause persistent infections, tend to codiverge with their host species over extended time periods and evolve slowly, broadly mirroring the division between RNA and DNA viruses. (*b*) The spatial (geographic) patterns exhibited by human RNA viruses follow distinct patterns that can be distinguished through the techniques of molecular phylogenetics. (*c*) Major transitions in the ecological history of human populations have likely had a profound impact on the evolution and emergence of viral infections, although a rapid turnover of genetic diversity means that only recent ecological events have left their signature in viral genome sequences.

POPULATION DYNAMICS AND GENETICS

Although the work described in this paper builds upon a variety of disciplines, from virology to ecology, and from phylogenetics to population genetics, its foundations lay with two research fields that have different intellectual histories but have converged on the same end point. The first is the epidemiological study of the population dynamics of infectious disease, set within the framework established by the seminal work of Anderson & May (4) and Nowak & May (87). The most fundamental unit of measurement in this field is the basic reproduction number, R_0, defined as the number of secondary cases produced when a pathogen is introduced into an entirely susceptible population. As individuals become infected and then recover or die, the proportion of susceptible hosts declines along with the number of secondary cases per infection, a quantity denoted R. If $R < 1$, as is the case for most viruses considered emergent and that result in only spillover infections, an infection will not cause a major epidemic (71). However, when $R > 1$, a new infection has the potential for widespread epidemic transmission. Consequently, understanding the conditions by which R evolves from less than to greater than unity is a key aspect in understanding disease emergence (5), and accurately measuring R itself can be thought of as the cornerstone of modern-day infectious disease epidemiology. Another epidemiological concept of importance for this review article is the critical community size

(CCS), which can be defined as the minimum number of individuals in a population required for a pathogen to persist in a population (i.e., to attain $R > 1$). As such, the CCS reflects the ongoing societal and ecological changes that have occurred within human populations since their origin.

The second discipline of importance for this paper is the branch of population genetics referred to as coalescent theory. The aim of coalescent theory is to infer key evolutionary processes, particularly aspects of population demography (i.e., rates of population growth and decline), from the distribution of branching (coalescent) events in phylogenetic trees of isolates sampled from within a single species (31, 67, 99). Although all such analyses inevitably entail a number of important assumptions, coalescent theory represents a powerful way by which to describe the processes that shape the genetic variation within viral populations and has led to major insights into viral population biology (72, 97). Most importantly, coalescent theory allows the estimation of R directly from gene sequence data, providing a natural link between the evolutionary analysis of viral genome sequences and the epidemiology of infectious disease (48, 97).

COMPARING RNA AND DNA VIRUSES

Perhaps the simplest and most powerful generality relating to the evolutionary history of human viruses is the division between viruses that frequently jump species boundaries and viruses that are passed on vertically between species over longer stretches of evolutionary time (**Table 1**). In general terms, this mirrors the division between RNA and DNA viruses: the former tend to (but not always) establish only acute infections in their hosts and evolve by a mechanism of cross-species transmission, and the latter often cause persistent (chronic) infections in their hosts and are more likely to have evolved by a process of long-term virus-host codivergence, in which the evolutionary history of the virus tracks that of its host species over many

Basic reproduction number (R_0): the number of secondary cases produced when a virus or other pathogen is newly introduced into an entirely susceptible population

Spillover infection: transfer of a pathogen from one species to another without sustained onward transmission in the new species

CCS: critical community size

Coalescent: population genetic theory that links the branching times on a phylogenetic tree of individuals sampled from a population with the demographic history of that population

Codivergence: the parallel diversification (including speciation) of a pathogen and its host, inferred when there is congruence between the phylogenetic trees and divergence times of each

Table 1 Major evolutionary and epidemiological properties of RNA and DNA viruses

Property	RNA virus-like	DNA viruses
Genome type	RNA, ssDNA	dsDNA
Genome size	Small	Small or large
Evolutionary rate	High (10^{-3} to 10^{-5} subs/site/year)	Low (10^{-7} to 10^{-9} subs/site/year)
Infection time	Generally acute	Generally persistent
Evolutionary mode	Cross-species transmission	Long-term codivergence
Transmission mode	Aerosol, body fluid, fecal-oral, vector-borne	Sexual, vertical
Critical community size	Large	Small
Virulence	Can be high	Often low

millions of years. The process of host jumping is of particular importance because it also the principal mechanism of viral emergence, with numerous examples, including HIV and SARS (23, 134), documented in recent years. Understanding the evolution of cross-species transmission is therefore fundamental to understanding the nature of viral emergence, although in many cases, including high-prevalent infections such as hepatitis C virus, is not clear which animal species have acted as the reservoir populations for viruses observed in humans. More generally, the ability of viruses to jump between hosts from different species can be considered a natural extension of the normal mode of transmission in most RNA viruses, in which viruses move horizontally among members of the same host species.

The central evolutionary division between host-jumping RNA viruses and codiverging DNA viruses correlates with a number of other key characteristics of these viral pathogens, particularly their mode of transmission, their virulence, and their rates of evolutionary change (**Table 1**). In brief, RNA viruses tend to be transmitted in a horizontal manner by aerosols, body fluids, fecal material, or vectors; can result in high virulence; and evolve rapidly. The end result of occupying this part of evolutionary parameter space is that acute RNA viruses often require large and well-connected host populations to survive: Any reduction in the number of susceptible hosts to below the CCS threshold results in $R < 1$ and hence in local extinc-

tion. In marked contrast, chronic DNA viruses tend to be transmitted either vertically (that is, between mother and offspring) or sexually; are often of lower or delayed virulence, as the virus is required to remain in the host for extended time periods to ensure transmission; and tend to evolve more slowly. The evolution of persistence in viral infections may therefore require a mode of transmission—vertical or sexual—that facilitates long-term codivergence. Early evidence for this fundamental ecological division was studies noting that indigenous Amerindian populations, which tend to be small, do not endemically carry acute RNA viral infections because these host populations are not of sufficient size to allow R to exceed unity (15). In contrast, the same Amerindian populations frequently carried persistent pathogen infections, presumably because these require a smaller CCS to sustain themselves at $R > 1$.

The root cause of the division between RNA and DNA viruses most likely lies with their different rates of evolutionary change, itself a function of the differing frequency of mutation in these two groups. RNA viruses are characterized by a high intrinsic mutation rate—estimated at up to one mutation per genome, per replication—resulting from replication with either RNA-dependent RNA polymerase (RNA viruses) or reverse transcriptase (retroviruses), both of which lack capacity for proofreading or error correction (32, 34, 77). Such frequent mutation, combined with rapid rates of replication, provides RNA viruses with high overall rates of evolutionary change, such that on average each nucleotide site fixes one mutation every 10^3 to 10^4 years [equivalent to nucleotide substitution rates of 10^{-3} to 10^{-4} substitutions per site per year (subs/site/year) (51, 63)]. Notably, recent work has shown that small, single-stranded (ss)DNA viruses evolve at rates broadly similar to those of their RNA counterparts, even though they replicate using host DNA polymerase (9, 36, 106, 107). Although the precise mechanistic basis for these high rates of evolutionary change is unknown, other aspects of viral genome architecture and ecology, such as genome size and replication speed, must also be implicated in setting overall rates of evolutionary change (38). The rapid evolutionary dynamics of RNA viruses (and ssDNA viruses) provide them with the capacity to quickly generate the genetic variation required to evade both innate and adaptive host immunity and so sustain their transmission while only infecting each individual host for a few days. In contrast, double-stranded (ds)DNA viruses, which replicate using DNA polymerases that come with added repair mechanisms, have far lower mutation rates (33) and thus evolve many logs more slowly than their RNA counterparts. Such low rates of nucleotide substitution, in the region of 10^{-7} to 10^{-9} subs/site/year, have been observed in large dsDNA viruses such as the herpesviruses (79) and the poxviruses (60), as well as small dsDNA viruses such as the papillomaviruses (12).

Although the division between RNA and DNA viruses is a powerful one, exceptions exist. In particular, a number of RNA viruses have been proposed to cause persistent infections, codiverge with their host species, and therefore evolve many orders of magnitude slower than expected given their reliance on RNA replication. The most convincing example is provided by the retrovirus *Simian foamy virus* (SFV). In this case the match between the phylogenies of the virus and its (primate) hosts is so strong that it argues for a long association between them (118). Assuming codivergence, the substitution rate of SFV is approximately 10^{-8} subs/site/year, many orders of magnitude

lower than that observed in other retroviruses, and close to that reported for human mitochondrial DNA (118). This low rate is mostly likely a reflection of low rates of replication, so that the virus is effectively latent within hosts (80), rather than a reduced error rate, as SFV still replicates using the standard retroviral reverse transcriptase and genetic variation as has been observed within infected animals (105).

The pattern of long-term virus-host codivergence coupled with low rates of evolutionary change has also been proposed for two groups of viruses that replicate using RNA-dependent RNA polymerase: the rodent hantaviruses and the primate flaviviruses GBV-A and GBV-C. In the former, there is seemingly strong congruence between the phylogeny of the virus and its rodent hosts (94). Indeed, aside from SFV, this constitutes the best evidence for long-term codivergence in an RNA virus (62). However, despite this phylogenetic pattern, the notion that the hantaviruses evolve anomalously slowly is in conflict with their replication via RNA polymerase and that rapid evolutionary dynamics have been observed in other hantaviruses (85). Similar reservations apply to GBV-A/C, although in this case the virus has only been isolated from a small number of primate species (22). Further, studies of GBV-C evolution in the short-term, including within single individuals, have revealed substitution rates that are within the normal range proposed for RNA viruses (83, 104). As such, whether the rodent hantaviruses and GBV-A/C truly codiverge with their host species, and thereby evolve anomalously slowly, is still open to debate.

Although the host species specificity of RNA viruses was originally thought to constitute a strong signal for long-term codivergence, work in recent years has demonstrated that a variety of other epidemiological and ecological processes produce viral phylogenies that superficially match those of their host species, wrongly leading to the conclusion that there has been extensive codivergence. This is perhaps best documented for the primate lentiviruses, the most famous member of which is HIV. Initially, studies of primate lentivirus biodiversity indicated

ssDNA: single-stranded DNA

dsDNA: double-stranded DNA

SFV: *Simian foamy virus*

that each primate species, specifically African cercopithecoid monkeys, had its own phylogenetically distinct lentivirus, such that there was widespread species specificity, in turn implying that these viruses codiverged with their primate hosts over millions of years (2, 11, 13).

Although the timescale of the evolutionary history of the primate lentiviruses is still unclear, particularly since the discovery of endogenous lentiviruses (64), there is little compelling evidence for codivergence. For example, there is a lack of congruence between the phylogenies of the viruses and their African green monkey hosts (131). This is more dramatically demonstrated with the hominoid primates. Here, three closely related species, humans, chimpanzees, and gorillas, all harbor closely related lentiviruses, which without further investigation might suggest codivergence because they share a common ancestor some six million years ago. However, detailed phylogenetic analyses have shown that all three represent recent cross-species transmission events (7, 65, 124, 136). Hence, it is dangerous to draw strong conclusions on the antiquity and mode of viral evolution simply from an examination of current species distributions.

There are a number of reasons why a process of cross-species transmission might be expected to produce phylogenies suggestive of codivergence more often than expected by chance alone. First, an important generality from studies of viral emergence is that the closer the host species involved, the more likely that viruses successfully jump between them (25, 57). This is at least in part because cellular environments, particularly cell receptors, are likely to be similar among closely related species, and viable receptor binding is a major requirement for successful viral emergence (8). As a consequence, humans are more likely to acquire successful new viral infections from other primates rather than any other group of animals, although exceptions are commonplace, and the extent of ecological contact between species also plays a major role in determining the likelihood of emergence. Within a specific group of animals such as the primates, this relationship may mean

that viruses are more able to jump between their immediate than their more distant relatives, producing a pattern of pure cross-species transmission that effectively mirrors that of codivergence. Such a process might explain the current biodiversity of the primate lentiviruses (21). Second, if closely related host species tend to live sympatrically, which is to be expected under a model of sympatric speciation, then a process of cross-species transmission is again predicted to generate phylogenies that match those of their host species. As viruses are sampled from increasingly large numbers of animal species, we will obtain more accurate estimates of the relative frequency of cross-species transmission versus codivergence.

THE SHALLOW GENETIC DIVERSITY OF HUMAN RNA VIRUSES

One of the most intriguing observations stemming from phylogenetic studies of the evolutionary history of human RNA viruses is that their sampled genetic diversity tends to have a recent common ancestry, dating back only a few hundred years at most. For example, although there have been suggestions of epidemics of dengue virus (DENV) in humans for a thousand years (50), the age of the genetic diversity within each viral serotype estimated using even the most sophisticated molecular clock techniques is far more recent, dating back a few hundred years at most (37, 122). Given the widespread notion that RNA viruses must have been associated with human populations for far longer time periods, such shallow diversity at first seems paradoxical (54). There are, however, good reasons to suspect that these molecular clock estimates are accurate.

First, a common misunderstanding in studies of the timescale of viral evolution is that estimates of the age of genetic diversity stemming from phylogenetic studies inform on the absolute age of the virus. They do not. Rather, estimates of the age of genetic diversity simply indicate when the sampled lineages last shared a common ancestor. For example, while molecular clock estimates place the sampled diversity of measles at a few hundred years at most (95), the true age of measles virus can only be estimated by determining the time of divergence of the measles virus lineage from its closest relative (which available data suggest may be rinderpest virus). Hence, although measles virus has likely imposed a burden on human populations since the time of the first cities (19), the genetic diversity currently circulating in this virus has arisen far more recently.

Second, under simple population genetic theory and a neutral evolutionary process, rapidly evolving RNA viruses with short generation times are not expected to harbor ancient genetic diversity. As a model calculation, consider an RNA virus with a generation time (g) of 4 days (rather like influenza A virus) and an effective population size (N_e) of 10,000, which may approximate population sizes during epidemic troughs. Under these parameters, the mean time to the most recent common ancestor (TMRCA) expected under a process of random genetic drift in a haploid population is given by $2N_e g = 20,000 \times 4$ days $= {\sim}200$ years (although with a large variance). That this value approximates those seen in many human RNA viruses suggests that a large proportion of their genetic diversity turns over in a largely neutral manner (46).

Conversely, it is also possible that selective sweeps, such as those for novel antigenic and drug resistance alleles, infrequently purge genetic diversity from viral populations, thereby greatly reducing times of shared common ancestry. This process has been frequently described for human influenza A virus, for which strong immune-mediated selection pressure coupled with a large and well-mixed population and a highly infectious pathogen result in frequent lineage turnover, particularly in the viral hemagglutinin protein (43, 44). However, whether widespread selective sweeps commonly occur in other RNA viruses is open to question. Indeed, given the disjunct spatial distributions exhibited by many RNA viruses and indicative of weak population mixing, the ability of natural selection to purge genetic diver-

sity on a global basis appears limited but clearly represents an important area for future study.

THE PHYLOGEOGRAPHIC PATTERNS OF HUMAN RNA VIRUSES

That RNA viruses generally exhibit rapid evolutionary dynamics and shallow genetic diversity in turn means that their current phylogeography is largely shaped by the movement and growth of human populations. Consequently, human demography plays a major role in shaping the genetic structure of viral populations. Further, these spatial patterns, as well as estimates of key epidemiological parameters, can be recovered through detailed phylogenetic analysis in a discipline called phylodynamics (48). The central tenet of the phylodynamics approach is that specific epidemiological processes, particularly rates of population growth and decline, the extent of population subdivision, and the strength and form of natural selection, are written into gene sequences and can be recovered using a suite of phylogenetic techniques, with the coalescent paramount among them.

Studies of human RNA viruses have revealed a number of important epidemiological generalities. In particular, a variety of different phylogeographic patterns are observed in human viruses (**Figure 1**), although these are closely connected such that the barriers between them are often fluid. Five such patterns are as follows.

- No clear spatial structure. There is complex and even random mixing among isolates sampled from different geographic regions, indicative of frequent viral traffic among localities.
- Wave-like transmission. Viruses move outward from a central starting point, thereby producing a relatively simple relationship between geographical and genetic distance. This is most elegantly exemplified by the measles virus (47).
- Source-sink (or core-satellite) models. One or a limited number of geographical areas act as a source population,

Selective sweep: the reduction of genetic variation within a population due to the recent and strong action of natural selection on an individual nucleotide site and any linked sites

Phylodynamics: how genetic variation within pathogen populations is modulated by host immunity, transmission bottlenecks and epidemic dynamics to determine the range of phylogenetic structures observed at scales from individual hosts to entire populations

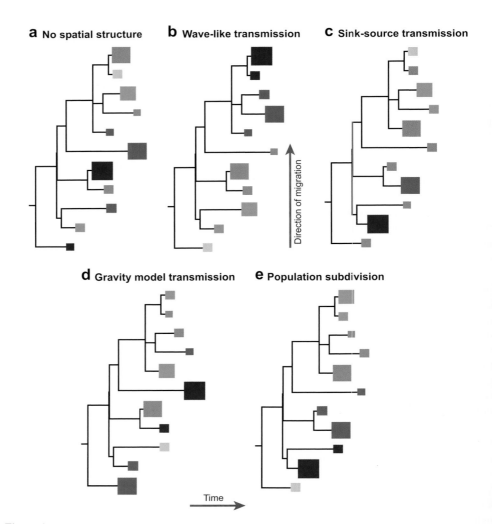

a No spatial structure

b Wave-like transmission

c Sink-source transmission

Direction of migration

d Gravity model transmission

e Population subdivision

Time

Figure 1

Schematic representation of the different phylogeographic patterns exhibited by human RNA viruses. Each population is represented by a different colored square, with the size of the square reflecting the size of the population. The darker the colors, the closer the populations are in space. The trees are also rooted, so that evolutionary time runs from left (oldest) to right (youngest). (*a*) No spatial structure, with a random mixing of populations. (*b*) Wave-like transmission, in which viruses move in a wave from the most distant to the nearest populations. (*c*) Source-sink transmission, in which the red populations act as the source for viruses seen in all other populations. (*d*) Gravity-like dynamics, in which viruses first move among the largest populations before diffusing to smaller populations. (*e*) Strong spatial (population) subdivision, with no clear evidence of migration among populations.

exporting viral lineages to other sink populations where viruses may only survive in the short-term (and perhaps with a strongly seasonal basis) and may generate transmission waves.

- Gravity-like dynamics. Patterns of viral transmission are driven by major population centers, which act as gravity attractors, perhaps following patterns of human work-flow (137).

■ Strong spatial subdivision. Different geographical localities are characterized by the presence of phylogenetically distinct viral isolates with little evidence of viral traffic among them. Although strong spatial subdivision is expected if a virus has been associated with humans for extended periods, it can also be established in only the recent past, given a combination of rapid evolution and strong founder effects, as seen in HIV (100).

A range of human viruses follow each of these patterns, reflecting the relative rates of viral gene flow, time of association with human populations, and the precise mode of transmission. It is also possible for single viruses to possess multiple phylogeographic patterns depending on the spatial scale under consideration. Relatively strong spatial subdivision is characteristic of the hepatitis C virus, in which genetic diversity is partitioned into a series of types and subtypes, each of which has a distinct geographical distribution and is often associated with particular risk groups (111). For example, populations of injecting drug users in industrialized nations are characterized by the presence of subtypes 1a and 3a, which have spread rapidly in these regions during the past 60 years (98, 111). In contrast, greater genetic diversity is present within African (types 1 and 2) and Asian (types 3 and 6) populations and may have been present for thousands of years. Conversely, the phylogeographic structure of human influenza A virus is characterized by the fluid movement of the virus among populations. On a global scale, influenza virus diversity is complex, with little spatial clustering, and reflects the frequent movement of viruses among localities (84). The global transmission of human influenza A virus might therefore follow a source-sink model. Under this scenario, viral transmission is highly seasonal in temperate regions, where influenza is strongly associated with the winter season, such that these geographical regions act as sink populations, with local viral extinction in the summer season.

Conversely, influenza has a more annual pattern of transmission in tropical regions, including Southeast Asia, which might act as global source populations (125).

Additional complexity is provided by observations that different spatial structures are apparent within individual localities, supporting both wave-like and gravity models. For example, in Brazil, influenza seems to move in a north-to-south traveling wave (3), whereas in the United States, influenza seems to satisfy the conditions of a gravity model, largely following the movement of people to and from their places of work (126). Given the obvious importance of socioeconomic factors in determining the spread of infectious diseases, gravity models are likely to apply to many human viral infections, although there are insufficient data to assess their general importance. Finally, wave-like transmission has also been proposed for the transmission of Ebola Zaire in Africa since 1976 (128), although the number of isolates currently available for analysis is limited. Crucially, all these patterns can be distinguished from gene sequence data using phylogenetic techniques (20, 101, 127); however, frequent recombination can greatly complicate lineage assignment (88), and additional epidemiological data, such as population sizes for gravity models, are often required. At an even more localized scale it may be possible to use molecular phylogenetics to determine which individuals in a population act as superspreaders (75).

The complexity of the phylogeographic patterns exhibited by human viruses, and how they vary by geographic scale, can be illustrated by DENV, the most common vector-borne viral infection of humans and an emerging pathogen for which gene sequence data are relatively abundant. The extensive genetic diversity in DENV has been described for over 20 years and comprises four highly divergent serotypes (DENV-1, -2, -3, and -4) that are divided into a series of phylogenetically defined subtypes (58, 102). However, more detailed phylogenetic analyses of DENV populations have revealed far more phylogeographic complexity.

Founder effect: the loss of genetic diversity when a population is formed in a new location by a small number of individuals from a larger population located elsewhere

First, DENV subtypes often have dif-fering spatial distributions, with some more widespread than others, indicating that both population subdivision and gene flow are important in the structuring genetic diversity in DENV. This is best documented in the case of DENV-2, of which two subtypes are apparently restricted to Southeast Asia and another to the Americas, whereas another subtype has a far wider distribution among tropical and subtropical localities (121). Whether these spatial patterns reflect different epidemic potentials (i.e., fitness) among subtypes is still uncertain (102).

Second, both serotypes and subtypes frequently cocirculate within the same geographic locality, which may lead to complex patterns of selective competition (141). For example, phylogenetic analyses have revealed a relatively high rate of lineage extinction and replacement within serotypes, generating periodic fluctuations in diversity that can have a major affect on population genetic structure (69, 113, 120, 133). Hence, it seems likely that the genetic structure of DENV populations is greatly influenced by the extent of cross-immunity among serotypes, both enhancing (42) and protective (1). Detailed phylogenetic analyses also hint at a possible distinction between endemic subtypes that have circulated within particular localities for extended time periods and epidemic subtypes that have spread rapidly through multiple populations. This is particularly apparent in DENV-1, for which one subtype has seemingly spread rapidly through continental regions of Southeast Asia, perhaps with wave-like dynamics, whereas islands of the Pacific and Melanesia are characterized by geographically distinct variants indicative of spatial subdivision over longer time periods (6).

Finally, the phylogenetic analyses of all four serotypes reveal that Southeast Asia harbors the greatest genetic diversity, with viral lineages sampled from this region often falling deep in phylogenetic trees. This suggests that Southeast Asia acts as a source population, generating the strains that then ignite epidemics elsewhere, including the Americas (103).

VIRUSES ARE POOR MARKERS OF HUMAN POPULATION HISTORY

Ever since the first gene sequences were generated for viral populations, attempts have been made to use these data to retrace the evolutionary history and demography of their human hosts. If the evolutionary history of a specific virus is characterized by strong virus-host co-divergence, then viruses may indeed represent valid phylogenetic markers, as elegantly shown in some animal populations (14). Similarly, it has also been proposed that viruses can provide valuable insights into host demographic history (14, 89). Further, because viruses, particularly those with RNA genomes, usually evolve faster than human nuclear and mitochondrial DNA, they will provide more phylogenetic resolution per nucleotide sequenced. As a dramatic case in point, it has been proposed that viruses constitute valid forensic tools for determining the ethnic background of identified cadavers (61). Although a number of viruses have been proposed as human population markers, three have proven most popular in this respect, JCV, GBV-C, and human papillomavirus (HPV). Similar proposals have also been made for various bacterial species, particularly *Helicobacter pylori* (74, 132).

Despite the obvious attractions, viruses make rather poor, and unnecessary, markers of human population history for a variety of reasons. First, the increased spread and reduced cost of human genome sequencing (78) mean that the use of viruses as surrogates is less important than it was only a few years ago. Second, with very few exceptions, all viruses to some extent move horizontally through populations (that is, without exclusive vertical transmission), a process that muddies the association between virus and host phylogeny. Although HPV is a sexually transmitted agent (and a common cause of cervical cancer), when human populations move the resultant admixture will necessarily reduce the extent of congruence between the virus and host phylogenies. Third, the evidence that human viruses carry a strong signal of host phylogeny is often debatable at best. For example, in the case of GBV-C, the evidence that this virus follows the history of human populations is largely based on the observation that viruses sampled from different human populations are phylogenetically distinct (92), without a robust demonstration that they accurately mirror human phylogenetic history and with a relatively small sample of human ethnic diversity. In these circumstances the use of viral gene sequences adds little to our understanding of human population history.

The uncertainties in the use of viruses as markers of human population history are reflected in the case of JCV, the virus perhaps most commonly used in this context. JCV-C is a small dsDNA virus that commonly infects human populations, attaining an estimated seroprevalence of 70%–90% globally (91). Initial studies suggested that the virus had a genetic structure compatible with the phylogenetic history of human populations, particularly that distinct viral populations (subtypes) were observed in individuals of African, Asian, and European origin (93, 117, 140). However, closer inspection reveals that the match between the virus and human phylogenies is weak at best, with evidence of extensive admixture, making the virus a worse marker than human mitochondrial DNA (108). For example, there is widespread mixing between Asian and European populations (i.e., European clade viruses contain JCV strains isolated from the Ainu, Nanai, and Koryak peoples of Asia), and the most divergent viruses (subtypes 1 and 4) were sampled from Europeans rather than Africans (68, 108). This overall phylogenetic incongruence notwithstanding, some evidence for a long antiquity of JCV in humans is provided by the grouping of some indigenous Asian and American populations (although viruses sampled from Native Americans are not monophyletic). Consequently, JCV represents a poor marker of human population

history, and the length of its association with humans is unknown. A similar uncertainty involves the rate of JCV evolution. If the theory of codivergence with human populations over a period of ~150,000 years is correct, then JCV evolves at a relatively low rate, in the realm of 10^{-7} subs/site/year, similar to large dsDNA viruses (53, 116). However, estimates of substitution rate based on the coalescent analysis of time-structured sequence data are far higher, never less than 10^{-6} subs/site/year, in which case the timescale of JCV evolution is far more recent (108). Which estimate correctly describes the evolutionary dynamics of JCV awaits further investigation.

MAJOR TRANSITIONS IN HUMAN ECOLOGY HAVE A PROFOUND INFLUENCE ON VIRAL EVOLUTION

The ultimate origins of RNA viruses are the source of much debate and little hard data (70); however, their wide species distributions indicate that RNA viruses have existed for millennia and that humans have suffered the burden of RNA viral infections throughout their evolutionary history. Indeed, it is reasonable to assume that the morbidity and mortality due to RNA viruses was one of the greatest challenges facing modern humans as they migrated from Africa and achieved global colonization. Similarly, as human ecology has changed through time, new niches for viral infection have been established. Broadly speaking, four such major transitions in the epidemiology of human viral infections can be postulated: (a) the establishment of farming; (b) the development of urbanization; (c) the rise of global travel; and (d) the modern human world, characterized by small-world dynamics, major changes in land use exemplified by extensive deforestation, and the occurrence of widespread immunodeficiency, together which might be thought of as the age of emergence.

In the case of the evolution of agriculture, the abandonment of the hunter-gatherer life style for the sedentary ways of farming brought humans in closer contact with animal species, increasing the likelihood of cross-species transmission and emergence, and also increased human population size, allowing acute RNA viruses to establish themselves for the first time in human populations (29). For example, it is possible that the ultimate origins of measles (humans) from rinderpest (cattle) occurred at this time. It should therefore not come as a surprise that paleopathological records indicate that the health of the first farmers was inferior to that of their hunter-gatherer ancestors (115). Similarly, the major effect of urbanization on the epidemiology of human viral infections is the increase of the number of susceptible hosts, as well as the contacts among them, thereby increasing R. Hence, the rise of urbanization may mark the point in time when acute RNA viruses were first able to sustain themselves in humans without continual replenishment from an animal reservoir population. A famous, and still highly illustrative, example is provided by measles virus, for which the CCS has been estimated between 250,000 and 500,000 (10). Such a large number of human hosts would not have been apparent until the rise of cities. However, despite the importance of ecological changes in disease emergence, there are few, if any, examples in which the molecular clock analysis of a human RNA virus has unequivocally demonstrated an ancestry that dates back to the rise of agriculture or of cities. Viral lineages turn over at a high rate, with a process of random genetic drift expected to result in more recent common ancestry.

Of all the major transitions in human disease ecology, perhaps that of most importance was the development of global travel, as this allowed the rapid dissemination of viruses to diverse geographical areas. It is the relative frequency of migration versus population isolation that is fundamental in shaping the phylogeographic structure of viral populations. Moreover, as widespread travel developed relatively recently, the history of human global travel is often written into viral genomes and frequently revealed in phylogenetic studies (55).

It is also likely that viruses spread quickly as soon as global trade networks were established. For example, although dengue is an archetypal tropical disease, with an ancestry that undoubtedly lies in the jungles of Africa or Asia, it was first described in Philadelphia in 1789, most likely a result of the slave trade bringing both infected hosts and vectors to the Americas (50). The influence of the slave trade on the global spread of human viral infections is demonstrated even more clearly in the case of *Yellow fever virus* (YFV). Some have claimed that yellow fever may have been resident in the New World long before European colonization, but recent molecular analyses provide compelling evidence that YFV moved from Africa to the Americas during the slave trade, with a subsequent east-to-west transmission wave across the Americas (18).

The way in which humans live today has greatly assisted the emergence and spread of new viral infections. A variety of ecological factors have facilitated the emergence and onward spread of new viral infections, including changes in agricultural practices and land use such as deforestation, famine, war, rapid global travel and the growth of mega-cities. Indeed, it is possible to find a direct ecological cause for virtually every modern emergent disease of humans (81). An important example is the epidemic of SARS coronavirus (SARS-CoV) that infected some 8000 people globally during 2003, resulting in a death toll of approximately 800. Detailed epidemiological investigations have revealed that the ultimate source of SARS-CoV is a group of related viruses that circulate widely in bat species (73). However, the spread of the virus in humans was most likely facilitated by exposure to secondarily infected civet cats on sale in markets in Southern China (49). The virus then spread rapidly in Hong Kong, assisted by local superspreaders, before disseminating on a wider scale by way of international air travel.

Another facet of modern human ecology that may greatly assist the emergence of human viral infections is widespread immunodeficiency. Current estimates place some

33 million people globally as carrying HIV, largely in sub-Saharan Africa, the majority of whom will go on to develop profound immunodeficiency manifest as full-blown AIDS (123). As well as having a major impact on morbidity and mortality, it is likely that such frequent immunodeficiency will allow the spread of other infectious agents that would have otherwise been cleared by healthy immune systems (129). For example, the global spread of AIDS has stimulated a resurgence in opportunistic pathogens such as *Mycobacterium tuberculosis*. Similarly, there is a growing body of evidence to suggest that normally acute RNA viral infections, such as influenza, present as persistent infections in the face of immunodeficiency (16, 40).

THE COMPLEX EVOLUTIONARY HISTORY OF HEPATITIS B VIRUS

As a coda to this paper, and to illustrate some of the topics described in detail above, I consider in more detail one human virus whose evolutionary history and phylogeography are remarkably complex—hepatitis B virus (HBV).

HBV (family *Hepadnaviridae*) is one of the most common and serious viral infections of humans. It is estimated that there are more than 350 million chronic carriers of the virus on a global basis, a significant percentage of whom will go on to develop serious liver diseases such as cirrhosis and hepatocellular carcinoma after many years of infection (135). At face value, this small dsDNA virus has all the attributes of one that has codiverged with human populations and other primate species over many millions of years (**Table 1**). HBV has a partially dsDNA genome, is frequently passed on by sexual transmission, such that it has a low CCS, and has been found in a diverse array of nonhuman primates, namely chimpanzees, gorillas, gibbons, orangutans, and woolly monkeys (a New World monkey). More divergent HBVs are found in some rodents and avian species. Not surprisingly, this phylogenetic background has led to suggestions that the evolutionary his-

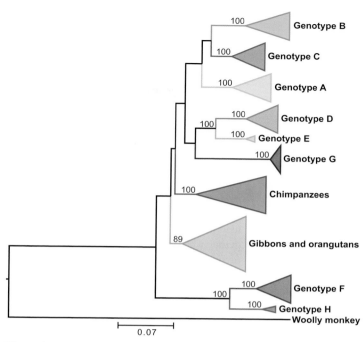

Figure 2

Evolutionary history of hepatitis B virus (HBV). The phylogenetic tree of 99 complete genome sequences of HBV was inferred using the maximum-likelihood method available in the PAUP* package (119) and assuming the best-fit model of nucleotide substitution determined by MODELTEST (96). For ease of representation, branches are collapsed and color-coded according to the human genotype (A–H) or host species from which they were sampled. Bootstrap values (>95%) are shown for key nodes, with a marked lack of phylogenetic resolution for the intergenotype branches. All horizontal branches are scaled according to the number of nucleotide substitutions per site, and the tree is rooted using the single sequence available from the New World woolly monkey.

tory of HBV is an ancient one, spanning millions of years (110).

The true age of HBV is still the subject of much debate; however, a closer inspection of available genome sequence data indicates that the evidence for long-term codivergence is weaker than might be expected. Although the virus is found in humans and our closest relatives, the phylogenetic pattern is evidently not that expected under simple codivergence. In particular, viruses sampled from nonhuman primates fall within the genetic diversity seen in humans, that is, the most divergent viruses are found in humans, not in other primates (**Figure 2**). Though this finding may in part be

HBV: hepatitis B virus

due to limited sampling coupled with a complex pattern of sequence evolution, the phylogenetic relationships of viruses sampled from apes better follow patterns of geographical distribution than of host phylogeny. Specifically, orangutans are more closely related to humans, chimpanzees, and gorillas than to gibbons, but the HBV isolates from gibbons and orangutans cluster closely together, reflecting their shared geographical distribution in Southeast Asia (114). In addition, the distribution of gibbon and orangutan HBV in Southeast Asia is strongly influenced by the precise place of sampling, with viruses sampled from particular localities (such as Borneo) grouping together whether they were taken from gibbons or orangutans (114). Such clustering by place over host species, which is mirrored by the phylogeographic patterns demonstrated by chimpanzees and gorillas in Africa, is strongly suggestive of recent cross-species transmission. Indeed, it is also striking that despite large-scale surveying, hepadnaviruses have yet to be detected in species of Old World monkeys (114). The occurrence of recombination events between human and ape viruses also argues for common viral gene flow among them (76, 138).

Overall, the most parsimonious hypothesis to be drawn from the phylogeny of the primate hepdnaviruses is that humans act as reservoir species for viruses found in other apes rather than vice versa, although far more data are needed to test this provocative hypothesis. Irrespective of the direction of viral emergence, the case of HBV again illustrates why species distributions alone should not be used to make inferences on the timescale and mode of viral evolution.

A lack of congruence between the phylogenetic histories of the host and virus is also seen in the genetic diversity of HBV in humans. Such diversity is strongly geographically structured, with a variety of genotypes (A–H), intergenotype recombinants, and subgenotypes exhibiting distinct geographical locations. Under the simple hypothesis of codivergence it might be expected that HBV isolates sampled from African patients would be the most divergent, reflecting the antiquity of human populations in this region. Unfortunately only a relatively small sample of African viruses have been analyzed to date, although the genetic diversity of HBV in Africa, most notably genotype E, is relatively shallow, suggestive of recent ancestry. In contrast, the most divergent human genotypes are observed in populations either from the New World (genotypes F and H) or from Asia (genotypes B and C), depending on the precise model of nucleotide substitution used in phylogenetic estimation (17). Detailed studies of the New World genotypes also indicate that HBV isolates sampled from Amerindians (genotype F) do not reflect the phylogenetic relationships of these host populations, again suggestive of more recent common ancestry (27). However, there are also intriguing phylogenetic relationships among HBV isolates sampled from Amerindian populations throughout North and South America (28), compatible with a far deep ancestry.

The notion that HBV has codiverged with animal populations over many millions of years is also incompatible with molecular clock estimates of the rate and timescale of viral evolution. The accurate estimation of nucleotide substitution rates in HBV is complicated by a number of factors, the most notable of which is the small size and overlapping nature of the HBV genome (~50% of the viral genome encodes more than a single protein product). This results in strong selective constraints, so that substitution rates measured in the short term may better reflect the intrinsic rate of mutation rather than the long-term rate of nucleotide substitution. Similarly, frequent recombination may complicate molecular clock estimates of divergence times (112, 142). Bearing such complexities in mind, most estimates of the rate of evolutionary change in HBV, estimated using a variety of techniques and data sets, are in the range of 10^{-5} subs/site/year and far higher than those observed in other dsDNA viruses (41, 52, 90). Although such a high rate of molecular evolution may seen anomalous for a DNA virus, it is expected given that HBV, like retroviruses, replicates using reverse transcriptase, although with rather lower replication rates (86). Consequently, HBV is expected to exhibit rapid evolutionary dynamics (142). Conversely, for the hepadnaviruses to have codiverged with their primate hosts would require rates of nucleotide substitution that are many logs lower, which is incompatible with the data (17). However, while there is currently little compelling evidence for codivergence in HBV, the true timescale of its evolutionary history remains a mystery.

CONCLUSIONS

The success of studies of viral evolution has been built on a combination of a growing database of gene and genome sequences coupled with advances in phylogenetic and coalescent methodology (35, 59), the latter having a profound effect on our ability to infer the most complex of evolutionary patterns. However, additional advances in both areas are required if we are to fully understand the intricacies of viral evolution. From a data perspective, it is essential that we sample viruses from as wide a range of geographical localities as possible, and not simply from the industrialized nations of the Western world. Similarly, to fully choose between different models of spatial structure, it will be necessary to obtain vastly more viral genome sequences from specific populations than are available. Thankfully, progress is being made in this respect for some viruses, chiefly influenza (45). Within the analytical arena the huge number of viral genome sequences now available, as well as those planned for the near future, requires major advantages in both phylogenetic methodology and computational power to decipher all the evolutionary signals contained within them (56). Likewise, new methods are required to infer the patterns and pathways of viral spread within complex transmission networks (130) and to accurately infer both spatial and temporal dynamics. Finally, it is essential that we investigate the evolutionary history and phylogeography of viruses in the context of all the other pathogens, both bacterial and viral, that simultaneously circulate within human populations. It is possible, if not likely, that the evolutionary and ecological interactions among pathogens are of central importance in shaping patterns of viral genetic diversity.

SUMMARY POINTS

1. The rapid rate of evolutionary change in RNA viruses means that the epidemiological and ecological processes that shape their diversity act on roughly the same timescale as mutations are fixed in viral populations.

2. There is a basic division between viruses that cause acute infections, frequently jump species boundaries, and evolve rapidly, and viruses that cause persistent infections, co-diverge with their host species over extended time periods, and evolve slowly (broadly mirroring the division between RNA and DNA viruses).

3. Preferential cross-species transmission among closely related host species and among species that live sympatrically can produce phylogenetic patterns that mirror patterns of long-term codivergence.

4. The genetic diversity of human RNA viruses tends to be shallow, indicative of a recent common ancestry of sampled isolates, and is most likely due to the rapid turnover of neutral genetic diversity.

5. The spatial patterns exhibited by human RNA viruses follow a number of clear transmission patterns, ranging from frequent migration to strong population subdivision, that can be distinguished through the application of molecular phylogenetics.

6. Major transitions in the ecological history of human populations have had a profound impact on the emergence and evolution of viral infections, although only recent events, mediated by the global travel of human hosts, have left their signature in viral genome sequences.

7. Despite their rapid evolution, viruses make relatively poor, and unnecessary, markers of human evolutionary history because of frequent admixture and limited phylogenetically informative variation.

FUTURE ISSUES

1. Do any RNA viruses codiverge with their host species over millions of years?

2. Do any RNA viruses evolve at the lower rates exhibited by dsDNA viruses?

3. What is the full range of phylogeographic patterns seen in human viruses, and what viruses confirm to each pattern?

4. How common are global selective sweeps in viral populations?

5. What is the timescale of HBV evolution? Have humans transmitted this virus to other primate species?

6. Can complex transmission networks be inferred using molecular phylogenetic methods?

7. What is the extent of viral biodiversity in human populations from nonindustrialized nations?

8. What are the extent and nature of the evolutionary and ecological interactions among the pathogens that circulate within individual populations?

DISCLOSURE STATEMENT

The authors are not aware of any biases that might be perceived as affecting the objectivity of this review.

LITERATURE CITED

1. Adams B, Holmes EC, Zhang C, Mammen MP Jr, Nimmannitya S, et al. 2006. Cross-protective immunity can account for the alternating epidemic pattern of dengue virus serotypes circulating in Bangkok. *Proc. Natl. Acad. Sci. USA* 103:14234–39

2. Allan JS, Short M, Taylor ME, Su S, Hirsch VM, et al. 1991. Species-specific diversity among simian immunodeficiency viruses from African green monkeys. *J. Virol.* 65:2816–28

3. Alonso WJ, Viboud C, Simonsen L, Hirano EW, Daufenbach LZ, et al. 2007. Seasonality of influenza in Brazil: a traveling wave from the Amazon to the subtropics. *Am. J. Epidemiol.* 165:1434–42

4. **Anderson RM, May RM. 1991.** *Infectious Diseases of Humans: Epidemiology and Control.* **Oxford: Oxford Univ. Press**

5. Antia R, Regoes RR, Koella JC, Bergstrom CT. 2003. The role of evolution in the emergence of infectious diseases. *Nature* 426:658–61

6. A-Nuegoonpipat A, Berlioz-Arthaud A, Chow V, Endy T, Lowry K, et al. 2004. Sustained transmission of dengue virus type 1 in the Pacific due to repeated introduction of different Asian genotypes. *Virology* 329:505–12

4. Seminal work on the mathematical epidemiology of infectious diseases.

7. Bailes E, Gao F, Bibollet-Ruche F, Courgnaud V, Peeters M, et al. 2003. Hybrid origin of SIV in chimpanzees. *Science* 300:1713

8. Baranowski E, Ruiz-Jarabo CM, Domingo E. 2001. Evolution of cell recognition by viruses. *Science* 292:1102–5

9. Barros de Freitas R, Durigon EL, Oliveira Dde S, Romano CM, Castro de Freitas M, et al. 2007. The "pressure pan" evolution of human erythrovirus B19 in the Amazon, Brazil. *Virology* 369:281–87

10. Bartlett MS. 1957. Measles periodicity and community size. *J. R. Stat. Soc. A* 120:48–70

11. Beer BE, Bailes E, Goeken R, Dapolito G, Coulibaly C, et al. 1999. Simian immunodeficiency virus (SIV) from sun-tailed monkeys (*Cercopithecus solatus*): evidence for host-dependent evolution of SIV within the *C. lhoesti* superspecies. *J. Virol.* 73:7734–44

12. Bernard HU. 1994. Coevolution of papillomaviruses with human populations. *Trends Microbiol.* 2:140–43

13. Bibollet-Ruche F, Bailes E, Gao F, Pourrut X, Barlow KL, et al. 2004. New simian immunodeficiency virus infecting De Brazza's monkeys (*Cercopithecus neglectus*): evidence for a *Cercopithecus* monkey virus clade. *J. Virol.* 78:7748–62

14. Biek R, Drummond AJ, Poss M. 2006. A virus reveals population structure and recent demographic history of its carnivore host. *Science* 311:538–41

15. Black FL. 1975. Infectious diseases in primitive societies. *Science* 187:515–18

16. Boivin G, Goyette N, Bernatchez H. 2002. Prolonged excretion of amantadine-resistant influenza A virus quasi species after cessation of antiviral therapy in an immunocompromised patient. *Clin. Infect. Dis.* 34:E23–25

17. Bollyky PL, Holmes EC. 1999. Reconstructing the complex evolutionary history of hepatitis B virus. *J. Mol. Evol.* 49:139–41

18. Bryant JE, Holmes EC, Barrett ADT. 2007. Out of Africa: a molecular perspective on the introduction of yellow fever virus into the Americas. *PLoS Pathog.* 3:e75

19. Carbone KM, Wolinsky JS. 2001. Mumps virus. In *Virology*, ed. BN Fields, DM Knipe, PM Howley, pp. 1381–400. Philadelphia: Lippincott, Williams & Wilkins

20. Carrington CVF, Foster JE, Pybus OG, Bennett SN, Holmes EC. 2005. Invasion and maintenance of dengue virus type 2 and type 4 in the Americas. *J. Virol.* 79:14680–87

21. Charleston MA, Robertson DL. 2002. Preferential host switching by primate lentiviruses can account for phylogenetic similarity with the primate phylogeny. *Syst. Biol.* 51:528–35

22. Charrel RN, de Micco P, de Lamballerie X. 1999. Phylogenetic analysis of GB viruses A and C: evidence for cospeciation between virus isolates and their primate hosts. *J. Gen. Virol.* 80:2329–35

23. Cleaveland S, Laurenson MK, Taylor LH. 2001. Diseases of humans and their domestic mammals: pathogen characteristics, host range and the risk of emergence. *Philos. Trans. R. Soc. London Ser. B* 356:991–99

24. Cottam EM, Haydon DT, Paton DJ, Gloster J, Wilesmith JW, et al. 2006. Molecular epidemiology of the foot-and-mouth disease virus outbreak in the United Kingdom in 2001. *J. Virol.* 80:11274–82

25. DeFilippis VR, Villarreal LP. 2000. An introduction to the evolutionary ecology of viruses. In *Viral Ecology*, ed. CJ Hurst, pp. 126–208. New York: Academic

26. de Oliveira T, Pybus OG, Rambaut A, Salemi M, Cassol S, et al. 2006. Molecular epidemiology: HIV-1 and HCV sequences from Libyan outbreak. *Nature* 444:836–37

27. Devesa M, Loureiro CL, Rivas Y, Monsalve F, Cardona N, et al. 2008. Subgenotype diversity of hepatitis B virus American genotype F in Amerindians from Venezuela and the general population of Colombia. *J. Med. Virol.* 80:20–26

28. Devesa M, Pujol FH. 2007. Hepatitis B virus genetic diversity in Latin America. *Virus Res.* 127:177–84

29. Dobson AP, Carper ER. 1996. Infectious diseases and human population history. *Bioscience* 46:115–26

30. Domingo E, Holland JJ. 1997. RNA virus mutations for fitness and survival. *Annu. Rev. Microbiol.* 51:151–78

31. Donnelly P, Tavaré S. 1995. Coalescents and genealogical structure under neutrality. *Annu. Rev. Genet.* 29:401–21

32. Drake JW. 1993. Rates of spontaneous mutation among RNA viruses. *Proc. Natl. Acad. Sci. USA* 90:4171–75

15. Classic study of the range of pathogens carried by Amerindian populations.

26. Elegant demonstration of the power of molecular epidemiology and phylogenetics in a forensic setting.

33. Drake JW, Charlesworth B, Charlesworth D, Crow JF. 1998. Rates of spontaneous mutation. *Genetics* 148:1667–86

34. Drake JW, Holland JJ. 1999. Mutation rates among RNA viruses. *Proc. Natl. Acad. Sci. USA* 96:13910–13

35. Drummond AJ, Ho SYW, Phillips MJ, Rambaut A. 2006. **Relaxed phylogenetics and dating with confidence.** *PLoS Biol.* **4:e88**

36. Duffy S, Holmes EC. 2008. Phylogenetic evidence for rapid rates of molecular evolution in the single-stranded begomovirus *Tomato yellow leaf curl virus* (TYLCV). *J. Virol.* 82:957–65

37. Dunham EJ, Holmes EC. 2007. Inferring the time-scale of dengue virus evolution under realistic models of DNA substitution. *J. Mol. Evol.* 64:656–61

38. Elena SF, Sanjuan R. 2005. Adaptive value of high mutation rates of RNA viruses: separating causes from consequences. *J. Virol.* 79:11555–58

39. Esposito JJ, Sammons SA, Frace AM, Osborne JD, Olsen-Rasmussen M, et al. 2006. Genome sequence diversity and clues to the evolution of variola (smallpox) virus. *Science* 313:807–12

40. Evans KD, Kline MK. 1995. Prolonged influenza A infection responsive to rimantadine therapy in a human immunodeficiency virus-infected child. *Pediatr. Infect. Dis. J.* 14:332–34

41. Fares MA, Holmes EC. 2002. A revised evolutionary history of hepatitis B virus (HBV). *J. Mol. Evol.* 54:807–14

42. Ferguson N, Anderson R, Gupta S. 1999. The effect of antibody-dependent enhancement on the transmission dynamics and persistence of multiple-strain pathogens. *Proc. Natl. Acad. Sci. USA* 96:790–94

43. Ferguson NM, Galvani AP, Bush RM. 2003. Ecological and immunological determinants of influenza evolution. *Nature* 422:428–33

44. Fitch WM, Leiter JME, Li X, Palese P. 1991. Positive Darwinian evolution in human influenza A viruses. *Proc. Natl. Acad. Sci. USA* 88:4270–74

45. Ghedin E, Sengamalay NA, Shumway M, Zaborsky J, Feldblyum T, et al. 2005. **Large-scale sequencing of human influenza reveals the dynamic nature of viral genome evolution.** *Nature* **437:1162–66**

46. Gojobori T, Moriyama EN, Kimura M. 1990. Molecular clock of viral evolution, and the neutral theory. *Proc. Natl. Acad. Sci. USA* 87:10015–18

47. Grenfell BT, Bjørnstad ON, Kappey J. 2001. Travelling waves and spatial hierarchies in measles epidemics. *Nature* 414:716–23

48. Grenfell BT, Pybus OG, Gog JR, Wood JLN, Daly JM, et al. 2004. Unifying the epidemiological and evolutionary dynamics of pathogens. *Science* 303:327–32

49. Guan Y, Zheng BJ, He YQ, Liu XL, Zhuang ZX, et al. 2003. Isolation and characterization of viruses related to the SARS coronavirus from animals in southern China. *Science* 302:276–78

50. Gubler DJ. 1997. Dengue and dengue hemorrhagic fever: its history and resurgence as a global public health problem. In *Dengue and Dengue Hemorrhagic Fever*, ed. DJ Gubler, G Kuno, pp. 1–22. New York: CABI Int.

51. Hanada K, Suzuki Y, Gojobori T. 2004. A large variation in the rates of synonymous substitution for RNA viruses and its relationship to a diversity of viral infection and transmission modes. *Mol. Biol. Evol.* 21:1074–80

52. Hannoun C, Horal P, Lindh M. 2000. Long-term mutation rates in the hepatitis B virus genome. *J. Gen. Virol.* 81:75–83

53. Hatwell JN, Sharp PM. 2000. Evolution of human polyomavirus JC. *J. Gen. Virol.* 81:1191–200

54. Holmes EC. 2003. Molecular clocks and the puzzle of RNA virus origins. *J. Virol.* 77:3893–97

55. Holmes EC. 2004. The phylogeography of human viruses. *Mol. Ecol.* 13:745–56

56. Holmes EC. 2007. Viral evolution in the genomic age. *PLoS Biol.* 5:e278

57. Holmes EC, Rambaut A. 2004. Viral evolution and the emergence of SARS coronavirus. *Philos. Trans. R. Soc. London Ser. B* 359:1059–65

58. Holmes EC, Twiddy SS. 2003. The origin, emergence and evolutionary genetics of dengue virus. *Infect. Genet. Evol.* 3:19–28

59. Huelsenbeck JP, Ronquist F, Nielsen R, Bollback JP. 2001. Bayesian inference of phylogeny and its impact on evolutionary biology. *Science* 294:2310–14

35. Shows how rates of nucleotide substitution can be estimated from RNA viruses and ssDNA viruses.

45. Demonstrates the power of the complete genome sequence analysis of influenza A virus.

60. Hughes AL. 2002. Origin and evolution of viral interleukin-10 and other DNA virus genes with vertebrate homologues. *J. Mol. Evol.* 54:90–101

61. Ikegaya H, Iwase H, Sugimoto C, Yogo Y. 2002. JC virus genotyping offers a new means of tracing the origins of unidentified cadavers. *Int. J. Legal. Med.* 116:242–45

62. Jackson AP, Charleston MA. 2004. A cophylogenetic perspective of RNA-virus evolution. *Mol. Biol. Evol.* 21:45–57

63. Jenkins GM, Rambaut A, Pybus OG, Holmes EC. 2002. Rates of molecular evolution in RNA viruses: a quantitative phylogenetic analysis. *J. Mol. Evol.* 54:152–61

64. Katzourakis A, Tristem M, Pybus OG, Gifford RJ. 2007. Discovery and analysis of the first endogenous lentivirus. *Proc. Natl. Acad. Sci. USA* 104:6261–65

65. **Keele BF, Van Heuverswyn F, Li Y, Bailes E, Takehisa J, et al. 2006. Chimpanzee reservoirs of pandemic and nonpandemic HIV-1.** *Science* **313:523–26**

66. Kilpatrick AM, Daszak P, Goodman SJ, Rogg H, Kramer LD, et al. 2006. Predicting pathogen introduction: West Nile virus spread to Galápagos. *Conserv. Biol.* 20:1224–31

67. Kingman JFC. 1982. On the genealogy of large populations. *J. Appl. Probab.* 19A:27–43

68. Kitchen A, Miyamoto MM, Mulligan CJ. 2007. Utility of DNA viruses for studying human host history: case study of JC virus. *Mol. Phylogenet. Evol.* 46:673–82

69. Klungthong C, Zhang C, Mammen MP Jr, Ubol S, Holmes EC. 2004. The molecular epidemiology of dengue virus serotype 4 in Bangkok, Thailand. *Virology* 329:168–79

70. Koonin EV, Senkevich TG, Dolja VV. 2006. The ancient Virus World and evolution of cells. *Biol. Direct.* 1:29

71. Kuiken T, Holmes EC, McCauley J, Rimmelzwaan GF, Williams CS, et al. 2006. Host species barriers to influenza virus infections. *Science* 312:394–97

72. Lemey P, Pybus OG, Wang B, Saksena NK, Salemi M, et al. 2003. Tracing the origin and history of the HIV-2 epidemic. *Proc. Natl. Acad. Sci. USA* 100:6588–92

73. Li W, Shi Z, Yu M, Ren W, Smith C, et al. 2005. Bats are natural reservoirs of SARS-like coronaviruses. *Science* 310:676–79

74. Linz B, Balloux F, Moodley Y, Manica A, Liu H, et al. 2007. An African origin for the intimate association between humans and *Helicobacter pylori*. *Nature* 445:915–18

75. Lloyd-Smith JO, Schreiber SJ, Kopp PE, Getz WM. 2005. Superspreading and the effect of individual variation on disease emergence. *Nature* 438:355–59

76. Magiorkinis EN, Magiorkinis GN, Paraskevis DN, Hatzakis AE. 2005. Re-analysis of a human hepatitis B virus (HBV) isolate from an East African wild born *Pan troglodytes schweinfurthii*: evidence for interspecies recombination between HBV infecting chimpanzee and human. *Gene* 349:165–71

77. Malpica JM, Fraile A, Moreno I, Obies CI, Drake JW, et al. 2002. The rate and character of spontaneous mutation in an RNA virus. *Genetics* 162:1505–11

78. Margulies M, Egholm M, Altman WE, Attiya S, Bader JS, et al. 2005. Genome sequencing in microfabricated high-density picolitre reactors. *Nature* 437:376–80

79. McGeoch DJ, Gatherer D. 2005. Integrating reptilian herpesviruses into the family *Herpesviridae*. *J. Virol.* 79:725–31

80. Meiering CD, Linial ML. 2001. Historical perspective of foamy virus epidemiology and infection. *Clin. Microbiol. Rev.* 14:165–76

81. **Morse SS. 1995. Factors in the emergence of infectious diseases.** *Emerg. Infect. Dis.* **1:7–15**

82. Moya A, Holmes EC, González-Candelas F. 2004. The population genetics and evolutionary epidemiology of RNA viruses. *Nat. Rev. Microbiol.* 2:279–87

83. Nakao H, Okomoto H, Fukuda M, Tsuda F, Mitsui T, et al. 1997. Mutation rate of GB virus C/hepatitis G virus over the entire genome and in subgenomic regions. *Virology* 233:43–50

84. Nelson MI, Simonsen L, Viboud C, Miller MA, Holmes EC. 2007. Phylogenetic analysis reveals the global migration of seasonal influenza A viruses. *PLoS Pathog.* 3:e131

85. Nichol ST, Spiropoulou CF, Morzunov S, Rollin PE, Ksiazek TG, et al. 1993. Genetic identification of a hantavirus associated with an outbreak of acute respiratory illness. *Science* 262:914–17

65. Important description of the biodiversity, evolution, and origins of HIV-1 from chimpanzee SIV.

81. Review of the ecological factors responsible for the emergence of infectious diseases.

86. Nowak MA, Bonhoeffer S, Hill AM, Boehme R, Thomas HC, et al. 1996. Viral dynamics in hepatitis B virus infection. *Proc. Natl. Acad. Sci. USA* 93:214–23

87. Nowak MA, May RM. 2000. *Virus Dynamics: Mathematical Principles of Immunology and Virology*. Oxford: Oxford Univ. Press

88. Oberste MS, Peñaranda S, Pallansch MA. 2004. RNA recombination plays a major role in genomic change during circulation of coxsackie B viruses. *J. Virol.* 78:2948–55

89. Ong C-K, Nee S, Rambaut A, Bernard H-U, Harvey PH. 1997. Elucidating the population histories and transmission dynamics of papillomaviruses using phylogenetic trees. *J. Mol. Evol.* 44:199–206

90. Orito E, Mizokami M, Ina Y, Moriyama EN, Kameshima N, et al. 1989. Host-independent evolution and a genetic classification of the hepadnavirus family based on nucleotide sequences. *Proc. Natl. Acad. Sci. USA* 86:7059–62

91. Padgett BL, Walker DL. 1973. Prevalence of antibodies in human sera against JC virus, an isolate from a case of progressive multifocal leukoencephalopathy. *J. Infect. Dis.* 127:467–70

92. Pavesi A. 2001. Origin and evolution of GBV-C/hepatitis G virus and relationships with ancient human migrations. *J. Mol. Evol.* 53:104–13

93. Pavesi A. 2005. Utility of JC polyomavirus in tracing the pattern of human migrations dating to prehistoric times. *J. Gen. Virol.* 86:1315–26

94. Plyusnin A, Morzunov SP. 2001. Virus evolution and genetic diversity of hantaviruses and their rodent hosts. *Curr. Top. Microbiol. Immunol.* 256:47–75

95. Pomeroy LW, Bjørnstad ON, Holmes EC. 2008. The evolutionary and epidemiological dynamics of the *Paramyxoviridae*. *J. Mol. Evol.* 66:98–106

96. Posada D, Crandall KA. 1998. MODELTEST: testing the model of DNA substitution. *Bioinformatics* 14:817–18

97. Pybus OG, Charleston MA, Gupta S, Rambaut A, Holmes EC, et al. 2001. The epidemic behaviour of the hepatitis C virus. *Science* 292:2323–25

98. Pybus OG, Cochrane A, Holmes EC, Simmonds P. 2005. The hepatitis C virus epidemic among injecting drug users. *Infect. Genet. Evol.* 5:131–39

99. Pybus OG, Rambaut A, Harvey PH. 2000. An integrated framework for the inference of viral population history from reconstructed genealogies. *Genetics* 155:1429–37

100. Rambaut A, Robertson DL, Pybus OG, Peeters M, Holmes EC. 2001. Phylogeny and the origin of HIV-1. *Nature* 410:1047–48

101. Real LA, Henderson JC, Biek R, Snaman J, Jack TL, et al. 2005. Unifying the spatial population dynamics and molecular evolution of epidemic rabies virus. *Proc. Natl. Acad. Sci. USA* 102:12107–11

102. Rico-Hesse R. 2003. Microevolution and virulence of dengue viruses. *Adv. Virus Res.* 59:315–41

103. Rico-Hesse R, Harrison LM, Salas RA, Tovar D, Nisalak A, et al. 1997. Origins of dengue type 2 viruses associated with increased pathogenicity in the Americas. *Virology* 230:244–51

104. Sarrazin C, Rüster B, Lee JH, Kronenberger B, Roth WK, et al. 2000. Prospective follow-up of patients with GBV-C/HGV infection: specific mutational patterns, clinical outcome, and genetic diversity. *J. Med. Virol.* 62:191–98

105. Schweizer M, Schleer H, Pietrek M, Liegibel J, Falcone V, et al. 1999. Genetic stability of foamy viruses: long-term study in an African green monkey population. *J. Virol.* 73:9256–65

106. Shackelton LA, Holmes EC. 2006. Phylogenetic evidence for the rapid evolution of human B19 erythrovirus. *J. Virol.* 80:3666–69

107. Shackelton LA, Parrish CR, Truyen U, Holmes EC. 2005. High rate of viral evolution associated with the emergence of canine parvoviruses. *Proc. Natl. Acad. Sci. USA* 102:379–84

108. Shackelton LA, Rambaut A, Pybus OG, Holmes EC. 2006. JC virus evolution and its association with human populations. *J. Virol.* 80:9928–33

109. Sharp PM. 2002. Origins of human virus diversity. *Cell* 108:305–12

110. Simmonds P. 2001. 2000 Fleming Lecture. The origin and evolution of hepatitis viruses in humans. *J. Gen. Virol.* 82:693–712

97. First demonstration of how key demographic processes in viral populations can be inferred from gene sequence data.

101. Demonstration of how complex spatial patterns can be inferred from the phylogenetic analysis of viral gene sequence data.

111. Simmonds P. 2004. Genetic diversity and evolution of hepatitis C virus—15 years on. *J. Gen. Virol.* 85:3173–88

112. Simmonds P, Midgley S. 2005. Recombination in the genesis and evolution of hepatitis B virus genotypes. *J. Virol.* 79:15467–76

113. Sittisombut N, Sistayanarain A, Cardosa MJ, Salminen M, Damrongdachakul S, et al. 1997. Possible occurrence of a genetic bottleneck in dengue serotype 2 viruses between the 1980 and 1987 epidemic seasons in Bangkok, Thailand. *Am. J. Trop. Hyg. Med.* 57:100–8

114. Starkman S, MacDonald DM, Lewis JCM, Holmes EC, Simmonds P. 2003. Geographic and species association of hepatitis B virus genotypes in nonhuman primates. *Virology* 314:381–93

115. Strassman BI, Dunbar RIM. 1999. Human evolution and disease: putting the stone age in perspective. In *Evolution in Health and Disease*, ed. SC Stearns, pp. 91–101. Oxford: Oxford Univ. Press

116. Sugimoto C, Hasegawa M, Kato A, Zheng HY, Ebihara H, et al. 2002. Evolution of human polyomavirus JC: implications for the population history of humans. *J. Mol. Evol.* 54:285–97

117. Sugimoto C, Kitamura T, Guo J, Al Ahdal MN, Shchelkunov SN, et al. 1997. Typing of urinary JC virus DNA offers a novel means of tracing human migrations. *Proc. Natl. Acad. Sci. USA* 94:9191–96

118. Switzer WM, Salemi M, Shanmugam V, Gao F, Cong M-E, et al. 2005. Ancient cospeciation of simian foamy viruses and primates. *Nature* 434:376–80

119. Swofford DL. 2003. PAUP*. *Phylogenetic Analysis Using Parsimony (*and other methods)*. Version 4. Sunderland, MA: Sinauer

120. Thu HM, Lowry K, Jian L, Hlaing T, Holmes EC, et al. 2005. Lineage extinction and replacement in dengue type 1 virus populations is due to stochastic events rather than natural selection. *Virology* 336:163–72

121. Twiddy SS, Farrar JF, Chau NV, Wills B, Gould EA, et al. 2002. Phylogenetic relationships and differential selection pressures among genotypes of dengue-2 virus. *Virology* 298:63–72

122. Twiddy SS, Holmes EC, Rambaut A. 2003. Inferring the rate and time-scale of dengue virus evolution. *Mol. Biol. Evol.* 20:122–29

123. UNAIDS. 2007. 2007 AIDS epidemic update. **http://www.unaids.org/en/KnowledgeCentre/ HIVData/EpiUpdate/EpiUpdArchive/2007default.asp**

124. Van Heuverswyn F, Li Y, Neel C, Bailes E, Keele BF, et al. 2006. Human immunodeficiency viruses: SIV infection in wild gorillas. *Nature* 444:164

125. Viboud C, Alonso WJ, Simonsen L. 2006. Influenza in tropical regions. *PLoS Med.* 3:e89

126. **Viboud C, Bjornstad ON, Smith DL, Simonsen L, Miller MA, et al. 2006. Synchrony, waves, and spatial hierarchies in the spread of influenza. *Science* 312:447–51**

127. Wallace RG, Hodac H, Lathrop RH, Fitch WM. 2007. A statistical phylogeography of influenza A H5N1. *Proc. Natl. Acad. Sci. USA* 104:4473–78

128. Walsh PD, Biek R, Real LA. 2005. Wave-like spread of Ebola Zaire. *PLoS Biol.* 3:e371

129. Weiss RA. 2001. The Leeuwenhoek lecture 2001. Animal origins of human infectious disease. *Philos. Trans. R. Soc. London Ser. B* 356:957–77

130. Welch D, Nicholls GK, Rodrigo A, Solomon W. 2005. Integrating genealogy and epidemiology: the ancestral infection and selection graph as a model for reconstructing host virus histories. *Theor. Popul. Biol.* 68:65–75

131. Wertheim JO, Worobey M. 2007. A challenge to the ancient origin of SIVagm based on African green monkey mitochondrial genomes. *PLoS Pathog.* 3:e95

132. Wirth T, Wang X, Linz B, Novick RP, Lum JK, et al. 2004. Distinguishing human ethnic groups by means of sequences from *Helicobacter pylori*: lessons from Ladakh. *Proc. Natl. Acad. Sci. USA* 101:4746–51

133. Wittke V, Robb TE, Thu HM, Nimmannitya S, Kalayanrooj S, et al. 2002. Extinction and rapid emergence of strains of dengue 3 virus during an interepidemic period. *Virology* 301:148–56

134. Woolhouse MEJ. 2002. Population biology of emerging and re-emerging pathogens. *Trends Microbiol.* 10:S3–S7

135. World Health Organization. 2000. *Hepatitis B. World Health Organization fact sheet 204.* **http://www.who. int/mediacentre/factsheets/fs204/en/**

126. Study of the transmission dynamics of influenza; demonstrates the utility of gravity models.

136. Worobey M, Santiago ML, Keele BF, Ndjango J-BN, Joy JB, et al. 2004. Origin of AIDS: contaminated polio vaccine theory refuted. *Nature* 428:820

137. Xia Y, Bjørnstad ON, Grenfell BT. 2004. Measles metapopulation dynamics: a gravity model for epidemiological coupling and dynamics. *Am. Nat.* 164:267–81

138. Yang J, Xi Q, Deng R, Wang J, Hou J, et al. 2007. Identification of interspecies recombination among hepadnaviruses infecting cross-species hosts. *J. Med. Virol.* 79:1741–50

139. Yeh S-H, Wang H-Y, Tsai C-Y, Kao C-L, Yang J-Y, et al. 2004. Characterization of severe acute respiratory syndrome coronavirus genomes in Taiwan: molecular epidemiology and genome evolution. *Proc. Natl. Acad. Sci. USA* 101:2542–47

140. Yogo Y, Sugimoto C, Zheng H-Y, Ikegaya H, Takasaka T, et al. 2004. JC virus genotyping offers a new paradigm in the study of human populations. *Rev. Med. Virol.* 14:179–91

141. Zhang C, Mammen MP Jr, Chinnawirotpisan P, Klungthong C, Rodpradit P, et al. 2005. Clade replacements in dengue virus serotypes 1 and 3 are associated with changing serotype prevalence. *J. Virol.* 79:15123–30

142. Zhou Y, Holmes EC. 2007. Bayesian estimates of the evolutionary rate and age of hepatitis B virus. *J. Mol. Evol.* 65:197–205

Population Structure of *Toxoplasma gondii*: Clonal Expansion Driven by Infrequent Recombination and Selective Sweeps

L. David Sibley[1] and James W. Ajioka[2]

[1]Department of Molecular Microbiology, Washington University School of Medicine, St. Louis, Missouri, 63130; email: sibley@wustl.edu

[2]Department of Pathology, University of Cambridge, Cambridge CB2 1QP, United Kingdom; email: ja131@cam.ac.uk

Annu. Rev. Microbiol. 2008. 62:329–51

First published online as a Review in Advance on June 10, 2008

The *Annual Review of Microbiology* is online at micro.annualreviews.org

This article's doi: 10.1146/annurev.micro.62.081307.162925

Key Words

parasite, genetics, virulence, global, transmission

Abstract

Toxoplasma gondii is among the most successful parasites. It is capable of infecting all warm-blooded animals and causing opportunistic disease in humans. *T. gondii* has a striking clonal population structure consisting of three predominant lineages in North America and Europe. Clonality is associated with the recent emergence of a monomorphic version of Chr1a, which drove a selective genetic sweep within the past 10,000 years. Strains from South America diverged from those in North America some 1–2 mya; recently, however, the monomorphic Chr1a has extended into regions of South America, where it is also associated with clonality. The recent spread of a few dominant lineages has dramatically shaped the population structure of *T. gondii* and has resulted in most lineages sharing a highly pathogenic nature. Understanding the factors that have shaped the population structure of *T. gondii* has implications for the emergence and transmission of human pathogens.

Contents

LIFE CYCLE AND TRANSMISSION

Toxoplasma gondii is a protozoan parasite with a complex life cycle involving sexual replication in members of the cat family (Felidae) and asexual propagation in a wide variety of warm-blooded hosts (**Figure 1**) (29). Three different invasive forms in the life cycle mediate survival and spread in the parasite's various hosts. Each invasive form is designed to accomplish different modes of transmission: Sporozoites are shed into the environment within resistant spores (oocysts), which are capable of causing oral infection in herbivorous animals; tachyzoites rapidly expand and disseminate within the naïve host; and bradyzoites infect long-lived host cells, where they are semidormant, survive in immunocompetent hosts, and assure oral transmission into carnivorous hosts (**Figure 1**).

Development in *T. gondii* is enormously flexible, as interconversion between these stages occurs readily (e.g., sporozoite to tachyzoite, tachyzoite to bradyzoite, and bradyzoite to tachyzoite). The entire life cycle can be regenerated from a single cloned organism, which can undergo differentiation to both male and female gametocytes (24, 95); this indicates that mating types are not predetermined. Hence, infection of a cat with a single isolate can give rise to progeny through a process of self-mating.

Transmission: the ability to pass infection from one susceptible host to another

Although meiosis does occur, there is little net gain in genetic diversity because both parental genotypes contribute identical alleles.

Sexual development (most likely due to co-evolution with its original host) only commences in enterocytes of the small intestine of the cat, where micro- and macrogametes develop intracellularly (33). Exflagellation of male gametes, followed by fertilization of female gametes (which remain intracellular), results in formation of a diploid zygote (43). Following further development, unsporulated oocysts are shed in the feces in high numbers (36). Once within the environment, oocysts undergo sporulation to yield two separate sporocysts containing four haploid sporozoites each (44). These tiny packets of infection can lie dormant in the environment for many months, where they resist destruction by physical forces (52) and common chemical disinfectants, making them a hazard for waterborne infection (4).

Ingestion of oocysts by a variety of warm-blooded animals, including herbivorous and omnivorous feeders, leads to asexual development. Nearly all orders of placental mammals (except baleen whales and insectivorous bats) are known hosts, and *T. gondii* also infects marsupials and birds (31). In these organisms, sporozoites first penetrate the small intestine and either develop and replicate within the enterocytes or cross the epithelial layer and basement membrane and thus gain access to the lamina propria, where they encounter a variety of resident cells including leukocytes (38, 109). Sporozoites can replicate in a variety of cells within the lamina propria and can also undergo a developmental switch to the fast-growing tachyzoite stage, which serves to rapidly expand parasite numbers and to disseminate the infection (109, 110).

Tachyzoites divide every 6–9 h by binary fission using a unique process called endodyogeny. Daughter cells form internally within the mother cell, re-creating a complete set of cytoskeletal and secretory organelles with each round of nuclear replication and division (111). Like the other invasion stages of the parasite, tachyzoites move by actin-based gliding

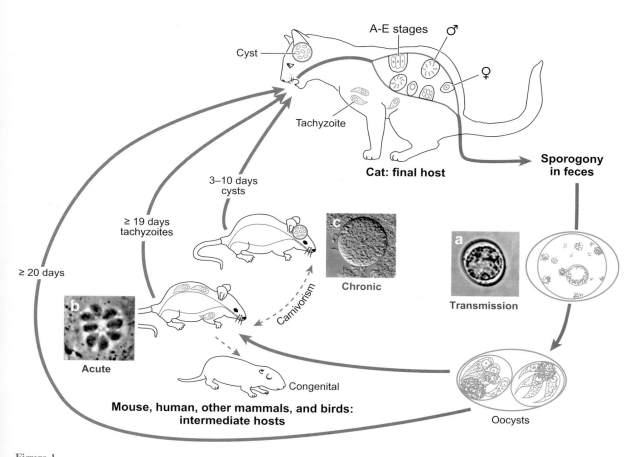

Figure 1

Life cycle depicting stages and modes of transmission of *Toxoplasma gondii*. Members of the cat family serve as definitive hosts, and sexual development only occurs within the small intestine, where micro- and macrogametes form within enterocytes following a round of mitotic replication (A-E stages). Fusion of the gametes yields a diploid zygote that is shed as a resistant spore (oocyst) in the feces. Meiosis occurs within the environment, yielding eight haploid progeny. Transmission (image *a* shows an unsporulated oocyst) occurs when oocysts contaminate food or water. Ingestion by a variety of warm-blooded hosts leads to acute infection, typified by fast-growing tachyzoites (image *b* shows a single vacuole with eight haploid parasites). Development of tissue cysts results in long-lasting chronic infection (image *c* shows a tissue cyst from mouse brain), which can also be transmitted by carnivorous feeding or scavenging. Adapted with permission from Reference 50.

motility, using this mechanism to gain entry into cells and to migrate through tissues (104). Active migration across biological barriers may contribute to dissemination into deep tissues (including the central nervous system) and into the placenta, where infections cause the most severe disease symptoms (6, 7). Primary infection with *T. gondii* in a naïve host evokes a strong immune response, and potent Th1 cytokine responses lead to control of infection through both innate (17) and adaptive immunity (99).

In response to immune pressure or environmental stresses, the parasite differentiates into bradyzoites (118). Bradyzoites survive for long periods in tissue cysts, which develop within long-lived differentiated cells such as neurons and muscle cells (118). The tissue cyst is an adaptation for chronicity and affords greater opportunity for transmission when ingested. Heteroxenous (i.e., two-host) life cycles of tissue-dwelling coccidians typically involve obligate cycling between the intermediate

MLEE: multilocus
enzyme
electrophoresis

RFLP: restriction
fragment length
polymorphism

Clonality: expansion
from a few progenitors
in the absence of
recombination to
create overrepresented
clusters of clones that
are related by a
star-shaped phylogeny

host (asexual phase) and definitive host (sexual phase), and the accidental ingestion of tissue cysts by an inappropriate host does not lead to transmission (29). Such obligatory two-host life cycles are common to all other tissue-dwelling coccidians, such as *Neospora caninum*, which infects the dog as a definitive host (30). *N. caninum* also infects a variety of ungulates that serve as intermediate hosts, yet this parasite is not transmissible by oral ingestion of tissues cysts between such intermediate hosts. In stark contrast, *T. gondii* has broken through this transmission barrier; ingestion of tissue cysts readily leads to infection in new hosts (112). Direct oral transmission by ingestion of tissue cysts effectively bypasses the need for genetic recombination in the cat and affords many new routes for dissemination through the food chain via carnivorous or omnivorous feeding or scavenging. In part, this adaptation may explain the wide host range of *T. gondii*.

CLONAL POPULATION STRUCTURE

Early studies of *T. gondii* strains emphasized the absence of major antigenic serotypes between isolates, which led to a widely held view that strains were homogeneous. With the advent of methods for multilocus enzyme electrophoresis (MLEE) and restriction fragment length polymorphism (RFLP) analysis, it quickly became apparent that this perception was inaccurate. MLEE analysis of strains from Europe (especially France) (26) and RFLP analysis of a collection of strains from North America and Europe (105) revealed a strikingly clonal population pattern (**Figure 2a**). Extension of these studies to a wider collection of strains from congenital human infections, AIDS patients, and animals (typically nonclinical infections) revealed that three major lineages predominate in North America and Europe (64). Highly similar or identical genotypes were sampled from different hosts in different geographic regions; the overall rate of recombinant strains (those exhibiting mixtures of the three clonal geno-

types) was less than 5%. There are no strong geographic differences between North America and Europe, and similar parasite genotypes were found in both animal and human infections. Remarkably, despite clonal propagation these three lineages are highly similar, differing by only a few percent at the DNA level for any given locus (64).

Many single-celled eukaryotes lack a defined sexual cycle, precluding genetic exchange by meiosis. Not surprisingly, clonality has also been described in parasites such as *Trypanosoma cruzi*, *Entamoeba histolytica*, and *Leishmania* spp. (116), which appear to lack sexual phases in their life cycles. Clonality has also been described in *Plasmodium falciparum* (115), whose obligatory sexual phase in the mosquito is required for human-to-human transmission (except in rare cases in which transmission occurs via blood transfusion). Some authors have questioned the clonality of *P. falciparum* in natural populations, except under rare conditions of transmission (22).

Although *T. gondii* is one of the few parasitic protozoa with a clearly defined sexual cycle, this mode of propagation evidently occurs only rarely in the wild. For example, given the number of alleles found at multiple loci sampled in the Howe et al. study (64), more than 1500 distinct genotypes would be expected to occur under assumptions of random mating and Hardy-Weinberg equilibrium. In contrast, fewer than 20 distinct genotypes were observed. There are two possible explanations for this striking pattern. First, intermediate hosts develop strong immunity to primary infection, thus limiting the chances for simultaneous infections with comparable tissue burdens of more than one strain (97). Combined with the short period in gametocyte formation in the cat (36), it is likely that relatively few cats are simultaneously infected with multiple strains, so the chances for genetic exchange are limited. Second, the ability of the parasite to bypass the cat in transmission by direct oral infection between intermediate hosts frees it of the necessity for sexual development, thus limiting opportunity for genetic exchange. The result is one of the

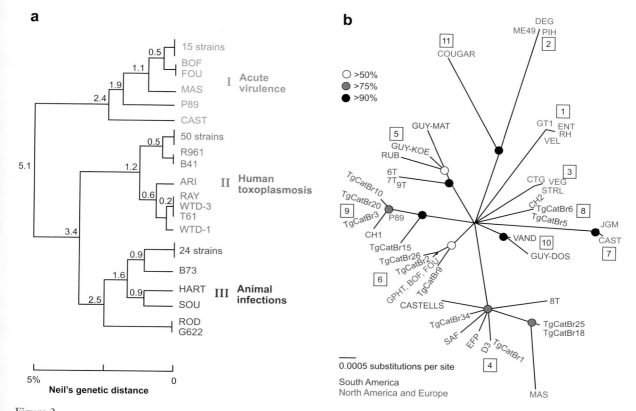

a

0.5 — 15 strains
BOF
1.1 — FOU
1.9 — MAS
2.4 — P89
CAST

I Acute virulence

0.5 — 50 strains
R961
B41
1.2 — ARI
0.6 — RAY
0.2 — WTD-3
T61
WTD-1

II Human toxoplasmosis

3.4

5.1

0.9 — 24 strains
B73
1.6 — HART
0.9 — SOU
2.5 — ROD
G622

III Animal infections

5%   0

Neil's genetic distance

b

DEG
ME49 / PIH
11   2
COUGAR

○ >50%
◉ >75%
● >90%

GUY-MAT
5   1
GUY-KOE   GT1 ENT
RUB   RH
6T   VEL
TgCatBr10   7T
9T   CTG VEG   3
TgCatBr20   STRL
9   CH2
TgCatBr3   P89   TgCatBr6 8
CH1   TgCatBr5
TgCatBr15   JGM
TgCatBr26   VAND 10   CAST
TgCatBr2   GUY-DOS   7
GPHT, BOF, FOU TgCatBr9
6   TgCatBr9
CASTELLS   8T
TgCatBr34   TgCatBr25
SAF   EFP D3 TgCatBr1   TgCatBr18
4
MAS

⎯⎯
0.0005 substitutions per site

South America
North America and Europe

Figure 2

Population structure of *Toxoplasma gondii* as inferred by phylogenetic studies. (*a*) Comparison of strains from human and animal sources from North America and Europe revealed that *T. gondii* consists of three highly clonal lineages. Analysis based on restriction fragment length polymorphism markers at six independent loci by neighbor-joining analysis and inferred genetic distance. Phenotypic differences are also strongly associated with specific lineages. Adapted with permission from Reference 64. (*b*) Phylogenetic analysis of strains from North America, Europe, and South America based on sequencing of introns. Eleven distinct haplogroups (in *boxes*) show strong geographic segregation between North and South America. Clonality predominates in North America (groups 1, 2, and 3 are equivalent to I, II, and III in panel *a*). Clonal groups also exist in South America (i.e., groups 8 and 9), although overall there is greater genetic diversity. Neighbor-joining distance analysis and bootstrap values are indicated by circles. Used with permission from Reference 74.

most striking examples of clonality known in parasites.

ORIGINS OF CLONALITY

Another striking oddity of *T. gondii* population biology is the low genetic heterogeneity among strains. In fact, early studies on the frequency of RFLPs among different lineages revealed only two distinct RFLP patterns (i.e., alleles) at each locus (11, 106). Further analysis based on the DNA sequences of genetically distinct loci revealed that this pattern is widely conserved (57).

With the advent of the whole-genome sequence for *T. gondii* (see **http://toxodb.org/toxo/**) and the assembly into chromosomes (discussed in detail below), it became possible to analyze such patterns on a genome-wide scale. Efforts to generate sequence polymorphism data by sequencing expressed sequence tags across members of the different lineages (1, 82, 85) proved enormously useful in this analysis. Mapping the single nucleotide polymorphisms (SNPs) common to a given strain type onto the assembled chromosomes provided a genome-wide analysis of the inheritance of alleles for each of the

SNP: single nucleotide polymorphism

three lineages (13). This study confirmed extensive biallelism and also led to the interpretation that the existing clonal groups were formed from just a few genetic crosses between genetically similar strains in the wild (13). This simple pattern of crosses left an indelible imprint on the genomes in the form of large blocks (haploblocks), which retain a signature of strain-specific alleles. In the absence of further extensive recombination, these blocks have been preserved within each lineage. After these founding events, the three lineages evidently expanded rapidly to populate many different hosts, including domestic and wild animals and humans. This finding reveals the striking effect that infrequent genetic recombination (or, to use the biological term, sex) has had on the population structure of *T. gondii*, despite the persistence of a highly clonal population structure.

The similarity of the clonal lineages is so extensive that it can be difficult to identify authentic mutations between members of the same genotype. Many differences detected by polymerase chain reaction or sequencing turn out to be errors, as the rates of Taq polymerase or sequencing errors are far higher (1 in 1000 bp) than the true rates of divergence within each lineage (perhaps less than 1 in 10,000 bp). High-fidelity sequencing of single-copy genetic regions comprising both antigens and introns was necessary to accurately estimate this divergence (112). The resulting data set reaffirmed the pattern that the clonal lineages were highly similar to each other and that the sequence divergence within each group was exceedingly low. Interpreting these data required a careful classification of SNPs as well as some theoretical calculations relating the rate of polymorphism to the time of the most recent common ancestor (MRCA). Analysis of the SNPs required that they first be placed into separate categories: (*a*) those that define allelic differences between the lineages and hence represent divergence since the last common ancestor of the parents of the clonal lineages (i.e., one locus, two allele polymorphisms with lineage-specific patterns), and (*b*) those that have arisen since the founding of the clonal lineages (i.e., unique to one

or more strains). Based on the assumption that mutations arise at a constant rate, it is possible to relate the extent of polymorphism to the length of time since a common ancestral origin (60). Several mutation rates that had previously been calculated for malaria (66, 98), as well as an estimate for introns in *T. gondii* (112), were used in this analysis. By analyzing only those mutations occurring within a given lineage, it was estimated that the MRCA since the common origin of the clonal lineages was on the order of 10,000 years (112).

The age of common ancestry of the *T. gondii* clonal lineages coincides closely with the dawn of agriculture and the domestication of companion animals and animals used for food. Several other human pathogens also show profound genetic bottlenecks related to this time frame (10,000–20,000 years ago), for example, *Mycobacterium tuberculosis* (16) and the emergence of *P. falciparum* out of Africa (71, 98). Such events have been attributed to the increased density of humans and hence increased opportunities for transmission between individuals, although in other cases a change in host from animal to human is thought to be responsible (120). However, as mentioned above, human-to-human transmission of *T. gondii* is rare. Newly created conditions of agriculture, which brought together for the first time companion animals (domestic cats), animals used for food (livestock), and pests (mice and rats) into close proximity with humans, may have dramatically increased opportunities for transmission. Transmission via a sylvatic cycle (i.e., between wild cats and mice) may have quickly moved into a domestic cycle (i.e., among domestic cats, pests, and food animals), exposing humans to a much higher risk of infection. Thus, the recent origin of the clonal lineages of *T. gondii* may be related to adaptation to domestic and/or companion animals.

Although most extant strains of *T. gondii* have a recent common ancestry, they populate hosts that have diverged over millions of years. Hence it is reasonable to conclude that the clonal strains have greatly expanded their host range (or displaced previous isolates) in the recent past. In the process of this expansion, the strains have propagated largely clonally. The unusual transmission of *T. gondii* by direct oral ingestion between intermediate hosts suggests that the acquisition of this trait may have been responsible for the expansion of the clonal lineages (112), driving a selective genetic sweep and carrying a majority of genes along via a hitchhiking effect (87). Enhanced oral transmission would simultaneously free these lineages from the sexual cycle, thus assuring clonal propagation. However, wider testing of strains revealed that oral transmission is found in many but not all exotic strains (54) and that it is not tightly associated with the origin of the clonal lineages (74), at least not as monitored in the laboratory mouse. Whether the clonal expansion of a few predominant lineages has been selected for by enhanced transmission among natural hosts (i.e., wild rodents) remains to be determined.

NORTH AMERICAN–SOUTH AMERICAN DIVERGENCE

Although the population structure of *T. gondii* in North America and Europe is well defined, with most strains fitting neatly into one of three predominant genotypes, exceptions to this rule have occasionally been found. Initially, these exceptions were classified as atypical or exotic strains because they had a mixture of alleles or entirely new alleles at specific loci (3, 64, 105). Studies of wider samples indicated that exotic genotypes may in fact be associated with more remote geographic areas (2, 3). For example, a collection of strains isolated from French Guyana proved to be highly unusual genotypically as well as highly pathogenic in immunocompetent individuals (20, 27). Further studies on strains from South America revealed that they were genetically distinct from strains in North America (32, 39, 45, 46, 75, 81).

There are a number of complications involved in interpreting many of the studies on South American strain diversity. First, because all the available genetic probes at that time were based on diversity seen in North America, a limited set of polymorphisms was used in analyzing strains from South America. Additionally, some genotyping studies relied on microsatellite markers (81), which are prone to homoplasy and thus are less reliable for genotyping and phylogenetic analyses. As such, many studies underestimate the level of genetic diversity in the South American strains or misclassify strains because of inappropriate markers, resulting in incorrect conclusions. For example, the idea that some clonal types seen in North America predominate in South America, such as highly virulent type I strains (35, 37, 91, 103, 107), is based on markers that do not completely capture the genetic diversity of South American strains and inappropriately group them with the clonal type I strains.

To provide a more balanced data set for analysis, introns were chosen on the basis of the assumption that they are selectively neutral and thus should acquire mutations at a constant rate. Moreover, a majority of these introns are within housekeeping genes, thus (hopefully) minimizing the chance that they carry mutations driven by selection at closely linked sites (in contrast, introns within antigenic genes may suffer from a strong hitchhiking effect). Analysis of introns from unlinked loci provided a strikingly different picture of the genetic diversity of strains from South America (74). First, polymorphic SNPs that were defined in clonal types of the North American strains were monomorphic in the South American strains (74). South American strains often varied in terms of which lineages they matched at particular SNPs, based on which allele was ancestral (i.e., shared by North and South American strains). Second, entirely new SNPs were observed in South American strains; again, these SNPs were biallelic and were invariably fixed in the North American strains (74). The simplest explanation for these patterns is that North and South American strains had diverged from a common ancestor, subsequently acquiring mutations that became fixed on their respective continent. This pattern is strikingly different from that observed between North America and Europe, where the populations are homogeneous.

What forces led to the separation of North American and South American strains and how are they maintained separately? Although we may never learn the complete explanation, an estimation of the MRCA between North and South American strains provides some clues. In this case, differences that define each region are factored into the age calculation, resulting in an estimate of 1–2 mya for their common ancestry (74). This conclusion refutes an earlier report suggesting that strains in South America were older than those strains seen in North America (81). This misconception was likely due to the fact that a recent genetic sweep has greatly restricted the genetic divergence of strains in North America (74, 112). Several events in biogeography may explain the North American–South American divergence of *T. gondii*. For example, this time frame roughly coincides with the reconnection of the Panamanian land bridge after a separation of more than 50 million years (86). It is estimated that at about the same time, members of the cat family migrated from North America to South America and subsequently underwent

a rapid speciation (69). Thus, it may be that *T. gondii* moved into South America in association with the cat migration, and then became geographically isolated and underwent genetic drift. Consistent with this model, South American strains of *T. gondii* show greater diversity than do North American strains.

Combining strains of *T. gondii* from regions of North America, Europe, and South America into a phylogenic analysis gives a picture of the population structure (**Figure 2b**) different from that discussed above. Instead of three clonal lineages, there are now 11 distinct haplogroups defined by polymorphism in introns. South American strains, although more diverse, show pockets of clonality (i.e., haplogroups 8, 9) (**Figure 2b**). When these same strains were analyzed using a Bayesian model for predicting population structure (41), similar haplogroups were identified (**Figure 3**). Structure analysis also provides an estimate of the founding populations that can collectively explain the current population structure, assuming admixture of the original genotypes. In this case, all existing *T. gondii* haplogroups are predicted to be

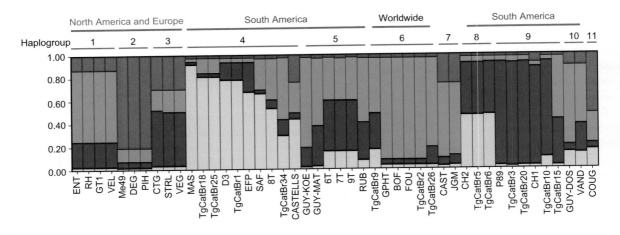

Figure 3

Population structure of *Toxoplasma gondii*. Analysis of the composition of *T. gondii* strains based on sequences of introns and analysis of polymorphisms (41) reveals the presence of 11 distinct haplogroups comprising mixtures of four distinct ancestral populations. Analysis was performed with K = 4 ancestral populations using a Bayesian statistical approach to infer population structure under the assumption of admixture (41). Colors represent contribution from ancestral populations. Distinct haplogroups show a strong geographic separation: Groups 1, 2, and 3 are found predominantly in Europe and North America, groups 4, 5, 9, and 10 are found primarily in South America, and group 6 is widespread in both Europe and South America. Strains are listed along the bottom. Used with permission from Reference 74.

derived from four original genotypes (74). Notably, only a few genetic crosses are necessary to generate the existing 11 haplogroups from these founder genotypes (74). Like the example listed above, in which only a few genetic crosses gave rise to the genetic diversity in North America (13), infrequent genetic exchanges followed by clonal expansion have contributed to the population structure as a whole. Thus, genetic recombination, although rare, has dramatically shaped the population structure of *T. gondii*.

FIXATION OF Chr1a AND GLOBAL SPREAD

Sequencing of the *T. gondii* genome led to a fortuitous finding that may help explain the recent ancestry of clonal lineages. Whole-genome sequencing of the type II strain ME49 at TIGR (**http://www.tigr.org/**) was paralleled by a Chr1-centric approach undertaken at the Sanger Institute using the type I strain RH. Comparison of the sequences of Chr1a and Chr1b, which comigrate on pulsed-field gel electrophoresis and hence were undertaken together, revealed a striking difference in the polymorphism rate between strains (**Figure 4a**). On Chr1a, SNPs were exceedingly rare (~1 in 10,000 bp) between type I and II, whereas they occurred at a frequency of ~1 in 100 bp on Chr1b (73). Based on the development of a set of polymorphic RFLP markers for genetic studies, this pattern of 1%–2% divergence—as seen on Chr1b—appears to hold for the entire genome (77). In stark contrast, the absence of SNPs on Chr1a suggests a common and recent origin. Resequencing of selected tracks from representative members of all three lineages confirmed that they share a common monomorphic version of Chr1a (**Figure 4b**) (73). Using similar methods to estimate the MRCA, it was demonstrated that the approximate age of the common Chr1a coincides with the origin of the clonal lineages (73). Even considering the relatively few genetic crosses that may have given rise to the three clonal lineages, this event is unlikely to have occurred by chance. Collectively, these re-

sults suggest that a unique combination of alleles among the genes on Chr1a was responsible for the recent and rapid emergence of the clonal lineages.

Although clonality is almost universal in *T. gondii* strains in North America, it is not as common in strains from South America. This observation raised the important question of whether South American strains also share the monomorphic Chr1a that typifies those in North America. Sequencing of similar blocks from representative members of the haplogroups that predominate in South America revealed an interesting pattern. Strains belonging to groups that exhibited clonality also showed evidence of the monomorphic Chr1a, whereas those with greater genetic diversity had a highly variable version of Chr1a (74). Other groups, such as haplogroup 6, showed clear patterns of recombination across Chr1a, suggesting that only portions of it have been retained in some highly successful lineages (74). Moreover, experimental crosses (discussed below) revealed that Chr1a readily undergoes recombination at a slightly higher rate than other chromosomes (77). The finding that acquisition of Chr1 is associated with clonal expansion, even in otherwise disparate genetic backgrounds, is further support for the idea that it contains genes that impart enhanced fitness. These patterns also demonstrate that despite their long-time separation, South American strains have undergone recent genetic exchange with strains from North America. One prediction of this finding is that strains from South America that contain the monomorphic Chr1a will show signature SNPs that define North American strains at other regions of their genomes. Genome-wide sequence information will be required to test this model and to reveal the extent of recombination that has occurred in these populations.

The striking pattern that emerges from the population structure of *T. gondii* is one of infrequent sexual recombination, followed by rapid expansion of select lineages that propagate asexually for extended periods. Rare sexual recombination, punctuated by periods of clonal expansion, has in effect fossilized these events,

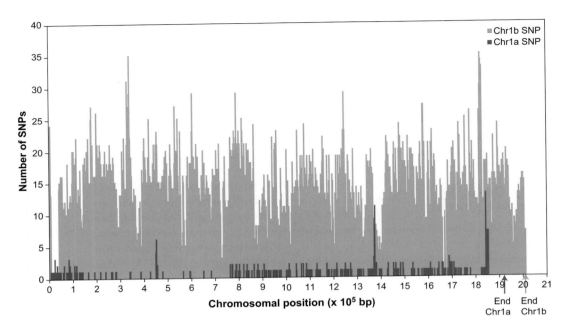

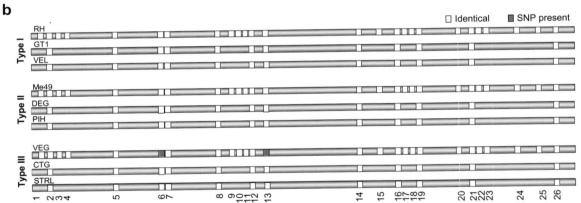

Figure 4

Chr1a is essentially identical among three *Toxoplasma gondii* lineages. Comparison from the whole-genome analysis of the type I RH strain by the Sanger Institute and the type I ME49 strain by TIGR. (*a*) Whole-genome sequencing revealed a markedly different abundance of single nucleotide polymorphisms (SNPs). Chr1b shows a SNP frequency of ~1 in 100 bp, similar to the rest of the genome. However, Chr1a shows a SNP frequency 100-fold lower across most of the chromosome. (*b*) Sequencing of selected blocks of Chr1a from the clonal lineages confirmed that they are virtually identical. The white boxes indicate sequenced regions. Used with permission from Reference 73.

making it possible to reconstruct history from the genomic record. *T. gondii* likely diverged from a common ancestry with *N. caninum* some ~10 mya [and likely more recently from *Hammondia* spp. (112)] (**Figure 5**). This event

may have been associated with a change in transmission, specifically the advent of direct oral transmission between intermediate hosts. Geographic isolation some 1–2 mya led to separate genotypes predominating in North and

South America (**Figure 5**). More recently, an event associated with the origin of Chr1a led to the rapid expansion of a few closely related strains in North America (**Figure 5**). The subsequent introgression of North American strains into South America has been associated with a gradual spread of the monomorphic Chr1 southward (**Figure 5a**). Although the expansion of haplogroups carrying this monomorphic Chr1 is a driving force in the recent evolution of *T. gondii*, the precise advantage this expansion confers is still unclear. The marked linkage disequilibrium and highly stratified population structure of *T. gondii* preclude the use of association studies to infer the genetic basis of phenotypes that may differ between clonal lineages. Experimental genetic approaches (described below) may address this question by comparing the relative fitness of progeny derived from crosses bearing the monomorphic Chr1a versus other variants.

GENOTYPES AND PATHOGENESIS

Human infection with *T. gondii* can occur by contamination of water with oocysts (4, 5, 8, 12)

or by ingestion of undercooked meat harboring tissue cysts (21, 23, 88). Toxoplasmosis typically takes the form of a short-lived acute infection characterized by lymphadenopathy, fever, and muscle weakness; infections are effectively controlled by the immune system and rarely require treatment (90). Conversion to bradyzoites

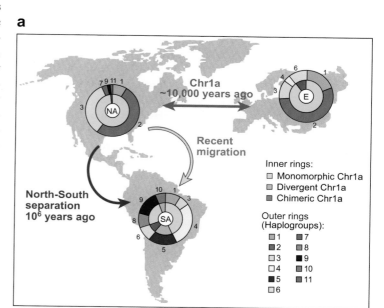

a

Figure 5

Ancestry of *Toxoplasma gondii* lineages. (*a*) Model for the recent spread of highly successful lineages of *T. gondii*. Separation of North American and South American lineages is estimated to have occurred 1–2 mya. Chr1a has a common ancestry of ~10,000 years and unites the predominant lineages that are abundant in North America and Europe. The more recent spread of Chr1a into South America is associated with increasingly clonal populations. Used with permission from Reference 74. (*b*) Estimation of the most recent common ancestry for key events in the history of *T. gondii* lineages. The split between North America and South America occurred ~10^6 years ago, and the appearance of monomorphic Chr1a (*yellow dots*) dates back ~10,000 years (73). The occurrence of Chr1a is not strictly correlated with virulence (VIR) in the mouse model, although this is a widespread trait (74). The origin of oral transmission between intermediate hosts (112) is predicted to have predated the split of North (*blue*) and South American (*red*) lineages but to have arisen after the last common ancestry with *Neospora*. Used with permission from Reference 74.

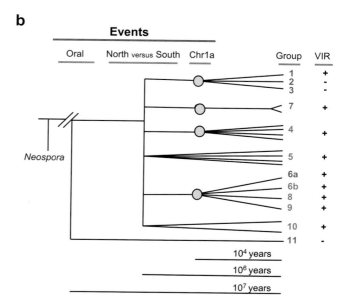

b

is thought to give rise to long-lived chronic infection that remains subclinical in healthy individuals. However, this state predisposes individuals to potential for reactivation when immune function becomes compromised (90). As such, toxoplasmosis is a severe cause of disease in patients with advanced HIV infection (83), patients undergoing cancer chemotherapy (67), or organ transplant patients (68). Notably, human-to-human infection is rare except in the case of congenital infection, which occurs when a naïve mother is infected during pregnancy (90). In such cases, severe symptoms can occur in the developing fetus depending on the time of transplacental infection (28). Additionally, ocular toxoplasmosis can occur as a result of congenital infection and it may manifest later in life (61, 62). Surprisingly, recent studies from southern Brazil indicate that newly acquired infections in healthy adults can give rise to recurrent severe ocular disease, forcing a revision of the previous idea that ocular toxoplasmosis was primarily a manifestation of congenital infection (55, 56, 108). Molecular analysis of strains from this region revealed that they differ from North American and European strains; the former fall primarily into haplogroup 4, which is predominant in South America (75). A study involving a small group of immunocompromised patients from the San Francisco area also showed evidence of mixed or atypical genotypes associated with ocular disease, although it is not known where they acquired their infections (58). Collectively, these studies imply that the genotype of the parasite may influence the severity of infection.

Analyzing the genetic makeup of strains that cause human infection is hampered by the fact that subclinical infections are not easily sampled, as the parasites are dormant in tissue. Hence, we lack the necessary information on the strains that cause baseline infections in humans to compare them against the strains isolated from disease-causing cases. In general, human-derived isolates are taken from clinically apparent cases, which may result in sampling bias. This problem has been partially alleviated by the development of serological tests that detect epitope-specific antibody responses (78). At present these tests are limited to distinguishing type II from type I or type III, and they do not contain sufficient diversity to accurately type strains that are atypical (such as those from South America or elsewhere).

Despite their numerous limitations, comparisons of studies performed in North America and Europe show that most human cases (congenital, transplant, and AIDS) are due to type II strains (3, 26, 63, 64). Type II strains have a high capacity to produce cysts in animal models and are frequently associated with infections in agricultural animals (89, 93). Thus, the prevalence of type II infections in humans may be due to more common exposure to strains of this genotype. Type III strains appear to be more common in animals, although in general they are not associated with disease (64). In contrast, type III strains are only seen in a few human cases, and these sufferers tend to have an underlying immunodeficient state. It is still unknown whether this characteristic of type III strains arises from lower tissue cyst burdens, a limited ability to cause infection, or a reduced capacity to cause human disease. Type I strains, although rare in animals, have shown increased prevalence in some cases of congenital infection (53) and in AIDS patients (76), suggesting that they are more likely to cause disease in humans. Collectively, there is an urgent need to expand the analysis of strains that cause human infection versus disease, preferably by utilizing an expanded array of peptides for serological testing of diverse strains.

DEFINING VIRULENCE GENES THROUGH GENETICS

Analyses of the population structure of *T. gondii* can make certain predictions about the history of lineages; however, they are less able to delineate the contribution of specific genes to the shaping of these patterns. Fortunately, *T. gondii* is equipped with excellent experimental genetic tools. Forward genetic analyses can potentially reveal the genes responsible for complex traits such as virulence and transmission and thus provide insight into the successful expansion of specific lineages. Identification of such traits relies on animal models, and yet the findings are highly relevant to human health.

Although early studies of *T. gondii* recognized that strains display vastly different levels of virulence in mice, as defined by survival of acute infection (49, 72, 97), subsequent molecular analyses revealed that they are genetically similar (11, 26). The observation that dramatic differences in mouse virulence correlated with limited genetic differences (105) set the stage for analysis of the genetic basis of virulence. However, this approach was not without obstacles, and several alternative models suggested that identifying the molecular basis of virulence might not be straightforward. First, morphological studies of meiosis in apicomplexans suggested that chiasmata did not occur between paired chromosomes, potentially limiting the utility of classical genetics (18, 19). Second, the trait of virulence in laboratory mice was largely attributable to passage history, which often involved repeated inoculation of tachyzoites in laboratory mice (51). Finally, it was suggested that virulence is not the result of specific alleles at a few dominant loci but rather the result of allelic combinations from many loci. In such a model, the contribution to virulence is heavily dependent on the genetic background rather than an intrinsic feature of the gene. Evidence for this model came from a cross between two relatively nonvirulent lineages (II and III), which gave rise to at least one clone with enhanced virulence in the mouse model (57). Although this finding demonstrated that recombination can result in enhanced virulence, it was not mutually exclusive with the alternative model (that alleles at some loci would confer virulence in an absolute sense, i.e., have a major impact irrespective of genetic background).

Fortunately, *T. gondii* has a number of useful features for performing genetic analysis of complex traits: (*a*) The parasite can be propagated indefinitely as a haploid stage using a variety of in vitro cell lines, (*b*) mutants are readily generated by chemical or genetic methods, (*c*) clones can be cryopreserved, and (*d*) recombinant lines can be phenotyped many times independently. An array of tools for forward and reverse genetics, analysis of cellular phenotypes, and animals models have been developed over the past two decades. These advances allowed the establishment of recombinant lines for genotypic and phenotypic analysis, construction of a genetic map, and assembly of the genome of *T. gondii*. Ultimately, such approaches enabled analysis of the genetic basis of complex traits, leading to the identification of genes that govern virulence as assessed in the mouse model. The key parameters of the genome, genetic tools, and important developments leading to this advance are summarized below.

Throughout most of its life cycle *T. gondii* is haploid, dividing asexually by a process of binary fission. Completion of the genome sequence revealed that the nuclear genome consists of 14 linear chromosomes and is ~65 Mb in size (**Figure 6a**) (see **http://toxodb.org/toxo/**). In addition, organellar genomes are found in the mitochondria and in a remnant endosymbiont called the apicoplast (**Figure 6a**). Expansion of mitochondrial repeats that are scattered around the nuclear genome (92) has hampered study of the intact mitochondrial genome. However, the mitochondrial genome in *T. gondii* may resemble that described in malaria as a 6-kb tandem repeat consisting of genes for cytochrome *c* oxidase subunits, cytochrome *b*, and fragmented ribosomal RNAs (70, 117). The apicoplast genome is a 35-kb circle that contains relatively few protein-encoding genes yet retains translation machinery including tandemly duplicated ribosomal RNAs and tRNAs (100). Both mitochondria (25) and apicoplast (42) genomes are inherited maternally and therefore provide useful markers for population structure. The highly interrelated nature of the different haplotypes of *T. gondii* was demonstrated by examination of the apicoplast genome, which revealed that only a few maternal sources contributed to the widely abundant genotypes that predominate in North America, Europe, and South America (74).

Haplotype: a combination of alleles at multiple loci that are inherited together, usually indicative of linkage disequilibrium or common ancestry

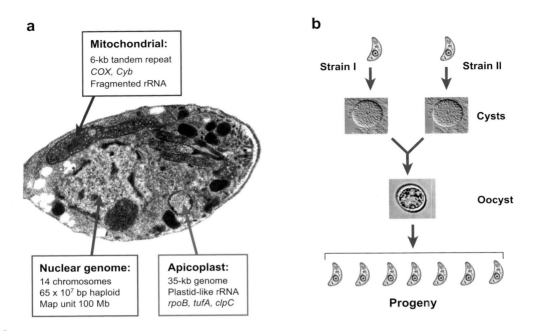

a

Mitochondrial:
6-kb tandem repeat
COX, Cyb
Fragmented rRNA

Nuclear genome:
14 chromosomes
65 x 10⁷ bp haploid
Map unit 100 Mb

Apicoplast:
35-kb genome
Plastid-like rRNA
rpoB, tufA, clpC

b

Strain I

Strain II

Cysts

Oocyst

Progeny

Figure 6

Genomes and genetic crosses in *Toxoplasma gondii*. (*a*) *T. gondii* contains three distinct genomes: mitochondrial, apicoplast, and nuclear. The mitochondrial genome is poorly known owing to the overrepresentation of mitochondria-like fragments in the nucleus (92). The features shown here are based on *Plasmodium gallinaceum* (70, 117). The apicoplast genome consists of a 35-kb circular genome. The nuclear genome consists of 14 linear chromosomes with a total size of ~65 Mb (77). (*b*) Schematic of a genetic cross in *T. gondii*. Chronic infections with distinct strains of *T. gondii* are induced in mice. Tissue cysts are co-fed to cats; they then shed oocysts (2N stage). Sporulation (i.e., meiosis) of oocysts yields haploid progeny, which can be cloned out for phenotypic and genotypic analysis. Further details on genetic crosses can be found at **http://toxomap.wustl.edu/**.

Pioneering work by Pfefferkorn and colleagues demonstrated that experimental genetic crosses between isolates of *T. gondii* were possible and that progeny were generated through a process of conventional meiosis (94–96) (**Figure 6*b***). Expansion of the initial drug-resistant markers to a wider set of RFLP markers, analyzed among a number of different crosses (see **http://toxomap.wustl.edu/**), was used to build a 5-cM linkage map that defined 14 distinct chromosomes (77). Moreover, this genetic map was used to assemble the 10X whole-genome shotgun sequence into a chromosome-centric view of the genome (see **http://toxodb.org/toxo/**). These studies demonstrated that crossovers are reasonably random and are sufficiently frequent to map phenotypes to within 300–500 kb with a single round of linkage analysis (77).

Completion of the whole-genome sequence and assembly of the above-mentioned genetic map (77) made it possible to address the phenotypic differences in virulence among strains using forward genetics. The clonal lineages of *T. gondii* resemble highly inbred lines, thus facilitating identification of genes that control the phenotypic differences among them. Notably, the clonal lineages differ markedly in acute virulence in the mouse model, as well as in several other traits related to pathogenesis. Following challenge with tachyzoites by needle inoculation, strains of the type I lineage are uniformly lethal in all strains of laboratory mice regardless of dose (65, 105). Moreover, the original isolate of the laboratory strain RH was virulent in mice at first isolation (101), and ~20 other type I strains have been isolated from animal and human infections, all of which share the same lethality (64). In contrast, type II and type III strains are relatively avirulent, with LD_{50} values of $>10^3$ in inbred mice and $>10^5$ in outbred mice (65, 105). Although the virulence of type II and type III strains increases with passage, it does not attain the level seen in type I strains (i.e., LD_{100} = a single viable organism). Ultimately, the concern about the stability of virulence was not a factor in mapping the genes responsible for this trait in *T. gondii*.

The application of classical genetics to analysis of acute virulence as typified by type I strains revealed that this trait is remarkably stable, heritable, and determined by relatively few genetic loci (113). Genetic crosses between the virulent type I and avirulent type III lineages were used to map and positionally clone the gene(s) responsible for this trait. Remarkably, the high virulence of the type I lineages proved to be largely (90%) due to a single allele at the *ROP18* locus, a gene encoding a rhoptry protein (ROP) (114).

Rhoptries are secretory organelles that discharge their contents at the time of cell invasion, contributing both to the formation of a tight junction between the parasite and host cell membranes and to the forming parasite-containing vacuole (10, 14). ROP proteins are also discharged into the host cell cytosol at the time of invasion (59), perhaps through small breaches in the membrane as it invaginates to accommodate the invading parasite. These ROP proteins have different destinations in the host cell, some trafficking to the host cell nucleus while others are targeted to the external surface of the parasite-containing vacuole (10, 14). Proteomic analysis of the rhoptries reveals a diverse array of proteins, some bearing kinase, phosphatases, and other functional domains (15), although their specific roles in parasite biology have not been defined for the most part.

Identification of ROP18 as the virulence-conferring gene was aided by several unusual features. The type I allele ($ROP18_I$) differs in two significant ways from the allele found in the type III strain ($ROP18_{III}$). First, $ROP18_I$ is remarkably polymorphic, differing in almost 15% of its residues—an order of magnitude more

divergent than typical proteins in the genome (114). Second, the expression level of ROP18 was reduced by more than 100-fold in the type III lineages, effectively generating a null background (114). Transfection of the $ROP18_I$ allele into this background resulted in a 4-log increase in lethality in outbred mice, a trait that is dependent on the serine-threonine kinase activity of ROP18 (114). This finding provides evidence that virulence of the type I strains is due to a specific allele unique to this lineage. Subsequent analysis of *T. gondii* strains from South America demonstrated that many of them also share the enhanced virulence of type I strains (**Figure 5b**) (74). This result implies that acute virulence is not a recent trait, as it predates the origin of the clonal lineages. Whether the enhanced virulence of other more divergent strains is also related to ROP18, or if it is due to a different mechanism, remains to be determined.

Independently, a study attempting to define genetic differences between type II and type III strains that influence pathogenicity in inbred mice also identified a role for the type II allele of ROP18 (102). However, in this instance ROP18 was one of five or more genetic loci that contribute quantitative effects to virulence difference; some of these loci enhance lethality in inbred mice, and others reduce it (102). Expression of the $ROP18_{II}$ allele also enhanced acute virulence in the type III lineage, albeit in the inbred mouse model. Collectively, these findings illustrate that the quantitative effects of a given gene can differ dramatically in the extent to which they affect a given multigenic trait, depending on the cross analyzed. This variation is a function of both the allele in question (and its inherent virulence potential) and the interaction it has with other alleles at different loci. In the case of type I virulence, $ROP18_I$ has an almost Mendelian (binary) contribution due to its high quantitative contribution to virulence in type I strains and all progeny of the I × III cross receiving this allele. On the other hand, $ROP18_{II}$ contributes much less to the overall virulence difference between type II and type III strains, which are more significantly affected by other genes. Both of these views are compatible with the quantitative nature of genetic contributions to complex traits (79, 80). Collectively, these studies also highlight the importance of rhoptry secretory proteins as effectors that are injected into the host cell during invasion and that have major influence on the outcome of infection (10, 14).

Although the contribution of genetic mapping to defining complex traits in *T. gondii* has seen great advances, it remains grounded in an artificial model, the laboratory mouse. Laboratory mice are susceptible relative to many other animals, including humans. One might ask: What is the significance of acute virulence in the laboratory mouse? Moreover, why would a highly successful parasite like *T. gondii* want to kill its host when this might disrupt opportunities for transmission? The answer to this apparent contradiction might lie in differences in susceptibility between laboratory mice and their wild ancestors. Studies performed in 1953 with the type I RH strain of *T. gondii* revealed that although this strain is uniformly lethal to all laboratory mice, White-footed mice (*Peromyscus leucopus*) occasionally survive and develop chronic infections (49). Moreover, rats are highly resistant to *T. gondii* and readily develop chronic infections characterized by high cyst burdens (34), including with type I strains, which are normally lethal in mice.

It has been argued that pathogens should evolve away from virulence, or the ability to inflict harm on one's host, in an effort to maximize survival. One example of this is *Mycobacterium tuberculosis*, a widespread human pathogen with relatively low virulence: It grows extremely slowly, subverts the immune system by its passivity, and shows evidence of coevolution in host responses to minimize pathology (9). However, not all pathogens show evidence of evolving toward lower virulence. In some cases this can be attributed to new genetic variants that arose by recombination, as in the case of *Cryptococcus gattii* (48), or by organisms that are newly extended outside of their former range [i.e., Ebola or Marburg virus infections in humans (120)]. However, even once established, not all stable host-pathogen relationships evolve away from virulence. Evolutionary theory and mathematical modeling reveal that pathogens do not evolve toward lower virulence, but rather toward maximum transmission (47). In the malaria parasite, selection favors higher parasitemia, and hence greater transmission rates, which results in higher virulence (84). Enhanced virulence is tolerated in this trade-off, provided that it does not cause adverse mortality to the host (47).

The $ROP18_1$ allele has been associated with enhanced growth of the parasite (40, 114), which may lead to higher parasite burdens and hence enhanced transmission. Such an adaptation might favor infection in relatively resistant hosts, which might otherwise clear infection and restrict opportunities for transmission. The inadvertent consequence of this adaptation may be the enhanced virulence typified by type I strains, as well as many lineages common in South America. It may not be coincidental that the only other region contributing to enhanced virulence in the type I lineage was a locus on Chr1a (114), which is associated with the rapid clonal expansion of *T. gondii* (73, 74). Further population and experimental genetic analyses may reveal the mechanism by which these factors interact to promote successful expansion of specific lineages of *T. gondii*.

SUMMARY POINTS

1. *T. gondii* has a highly unusual population structure that is dominated by a few clonal lineages. In North America and Europe, three highly clonal lineages predominate; these lineages evolved from a common ancestor within the past ~10,000 years.

2. Strains in South America are genetically distinct and evolved away from those in North America $\sim 10^6$ years ago. Despite having a more diverse population structure, South American strains still show pockets of clonality, which is associated with more recent mixing with North American strains.

3. Clonality is strongly associated with the appearance of a monomorphic Chr1a that dates to within the past 10,000 years. Since then, relatively few genetic exchanges in the wild have dramatically affected population structure.

4. The striking genetic sweep associated with Chr1a is suggestive of a major adaptive trait that simultaneously leads to the emergence of a few dominant lineages and to the apparent suppression of recombination.

5. Although genetic differences between the clonal lineages are small, they underlie large phenotypic differences in virulence and infectivity. Forward genetic analysis is an effective means of identifying the genetic traits responsible for virulence.

FUTURE ISSUES

1. Studies of genetic diversity in North America, Europe, and South America have revealed a remarkable pattern of clonality in *T. gondii*. Researchers formerly thought that this represented the global population structure, but recent studies from South America have challenged this view and revealed a far more complex population structure. We do not yet know the makeup of parasite populations in Africa, Asia, and other parts of the globe where *T. gondii* is abundant. If the evolution of *T. gondii* has paralleled its feline hosts, it might be expected to show an equally complex evolutionary history, which may still be grounded in geographic differences.

2. Modern-day patterns of immigration of humans as well as transport of livestock are predicted to affect such patterns and may explain the recent spread of lineages harboring a monomorphic version of Chr1a.

3. Understanding the selective advantage of genes on Chr1a may hold the secret to the success of *T. gondii* as an abundant parasite of animals and humans. Recent analyses have stressed the importance of sequence-based genotyping across multiple loci and such analyses need to be expanded to develop genome-wide haplotype maps.

4. Analysis of the pattern of interactions among different loci within distinct lineages may reveal further evidence of genetic sweeps, selective loci, and adaptations for specific niches. Distinct lineages of *T. gondii* have different biological traits including virulence, transmission, and infectivity in humans.

5. Further analysis of the genetic structure of populations will aid us in understanding the mechanism(s) of the evolution of virulence as well as the underlying molecular basis of pathogenesis.

DISCLOSURE STATEMENT

The authors are not aware of any biases that might be perceived as affecting the objectivity of this review.

ACKNOWLEDGMENTS

The Wellcome Trust and the National Institutes of Health provided support for work in the authors' laboratories. Dedicated to Frank Verley, whose tremendous enthusiasm as a teacher provided an exciting introduction to population genetics. We are also grateful for the many contributions of Rubens Bellfort, Matt Berriman, John Boothroyd, Jon Boyle, Marie Laure Dardé, J.P. Dubey, Jack Frenkel, Michael Grigg, Krys Kelly, Ian Paulsen, Elmer Pfefferkorn, David Roos, Ben Rosenthal, Jack Remington, Claudio Silveira, Chunlei Su, Michael White, John Wootton, and members of our laboratories.

LITERATURE CITED

1. Ajioka JA, Boothroyd JC, Brunk BP, Hehl A, Hillier L, et al. 1998. Gene discovery by EST sequencing in *Toxoplasma gondii* reveals sequences restricted to the *Apicomplexa*. *Genet. Res.* 8:18–28

2. Ajzenberg D, Bañuls AL, Su C, Dumètre A, Demar M, et al. 2004. Genetic diversity, clonality and sexuality in *Toxoplasma gondii*. *Int. J. Parasitol.* 34:1185–96

3. Ajzenberg D, Cogné N, Paris L, Bessieres MH, Thulliez P, et al. 2002. Genotype of 86 *Toxoplasma gondii* isolates associated with human congenital toxoplasmosis and correlation with clinical findings. *J. Infect. Dis.* 186:684–89

4. Aramini JJ, Stephen C, Dubey JP, Engelstoft C, Schwantje H, Ribble CS. 1999. Potential contamination of drinking water with *Toxoplasma gondii* oocysts. *Epidemiol. Infect.* 122:305–15

5. Bahia-Oliveira LM, Jones JL, Azevedo-Silva J, Alves CC, Orefice F, Addiss DG. 2003. Highly endemic, waterborne toxoplasmosis in north Rio de Janeiro state, Brazil. *Emerg. Infect. Dis.* 9:55–62

6. Barragan A, Sibley LD. 2002. Transepithelial migration of *Toxoplasma gondii* is linked to parasite motility and virulence. *J. Exp. Med.* 195:1625–33

7. Barragan A, Sibley LD. 2003. Migration of *Toxoplasma gondii* across biological barriers. *Trends Microbiol.* 11:426–30

8. Benenson MW, Takafuji ET, Lemon SM, Greenup RL, Sulzer AJ. 1982. Oocyst-transmissed toxoplasmosis associated with ingestion of contaminated water. *N. Engl. J. Med.* 307:666–69

9. Blaser MJ, Kirschner DE. 2007. The equilibria that allow bacterial persistence in human hosts. *Nature* 449:843–49

10. Boothroyd JC, Dubremetz JF. 2008. Kiss and spit: the dual roles of *Toxoplasma* rhoptries. *Nat. Rev. Microbiol.* 6:79–88

11. Boothroyd JC, LeBlanc AJ, Sibley LD. 1993. Allelic polymorphism in *Toxoplasma gondii*: implications for interstrain mating. In *Toxoplasmosis*, ed. J Smith, Ser. H 78:3–8. Springer-Verlag

12. Bowie WR, King AS, Werker DH, Issac-Renton JL, Eng SB, Marion SA. 1997. Outbreak of toxoplasmosis associated with municipal drinking water. The BC *Toxoplasma* investigation team. *Lancet* 19:173–77

13. **Boyle JP, Rajasekar B, Saeij JPJ, Ajioka JW, Berriman M, et al. 2006. Just one cross appears capable of dramatically altering the population biology of a eukaryotic pathogen like *Toxoplasma gondii*. *Proc. Natl. Acad. Sci. USA* 103:10514–19**

14. Bradley PJ, Sibley LD. 2007. Rhoptries: an arsenal of secreted virulence factors. *Curr. Opin. Microbiol.* 10:582–87

15. Bradley PJ, Ward C, Cheng SJ, Alexander DL, Coller S, et al. 2005. Proteomic analysis of rhoptry organelles reveals many novel constituents for host-parasite interactions in *T. gondii* . *J. Biol. Chem.* 280:34245–58

16. Brosch R, Gordon SV, Marmiesse M, Brodin P, Buchrieser C, et al. 2002. A new evolutionary scenario for the *Mycobacterium tuberculosis* complex. *Proc. Natl. Acad. Sci. USA* 99:3684–89

17. Buzoni-Gatel D, Kasper LH. 2007. Innate immunity in *Toxoplasma gondii* infection. See Ref. 119, pp. 593–607

18. Canning EU, Anwar M. 1968. Studies on meiotic division in coccidial and malarial parasites. *J. Protozool.* 15:290–98

13. Demonstrated that biallelism in *T. gondii* can be attributed to the origin of the three lineages from just a few genetic crosses in the wild that occurred between highly similar strains.

19. Canning EU, Morgan K. 1975. DNA synthesis, reduction, and elimination during the life cycles of the eimeriine coccidian, *Eimeria tenella*, and the haemogregarine, *Hepatozooan domerguei. Exp. Parasitol.* 38:217–27

20. Carme B, Bissuel F, Ajzenberg D, Bouyne R, Aznar C, et al. 2002. Severe acquired toxoplasmosis in immunocompetent adult patients in French Guyana. *J. Clin. Microbiol.* 40:4037–44

21. Choi W, Nam H, Kwak N, Huh W, Kim Y, et al. 1997. Foodborne outbreaks of human toxoplasmosis. *J. Infect. Dis.* 175:1280–82

22. Conway DJ, Roper C, Oduola AM, Armot DE, Kremsner P, et al. 1999. High recombination rate in populations of *Plasmodium falciparum. Proc. Natl. Acad. Sci. USA* 96:4506–11

23. Cook AJ, Gilbert RE, Buffolano W, Zufferey J, Petersen E, et al. 2000. Sources of *Toxoplasma* infection in pregnant women: European multicentre case-control study. European Research Network on Congenital Toxoplasmosis. *BMJ* 321:142–47

24. Cornelissen AWCA, Overdulve JP. 1985. Sex determination and sex differentiation in *Coccidia*: gametogony and oocyst production after monoclonal infection of cats with free-living and intermediate host stages of *Isospora (Toxoplasma) gondii. Parasitology* 90:35–44

25. Creasey AM, Ranford-Cartwright LC, Moore DJ, Williamson DH, Wilson RJM, et al. 1993. Uniparental inheritance of the mitochondrial gene cytochrome *b* in *Plasmodium falciparum. Curr. Genet.* 23:360–64

26. Dardé ML, Bouteille B, Pestre-Alexandre M. 1992. Isoenzyme analysis of 35 *Toxoplasma gondii* isolates and the biological and epidemiological implications. *J. Parasitol.* 78:786–94

27. Dardé ML, Villena I, Pinon JM, Beguinot I. 1998. Severe toxoplasmosis caused by a *Toxoplasma gondii* strain with a new isotype acquired in French Guyana. *J. Clin. Microbiol.* 36:324

28. Desmonts G, Couvreur J. 1974. Congenital toxoplasmosis: a prospective study of 378 pregnancies. *N. Engl. J. Med.* 290:1110–16

29. Dubey JP. 1977. *Toxoplasma, Hammondia, Besniotia, Sarcocystis*, and other tissue cyst–forming coccidia of man and animals. In *Parasitic Protozoa*, ed. JP Kreier, pp. 101–237. New York: Academic

30. Dubey JP. 2003. Review of *Neospora caninum* and neosporosis in animals. *Kor. J. Parasitol.* 41:1–16

31. Dubey JP, Beattie CP. 1988. *Toxoplasmosis of Animals and Man*. Boca Raton, FL: CRC Press

32. Dubey JP, Cortes-Vecino JA, Vargas-Duarte JJ, Sundar N, Velmurugan GV, et al. 2007. Prevalence of *Toxoplasma gondii* in dogs from Colombia, South America and genetic characterization of *T. gondii* isolates. *Vet. Parasitol.* 145:45–50

33. Dubey JP, Frenkel JF. 1972. Cyst-induced toxoplasmosis in cats. *J. Protozool.* 19:155–77

34. Dubey JP, Frenkel JK. 1998. Toxoplasmosis of rats: a review, with considerations of their value as an animal model and their possible role in epidemiology. *Vet. Parasitol.* 77:1–32

35. Dubey JP, Graham DH, Blackston CR, Lehmann T, Gennari SM, et al. 2002. Biological and genetic characterisation of *Toxoplasma gondii* isolates from chickens (*Gallus domesticus*) from São Paulo, Brazil: unexpected findings. *Int. J. Parasitol.* 32:99–105

36. Dubey JP, Miller NL, Frenkel JK. 1970. The *Toxoplasma gondii* oocyst from cat feces. *J. Exp. Med.* 132:636–62

37. Dubey JP, Navarro IT, Sreekumar C, Dahl E, Freire RL, et al. 2004. *Toxoplasma gondii* infections in cats from Paraná, Brazil: seroprevalence, tissue distribution, and biologic and genetic characterization of isolates. *J. Parasitol.* 90:721–26

38. Dubey JP, Speer CA, Shen SK, Kwok OCH, Blixt JA. 1997. Oocyst-induced murine toxoplasmosis: life cycle, pathogenicity, and stage conversion in mice fed *Toxoplasma gondii* oocysts. *J. Parasitol.* 83:870–82

39. Dubey JP, Sundar N, Gennari SM, Minervino AH, Farias NA, et al. 2007. Biologic and genetic comparison of *Toxoplasma gondii* isolates in free-range chickens from the northern Pará state and the southern state Rio Grande do Sul, Brazil, revealed highly diverse and distinct parasite populations. *Vet. Parasitol.* 143:182–88

40. El Hajj H, Lebrun M, Arold ST, Vial H, Labesse G, Dubremetz JF. 2007. ROP18 is a rhoptry kinase controlling the intracellular proliferation of *Toxoplasma gondii. PLoS Pathog.* 3:e14

41. Falush D, Stephens M, Pritchard JK. 2003. Inference of population structure using multilocus genotype data: linked loci and correlated allele frequencies. *Genetics* 164:1567–87

42. Ferguson DJ, Henriquez FL, Kirisits MJ, Muench SP, Prigge ST, et al. 2005. Maternal inheritance and stage-specific variation of the apicoplast in *Toxoplasma gondii* during development in the intermediate and definitive host. *Eukaryot. Cell* 4:814–26

43. Ferguson DJ, Hutchison WM, Siim JC. 1975. The ultrastructural development of the macrogamete and formation of the oocyst wall of *Toxoplasma gondii*. *Acta Pathol. Microbiol. Scand.* 83:491–505

44. Ferguson DJP, Birch-Andersen A, Siim JC, Hutchison WM. 1979. Ultrastructural studies of the sporulation of oocysts of *Toxoplasma gondii*. *Acta Pathol. Microbiol. Scand.* 87:183–90

45. Ferreira AM, Vitor RW, Gazzinelli RT, Melo MN. 2006. Genetic analysis of natural recombinant Brazilian *Toxoplasma gondii* strains by multilocus PCR-RFLP. *Infect. Genet. Evol.* 6:22–31

46. Ferreira AM, Vitor RWA, Carneiro ACAV, Brandão GP, Melo MN. 2004. Genetic variability of Brazilian *Toxoplasma gondii* strains detected by random amplified polymorphic DNA–polymerase chain reaction (RAPD–PCR) and simple sequence repeat—anchored PCR (SSR–PCR). *Infect. Genet. Evol.* 4:131–42

47. Frank SA. 1996. Models of parasite virulence. *Q. Rev. Biol.* 71:37–78

48. Fraser JA, Giles SS, Wenik EC, Geunes-Boyer SG, Wright JR, et al. 2005. Same-sex mating and the origin of the Vancouver Island *Cryptococcus gattii* outbreak. *Nature* 437:1360–64

49. Frenkel JK. 1953. Host, strain and treatment variation as factors in the pathogenesis of toxoplasmosis. *Am. J. Trop. Med. Hyg.* 2:390–415

50. Frenkel JK. 1973. *Toxoplasma* in and around us. *Biol. Sci.* 23:343–52

51. Frenkel JK, Ambroise-Thomas P. 1997. Genomic drift of *Toxoplasma gondii*. *Parasitol. Res.* 83:1–5

52. Frenkel JK, Ruiz A, Chinchilla M. 1975. Soil survival of *Toxoplasma* oocysts in Kansas and Costa Rica. *Am. J. Trop. Med. Hyg.* 24:439–43

53. Fuentes I, Rubio JM, Ramírez C, Alvar J. 2001. Genotypic characterization of *Toxoplasma gondii* strains associated with human toxoplasmosis in Spain: direct analysis from clinical samples. *J. Clin. Microbiol.* 39:1566–70

54. Fux B, Nawas J, Khan A, Gill DB, Su C, Sibley LD. 2007. *Toxoplasma gondii* strains defective in oral transmission are also defective in developmental stage differentiation. *Infect. Immun.* 75:2580–90

55. Gilbert RE, Stanford MR. 2000. Is ocular toxoplasmosis caused by prenatal or postnatal infection? *Br. J. Ophthalmol.* 84:224–26

56. Glasner PD, Silveira C, Kruszon-Moran D, Martins MC, Burnier M, et al. 1992. An unusually high prevalence of ocular toxoplasmosis in southern Brazil. *Am. J. Opthalmol.* 114:136–44

57. Grigg ME, Bonnefoy S, Hehl AB, Suzuki Y, Boothroyd JC. 2001. Success and virulence in *Toxoplasma* as the result of sexual recombination between two distinct ancestries. *Science* 294:161–65

58. Grigg ME, Ganatra J, Boothroyd JC, Margolis TP. 2001. Unusual abundance of atypical strains associated with human ocular toxoplasmosis. *J. Infect. Dis.* 184:633–39

59. Håkansson S, Charron AJ, Sibley LD. 2001. *Toxoplasma* evacuoles: a two-step process of secretion and fusion forms the parasitophorous vacuole. *EMBO J.* 20:3132–44

60. Hartl D, Clark AG. 1997. *Principles of Population Genetics*. Sunderland, MA: Sinauer. 3rd ed.

61. Holland GN. 2003. Ocular toxoplasmosis: a global reassessment. Part I: Epidemiology and course of disease. *Am. J. Ophthalmol.* 136:973–88

62. Holland GN. 2004. Ocular toxoplasmosis: a global reassessment. Part II: Disease manifestations and management. *Am. J. Ophthalmol.* 137:1–17

63. Howe DK, Honoré S, Derouin F, Sibley LD. 1997. Determination of genotypes of *Toxoplasma gondii* strains isolated from patients with toxoplasmosis. *J. Clin. Microbiol.* 35:1411–14

64. Howe DK, Sibley LD. 1995. *Toxoplasma gondii* comprises three clonal lineages: correlation of parasite genotype with human disease. *J. Infect. Dis.* 172:1561–66

65. Howe DK, Summers BC, Sibley LD. 1996. Acute virulence in mice is associated with markers on chromosome VIII in *Toxoplasma gondii*. *Infect. Immun.* 64:5193–98

66. Hughes AL, Verra F. 2001. A very large long-term effective population size in the virulent human malaria parasite *Plasmodium falciparum*. *Proc. R. Soc. London Biol. Sci. Ser. B* 268:1855–60

67. Israelski DM, Remington JS. 1993. Toxoplasmosis in patients with cancer. *Clin. Infect. Dis.* 17(Suppl. 2):S423–35

68. Israelski DM, Remington JS. 1993. Toxoplasmosis in the non-AIDS immunocompromised host. *Curr. Clin. Top. Infect. Dis.* 13:322–56

64. The dominance of just three clonal lineages in North America and Europe was demonstrated.

69. Johnson WE, Eizirik E, Pecon-Slattery J, Murphy WJ, Auntunes A, et al. 2006. The late Miocene radiation of modern Felidae: a genetic assessment. *Science* 311:73–77

70. Joseph JT, Aldritt SM, Unnasch T, Puijalon O, Wirth DF. 1989. Characterization of a 68. conserved extrachromosomal element isolated from the avian malarial parasite *Plasmodium gallinaceum*. *Mol. Cell. Biol.* 9:3621–29

71. Joy DA, Feng X, Mu J, Furuya T, Chotivanich K, et al. 2003. Early origin and recent expansion of *Plasmodium falciparum*. *Science* 300:318–21

72. Kaufman HE, Melton ML, Remington J, Jacobs L. 1959. Strain differences of *Toxoplasma gondii*. *J. Parasitol.* 45:189–90

73. Khan A, Bohme U, Kelly KA, Adlem E, Brooks K, et al. 2006. Common inheritance of chromosome Ia associated with clonal expansion of *Toxoplasma gondii*. *Genet. Res.* 16:1119–25

74. Khan A, Fux B, Su C, Dubey JP, Darde ML, et al. 2007. Recent transcontinental sweep of *Toxoplasma gondii* driven by a single monomorphic chromosome. *Proc. Natl. Acad. Sci. USA* 104:14872–77

75. Khan A, Jordan C, Muccioli C, Vallochi AL, Rizzo LV, et al. 2006. Genetic divergence of *Toxoplasma gondii* strains associated with ocular toxoplasmosis Brazil. *Emerg. Infect. Dis.* 12:942–49

76. Khan A, Su C, German M, Storch GA, Clifford D, Sibley LD. 2005. Genotyping of *Toxoplasma gondii* strains from immunocompromised patients reveals high prevalence of type I strains. *J. Clin. Microbiol.* 43:5881–87

77. Khan A, Taylor S, Su C, Mackey AJ, Boyle J, et al. 2005. Composite genome map and recombination parameters derived from three archetypal lineages of *Toxoplasma gondii*. *Nucleic Acids Res.* 33:2980–92

78. Kong JT, Grigg ME, Uyetake L, Parmley SF, Boothroyd JC. 2003. Serotyping of *Toxoplasma gondii* infections in humans using synthetic peptides. *J. Infect. Dis.* 187:1484–95

79. Lander E, Kruglyak L. 1995. Genetic dissection of complex traits: guidelines for interpreting and reporting linkage results. *Nat. Genet.* 11:241–47

80. Lander ES, Botstein D. 1989. Mapping Mendelian factors underlying quantitative traits using RFLP linkage maps. *Genetics* 121:185–99

81. Lehmann T, Marcet PL, Graham DH, Dahl ER, Dubey JP. 2006. Globalization and the population structure of *Toxoplasma gondii*. *Proc. Natl. Acad. Sci. USA* 103:11423–28

82. Li L, Brunk BP, Kissinger JC, Pape D, Tang K, et al. 2003. Gene discovery in the *Apicomplexa* as revealed by EST sequencing and assembly of a comparative gene database. *Genet. Res.* 13:443–54

83. Luft BJ, Remington JS. 1992. Toxoplasmic encephalitis in AIDS. *Clin. Infect. Dis.* 15:211–22

84. Mackinnon MJ, Read AF. 2004. Virulence in malaria: an evolutionary viewpoint. *Philos. Trans. R. Soc. London B. Biol. Sci.* 359:965–86

85. Manger ID, Adrian H, Parmley S, Sibley LD, Marra M, et al. 1998. Expressed sequence tag analysis of the bradyzoite stage of *Toxoplasma gondii*: identification of developmentally regulated genes. *Infect. Immun.* 66:1632–37

86. Marchall LG, Webb SD, Sepkoski JJ, Raup DM. 1982. Mammalian evolution and the great American interchange. *Science* 215:1351–57

87. Maynard Smith J, Haigh J. 1974. The hitch-hiking effect of a favourable gene. *Genet. Res.* 23:23–35

88. Mead PS, Slutsker L, Dietz V, McCaig LF, Bresee JS, et al. 1999. Food-related illness and death in the United States. *Emerg. Infect. Dis.* 5:607–25

89. Mondragon R, Howe DK, Dubey JP, Sibley LD. 1998. Genotypic analysis of *Toxoplasma gondii* isolates in pigs. *J. Parasitol.* 84:639–41

90. Montoya JG, Liesenfled O. 2004. Toxoplasmosis. *Lancet* 363:1965–76

91. Moura L, Bahia-Oliveira LMG, Wada MA, Jones JL, Tuboi SH, et al. 2006. Waterborne toxoplasmosis, Brazil, from field to gene. *Emerg. Infect. Dis.* 12:326–29

92. Ossorio PN, Sibley LD, Boothroyd JC. 1991. Mitochondrial-like DNA sequences flanked by direct and inverted repeats in the nuclear genome of *Toxoplasma gondii*. *J. Mol. Biol.* 222:525–36

93. Owen MR, Trees AJ. 1999. Genotyping of *Toxoplasma gondii* associated with abortion in sheep. *J. Parasitol.* 85:382–84

74. Demonstrated the recent origin and genetic sweep of a single monomorphic Chr1a in *T. gondii* and provided improved methods for population analysis.

77. Described the construction of a genetic linkage map of *T. gondii* and provided key genetic parameters that were used to assemble the genome and map complex phenotypes.

78. Demonstrated the feasibility of serotyping infections based on allele-specific peptides.

95. Provided key experimental data that there is no predefined mating type in *T. gondii*, hence a single strain can complete the entire life cycle.

96. Provided the first example of genetic crosses between strains of *T. gondii* and formed the foundation for current genetics in this system.

102. Mapped the genetic basis of differences in pathogenicity between parasite genotypes II and III and identified several loci including ROP18 as contributing to virulence.

105. First established the link to acute virulence with a particular clonal genotype.

112. Defined key transmission differences in the life cycle of *T. gondii* as newly emergent and related to the clonal population structure.

114. Defined the basis for differences in acute pathogenicity in mice between parasite genotypes I and III using classical genetics and identified the gene ROP18 as the key determinant of virulence.

94. Pfefferkorn ER, Pfefferkorn LC. 1979. Quantitative studies of the mutagenesis of *Toxoplasma gondii*. *J. Parasitol.* 65:363–70

95. **Pfefferkorn ER, Pfefferkorn LC, Colby ED. 1977. Development of gametes and oocysts in cats fed cysts derived from cloned trophozoites of *Toxoplasma gondii*. *J. Parasitol.* 63:158–59**

96. **Pfefferkorn LC, Pfefferkorn ER. 1980. *Toxoplasma gondii*: genetic recombination between drug resistant mutants. *Exp. Parasitol.* 50:305–16**

97. Reikvam S, Lorentzen-Styr AM. 1976. Virulence of different strains of *Toxoplasma gondii* and host response in mice. *Nature* 261:508–9

98. Rich SM, Licht MC, Hudson RR, Ayala FJ. 1998. Malaria's Eve: evidence for a recent population bottleneck throughout the world populations of *Plasmodium falciparum*. *Proc. Natl. Acad. Sci. USA* 95:4425–30

99. Roberts CW, Gazzinelli RT, Khan IA, Nowakowska D, Esquivel A, Mcleod R. 2007. Adaptive immunity and genetics of the host immune response. See Ref. 119, pp. 610–720

100. Roos DS, Crawford MJ, Donald RGK, Kissinger JC, Klimczak LJ, Striepen B. 1999. Origin, targeting, and function of the apicomplexan plastid. *Curr. Opin. Microbiol.* 2:426–32

101. Sabin AB. 1941. Toxoplasmic encephalitis in children. *JAMA* 116:801–7

102. **Saeij JPJ, Boyle JP, Coller S, Taylor S, Sibley LD, et al. 2006. Polymorphic secreted kinases are key virulence factors in toxoplasmosis. *Science* 314:1780–83**

103. Santos CBA, Carvalho ÂCFB, Ragozo AMA, Soares RM, Amaku M, et al. 2005. First isolation and molecular characterization of *Toxoplasma gondii* from finishing pigs from São Paulo State, Brazil. *Vet. Parasitol.* 131:207–11

104. Sibley LD. 2004. Invasion strategies of intracellular parasites. *Science* 304:248–53

105. **Sibley LD, Boothroyd JC. 1992. Virulent strains of *Toxoplasma gondii* comprise a single clonal lineage. *Nature* 359:82–85**

106. Sibley LD, LeBlanc AJ, Pfefferkorn ER, Boothroyd JC. 1992. Generation of a restriction fragment length polymorphism linkage map for *Toxoplasma gondii*. *Genetics* 132:1003–15

107. Silva AV, Pezerico SB, Lima VY, d'Arc Moretti L, Pinheiro JP, et al. 2005. Genotyping of *Toxoplasma gondii* strains isolated from dogs with neurological signs. *Vet. Parasitol.* 127:23–27

108. Silveira C, Belfort R Jr, Muccioli C, Abreu MT, Martins MC, et al. 2001. A follow-up study of *Toxoplasma gondii* infection in southern Brazil. *Am. J. Ophthalmol.* 131:351–54

109. Speer CA, Dubey JP. 1998. Ultrastructure of early stages of infections in mice fed *Toxoplasma gondii* oocysts. *Parasitology* 116:35–42

110. Speer CA, Dubey JP, Blixt JA, Prokop K. 1997. Time lapse video microscopy and ultrastructure of penetrating sporozoites, types 1 and 2 parasitophorous vacuoles, and the transformation of sporozoites to tachyzoites of the VEG strain of *Toxoplasma gondii*. *J. Parasitol.* 83:565–74

111. Striepen B, Jordan CN, Reiff S, van Dooren GG. 2007. Building the perfect parasite: cell division in *Apicomplexa*. *PloS Pathog.* 3:691–98

112. **Su C, Evans D, Cole RH, Kissinger JC, Ajioka JW, Sibley LD. 2003. Recent expansion of *Toxoplasma* through enhanced oral transmission. *Science* 299:414–16**

113. Su C, Howe DK, Dubey JP, Ajioka JW, Sibley LD. 2002. Identification of quantitative trait loci controlling acute virulence in *Toxoplasma gondii*. *Proc. Natl. Acad. Sci. USA* 99:10753–58

114. **Taylor S, Barragan A, Su C, Fux B, Fentress SJ, et al. 2006. A secreted serine-threonine kinase determines virulence in the eukaryotic pathogen *Toxoplasma gondii*. *Science* 314:1776–80**

115. Tibayrenc M, Kjellberg F, Araud J, Oury B, Breniere SF, et al. 1991. Are eukaryotic microorganisms clonal or sexual? A population genetics vantage. *Proc. Natl. Acad. Sci. USA* 88:5129–33

116. Tibayrenc M, Kjellberg F, Ayala FJ. 1990. A clonal theory of parasitic protozoa: the population structures of *Entamoeba*, *Giardia*, *Leishmania*, *Naegleria*, *Plasmodium*, *Trichamonas*, and *Trypanosoma* and their medical and taxonomic consequences. *Proc. Natl. Acad. Sci. USA* 87:2414–18

117. Vaidya AB, Akella R, Suplick K. 1989. Sequences similar to genes for two mitochondrial proteins and portions of ribosomal RNA in tandemly arrayed 6-kb-pair DNA of a malarial parasite. *Mol. Biochem. Parasitol.* 35:97–107

118. Weiss LM, Kim K. 2007. Bradyzoite development. See Ref. 119, pp. 341–66

119. Weiss LM, Kim K, eds. 2007. Toxoplasma gondii, *the Model Apicomplexan: Perspectives and Methods.* London: Academic/Elsevier
120. Wolfe ND, Dunavan CP, Diamond J. 2007. Origins of major human infectious diseases. *Nature* 447:279–83

RELATED RESOURCES

http://toxodb.org/toxo/
http://toxomap.wustl.edu/
http://toxohapmap.wustl.edu/

Peptide Release on the Ribosome: Mechanism and Implications for Translational Control

Elaine M. Youngman, Megan E. McDonald, and Rachel Green

Howard Hughes Medical Institute, Department of Molecular Biology and Genetics, Johns Hopkins University School of Medicine, Baltimore, Maryland 21205; email: ragreen@jhmi.edu

Annu. Rev. Microbiol. 2008. 62:353–73

First published online as a Review in Advance on June 10, 2008

The *Annual Review of Microbiology* is online at micro.annualreviews.org

This article's doi: 10.1146/annurev.micro.61.080706.093323

Key Words

protein synthesis, translation termination, release factors, induced fit

Abstract

Peptide release, the reaction that hydrolyzes a completed protein from the peptidyl-tRNA upon completion of translation, is catalyzed in the active site of the large subunit of the ribosome and requires a class I release factor protein. The ribosome and release factor protein cooperate to accomplish two tasks: recognition of the stop codon and catalysis of peptidyl-tRNA hydrolysis. Although many fundamental questions remain, substantial progress has been made in the past several years. This review summarizes those advances and presents current models for the mechanisms of stop codon specificity and catalysis of peptide release. Finally, we discuss how these views fit into a larger emerging theme in the translation field: the importance of induced fit and conformational changes for progression through the translation cycle.

Contents

INTRODUCTION TO TRANSLATION TERMINATION

RFs: release factor proteins

Class I release factor: RF1 and RF2 in bacteria, eRF1 in eukaryotes. Proteins that bind the stop codon and directly participate in catalysis of peptide release

Class II release factor: RF3 in bacteria, eRF3 in eukaryotes. These proteins are GTPases that facilitate translation termination, although their functions are distinct in bacteria and eukaryotes

Posttermination complex: a ribosome complex where peptidyl-tRNA hydrolysis has occurred, leaving a deacylated tRNA in the P site and a class I RF bound in the A site

The completion of a polypeptide chain is signaled in all organisms by the presence of one of three nearly universal stop codons in the ribosomal A site, marking the end of translation elongation. Unlike sense codons, which are decoded by aminoacyl-tRNAs (aa-tRNAs), stop codons are decoded by release factor proteins (RFs). Ultimately, stop codon recognition by RFs results in hydrolysis of the completed polypeptide chain from the P-site tRNA. Although the existence of protein accessory factors required for the termination of translation was demonstrated 30 years ago (7), termination is less well studied than elongation, and mechanistic understanding of the process is only beginning to emerge.

Two types of accessory factors are required for efficient termination. Class I RFs directly recognize the stop codon and participate in catalysis of the peptide release reaction. To decode the three stop codons (UAG, UAA, and UGA), bacteria possess two class I RFs (RF1 and RF2) with overlapping codon specificities (69), whereas eukaryotes and archaea have only one class I RF (eRF1), called omnipotent because it decodes all three stop codons (34).

Sequence analysis of the class I RFs indicates that the bacterial and eukaryotic/archaeal proteins evolved independently, with no homology at the sequence level beyond a GGQ tripeptide motif conserved in all known factors (18). Class II RFs are GTPases in all organisms but play distinct roles in bacteria and eukaryotes, consistent with the notion that they evolved to facilitate the actions of wholly different class I RFs.

In bacteria, the class II RF (RF3) functions in a step after peptidyl-tRNA hydrolysis to initiate disassembly of the posttermination complex. After the peptide release reaction has been catalyzed by a class I RF, the posttermination ribosome complex, which contains a deacylated tRNA in the P site and a class I RF in the A site, acts as a guanine nucleotide exchange factor (GEF), promoting the exchange of GDP for GTP on RF3 (92). RF3·GTP binds with high affinity to the ribosome complex and leads to the dissociation of the class I RF from the ribosome (17, 93). Recent structural data indicate that RF3·GTP binding promotes the dissociation of RF1 or RF2 not because they have overlapping binding sites but because RF3·GTP binding promotes conformational changes in the decoding and GTPase-associated centers of the ribosome that are incompatible with continued binding of RF1 and RF2 (19). RF3 in bacteria therefore accelerates the dissociation of class I RFs from ribosomal complexes once the peptide release reaction is complete. GTP hydrolysis by and dissociation of RF3 then leave the posttermination ribosome complex ready for disassembly by ribosome recycling factor (RRF) and other factors. RF3 deletion is not lethal in bacteria (22, 43), perhaps because the rate of class I RF recycling is accelerated only about fivefold in the presence of RF3·GTP (92).

The function of the eukaryotic class II RF (eRF3) in termination is distinct. Biochemical studies suggest that eRF1 and eRF3 are associated in the cytoplasm of the cell, where eRF1 effectively acts as a GTP dissociation inhibitor (TDI), preferentially stabilizing the GTP·eRF1·eRF3 ternary complex (55). It is this complex that interacts with the pretermination ribosome complex carrying a stop codon in the A site. Binding of the ternary complex appears to induce conformational changes in the ribosome complex that are independent of GTP hydrolysis or peptide release, because they are manifested in toeprinting experiments with a nonhydrolyzable GTP analog (GMPPNP) or with a catalytically inactive eRF1 variant (1). Optimal rates of peptide release catalyzed by eRF1 depend on the presence of eRF3 and GTP, although slow release is observed even in their absence. The dependence of peptide release on a hydrolyzable GTP argues that GTP hydrolysis precedes catalysis of peptide release, although the exact timing of these events has not yet been determined. This topic of considerable interest is not the primary focus of this review.

From a broad perspective, termination involves tasks similar to those encountered by aa-tRNAs during elongation: In each case, the codon presented in the small ribosomal subunit A site must be recognized, and subsequently a nucleophilic attack must be catalyzed in the distantly located active site of the large subunit. RFs and aa-tRNAs must interact with the same underlying ribosomal machinery to perform these tasks, and the idea that RFs are functional mimics of tRNA molecules is an often-repeated theme in the literature first presented over a decade ago (44). However, although both elongation and termination are facilitated by bifunctional molecules (tRNAs and class I RFs) interacting with the same conserved ribosome functional centers, obvious distinctions exist between these events. First, the codons presented during elongation are decoded by aa-tRNAs through RNA–RNA base-pairing interactions, whereas stop codons are decoded during termination by protein RFs, a process that depends on protein-RNA interactions that are currently uncharacterized. In addition, the attacking nucleophile during elongation is the primary amine of the incoming aa-tRNA and the result of the reaction is peptidyl transfer, whereas during termination the nucleophile is a water molecule and the reaction results in peptidyl-tRNA hydrolysis.

A great deal of progress has been made in understanding the mechanisms that underlie elongation, such as the mechanism of ribosomal catalysis of peptidyl transfer, communication between the two ribosomal subunits, and the role of ribosomal motions in the various steps of the elongation cycle. Given the similarities between peptide bond formation and peptide release, this progress will also inform future studies of termination. Study of elongation mechanisms has been advanced by an expansion of available technologies, from the development of modern rapid kinetic approaches and fluorescence-based reporters for events along the elongation pathway to the determination of high-resolution crystal structures of biologically meaningful elongation complexes. The toolkit available for the study of release lags behind substantially. As has been the case for other ribosome–accessory factor complexes, high-resolution structures of ribosome-bound RFs have been difficult to obtain (54), and structures of isolated RFs do not mimic the ribosome-bound situation (75, 83). Rapid kinetic assays are available but incompletely characterized, and knowledge of any kinetic intermediates that exist in the reaction pathway is lacking. As a result, the fundamental questions associated with peptide release, i.e., how RFs specifically recognize stop codons, and how the RF-ribosome complex then catalyzes the hydrolysis reaction, remain incompletely understood.

This review summarizes recent inroads into addressing these questions and discusses some key experiments that remain. From a broader perspective, the big picture that has emerged in recent years in the ribosome field is that the ribosome is a remarkable macromolecular machine. Life depends on fast and accurate protein synthesis—in molecular terms, this means quick decisions about right and wrong substrates. The underlying design principles that accomplish this fundamental task during each step of the translation cycle, including termination, are likely at the heart of not just everyday protein synthesis, but its regulation as well. We conclude with a discussion of this emerging theme of switch-like conformational behavior in the ribosome and how regulatory mechanisms may directly act on these ubiquitous switches.

CLASS I RELEASE FACTORS AS tRNA MIMICS

Although high-resolution structural information remains unavailable for the full pretermination complex (i.e., a class I RF bound to a stop-codon-programmed ribosome), a variety of structural approaches have led to significant advances in our understanding of RF-ribosome interactions. The similarity between the basic function of class I RFs and aa-tRNAs leads to obvious predictions that the class I RFs are functional and thus structural tRNA mimics (44). Here we discuss biochemical and structural data that validate this model.

At the simplest level, if the class I RF is a tRNA mimic, it must bind at some stage of the termination cycle to the ribosome in the A site where the tRNA normally binds during elongation. Specific regions of class I RFs have been identified as likely playing key roles in the two broad functions of these factors. Interactions with stop codons in the decoding center (DC) involve the PA(V)T/SPF motif in bacteria (known as the peptide anticodon) and, more speculatively, NIKS in eukaryotes. Promotion of peptide hydrolysis in the peptidyl transferase center (PTC) is linked to the GGQ domain in both kingdoms. These regions were identified as crucial functional elements in both genetic and biochemical experiments, and a number of structural approaches place them in the appropriate ribosome functional centers to perform these tasks. Supporting the idea that RFs directly bind and decode the stop codon in the small-subunit A site, both bacterial (5, 76) and eukaryotic RFs (9) can be cross-linked to termination complexes bearing reactive 4-thio-U residues in or immediately adjacent to the stop codon. Further experiments mapped the cross-link site between a 4-thio-U at the first position of a stop codon and eRF1 to a KSR tripeptide (amino acids 63–65) that overlaps the highly conserved NIKS motif (amino acids 61–64) (11).

Further, directed hydroxyl radical footprinting experiments using Fe-BABE tethered to bacterial RF1 or RF2 at single cysteine residues placed the PA(V)T/SPF tripeptide anticodon motif in the small subunit near the DC and the GGQ motif in the large subunit proximal to the PTC, the A loop, and the GTPase-associated center (66, 86). Additional strong support for the idea that RFs are functional mimics of tRNAs comes from recent low-resolution structures of ribosome-bound bacterial RFs: Both cryo-EM structures (33, 59, 60) and 6 Å resolution crystal structures (54) of RF1 and RF2 bound to stop-codon-programmed ribosomes reveal an extended class I RF molecule binding to the ribosome, easily spanning the two key functional centers. Although no ribosome-bound structures exist for the eukaryotic system, high-resolution structural data on the isolated class I RF reveal two functionally important lobes that could similarly span this same distance (75). Indeed, a glance at the superimposition of the class I RF from the low-resolution bacterial structure on an A-site-bound tRNA leaves no doubt that the functional ends of the RF (PVT and GGQ) are equivalent to the functional ends of the tRNA (the anticodon and CCA) (**Figure 1**).

Although high-resolution views of the interactions between class I RFs and ribosome functional centers are not yet available, several high-resolution structures of class I RFs in isolation have been solved, including both bacterial and eukaryotic representatives (73, 75, 83; Accession no. 1ZBT). These structures immediately reveal that the near-complete absence of homology between bacterial and eukaryotic RFs extends beyond the sequence level: The tertiary structure of these factors is also distinct. For example, although both bacterial and eukaryotic class I RFs present the conserved and functionally critical GGQ motif at the tip of an extended loop, the arrangements of secondary structure elements that make up their overall structure are entirely different (**Figure 2**). Despite these differences, both bacterial and eukaryotic RFs

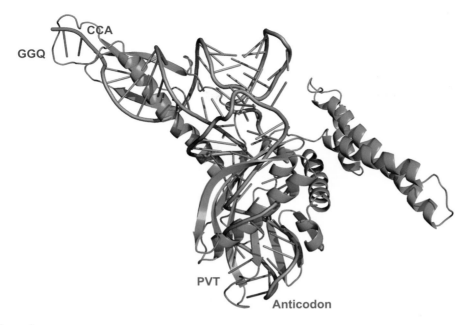

Figure 1

Class I release factors (RFs) are tRNA mimics. The overlaid structures of A-site-bound *Thermus thermophilus* RF1 and tRNA^Phe demonstrate that, although class I RFs are not strict structural mimics of a tRNA molecule, their functional ends (GGQ and PVT) interact with the same ribosome centers as the functional ends (CCA and the anticodon loop) of an A-site tRNA. Structures were overlaid by aligning the backbones of ribosomal proteins from the ribosome-bound RF1 structure (PDB ID 2B64) and the *T. thermophilus* 70S ribosome structure with bound mRNA and tRNAs (91) (PDB ID 1JGQ). The ribosome-bound RF1 structure shown here and throughout the figures is a model derived from the Cα positions of ribosome-bound RF1 using Quanta and O (30) to model side chain coordinates. All structure figures were made using PyMol (**http://pymol.sourceforge.net/**). Adapted from Reference 54, with permission from Elsevier.

are composed of multiple globular domains, and in both cases it has been proposed that flexibility of the domains relative to one another is important for termination function. In neither case can the class I RF conformation observed in isolated structures be directly accommodated into the presumed relevant functional position on the ribosome, leaving open the question of how the RF binds the ribosome in its functionally relevant conformation from solution.

Bacterial RF1 and RF2 are composed of four domains: The N-terminal domain 1 can be removed from RF1 without any impact on peptide release activity but is required for the downstream GTP-dependent acceleration of class I RF dissociation from the ribosome by

RF3 (45). Domain 2 contains the PA(V)T/SPF motif responsible for codon specificity, packing against domain 4 to form the rigid central core of the molecule, and domain 3 contains the GGQ motif important for catalysis. In crystal structures of *Thermotoga maritima* RF1 (73) and *Escherichia coli* RF2 (83), the structure of each of these individual domains is highly similar, as expected from the high degree of sequence homology between RF1 and RF2. However, RF1 and RF2 are in dramatically different conformations when in isolation compared with their ribosome-bound states. In isolated crystal structures, there is a large interaction surface between domain 3 and domains 2 and 4, resulting in a closed

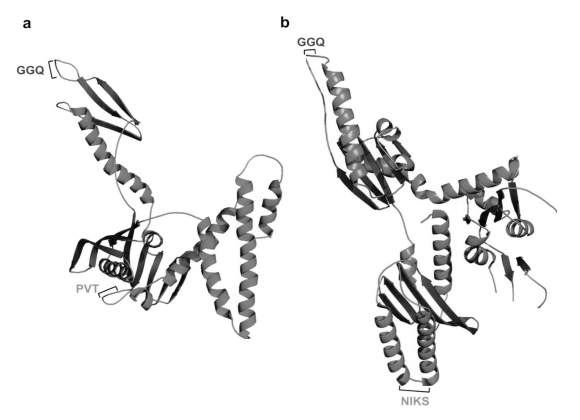

a GGQ   PVT

b GGQ   NIKS

Figure 2

The structures of bacterial and eukaryotic class I release factors (RFs) are not conserved. Although they perform equivalent functions, the overall structures of (*a*) *Thermus thermophilus* RF1 modeled from a ribosome-bound structure in an open conformation and (*b*) human eRF1 are not related (PDB ID 1DT9).

conformation where the PA(V)T/SPF anti-codon motif and GGQ catalytic motifs are separated by less than 25 Å, insufficient to span the 75 Å between the decoding and catalytic centers on the ribosome. The transition to the extended conformation of RF1 and RF2 seen on the ribosome is accommodated by dramatic movement of domain 3 away from the core of the molecule, with the loop between domains 2 and 3 acting as a hinge (**Figure 3**). In its extended form on the ribosome, domains 2, 3, and 4 occupy roughly the same space as an A-site tRNA (**Figure 1**). Domain 1 extends beyond the dimensions of a tRNA and interacts with the L11/helix 44 region of the large subunit, although the conformation of domain 1 and interactions with the

ribosome are somewhat different for RF1 and RF2 (54).

Human eRF1 is composed of three domains arranged roughly into a Y shape, with each domain projecting away from the core of the molecule (75). The N domain contains the NIKS loop, thought to interact with the stop codon; the M domain contains the catalytically critical GGQ motif presented on a loop at the end of a long α-helix that extends away from the body of the protein; and the C domain contains residues required for interaction with eRF3 (26, 42). The distance between the GGQ motif at the tip of domain M and the region thought to be responsible for codon interaction at the tip of domain N, between 80 and 100 Å, is not

a   b

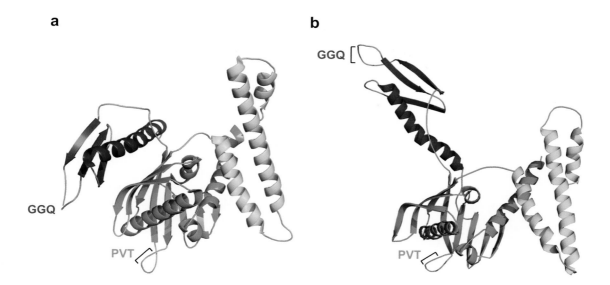

Figure 3

A comparison of the closed and open conformations of bacterial class I release factors (RFs). The closed conformation (*a*) is seen in all structures of bacterial class I RFs in isolation, whereas the open conformation (*b*) is observed in all ribosome-bound structures. Shown here as representative examples are (*a*) the crystal structure of *Thermotoga maritima* RF1 (73) and (*b*) the ribosome-bound crystal structure of *Thermus thermophilus* RF1 (54). Yellow, domain 1; red, domain 3; blue, domain 2/4.

exactly compatible with the requirement for simultaneous interaction with the PTC and DC, respectively, which are separated by ~75 Å. The difference in these two distance measurements could be accounted for in part by flexibility of the GGQ loop, demonstrated by recent NMR evidence (29).

CATALYSIS OF PEPTIDE RELEASE (AND INDUCED FIT)

At the conclusion of translation elongation the completed polypeptide chain remains esterified at its C terminus to the P-site tRNA. In peptide release, nucleophilic attack of the ester carbon by water releases the finished protein from the P-site tRNA. Peptide release is related to peptide bond formation, differing only in the nature of the attacking nucleophile (the primary amine of the incoming aa-tRNA in the case of peptide bond formation and water in the case of peptide release) and in the requirement for a protein accessory factor in the case of peptide release. Both reactions are catalyzed in the

same active site on the large ribosomal subunit and begin with the same ground state complex where the labile ester linkage of the peptidyl-tRNA is poised in the PTC.

Important features of this ground state have been deciphered from experiments aimed at understanding the mechanism of peptide bond formation, and these conclusions broadly apply to peptide release. Structurally, when the large-subunit A site is empty, the electrophilic carbon of the P-site tRNA is protected from attack by two mechanisms. First, access to the labile ester is sterically blocked on either side by 23S rRNA nucleotides. Second, the P-site substrate is rotated away from the A-site side. These details ensure that the key electrophilic carbon is not optimally oriented for the requisite in-line attack (**Figure 4**), whether the nucleophile is a tRNA-esterified amino acid or water. During peptidyl transfer, when an aa-tRNA substrate with an intact $C_{74}C_{75}A_{76}$ terminus binds in the A site, interactions with the 23S rRNA A loop lead to extensive conformational rearrangement of the PTC that ends in an

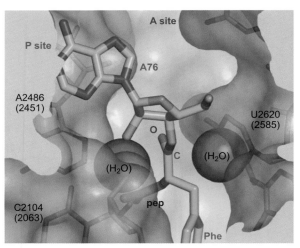

Figure 4

P-site-bound peptidyl-tRNA is protected from nucleophilic attack by water in the absence of an A-site ligand. Shown is a surface representation of large-subunit rRNA residues surrounding the peptidyl-tRNA mimic CCA-phenylalanine-caproic acid-biotin from a crystal structure with *Haloarcula marismortui* ribosomes (numbers in parentheses are *Escherichia coli* numbering). A sphere with the radius of a water molecule is shown at what would be the optimum angle for in-line attack of the ester carbon. Reprinted by permission from Macmillan Publishers Ltd: Nature (Reference 68), copyright 2005.

induced state in which the nucleophilic amine is well-positioned for attack on the ester carbonyl carbon (68). The structural data correlate well with biochemical data highlighting the importance of the interaction between the A loop and the A-site tRNA for maximal rates of catalysis. The minimal A-site substrate puromycin contains only the terminal adenosine of the CCA end, and rates of peptide bond formation with this substrate are sensitive to mutation of active site nucleotides, the length of the peptidyl moiety in the P site, and changes in pH (6, 32, 89). In contrast, intact aa-tRNA substrates are refractive to these same perturbations (3, 89), and addition of the C75 equivalent to puromycin (CPm) is sufficient to confer robust behavior to this minimal substrate (2, 6). Induced fit thus explains in structural terms a key contributor to control of catalysis (protection of the peptidyl-tRNA) and thus to catalysis itself.

Given the similarities between peptide bond formation and peptide release, one obvious approach is to ask to what extent the same strategies might be employed in catalysis of peptide

release. As a first clue, studies in the 1970s (and again more recently) showed that binding of a deacylated tRNA in the A site results in a large (up to 300-fold in some circumstances) increase in the rate of peptidyl hydrolysis, supporting the idea that engagement of the A loop by the CCA end of an A-site tRNA induces a catalytically active state (8, 72, 93). It is a priori clear that one function of RFs must be to induce the active conformation of the PTC, and recent studies helped to define the contributions of an induced-fit component to the estimated $2 \cdot 10^4$ in release catalysis (72). It is not yet clear whether, like tRNAs, they accomplish this through interaction with the ribosomal A loop or through a distinct mechanism, or to what extent the final induced state for peptide release resembles that of peptide bond formation.

Beyond induced fit, other contributors to catalysis of PT have been identified. In the most current model for peptide bond formation (61), the A76 2'OH of the P-site tRNA, which is within hydrogen-bonding distance of the attacking amine in prereaction crystal structures, participates in a proton shuttle mechanism that ultimately results in loss of a proton from the attacking amine and donation of a proton to the leaving 3'OH of the P-site tRNA. This mechanism invokes a six-membered transition state with an oxyanion that appears in crystal structures to be stabilized by a bound water molecule, which in turn is coordinated by PTC nucleotides (67). Consistent with this model, removal of the terminal 2'OH from the P-site tRNA results in a dramatic reduction in the rate constant for PT (85): on the order of 10^4-fold of a total rate acceleration of approximately 10^5-fold (74). Similar conclusions were reached in molecular dynamics simulations of the reaction (71, 77). Although equivalent data are not yet available for any eukaryotic system, the high degree of conservation of the elements involved suggests that similar mechanisms will control catalysis there as well.

Other potential contributions to catalysis by class I RFs have been evaluated in recent mutational and computational studies. Because water is an inherently poorer nucleophile than

Induced fit: a model for enzyme specificity; proposes that correct substrates induce structural changes in the enzyme that make it a better catalyst

the primary amine involved in peptide bond formation, catalysis of peptide release might require more active modes of catalysis than peptide bond formation, such as a general acid/base mechanism. Unlike the amino acid of the PT reaction, water is not tethered to the active site by covalent bonds, thus immediately suggesting that careful packing of the active site might be important for optimal catalysis. Consistent with these ideas, the release reaction is indeed more sensitive to mutation of PTC rRNA nucleotides than is peptide bond formation (89). Mutation of the active site nucleotides closest to the presumed center of chemistry led to defects in peptide release up to 300-fold in the case of mutations of A2602, whereas none of these mutations had an impact on the rate of peptide bond formation (15, 56, 89).

On the RF, similarity to the universally conserved CCA ends of tRNAs led to the early proposal that the universally conserved GGQ motif would interact with the PTC (18). Subsequently, the various structural studies noted above confirmed this prediction. Substantial attention has therefore been paid to this motif, and numerous experimental approaches have highlighted its importance for release function in vitro and in vivo both in bacteria and eukaryotes (18, 45, 70, 72, 75, 93). Surprisingly, however, mutation of either of the two glycine residues dramatically impairs peptide release (72, 93), but mutation of the GGQ glutamine has much less dramatic consequences (45, 72, 93) (**Table 1**). These data call into question early models suggesting that the conserved glutamine is critical for coordination of the attacking water molecule (75): If it were critical, one would expect mutation of the glutamine to result in larger deficiencies in peptide release. Recent nucleophile partitioning experiments from our own lab indicate that the conserved glutamine residue is important in excluding other nucleophiles from the active site (72), consistent with the idea that careful packing of this catalytic center is important to optimal RF function. Extensive mutagenesis of conserved amino acids in the GGQ domain of *E. coli* RF1 failed to find any residues other than the GGQ

Table 1 Summary of the effects of mutations in the conserved GGQ motif on class I RF activity

RF type	Mutant	Measurement	Rate/endpoint	Reference
eRF1	WT	k_{cat}	$7.8 \cdot 10^{-3}$ s^{-1}	(64)
eRF1	GGG	Endpoint	60%	(70)
RF1	WT	k_{cat}	0.5 s^{-1}	(93)
			0.65 s^{-1}	(72)
RF1	GAQ	k_{cat}	$11.3 \cdot 10^{-5}$ s^{-1}	(93)
RF1	GAQ	k_{cat}	$1.5 \cdot 10^{-4}$ s^{-1}	(72)
RF1	GAQ	Predicted k_{cat}	$\sim 5 \cdot 10^{-5}$ s^{-1} [a]	(78)
RF1	GGA	k_{cat}	0.18 s^{-1}	(72)
RF1	GGA	Predicted k_{cat}	$\sim 7 \cdot 10^{-4}$ s^{-1} [a]	(78)
RF1	AGQ	k_{cat}	$6.3 \cdot 10^{-4}$ s^{-1}	(72)
RF1	WT	k_{cat}/K_m	49 μM^{-1} s^{-1}	(45)
RF1	GGA	k_{cat}/K_m	10.2 μM^{-1} s^{-1}	(45)
RF2	WT	k_{cat}	1.5 s^{-1}	(16)
RF2	GAQ	k_{cat}	$7.7 \cdot 10^{-5}$ s^{-1}	(93)
RF2	GGA	k_{cat}	$1.7 \cdot 10^{-2}$ s^{-1}	(93)
RF2 (T246A)	WT	k_{cat}/K_m	62.5 μM^{-1} s^{-1}	(45)
RF2 (T246A)	GGA	k_{cat}/K_m	13.1 μM^{-1} s^{-1}	(45)

[a]Calculated from fold reductions reported in Reference 78, based on a WT RF1 rate constant of 0.5 s^{-1}.
Abbreviations: RF1, RF2, bacterial class I release factor; eRF1, eukaryotic class I release factor; WT, wild type.

glycines whose mutation caused dramatic reductions in the rate constant for peptide release (72), leaving the hypothesis for general acid/base contributions from the RF unlikely. Although careful kinetic analysis of the equivalent mutations in eRF1 has not yet been performed, initial experiments support the model derived from the bacterial system: Mutation of either glycine yields eRF1 with no detectable release activity, whereas mutation of the glutamine to any of several amino acids results in only a partial loss of activity in endpoint assays (18, 70).

Beyond conservation of the GGQ tripeptide sequence, posttranslational modification of the glutamine to N^5-methylglutamine appears to be conserved in both bacteria and eukaryotes (14, 23, 24, 57). This methylation is accomplished by nonessential genes in both *E. coli* (PrmC) and budding yeast (Mtq2p) (23, 24, 57). A model for catalysis based on a molecular dynamics approach similar to one successfully used to model peptide bond formation (78) concluded that the carbonyl group of the glutamine side chain acts to position a water molecule in the active site pocket, and that methylation of the glutamine contributes critical packing interactions that orient the side chain. This study further suggested that the GGA mutant is relatively active because an extra water molecule positioned by the protein backbone occupies the space normally occupied by the carbonyl oxygen. The model derived from this simulation additionally proposes that, as in PT, the RF acts in part by promoting the induced state of the PTC, and that the 2′OH of the P-site tRNA acts as a proton shuttle. In agreement with this model, recent unpublished work from our own lab indicates that removal of the terminal 2′OH from the P-site tRNA nearly abolishes the release activity of *E. coli* RF1 (J. Brunelle, J. Shaw, E. Youngman & R. Green, manuscript in preparation). This computational model, however, likely overestimates the contribution of the methyl group to catalysis, as overexpressed bacterial RFs are severely undermethylated and yet generally have nearly wild-type activity (14).

The combination of these data suggests that the mechanism for peptide release is strikingly similar to that of peptide bond formation: Class I RFs act as tRNA mimics to activate the PTC for catalysis in a mechanism that uses the A76 2′OH of the P-site tRNA in a proton shuttle. The attacking water molecule is positioned by the GGQme in the context of a tightly folded pocket created by rRNA residues and the remainder of the GGQ loop.

SPECIFICITY OF PEPTIDE RELEASE (AND INDUCED FIT)

In vivo, bacterial RFs catalyze erroneous peptide release at less than 1 in 10^5 sense codons (31), about as accurate as tRNA-mediated decoding of sense codons during elongation, which has an error rate of 1 incorrect amino acid incorporated per 10^5–10^6 codons (36). The question of how RFs terminate translation with such high fidelity can be divided into several components. First, what are the molecular interactions between RFs and stop codons that facilitate codon-specific binding of the factors? Second, are thermodynamic differences in the affinity of RFs for sense and nonsense codons sufficient to explain the accuracy of termination, or are kinetic mechanisms also involved? Third, does the ribosome itself play a role in achieving the specificity of termination?

In contrast to the better-studied example of tRNA selection, in which the molecular interactions that ultimately allow the decoding of sense codons by aa-tRNAs are RNA-RNA Watson-Crick-like interactions between the codon and anticodon, the protein-RNA interactions involved in RF recognition of stop codons remain unknown. Genetic experiments in bacteria that sought a swap of specificity between UAG-specific RF1 and UGA-specific RF2 identified a motif in each factor, PA(V)T in RF1 and SPF in RF2, that is at least partially responsible for codon specificity in these factors and is referred to as the tripeptide anticodon (28). Structural studies discussed above confirmed that these sequence motifs are indeed near the stop codon in RF-ribosome complexes, although no structures of sufficient resolution are yet available to discern the details of this interaction. In the eukaryotic system, it has been more challenging to narrow the region of eRF1 responsible for stop codon recognition down to a few amino acids. From a conceptual standpoint, this challenge is different from that faced by the two distinct factors used in bacterial systems. Each bacterial factor is degenerate at only a single position, i.e., RF1 recognizes U in the first position, A in the second position, and either purine at the third position. In contrast, eRF1 must recognize U in all three stop codons, and then either AA, AG, or GA at the second and third positions, but not the GG combination, meaning that simply allowing only purines in the second and third positions is not sufficient (46). Although a conserved NIKS motif is important for eRF1 function, and cross-linking experiments place it near the stop codon (11), this motif is likely responsible for recognizing what is common to all three stop codons (the first codon position U) rather than the subsequent purine residues. Comprehensive sequence analysis across evolutionary space that includes organisms in which recoding of certain stop codons has occurred (i.e., where eRF1 has narrowed specificity) has revealed key amino acids likely to be important for distinguishing among stop and sense codons (38). Analysis of these sites will be critical in determining how this molecular challenge was solved by evolution (37, 65).

Apart from structural knowledge of the molecular interactions between RFs and stop codons, a full accounting for the observed fidelity of termination in vivo will also require detailed knowledge of both the thermodynamic and kinetic parameters of the system. In the simplest model, the specificity of peptide release could result entirely from differences in the affinity of RFs for stop- versus sense-codon-programmed ribosome complexes, and knowledge of the energetics of the interactions between RFs and stop codons would be sufficient to account for observed accuracy (a thermodynamic discrimination model). Alternatively, kinetic mechanisms could also be involved, and a

priori could include a two-step discrimination process that depends on the GTP hydrolysis activity of the class II RF. Such models would parallel current models for discrimination during tRNA selection and the essential role of GTP hydrolysis by EF-Tu in this process (62).

To distinguish among these possibilities, Freistroffer et al. (16) exhaustively measured the kinetic parameters k_{cat} and $K_{1/2}$ for peptide release by both E. coli RF1 and RF2 on their cognate stop codons as well as on the set of near-stop sense codons (differing at one position from the cognate set). These experiments were performed in both the presence and absence of the GTPase RF3. In no case did addition of RF3 enhance discrimination against near-stop codons, dismissing potential parallels to the kinetic proofreading mechanism utilized during tRNA selection. As expected, there are substantial differences in the $K_{1/2}$ of class I RFs for stop versus near-stop codons that likely reflect differences in affinity of the RF for these complexes. However, the $K_{1/2}$ values for peptide release on the set of near-stop codons are relatively uniform, suggesting that the energetics of these near-cognate interactions are not primarily driven by interactions of the RF with specific nucleotides of the codon, but rather by more general features of these complexes. This same study further noted that discrimination against near-stop codons also relies on differences in k_{cat} (16). Thus, even when near-stop complexes are completely bound by RF, catalysis is not rapid. These data argue that binding of RFs to stop codons is different from baseline binding to near-stop codons, inducing a particular activated state of the ribosome complex.

A recent study from our own lab provided further structural evidence for this induced-fit mechanism based on differences in release sensitivity of stop and near-stop codon ribosome complexes to the DC-binding aminoglycoside paromomycin (90). Finally, cross-linking experiments in a eukaryotic system have also provided support for models invoking a codon-independent binding step of the class I RF to the ribosome followed by rearrangement to a stop-codon-specific state that is fully active for peptide release (10). These data together are reminiscent of the mechanism of tRNA selection during elongation, in which initial codon-independent binding is followed by a codon recognition step and induced-fit rearrangements that ultimately promote downstream peptide bond formation (62).

The proposed induced-fit mechanism argues that conformational changes originate in the DC on stop codon recognition by the RF and ultimately lead to a lowering of the energy of the transition state for peptide release, a reaction catalyzed 75 Å away in the PTC. One possibility is that conformational changes are transduced to the large subunit through the RF itself. Genetic screens have identified mutations in domains 2 and 3 of the bacterial class I RF—both distant from the tripeptide anticodon motif—that can alter the specificity of peptide release (27, 79, 88). In the simplest case,

there are RF2 variants that function as the sole class I RF in bacteria, thus recognizing all three stop codons with some efficiency; these are called omnipotent suppressors (27, 79). These general up variants typically carry a charge change (E/D to K) at positions in domain 2 reasonably near the interface with domain 3 (**Figure 5**). It is speculated that these mutations impact interdomain movements (33, 60) and may be analogous to the Hirsh suppressor tRNA, which bears a mutation in the D arm—distant from the anticodon—and shows increased levels of miscoding at near-cognate codons (25). In the same way that deciphering the mechanism by which the Hirsh mutation alters the accuracy of tRNA selection provided insight into the role of tRNAs in induced fit for maintaining the fidelity of tRNA selection (12), thorough analysis of these RF mutations should reveal new insights into the mechanism

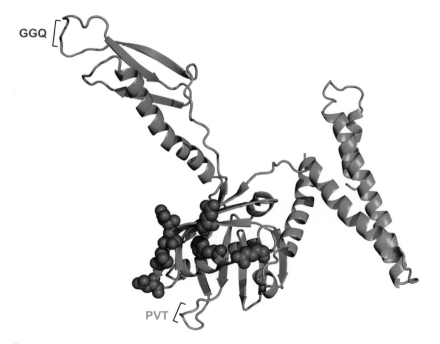

Figure 5

Charge-variant mutants that reduce the accuracy of peptide release by *Escherichia coli* RF2. Mutations in RF2 isolated by Ito and colleagues (27, 79) that can complement the growth of a strain lacking both RF1 and RF2 mapped onto the ribosome-bound *Thermus thermophilus* RF1 structure (54). The mutations, all Glu to Lys variants, are located in domain 2 distant from the PAT anticodon motif.

of action of class I RFs in peptide release. Early work in our lab indicates that these mutations do not simply bind more tightly to ribosome complexes programmed with sense codons, but actually result in an increase in the observed rate constant of peptide release (C. Louie, E. Youngman & R. Green, unpublished data). Computational approaches have suggested that protonation of histidine residues in the hinge regions of bacterial RFs may promote critical conformational changes (40), and it is possible that these charge-variant RFs similarly affect this same equilibrium in the RF.

CONFORMATIONAL CHANGES AND TRANSLATION TERMINATION

Kinetic and structural evidence discussed throughout this review makes a strong argument that conformational changes in the ribosome, RFs, or likely both play important roles in multiple steps of translation termination. On the one hand, conformational rearrangements within the ribosome have been described in some detail on the basis of cryo-EM and higher-resolution X-ray structures. These structural states may be key for understanding termination as well, though current structural data on termination complexes cannot yet decipher the key rearrangements. On the other hand, there is also much support for the idea that conformational rearrangements in the RFs themselves may be critical. The observation that the structures of RF1 and RF2 are dramatically different in isolation and on the ribosome immediately leads to the model that bacterial RFs exist in a closed conformation in solution and rearrange to their open, active form upon ribosome binding and, presumably, stop codon recognition. This model is appealing in that it offers a structural basis for the mechanism of specificity of peptide release and for RF-mediated activation of the PTC. As mentioned above, a molecular modeling approach argued that protonation of conserved histidine residues in a negative electrostatic environment could drive the transition from the closed to open form (40).

Other results, however, call such large-scale conformational changes into question. In a crystal structure of *E. coli* RF1 in complex with the methyltransferase (PrmC) responsible for methylation of the GGQ glutamine, PrmC makes essential contacts both with the GGQ-containing domain 3 and with the PAT-containing domain 2 of RF1 in its closed conformation (21). A simple model, then, argues that the closed conformation of bacterial class I RFs exists to allow interaction with the methyltransferase and does not play an important role in the termination function. Small-angle X-ray scattering (SAXS) data collected from *E. coli* RF1 in solution are more consistent with the open conformation observed on the ribosome than the closed conformation seen in crystal structures (82), although similar experiments with the *T. thermophilus* RF2 argued that most of the population (80%) was in the closed conformation in solution (94). These differences may reflect alternative equilibria for factors from mesophilic and thermophilic bacteria. What the equilibrium between these two states is, and whether the closed conformation has any affinity for the ribosome, remain important unresolved questions.

In addition to these rather large-scale rearrangements of the RF, smaller-scale conformational changes may also play an important role in interactions with the ribosome. A recent NMR study of the GGQ-containing M domain of human eRF1 revealed a large degree of conformational flexibility in the GGQ loop (as well as in the loop at the C-terminal end of α helix 8) (29). This C-terminal loop could come into contact with the NIKS-containing N domain in the intact protein, and the authors suggest that conformational changes could be transmitted upon stop codon recognition via either the N domain or ribosomal elements to this C-terminal loop, and subsequently through the more rigid α helix 8 to the GGQ loop, resulting in a catalysis-competent conformation.

Finally, downstream of peptide release itself, other conformational rearrangements are likely involved in RF3-mediated dissociation of class I RFs from the ribosome following peptide release. The observation that deacylated tRNA in the P site stimulates the GTPase activity of RF3 by 15-fold relative to peptidyl-tRNA suggested that release of the peptide from P-site tRNA, and subsequent movement of the CCA end of this tRNA to the E site, could promote an alternate conformational state of the ribosome. Indeed, it has previously been postulated that movement of deacylated P-site tRNA to the hybrid P/E configuration unlocks the ribosome such that binding of the GTPase EF-G during elongation promotes a counter-clockwise rotation of the small subunit relative to the large subunit (80). Cryo-EM structures recently showed rotational movements of ribosome subunits in posttermination ribosomes bound by RF3, and the authors propose that these rearrangements underlie the mechanism for the RF3-mediated increase in the rate of class I RF dissociation from the ribosome (19).

SWITCHES AND REGULATION

When integrated into the larger picture of the mechanism of translation on the ribosome, recent advances in termination add to an emerging common theme: Ribosome function is specified by conformational changes that directly impact the rates of many steps throughout the translation cycle. During elongation, conformational switches in the small-subunit DC are triggered by recognition of cognate codon:anticodon helices, thus facilitating subsequent steps in the tRNA selection process (GTPase activation and tRNA accommodation). Similarly, conformational switches in the large-subunit PTC are triggered by the CCA end of the EF-Tu-deposited A-site tRNA, promoting efficient catalysis in this active site. Here we have described phenomenologically similar events that take place during the termination step of protein synthesis. These examples highlight a repeating theme of switch-like behavior in ribosome functional centers. The ribosome has evolved elegant molecular mechanisms for substrate selection during protein synthesis, thus ensuring not only efficient but high-fidelity protein synthesis.

It is clear that such switch-like behavior is critical to everyday protein synthesis. In addition, it is easy to imagine that there might be regulatory mechanisms that take advantage of this fundamental behavior. Most discussions of translational control focus on regulatory events that target the initiation steps of protein synthesis in bacteria and eukaryotes, which often simply prevent the ribosome from accessing the mRNA of interest by physical obstruction. This barrier is most commonly accomplished in bacteria by sequestering the Shine-Dalgarno initiation sequence and in eukaryotes by disrupting interaction with the mRNA cap that is essential for AUG start site recognition. Because these regulatory events are not ribosome targeted we will not consider them further here.

More downstream events in the initiation process, beyond cap recognition and scanning, appear to depend on conformational switches of the ribosome itself. For example, conformational rearrangements in the 43S initiation complex in eukaryotes have been proposed to be key for triggering GTP hydrolysis by eIF2 and release of eIF1 on specific recognition of AUG start codons (41, 52). Such conformational switches regulating the fidelity of initiation may be related to those that occur during elongation.

While much regulation appears to target translation initiation, there are scattered examples in the literature of regulation during elongation, though the molecular mechanisms at play are far from understood. The most well-characterized cases occur in bacteria, in which specific peptide sequences, often in conjunction with ribosome-binding small molecules, are critical for mediating stalling during elongation or termination. In one well-studied example, accessibility of the Shine-Dalgarno sequence of the *secA* gene is controlled by ribosome stalling at a 17-meric amino acid sequence near the C terminus of the upstream open reading frame *secM* (48, 49). The stalling event occurs with Pro-tRNA^{Pro} bound in the A site but is unreactive toward the P-site peptidyl-tRNA (47). Stalling seems to require communication between the peptide in the exit tunnel and the active site, because mutations of ribosomal protein L22 that widen the tunnel lead to defective stalling (49) that, remarkably, can be rescued by mutations in the nascent peptide thought to alter its conformation in the tunnel (87).

In the *E. coli tna* operon, stalling at the upstream leader peptide *tnaC* controls mRNA levels through a transcription attenuation mechanism; in this case, ribosomes stall in the presence of high concentrations of free tryptophan at the UGA stop codon with RF2 bound but inactive. Tryptophan appears to bind in the vicinity of the large-subunit A site, and the stalled peptide is inactive not only in peptide release but also in peptide bond formation with puromycin, suggesting a general inactivation of the catalytic center (20). As with *secM*, stalling appears to depend on interactions between specific, critically positioned amino acids and the exit tunnel (20). The shared features of these and other examples (39, 58, 84) highlight a potential communication network that is triggered in the exit channel and leads to shutdown of the catalytic center of the large ribosomal subunit. It is appealing to think that the inactivation of catalysis in this case depends on a reversal of the same conformational rearrangements that we previously described as essential for generally activating the PTC.

As one might suspect, communication from the exit channel seems to propagate in two directions—into the catalytic core of the ribosome and out to the cytoplasmic environment. This point has been beautifully demonstrated through studies on the SRP and its engagement with the ribosome. In this system, affinity of the SRP for the ribosome is modulated by nascent peptide still fully contained within the exit channel (4). In this case, signaling to the exterior of the ribosome by the nascent peptide appears to depend on interactions with the large-subunit protein L23, which extends a domain into the exit tunnel past the point of maximal constriction as discussed above. Although likely a distinct signal transduction pathway, the idea is the same: Long-range communication through conformational rearrangements of the ribosome leads to functional consequences.

Again, this well-characterized example is unlikely to be an isolated one, and we anticipate seeing many cases in which conformational changes triggered by interactions in the exit channel have key effects on engagement with downstream, external factors.

How ubiquitous might such regulation be? The observation that ribosome function depends so heavily on switches is not surprising. The use of induced-fit mechanisms for careful substrate selection, first proposed in 1958 (35), has now been documented for a wide range of polymerases and other enzymes. The extent to which these conformational switches are used for regulatory purposes is less well understood. We propose that ribosome switches described in this review are at the heart of much regulation. Upstream open reading frames are more common than previously suspected (63), and these present us with potential systems for understanding ribosome-targeted regulation. 3' untranslated regions in eukaryotes, implicated in many systems as essential for postinitiation translation control, feature complex RNA structure and multiple binding sites for proteins and microRNAs that may act together or separately (81). The effects of this targeting could be mediated through direct communication to the core of the ribosome, or through engagement of the peptide as it emerges from the exit channel. MicroRNAs and RNA-binding proteins proposed to regulate translation at a postinitiation step (50, 51, 53) may well directly affect ribosome function rather than other downstream events. Similarly, it seems reasonable to suspect that key components in the nonsense-mediated decay (NMD) pathway affect ribosome function directly via known interactions with the eukaryotic RFs (13) before the degradative machinery is recruited. The direct manipulation of core machinery and its basal movements is an appealing molecular mechanism for achieving regulation, and one that was already in place as higher-order gene regulation evolved. If one accepts such arguments, the elegant mechanisms that evolved to make proteins efficiently and with high fidelity are what allow for sophisticated gene regulation.

CONCLUDING REMARKS

This review summarizes recent advances in our understanding of the molecular mechanisms underlying translation termination in bacterial and eukaryotic systems. Insights have come from biochemical, genetic, and structural approaches and from an integration of their various strengths. A considerable complication in studying termination (or elongation) is the competitive nature of these processes: Genetic screens that look for nonsense suppression may yield mutations in either the termination or the elongation machinery. Such mutants, however, particularly when combined with high-quality structural information, provide excellent substrates for well-behaved in vitro assays designed to remove these complications. In the future we anticipate that these types of experiments will yield significant insights into the mechanism of translation termination and, further, that this detailed understanding will provide a launching point for the study of translation regulation in these systems.

SUMMARY POINTS

1. RFs are bifunctional molecules that, like tRNAs, have an end that recognizes stop codons in the small-subunit DC and another that reaches into the large-subunit PTC to participate in catalysis.

2. Several mechanisms appear to contribute to peptide release: general activation of the PTC by class I RFs, positioning of the attacking water molecule by the tight pocket formed by the RF GGQ loop and active site nucleotides, and contribution of the 2'OH of the P-site tRNA A76 to a proton shuttle mechanism similar to the one proposed to be important for catalysis of peptidyl transfer.

3. Both thermodynamic and kinetic mechanisms appear to contribute to the specificity of peptide release. Class I RFs have a much lower affinity for ribosomes programmed with sense codons rather than stop codons, and even when bound they have kinetic deficiencies in peptide release.

4. Conformational changes in the ribosome, RF, or both contribute to the mechanism of peptide release, although the specifics of these rearrangements are not known.

5. Switch-like movement between different ribosome conformational states is critical to progression through many of the steps in the translation cycle. We propose that molecules involved in the regulation of translation have evolved to take advantage of this fundamental property and thus function by altering the accessibility of or equilibrium between these various states.

FUTURE ISSUES

1. What are the molecular interactions between RFs, stop codons, and the ribosome that are critical for peptide release? A combination of structural and mutagenesis approaches should provide insight into this question in the near future.

2. What is the full mechanism for catalysis of peptide release? While details differ, current assays for peptide release generally use a saturating concentration of a class I RF to catalyze release of a peptidyl-tRNA in the P site. No experiments have yet addressed whether the chemistry of the reaction is rate limiting, and the existence of RF mutants at sites distant from the catalytic center with faster rate constants for peptide release than for the wild-type factor suggests that some step prior to release is rate limiting. Thorough investigation of the mechanism of peptide release, including experiments such as pH-rate profiles, will require an assay that is rate limited by chemistry.

3. Are there kinetic intermediates in the peptide release reaction pathway? The reaction pathway culminating in peptide bond formation, where aa-tRNAs are loaded into the A site by EF-Tu, codon specificity is examined, and the aa-tRNA is accommodated into the PTC, occurs in a number of kinetically discrete steps. Analysis of whether such kinetic intermediates also exist in the peptide release reaction pathway, in particular an equivalent to the accommodation step, requires development of appropriate assays.

4. What is the role of ribosomal residues in the specificity of peptide release? Although a number of rRNA mutations with suppressor phenotypes have been isolated, the contribution of these nucleotides has not been evaluated using in vitro assays that eliminate problematic competition from near-cognate tRNAs.

5. Do external factors (such as NMD pathway components) impinge on the efficiency of peptide release for regulatory purposes?

DISCLOSURE STATEMENT

The authors are not aware of any biases that might be perceived as affecting the objectivity of this review.

ACKNOWLEDGMENTS

We wish to thank J. Shaw and H. Zaher for comments on the manuscript, and we are especially indebted to M. Bianchet for assistance with modeling the ribosome-bound RF structure. Work in the Green laboratory is funded by NIH and HHMI.

LITERATURE CITED

1. Alkalaeva EZ, Pisarev AV, Frolova LY, Kisselev LL, Pestova TV. 2006. In vitro reconstitution of eukaryotic translation reveals cooperativity between release factors eRF1 and eRF3. *Cell* 125:1125–36
2. Beringer M, Rodnina MV. 2007. Importance of tRNA interactions with 23S rRNA for peptide bond formation on the ribosome: studies with substrate analogs. *Biol. Chem.* 388:687–91
3. Bieling P, Beringer M, Adio S, Rodnina MV. 2006. Peptide bond formation does not involve acid-base catalysis by ribosomal residues. *Nat. Struct. Mol. Biol.* 13:423–28
4. Bornemann T, Jöckel J, Rodnina MV, Wintermeyer W. 2008. Signal sequence-independent mechanisms targeting of ribosomes containing short nascent peptides within the exit tunnel. *Nat. Struct. Mol. Biol.* 15:494–99
5. Brown CM, Tate WP. 1994. Direct recognition of mRNA stop signals by *Escherichia coli* polypeptide chain release factor two. *J. Biol. Chem.* 269:33164–70

6. Brunelle JL, Youngman EM, Sharma D, Green R. 2006. The interaction between C75 of tRNA and the A loop of the ribosome stimulates peptidyl transferase activity. *RNA* 12:33–39

7. Capecchi MR. 1967. Polypeptide chain termination in vitro: isolation of a release factor. *Proc. Natl. Acad. Sci. USA* 58:1144–51

8. Caskey CT, Beaudet AL, Scolnick EM, Rosman M. 1971. Hydrolysis of fMet-tRNA by peptidyl transferase. *Proc. Natl. Acad. Sci. USA* 68:3163–67

9. Chavatte L, Frolova L, Kisselev L, Favre A. 2001. The polypeptide chain release factor eRF1 specifically contacts the s(4)UGA stop codon located in the A site of eukaryotic ribosomes. *Eur. J. Biochem.* 268:2896–904

10. Chavatte L, Frolova L, Laugaa P, Kisselev L, Favre A. 2003. Stop codons and UGG promote efficient binding of the polypeptide release factor eRF1 to the ribosomal A site. *J. Mol. Biol.* 331:745–58

11. Chavatte L, Seit-Nebi A, Dubovaya V, Favre A. 2002. The invariant uridine of stop codons contacts the conserved NIKSR loop of human eRF1 in the ribosome. *EMBO J.* 21:5302–11

12. Cochella L, Green R. 2005. An active role for tRNA in decoding beyond codon:anticodon pairing. *Science* 308:1178–80

13. Czaplinski K, Ruiz-Echevarria MJ, Paushkin SV, Han X, Weng Y, et al. 1998. The surveillance complex interacts with the translation release factors to enhance termination and degrade aberrant mRNAs. *Genes Dev.* 12:1665–77

14. Dincbas-Renqvist V, Engstrom A, Mora L, Heurgue-Hamard V, Buckingham R, Ehrenberg M. 2000. A post-translational modification in the GGQ motif of RF2 from *Escherichia coli* stimulates termination of translation. *EMBO J.* 19:6900–7

15. Feinberg JS, Joseph S. 2006. A conserved base-pair between tRNA and 23 S rRNA in the peptidyl transferase center is important for peptide release. *J. Mol. Biol.* 364:1010–20

16. Freistroffer DV, Kwiatkowski M, Buckingham RH, Ehrenberg M. 2000. The accuracy of codon recognition by polypeptide release factors. *Proc. Natl. Acad. Sci. USA* 97:2046–51

17. Freistroffer DV, Pavlov MY, MacDougall J, Buckingham RH, Ehrenberg M. 1997. Release factor RF3 in *E. coli* accelerates the dissociation of release factors RF1 and RF2 from the ribosome in a GTP-dependent manner. *EMBO J.* 16:4126–33

18. Frolova LY, Tsivkovskii RY, Sivolobova GF, Oparina NY, Serpinsky OI, et al. 1999. Mutations in the highly conserved GGQ motif of class 1 polypeptide release factors abolish ability of human eRF1 to trigger peptidyl-tRNA hydrolysis. *RNA* 5:1014–20

19. Gao H, Zhou Z, Rawat U, Huang C, Bouakaz L, et al. 2007. RF3 induces ribosomal conformational changes responsible for dissociation of class I release factors. *Cell* 129:929–41

20. Gong F, Ito K, Nakamura Y, Yanofsky C. 2001. The mechanism of tryptophan induction of tryptophanase operon expression: Tryptophan inhibits release factor-mediated cleavage of TnaC-peptidyl-tRNA(Pro). *Proc. Natl. Acad. Sci. USA* 98:8997–9001

21. Graille M, Heurgue-Hamard V, Champ S, Mora L, Scrima N, et al. 2005. Molecular basis for bacterial class I release factor methylation by PrmC. *Mol. Cell* 20:917–27

22. Grentzmann G, Brechemier-Baey D, Heurgue V, Mora L, Buckingham RH. 1994. Localization and characterization of the gene encoding release factor RF3 in *Escherichia coli*. *Proc. Natl. Acad. Sci. USA* 91:5848–52

23. Heurgue-Hamard V, Champ S, Engstrom A, Ehrenberg M, Buckingham RH. 2002. The hemK gene in *Escherichia coli* encodes the N(5)-glutamine methyltransferase that modifies peptide release factors. *EMBO J.* 21:769–78

24. Heurgue-Hamard V, Champ S, Mora L, Merkulova-Rainon T, Kisselev LL, Buckingham RH. 2005. The glutamine residue of the conserved GGQ motif in *Saccharomyces cerevisiae* release factor eRF1 is methylated by the product of the YDR140w gene. *J. Biol. Chem.* 280:2439–45

25. Hirsh D. 1971. Tryptophan transfer RNA as the UGA suppressor. *J. Mol. Biol.* 58:439–58

26. Ito K, Ebihara K, Nakamura Y. 1998. The stretch of C-terminal acidic amino acids of translational release factor eRF1 is a primary binding site for eRF3 of fission yeast. *RNA* 4:958–72

27. Ito K, Uno M, Nakamura Y. 1998. Single amino acid substitution in prokaryote polypeptide release factor 2 permits it to terminate translation at all three stop codons. *Proc. Natl. Acad. Sci. USA* 95:8165–69

28. Ito K, Uno M, Nakamura Y. 2000. A tripeptide 'anticodon' deciphers stop codons in messenger RNA. *Nature* 403:680–84

29. Ivanova EV, Kolosov PM, Birdsall B, Kelly G, Pastore A, et al. 2007. Eukaryotic class 1 translation termination factor eRF1—the NMR structure and dynamics of the middle domain involved in triggering ribosome-dependent peptidyl-tRNA hydrolysis. *FEBS J.* 274:4223–37

30. Jones T, Zou J-Y, Cowan S, Kjeldgaard M. 1991. Improved methods for building protein models in electron density maps and the location of errors in these models. *Acta Crystallogr. A* 47:110–19

31. Jorgensen F, Adamski FM, Tate WP, Kurland CG. 1993. Release factor-dependent false stops are infrequent in *Escherichia coli. J. Mol. Biol.* 230:41–50

32. Katunin VI, Muth GW, Strobel SA, Wintermeyer W, Rodnina MV. 2002. Important contribution to catalysis of peptide bond formation by a single ionizing group within the ribosome. *Mol. Cell* 10:339–46

33. Klaholz BP, Pape T, Zavialov AV, Myasnikov AG, Orlova EV, et al. 2003. Structure of the *Escherichia coli* ribosomal termination complex with release factor 2. *Nature* 421:90–94

34. Konecki DS, Aune KC, Tate W, Caskey CT. 1977. Characterization of reticulocyte release factor. *J. Biol. Chem.* 252:4514–20

35. Koshland DE. 1958. Application of a theory of enzyme specificity to protein synthesis. *Proc. Natl. Acad. Sci. USA* 44:98–104

36. Kramer EB, Farabaugh PJ. 2007. The frequency of translational misreading errors in *E. coli* is largely determined by tRNA competition. *RNA* 13:87–96

37. Lekomtsev S, Kolosov P, Bidou L, Frolova L, Rousset JP, Kisselev L. 2007. Different modes of stop codon restriction by the *Stylonychia* and *Paramecium* eRF1 translation termination factors. *Proc. Natl. Acad. Sci. USA* 104:10824–29

38. Liang H, Wong JY, Bao Q, Cavalcanti AR, Landweber LF. 2005. Decoding the decoding region: analysis of eukaryotic release factor (eRF1) stop codon-binding residues. *J. Mol. Evol.* 60:337–44

39. Lovett PS, Rogers EJ. 1996. Ribosome regulation by the nascent peptide. *Microbiol. Rev.* 60:366–85

40. Ma B, Nussinov R. 2004. Release factors eRF1 and RF2: a universal mechanism controls the large conformational changes. *J. Biol. Chem.* 279:53875–85

41. Maag D, Fekete CA, Gryczynski Z, Lorsch JR. 2005. A conformational change in the eukaryotic translation preinitiation complex and release of eIF1 signal recognition of the start codon. *Mol. Cell* 17:265–75

42. Merkulova TI, Frolova LY, Lazar M, Camonis J, Kisselev LL. 1999. C-terminal domains of human translation termination factors eRF1 and eRF3 mediate their in vivo interaction. *FEBS Lett.* 443:41–47

43. Mikuni O, Ito K, Moffat J, Matsumura K, McCaughan K, et al. 1994. Identification of the prfC gene, which encodes peptide-chain-release factor 3 of *Escherichia coli. Proc. Natl. Acad. Sci. USA* 91:5798–802

44. Moffat JG, Tate WP. 1994. A single proteolytic cleavage in release factor 2 stabilizes ribosome binding and abolishes peptidyl-tRNA hydrolysis activity. *J. Biol. Chem.* 269:18899–903

45. Mora L, Zavialov A, Ehrenberg M, Buckingham RH. 2003. Stop codon recognition and interactions with peptide release factor RF3 of truncated and chimeric RF1 and RF2 from *Escherichia coli. Mol. Microbiol.* 50:1467–76

46. Muramatsu T, Heckmann K, Kitanaka C, Kuchino Y. 2001. Molecular mechanism of stop codon recognition by eRF1: a wobble hypothesis for peptide anticodons. *FEBS Lett.* 488:105–9

47. Muto H, Nakatogawa H, Ito K. 2006. Genetically encoded but nonpolypeptide prolyl-tRNA functions in the A site for SecM-mediated ribosomal stall. *Mol. Cell* 22:545–52

48. Nakatogawa H, Ito K. 2001. Secretion monitor, SecM, undergoes self-translation arrest in the cytosol. *Mol. Cell* 7:185–92

49. Nakatogawa H, Ito K. 2002. The ribosomal exit tunnel functions as a discriminating gate. *Cell* 108:629–36

50. Nottrott S, Simard MJ, Richter JD. 2006. Human let-7a miRNA blocks protein production on actively translating polyribosomes. *Nat. Struct. Mol. Biol.* 13:1108–14

51. Olsen PH, Ambros V. 1999. The lin-4 regulatory RNA controls developmental timing in *Caenorhabditis elegans* by blocking LIN-14 protein synthesis after the initiation of translation. *Dev. Biol.* 216:671–80

52. Passmore LA, Schmeing TM, Maag D, Applefield DJ, Acker MG, et al. 2007. The eukaryotic translation initiation factors eIF1 and eIF1A induce an open conformation of the 40S ribosome. *Mol. Cell* 26:41–50

53. Petersen CP, Bordeleau ME, Pelletier J, Sharp PA. 2006. Short RNAs repress translation after initiation in mammalian cells. *Mol. Cell* 21:533–42

54. Petry S, Brodersen DE, Murphy FV 4th, Dunham CM, Selmer M, et al. 2005. Crystal structures of the ribosome in complex with release factors RF1 and RF2 bound to a cognate stop codon. *Cell* 123:1255–66

55. Pisareva VP, Pisarev AV, Hellen CU, Rodnina MV, Pestova TV. 2006. Kinetic analysis of interaction of eukaryotic release factor 3 with guanine nucleotides. *J. Biol. Chem.* 281:40224–35

56. Polacek N, Gomez MJ, Ito K, Xiong L, Nakamura Y, Mankin A. 2003. The critical role of the universally conserved A2602 of 23S ribosomal RNA in the release of the nascent peptide during translation termination. *Mol. Cell* 11:103–12

57. Polevoda B, Span L, Sherman F. 2006. The yeast translation release factors Mrf1p and Sup45p (eRF1) are methylated, respectively, by the methyltransferases Mtq1p and Mtq2p. *J. Biol. Chem.* 281:2562–71

58. Raney A, Law GL, Mize GJ, Morris DR. 2002. Regulated translation termination at the upstream open reading frame in *S*-adenosylmethionine decarboxylase mRNA. *J. Biol. Chem.* 277:5988–94

59. Rawat U, Gao H, Zavialov A, Gursky R, Ehrenberg M, Frank J. 2006. Interactions of the release factor RF1 with the ribosome as revealed by cryo-EM. *J. Mol. Biol.* 357:1144–53

60. Rawat UB, Zavialov AV, Sengupta J, Valle M, Grassucci RA, et al. 2003. A cryo-electron microscopic study of ribosome-bound termination factor RF2. *Nature* 421:87–90

61. Rodnina MV, Beringer M, Wintermeyer W. 2006. Mechanism of peptide bond formation on the ribosome. *Q. Rev. Biophys.* 39:203–25

62. Rodnina MV, Wintermeyer W. 2001. Fidelity of aminoacyl-tRNA selection on the ribosome: kinetic and structural mechanisms. *Annu. Rev. Biochem.* 70:415–35

63. Sachs MS, Geballe AP. 2006. Downstream control of upstream open reading frames. *Genes Dev.* 20:915–21

64. Salas-Marco J, Bedwell DM. 2004. GTP hydrolysis by eRF3 facilitates stop codon decoding during eukaryotic translation termination. *Mol. Cell. Biol.* 24:7769–78

65. Salas-Marco J, Fan-Minogue H, Kallmeyer AK, Klobutcher LA, Farabaugh PJ, Bedwell DM. 2006. Distinct paths to stop codon reassignment by the variant-code organisms *Tetrahymena* and *Euplotes*. *Mol. Cell. Biol.* 26:438–47

66. Scarlett DJ, McCaughan KK, Wilson DN, Tate WP. 2003. Mapping functionally important motifs SPF and GGQ of the decoding release factor RF2 to the *Escherichia coli* ribosome by hydroxyl radical footprinting. Implications for macromolecular mimicry and structural changes in RF2. *J. Biol. Chem.* 278:15095–104

67. Schmeing TM, Huang KS, Kitchen DE, Strobel SA, Steitz TA. 2005. Structural insights into the roles of water and the 2′ hydroxyl of the P site tRNA in the peptidyl transferase reaction. *Mol. Cell* 20:437–48

68. Schmeing TM, Huang KS, Strobel SA, Steitz TA. 2005. An induced-fit mechanism to promote peptide bond formation and exclude hydrolysis of peptidyl-tRNA. *Nature* 438:520–24

69. Scolnick E, Tompkins R, Caskey T, Nirenberg M. 1968. Release factors differing in specificity for terminator codons. *Proc. Natl. Acad. Sci. USA* 61:768–74

70. Seit-Nebi A, Frolova L, Justesen J, Kisselev L. 2001. Class-1 translation termination factors: Invariant GGQ minidomain is essential for release activity and ribosome binding but not for stop codon recognition. *Nucleic Acids Res.* 29:3982–87

71. Sharma PK, Xiang Y, Kato M, Warshel A. 2005. What are the roles of substrate-assisted catalysis and proximity effects in peptide bond formation by the ribosome? *Biochemistry* 44:11307–14

72. Shaw JJ, Green R. 2007. Two distinct components of release factor function uncovered by nucleophile partitioning analysis. *Mol. Cell* 28:458–67

73. Shin DH, Brandsen J, Jancarik J, Yokota H, Kim R, Kim SH. 2004. Structural analyses of peptide release factor 1 from *Thermotoga maritima* reveal domain flexibility required for its interaction with the ribosome. *J. Mol. Biol.* 341:227–39

74. Sievers A, Beringer M, Rodnina MV, Wolfenden R. 2004. The ribosome as an entropy trap. *Proc. Natl. Acad. Sci. USA* 101:7897–901

75. Song H, Mugnier P, Das AK, Webb HM, Evans DR, et al. 2000. The crystal structure of human eukaryotic release factor eRF1—mechanism of stop codon recognition and peptidyl-tRNA hydrolysis. *Cell* 100:311–21

76. Tate W, Greuer B, Brimacombe R. 1990. Codon recognition in polypeptide chain termination: site directed crosslinking of termination codon to *Escherichia coli* release factor 2. *Nucleic Acids Res.* 18:6537–44

77. Trobro S, Aqvist J. 2005. Mechanism of peptide bond synthesis on the ribosome. *Proc. Natl. Acad. Sci. USA* 102:12395–400

78. Trobro S, Aqvist J. 2007. A model for how ribosomal release factors induce peptidyl-tRNA cleavage in termination of protein synthesis. *Mol. Cell* 27:758–66

79. Uno M, Ito K, Nakamura Y. 2002. Polypeptide release at sense and noncognate stop codons by localized charge-exchange alterations in translational release factors. *Proc. Natl. Acad. Sci. USA* 99:1819–24

80. Valle M, Zavialov A, Sengupta J, Rawat U, Ehrenberg M, Frank J. 2003. Locking and unlocking of ribosomal motions. *Cell* 114:123–34

81. Vella MC, Choi EY, Lin SY, Reinert K, Slack FJ. 2004. The *C. elegans* microRNA let-7 binds to imperfect let-7 complementary sites from the lin-41 3′UTR. *Genes Dev.* 18:132–37

82. Vestergaard B, Sanyal S, Roessle M, Mora L, Buckingham RH, et al. 2005. The SAXS solution structure of RF1 differs from its crystal structure and is similar to its ribosome bound cryo-EM structure. *Mol. Cell* 20:929–38

83. Vestergaard B, Van LB, Andersen GR, Nyborg J, Buckingham RH, Kjeldgaard M. 2001. Bacterial polypeptide release factor RF2 is structurally distinct from eukaryotic eRF1. *Mol. Cell* 8:1375–82

84. Wang Z, Fang P, Sachs MS. 1998. The evolutionarily conserved eukaryotic arginine attenuator peptide regulates the movement of ribosomes that have translated it. *Mol. Cell. Biol.* 18:7528–36

85. Weinger JS, Parnell KM, Dorner S, Green R, Strobel SA. 2004. Substrate-assisted catalysis of peptide bond formation by the ribosome. *Nat. Struct. Biol.* 11:1101–6

86. Wilson KS, Ito K, Noller HF, Nakamura Y. 2000. Functional sites of interaction between release factor RF1 and the ribosome. *Nat. Struct. Biol.* 7:866–70

87. Woolhead CA, Johnson AE, Bernstein HD. 2006. Translation arrest requires two-way communication between a nascent polypeptide and the ribosome. *Mol. Cell* 22:587–98

88. Yoshimura K, Ito K, Nakamura Y. 1999. Amber (UAG) suppressors affected in UGA/UAA-specific polypeptide release factor 2 of bacteria: genetic prediction of initial binding to ribosome preceding stop codon recognition. *Genes Cells* 4:253–66

89. Youngman EM, Brunelle JL, Kochaniak AB, Green R. 2004. The active site of the ribosome is composed of two layers of conserved nucleotides with distinct roles in peptide bond formation and peptide release. *Cell* 117:589–99

90. Youngman EM, He SL, Nikstad LJ, Green R. 2007. Stop codon recognition by release factors induces structural rearrangement of the ribosomal decoding center that is productive for peptide release. *Mol. Cell* 28:533–43

91. Yusupova GZ, Yusupov MM, Cate JH, Noller HF. 2001. The path of messenger RNA through the ribosome. *Cell* 106:233–41

92. Zavialov AV, Buckingham RH, Ehrenberg M. 2001. A posttermination ribosomal complex is the guanine nucleotide exchange factor for peptide release factor RF3. *Cell* 107:115–24

93. Zavialov AV, Mora L, Buckingham RH, Ehrenberg M. 2002. Release of peptide promoted by the GGQ motif of class 1 release factors regulates the GTPase activity of RF3. *Mol. Cell* 10:789–98

94. Zoldak G, Redecke L, Svergun DI, Konarev PV, Voertler CS, et al. 2007. Release factors 2 from *Escherichia coli* and *Thermus thermophilus*: structural, spectroscopic and microcalorimetric studies. *Nucleic Acids Res.* 35:1343–53

Rules of Engagement: Interspecies Interactions that Regulate Microbial Communities

Ainslie E.F. Little,[1] Courtney J. Robinson,[1] S. Brook Peterson,[1] Kenneth F. Raffa,[3] and Jo Handelsman[1,2]

[1]Department of Bacteriology, [2]Department of Plant Pathology, and [3]Department of Entomology, University of Wisconsin, Madison, Wisconsin, 53706; email: alittle@wisc.edu, cjr@plantpath.wisc.edu, snowbp@u.washington.edu, raffa@entomology.wisc.edu, joh@plantpath.wisc.edu

Annu. Rev. Microbiol. 2008. 62:375–401

First published online as a Review in Advance on June 10, 2008

The *Annual Review of Microbiology* is online at micro.annualreviews.org

This article's doi: 10.1146/annurev.micro.030608.101423

Key Words

microbial ecology, community ecology, community genetics, symbiosis, metagenomics

Abstract

Microbial communities comprise an interwoven matrix of biological diversity modified by physical and chemical variation over space and time. Although these communities are the major drivers of biosphere processes, relatively little is known about their structure and function, and predictive modeling is limited by a dearth of comprehensive ecological principles that describe microbial community processes. Here we discuss working definitions of central ecological terms that have been used in various fashions in microbial ecology, provide a framework by focusing on different types of interactions within communities, review the status of the interface between evolutionary and ecological study, and highlight important similarities and differences between macro- and microbial ecology. We describe current approaches to study microbial ecology and progress toward predictive modeling.

Contents

INTRODUCTION

Biology in the twentieth century was dominated by simplification and order. This was driven by the desire to improve experimental controls and resulted in a landscape of intellectual frameworks unified by reductionism. Study of systems was replaced by study of parts, organism with cells, cells with genes and proteins, and genes and proteins with their atoms. The scientific triumphs were many and the practical outcomes—vaccines, antibiotics, and high-yielding crops—transformed human health and food security. But the cost was a reduction of emphasis, training, and vision in systems-level biology, and with that a reduced ability to address some of the most important current environmental and health challenges.

Community: an assemblage of populations occupying a given area at the same time

Diversity: the richness and evenness of species within a community

As the twentieth century drew to a close, we were confronted by new challenges that rekindled widespread interest and identified the need to understand systems-level biology. Certain human diseases emerged whose origins were understood from landscape-level events that did not fit neatly into a reductionist scheme. Similarly, global climate change, and its underlying human causation, was recognized as a reality, and any realistic solutions required study of interconnecting spheres of society and the biosphere. These events and others like them arrived just as powerful new methods in microbiology emerged to open the way for a renaissance of ecology in general and microbial community ecology in particular. Although the need for systems biology has always been apparent to ecologists, who can offer many examples of ecosystems in which studying a binary interaction led to an erroneous conclusion that was corrected only by introducing more complexity into the model, the change in perspective was a surprise to much of the microbiology community (158).

Over the past century of microbiology, the emphasis on the study of microbes in pure culture has isolated microorganisms from their communities and focused on their behavior in the biologically simple environments of the petri dish and test tube. Although simple model systems have driven an explosion of knowledge in cellular processes and host-microbe interactions over the past two decades, the reality of natural communities demands that we direct attention to complex assemblages as well. Global microbial diversity is enormous, likely representing 10^7 species, of which only 0.01% to 0.1% are known (33, 34, 50, 60, 143). Microbial communities can be complex, with high species richness and unevenness, and their structures are continually influenced by changing biological, chemical, and physical factors. Most microorganisms do not submit readily to growth in the laboratory, leaving microbiologists to either concentrate on the subsets that do perform well under artificial conditions or grope for other methods to describe the species that compose natural communities. Therefore, the structure of most microbial communities has been difficult to illuminate.

Describing community structure is often a prelude to understanding community function, which has been similarly difficult to elucidate (76, 142, 144). One of the barriers confronting microbial ecologists is the lack of ecological principles that provide the foundation for predictive models. Broadly based, validated principles derived from systems that can be manipulated experimentally would allow for predictions regarding behavior of communities that are less tractable for study. Many of these principles can be borrowed from macroecology, although some need to be reformatted to fit the microbial lifestyle. In this review, we explore the internal processes of microbial communities in an effort to begin to define the principles that underpin ecological and evolutionary patterns of microbial communities. Defining these principles is necessary to enable predictive modeling of ecological dynamics of microbial communities. In addition to their importance to fundamental understanding of the biosphere, predictive models have numerous practical applications. For example, they can provide guidance to strategies for manipulating communities on plant or animal surfaces to suppress pathogens, maintaining community integrity when applying chemicals such as antibiotics or pesticides that could destabilize communities, or successfully introducing a beneficial microorganism such as a biocontrol agent in agriculture or a probiotic in veterinary and human medicine.

ECOLOGICAL PROPERTIES OF MICROBIAL COMMUNITIES

The properties of microbial communities can be divided into two categories: structural and functional. Structural properties describe how the community varies and what it looks like—the types and numbers of members across a range of environments. Functional properties define the community's behavior—how the community processes substrates, interacts with

forces in its environment, and responds to perturbations such as invasion.

Structural Properties

One of the simplest ways to characterize a community is to list its members (composition) and to tabulate the total number of members (richness). To answer such questions as How many different species are there in a given community? and What are they? seems easy and straightforward, but answering them is challenging when they are applied to microbial communities (177). The challenges derive from both biological and statistical issues. Enumeration by culturing limits the description to those members that can be cultured, which constitute the minority (often less than 1%) in most communities. Molecular methods present culture-independent alternatives, which capture far more richness than does culturing (32). Sequence analysis of the 16S rRNA gene is the dominant method of determining identity and phylogenetic relatedness of microorganisms (32), although other genes, such as $rpoB$, may provide greater resolution in phylogenetic associations at the species and subspecies levels (21). To avoid the inherent difficulty in sampling every species in every community, macro- and microbial ecologists often use estimates of richness based on samples of the communities; however, these estimates can vary depending on the estimator chosen and the type of data analyzed (8, 177; see Reference 32 for a review of the difficulties of quantifying properties in microbial communities, the statistics used in the analyses, as well as recent accomplishments in the area). Estimates of richness based on gene sequence relatedness are often calculated using software such as EstimateS and DOTUR (26, 174). DOTUR, for example, has been used for a variety of genes and environments to assign sequences to operational taxonomic units based on phylogenetic distances and to calculate richness estimates at different degrees of phylogenetic resolution (17, 30, 46, 56, 96, 175, 201).

Diversity indices take into account both species richness and evenness of distribution of

Community structure: the composition (membership) of a community and the abundance of individual members species (i.e., abundance of individuals) (8). The diversity of a community is difficult to interpret on its own but can be valuable when used to compare communities. Community structure, like diversity, is an attribute that is most useful when analyzed for comparative purposes. Community structure incorporates both the composition of the community and the abundance of individual members. Diversity and structure measure different aspects of community characteristics, so they can vary independently of each other (78). In soil communities, the presence of a plant community, the introduction of various transgenes into a tree community, and rhizomediation influence composition and structure, but not diversity (40, 111, 214). However, in other situations structural changes influence diversity as well.

In addition to the structure of the entire community, understanding the structure of assemblages within a community can also be important. For example, Perez & Sommaruga (148) monitored the response of the Betaproteobacteria and *Actinobacteria* populations within a lake water community to solar radiation and dissolved organic matter from lake, algal, or soil sources. Functionally defined assemblages, such as guilds, are also of interest because of the activities of community members. In another recent study, the structure of a methanotroph assemblage within a rice field community was monitored using terminal restriction-length polymorphism (T-RFLP) analysis, *pmoA* gene analysis, and stable isotope probing (181). The results indicated that the activity and structure of the methanotroph assemblage (composed of type I and II methanotrophs) fluctuated over time and with CH_4 availability and that type I and type II methanotrophs occupied two different niches within the rice field ecosystem. Other guilds such as ammonium-oxidizers, methanogens, and iron-reducers have been analyzed for structural properties (24, 77, 184). Similarity of community structure can be calculated using a number of computer programs such as S-LIBSHUFF, SONS, TreeClimber, and Unifrac and techniques such as analysis of molecular variance (AMOVA) and homogeneity of molecular variance (HOMOVA) (123, 124, 128, 173, 175, 176, 178). The differences between the hypotheses tested by each of these tools are discussed elsewhere (173).

Robustness

Community robustness is the ability of the community to maintain its functional and structural integrity in the face of potential perturbations (8). This is consistent with other uses of robustness in engineering and statistics that pertain to the heartiness of a system and its ability to function under various, often adverse, conditions (62). Just like complex engineered systems, biological systems, such as cells, tissues, organs, and ecological webs, are composed of diverse and often multifunctional components (100). Systems that maintain their function, characteristic behavior, or some other property despite internal and external perturbations and adapt to their environments are robust (100, 101, 187). Although robustness is a characteristic of all biological systems, it is a relative property that depends on the perturbation and the behavior monitored (187). For example, cancer cells that establish in the human body are particularly robust against the host's defenses, but perturbations against which they are weakly robust offer promising therapies (101).

We use structural robustness, similar to ecosystem stability (although ecosystem stability does not always refer to structure), to describe the constancy in community structure over time (temporal stability), the ability to resist change following a perturbation (resistance), and the return to its native structure following a change to structure (resilience). Components of robustness are often studied individually. Temporal stability, though not always referred to as such, is studied far more often than the other components of robustness. For example, Kikuchi & Graf (99) recently reported that populations in the microbial community of leech crops, comprising *Aeromonas veronii* bv. sobria and a *Rikenella*-like organism, fluctuated within 6 h to 14 days after blood

Individual: smallest unit within a population; for prokaryotes, it is usually a single cell

Guild: a group of species that occupy a common niche in a given community, characterized by exploitation of environmental resources in the same way

Population: all the individuals of one species in a given area at the same time

Robustness: description of a community's temporal stability and resistance and resilience to perturbations that challenge structure (structural robustness) or function (functional robustness)

feeding. The population size of both members initially increased following feeding, but the abundance of *A. veronii* decreased 4 days after feeding while the abundance of the *Rikenella*-like species remained constant over the timescale studied (99). Temporal stability has also been examined in rice field soil, cabbage white butterfly midguts, and a number of aquatic communities (2, 20, 23, 94, 98, 113, 135, 137; C. Robinson, P. Schloss, K. Raffa & J. Handelsman, manuscript submitted).

Functional robustness refers to the ability of a community to maintain a particular activity despite perturbation, which, unlike structural robustness, is not necessarily linked to composition, although links between structure and function have been established many times (1, 22, 24, 28, 61, 181, 184). Saison et al. (171) showed that soil communities exhibited functional and structural resilience to low, but not high, levels of winery compost. In another study, Yannarell et al. (211) found that nitrogen fixation returned to normal levels in a Bahamian microbial mat following a Category Four hurricane, despite a shift in community structure.

At the intersection of functional and structural robustness is the years-old diversity-stability debate. Briefly, the debate questions whether increased community diversity increases or decreases community stability (see References 89 and 131 for reviews of the diversity-stability debate). Macro- and, to a lesser extent, microbial ecology have long sought a common rule that governs the relationship between diversity of a given community and its ecosystem stability, with stability often measured by a specific activity or function. Several years ago, McGrady-Steed et al. (133) found that as diversity increased, decomposition increased, but carbon dioxide uptake decreased in an aquatic microcosm that contained bacteria, protists, and metazoans. They also noted that resistance to invasion increased as the abundance of certain members increased.

More recently, Girvan et al. (67) examined the temporal structural stability and the functional resilience of two soil communities of different diversities that had been per-

turbed with benzene or copper using denaturing gradient gel electrophoresis and monitoring of broadscale (mineralization of ^{14}C-labeled wheat shoot) and narrow-scale (mineralization of ^{14}C-labeled 2,4-dichlorophenol) functions for 9 weeks. Temporal shifts in structure were observed in all soils, although copper-treated and control soils were consistently more similar to each other than either was to the benzene-treated soil (67). This indicates that in some systems the source of the perturbation may be more important than community diversity in structural temporal stability. Benzene perturbation reduced the ability of both communities to perform the narrow niche function; however, the more diverse community reacquired the function by week 9 of the experiment, thereby exhibiting functional resilience (67). Copper treatment increased broadscale function but initially reduced narrow niche function for both communities before they both recovered (67). These findings indicate that diversity and source of perturbation may be important for functional robustness. The results also suggest that soil exhibits functional resilience owing in part to functional redundancy, but that greater diversity in soil communities may also lead to greater resistance and functional stability. The positive association between community diversity and stability is consistent with what has been observed in some macroecological systems such as grassplots (195). More study of microbial community robustness is needed to establish the rules of engagement, i.e., governing principles and predictive models.

INTERACTIONS WITHIN MICROBIAL COMMUNITIES

A first step toward understanding the nature of ecological interactions in natural microbial communities is cataloging mechanisms by which microorganisms interact. Symbiotic interactions can be divided into three overlapping categories, which exist in a continuum from parasitic to mutualistic (**Figure 1**). Parasites (Greek, *para*, "near," and *sitos*, "food") are organisms that live on or in another organism

Protagonistic	Benign	Antagonistic

Mutualists	Commensals	Competitors, predators, and parasites

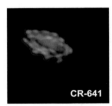

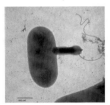

Figure 1

Continuum of interspecific interactions that occur in microbial communities. (*Left*) The beneficial relationships that occur between mutualists such as the phototrophic consortia (shown below) comprising central rod-shaped bacteria in the family *Comamonadaceae* (*red*) and green epibiotic sulfur bacteria (*blue*) (reprinted from Reference 95). Commensal, or benign, relationships are depicted in the middle of the continuum by an electron micrograph of the microbial community of the cabbage white butterfly. (*Right*) Antagonistic relationships are illustrated by an electron micrograph of *Bdellovibrio* attacking *Shewanella oneidsis* (photo credit: Robert Chamberlain, Wayne Rickoll, and Mark Martin).

and obtain all or part of their necessary nutrients at the expense of their host. Commensalism (Latin, *com*, "together," and *mensa*, "table") includes relationships in which one partner derives benefit from the other and the other partner neither is harmed nor benefits from the association. Mutualism (Latin, *mutualis*, "reciprocal") is an association in which both organisms derive benefit from one another. In addition to specific symbiotic interactions, microorganisms can interact antagonistically with other microorganisms via competition for a common resource or via predation of one organism upon another. The mechanisms that dictate interactions among microorganisms are largely responsible for the properties of the community as a whole. Dissecting the binary and tertiary interactions among community members is one essential component to understanding the properties of the community.

Exploitative Competition

As in interactions between macroorganisms, exploitative competition, or competition for nutrients and space, plays an important role in shaping microbial interactions. In eukaryote-associated microbial communities, such as in the human gut, competition for nutrients and space from resident microflora is thought to present one barrier to infection by pathogens, a phenomenon called the barrier effect (74). In cabbage white butterfly larvae, changes in the resident microflora community structure resulting from treatment with antibiotics facilitates invasion by a nonresident, in support of the role of competition from residents in normally preventing invasion (C. Robinson, Y. Ramos, K. Raffa & J. Handelsman, unpublished observations). Probiotic bacteria, such as some *Bifidobacterium* and *Lactobacillus* species, are thought to exert positive effects on host health in part via competitive interactions with pathogenic bacteria for space and nutrients, and also through interference competition by production of toxic compounds (161). In vitro, binding of some probiotic bacteria to cultured epithelial cells can prevent binding of pathogens, in support of the hypothesis that bacteria successfully compete for space (73, 112). Competition for nutrients among bacterial species from functional groups having different nutritional requirements can be important in structuring microbial communities in nutritionally heterogeneous environments. For example, in microcosms containing picoplanktonic cyanobacteria and heterotrophic bacteria along crossed gradients of glucose and phosphate, the cyanobacteria positively responded to increased phosphate only when glucose was low, presumably because of increased competition from the heterotrophs when organic carbon was supplied (47).

Exploitative and interference competition. The aspect of microbial interactions that has arguably received the most attention is the ability of some microorganisms to produce compounds that, at least in laboratory studies, directly antagonize other microorganisms. It has been assumed that organisms produce these compounds as a means of chemical warfare, providing a competitive edge to the producers by directly inhibiting growth or killing off potential competitors, a form of interference competition. One example in which evidence supports this proposition is in the rhizosphere, where antibiotic production by a number of bacteria contributes to their ability to protect plants from particular pathogens (27, 157). The antibiotics these bacteria produce in vitro have been detected in the rhizosphere (12), and mutants deficient in antibiotic production often exhibit a reduced ability to protect the plant from the pathogen (97, 183, 192) or a reduced fitness in the rhizosphere (129, 153). Additionally, production of the peptide antibiotic trifolitoxin by some strains of *Rhizobium etli* strains contributes to their competitive ability in the rhizosphere, leading to increased occupation of root nodules by producing strains (166, 167). In addition to the rhizosphere systems, evidence from a number of invertebrate-microbe interactions supports the role of antibiotics in antagonistic relationships within communities. For example, antibiotics produced by *Actinomycetes* associated with leafcutter ants protect the ants' fungal gardens from parasitism by another fungus (31, 121). Similarly, larvae of some crustaceans, beewolves, and bark beetles rely on production of antifungal compounds by bac-

terial symbionts to avoid infection by fungal pathogens (19, 65, 66, 93).

In many cases, however, antibiotic production in vitro has not been demonstrated to result in antagonism in situ, leading to speculation that antibiotics play roles other than as growth inhibitors (39). At subinhibitory concentrations, structurally diverse antibiotics affect transcription of many bacterial genes not necessarily associated with stress responses, suggesting that antibiotics may function as signaling molecules in the environment when produced at low concentrations (39).

Interference competition via signal disruption. Some mechanisms of interference competition between microorganisms are independent of antibiotic production, such as disruption of signaling cascades. Diverse bacteria degrade acyl-homoserine lactone signal molecules (110, 119, 198–200, 210), and the rapid turnover rate of acyl-homoserines in nonsterile soil suggests this is a common bacterial trait at least in that environment (204). Signaling by small peptides can also be disrupted; for example, siamycin, a secondary metabolite produced by a soil *Streptomyces* strain, inhibits signaling by the gelatinase biosynthesis-activating pheromone of *Enterococcus faecalis* (138). The ability to interfere with signaling would provide a competitive advantage if competitive determinants, such as antimicrobial toxin production, were regulated by signaling, as has been suggested (3). However, a clear link between the disruption of signaling and competitive advantage has yet to be established.

Predation

In macroecological systems, predation, or the consumption of one organism by another, is frequently a key stabilizer of community structure (52, 53, 82). The top predator often regulates abundance of other species that in turn regulate other species, providing a cascade of effects that have a sweeping influence. Often, this effect on the community far exceeds the numerical representation of the predator and can transform entire landscapes, which defines the predator as a keystone species (145, 165).

Predation of bacteria by microbial eukaryotes and bacteriophages provides a key link between microbial and macroorganismal food webs (25, 150) and has global effects on bacterial community structure and composition in many environments. In both freshwater and marine habitats, predation is a leading cause of bacterial mortality (150, 191). In some aquatic environments, top-down control by predation also appears to regulate bacterial population sizes (150). Additionally, predation has been suggested as an influence on bacterial species richness and evenness (150, 213), selectively limiting population sizes of some readily culturable aquatic bacteria that rapidly proliferate when grazing pressure is experimentally reduced (7). In soil, predation by protozoa similarly limits bacterial population sizes and can influence bacterial community composition and structure (25, 137). Difficulties associated with quantification of bacteriophage in soil have hampered efforts to ascertain the role of predation by phage in these ecosystems (5, 207). In engineered microbial communities, predation pressure can influence bacterial population structure by selecting for strains adapted to defend against predator attack (107). Manipulation of microbial communities via phage that prey specifically on lineages of bacteria could be used as a tool to study community dynamics, given the strength of phage selective pressures in some communities (15).

Bacteria or archaea that prey on other bacteria or archaea appear to be relatively rare compared with their eukaryotic counterparts, but examples of each have been described. *Bdellovibrio*-like organisms are obligate bacterial predators, which now appear to be more diverse and widespread than previously recognized (36, 155). Most of these organisms penetrate the outer membrane and wall of their prey and replicate in the cytoplasm, thereby killing the host. However, some, including the *Micavibrio* sp. (Alphaproteobacteria), attach to the outside of prey cells and replicate epibiotically (37). The range of prey

organisms targeted by different *Bdellovibrio*-like predators varies but typically includes only a limited number of species (90, 154, 168). A different predatory strategy is employed by a second group of bacterial predators, the myxobacteria (Deltaproteobacteria). Populations of myxobacteria exhibit cooperative, surface-associated motility and collectively subsume prey organisms they encounter by producing secreted and cell-associated degradative enzymes (162). However, unlike *Bdellovibrio*-like organisms, myxobacteria can also obtain nutrients by degrading macromolecules instead of live prey; genome sequencing of *Myxococcus xanthus* suggests that prey bacteria serve as a source of branched-chain amino acids, which the predator does not have the capacity to synthesize (71). In addition to many proteases and cell wall–degrading enzymes, myxobacteria also produce diverse secondary metabolites, which may also play a role in predation or may mediate competition with other species or other myxobacterial strains (55, 71). Finally, a number of gram-negative bacteria release membrane vesicles containing hydrolytic enzymes that can fuse with and lyse other bacteria (91, 117). Membrane-vesicle-mediated lysis may help to extract nutrients from target cells, although this has not been established empirically.

Parasitism

The only prokaryote thought to parasitize another prokaryote is *Nanoarchaeum equitans*, isolated from hydrothermal vents. *N. equitans* is small in physical stature and has a tiny, compact genome, less than 0.5 Mb, predicted to contain a 95% coding sequence (87, 132). This species is completely dependent on its host, the larger archaeon, *Ignicoccus hospitalis* (87, 147). The genome of *N. equitans* lacks many key metabolic functions, including genes encoding glycolysis and trichloroacetic acid cycle enzymes, as well as most amino acid and lipid biosynthesis pathways, indicating that it must obtain many nutrients and metabolites from its host (205). The host species, however, can be found in a free-living state, and association with *N. equitans* appears to take a toll on its fitness, thus suggesting that the smaller associate is a parasite (87, 147).

Mutualistic and Commensal Interactions

At least as prevalent among microorganisms as the antagonistic interactions described above are interactions in which both partners benefit (mutualism) or in which one partner benefits with no apparent effect on the other (commensalism). Purely commensal relationships may not exist; perhaps we simply have not discovered the benefit to the second partner. More likely, those organisms are either beneficial or harmful to their hosts, depending on the community dynamics in the niche, but researchers have yet to delimit and quantify the costs and benefits exchanged between the host and symbiont. For example, many microbes in the human gut historically termed commensal are now recognized as critical factors in gut and immunity development, nutrient uptake, and homeostasis of the system (85, 86, 159). Furthermore, relationships may be context dependent, that is, an organism could be beneficial under certain conditions and commensal under others.

Obligate associations. In obligate mutualisms each partner depends on the other for survival and reproduction. One particularly elegant obligate microbial association is the phototrophic consortium detected in many freshwater habitats (68). In these assemblages, a central motile, nonphotosynthetic Betaproteobacterium from the family *Comamonadaceae* is surrounded by green sulfur bacteria in an organized structure (59, 95). The epibiotic sulfur bacteria are thought to benefit from the motility provided by the central bacterium, enabling the consortia to chemotax toward sulfide (68). The central *Comamonadaceae* may benefit from carbon secreted by the sulfur bacterium during photoautotrophic growth (68, 69). Additionally, the partners appear to coordinate behaviors via as yet unidentified signal exchange. For example, the consortia chemotax toward sulfide and

the organic compound 2-oxoglutarate only in light, which the motile central organism itself cannot detect (69). The consortia also move preferentially to the optimal light wavelength absorbed by the nonmobile photosynthetic bacteria (58, 59).

Several obligate associations between microbial species occur within the context of a eukaryotic host. For example, in the glassy-winged sharpshooter, *Homalodisca coagulata*, metagenomic analysis of its microbial symbionts revealed the presence of complementary amino acid, vitamin, and cofactor biosynthetic pathways in two microbial symbionts, *Baumannia cicadellinicola* and *Sulcia muelleri*. Both symbionts are required to sustain the sharpshooter, which feeds on the amino acid–poor diet of plant xylem (209). Several species from another group of plant sap-feeding insects, the mealybugs, harbor not only multiple symbionts, but one symbiont, a Gammaproteobacterium, is housed inside the second symbiont, a Betaproteobacterium (203). The functions provided to the host by each symbiont have not yet been identified, but the associations appear to be stably maintained and vertically transmitted, as reflected by cospeciation in symbiont phylogenies (190).

Facultative associations. Conditions under which one or both species of a mutualism or commensalism survive and maintain populations in the absence of the other partner are called facultative. Many instances of facultative mutualism in microbe-microbe relationships involve the exchange or sharing of nutritional resources. For example, metabolic cooperation can result from complementary degradative capabilities or from the ability of one organism to make use of byproducts generated by another. In the human oral cavity, metabolic cooperation plays a key role in structuring the complex, multispecies biofilm formed on tooth surfaces. The late successional stage colonizer *Porphymonas gingivalis* benefits the earlier colonizer *Fusobacterium nucleatum* by activating a host protease, plasmin, which *F. nucleatum* subsequently captures and uses to obtain nutrients (35). Another

Coevolution:
reciprocal genetic
changes in two or
more species in
response to each other

facultative commensalism is in plant root exudate, where peptidoglycan from the cell wall of *Bacillus cereus* rhizosphere isolates provides carbon to sustain the growth of *Flavobacterium* and *Chryseobacterium* species, which is otherwise carbon limited (151, 152), without impacting the growth of *B. cereus*. In some cases, metabolic cooperation results from the ability of one organism to alleviate the effects of a toxin on another organism. For example, in a model system to evaluate effects of mixed organic waste on organisms important for detoxification, the *p*-cresol-degrading organism *Pseudomonas putida* DMP1 protected the *p*-cresol-sensitive strain *Pseudomonas* sp. strain GJ1, which could then degrade a second common waste compound, 2-chloroethanol (29).

Syntrophy. The hallmark of syntrophic interactions, which can be obligate or facultative, is the coupling of metabolic processes in two organisms, typically by transfer of electrons between the organisms by hydrogen or other carriers, which facilitates metabolisms that would otherwise be thermodynamically unfavorable. Under methanogenic conditions, syntrophy appears to facilitate a number of the intermediate transformations between primary fermentation of complex organic matter and eventual production of methane (172). In methane-rich marine sediments, syntrophy between archaea thought to perform reverse methanogenesis and sulfate-reducing bacteria plays a role in mediating methane oxidation, an important control of the flux of this potent greenhouse gas (10, 75). Degradation of some xenobiotic compounds also relies on syntrophy. For example, interspecies hydrogen transfer from a sulfate-reducing organism facilitates tetrachloroethene dehalorespiration by another organism (48). Similarly, vinyl chloride dechlorination by *Methanosarcina* spp. also requires interspecies hydrogen transfer by a syntrophic partner organism (80).

Coaggregation and multispecies biofilm formation. Many beneficial interactions between microorganisms require the partners to be maintained in close proximity, which is often achieved by the formation of multispecies biofilms or aggregates. In some cases, most notably in the oral cavity of vertebrates, development of complex communities results from specific, receptor-mediated interactions between pairs of organisms, known as coaggregation (109). Early colonizers to tooth surfaces, such as *Streptococcus gordonii* and other viridans streptococci, can bind a variety of host molecules and subsequently facilitate colonization by the second-stage species through coaggregation with specific partners (105). *F. nucleatum*, the most abundant gram-negative bacterium in mouths of healthy people, is thought to serve as a bridge between these early and subsequent late colonizers because of its ability to coaggregate with many species from both classes (104).

EVOLUTION IN MICROBIAL COMMUNITIES

The intersection of ecology and evolution is important to our understanding of communities but has not been sufficiently studied to produce a cohesive framework. Antagonistic and mutualistic behaviors have evolved as adaptations to life in a community. Organisms exploit or compete with each other for resources, leading to the grand diversity of ecological mechanisms in the biological world. Individual species evolve in the context of a community, resulting in coevolution, and the community evolves as a composite of many species. Identifying the selection pressures that favor certain interactions is the key to deriving an evolutionary understanding of microbial communities. And perhaps there is a larger conceptual framework to be developed that will describe microbial community evolution, with the entire community as the unit upon which selection acts.

In many ways, the evolution of prokaryotes and eukaryotes is similar. Natural selection and genetic drift operate on population-level genetic variation caused by mutation and gene flow. Together these processes alter the genetic composition of populations and directly and indirectly affect the species interactions that dictate community ecology. However, some evolutionary processes play different roles in prokaryotic and eukaryotic populations, and these contrasts are particularly important to consider in the context of microbial community dynamics.

Genetic variation is the target on which selection acts, whereas ecological forces, including biotic factors such as species interactions (competition, parasitism, mutualism), are the agents of selection. Thus, the two fields of study, population genetics and community ecology, seem inevitably coupled through evolution.

Most ecological and evolutionary theory has been developed on the basis of observations made in eukaryotic organisms. Natural selection, developed with plants and animals, and Mendelian genetics, originating from studies of plants, were integrated to form the modern synthesis, which is the basis of current evolutionary theory, but none of the major leaders in the development of the modern synthesis (Fisher, Dobzhansky, Haldane, Wright, Huxley, Mayr, Rensch, Simpson, and Stebbins) focused on prokaryotes. Consequently, models for understanding adaptation, evolution, and speciation in prokaryotic biology were not developed until the early 1980s (114). Advances in molecular biology have propelled the expansion of prokaryotic models for evolution over the past 20 years, from which two major differences between prokaryotic and eukaryotic evolution have emerged: the frequency of recombination and the phylogenetic breadth among which genetic materials can be exchanged. Intragenomic processes such as recombination are likely to have the greatest influence in short-term changes in a community, leading to population adaptation to changing conditions or new metabolic opportunities. Intergenomic processes such as horizontal gene transfer have profound effects on the long-term evolution of communities, possibly leading to the formation of new species (41, 84).

Although a superficial examination of evolutionary processes in prokaryotes and eukaryotes reveals stark differences, deeper examination might unite them. Clonal eukaryotes, for example, may be governed by similar principles as prokaryotes. More significantly, the same forces may regulate hybridization between plant species and interspecies gene transfer in prokaryotes and the resultant afront to the integrity of the species (41, 84).

Intragenomic alterations. Sequential point mutations and gene rearrangements can lead to adjustments in the genotypic and, consequently, phenotypic content of population members. Mutations or alterations that are selected typically improve the fitness of an organism in its current ecological niche, which is necessary to maintain interspecific interactions such as competition, predation, and mutualism. Several hypotheses regarding microbial fitness are readily testable. For example, improvements in fitness over thousands of generations under glucose-limited conditions have been measured (146). An interesting and intensely studied pattern resulting from closely evolving interactions is the coevolutionary process, in which genetically based adaptations in one species invoke reciprocal genetic changes in populations of its partner species (e.g., competitor, parasite) or guilds of species (193, 194).

Horizontal gene transfer. The transfer of genetic information between species is a central mechanism of generating genetic variation in microbial communities. For example, multilocus sequence typing data and proteomic and comparative genomics data indicate that bacterial species in acid mine biofilm communities exchange large (up to hundreds of kilobases) regions of DNA as well as smaller sections that may play a role in resistance to phage (122, 197). Events of horizontal gene transfer can be detected through phylogenetics, by seeking atypical distributions of genes across organisms, or through phylogeny-independent methods that examine genes that appear aberrant in their current genomic context. Complete genome sequencing has arguably been the most important factor in unveiling instances of horizontal gene transfer, illustrating the impact of horizontal gene transfer on bacterial evolution (6, 10, 13). Perhaps one of the most dramatic impacts of horizontal transfer of genetic information by accessory genetic elements and vectors of genes (e.g., viruses) is ecological. Horizontal gene transfer can enable a microbe to rapidly expand and/or alter its ecological niche, making this genetic process important when considering evolution in microbial communities through deep time (i.e. an evolutionary timescale rather than an ecological timescale). Horizontal gene transfer has also been proposed to contribute to speciation, which is a critical aspect of community function and evolution (41, 84).

Ecological processes, in turn, affect evolution by providing opportunities for interspecies gene transfer and providing selection pressure. The architecture of communities, which dictates proximity of cells of different species, affects the probability of gene exchange across wide phylogenetic distances. The physical and biological features of the community will affect the susceptibility of cells to transformation or transduction, thereby affecting the frequency of gene transfer. The ecological processes and characteristics of the community create the selection pressures that determine the direction of change in frequencies of certain genotypes.

When entire genes or groups of genes are transferred between individuals, especially distantly related individuals, a trait can rapidly sweep through a population under appropriate selection pressure. A contemporary example of this is the rapid spread of antibiotic resistance in bacterial populations. In other instances, the changes can lead to rapid lineage diversification (164). In some cases these changes, especially those that involve the metabolic repertoire, enable the recipient of horizontal gene transfer to invade and adapt in a new ecological niche. Classic examples of niche-altering gene acquisitions include the *lac* operon by *Escherichia coli* and pathogenicity islands by *Salmonella* sp. Changes that enable an organism to invade a new niche(s) have strong implications at the community level: They have the potential to alter interactions between species and the structure, diversity, and robustness of communities. Ultimately, the ecological selection pressures that drive microbial evolution are major contributors to the emergent structure and function of the community.

APPROACHES TO THE STUDY OF MICROBIAL COMMUNITIES

Community ecology, as it pertains to microbes, remains in its infancy. Most studies of microbe-microbe and host-microbe interactions have extracted the organisms from their native community and studied them as binary interactions; indeed, this is simpler to understand and a necessary step toward understanding interspecific interactions in a community context (**Figure 2**). However, the information gained from such studies can be inadequate or misleading, as has been shown repeatedly in macroecological research (18, 120, 182, 185), and as such, results obtained from those studies must be interpreted with caution. To this end, we have parsed microbial ecology studies of communities into four groups on the basis of the questions being asked and presented them chronologically in terms of the order in which questions must be addressed and answered to generate further information on microbial community characteristics and processes. Each question is illustrated with classic experiments and recent technological advances that have enabled their investigation (**Figure 3**).

Who Is Present in the Community?

The first question a community ecologist asks is, Who makes up this community?, and this is indeed a good starting point. But with microorganisms, this is not a trivial question, nor has it been easy to identify the diversity of species found in various communities. Two central techniques are used to identify microbial phylotypes within community samples: One technique relies on culturing, and the other is culture independent.

	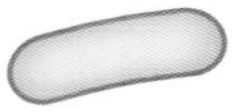 Individual	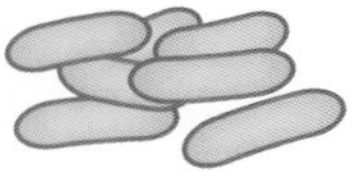 Population	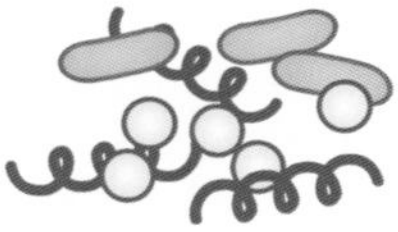 Community
Ecology	**Physiology:** Differential gene expression in response to change	**Demographics:** Birth, death, immigration, emigration	**Community ecology:** Interspecific interactions that shape community structure and function
Genomics	Fine-scale mapping of individual genomes	**Population genomics:** Comparative genomic analyses to assess variation	**Metagenomics:** Genetic potential of collective members of community
Genetics	**Bacterial genetics:** Role of genes under various conditions	**Population genetics:** Allele frequency distribution	**Community genetics:** Interplay between genetic composition of community and ecological community properties

Figure 2

Progression from studies on the individual scale to studies on the community scale.

Culture-based methods. Koch's discovery that bacteria could be isolated and grown in pure culture on solid artificial medium enabled the discipline bacteriology to develop. Growing bacteria in pure culture provided morphological and physiological data, which together provided the basis to identify bacterial species. More advanced culture techniques incorporated various nutrients and abiotic conditions that closely mimicked the environments from which the samples were isolated. However, recent estimates indicate that less than 1% of the membership of many communities is culturable, making it necessary to assess the identity of the as yet uncultured organisms to generate a complete list of community members.

Culture-independent methods. One molecular approach that provides a powerful complement to culture-dependent techniques is the amplification of 16S rRNA gene sequences directly from environmental samples, such as soil, using PCR and universal or domain-specific primers, which is usually followed by clone library construction. Clones are then screened to analyze sequence differences (11), or re-

striction fragment length polymorphisms (136), which are then used to identify species. To date, 16S rRNA gene amplification and identification remain the most reliable tool to describe prokaryotic species. Because the gene is universal and can therefore be used as an identifier for any bacterial or archaeal species, it accounts for both culturable and nonculturable prokaryotic organisms, and it has phylogenetic meaning. There are, however, limitations to the 16S rRNA gene approach to phylogeny. Interspecies gene transfer muddies phylogenetic assessments derived from 16S rRNA. If organisms are hybrids with fragments of DNA of different organismal origins, then what is a species? Should they be defined by the 16S rRNA gene affiliation or by a census of functional genes? Prokaryote phylogeny is an emergent field and the species concept will be one to grapple with in the future.

The universal primers used to assess entire communities may not detect all species. Recent work suggests significant differences in the groups whose 16S rRNA genes are amplified when the universal primers are replaced by miniprimers with very broad specificity (88). All methods impose bias and it is likely that

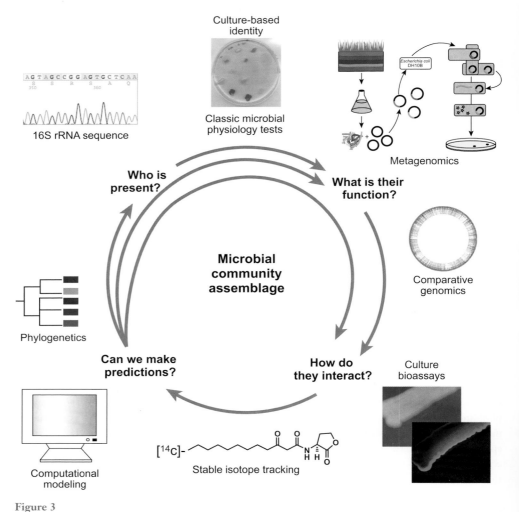

Figure 3

Four groups of questions in microbial ecology and some techniques used to address them.

further development of phylogenetic tools will reveal greater diversity and perhaps groups of organisms that are not suspected from current surveys.

What Are the Functions of Individual Organisms?

After identifying which organism(s) is present within a community, the next challenge for microbial ecologists is to identify which organism demonstrates each of the various metabolic processes in its native community.

Classic in vitro microbial physiology. Up until the last 15 years of the twentieth century, pure-culture experimental setup, as described above, was the central method of associating organisms with metabolic processes. In an effort to understand metabolic processes within the context of a microbe's native community, it was typical to inoculate enrichment cultures with samples from the natural environment of interest, determine which bacteria grow, and then make inferences on the basis of substrate use about the microbe's activity in its native community.

Metagenomics. The diversity of the as yet uncultturable members of microbial communities is vast compared with that of the culturable members (9, 13, 14, 108, 118, 130, 156). To capture and study the functional diversity of these organisms, a new field designated metagenomics has sprung to life. Metagenomics is the culture-independent analysis of genomes from an assemblage of microorganisms. A metagenomic analysis entails extracting DNA directly from soil, cloning it into a culturable host bacterium, and analyzing it (169, 186). This method has recently been used for massive capture and sequencing of DNA from the Sargasso Sea (202), acid mine drainage (197), a Minnesota soil sample (196), and a global ocean survey (170). DeLong et al. (42) applied metagenomics to planktonic microbial communities in the North Pacific Subtropical Gyre, in which they identified stratified microbial communities through comparative genomics delineated by taxonomic zonation and functional and metabolic potential. Through detailed analysis of the genes in each stratum, they inferred the photosynthetic activity at various depths in the ocean. They also found a surprisingly high frequency of cyanophage-infected cells (up to 12%), which likely structure the planktonic community via predation.

Functional metagenomics. Entire phyla in soil are known only by their 16S rRNA gene signatures, with nothing known about their physiology, genetics, or role in the soil community. Most work in metagenomics is driven by sequence analysis, but this work is limited by the ability to recognize gene function on the basis of sequence alone. Because many of the genes isolated from the environment have no significant similarity to genes of known function, an alternative approach is to search for genes of a particular function by functional metagenomics. In functional metagenomics, genes are sought that confer a function of interest on a host bacterium. This method requires that the genes be expressed, but it does not require that their functions be recognizable by sequence (79, 83, 102, 103, 126, 188, 208).

Stable isotope probing. Stable isotope probing (SIP) involves introducing a stable isotope-labeled substrate into the community and tracking the movement of the substrate by extracting diagnostic molecules (e.g., lipids and nucleic acids) to determine which molecules have incorporated the substrate. Stable isotope ratios have been used by ecologists to follow resource use through trophic levels for decades (43, 44, 179). More recently, stable isotopes have become a tool used by microbial ecologists to track the movement of substrates through microbial communities and identify which community members utilize which substrates. The main advantages of SIP are that it does not rely on culturability, and it allows direct observation of substrate movement with minimal disruption of the natural environment and community. SIP in microbial communities has been recently reviewed elsewhere (106).

Single-cell analyses. Fully comprehending community function necessitates understanding the function and activities at all levels, including that of individual members. Individuals within a population vary in levels of expression of certain genes and growth rates (92). For example, using flow cells and laser scanning, Strovas et al. (189) revealed that individuals within *Methylobacterium extorquens* grown continuously vary in cell size at division, division time, and growth rate, and they respond differently to a substrate shift. In addition to explaining more about individual members of a community, single-cell analyses should also prove useful for studying rare and uncultured organisms. Although used to amplify low-quantity DNA for metagenomic studies, multiple displacement amplification, a technique that involves random primers and φ29 DNA polymerase), can also be used to amplify whole genomes of single cells (72, 81, 139). Multiple displacement amplification combined with technologies for capturing individual cells, such as microfluids, may lead to situations in which the genomes of rare members could be analyzed to determine the potential function of the member within its community.

Metagenomics: culture-independent analysis of DNA extracted directly from communities in environmental samples

How Do Organisms Interact?

Interactions are the fulcrum of communities. Studying them is essential to understanding community function, but that study is challenging. Lessons from study of cultured organisms can guide methodological choices. Culturing has been, by far, the most impactful method introduced into microbiology since the microscope. The study of organisms in pure culture has produced the staggering depth and breadth of current knowledge about cellular microbiology. Likewise, coculture can be used to study interactions in a controlled environment in which variables can be manipulated. Bacterial genetics has been similarly influential, providing experimental precision and rigor that elevate associations to causal relationships. The adaptation of cellular genetics to study interactions among organisms is likely to yield surprises and provide a foundation for community ecology principles.

Coculture experiments. Although the trend in the twentieth century was toward reduction, which meant studying microbes in isolation or sometimes pairs, studying organismal interactions in their native communities, or in assemblages that more closely resemble native communities, is not a recent idea. In 1895, Winogradsky isolated *Clostridium pasteurianum*, an organism that fixes free nitrogen from the air (163, 197). But through a series of pure-culture and coculture experiments, it became clear to Winogradsky that *C. pasteurianum*, a strict anaerobe, could only fix nitrogen under aerobic conditions. If *C. pasteurianum* was grown in close association with an extreme aerobe that essentially creates an anaerobic environment for *C. pasteurianum*, nitrogen fixation was restored. Studying microbial assemblages in culture remains an important method to investigate organismal interactions. In 1993 Gilbert and colleagues (64) reported that the application of *Bacillus cereus* UW85, an antibiotic-producing strain used for biological control, to the soybean rhizosphere increased the abundance of bacteria from the *Cytophaga-Flavobacterium* (CF) group. More recently,

Peterson et al. (152) used coculturing to determine that the commensal relationship between *B. cereus* UW85 and the bacteria from the CF group is mediated by a *B. cereus* peptidoglycan. Although coculture experiments are invaluable to our understanding of interspecies interactions, a main restriction is that they are limited to a small number of species interacting together in liquid or on a plate.

Bacterial genetics. Just as bacterial genetics has provided profound insight into the function of organisms in pure culture, it can lead to an understanding of species interactions in communities. The foundation of genetics is construction of random mutations followed by mutant hunts. The randomness coupled with assessment of phenotype produces a minimally biased approach in which the genes required for a certain function or process are identified. In this age of gene arrays and genomics, mutant hunts have been replaced by more-directed methods, but broad searches remain critical to expanding our knowledge beyond the human imagination. Targeted approaches typically require the researcher to make educated predictions about the nature or type of genes involved in a process. Broad mutant hunts enable the bacteria to answer that question. Historically, the answers have been surprising, and many of the genes identified would not have been predicted to be involved in the process on the basis of sequence alone.

The community context presents new challenges for bacterial genetics. Identifying genes involved in community function requires complex screens that will be difficult to apply to large collections of mutants, but the search will be worthwhile. In addition, studies of targeted mutants in communities can be revealing. For example, the study of genes involved in quorum sensing in pure culture and in a community provides very different insights. The ubiquitous presence of genes involved in quorum-sensing indicates that density-dependent cell-cell communication is a common mode of bacterial communication (6, 127, 134). Demonstration of signal exchange between strains in a community in a caterpillar gut provided surprising evidence of stability of quorum-sensing signal in a high pH environment (B. Borlee, G. Geske, C. Robinson, H. Blackwell & J. Handelsman, manuscript in revision). Genes that code for antibacterial compounds are found in several environmental isolates (16, 45, 51, 116). The role of antimicrobials in nature is controversial (16, 38, 45, 51, 116, 212). Some studies, however, have shown that producing these compounds yields a competitive advantage for bacteria (57, 63). An antibacterial protein produced by a marine bacterium is important for its competition with other marine bacteria for the formation of biofilms as well as dispersal of cells from biofilms (125, 160).

Can We Make Predictions at the Community Scale?

Perhaps the most important implication of information on ecological community principles and dynamics from both an application and a theoretical standpoint is the potential to build predictive models. By defining fundamental principles of microbial community ecology derived from phylogenetically and ecologically disparate systems amenable to experimental manipulation, we can generate models to predict information about other communities that are less tractable for study. Much of the baseline research needed to generate such models falls into the categories described above, namely, who is present in a community, what is their function, and how do they interact with one another. Advances in molecular biology, and computational biology in particular, have allowed development in predictive ecological modeling that is poised for application to microbial communities.

Integrative modeling. In parallel with the recurring theme of moving from reductionist science to systems-based biology, many current approaches toward understanding properties of microbial communities include integrating biochemistry, thermodynamics, metabolite transport and utilization, metagenomic sequencing,

regulatory and metabolic network analysis, and comparative and evolutionary genomics. On the mechanistic scale, several groups aim to predict microbially mediated metabolic activities in specific environments using microarrays and to predict a microbe's behavior and lifestyle directly from its genomic sequence (4). On the ecological scale, ecological niche modeling packages can output predictions of geographic ranges for species on the basis of current species records and layers of environmental data (e.g., GARP, Genetic Algorithm for Rule-set Production).

Community genetics. An integrative field of study becoming increasingly important in generating predictions about how microbial community members interact is community genetics, which involves coupling changes in genetic distributions with species interactions and community structure (49, 206). A few studies of macroecological systems have generated predictions. For example, Whitham and colleagues found that genetic variation in plants affects diverse communities of insect herbivores, birds, and fungi (206). Other studies have shown that population dynamics and trophic interactions affect the rate at which pests evolve resistance to genetically engineered crops. Experimental results from a predator-prey system (rotifers that consume algae) provided the basis for building models for ecological and evolutionary hypotheses about the predator-prey cycles. These studies in situations involving rapid evolution model successfully predicted ecological consequences (180).

There has been little application of community genetics to make predictions of ecological phenomena in predominantly microbial communities, but it will be a fruitful avenue of research. Community genetics in microbial communities has the special power of using constructed, defined mutants that can be introduced into a community. The behavior of the community in the presence of the mutant and wild type can be compared, providing insight into the role of a single gene in a population of a species and in the community

Community genetics: studies of the interplay between community genetic composition and community structure and function context. Just as bacterial genetics brought power and precision to the dissection of bacterial cell processes in the twentieth century, the same approaches will transform our understanding of community processes in the twenty-first century.

SUMMARY POINTS

1. An important feature of microbial communities is robustness, which has structural and functional components.

2. Understanding microbial community ecology necessitates study of evolutionary mechanisms that underlie community structure.

3. Recent advances in molecular biology provide a means to address questions about microbial communities, including both culturable and as yet uncultured members.

4. Microbial communities offer a unique opportunity to apply community genetics in a manner that is difficult in most macroorganism communities: Introduction of defined mutants into communities will advance our understanding of the interplay between community genetic composition and community structure and function under various selection pressures.

5. The approaches described here will contribute to the inputs needed to build and test predictive models that will elucidate principles that govern community interactions, providing a set of rules of engagement among community members that dictate community structure and function.

DISCLOSURE STATEMENT

The authors are not aware of any biases that might be perceived as affecting the objectivity of this review.

ACKNOWLEDGMENTS

We are grateful for support from an Interdisciplinary Hatch Project from the University of Wisconsin College of Agricultural and Life Sciences and the Howard Hughes Medical Institute. We thank Karen Cloud-Hansen for helpful discussions of an earlier draft of the manuscript.

LITERATURE CITED

1. Adamczyk J, Hesselsoe M, Iversen N, Horn M, Lehner A, et al. 2003. The isotope array, a new tool that employs substrate-mediated labeling of rRNA for determination of microbial community structure and function. *Appl. Environ. Microbiol.* 69:6875–87

2. Alonso-Saez L, Balague V, Sa EL, Sanchez O, Gonzalez JM, et al. 2007. Seasonality in bacterial diversity in north-west Mediterranean coastal waters: assessment through clone libraries, fingerprinting and FISH. *FEMS Microbiol. Ecol.* 60:98–112

3. An D, Danhorn T, Fuqua C, Parsek MR. 2006. Quorum sensing and motility mediate interactions between *Pseudomonas aeruginosa* and *Agrobacterium tumefaciens* in biofilm cocultures. *Proc. Natl. Acad. Sci. USA* 103:3828–33

4. Antonovics J. 1992. Towards community genetics. In *Plant Resistance to Herbivores and Pathogens: Ecology, Evolution, and Genetics*, ed. RS Fritz, EL Simms, pp. 426–49. Chicago: Univ. Chicago Press

5. Ashelford KE, Day MJ, Fry JC. 2003. Elevated abundance of bacteriophage infecting bacteria in soil. *Appl. Environ. Microbiol.* 69:285–89

6. Bassler BL. 1999. How bacteria talk to each other: regulation of gene expression by quorum sensing. *Curr. Opin. Microbiol.* 2:582–87

7. Beardsley C, Pernthaler J, Wosniok W, Amann R. 2003. Are readily culturable bacteria in coastal North Sea waters suppressed by selective grazing mortality? *Appl. Environ. Microbiol.* 69:2624–30

8. Begon M, Harper J, Townsend C. 1990. *Ecology: Individuals, Populations, and Communities.* Cambridge, MA: Blackwell. 945 pp.

9. Bintrim SB, Donohue TJ, Handelsman J, Roberts GP, Goodman RM. 1997. Molecular phylogeny of Archaea from soil. *Proc. Natl. Acad. Sci. USA* 94:277–82

10. Boetius A, Ravenschlag K, Schubert CJ, Rickert D, Widdel F, et al. 2000. A marine microbial consortium apparently mediating anaerobic oxidation of methane. *Nature* 407:623–26

11. Bond PL, Hugenholtz P, Keller J, Blackall LL. 1995. Bacterial community structures of phosphate-removing and nonphosphate-removing activated sludges from sequencing batch reactors. *Appl. Environ. Microbiol.* 61:1910–16

12. Bonsall RF, Weller DM, Thomashow LS. 1997. Quantification of 2,4-diacetylphloroglucinol produced by fluorescent *Pseudomonas* spp. in vitro and in the rhizosphere of wheat. *Appl. Environ. Microbiol.* 63:951–55

13. Borneman J, Skroch PW, O'Sullivan KM, Palus JA, Rumjanek NG, et al. 1996. Molecular microbial diversity of an agricultural soil in Wisconsin. *Appl. Environ. Microbiol.* 62:1935–43

14. Borneman J, Triplett EW. 1997. Molecular microbial diversity in soils from eastern Amazonia: evidence for unusual microorganisms and microbial population shifts associated with deforestation. *Appl. Environ. Microbiol.* 63:2647–53

15. Breitbart M, Wegley L, Leeds S, Schoenfeld T, Rohwer F. 2004. Phage community dynamics in hot springs. *Appl. Environ. Microbiol.* 70:1633–40

16. Brinkhoff T, Bach G, Heidorn T, Liang L, Schlingloff A, Simon M. 2004. Antibiotic production by a Roseobacter clade-affiliated species from the German Wadden Sea and its antagonistic effects on indigenous isolates. *Appl. Environ. Microbiol.* 70:2560–65

17. Brodie EL, DeSantis TZ, Parker JPM, Zubietta IX, Piceno YM, Andersen GL. 2007. Urban aerosols harbor diverse and dynamic bacterial populations. *Proc. Natl. Acad. Sci. USA* 104:299–304

18. Bronstein J, Barbosa P. 2002. Multi-trophic/multi-species interactions: the role of nonmutualists in shaping and mediating mutualisms. In *Multitrophic Interactions*, ed. T Tscharntke, BA Hawkins, pp. 44–66. Cambridge, UK: Cambridge Univ. Press

19. Cardoza YJ, Klepzig KD, Raffa KF. 2006. Bacteria in oral secretions of an endophytic insect inhibit antagonistic fungi. *Ecol. Entomol.* 31:636–45

20. Carrino-Kyker SR, Swanson AK. 2008. Temporal and spatial patterns of eukaryotic and bacterial communities found in vernal pools. *Appl. Environ. Microbiol.* 74:2554–57

21. Case RJ, Boucher Y, Dahllof I, Holmstrom C, Doolittle WF, Kjelleberg S. 2007. Use of 16S rRNA and *rpoB* genes as molecular markers for microbial ecology studies. *Appl. Environ. Microbiol.* 73:278–88

22. Chandler DP, Jarrell AE, Roden ER, Golova J, Chernov B, et al. 2006. Suspension array analysis of 16S rRNA from Fe- and SO42-reducing bacteria in uranium-contaminated sediments undergoing bioremediation. *Appl. Environ. Microbiol.* 72:4672–87

23. Chenier MR, Beaumier D, Fortin N, Roy R, Driscoll BT, et al. 2006. Influence of nutrient inputs, hexadecane, and temporal variations on denitrification and community composition of river biofilms. *Appl. Environ. Microbiol.* 72:575–84

24. Chu H, Fujii T, Morimoto S, Lin X, Yagi K, et al. 2007. Community structure of ammonia-oxidizing bacteria under long-term application of mineral fertilizer and organic manure in a sandy loam soil. *Appl. Environ. Microbiol.* 73:485–91

25. Clarholm M. 2002. Bacteria and protozoa as integral components of the forest ecosystem—their role in creating a naturally varied soil fertility. *Antonie van Leeuwenhoek* 81:309–18

26. Colwell RK. 2005. EstimateS Ver. 7.5: statistical estimation of species richness and shared species from samples. (Software and User's Guide) **http://viceroy.eeb.uconn.edu/EstimateSPages/EstimateSSupport.htm**

27. Compant S, Duffy B, Nowak J, Clement C, Barka EA. 2005. Use of plant growth-promoting bacteria for biocontrol of plant diseases: principles, mechanisms of action, and future prospects. *Appl. Environ. Microbiol.* 71:4951–59

28. Cottrell MT, Kirchman DL. 2000. Natural assemblages of marine Proteobacteria and members of the Cytophaga-Flavobacter cluster consuming low- and high-molecular-weight dissolved organic matter. *Appl. Environ. Microbiol.* 66:1692–97

29. Cowan SE, Gilbert E, Liepmann D, Keasling JD. 2000. Commensal interactions in a dual-species biofilm exposed to mixed organic compounds. *Appl. Environ. Microbiol.* 66:4481–85

30. Cox CR, Gilmore MS. 2007. Native microbial colonization of *Drosophila melanogaster* and its use as a model of *Enterococcus faecalis* pathogenesis. *Infect. Immun.* 75:1565–76

31. Currie CR, Poulsen M, Mendenhall J, Boomsma JJ, Billen J. 2006. Coevolved crypts and exocrine glands support mutualistic bacteria in fungus-growing ants. *Science* 311:81–83

32. Curtis TP, Head IM, Lunn M, Woodcock S, Schloss PD, Sloan WT. 2006. What is the extent of prokaryotic diversity? *Philos. Trans. R. Soc. London Ser. B* 361:2023–37

33. Curtis TP, Sloan WT. 2004. Prokaryotic diversity and its limits: microbial community structure in nature and implications for microbial ecology. *Curr. Opin. Microbiol.* 7:221–26

34. Curtis TP, Sloan WT, Scannell JW. 2002. Estimating prokaryotic diversity and its limits. *Proc. Natl. Acad. Sci. USA* 99:10494–99

35. Darenfed H, Grenier D, Mayrand D. 1999. Acquisition of plasmin activity by *Fusobacterium nucleatum* subsp. *nucleatum* and potential contribution to tissue destruction during periodontitis. *Infect. Immun.* 67:6439–44

36. Davidov Y, Friedjung A, Jurkevitch E. 2006. Structure analysis of a soil community of predatory bacteria using culture-dependent and culture-independent methods reveals a hitherto undetected diversity of *Bdellovibrio*-and-like organisms. *Environ. Microbiol.* 8:1667–73

37. Davidov Y, Huchon D, Koval SF, Jurkevitch E. 2006. A new alpha-proteobacterial clade of *Bdellovibrio*-like predators: implications for the mitochondrial endosymbiotic theory. *Environ. Microbiol.* 8:2179–88

38. Davies J. 2006. Are antibiotics naturally antibiotics? *J. Ind. Microbiol. Biotechnol.* 33:496–99

39. Davies J, Spiegelman GB, Yim G. 2006. The world of subinhibitory antibiotic concentrations. *Antimicrobials/Genomics* 9:445–53

40. de Carcer DA, Martin M, Karlson U, Rivilla R. 2007. Changes in bacterial populations and in biphenyl dioxygenase gene diversity in a polychlorinated biphenyl-polluted soil after introduction of willow trees for rhizoremediation. *Appl. Environ. Microbiol.* 73:6224–32

41. de la Cruz F, Davies J. 2000. Horizontal gene transfer and the origin of species: lessons from bacteria. *Trends Microbiol.* 8:128–33

42. DeLong EF, Preston CM, Mincer T, Rich V, Hallam SJ, et al. 2006. Community genomics among stratified microbial assemblages in the ocean's interior. *Science* 311:496–503

43. Deniro MJ, Epstein S. 1978. Influence of diet on distribution of carbon isotopes in animals. *Geochim. Cosmochim. Acta* 42:495–506

44. Deniro MJ, Epstein S. 1981. Influence of diet on the distribution of nitrogen isotopes in animals. *Geochim. Cosmochim. Acta* 45:341–51

45. Derzelle S, Duchaud E, Kunst F, Danchin A, Bertin P. 2002. Identification, characterization, and regulation of a cluster of genes involved in carbapenem biosynthesis in *Photorhabdus luminescens. Appl. Environ. Microbiol.* 68:3780–89

46. Diaz PI, Chalmers NI, Rickard AH, Kong C, Milburn CL, et al. 2006. Molecular characterization of subject-specific oral microflora during initial colonization of enamel. *Appl. Environ. Microbiol.* 72:2837–48

47. Drakare S. 2002. Competition between picoplanktonic cyanobacteria and heterotrophic bacteria along crossed gradients of glucose and phosphate. *Microb. Ecol.* 44:327–35

48. Drzyzga O, Gottschal JC. 2002. Tetrachloroethene dehalorespiration and growth of *Desulfitobacterium frappieri* TCE1 in strict dependence on the activity of *Desulfovibrio fructosivorans. Appl. Environ. Microbiol.* 68:642–49

49. Dungey HS, Potts BM, Whitham TG, Li HF. 2000. Plant genetics affects arthropod community richness and composition: evidence from a synthetic eucalypt hybrid population. *Evol. Int. J. Org. Evol.* 54:1938–46

50. Dykhuizen DE. 1998. Santa Rosalia revisited: Why are there so many species of bacteria? *Antonie van Leeuwenhoek* 73:25–33

51. Emmert EAB, Klimowicz AK, Thomas MG, Handelsman J. 2004. Genetics of zwittermicin A production by *Bacillus cereus*. *Appl. Environ. Microbiol.* 70:104–13

52. Estes JA. 1996. Predators and ecosystem management. *Wildl. Soc. Bull.* 24:390–96

53. Estes JA, Crooks K, Holt R. 2001. Predators, ecological role of. In *Encyclopedia of Biodiversity*, ed. SA Levin, 4:857–78. San Diego, CA: Academic Press

54. Feil EJ, Holmes EC, Bessen DE, Chan MS, Day NP, et al. 2001. Recombination within natural populations of pathogenic bacteria: short-term empirical estimates and long-term phylogenetic consequences. *Proc. Natl. Acad. Sci. USA* 98:182–87

55. Fiegna F, Velicer GJ. 2005. Exploitative and hierarchical antagonism in a cooperative bacterium. *PLoS Biol.* 3:e370

56. Francis CA, Roberts KJ, Beman JM, Santoro AE, Oakley BB. 2005. Ubiquity and diversity of ammonia-oxidizing archaea in water columns and sediments of the ocean. *Proc. Natl. Acad. Sci. USA* 102:14683–88

57. Franks A, Egan S, Holmstrom C, James S, Lappin-Scott H, Kjelleberg S. 2006. Inhibition of fungal colonization by *Pseudoalteromonas tunicata* provides a competitive advantage during surface colonization. *Appl. Environ. Microbiol.* 72:6079–87

58. Frostl JM, Overmann J. 1998. Physiology and tactic response of the phototrophic consortium "Chlorochromatium aggregatum." *Arch. Microbiol.* 169:129–35

59. Frostl JM, Overmann J. 2000. Phylogenetic affiliation of the bacteria that constitute phototrophic consortia. *Arch. Microbiol.* 174:50–58

60. Gans J, Wolinsky M, Dunbar J. 2005. Computational improvements reveal great bacterial diversity and high metal toxicity in soil. *Science* 309:1387–90

61. Gentile ME, Lynn Nyman J, Criddle CS. 2007. Correlation of patterns of denitrification instability in replicated bioreactor communities with shifts in the relative abundance and the denitrification patterns of specific populations. *ISME J.* 1:714–28

62. Geraci A. 1991. *IEEE Standard Computer Dictionary: Compilation of IEEE Standard Computer Glossaries*. New York: Inst. Elect. Electron. Eng. Inc.

63. Giddens SR, Houliston GJ, Mahanty HK. 2003. The influence of antibiotic production and pre-emptive colonization on the population dynamics of *Pantoea agglomerans* (*Erwinia herbicola*) Eh1087 and *Erwinia amylovora* in planta. *Environ. Microbiol.* 5:1016–21

64. Gilbert G, Parke JL, Clayton MK, Handelsman J. 1993. Effects of an introduced bacterium on bacterial communities on roots. *Ecology* 74:840–54

65. Gil-Turnes MS, Fenical W. 1992. Embryos of *Homarus americanus* are protected by epibiotic bacteria. *Biol. Bull.* 182:105–8

66. Gil-Turnes MS, Hay ME, Fenical W. 1989. Symbiotic marine bacteria chemically defend crustacean embryos from a pathogenic fungus. *Science* 246:116–18

67. Girvan MS, Campbell CD, Killham K, Prosser JI, Glover LA. 2005. Bacterial diversity promotes community stability and functional resilience after perturbation. *Environ. Microbiol.* 7:301–13

68. Glaeser J, Overmann J. 2003. Characterization and in situ carbon metabolism of phototrophic consortia. *Appl. Environ. Microbiol.* 69:3739–50

69. Glaeser J, Overmann J. 2003. The significance of organic carbon compounds for in situ metabolism and chemotaxis of phototrophic consortia. *Environ. Microbiol.* 5:1053–63

70. Gogarten JP, Doolittle WF, Lawrence JG. 2002. Prokaryotic evolution in light of gene transfer. *Mol. Biol. Evol.* 19:2226–38

71. Goldman BS, Nierman WC, Kaiser D, Slater SC, Durkin AS, et al. 2006. Evolution of sensory complexity recorded in a myxobacterial genome. *Proc. Natl. Acad. Sci. USA* 103:15200–5

72. Gonzalez JM, Portillo MC, Saiz-Jimenez C. 2005. Multiple displacement amplification as a prepolymerase chain reaction (pre-PCR) to process difficult to amplify samples and low copy number sequences from natural environments. *Environ. Microbiol.* 7:1024–28

73. Gopal PK, Prasad J, Smart J, Gill HS. 2001. In vitro adherence properties of *Lactobacillus rhamnosus* DR20 and *Bifidobacterium lactis* DR10 strains and their antagonistic activity against an enterotoxigenic *Escherichia coli*. *Int. J. Food Microbiol.* 67:207–16

74. Guarner F, Malagelada JR. 2003. Gut flora in health and disease. *Lancet* 361:512–19

75. Hallam SJ, Putnam N, Preston CM, Detter JC, Rokhsar D, et al. 2004. Reverse methanogenesis: testing the hypothesis with environmental genomics. *Science* 305:1457–62

76. Handelsman J. 2004. Metagenomics: application of genomics to uncultured microorganisms. *Microbiol. Mol. Biol. Rev.* 68:669–85

77. Hansel CM, Fendorf S, Jardine PM, Francis CA. 2008. Changes in bacterial and archaeal community structure and functional diversity along a geochemically variable soil profile. *Appl. Environ. Microbiol.* 74:1620–33

78. Hartmann M, Widmer F. 2006. Community structure analyses are more sensitive to differences in soil bacterial communities than anonymous diversity indices. *Appl. Environ. Microbiol.* 72:7804–12

79. Healy FG, Ray RM, Aldrich HC, Wilkie AC, Ingram LO, Shanmugam KT. 1995. Direct isolation of functional genes encoding cellulases from the microbial consortia in a thermophilic, anaerobic digester maintained on lignocellulose. *Appl. Microbiol. Biotechnol.* 43:667–74

80. Heimann AC, Batstone DJ, Jakobsen R. 2006. *Methanosarcina* spp. drive vinyl chloride dechlorination via interspecies hydrogen transfer. *Appl. Environ. Microbiol.* 72:2942–49

81. Hellani A, Coskun S, Sakati N, Benkhalifa M, Al-Odaib A, Ozand P. 2004. Multiple displacement amplification on single cell and possible preimplantation genetic diagnosis applications. *Fertil. Steril.* 82:S28

82. Henke SE, Bryant FC. 1999. Effects of coyote removal on the faunal community in western Texas. *J. Wildl. Manag.* 63:1066–81

83. Henne ADR, Schmitz RA, Gottschalk G. 1999. Construction of environmental DNA libraries in *Escherichia coli* and screening for the presence of genes conferring utilization of 4-hydroxybutyrate. *Appl. Environ. Microbiol.* 65:3901–7

84. Hoffmeister M, Martin W. 2003. Interspecific evolution: microbial symbiosis, endosymbiosis and gene transfer. *Environ. Microbiol.* 5:641–49

85. Hooper LV. 2004. Bacterial contributions to mammalian gut development. *Trends Microbiol.* 12:129–34

86. Hooper LV, Midtvedt T, Gordon JI. 2002. How host-microbial interactions shape the nutrient environment of the mammalian intestine. *Annu. Rev. Nutr.* 22:283–307

87. Huber H, Hohn MJ, Rachel R, Fuchs T, Wimmer VC, Stetter KO. 2002. A new phylum of Archaea represented by a nanosized hyperthermophilic symbiont. *Nature* 417:63–67

88. Isenbarger TA, Finney M, Ríos-Velázquez C, Handelsman J, Ruvkun G. 2008. Miniprimer PCR, a new lens for viewing the microbial world. *Appl. Environ. Microbiol.* 74:840–49

89. Ives AR, Carpenter SR. 2007. Stability and diversity of ecosystems. *Science* 317:58–62

90. Jurkevitch E, Minz D, Ramati B, Barel G. 2000. Prey range characterization, ribotyping, and diversity of soil and rhizosphere *Bdellovibrio* spp. isolated on phytopathogenic bacteria. *Appl. Environ. Microbiol.* 66:2365–71

91. Kadurugamuwa JL, Beveridge TJ. 1996. Bacteriolytic effect of membrane vesicles from *Pseudomonas aeruginosa* on other bacteria including pathogens: conceptually new antibiotics. *J. Bacteriol.* 178:2767–74

92. Kaern M, Elston TC, Blake WJ, Collins JJ. 2005. Stochasticity in gene expression: from theories to phenotypes. *Nat. Rev. Genet.* 6:451–64

93. Kaltenpoth M, Gottler W, Herzner G, Strohm E. 2005. Symbiotic bacteria protect wasp larvae from fungal infestation. *Curr. Biol.* 15:475–79

94. Kan J, Suzuki MT, Wang K, Evans SE, Chen F. 2007. High temporal but low spatial heterogeneity of bacterioplankton in the Chesapeake Bay. *Appl. Environ. Microbiol.* 73:6776–89

95. Kanzler BEM, Pfannes KR, Vogl K, Overmann J. 2005. Molecular characterization of the nonphotosynthetic partner bacterium in the consortium "Chlorochromatium aggregatum." *Appl. Environ. Microbiol.* 71:7434–41

96. Katayama T, Tanaka M, Moriizumi J, Nakamura T, Brouchkov A, et al. 2007. Phylogenetic analysis of bacteria preserved in a permafrost ice wedge for 25,000 years. *Appl. Environ. Microbiol.* 73:2360–63

97. Keel C, Schnider U, Maurhofer M, Voisard C, Laville J, et al. 1992. Suppression of root diseases by *Pseudomonas fluorescens* CHA 0: importance of the bacterial secondary metabolite 2,4-diacetylphloroglucinol. *Mol. Plant-Microbe Interact.* 5:4–13

98. Kent AD, Jones SE, Lauster GH, Graham JM, Newton RJ, McMahon KD. 2006. Experimental manipulations of microbial food web interactions in a humic lake: shifting biological drivers of bacterial community structure. *Environ. Microbiol.* 8:1448–59

99. Kikuchi Y, Graf J. 2007. Spatial and temporal population dynamics of a naturally occurring two-species microbial community inside the digestive tract of the medicinal leech. *Appl. Environ. Microbiol.* 73:1984–91

100. Kitano H. 2002. Computational systems biology. *Nature* 420:206–10

101. Kitano H. 2003. Cancer robustness: tumour tactics. *Nature* 426:125

102. Knietsch A, Bowien S, Whited G, Gottschalk G, Daniel R. 2003. Identification and characterization of coenzyme B-12-dependent glycerol dehydratase- and diol dehydratase-encoding genes from metagenomic DNA libraries derived from enrichment cultures. *Appl. Environ. Microbiol.* 69:3048–60

103. Knietsch A, Waschkowitz T, Bowien S, Henne A, Daniel R. 2003. Metagenomes of complex microbial consortia derived from different soils as sources for novel genes conferring formation of carbonyls from short-chain polyols on *Escherichia coli*. *J. Mol. Microbiol. Biotechnol.* 5:46–56

104. Kolenbrander PE, Andersen RN, Blehert DS, Egland PG, Foster JS, Palmer RJ Jr. 2002. Communication among oral bacteria. *Microbiol. Mol. Biol. Rev.* 66:486–505

105. Kolenbrander PE, Andersen RN, Moore LV. 1990. Intrageneric coaggregation among strains of human oral bacteria: potential role in primary colonization of the tooth surface. *Appl. Environ. Microbiol.* 56:3890–94

106. Kreuzer-Martin HW. 2007. Stable isotope probing: linking functional activity to specific members of microbial communities. *Soil Sci. Soc. Am. J.* 71:611–19

107. Kunin V, He S, Warnecke F, Peterson SB, Garcia Martin H, et al. 2008. A bacterial metapopulation adapts locally to phage predation despite global dispersal. *Genome Res.* 18:293–97

108. Kuske CR, Ticknor LO, Miller ME, Dunbar JM, Davis JA, et al. 2002. Comparison of soil bacterial communities in rhizospheres of three plant species and the interspaces in an arid grassland. *Appl. Environ. Microbiol.* 68:1854–63

109. Lamont RJ, El-Sabaeny A, Park Y, Cook GS, Costerton JW, Demuth DR. 2002. Role of the *Streptococcus gordonii* SspB protein in the development of *Porphyromonas gingivalis* biofilms on streptococcal substrates. *Microbiology* 148:1627–36

110. Leadbetter JR, Greenberg EP. 2000. Metabolism of acyl-homoserine lactone quorum-sensing signals by *Variovorax paradoxus*. *J. Bacteriol.* 182:6921–26

111. LeBlanc PM, Hamelin RC, Filion M. 2007. Alteration of soil rhizosphere communities following genetic transformation of white spruce. *Appl. Environ. Microbiol.* 73:4128–34

112. Lee YK, Puong KY. 2002. Competition for adhesion between probiotics and human gastrointestinal pathogens in the presence of carbohydrate. *Br. J. Nutr.* 88(Suppl. 1):S101–8

113. Lepere C, Boucher D, Jardillier L, Domaizon I, Debroas D. 2006. Succession and regulation factors of small eukaryote community composition in a lacustrine ecosystem (Lake Pavin). *Appl. Environ. Microbiol.* 72:2971–81

114. Levin BR. 1981. Periodic selection, infectious gene exchange, and the genetic structure of *E. coli* populations. *Genetics* 99:1–23

115. Levin BR, Bergstrom CT. 2000. Bacteria are different: observations, interpretations, speculations, and opinions about the mechanisms of adaptive evolution in prokaryotes. *Proc. Natl. Acad. Sci. USA* 97:6981–85

116. Li J, Beatty PK, Shah S, Jensen SE. 2007. Use of PCR-targeted mutagenesis to disrupt production of fusaricidin-type antifungal antibiotics in *Paenibacillus polymyxa*. *Appl. Environ. Microbiol.* 73:3480–89

117. Li Z, Clarke AJ, Beveridge TJ. 1998. Gram-negative bacteria produce membrane vesicles which are capable of killing other bacteria. *J. Bacteriol.* 180:5478–83

118. Liles MR, Manske BF, Bintrim SB, Handelsman J, Goodman RM. 2003. A census of rRNA genes and linked genomic sequences within a soil metagenomic library. *Appl. Environ. Microbiol.* 69:2684–91

119. Lin Y-H, Xu J-L, Hu J, Wang L-H, Ong SL, et al. 2003. Acyl-homoserine lactone acylase from *Ralstonia* strain XJ12B represents a novel and potent class of quorum-quenching enzymes. *Mol. Microbiol.* 47:849–60

120. Little AEF, Currie CR. 2008. Black yeast symbionts compromise the efficiency of antibiotic defenses in fungus-growing ants. *Ecology* 89:1216–22

121. Little AEF, Murakami T, Mueller UG, Currie CR. 2006. Defending against parasites: Fungus-growing ants combine specialized behaviours and microbial symbionts to protect their fungus gardens. *Biol. Lett.* 2:12–16

122. Lo I, Denef VJ, Verberkmoes NC, Shah MB, Goltsman D, et al. 2007. Strain-resolved community proteomics reveals recombining genomes of acidophilic bacteria. *Nature* 446:537–41

123. Lozupone C, Knight R. 2005. UniFrac: a new phylogenetic method for comparing microbial communities. *Appl. Environ. Microbiol.* 71:8228–35

124. Lozupone CA, Hamady M, Kelley ST, Knight R. 2007. Quantitative and qualitative beta diversity measures lead to different insights into factors that structure microbial communities. *Appl. Environ. Microbiol.* 73:1576–85

125. Mai-Prochnow A, Evans F, Dalisay-Saludes D, Stelzer S, Egan S, et al. 2004. Biofilm development and cell death in the marine bacterium *Pseudoalteromonas tunicata*. *Appl. Environ. Microbiol.* 70:3232–38

126. Majernik A, Gottschalk G, Daniel R. 2001. Screening of environmental DNA libraries for the presence of genes conferring Na(+)(Li(+))/H(+) antiporter activity on *Escherichia coli*: characterization of the recovered genes and the corresponding gene products. *J. Bacteriol.* 183:6645–53

127. Manefield M, Turner SL. 2002. Quorum sensing in context: out of molecular biology and into microbial ecology. *Microbiol.* 148:3762–64

128. Martin AP. 2002. Phylogenetic approaches for describing and comparing the diversity of microbial communities. *Appl. Environ. Microbiol.* 68:3673–82

129. Mazzola M, Cook RJ, Thomashow LS, Weller DM, Pierson LS 3rd. 1992. Contribution of phenazine antibiotic biosynthesis to the ecological competence of fluorescent pseudomonads in soil habitats. *Appl. Environ. Microbiol.* 58:2616–24

130. McCaig AE, Grayston SJ, Prosser JI, Glover LA. 2001. Impact of cultivation on characterisation of species composition of soil bacterial communities. *FEMS Microbiol. Ecol.* 35:37–48

131. McCann KS. 2000. The diversity-stability debate. *Nature* 405:228–33

132. McCliment EA, Voglesonger KM, O'Day PA, Dunn EE, Holloway JR, Cary SC. 2006. Colonization of nascent, deep-sea hydrothermal vents by a novel archaeal and nanoarchaeal assemblage. *Environ. Microbiol.* 8:114–25

133. McGrady-Steed J, Harris PM, Morin PJ. 1997. Biodiversity regulates ecosystem predictability. *Nature* 390:162–65

134. Miller MB, Bassler BL. 2001. Quorum sensing in bacteria. *Annu. Rev. Microbiol.* 55:165–99

135. Moss JA, Nocker A, Lepo JE, Snyder RA. 2006. Stability and change in estuarine biofilm bacterial community diversity. *Appl. Environ. Microbiol.* 72:5679–88

136. Moyer CL, Dobbs FC, Karl DM. 1994. Estimation of diversity and community structure through restriction fragment length polymorphism distribution analysis of bacterial 16S rRNA genes from a microbial mat at an active, hydrothermal vent system, Loihi Seamount, Hawaii. *Appl. Environ. Microbiol.* 60:871–79

137. Murase J, Noll M, Frenzel P. 2006. Impact of protists on the activity and structure of the bacterial community in a rice field soil. *Appl. Environ. Microbiol.* 72:5436–44

138. Nakayama J, Tanaka E, Kariyama R, Nagata K, Nishiguchi K, et al. 2007. Siamycin attenuates fsr quorum sensing mediated by a gelatinase biosynthesis-activating pheromone in *Enterococcus faecalis*. *J. Bacteriol.* 189:1358–65

139. Neufeld JD, Chen Y, Dumont MG, Murrell JC. 2008. Marine methylotrophs revealed by stable-isotope probing, multiple displacement amplification and metagenomics. *Environ. Microbiol.* 10:1526–35

140. Neuhauser C, Andow DA, Heimpel GE, May G, Shaw RG, Wagenius S. 2003. Community genetics: expanding the synthesis of ecology and genetics. *Ecology* 84:545–58

141. Ochman H, Lawrence JG, Groisman EA. 2000. Lateral gene transfer and the nature of bacterial innovation. *Nature* 405:299–304

142. Pace NR. 1995. Opening the door onto the natural microbial world: molecular microbial ecology. *Harvey Lect.* 91:59–78

143. Pace NR. 1997. A molecular view of microbial diversity and the biosphere. *Science* 276:734–40

144. Pace NR, Stahl DA, Lane DJ, Olsen GJ. 1985. Analyzing natural microbial populations by rRNA sequences. *ASM News* 51:4–12

145. Paine RT. 1969. A note on trophic complexity and community stability. *Am. Nat.* 103:91–93

146. Papadopoulos D, Schneider D, Meier-Eiss J, Arber W, Lenski RE, Blot M. 1999. Genomic evolution during a 10,000-generation experiment with bacteria. *Proc. Natl. Acad. Sci. USA* 96:3807–12

147. Paper W, Jahn U, Hohn MJ, Kronner M, Nather DJ, et al. 2007. *Ignicoccus hospitalis* sp. nov., the host of 'Nanoarchaeum equitans.' *Int. J. Syst. Evol. Microbiol.* 57:803–8

148. Perez MT, Sommaruga R. 2007. Interactive effects of solar radiation and dissolved organic matter on bacterial activity and community structure. *Environ. Microbiol.* 9:2200–10

149. Perna NT, Plunkett G 3rd, Burland V, Mau B, Glasner JD, et al. 2001. Genome sequence of entero-haemorrhagic *Escherichia coli* O157:H7. *Nature* 409:529–33

150. Pernthaler J. 2005. Predation on prokaryotes in the water column and its ecological implications. *Nat. Rev. Microbiol.* 3:537–46

151. Peterson SB. 2008. *Interactions between rhizosphere bacteria.* PhD thesis. Univ. Wis.-Madison

152. Peterson SB, Dunn AK, Klimowicz AK, Handelsman J. 2006. Peptidoglycan from *Bacillus cereus* mediates commensalism with rhizosphere bacteria from the Cytophaga-Flavobacterium group. *Appl. Environ. Microbiol.* 72:5421–27

153. Pierson LS, Pierson EA. 1996. Phenazine antibiotic production in *Pseudomonas aureofaciens*: role in rhizosphere ecology and pathogen suppression. *FEMS Microbiol. Lett.* 136:101–8

154. Pineiro SA, Sahaniuk GE, Romberg E, Williams HN. 2004. Predation pattern and phylogenetic analysis of *Bdellovibrionaceae* from the Great Salt Lake, Utah. *Curr. Microbiol.* 48:113–17

155. Pineiro SA, Stine OC, Chauhan A, Steyert SR, Smith R, Williams HN. 2007. Global survey of diversity among environmental saltwater *Bacteriovoracaceae*. *Environ. Microbiol.* 9:2441–50

156. Quaiser A, Ochsenreiter T, Klenk HP, Kletzin A, Treusch AH, et al. 2002. First insight into the genome of an uncultivated crenarchaeote from soil. *Environ. Microbiol.* 4:603–11

157. Raaijmakers JM, Vlami M, de Souza JT. 2002. Antibiotic production by bacterial biocontrol agents. *Antonie van Leeuwenhoek* 81:537–47

158. Raffa KF. 2004. Transgenic resistance in short-rotation plantation trees. In *The Bioengineered Forest: Challenges for Science and Society*, ed. SH Strauss, HD Bradshaw, pp. 208–27. Washington, DC: RFF Press

159. Rakoff-Nahoum S, Paglino J, Eslami-Varzaneh F, Edberg S, Medzhitov R. 2004. Recognition of commensal microflora by Toll-like receptors is required for intestinal homeostasis. *Cell* 118:229–41

160. Rao D, Webb JS, Kjelleberg S. 2005. Competitive interactions in mixed-species biofilms containing the marine bacterium *Pseudoalteromonas tunicata*. *Appl. Environ. Microbiol.* 71:1729–36

161. Rastall RA, Gibson GR, Gill HS, Guarner F, Klaenhammer TR, et al. 2005. Modulation of the microbial ecology of the human colon by probiotics, prebiotics and synbiotics to enhance human health: an overview of enabling science and potential applications. *FEMS Microbiol. Ecol.* 52:145–52

162. Reichenbach H. 1999. The ecology of the myxobacteria. *Environ. Microbiol.* 1:15–21

163. Rhee SK, Liu XD, Wu LY, Chong SC, Wan XF, Zhou JZ. 2004. Detection of genes involved in biodegradation and biotransformation in microbial communities by using 50-mer oligonucleotide microarrays. *Appl. Environ. Microbiol.* 70:4303–17

164. Riley MS, Cooper VS, Lenski RE, Forney LJ, Marsh TL. 2001. Rapid phenotypic change and diversification of a soil bacterium during 1000 generations of experimental evolution. *Microbiology* 147:995–1006

165. Ripple WJ, Beschtab RL. 2003. Wolf reintroduction, predation risk, and cottonwood recovery in Yellowstone National Park. *For. Ecol. Man.* 184:299–313

166. Robleto EA, Borneman J, Triplett EW. 1998. Effects of bacterial antibiotic production on rhizosphere microbial communities from a culture-independent perspective. *Appl. Environ. Microbiol.* 64:5020–22

167. Robleto EA, Scupham AJ, Triplett EW. 1997. Trifolitoxin production in *Rhizobium etli* strain CE3 increases competitiveness for rhizosphere. *Mol. Plant-Microbe Interact.* 10:228–33

168. Rogosky AM, Moak PL, Emmert EA. 2006. Differential predation by *Bdellovibrio bacteriovorus* 109J. *Curr. Microbiol.* 52:81–85

169. Rondon MR, August PR, Bettermann AD, Brady SF, Grossman TH, et al. 2000. Cloning the soil metagenome: a strategy for accessing the genetic and functional diversity of uncultured microorganisms. *Appl. Environ. Microbiol.* 66:2541–47

170. Rusch DB, Halpern AL, Sutton G, Heidelberg KB, Williamson S, et al. 2007. The Sorcerer II global ocean sampling expedition: northwest Atlantic through eastern tropical Pacific. *PLoS Biol.* 5:e77

171. Saison C, Degrange V, Oliver R, Millard P, Commeaux C, et al. 2006. Alteration and resilience of the soil microbial community following compost amendment: effects of compost level and compost-borne microbial community. *Environ. Microbiol.* 8:247–57

172. Schink B. 2002. Synergistic interactions in the microbial world. *Antonie van Leeuwenhoek* 81:257–61

173. Schloss PD. 2008. Evaluating different approaches that test whether microbial communities have the same structure. *ISME J.* 2:265–75

174. Schloss PD, Handelsman J. 2005. Introducing DOTUR, a computer program for defining operational taxonomic units and estimating species richness. *Appl. Environ. Microbiol.* 71:1501–6

175. Schloss PD, Handelsman J. 2006. Introducing SONS, a tool for operational taxonomic unit-based comparisons of microbial community memberships and structures. *Appl. Environ. Microbiol.* 72:6773–79

176. Schloss PD, Handelsman J. 2006. Introducing TreeClimber, a test to compare microbial community structures. *Appl. Environ. Microbiol.* 72:2379–84

177. Schloss PD, Handelsman J. 2007. The last word: books as a statistical metaphor for microbial communities. *Annu. Rev. Microbiol.* 61:23–34

178. Schloss PD, Larget BR, Handelsman J. 2004. Integration of microbial ecology and statistics: a test to compare gene libraries. *Appl. Environ. Microbiol.* 70:5485–92

179. Schoeninger MJ, Deniro MJ, Tauber H. 1983. Stable nitrogen isotope ratios of bone collagen reflect marine and terrestrial components of prehistoric human diet. *Science* 220:1381–83

180. Shertzer KW, Ellner SP, Fussmann GF, Hairston NG. 2002. Predator-prey cycles in an aquatic microcosm: testing hypotheses of mechanism. *J. Anim. Ecol.* 71:802–15

181. Shrestha M, Abraham W-R, Shrestha PM, Noll M, Conrad R. 2008. Activity and composition of methanotrophic bacterial communities in planted rice soil studied by flux measurements, analyses of pmoA gene and stable isotope probing of phospholipid fatty acids. *Environ. Microbiol.* 10:400–12

182. Sih A, Crowley P, McPeek M, Petranka J, Strohmeier K. 1985. Predation, competition, and prey communities—a review of field experiments. *Annu. Rev. Ecol. Syst.* 16:269–311

183. Silo-Suh LA, Lethbridge BJ, Raffel SJ, He H, Clardy J, Handelsman J. 1994. Biological activities of two fungistatic antibiotics produced by *Bacillus cereus* UW85. *Appl. Environ. Microbiol.* 60:2023–30

184. Smith JM, Castro H, Ogram A. 2007. Structure and function of methanogens along a short-term restoration chronosequence in the Florida Everglades. *Appl. Environ. Microbiol.* 73:4135–41

185. Stanton ML. 2003. Interacting guilds: moving beyond the pairwise perspective on mutualisms. *Am. Nat.* 162:S10–23

186. Stein JL, Marsh TL, Wu KY, Shizuya H, DeLong EF. 1996. Characterization of uncultivated prokaryotes: isolation and analysis of a 40-kilobase-pair genome fragment from a planktonic marine archaeon. *J. Bacteriol.* 178:591–99

187. Stelling J, Sauer U, Szallasi Z, Doyle FJ III, Doyle J. 2004. Robustness of cellular functions. *Cell* 118:675–85

188. Streit WR, Daniel R, Jaeger K-E. 2004. Prospecting for biocatalysts and drugs in the genomes of noncultured microorganisms. *Curr. Opin. Biotechnol.* 15:285–90

189. Strovas TJ, Sauter LM, Guo X, Lidstrom ME. 2007. Cell-to-cell heterogeneity in growth rate and gene expression in *Methylobacterium extorquens* AM1. *J. Bacteriol.* 189:7127–33

190. Thao ML, Gullan PJ, Baumann P. 2002. Secondary (gamma-Proteobacteria) endosymbionts infect the primary (beta-Proteobacteria) endosymbionts of mealybugs multiple times and coevolve with their hosts. *Appl. Environ. Microbiol.* 68:3190–97

191. Thingstad TF. 2000. Elements of a theory for the mechanisms controlling abundance, diversity, and biogeochemical role of lytic bacterial viruses in aquatic systems. *Limnol. Oceanogr.* 45:1320–28

192. Thomashow LS, Weller DM. 1988. Role of a phenazine antibiotic from Pseudomonas fluorescens in biological control of Gaeumannomyces graminis var. tritici. *J. Bacteriol.* 170:3499–508

193. Thompson JN. 1994. *The Coevolutionary Process*. Chicago: Univ. Chicago Press. 376 pp.

194. Thompson JN. 2005. Coevolution: the geographic mosaic of coevolutionary arms races. *Curr. Biol.* 15:R992–94

195. Tilman D, Wedin D, Knops J. 1996. Productivity and sustainability influenced by biodiversity in grassland ecosystems. *Nature* 379:718–20

196. Tringe SG, von Mering C, Kobayashi A, Salamov AA, Chen K, et al. 2005. Comparative metagenomics of microbial communities. *Science* 308:554–57

197. Tyson GW, Chapman J, Hugenholtz P, Allen EE, Ram RJ, et al. 2004. Community structure and metabolism through reconstruction of microbial genomes from the environment. *Nature* 428:37–43

198. Uroz S, Chhabra SR, Camara M, Williams P, Oger P, Dessaux Y. 2005. *N*-acylhomoserine lactone quorum-sensing molecules are modified and degraded by *Rhodococcus erythropolis* W2 by both amidolytic and novel oxidoreductase activities. *Microbiology* 151:3313–22

199. Uroz S, D'Angelo-Picard C, Carlier A, Elasri M, Sicot C, et al. 2003. Novel bacteria degrading *N*-acylhomoserine lactones and their use as quenchers of quorum-sensing-regulated functions of plant-pathogenic bacteria. *Microbiology* 149:1981–99

200. Uroz S, Oger P, Chhabra SR, Camara M, Williams P, Dessaux Y. 2007. *N*-acylhomoserine lactones are degraded via an amidolytic activity in *Comamonas* sp. strain D1. *Arch. Microbiol.* 187:249–56

201. Vasanthakumar A, Delalibera I, Handelsman J, Klepzig KD, Schloss PD, Raffa KF. 2006. Characterization of gut-associated bacteria in larvae and adults of the southern pine beetle, *Dendroctonus frontalis* Zimmermann. *Environ. Entomol.* 35:1710–17

202. Venter JC, Remington K, Heidelberg JF, Halpern AL, Rusch D, et al. 2004. Environmental genome shotgun sequencing of the Sargasso Sea. *Science* 304:66–74

203. von Dohlen CD, Kohler S, Alsop ST, McManus WR. 2001. Mealybug [beta]-proteobacterial endosymbionts contain [gamma]-proteobacterial symbionts. *Nature* 412:433–36

204. Wang Y-J, Leadbetter JR. 2005. Rapid acyl-homoserine lactone quorum signal biodegradation in diverse soils. *Appl. Environ. Microbiol.* 71:1291–99

205. Waters E, Hohn MJ, Ahel I, Graham DE, Adams MD, et al. 2003. The genome of *Nanoarchaeum equitans*: insights into early archaeal evolution and derived parasitism. *Proc. Natl. Acad. Sci. USA* 100:12984–88

206. Whitham TG, Young WP, Martinsen GD, Gehring CA, Schweitzer JA, et al. 2003. Community and ecosystem genetics: a consequence of the extended phenotype. *Ecology* 84:559–73

207. Williamson KE, Radosevich M, Wommack KE. 2005. Abundance and diversity of viruses in six Delaware soils. *Appl. Environ. Microbiol.* 71:3119–25

208. Winogradsky S. 1895. Researches sur l'assimilation de l'azote libre de l'atmosphere par les microbes. *Arch. Sci. Biol.* 3:297–352

209. Wu D, Daugherty SC, Van Aken SE, Pai GH, Watkins KL, et al. 2006. Metabolic complementarity and genomics of the dual bacterial symbiosis of sharpshooters. *PLoS Biol.* 4:e188

210. Yang F, Wang L-H, Wang J, Dong Y-H, Hu JY, Zhang L-H. 2005. Quorum quenching enzyme activity is widely conserved in the sera of mammalian species. *FEBS Lett.* 579:3713–17

211. Yannarell AC, Steppe TF, Paerl HW. 2007. Disturbance and recovery of microbial community structure and function following Hurricane Frances. *Environ. Microbiol.* 9:576–83

212. Yim G, Wang HH, Davies J. 2007. Antibiotics as signalling molecules. *Philos. Trans. R. Soc. London Ser. B* 362:1195–200

213. Zhang R, Weinbauer MG, Qian P-Y. 2007. Viruses and flagellates sustain apparent richness and reduce biomass accumulation of bacterioplankton in coastal marine waters. *Environ. Microbiol.* 9:3008–18

214. Zul D, Denzel S, Kotz A, Overmann J. 2007. Effects of plant biomass, plant diversity, and water content on bacterial communities in soil lysimeters: implications for the determinants of bacterial diversity. *Appl. Environ. Microbiol.* 73:6916–29

Host Restriction of Avian Influenza Viruses at the Level of the Ribonucleoproteins

Nadia Naffakh, Andru Tomoiu,
Marie-Anne Rameix-Welti, and Sylvie van der Werf

Unité de Génétique Moléculaire des Virus Respiratoires, URA CNRS 3015, Institut Pasteur, Paris, 75015 France; email: svdwerf@pasteur.fr

Annu. Rev. Microbiol. 2008. 62:403–424

The *Annual Review of Microbiology* is online at micro.annualreviews.org

This article's doi:
10.1146/annurev.micro.62.081307.162746

Key Words

evolution, RNA-dependent RNA polymerase, interspecies transmission, adaptation, host factors

Abstract

Although transmission of avian influenza viruses to mammals, particularly humans, has been repeatedly documented, adaptation and sustained transmission in the new host is a rare event that in the case of humans may result in pandemics. Host restriction involves multiple genetic determinants. Among the known determinants of host range, key determinants have been identified on the genes coding for the nucleoprotein and polymerase proteins that, together with the viral RNA segments, form the ribonucleoproteins (RNPs). The RNP genes form host-specific lineages and harbor host-associated genetic signatures. The functional significance of these determinants has been studied by reassortment and reverse genetics experiments, underlining the influence of the global genetic context. In some instances the molecular mechanisms have been approached, pointing to the importance of the polymerase activity and interaction with cellular host factors. Better knowledge of determinants of host restriction will allow monitoring of the pandemic potential of avian influenza viruses.

Contents

INTRODUCTION

Influenza A viruses are major pathogens that have considerable impact on human health during yearly epidemics and have caused particularly high morbidity and mortality during the pandemics that occurred in 1918, 1957, and 1968. Influenza A viruses also have a major impact on the health of pigs and horses, in which they cause respiratory disease, as well as poultry, in which some highly pathogenic strains are responsible for systemic disease with high mortality rates. Wild aquatic birds are the reservoir of all known subtypes of influenza A viruses, which have two viral surface glycoproteins, the hemagglutinin (HA) (H1–

H16) and neuraminidase (NA) (N1–N9). In aquatic birds the virus usually causes little or no disease. It is acquired mainly by oral infection, replicates primarily in the intestinal tract, and is excreted in large quantities in the feces (91, 92). Avian influenza viruses from aquatic birds are occasionally transmitted to other bird species, particularly poultry, and to aquatic (seals, dolphins, whales) or terrestrial mammals (horses, pigs, mink). In humans, cases of infection by a number of avian influenza viruses transmitted mainly from poultry have been documented (85). These incidents were responsible for conjunctivitis or mild respiratory disease (e.g., H7N7, H9N2). In the case of the highly pathogenic avian influenza (HPAI) H5N1 viruses, severe disease associated with acute respiratory distress and a high fatality rate was observed in 1997 in Hong Kong (18 cases and 6 deaths) (12) and more recently since late 2003 in the widespread circulation of H5N1 viruses in Southeast Asia, the Middle East, Africa, and Europe (348 cases and 216 deaths as of January 3, 2008; http://www.who.int/en/). However, only certain subtypes have established in mammals, such as the H1N1, H2N2, and H3N2 viruses in humans. This observation raises the question of the mechanisms and molecular determinants involved in the restriction of transmission and adaptation of avian influenza viruses to new hosts. Given the intimate interactions with host components throughout virus multiplication, these are likely to be multigenic traits.

Influenza A viruses, members of the *Orthomyxoviridae* family, are enveloped negative-stranded RNA viruses with a segmented genome (70). The envelope contains the two surface glycoproteins, HA and NA, as well as the M2 protein. The matrix protein M1 is underlying the envelope. Within the virus particle, the eight viral RNA (vRNA) segments are present in the form of helicoidal and pseudo-circular ribonucleoproteins (vRNPs) (16). The vRNAs are associated with the nucleoprotein (NP), and their partially complementary extremities are bound to the heterotrimeric polymerase complex made of the PB1, PB2, and

---

HA: hemagglutinin

NA: neuraminidase

HPAI: highly pathogenic avian influenza

vRNA: viral RNA

RNP: ribonucleoprotein

NP: nucleoprotein

PA proteins (**Figure 1**) (3, 59). An additional structural protein NEP (nuclear export protein) is present in small amounts in the viral particle. The virus genome further encodes two nonstructural proteins, PB1-F2 and NS1 (70). Upon infection, the virus attaches to the cells via binding of the HA to sialic acids (SAs) (**Figure 2**). Receptor binding is followed by internalization of the viral particle by endocytosis. Upon exposure to low pH in the endosome, the HA undergoes a conformational change that results in fusion of the virus envelope and membrane of the endosome (70). Upon fusion, free vRNPs are released into the cytoplasm and transported to the nucleus, where transcription and replication of the vRNAs take place (64).

The initiation of mRNA synthesis by the associated polymerase complex involves a cap-snatching mechanism through binding of capped cellular premessenger RNAs by the PB2 protein and cleavage by the endonuclease activity of PB1. The PB1 protein, which is the RNA-dependent RNA polymerase, is responsible for elongation. Termination of mRNA synthesis and polyadenylation occur at an oligoU sequence located close to the 5' extremity of the vRNA. The viral mRNAs are translated by the cellular translation machinery. In the case of the M and NS segments splicing is required for the synthesis of the M2 and NEP proteins, respectively, whereas the unspliced messengers are translated into the M1 and NS1 proteins. The vRNPs are also used by the trimeric polymerase complex as templates for vRNA replication. In this case initiation is primer independent and results in the synthesis of colinear complementary copies of the vRNAs, or cRNAs. These in turn serve as templates for the synthesis of new vRNA molecules. Although PA is required along with PB2 and PB1 for both the transcription and replication processes, its exact function is not clear at present. Newly synthesized vRNPs may serve as templates for new rounds of transcription or replication, or they may exit the nucleus to be incorporated into progeny virions. Mechanisms that dictate the fate of vRNPs have not been fully elucidated. NP may favor replication over transcrip-

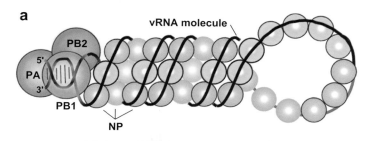

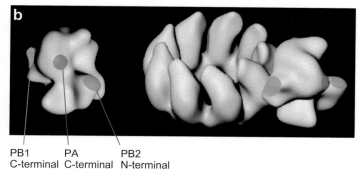

PB1   PA   PB2
C-terminal C-terminal N-terminal

Figure 1

Structural organization of influenza virus ribonucleoproteins (RNPs). (*a*) Schematic model of RNP organization. The single-stranded viral RNA (vRNA) molecule (*black line*) is coiled into a hairpin structure associated with nucleoprotein (NP) (*yellow spheres*). A short region of duplex vRNA (formed between the 5' and 3' ends) constitutes the binding site for the heterotrimeric RNA-dependent RNA polymerase. From Reference 73, with permission. (*b*) 3D model for the influenza virus polymerase. Electron microscopy images of the polymerase complex from isolated RNPs were used for 3D reconstruction. (*Left*) Model of the isolated polymerase. The colored areas indicate the location of specific subunit domains as deduced from 3D reconstruction of polymerase anti-PB2 and anti-PA monoclonal complexes, and of a tandem affinity purification-tagged PB1 polymerase. (*Right*) Position of the polymerase subunits with regard to the NP ring. From Reference 3, with permission.

tion (73), and phosphorylation of M1 and possibly of NEP and NP may determine vRNP export (18).

Exit of the vRNPs from the nucleus requires their association with M1, which in turn interacts through its C terminus with NEP, responsible for interactions with the cellular hCRM1 nuclear export machinery (18). Interaction with M1 prevents the nuclear reentry of the exported RNPs and promotes their targeting to the virion assembly sites at the apical side of polarized cells, where the envelope proteins are inserted and budding of the viral particles takes place. How a complete set of eight vRNPs is packaged into the newly formed virions is not

NEP: nuclear export protein

SA: sialic acid

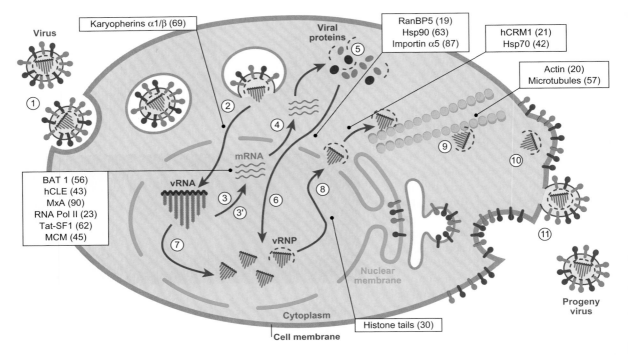

Figure 2

Schematic representation of the influenza virus life cycle. Following receptor binding and endocytosis (step 1), fusion of the viral envelope and endosome membrane results in cytoplasmic release of the free viral ribonucleoproteins (vRNPs) (step 2). These are transported to the nucleus, where primary transcription of the viral RNAs (vRNAs) takes place (step 3). Resulting mRNAs are exported (step 4) and translated by the cellular machinery into viral proteins (step 5). Proteins PB1, PB2, and PA are imported to the nucleus and assemble into viral polymerase complexes (step 6), which ensure secondary transcription (step 3′) and replication (step 7) of vRNAs. Newly formed vRNPs are exported to the cytoplasm (step 8) and transported to the cell membrane, probably through interactions with the cytoskeleton (step 9). Virion assembly (step 10) occurs at the apical side of polarized cells, where budding of progeny viral particles takes place (step 11). Known cellular protein partners of the RNP components are grouped according to the step of the viral life cycle at which they could possibly be involved. These include cellular factors involved in cellular transcription (such as RNA Pol II, BAT1, and hCLE), nuclear import (such as RanBP5), nuclear export (such as hCRM1), and intracytoplasmic trafficking (such as actin).

fully understood, although the identification of packaging signals located at the ends of each vRNA segment as well as the observation of the organization of the eight RNPs in budding particles indicated that the vRNPs are incorporated according to a specific mechanism (27, 67). The sialidase activity of the NA, which cleaves the SAs present on the cell membrane and viral envelope proteins, ensures proper release of the progeny virions.

Finally, cleavage of the HA is required for infectivity of viral particles. Cleavage at an Arg residue involves extracellular trypsin-like proteases that are secreted at the site of viral replication, i.e., the respiratory or intestinal epithe-

lium. In the case of HPAI viruses of the H5 or H7 subtype, a stretch of basic residues at the cleavage site is recognized by furin-like proteases in the *trans*-Golgi (31). Release of fully infectious virions from the cells upon intracellular cleavage of the HA thus accounts for the potential for systemic spread and ability of the H5 and H7 HPAI viruses to replicate in multiple organs. Here we focus on the components of the RNPs and review the genetic determinants of host range that have been identified on the RNP genes as well as our current knowledge about their molecular functions and the corresponding mechanisms involved in the control of transmission and adaptation to new hosts.

Host range: the range of species in which the virus replicates and is transmitted efficiently

HOST-SPECIFIC EVOLUTION OF INFLUENZA RNP GENES

Evolution of influenza viruses is driven by two types of mechanisms: the accumulation of point mutations during the replication of their genomes because of the low fidelity of the viral polymerase, and the reassortment of viral gene segments that can occur in cells infected with two different influenza viruses. Evolutionary rates determined for internal proteins are much lower than those for the HA and NA genes, which is probably related to differences in the selective pressure exerted by the host immune system. However, phylogenetic analyses demonstrate that influenza A RNP genes have evolved into divergent host-specific lineages. An alignment of all available sequences for each internal protein reveals that at specific positions some amino acids appear to be characteristic of the species in which the virus was isolated and thus could be involved in the determination of host range.

Host-Specific Lineages

Phylogenetic analyses of RNP gene sequences from a variety of avian and mammalian influenza viruses result in trees with two major branches, one corresponding to classical swine and human viruses and the other to avian and avian-like viruses including equine and H13 gull viruses (**Figure 3**). The sequence of the 1918 pandemic influenza virus appears similar to the common ancestor of the mammalian clade for each of the PB1, PB2, PA, and NP genes (74, 88). Phylogenetic amino acid trees share the same topology as nucleotide trees, but they reveal that the rate of evolution in the natural avian host reservoir is much slower than the rate of evolution in the mammalian host reservoir (32). Whereas at the nucleotide level the evolutionary rates reported for avian and mammalian influenza viruses are in the same range ($\sim 10^{-3}$ substitutions per site per year) (11, 33), the synonymous/nonsynonymous substitution rate ratios (S/N) are higher for avian viruses. For example, for the NP gene, the average S/N

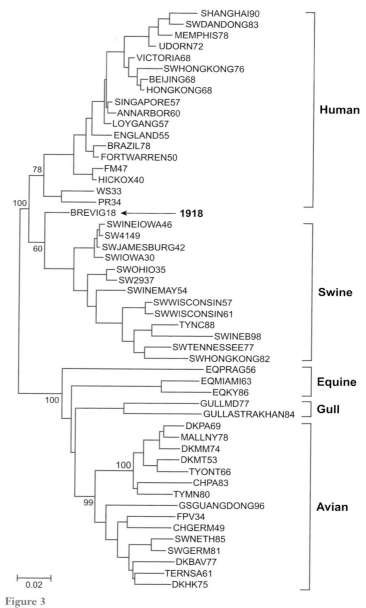

Figure 3

Nucleotide phylogenetic tree of influenza NP gene sequences. The tree was constructed using neighbor-joining, with the proportion of nucleotide differences as the distance measure. Bootstrap values are given for selected nodes of the tree, and a distance bar is shown below the tree. The major phylogenetic groups are highlighted, and the position of the sequence from the 1918 pandemic strain is indicated by an arrow. Reproduced from Reference 74, with permission.

ratio was estimated at 15.2 within the avian clade and 3.9 within the mammalian clade (75). The apparent elimination of genes with nonsynonymous mutations is generally thought to reflect an optimal adaptation of influenza A viruses to their natural avian host. In contrast, the slow but steady accumulation of amino acid changes over time in mammals seems to be the result of a selective pressure on the virus to adapt to a new host. The amino acid sequence of all genes from the influenza pandemic virus of 1918 is avian like, which suggests that the human and classical swine H1N1 lineages derive from a common avian ancestor (75). However, the nucleotide sequences, especially the NP gene sequence of the 1918 virus, are divergent from known sequences from wild bird strains. Such a divergence suggests that, before emerging in humans, the avian ancestor of the 1918 pandemic virus had evolved for several years in an animal species distinct from the wild birds sampled so far. Whether this intermediate host(s) is avian or mammal is still unknown and will probably be difficult to determine unless sequences for precursors of the 1918 virus become available.

When genes of avian origin were recently introduced into a mammalian host, their sequence in early isolates from the new lineage was similar to those from contemporary avian genes but showed a rapid accumulation of mutations in later isolates. This was the case when an avian virus PB1 gene was introduced together with avian HA/NA genes into the human virus gene pool by genetic reassortment in 1957 and again in 1968 (46). As a result, the PB1 gene tree has distinct features: H1N1 human viruses are joined with classical swine viruses, whereas H2N2 (1957–1968) and H3N2 (post-1968) human viruses form separate sublineages within the avian viruses lineage. The average S/N ratio in the PB1 gene was 10.9 for human H3N2 viruses and 18.9 for avian viruses (11). The PB2, PA, and NP segments were not affected by the 1957 and 1968 reassortments or by any other reassortment. In the case of PB1, stronger functional constraints may be limit-

ing the evolutionary rate, thus limiting the evolutionary divergence between host-specific lineages. Indeed, the evolutionary rate for PB1 of avian influenza viruses is estimated to be 2.86, compared with 3.15, 3.48, and 3.17×10^{-3} substitutions per site per year for the PB2, PA, and NP genes, respectively (11). The PB2, PA, and NP genes probably undergo a more significant host adaptation that prevents them from becoming integrated into virus gene pools of a distinct host lineage. The PB2, PA, and NP proteins share similar evolutionary pathways, suggesting that they are coevolving as a result of the strong physical and functional interactions they have with each other. It is likely that coevolution of the RNP proteins is an additional factor restraining the corresponding genomic segments from forming interspecies reassortants. Even among the avian virus gene pool wherein genetic reassortment of core protein genes seems to be frequent (38, 53), large-scale sequence analysis has revealed virus families showing conserved combinations of core protein motifs (68).

In 1979, an H1N1 avian influenza virus established a new lineage in European swine. Phylogenetic analysis of several genes, including the NP gene, indicated that the evolutionary rates of both the coding and the noncoding changes of the early H1N1 swine strains were up to 54% higher than those of classical swine or human viruses (52). These increased evolutionary rates led to the hypothesis that mutator mutations on the viral polymerase, increasing the error rate and allowing the generation of a broader spectrum of variants (83), could be involved in early steps of the adaptation to a new host and be later counterselected. A comparison of the error rates of an early European H1N1 swine virus and a well-established avian virus did not confirm this hypothesis (81). Darwinian selection is still thought to be the major mechanism contributing to early adaptation and high evolutionary rates (24).

More recently, in 1997 in Hong Kong, a highly pathogenic H5N1 virus that circulated in chickens was responsible for respiratory disease in at least 18 persons (13, 86). Each case appeared to result from independent transmission of the virus from poultry to humans, and in the absence of efficient human-to-human transmission, no new H5N1 lineage was established in the human population. The rates of evolution for the PB1, PB2, PA, and NP genes in poultry were higher than those observed in influenza viruses from wild aquatic birds, although no difference was observed for the HA gene (95). These findings suggested that the Hong Kong H5N1 isolates could be reassortants whose HA gene was well adapted to domestic poultry, and the rest of the genome was arising from a wild bird. Phylogenetic analyses revealed that the internal genes from a quail H9N2 virus (A/Qa/HK/G1/97) were similar to those of the H5N1 influenza viruses, suggesting that the quail may have been the internal gene donor (36). There is growing evidence that influenza viruses in poultry evolve at rates reported for mammals (82), which suggests that these species must not be considered as natural hosts of influenza viruses, but as distinct hosts and possibly as intermediate hosts in zoonotic transmission. Additional information revealed by the phylogeny of the Hong Kong 1997 H5N1 viruses is that their internal genes show great similarity to those of viruses of the H9N2 subtype, which were responsible for two cases of human infection in Hong Kong in 1999 (50). This similarity strongly suggests that the internal genes of influenza A viruses contain determinants of interspecies transmission.

Host-Associated Genetic Signatures

The number of influenza virus sequences available has increased significantly during the past years, allowing computational approaches to be developed to identify amino acid residues that are characteristic of the human or avian origin of the sequence. Of the 52 host-associated genetic signatures identified by Chen et al. (8), 35 are located in the RNP (2, 8, 10, and 15 in PB1, PB2, PA, and NP, respectively) (**Figure 4**). The nature of these specific residues has been ex-

amined in viruses of avian origin that crossed the species barrier and infected humans, the 1918 pandemic virus on the one hand and recent H5N1, H9N2, and H7N7 viruses on the other hand. In the 1918 virus, 12 of 35 species-associated residues are human like (8, 74, 88); of 21 avian viruses isolated from humans, 19 (90.5%) show a human-like residue at one or more of the 35 positions (8).

A particularly remarkable host-associated genetic signature is located at residue 627 of PB2. Human viruses generally have a Lys residue (rarely an Arg) at this position, whereas most avian viruses have a Glu. The only human isolates showing a Glu at position 627 correspond to sporadic cases of infection by an avian virus. Of the 43 recent H5N1 human isolates, 23 have the Glu627Lys substitution (79). During an H7N7 virus outbreak in the Netherlands in 2003, the same PB2-Glu627Lys substitution was found in a virus isolated from a human case of fatal pneumonia, but not in viruses isolated from human cases of conjunctivitis or from chicken (26). The 1918 pandemic virus has a Lys residue at position 627, although the amino acid sequence of PB2 is avian like (88). Finally, the Glu627Lys substitution was found in six of seven H5N1 viruses isolated from tigers in Thailand in 2004 (2, 47).

Together these observations suggest that the presence of a Lys at position 627 of PB2 is a strong determinant of adaptation to mammals. Equine viruses and swine viruses from avian-like lineages retain a Glu at position 627 of PB2 (79), suggesting that the selective pressure for this residue in horses and pigs is limited. The only avian isolates showing a Lys at position 627 are HPAI H5N1 viruses isolated from dead wild waterfowl around Qinghai Lake in western China in 2005 (10, 51) or from genetically closely related H5N1 viruses isolated subsequently from wild birds in Mongolia and Russia (9). Phylogenetic analyses indicated that the viruses causing the outbreak derive from reassortants between H5N1 viruses of various genotypes that circulated in poultry in southern China from 2003 to 2005.

Zoonotic transmission: transmission of an infectious agent from animals to humans

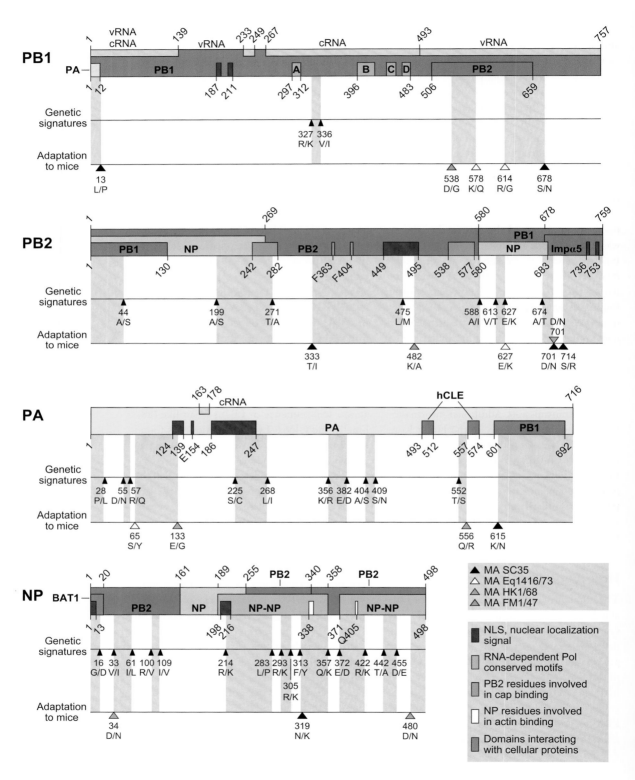

As pointed to by Taubenberger et al. (88), another remarkable residue is PB1 amino acid 375, which is a Ser in most human viruses and an Asn in most avian isolates. Whereas the sequences from the three human pandemic PB1 proteins differ from the avian consensus sequence by only four to seven amino acids, only one of these differences is common to all three pandemic viruses: the substitution of the Asn at position 375 by a Ser. A Gly375Ser substitution has also been observed upon introduction of a swine virus into humans (46). Taken together these observations support a role for PB1 residue 375 in adaptation to mammals. However, PB1 residue 375 cannot be defined as a host-specific genetic signature sensu stricto, as it is an Asn in some human H3N2 viruses and a Ser in a significant proportion of avian viruses (88). Compensatory mutations in PB1 or in a distinct viral protein may have occurred in these isolates.

A full understanding of the genetic basis of host specificity of influenza viruses requires a more global view of genetic diversity than the one revealed by single-amino-acid differences. Obenauer et al. (68) have developed a method for the identification of more complex signatures composed of several amino acids (proteotypes) that might provide a higher-resolution view of protein variability and coevolution. Another approach, developed by Sampath et al. (77), is to use reverse transcription PCR/electrospray ionization mass spectrometry technology for a global characterization of viral genotypes, with an accuracy high enough to track the evolution of influenza viruses and detect the emergence of novel viral genotypes.

GENETIC ANALYSES OF HOST RANGE DETERMINANTS ON INFLUENZA RNPs

Host Range of Reassortant or Mutant Viruses

SGR: single gene reassortant

The first evidence that influenza RNPs harbor determinants for host range came in 1977 from Almond's (1) work on the avian FPV-Rostock and FPV-Dobson viruses, which are able and unable to form plaques on mammalian BHK cells, respectively. The analysis of reassortant viruses revealed that the extended host range of FPV-Dobson was conferred by the PB2 segment (1). In the mid-1980s, various strains were used to generate avian-human reassortants upon coinfection of cultured cells. Influenza A viruses possessing the six internal genes of an avian virus and the HA and NA genes of a human virus (6:2 reassortants) appeared consistently restricted for replication in squirrel monkeys, as were the avian parental viruses (7, 14, 41, 80, 89). To investigate the contribution of each internal gene to host restriction, single gene reassortant (SGR) viruses bearing one avian internal gene and all the remaining genes from a human virus were generated, and their ability to replicate in vitro and/or in vivo was evaluated.

Reassortant viruses containing the NP gene from the avian A/Mallard/NY/6750/78 virus and the remaining genes from the human A/Udorn/307/72 virus appeared restricted in growth in the respiratory tract of squirrel monkeys, as were the 6:2 reassortants having six avian influenza internal genes (89). Similar results were reported

---

Figure 4

Functional domains of the ribonucleoprotein (RNP) subunits and residues possibly involved in host adaptation. This representation was adapted from Reference 4 with permission and completed with additional information. Residues defined by Chen et al. (8) as host-specific genetic signatures are represented by thin black triangles. The residue number and nature (amino acid found in avian/human viruses) are indicated. Residues possibly involved in mice adaptation are represented by large triangles with the following color code: black, A/Seal/Massachussets/1/80-derived mouse-adapted (MA) SC35 virus (28); white, MA A/Equine/London/1416/73 (Eq1416/73) virus (79); gray, MA A/Hong Kong/1/68 (HK1/68) virus (6); and orange, MA A/Fort Mouth/1/47 (FM1/47) virus (5). The residue number and nature (amino acid found in the parental/MA virus) are indicated.

Gene constellation:
unique set of genomic
segments as applied to
segmented viral
genomes

Reverse genetics:
production of
infectious viruses from
cloned cDNAs

for A/Pintail/Alberta/119/79 × A/Washington/897/80 reassortant viruses (80), suggesting that the NP gene was the major determinant of attenuation of avian-human A/Mallard/78 or A/Pintail/79 reassortant viruses in monkeys. Clements et al. (15) isolated all six potential A/Mallard/78 × A/Los Angeles/2/87 SGR viruses and tested their ability to replicate in squirrel monkeys and in adult human volunteers. Contribution of the A/Mallard/78-derived NP gene to the attenuation in monkeys was confirmed. However, the NP gene did not contribute to the attenuation in humans, underlying the fact that studies of SGRs conducted with monkeys would not necessarily lead to the identification of determinants of attenuation of influenza A viruses in humans.

Discordant results were also obtained depending on the strains used to generate human-avian reassortant viruses, which led to the idea that host range was a multigenic character, possibly determined by various constellations of gene segments encoding the internal proteins. For instance, the SGR virus that derived its PB2 gene from the avian A/Mallard/78 virus was attenuated in monkeys when the remaining genes derived from the human A/Los Angeles/87 virus (15), but not when the remaining genes derived from the human A/Udorn/72 virus (89). Moreover, the A/Mallard/78 × A/Los Angeles/87 PB2 SGR showed an unusual in vitro host range phenotype (15). This reassortant showed efficient replication in primary chicken kidney cells and restricted replication in mammalian MDCK cells, whereas the human and avian viruses each replicated efficiently in both primary chicken kidney and MDCK cells. Further study of this virus and of phenotypic revertants that had recovered the ability to replicate efficiently in MDCK cells and in the respiratory tract of squirrel monkeys allowed Subbarao et al. (84) to determine that a single Glu to Lys substitution at PB2 residue 627 was responsible for the extended host range phenotype (Table 1). This and the finding that PB2 amino acid 627 was a host-associated genetic signature (generally a Lys in human viruses, a Glu in avian viruses) strongly suggested that this residue was a major host range determinant of influenza A viruses. Still, it was unclear whether the avian PB2 gene itself or the specific constellation of the A/Mallard/78 × A/Los Angeles/87 SGR virus specified the restriction of viral replication in MDCK cells, in monkeys, and in humans.

The analysis of the contribution of specific amino acid substitutions or gene constellations to host range was facilitated when a plasmid-based reverse genetics system became available for influenza A viruses in 1999 (25, 66). In the context of the emergence of avian H5N1 viruses with a pandemic potential, reverse genetics was immediately and intensively used to document the genetic features that allow these viruses to cross the species barrier from birds to mammals. The first published study examined the molecular basis for the difference in virulence in mice shown by two H5N1 human isolates from the 1997 outbreak in Hong Kong (Gs/Gd genotype) (39). All possible SGR derived from A/Hong Kong/483/97 (HK483/97, highly pathogenic) and A/Hong Kong/486/97 (HK486/97, mildly pathogenic) were produced. The PB2 residue 627 was responsible for the difference in virulence between the two viruses, although the HA gene and, to a lesser extent, the NA gene also contributed to the high virulence of the HK483/97 virus (39).

The same strategy was used to compare a recent human H5N1 isolate from genotype Z (A/Vietnam/1203/04 or VN1203/04) with a nonpathogenic precursor (76), and two closely related H7N7 viruses isolated in the Netherlands in 2003 from a human case of fatal pneumonia (A/Netherlands/219/03 or NL219/03) and from a case of conjunctivitis (58). The VN1203/04 and NL219/03 viruses differed from their counterparts by only 31 and 14 amino acid differences, 12 and 6 of which involved the RNP, respectively. The determinants of VN1203/04 virulence in mice mapped to the polymerase complex but not to the HA, NA, or NS genes. The major determinant

412   *Naffakh et al.*

Table 1 Summary of the effects of a Glu to Lys substitution at PB2 residue 627 on influenza virus growth properties, as observed experimentally

Effect of Glu627Lys in PB2	High effect[a]	Null or *moderate* effect[a]	Reference(s)
Increased polymerase activity in mammalian cultured cells (37°C)	Mal6750/78 (H2N2), FPV34 (H7N1) Mal237/83 (H1N1)	PA908/97 (H3N2) HK156/97(H5N1)	(60) (48) (48)
	SC35 (H7N7)		(28)
Increased replication of the virus in mammalian cultured cells (37°C)	Mal6750/78xLA2/87 PB2-SGR (H3N2)[b] HK486/97 (H5N1)		(84) (78)
Systemic replication and increased virulence in mice	HK486/97 (H5N1) VN1203/04 (H5N1) NL219/03 (H7N7) Eq1416/73 (H7N7)	*VN1204/04 (H5N1)*[c]	(78) (54, 76) (58) (79)
Abrogation of polymerase cold sensitivity (at 33°C versus 37°C) in mammalian cultured cells	Mal6750/78 (H2N2), FPV34 (H7N1) Mal237/83 (H1N1)	PA908/97 (H3N2)[d] HK156/97(H5N1)[d]	(55) (48) (48)
	VN1204/04 (H5N1)		(40)
Increased ability of the virus to replicate in the upper respiratory tract of mice	Mal6750/78 (H2N2) VN1204/04 (H5N1) VN1203/04 (H5N1) Ck5/2003 (H5N1) Dk18/2003 (H5N1)		(40) (40) (40) (40) (40)

[a]Mal6750/78: A/Mallard/New York/6750/78; FPV34: A/FPV/Rostock/34; Mal237/83: A/Mallard/Marquenterre/MZ237/83; PA908/97: A/Paris/908/97; HK156/97: A/Hong Kong/156/97; SC35: derived from A/Seal/Massachussetts/1/80 by serial passages in chicken embryo cells; NL219/03: A/Netherlands/219/2003; LA2/87: A/Los Angeles/2/87; HK486/97: A/Hong Kong/486/97; VN1203/04: A/Vietnam/1203/04; VN1204/04: A/Vietnam/1204/04; Eq1416/73: A/equine/London/1416/73; Ck5/2003: A/Chicken/Vietnam/NCVD5/2003; Dk18/2003: A/muscovy duck/Vietnam/NCVD18/2003. The subtype of the viruses is indicated in parenthesis; the H5 and H7 viruses have a series of basic amino acids at the cleavage site of the HA.
[b]PB2-SGR: single-gene reassortant bearing the PB2 segment from Mal6750/78 and the remaining segments from LA2/87.
[c]In this particular study, PB2 residue 627 of VN1204/04 was not mutated experimentally from Glu to Lys. The moderate effect reported here indicates that VN1204/04 (PB2-627-Glu) was only slightly less pathogenic than the closely related VN1203/04 virus (PB2-627-Lys) in mice. The two viruses show seven amino acid differences in total.
[d]Unlike the Mal237/83 polymerase, the PA908/97 and HK156/97 polymerases show no cold sensitivity whether PB2 residue 627 is a Glu or a Lys.

of pathogenicity of the NL219/03 was the PB2 gene. Both the VN1203/04 and NL219/03 viruses had a Lys instead of a Glu at position 627 of PB2, which determined the outcome of infection in mice in both cases (58, 76). Reverse genetics thus confirmed that PB2 amino acid 627 could, within various genetic contexts, be a major determinant of host range and pathogenicity (**Table 1**).

However, the presence of a Glu or a Lys at PB2 residue 627 did not strictly correlate with virulence in animal models. A number of H5N1 viruses isolated in Hong Kong in 1997 presenting a Glu at PB2 residue 627 were lethal for mice (29, 44) or ferrets (34, 96). The

A/Vietnam/1204/04 virus, which showed seven amino acid differences with the VN1203/04 virus, including a Glu instead of a Lys at PB2 residue 627, had an LD_{50} increase only 40-fold compared with that of VN1203/04 and was still highly pathogenic (54) (**Table 1**). In humans, H5N1 infection primarily affects the lower respiratory tract, and although there is some evidence for extrapulmonary dissemination of the virus, the broad tissue tropism and high viral titers in multiple organs found in mice or ferrets have not been observed. The clinical outcome ranges from mild respiratory symptoms to severe pneumonia and death, and the presence of a Glu or a Lys at PB2 residue 627 does not strictly correlate with the severity of disease in humans (29). Overall, these observations indicate that the ability of PB2 residue 627 to modulate virulence is dependent on additional features of the viral genome, in agreement with the concept of the pathogenicity of influenza viruses as multigenic. The important role played by the C-terminal domain of PB2 is nevertheless underlined by the fact that in the presence of a Glu at residue 627 of PB2, the ability of the avian strain A/FPV/Dobson-4H/27 to replicate in mouse cells was determined by the region encoding amino acids 362 to 581 of PB2 (93), and the pathogenicity of a duck H5N1 influenza virus in mice was dramatically increased by the Asp to Asn substitution at PB2 residue 701 (49).

Genetic Changes Associated with Experimental Adaptation to a New Host

Influenza A viruses usually do not replicate efficiently in mice. Adaptation by serial passages in mouse lung has been reported for several viruses, including the human A/Hong Kong/1/68 virus (HK1/68) (6), the avian-like H7N7 virus (SC35) derived from the A/Seal/Massachussets/1/80 isolate (28), and the H7N7 equine A/Equine/London/1416/73 virus (Eq1416/73) (79). In all cases, the experimental process resulted in the selection of highly virulent variants that could kill mice at low doses. The LD_{50} of the mouse-adapted (MA) viruses was 10^2–10^3 pfu (particle forming unit), compared with 10^6–10^8 for the parental viruses. High virulence in mice correlated with an increased multiplication potential in mammalian cell lines, the MA SC35 and HK1/68 viruses growing to titers 2–4 logs higher than the parental viruses. Sequence analysis revealed that the MA variants had acquired mutations in multiple genes (**Figure 4**). Eleven amino acid substitutions involving all viral proteins except PB1 and NEP were detected in the genome of the most virulent clonal isolate of the MA HK1/68 virus (6). Four amino acid substitutions were detected in the MA Eq1416/73 virus,

including two in PB1, one in PB2, and one in PA. Of these, the Glu to Lys substitution at PB2 residue 627 was most important for the virulence in mice (79) (**Table 1**). The MA and parent SC35 viruses differed by seven amino acid substitutions in the RNP, one in the HA, and one in the NA (28).

Reverse genetics was used to determine which mutations conferred increased virulence in mice to the MA SC35 virus. A comparison of the LD_{50} in mice of reassortants between the MA and parent SC35 viruses indicated that each mutation in the RNP contributed to virulence. For example, the PB2-Arg714Ser substitution was sufficient to convert the highly virulent MA SC35 virus to a nonpathogenic virus with an $LD_{50} > 10^6$. The reciprocal PB2-Ser714Arg substitution had no effect when introduced alone in the genome of the SC35 parental virus. However, in combination with the PB2-Asp701Asn substitution, it induced a 3 log reduction of the LD_{50}, indicating a strong additive effect of the two substitutions in PB2. To a lesser extent, the Asn319Lys substitution in NP also had an additive effect when combined with the PB2-Ser714Arg or PB2-Asp701Asn substitution.

Noticeably, the PB2-Asp701Asn substitution was detected not only in the MA SC35 virus but also in the MA HK1/68 virus (6), and in duck H5N1 isolates, which have a natural capacity to infect mice (49). Such a repeated and independent occurrence of a common mutation in adapted variants is called convergent evolution and is a strong indicator of an adaptive significance of the mutation. Convergent evolution of the RNP is also illustrated by changes at PB2-701, PB2-714, PA-615, and NP-319, observed in the MA SC35 virus, as well as in many human H5N1 isolates that show high pathogenicity for mice and humans (28). Taken together, the data from experimental adaptation of influenza viruses to mice strongly suggest that mutations in the RNP confer increased replicative fitness and contribute to higher virulence in mice, although the functional role of these mutations still remains to be understood.

MOLECULAR FUNCTIONS AND MECHANISMS INVOLVED IN HOST RANGE DETERMINATION BY INFLUENZA RNPs

Although genetic analyses hint at a major role of RNP components in host restriction, the molecular functions and mechanisms involved have begun to be unraveled only recently. It is generally assumed that interactions among viral proteins and host factors play a key role in viral fitness and pathogenicity, and that adaptive mutations of the RNP lead to an optimized interaction with host factors. Once the virus has been transmitted to a new species, it is confronted with changes in the cellular microenvironment that might influence every step of the replicative cycle. The absence of a cellular factor upon which the virus used to depend, the presence of an inhibiting factor that has to be circumvented, or a lower or higher concentration of, or affinity for, such a positive or a negative factor will alter the ability of the virus to replicate. A growing amount of data indicate that trafficking and polymerase activity of RNPs indeed require the participation of host factors (4, 23). A correlation between the levels of polymerase activity in cultured mammalian cells and viral multiplication in mammalian hosts has been repeatedly observed. No particular host protein is known to be involved in the host range of influenza A viruses so far, but a number of hypotheses can be formulated on the basis of our current knowledge of the molecular biology of these viruses.

Polymerase Activity: The Major RNP Function Involved

The genetic system described by Pleschka et al. (71) for the in vivo reconstitution of functional influenza RNPs was instrumental in establishing a relationship between the levels of polymerase activity and viral multiplication in a mammalian host. Reconstitution of vRNPs is achieved by transfecting into cultured cells (usually 293T, COS-1, or Vero cells) a plasmid that directs the expression of a virus-like reporter RNA together with plas-

mids from which the four core proteins are expressed. When this system was used, RNPs from the A/Mallard/Marquenterre/MZ237/83 (Mal237/83) avian influenza isolate ensured transcription/replication of a viral-like RNA in mammalian cells about 10-fold less efficiently than did RNPs from the human A/Paris/908/97 (PA908/97) isolate (48). Consistently, the RNPs derived from the highly pathogenic VN1203/04 and SC35M viruses showed a 3.5- and 6-fold-higher activity, respectively, than their low-pathogenic counterparts (28, 58, 76). This enhanced polymerase activity most likely contributed to their increased ability to spread rapidly and cause disease in mice. A variant of the A/Puerto Rico/8/34 strain that is unusually virulent in mice carrying a functional *Mx1* resistance gene (hvPR8) also has a higher polymerase activity than a standard PR8 strain does (35). This observation and the facts that (*a*) the polymerase genes (in addition to the HA and NA genes) contributed to the virulence of reassortant hvPR8 x PR8 viruses, and that (*b*) interferon, which induces transcription of the *Mx1* gene, protected *Mx1+/+* mice if administered before viral infection suggest that high polymerase activity can contribute to virulence by helping to outcompete the antiviral response of the infected host.

Low temperature (33°C) reduced further viral replication and polymerase activity of avian influenza viruses in mammalian cells, whereas human viruses did not exhibit sensitivity to cold. The presence of a Glu instead of a Lys at PB2 residue 627 was the major determinant of the decline of avian-derived RNPs polymerase activity at 33°C (48, 55) (**Table 1**). Accordingly, reverse genetics experiments demonstrated that the Glu627Lys substitution conferred human H5N1 isolates the ability to grow efficiently at 33°C on diverse human cell lines (40). Moreover, H5N1 viruses with a Lys at position 627 of PB2 replicated in nasal turbinates as well as in lungs of infected mice, whereas viruses with a Glu at this position replicated poorly in nasal turbinates. These data suggest that cold sensitivity of the polymerase of avian influenza

Viral fitness: inherent ability of the virus to replicate in a given environment

viruses, which usually replicate in the intestinal tract of birds with a body temperature of approximately 41°C, could limit their growth in the upper respiratory tract of humans and prevent human-to-human transmission.

The data obtained in transient polymerase assays on reconstituted RNPs reinforce the concept of gene constellation. For example, the effect of PB2 amino acid 627 on polymerase activity is modulated by the genetic background. Changing PB2 residue 627 from a Glu to a Lys did not increase the transcription/replication activity of RNPs derived from the A/Hong Kong/156/97 virus (HK156/97, a human H5N1 isolate) as it did for RNPs from the Mal237/83 avian isolate when reconstituted in human cells (48) (**Table 1**). Experiments with heterospecific HK156/97-Mal237/83 RNPs suggested that molecular determinants in both the PB2 and PA proteins of HK156 compensated for the presence of a Glu at residue 627 of PB2 and contributed synergistically to the high levels of polymerase activity of HK156/97 RNPs in human cells (48). In the genetic background of SC35 virus-derived RNPs, changing the PB2 residue 627 from a Glu to a Lys led to a 20-fold increase in polymerase activity (28) (**Table 1**), although this substitution was not selected during the process of adaptation to mice, leading to the SC35M virus. Noticeably, among the substitutions observed in SC35M, those on the PB1 gene associated with the highest increase in polymerase activity of reconstituted RNPs did not confer virulence when tested alone in the context of infectious recombinant viruses (28).

These observations suggest that optimum polymerase activity, depending on a combination of upregulating and downregulating determinants, is probably required for efficient viral replication in a mammalian host. Reported gene constellation effects include the need for a match between PB1 and PA to achieve high polymerase activity of chimeric SC35-SC35M reconstituted RNPs (28), and the need for a match between PB2 and NP to achieve high polymerase activity of RNPs reconstituted from

a mixture of proteins derived from human and avian influenza viruses (60).

Possible Host-Restriction Mechanisms Related to Polymerase Activity

The genetic variations observed upon introduction of RNP genes into a new species, such as the host-associated genetic signatures or the adaptive mutations to mice, are distributed all along the sequences of the PB1, PB2, PA, and NP genes (**Figure 4**). Some of them map to functional domains involved in nuclear localization of the RNP components. Following export of newly transcribed viral mRNA, nuclear import and concurrent assembly of neosynthesized polymerase complex subunits rely on molecular interactions with host proteins such as importins (19, 69, 87) and chaperone Hsp90 (63). A recent structural study of a C-terminal domain of PB2 complexed with human importin α5 revealed that the bipartite NLS formed by amino acids 736–755 can interact with amino acid Asp701 (87), a residue previously found to mediate the adaptation of avian viruses to mice (28, 49). An attractive hypothesis would be that the host range alteration caused by the Asp701Asn substitution could be mechanistically explained by enhanced importin binding and, in turn, enhanced trimeric polymerase assembly in murine cells.

After final assembly in the nucleus, RNPs associate with the histone tails of nucleosome particles, most probably via the NP (30). This interaction was suggested to influence the chromatin structure and transcription pattern in infected cells. Viral transcription is critically dependent on ongoing cellular transcription (22), particularly on activities linked to the cellular DNA-dependent RNA polymerase II (Pol II). The viral polymerase uses short 5'-capped RNA fragments, derived from cellular Pol II transcripts, as primers for viral mRNA synthesis. Moreover, some viral transcripts require splicing, and because influenza virus does not encode splicing machinery, it is dependent on host splicing, an activity also related to Pol II transcription. Influenza RNPs bind the phosphorylated C-terminal domain of initiating Pol II (23), and it was proposed that this interaction provides the viral polymerase with high concentrations of capped RNA substrate for endonucleolytic cleavage. Among the other known interacting proteins of the host that could be involved in regulating viral transcription and replication, cellular splicing factor BAT1 (also known as UAP56/RAF-2p48/NPI-5) and transcription elongation factor Tat-SF1, and possibly nucleophosmin, stimulate vRNA synthesis by facilitating the formation of NP-RNA complexes (56, 62, 97). Last, the PA subunit interacts with hCLE, a putative transcriptional activator (43), and MCM, a putative DNA replication fork helicase (45). The full significance of the above-mentioned interactions still remains to be determined. In particular, their potential impact on host range has not been investigated so far.

As illustrated in **Figure 4**, a number of host-associated genetic signatures and adaptive mutations to mice map to domains of interaction between viral components of the RNP. Residue 627 of PB2, a major determinant of host range, is located within a region that promotes binding of PB2 to both PB1 and NP (72). The RNPs derived from an avian influenza isolate, which showed low levels of polymerase activity in human cells compared with RNPs of a human isolate, also showed reduced NP binding to the PB1-PB2-PA complex and to the isolated PB2 subunit (48). Both defects were restored when PB2 residue 627 was changed to a Lys, suggesting that PB2 might play a pivotal role in molecular interactions involving both the viral NP and cellular proteins. On the other hand, NP residue 319, which contributed to the extended host range of the SC35M virus (28), was located at the surface of the molecule in the recently solved 3D structure (94), and Gabriel et al. (28) proposed that it could mediate binding to the polymerase complex. Overall, the available data suggest that adaptive mutations on the RNP modulate a complex interplay between viral and host proteins. They might alter interactions with host factors directly by modifying their binding site, or indirectly by altering the association with other subunits and/or by inducing conformational changes in other subunits of the RNP.

Other Functions and Mechanisms Possibly Involved

RNP involvement in the viral replication cycle goes beyond nuclear import, assembly, and polymerase activity, i.e., the steps leading to transcription and replication of the viral genome. Nuclear export, cytoplasmic retention, and transport of the RNPs and viral particle assembly and egress are some of the key stages at which molecular interactions between RNPs and cellular components can also in theory determine host range (**Figure 2**). The NP protein seems to function as an adaptor molecule between the RNP and host cell pathways. The nuclear export of newly formed RNPs relies on the M1 and NEP viral proteins as well as the cellular hCRM1 nuclear export pathway. Interactions of NP with hCRM1 and of RNPs with hsp70 were suggested to participate in RNP export (21, 42). Once in the cytoplasm, association of RNPs with F-actin stress fibers via the NP could participate in the cytoplasmic retention of the viral complexes (20). Moreover, microtubule networks have been suggested to help polarize transport of the RNPs from the nuclear membrane to the apical side of the cell membrane (57). A defect in any of the mechanisms mentioned above could be responsible for a reduction of virion production. Besides, still poorly defined activities of the RNP components, such as the induction of proteolysis by PA, which varies in intensity depending on the avian or human origin of the virus (61), might be a determinant for host range.

A driving force of evolution in a new host is the pressure exerted by the host immune response. Influenza NP protein binds to the antiviral interferon-induced human MxA protein (90). Although the mechanism of action of MxA is not fully understood, this binding likely prevents incoming vRNPs from being transported into the nucleus (37). Some of the adaptive mutations observed on RNP genes when an avian influenza virus is introduced into a mammalian host might correspond to escape mutations from the innate immune system of the new host.

CONCLUSION

Successful establishment of avian influenza viruses in new hosts, particularly in humans, is a multistep process that requires (*a*) transmission from the avian reservoir to the new host; (*b*) efficient replication into the new host; and (*c*) an ability to establish a sustained chain of transmission and to efficiently compete with related viruses previously circulating in that new host. The probability of each event to occur may be influenced by numerous factors. For instance, transmission from the avian reservoir to humans will be determined by the density of the avian and human populations and by the number of contacts between the two populations. In this respect, the introduction of avian influenza viruses from the wild bird reservoir into domestic species results in an amplification of both the virus population and the number of contacts with humans. Whereas occasional transmission of avian influenza viruses from poultry to humans has been repeatedly documented, adaptation and sustained transmission in humans resulting in the initiation of pandemics is fortunately a rare event. This is consistent with the fact that multiple host range determinants and/or a specific combination thereof seem to be involved.

Efficient replication as well as transmission in the new host is determined mainly by the ability of the virus to interact with host components at each step of viral multiplication and to counteract the host response. In addition to the multiple determinants harbored by the RNPs described above, determinants of host restriction have been identified on several other viral components. Major determinants are related to the receptor specificity of the surface glycoproteins (65). Determinants of pathogenicity have been identified on the NS1 protein (65), which among other functions is an interferon antagonist, and on the PB1-F2 protein (17), which induces apoptosis of the infected cell. To what extent features of NS1 and PB1-F2 are related to host restriction or merely contribute to the virulence of the viruses according to their hosts requires further investigations.

The way the RNP interacts with the host seems to define its capacity to assume essential functions for viral replication and thus determine whether the virus is successful in establishing an infection. The acquisition of host range determinants results from the intrinsic genetic variability of influenza viruses by both mutations and reassortment events that occur at each replication cycle in the original and new host. In that respect, mutations on the RNPs that result in increased replication rates could be a prerequisite for the generation of increased genetic diversity from which variants with increased adaptation and transmission potential for the new host may be selected. In addition, viruses with increased replication potential are more likely to outcompete and escape efficiently the host innate and immune responses. This may account at least in part for initial virulence for the new host as seen for instance with the pandemic virus of 1918.

Analysis of the molecular mechanisms underlying host specificity has been facilitated in the past decade by the development of appropriate in vitro systems and of reverse genetics for influenza viruses. Although in recent years our knowledge of the molecular biology of influenza RNPs has improved, our understanding of the structure/function relationships within the RNPs and of the functional significance of their interactions with host components is still deficient. In light of the pandemic threat posed by the widely circulating HPAI H5N1 viruses with known potential to transmit to humans, a better knowledge of the determinants of host range and in particular of those present on the RNPs that confer increased replication rates is paramount.

SUMMARY POINTS

1. Influenza A RNP genes have evolved into divergent host-specific lineages with evolutionary rates similar at the nucleotide level for avian and mammalian viruses but higher rates of accumulation of amino acid changes over time for mammalian viruses, indicative of selective pressure to adapt to the new host.

2. The RNP genes contain important determinants of host range such as PB2 residue 627. Their effect can be modulated by the genetic background and gene constellations.

3. Determinants of host range map to domains of interaction between viral components of the RNP and/or are likely involved in interactions with cellular host factors. Current evidence indicates that they could contribute to increased replicative fitness according to the host.

FUTURE ISSUES

1. A better knowledge of the structure/function relationships of RNPs and of their interactions with host components such as host transcription machinery is required for a better understanding of the mechanisms underlying host specificity and the rational design of antiviral drugs against viruses with zoonotic potential.

2. Beyond the known functions of RNPs for transcription and replication, a better understanding is needed of the role of RNPs during the early and late steps of viral multiplication and their impact on virulence and escape of host defense mechanisms.

3. Further global characterization of viral genotypes should help investigators understand the impact of genetic diversity and mechanisms that result in the selection and adaptation of influenza viruses upon interspecies transmission and establishment of sustained spread within the new host.

DISCLOSURE STATEMENT

The authors are not aware of any biases that might be perceived as affecting the objectivity of this review.

ACKNOWLEDGMENTS

We thank Juan Ortín, Paul Digard, Jeffery Taubenberger, and Florence Baudin for kindly allowing us to use their illustrations, and Sandie Munier for critical reading of the manuscript. This work was supported in part by grants from the European Vigilance Network for the Management of Antiviral Drug Resistance (VIRGIL, LSHM-CT-2004-503359), and the European FLUINNATE program (SP5B-CT-2006-044161). A. Tomoiu was supported by a fellowship from the Région Ile-de-France. M.A. Rameix-Welti was supported by a fellowship from the Comité des Maladies Infectieuses et Tropicales (Institut Pasteur).

LITERATURE CITED

1. Almond JW. 1977. A single gene determines the host range of influenza virus. *Nature* 270:617–18
2. Amonsin A, Payungporn S, Theamboonlers A, Thanawongnuwech R, Suradhat S, et al. 2006. Genetic characterization of H5N1 influenza A viruses isolated from zoo tigers in Thailand. *Virology* 344:480–91
3. Area E, Martin-Benito J, Gastaminza P, Torreira E, Valpuesta JM, et al. 2004. 3D structure of the influenza virus polymerase complex: localization of subunit domains. *Proc. Natl. Acad. Sci. USA* 101:308–13
4. Boulo S, Akarsu H, Ruigrok RW, Baudin F. 2007. Nuclear traffic of influenza virus proteins and ribonucleoprotein complexes. *Virus Res.* 124:12–21
5. Brown EG, Bailly JE. 1999. Genetic analysis of mouse-adapted influenza A virus identifies roles for the NA, PB1, and PB2 genes in virulence. *Virus Res.* 61:63–76
6. Brown EG, Liu H, Kit LC, Baird S, Nesrallah M. 2001. Pattern of mutation in the genome of influenza A virus on adaptation to increased virulence in the mouse lung: identification of functional themes. *Proc. Natl. Acad. Sci. USA* 98:6883–88
7. Buckler-White AJ, Naeve CW, Murphy BR. 1986. Characterization of a gene coding for M proteins which is involved in host range restriction of an avian influenza A virus in monkeys. *J. Virol.* 57:697–700
8. Chen GW, Chang SC, Mok CK, Lo YL, Kung YN, et al. 2006. Genomic signatures of human versus avian influenza A viruses. *Emerg. Infect. Dis.* 12:1353–60
9. Chen H, Li Y, Li Z, Shi J, Shinya K, et al. 2006. Properties and dissemination of H5N1 viruses isolated during an influenza outbreak in migratory waterfowl in western China. *J. Virol.* 80:5976–83
10. Chen H, Smith GJ, Zhang SY, Qin K, Wang J, et al. 2005. Avian flu: H5N1 virus outbreak in migratory waterfowl. *Nature* 436:191–92
11. Chen R, Holmes EC. 2006. Avian influenza virus exhibits rapid evolutionary dynamics. *Mol. Biol. Evol.* 23:2336–41
12. Claas EC, de Jong JC, van Beek R, Rimmelzwaan GF, Osterhaus AD. 1998. Human influenza virus A/HongKong/156/97 (H5N1) infection. *Vaccine* 16:977–78
13. Claas EC, Osterhaus AD, van Beek R, De Jong JC, Rimmelzwaan GF, et al. 1998. Human influenza A H5N1 virus related to a highly pathogenic avian influenza virus. *Lancet* 351:472–77
14. Clements ML, Snyder MH, Buckler-White AJ, Tierney EL, London WT, Murphy BR. 1986. Evaluation of avian-human reassortant influenza A/Washington/897/80 x A/Pintail/119/79 virus in monkeys and adult volunteers. *J. Clin. Microbiol.* 24:47–51
15. Clements ML, Subbarao EK, Fries LF, Karron RA, London WT, Murphy BR. 1992. Use of single-gene reassortant viruses to study the role of avian influenza A virus genes in attenuation of wild-type human influenza A virus for squirrel monkeys and adult human volunteers. *J. Clin. Microbiol.* 30:655–62
16. Compans RW, Content J, Duesberg PH. 1972. Structure of the ribonucleoprotein of influenza virus. *J. Virol.* 10:795–800
17. Conenello GM, Zamarin D, Perrone LA, Tumpey T, Palese P. 2007. A single mutation in the PB1-F2 of H5N1 (HK/97) and 1918 influenza A viruses contributes to increased virulence. *PLoS Pathog.* 3:1414–21
18. Cros JF, Palese P. 2003. Trafficking of viral genomic RNA into and out of the nucleus: influenza, Thogoto and Borna disease viruses. *Virus Res.* 95:3–12
19. Deng T, Engelhardt OG, Thomas B, Akoulitchev AV, Brownlee GG, Fodor E. 2006. Role of ran binding protein 5 in nuclear import and assembly of the influenza virus RNA polymerase complex. *J. Virol.* 80:11911–19
20. Digard P, Elton D, Bishop K, Medcalf E, Weeds A, Pope B. 1999. Modulation of nuclear localization of the influenza virus nucleoprotein through interaction with actin filaments. *J. Virol.* 73:2222–31
21. Elton D, Simpson-Holley M, Archer K, Medcalf L, Hallam R, et al. 2001. Interaction of the influenza virus nucleoprotein with the cellular CRM1-mediated nuclear export pathway. *J. Virol.* 75:408–19
22. Engelhardt OG, Fodor E. 2006. Functional association between viral and cellular transcription during influenza virus infection. *Rev. Med. Virol.* 16:329–45
23. **Engelhardt OG, Smith M, Fodor E. 2005. Association of the influenza A virus RNA-dependent RNA polymerase with cellular RNA polymerase II. *J. Virol.* 79:5812–18**
24. Fitch WM, Leiter JM, Li XQ, Palese P. 1991. Positive Darwinian evolution in human influenza A viruses. *Proc. Natl. Acad. Sci. USA* 88:4270–74

23. Evidence for a physical association between influenza virus polymerase and the large subunit of RNA polymerase II, provided by biochemical and imaging analyses.

25. Fodor E, Devenish L, Engelhardt OG, Palese P, Brownlee GG, Garcia-Sastre A. 1999. Rescue of influenza A virus from recombinant DNA. *J. Virol.* 73:9679–82

26. Fouchier RA, Schneeberger PM, Rozendaal FW, Broekman JM, Kemink SA, et al. 2004. Avian influenza A virus (H7N7) associated with human conjunctivitis and a fatal case of acute respiratory distress syndrome. *Proc. Natl. Acad. Sci. USA* 101:1356–61

27. Fujii Y, Goto H, Watanabe T, Yoshida T, Kawaoka Y. 2003. Selective incorporation of influenza virus RNA segments into virions. *Proc. Natl. Acad. Sci. USA* 100:2002–7

28. Gabriel G, Dauber B, Wolff T, Planz O, Klenk HD, Stech J. 2005. The viral polymerase mediates adaptation of an avian influenza virus to a mammalian host. *Proc. Natl. Acad. Sci. USA* 102:18590–95

29. Gao P, Watanabe S, Ito T, Goto H, Wells K, et al. 1999. Biological heterogeneity, including systemic replication in mice, of H5N1 influenza A virus isolates from humans in Hong Kong. *J. Virol.* 73:3184–89

30. Garcia-Robles I, Akarsu H, Muller CW, Ruigrok RW, Baudin F. 2005. Interaction of influenza virus proteins with nucleosomes. *Virology* 332:329–36

31. Garten W, Klenk HD. 1999. Understanding influenza virus pathogenicity. *Trends Microbiol.* 7:99–100

32. Gorman OT, Bean WJ, Kawaoka Y, Donatelli I, Guo YJ, Webster RG. 1991. Evolution of influenza A virus nucleoprotein genes: implications for the origins of H1N1 human and classical swine viruses. *J. Virol.* 65:3704–14

33. Gorman OT, Donis RO, Kawaoka Y, Webster RG. 1990. Evolution of influenza A virus PB2 genes: implications for evolution of the ribonucleoprotein complex and origin of human influenza A virus. *J. Virol.* 64:4893–902

34. Govorkova EA, Rehg JE, Krauss S, Yen HL, Guan Y, et al. 2005. Lethality to ferrets of H5N1 influenza viruses isolated from humans and poultry in 2004. *J. Virol.* 79:2191–98

35. Grimm D, Staeheli P, Hufbauer M, Koerner I, Martinez-Sobrido L, et al. 2007. Replication fitness determines high virulence of influenza A virus in mice carrying functional Mx1 resistance gene. *Proc. Natl. Acad. Sci. USA* 104:6806–11

36. Guan Y, Shortridge KF, Krauss S, Webster RG. 1999. Molecular characterization of H9N2 influenza viruses: Were they the donors of the "internal" genes of H5N1 viruses in Hong Kong? *Proc. Natl. Acad. Sci. USA* 96:9363–67

37. Haller O, Staeheli P, Kochs G. 2007. Interferon-induced Mx proteins in antiviral host defense. *Biochimie* 89:812–18

38. Hatchette TF, Walker D, Johnson C, Baker A, Pryor SP, Webster RG. 2004. Influenza A viruses in feral Canadian ducks: extensive reassortment in nature. *J. Gen. Virol.* 85:2327–37

39. Hatta M, Gao P, Halfmann P, Kawaoka Y. 2001. Molecular basis for high virulence of Hong Kong H5N1 influenza A viruses. *Science* 293:1840–42

40. Hatta M, Hatta Y, Kim JH, Watanabe S, Shinya K, et al. 2007. Growth of H5N1 influenza A viruses in the upper respiratory tracts of mice. *PLoS Pathog.* 3:1374–79

41. Hinshaw VS, Webster RG, Naeve CW, Murphy BR. 1983. Altered tissue tropism of human-avian reassortant influenza viruses. *Virology* 128:260–63

42. Hirayama E, Atagi H, Hiraki A, Kim J. 2004. Heat shock protein 70 is related to thermal inhibition of nuclear export of the influenza virus ribonucleoprotein complex. *J. Virol.* 78:1263–70

43. Huarte M, Sanz-Ezquerro JJ, Roncal F, Ortin J, Nieto A. 2001. PA subunit from influenza virus polymerase complex interacts with a cellular protein with homology to a family of transcriptional activators. *J. Virol.* 75:8597–604

44. Katz JM, Lu X, Tumpey TM, Smith CB, Shaw MW, Subbarao K. 2000. Molecular correlates of influenza A H5N1 virus pathogenesis in mice. *J. Virol.* 74:10807–10

45. Kawaguchi A, Nagata K. 2007. De novo replication of the influenza virus RNA genome is regulated by DNA replicative helicase, MCM. *EMBO J.* 26:4566–75

46. Kawaoka Y, Krauss S, Webster RG. 1989. Avian-to-human transmission of the PB1 gene of influenza A viruses in the 1957 and 1968 pandemics. *J. Virol.* 63:4603–8

47. Keawcharoen J, Oraveerakul K, Kuiken T, Fouchier RA, Amonsin A, et al. 2004. Avian influenza H5N1 in tigers and leopards. *Emerg. Infect. Dis.* 10:2189–91

25. Initial description of a plasmid-based reverse genetics system for influenza viruses.

28. Analysis of the genetic changes associated with experimental adaptation of an avian-derived H7N7 virus to mice, using reverse genetics.

39. Identification of the genetic differences between two human influenza A H5N1 isolates responsible for their low and high virulence in mice, using reverse genetics.

48. Labadie K, Dos Santos Afonso E, Rameix-Welti MA, van der Werf S, Naffakh N. 2007. Host-range determinants on the PB2 protein of influenza A viruses control the interaction between the viral polymerase and nucleoprotein in human cells. *Virology* 362:271–82

49. Li Z, Chen H, Jiao P, Deng G, Tian G, et al. 2005. Molecular basis of replication of duck H5N1 influenza viruses in a mammalian mouse model. *J. Virol.* 79:12058–64

50. Lin YP, Shaw M, Gregory V, Cameron K, Lim W, et al. 2000. Avian-to-human transmission of H9N2 subtype influenza A viruses: relationship between H9N2 and H5N1 human isolates. *Proc. Natl. Acad. Sci. USA* 97:9654–58

51. Liu J, Xiao H, Lei F, Zhu Q, Qin K, et al. 2005. Highly pathogenic H5N1 influenza virus infection in migratory birds. *Science* 309:1206

52. Ludwig S, Stitz L, Planz O, Van H, Fitch WM, Scholtissek C. 1995. European swine virus as a possible source for the next influenza pandemic? *Virology* 212:555–61

53. Macken CA, Webby RJ, Bruno WJ. 2006. Genotype turnover by reassortment of replication complex genes from avian influenza A virus. *J. Gen. Virol.* 87:2803–15

54. Maines TR, Lu XH, Erb SM, Edwards L, Guarner J, et al. 2005. Avian influenza (H5N1) viruses isolated from humans in Asia in 2004 exhibit increased virulence in mammals. *J. Virol.* 79:11788–800

55. **Massin P, van der Werf S, Naffakh N. 2001. Residue 627 of PB2 is a determinant of cold sensitivity in RNA replication of avian influenza viruses. *J. Virol.* 75:5398–404**

56. Momose F, Basler CF, O'Neill RE, Iwamatsu A, Palese P, Nagata K. 2001. Cellular splicing factor RAF-2p48/NPI-5/BAT1/UAP56 interacts with the influenza virus nucleoprotein and enhances viral RNA synthesis. *J. Virol.* 75:1899–908

57. Momose F, Kikuchi Y, Komase K, Morikawa Y. 2007. Visualization of microtubule-mediated transport of influenza viral progeny ribonucleoprotein. *Microbes Infect.* 9:1422–33

58. **Munster VJ, de Wit E, van Riel D, Beyer WE, Rimmelzwaan GF, et al. 2007. The molecular basis of the pathogenicity of the Dutch highly pathogenic human influenza A H7N7 viruses. *J. Infect. Dis.* 196:258–65**

59. Murti KG, Webster RG, Jones IM. 1988. Localization of RNA polymerases on influenza viral ribonucleoproteins by immunogold labeling. *Virology* 164:562–66

60. Naffakh N, Massin P, Escriou N, Crescenzo-Chaigne B, van der Werf S. 2000. Genetic analysis of the compatibility between polymerase proteins from human and avian strains of influenza A viruses. *J. Gen. Virol.* 81:1283–91

61. Naffakh N, Massin P, van der Werf S. 2001. The transcription/replication activity of the polymerase of influenza A viruses is not correlated with the level of proteolysis induced by the PA subunit. *Virology* 285:244–52

62. Naito T, Kiyasu Y, Sugiyama K, Kimura A, Nakano R, et al. 2007. An influenza virus replicon system in yeast identified Tat-SF1 as a stimulatory host factor for viral RNA synthesis. *Proc. Natl. Acad. Sci. USA* 104:18235–40

63. Naito T, Momose F, Kawaguchi A, Nagata K. 2007. Involvement of Hsp90 in assembly and nuclear import of influenza virus RNA polymerase subunits. *J. Virol.* 81:1339–49

64. Neumann G, Brownlee GG, Fodor E, Kawaoka Y. 2004. Orthomyxovirus replication, transcription, and polyadenylation. *Curr. Top. Microbiol. Immunol.* 283:121–43

65. Neumann G, Kawaoka Y. 2006. Host range restriction and pathogenicity in the context of influenza pandemic. *Emerg. Infect. Dis.* 12:881–86

66. **Neumann G, Watanabe T, Ito H, Watanabe S, Goto H, et al. 1999. Generation of influenza A viruses entirely from cloned cDNAs. *Proc. Natl. Acad. Sci. USA* 96:9345–50**

67. Noda T, Sagara H, Yen A, Takada A, Kida H, et al. 2006. Architecture of ribonucleoprotein complexes in influenza A virus particles. *Nature* 439:490–92

68. **Obenauer JC, Denson J, Mehta PK, Su X, Mukatira S, et al. 2006. Large-scale sequence analysis of avian influenza isolates. *Science* 311:1576–80**

69. O'Neill RE, Jaskunas R, Blobel G, Palese P, Moroianu J. 1995. Nuclear import of influenza virus RNA can be mediated by viral nucleoprotein and transport factors required for protein import. *J. Biol. Chem.* 270:22701–4

55. Evidence that the nature of PB2 residue 627 can determine the level of activity of avian influenza virus–derived polymerases at low temperature.

58. Identification of PB2 residue 627 as the major determinant of the pathogenicity in mice of an influenza A H7N7 virus responsible for a human fatal case.

66. Initial description of a plasmid-based reverse genetics system for influenza viruses.

68. Large-scale sequencing of avian influenza viruses and analysis of the variability of viral proteins by proteotyping, i.e., identification of unique amino acid signatures.

70. Palese P, Shaw M. 2007. Orthomyxoviridae: the viruses and their replication. In *Fields Virology*, ed. D Knipe, P Howley, D Griffin, R Lamb, M Martin, et al., 2:1647–89. Philadelphia/New York: Lippincott Williams & Wilkins. 5th ed.

71. Pleschka S, Jaskunas R, Engelhardt OG, Zurcher T, Palese P, Garcia-Sastre A. 1996. A plasmid-based reverse genetics system for influenza A virus. *J. Virol.* 70:4188–92

72. Poole E, Elton D, Medcalf L, Digard P. 2004. Functional domains of the influenza A virus PB2 protein: identification of NP and PB1-binding sites. *Virology* 30:120–33

73. Portela A, Digard P. 2002. The influenza virus nucleoprotein: a multifunctional RNA-binding protein pivotal to virus replication. *J. Gen. Virol.* 83:723–34

74. Reid AH, Fanning TG, Janczewski TA, Lourens RM, Taubenberger JK. 2004. Novel origin of the 1918 pandemic influenza virus nucleoprotein gene. *J. Virol.* 78:12462–70

75. Reid AH, Taubenberger JK, Fanning TG. 2004. Evidence of an absence: the genetic origins of the 1918 pandemic influenza virus. *Nat. Rev. Microbiol.* 2:909–14

76. Salomon R, Franks J, Govorkova EA, Ilyushina NA, Yen HL, et al. 2006. The polymerase complex genes contribute to the high virulence of the human H5N1 influenza virus isolate A/Vietnam/1203/04. *J. Exp. Med.* 203:689–97

77. Sampath R, Hall TA, Massire C, Li F, Blyn LB, et al. 2007. Rapid identification of emerging infectious agents using PCR and electrospray ionization mass spectrometry. *Ann. N. Y. Acad. Sci.* 1102:109–20

78. Shinya K, Hamm S, Hatta M, Ito H, Ito T, Kawaoka Y. 2004. PB2 amino acid at position 627 affects replicative efficiency, but not cell tropism, of Hong Kong H5N1 influenza A viruses in mice. *Virology* 320:258–66

79. Shinya K, Watanabe S, Ito T, Kasai N, Kawaoka Y. 2007. Adaptation of an H7N7 equine influenza A virus in mice. *J. Gen. Virol.* 88:547–53

80. Snyder MH, Buckler-White AJ, London WT, Tierney EL, Murphy BR. 1987. The avian influenza virus nucleoprotein gene and a specific constellation of avian and human virus polymerase genes each specify attenuation of avian-human influenza A/Pintail/79 reassortant viruses for monkeys. *J. Virol.* 61:2857–63

81. Stech J, Xiong X, Scholtissek C, Webster RG. 1999. Independence of evolutionary and mutational rates after transmission of avian influenza viruses to swine. *J. Virol.* 73:1878–84

82. Suarez DL. 2000. Evolution of avian influenza viruses. *Vet. Microbiol.* 74:15–27

83. Suarez P, Valcarcel J, Ortin J. 1992. Heterogeneity of the mutation rates of influenza A viruses: isolation of mutator mutants. *J. Virol.* 66:2491–94

84. Subbarao EK, London W, Murphy BR. 1993. A single amino acid in the PB2 gene of influenza A virus is a determinant of host range. *J. Virol.* 67:1761–64

85. Subbarao K, Katz J. 2000. Avian influenza viruses infecting humans. *Cell Mol. Life. Sci.* 57:1770–84

86. Subbarao K, Klimov A, Katz J, Regnery H, Lim W, et al. 1998. Characterization of an avian influenza A (H5N1) virus isolated from a child with a fatal respiratory illness. *Science* 279:393–96

87. Tarendeau F, Boudet J, Guilligay D, Mas PJ, Bougault CM, et al. 2007. Structure and nuclear import function of the C-terminal domain of influenza virus polymerase PB2 subunit. *Nat. Struct. Mol. Biol.* 14:229–33

88. Taubenberger JK, Reid AH, Lourens RM, Wang R, Jin G, Fanning TG. 2005. Characterization of the 1918 influenza virus polymerase genes. *Nature* 437:889–93

89. Tian SF, Buckler-White AJ, London WT, Reck LJ, Chanock RM, Murphy BR. 1985. Nucleoprotein and membrane protein genes are associated with restriction of replication of influenza A/Mallard/NY/78 virus and its reassortants in squirrel monkey respiratory tract. *J. Virol.* 53:771–75

90. Turan K, Mibayashi M, Sugiyama K, Saito S, Numajiri A, Nagata K. 2004. Nuclear MxA proteins form a complex with influenza virus NP and inhibit the transcription of the engineered influenza virus genome. *Nucleic Acids Res.* 32:643–52

91. Webster RG, Bean WJ, Gorman OT, Chambers TM, Kawaoka Y. 1992. Evolution and ecology of influenza A viruses. *Microbiol. Rev.* 56:152–79

92. Wright P, Neumann G, Kawaoka Y. 2007. Orthomyxoviruses. In *Fields Virology*, ed. D Knipe, P Howley, D Griffin, R Lamb, M Martin, et al., 2:1691–739. Philadelphia/New York: Lippincott Williams & Wilkins. 5th ed.

84. Initial identification of PB2 residue 627 as a determinant of the ability of an avian-human reassortant influenza A virus to replicate on mammalian cells.

88. Sequence and phylogenetic analyses of the polymerase genes of the influenza virus responsible for the severe pandemic of 1918.

93. Yao Y, Mingay LJ, McCauley JW, Barclay WS. 2001. Sequences in influenza A virus PB2 protein that determine productive infection for an avian influenza virus in mouse and human cell lines. *J. Virol.* 75:5410–15

94. Ye Q, Krug RM, Tao YJ. 2006. The mechanism by which influenza A virus nucleoprotein forms oligomers and binds RNA. *Nature* 444:1078–82

95. Zhou NN, Shortridge KF, Claas EC, Krauss SL, Webster RG. 1999. Rapid evolution of H5N1 influenza viruses in chickens in Hong Kong. *J. Virol.* 73:3366–74

96. Zitzow LA, Rowe T, Morken T, Shieh WJ, Zaki S, Katz JM. 2002. Pathogenesis of avian influenza A (H5N1) viruses in ferrets. *J. Virol.* 76:4420–29

97. Mayer D, Molawi K, Martinez-Sobrido L, Ghanem A, Thomas S, et al. 2007. Identification of cellular interaction partners of the influenza virus ribonucleoprotein complex and polymerase complex using proteomic-based approaches. *J. Proteome Res.* 6:672–82

Cell Biology of HIV-1 Infection of Macrophages

Carol A. Carter and Lorna S. Ehrlich

Department of Molecular Genetics and Microbiology, Stony Brook University, Stony Brook, New York 11794; email: lsehrlich@ms.cc.sunysb.edu, ccarter@ms.cc.sunysb.edu

Annu. Rev. Microbiol. 2008. 62:425–443

The *Annual Review of Microbiology* is online at micro.annualreviews.org

This article's doi:
10.1146/annurev.micro.62.081307.162758

Key Words

Gag, MVBs, MDM, retrovirus, ESCRT

Abstract

HIV infection of macrophages is a critically important component of viral pathogenesis and progression to AIDS. Although the virus follows the same life cycle in macrophages and T lymphocytes, several aspects of the virus-host relationship are unique to macrophage infection. Examples of these are the long-term persistence of productive infection, sustained by the absence of cell death, and the ability of progeny virus to bud into and accumulate in endocytic compartments designated multivesicular bodies (MVBs). Recently, the hypothesis that viral exploitation of the macrophage endocytic machinery is responsible for perpetuating the chronic state of infection unique to this cell type has been challenged in several independent studies employing a variety of experimental strategies. This review examines the evidence supporting and refuting the canonical hypothesis and highlights recently identified cellular factors that may contribute to the unique aspects of the HIV-macrophage interaction.

Contents

INTRODUCTION

HIV-1 can replicate in many cell types; however, the two major cellular reservoirs in the natural host are latently infected resting CD4+ T cells and macrophages. The macrophage is widely recognized as the earlier cellular target of HIV-1 (68), with the subsequent infection of T cells and the ensuing T cell depletion resulting in the immunopathogenesis that characterizes AIDS. In contrast to T cells, macrophages are more resistant to the cytopathic effects of the virus and better able to evade the defensive action of the infected individual's immune system (68, 92). As a result, the HIV-1-infected macrophage produces and harbors the virus for a longer period. With the ability of this cell type to cross the blood-tissue barrier, an infected macrophage cell is a potent agent for delivery of HIV-1 to all tissues and organs, including the brain (68). Thus, the HIV-1-infected macrophage is of critical importance in the

pathogenesis of HIV because it is a major contributor to early-stage viral transmission, persistence, and virus dissemination throughout the body of the host.

Macrophages accumulate replication-competent HIV-1 for prolonged periods, even in patients receiving highly active antiretroviral treatment (HAART). It is widely assumed that the basis for the longevity of the HIV-1-infected macrophage lies in the nature of the virus-host interaction (21, 81). A long-held view has been that HIV-1 protein trafficking and virus release from T cells and macrophages follow different paradigms. Until recently, the consensus held that viral assembly events occur mainly at the plasma membrane in T cells and in most nonhematopoietic cell types but, in macrophages, took place in compartments of the endocytic pathway. Longevity in the macrophage could therefore result from storage of assembled progeny virus within specialized endosomes. However, the hypothesis that assembly and budding site selection occur differently in macrophages and T cells is currently controversial. The recent identification of a plethora of cellular proteins involved in HIV-1 Gag (group-specific antigen) trafficking (25) that might contribute to cell-type-dependent similarities and differences prompts a reexamination of the cell biology of viral assembly in these cells. In this review, we discuss the unique aspects of viral infection in macrophages, with particular emphasis on reexamining the evidence for the hypothesis that the mechanism of virus release from macrophages is distinct from that in T cells and other model systems. A conscientious effort has been made to acknowledge the contributions of the many laboratories that have published reports in this field. We apologize to those whose work may not have been cited because of oversight and space limitations.

Experimental Models for Tissue Macrophages

HIV-1 and other members of the lentiviral subgroup in the retroviral family infect cells

Macrophage: relatively long-lived phagocytic cells involved in antigen presentation and other immune surveillance functions in mammalian tissues

Group-specific antigen (Gag): the structural precursor polyprotein encoded by the *gag* gene of all retroviruses

of the mononuclear phagocyte lineage at different stages of their differentiation (7). The investigation of these tissue macrophages is often difficult because of their limited accessibility and inefficient recovery. Thus, most in vitro studies of infection utilize monocyte-derived macrophages (MDMs). The cells obtained by this method are highly variable owing to heterogeneity inherent in patient sources and the lack of uniformity in the experimental protocol employed by different investigators. A number of cell lines, mostly Mono Mac-1 and Mac-6, THP-1, U937, HL-60, and their derivative chronically infected counterparts (e.g., U1 and OM-10.1 cell lines), have complemented the use of MDMs. Although MDMs more accurately reflect HIV-1 infection of macrophages in vivo, cell lines offer the advantage of high accessibility, ease of standardization, unlimited expansion, and reasonably stable genetic and growth properties. Most importantly, several aspects of replication unique to macrophage infection in vivo, such as its remarkable longevity and fusogenic properties, have been reproduced in both MDMs and mononuclear phagocytic cell lines. Ironically, however, emerging studies suggest that neither MDMs nor macrophagic cell lines may recapitulate the complexity of the viral assembly and release events thought to underlie some of the unique aspects of macrophage versus T cell infection.

Modulation of Macrophage Gene Expression Following Infection

Several independent microarray studies document changes in host gene expression following infection of hematopoietic cells with HIV-1 or transfection of individual HIV-1 genes (23). Most of the genes identified as differentially up- or downregulated following HIV-1 infection of macrophages that have been independently validated by an alternative method subsequent to microarray-based detection function in signaling transcriptional program changes or in cell cycle regulation. Relative to CD4+ T cells and other cells, in macrophages there may be a preferential modulation of specific cellular factors for signal pathway activation of the host immune/inflammatory responses, as indicated by differential upregulation of signal transducer and activator of transcription 5A (STAT 5A); protein kinases [e.g., extracellular regulated kinase (ERK), double-stranded RNA-activated protein kinase (PKR)]; and proteins involved in immune responses [e.g., interferon (IFN)-induced protein with tetratricopeptide repeat 4 (IFIT4)]. Certain chemokines, such as CCL2 and CCL8, proposed to enhance virus dissemination by promoting recruitment of target CD4+ T cells and macrophages to sites of infection, also appear to be differentially upregulated in HIV-infected or gp120-exposed macrophages.

As noted above, macrophages support high levels of viral replication even though they are nondividing differentiated cells. Whereas regulatory viral protein (Vpr)-induced cell cycle arrest by p21-mediated upregulation occurs in both T cells and MDMs (91), modulation of some of the genes that promote cell cycle transition and arrest, such as the human 14-3-3 epsilon isoform YWHAE, occurs differentially in HIV-1-infected macrophages versus T cells (9). Recent studies using circulating monocytes from HIV-infected patients in which antiapoptosis genes were identified as upregulated also support the notion that differential modulation of gene expression is a mechanism contributing to the greater survival of macrophages versus T cells following HIV infection (23).

HIV Life Cycle: Unique Aspects of Macrophage Infection

Entry of HIV-1 into target cells takes place in two sequential steps. The first step involves the high-affinity binding of the viral surface glycoprotein gp120 to the CD4 receptor protein, which is present on both macrophages and T cells. The second step, promoted through the gp120-CD4 interaction, engages coreceptor molecules, differentially expressed on the two cell types, to mediate fusion of the viral envelope with the cell membrane. These coreceptor proteins belong to the superfamily of

Monocyte-derived macrophage (MDM): experimental model system for study of macrophages in vitro

PI(4,5)P$_2$:
phosphatidylinositol-
4,5-bisphosphate

G-protein-coupled receptors that normally bind chemokines, small molecules involved in inflammatory responses (1). Until recently, HIV-1 target cell tropism (macrophage or T cell) was considered synonymous with coreceptor selectivity (CCR5 or CXCR4, respectively). However, new data emerging from studies with primary isolates indicate that the relationship is much more complex. Whereas most CD4$^+$-transformed T cell lines express only CXCR4, the macrophages and primary lymphocytes that serve as the main cellular targets for infection in vivo express both coreceptor types on their surface. Rather than coreceptor expression per se, viral entry appears to be governed by viral strain and cell-specific utilization of the coreceptors, with the viral strain specificity determined by distinct characteristics of Env (26). Additional macrophage-specific membrane interactions that follow the unique receptor/coreceptor binding include those with host proteins annexin 2, p21, and α-v-integrin (4, 96). Whether entry through the CCR5 versus CXCR4 coreceptor influences downstream replication events is not known. However, in both cases the ribonucleoprotein complex or viral core that is released into the cytoplasm initiates the next stage of the infection. In this next stage, the viral RNA genome in the viral core structure is reverse transcribed into DNA by the virus-encoded reverse transcriptase enzyme and then established in the target cell by viral enzyme-mediated integration into host chromatin.

Whereas T cells must be activated before they can be productively infected, a process that includes cell division, productive infection can be established in macrophages even though they are not dividing. Jacque & Stevenson (37) recently determined that HeLa cells as well as MDMs lacking emerin, an inner nuclear membrane protein that bridges the interface between the inner nuclear envelope and chromatin, were unable to support productive infection, but this is controversial (52, 83). Several structural components of the viral core are also required specifically to deliver the so-called preintegration complex (PIC) across the nu-

clear membrane and to establish it in the nucleus. Although all members of the family *Retroviridae* encode the *gag* and *pol* genes, which determine this function, only the gene products encoded in members of the lentiviral subgroup, which includes HIV, possess the ability to direct entry into the nucleus of a nondividing cell such as the macrophage. It is generally believed that the nuclear membrane of nondividing cells is a barrier to the transport of large complexes, such as the PIC, and that the barrier is removed when the membrane breaks down during cell division. Some investigators have challenged this view (38, 101). The core constituents, matrix (MA), nucleocapsid (NC), integrase (IN), and Vpr, direct this function; their precise roles are highly controversial and apparently redundant (17). MA and NC are *gag* related, IN (integrase) is *pol* related, and Vpr is uniquely encoded by the primate lentiviruses. Emerging evidence suggests that capsid (CA), another *gag*-related component, also plays a role in cell cycle–independent infection, as the ability to infect primary macrophages was lost following mutation of a site in the CA protein (102).

Later in infection, transcription from the integrated DNA copy of the viral RNA genome marks the productive phase of the virus life cycle. Early studies pioneered by Wills & Craven (99) demonstrated that the structural precursor polyprotein (Gag) encoded by the *gag* gene is both necessary and sufficient for assembly and budding of immature virus particles. Gag is synthesized on free polysomes as a linear arrangement of structural domains with intervening spacer peptides, for example, as (N)-MA-CA-p2-NC-p1-p6-(C); cotranslationally myristylated at the N terminus of MA; and posttranslationally phosphorylated and ubiquitinated at various sites within the domains (47, 53, 64). Gag bears determinants for interactions with phospholipids, RNA, and cellular proteins, as well as determinants that target it to specific membrane compartments in the host cell. The N-terminal matrix domain in Gag bears determinants for binding to phosphatidylinositol-4,5-bisphosphate [PI(4,5)P$_2$], a phospholipid enriched in the plasma membrane (60, 78, 105).

Sequences in the MA domain of Gag are docking sites for the delta subunit of the AP-3 adaptor complex (15) and for annexin 2 (77) that result in the targeting of Gag to internal membrane compartments. The CA domain contains a di-leucine motif that also promotes Gag association with internal vesicles (42). The p6 region in Gag bears the so-called late (L) domain that drives interactions with cellular factors belonging to the endocytic trafficking machinery, specifically Tsg101 (20, 45, 93) and Alix/AIP-1 (44, 85, 94). The NC region also interacts with Alix/AIP-1 and may play an as yet undefined role in virus release (70). Thus, it is possible that Gag is targeted to different membrane compartments in macrophages and T cells through interplay of viral and cellular determinants.

During the final stages of viral assembly in the natural setting, the viral encoded protease cleaves sites between the domains of the immature Gag precursor assemblage, releasing mature MA, CA, NC, and p6 products that undergo morphogenetic rearrangement within the particle to form mature, infectious virions. **Figure 1** summarizes schematically the replication steps that exhibit cell-type-dependent features in macrophages and T cells. Several excellent reviews of *Retrovirus* replication, with emphasis on particular aspects of the Gag assembly process, have been published in recent years (10, 35, 39, 51, 74).

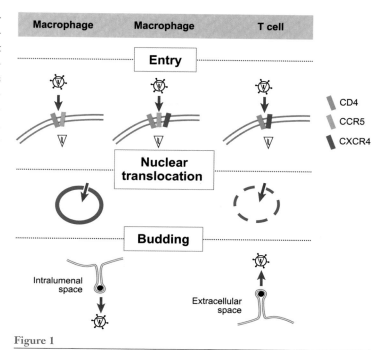

Figure 1

Macrophage- and T-cell-specific steps in HIV-1 infection. Entry into macrophages and T cells occurs by receptor-mediated fusion. The CD4 receptor (*green bar*) is required on both cell types. In addition, entry into macrophages requires interaction with the CCR5 (*orange bar*) or the CXCR4 (*blue bar*) coreceptor; entry into T cells occurs through the CXCR4 coreceptor. Nuclear translocation can occur in nondividing cells such as the macrophage where the nuclear membrane is intact (*solid line*), but it requires cell division (*dashed line*) in other cell types. Budding occurs predominantly at the plasma membrane of T cells, releasing virus into the extracellular space but mainly into cytoplasmic vacuoles in macrophages infected in vitro. Recent studies suggest that the latter occurrence may be less frequent in vivo.

Host Response

Differences between HIV-1 infection in macrophage and T cells may be due to differences in host factors interacting with the various determinants in Gag. L domain mutants produced in T cell lines and primary lymphocytes are released from the cell but exhibit a defect in virion-virion detachment, such that chains of tethered particles are formed. In contrast, release of L domain mutants from MDMs is markedly inhibited (11, 13). IFN, which influences this late event (14, 22, 27, 69), elicited different effects in the two cell types. HIV-1 replication is inhibited by IFN in both cell types, but in T cells treatment was accompanied by inhibition of viral antigen release and accumulation of virus particles at the surface of HIV-producing cell lines, indicating a blockade of a late event (59). In contrast, macrophages showed no evidence of active transcription, viral protein production, or transcriptase activity, indicative of an earlier block (32, 66, 96). Although IFN treatment is highly effective, the endogenous IFN response against infection is impotent, as macrophages express an ISG15-specific isopeptidase (ISGylation deconjugating protease), Ubp43 (UPS18), that inhibits type I IFN signaling (104). Inhibition of this antiviral response might contribute to the longevity of the virus-host interaction in the macrophage.

Clearly, cellular factors unique to macrophages modulate manifestations of both the infection and the host response.

Assembly in T Cells and Macrophages: Evolution of the Current Paradigms

Although HIV can establish a latent infection in a small population of macrophages (5), most macrophages produce copious amounts of virus (62, 95). This implies that mechanisms exist for efficient virus release, even though, unlike T cells, macrophages remain intact and functional. One of the historically defined features of HIV replication in macrophages is the budding into and accumulation of viral particles in internal cytoplasmic vacuoles of MDMs infected in vitro (63). In T cells and most non-hematopoietic cell types, HIV-1 assembles on the plasma membrane and progeny virus is released from this site (**Figure 2**). In contrast, early studies showed that, in infected MDMs, HIV-1 assembly takes place within structures that appeared to be intracellular vesicles (63). These compartments were later identified in other cells as late endosomes/multivesicular bodies (LEs/MVBs) mainly on the basis of their association with the tetraspanin protein CD63 (56, 65, 72, 82).

In the macrophage, specialized LEs/MVBs (MCII compartments) can traffic to the cell surface, fuse with the plasma membrane, and release their contents into the extracellular environment as exosomes via the so-called exosomal pathway (84). This outcome was thought to be unique to this cell type and some other hematopoietic cell types; thus, viral assembly

and budding appeared to occur through a process that is constitutive in these cells. This suggested that viral release was a natural outcome of infection of a cell type in which the exosomal pathway is active. The inaccessibility of intracellularly assembled virus to immune surveillance prior to release has been assumed to allow the macrophage to serve as an HIV-1 reservoir and as a potent disseminator of the virus to all tissues, including the central nervous system, as the macrophage has the ability to cross the blood-brain barrier.

In the past few years, however, several laboratories have reported detection of HIV-1 virions or the virus-like particles (VLPs) assembled by Gag in the LEs/MVBs of nonhematopoietic cells (56, 82). This led to the notion that these compartments might gate viral particle release (56). Thus, it appeared that LEs/MVBs played an important role in both macrophages and T cells, and that viral assembly in endocytic vesicles was not strictly cell-type-dependent. Indeed, Gould et al. (28) dubbed this model the Trojan exosome hypothesis (**Figure 3**) to highlight the notion that virus in LE/MVB compartments would not only escape degradation but might also evade immune system recognition and antiviral drugs following release into the extracellular environment. gp120 is normally exposed on the surface of cells and particles assembled at the plasma membrane but would not be exposed if virus was released as exosomes. Implicit in this hypothesis is that the virus possesses the means to thwart the natural sorting process that would otherwise deliver it to a degradative compartment; LEs/MVBs normally fuse with or mature into such compartments. It has been suggested that viral replication can interfere with trafficking of internalized cell surface receptors to degradative compartments (90), a process that requires Tsg101 (2, 43). Another corollary of the hypothesis is that HIV-1 can exploit the pathway even in cells where the exosomal machinery does not function constitutively. The in vivo relevance of the Trojan exosome hypothesis was recently challenged in an ultrastructural study of sites of HIV assembly in macrophages

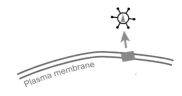

Figure 2

Gag assemblages (*green rectangle*) accumulate on the plasma membrane of T cells and nonhematopoietic cells and bud from there into the extracellular space.

infected in vitro and in vivo (62). Whereas the virus was regularly detected in LEs/MVBs in MDMs infected in culture, virus particle association with this compartment was infrequent in vivo, i.e., in specimens obtained from HIV-1-infected individuals.

Given the considerable mechanistic and physiological constraints on the concept of viral release as classically defined exosomes, the Trojan exosome hypothesis was subsequently modified to encompass the possibility that physical association of markers of internal vesicles with viral particles indicates not only exosomal release but also budding from LE/MVB-derived membrane microdomains at the cell surface. HIV was suggested to bud through discrete plasma membrane microdomains enriched in endosomal markers, lipid rafts, and tetraspanins (3) (**Figure 4**). Nydegger et al. (57) found tetraspanin-enriched microdomains (TEMs), or endosome-like domains, containing the tetraspanins CD9, CD63, CD81, and CD82 on the surface of HeLa cells and observed that HIV-1 Gag accumulated at these surface TEMs together with gp120 and with Tsg101 and Vps28, two components of ESCRT-1 (endosomal sorting complex required for transport 1), which is part of the cellular machinery critical for budding.

It has been suggested that both the ultrastructural and the genetic evidence supporting localization of Gag to the LE/MVB merely reflects internalization of released virions or adventitious endocytosis of the Gag precursor. These are tenable explanations given that (a) endocytosis of cellular cargo is a natural function of Tsg101. Tsg101 normally regulates endocytic trafficking through interactions with Pro-Thr/Ser-Ala-Pro (PT/SAP) motifs in proteins. With the cellular protein Hrs (which contains a Pro-Ser-Ala-Pro motif) as a binding partner, Tsg101 facilitates delivery of internalized membrane receptors to degradative compartments in response to ligand-stimulated receptor downregulation (2, 43). Tsg101 recognizes ubiquitinated cargo (41) and ubiquitination is the signal for this sorting. Tsg101 promotes endocytosis of Gag (24, 31, 54). Removal

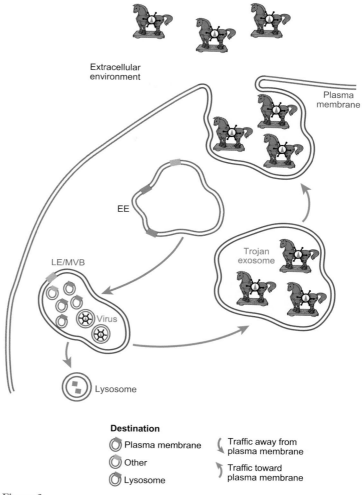

Figure 3

The Trojan exosome model. The model hypothesizes that HIV, on the cytoplasmic face of the late endosome/multivesicular body (LE/MVB) membrane, is internalized in intralumenal vesicles following membrane invagination and fission. LEs/MVBs, derived from the early endosome (EE), function constitutively in the sorting and trafficking of cellular cargo to the plasma membrane, other intracellular organelles, and the lysosome. It is proposed that the assemblage sequestered in the intralumenal vesicles (Trojan exosomes) is released into the extracellular environment when LE/MVBs are transported to the cell surface and fuse with the plasma membrane.

of the p6 region in Gag that binds Tsg101 inhibited virus release but did not prevent LE/MVB targeting of wild-type Gag in macrophages (61). (b) Abolishing LE motility altered the subcellular localization of Gag in primary macrophages but did not prevent particle release (40). These

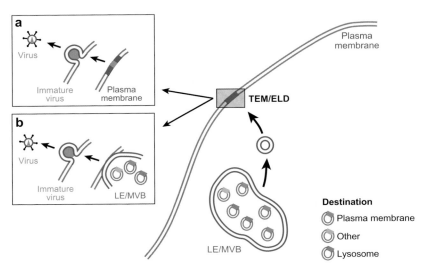

Figure 4

Tetraspanin-enriched microdomains (TEMs), or endosome-like domains (ELDs), as sites of virus budding. The plasma membrane (PM) is proposed to possess stable microdomains containing the natural tetraspanin components of late endosome/multivesicular body (LE/MVB) compartments and other endocytic membrane marker proteins. These budding microdomains may derive from constitutive membrane biogenesis or repair (*inset a*) and/or from natural LE/MVB-PM fusion events (*inset b*).

Endosomal sorting complex required for transport (ESCRT): three to four evolutionarily conserved complexes (ESCRT-0, -I, -II, and -III) that function in a coordinated manner in endosomal trafficking pathways

TEM: tetraspanin-enriched microdomain

investigators reported detection of newly synthesized Gag on the plasma membrane of the macrophages prior to its accumulation in endosomes. (*c*) Using a biarsenical labeling method to dynamically observe HIV-1 Gag within live HeLa, Mel JuSo, and Jurkat T cells, Rudner et al. (76) showed that Gag concentrates in specific plasma membrane areas rapidly after translation.

However, there is as yet no facile resolution of this controversy. The possibility that Gag association with LE/MVBs merely reflects internalized dead-end Gag is countered by evidence that (*a*) Gag targets the LE/MVB compartment before delivery to the plasma membrane even in cell types where virus assembly takes place predominantly at the plasma membrane (30, 67). Perlman & Resh (67) obtained this result using both the biarsenical labeling procedure employed by Rudner et al. (76) and synchronized Gag expression. (*b*) Gag does not colocalize with the endosomal marker EEA1, suggesting that Gag is not sorted to the LE/MVB compartment from the early endosomes, as

would be expected for typical endocytic cargo. Rather, Gag associates with a subset of early endosomes that are Hrs and Tsg101 positive (2, 24). (*c*) Antibody-coated beads phagocytosed by macrophages did not colocalize with Gag until several hours later, suggesting that Gag detected in the LE/MVB did not originate from phagocytosed virus particles (61). (*d*) Particles assembled from Gag lacking the L domain are detected tethered to the intralumenal membrane of intracellular vesicles in macrophages following electron microscopy analyses (11), not tethered to both peripheral and internal membranes, as might be expected if some of the Gag assemblages were internalized from the cell periphery. (*e*) Poole et al. (71) concluded that Gag colocalizes early after expression with genomic RNA in the perinuclear region and that the colocalized RNA and protein are transported through the cytoplasm to the plasma membrane.

On balance, these apparently conflicting observations most likely indicate that Gag association with LEs/MVBs is subject to regulation by a variety of factors in the cellular environment. In this regard, an idea introduced in an early study of enveloped virus budding site localization seems particularly pertinent here. Rodriguez-Boulan et al. (75) observed that vesicular stomatitis virus (VSV), which normally buds from the basolateral surface of polarized epithelial Madin-Darby canine kidney (MDCK) cells, frequently budded into intracellular vesicles in isolated suspended MDCK cells but rarely was detected in internal compartments in confluent monolayers. On the basis of their observation, they suggested that interaction with a substrate or another cell might suffice as a signal to trigger the return of the plasma membrane components responsible for polarized budding at the cell surface. This hypothesis offers a potentially unifying resolution that is amenable to experimental analysis.

Cell-Type Determinants of Targeting and Release

The MA domain in Gag bears determinants of membrane binding and targeting; however, sequences in the other domains also influence Gag membrane localization. These determinants exhibit cell-type dependency. Ono & Freed (61) observed that Gag proteins with mutations in the highly basic domain of MA or in residues 84 to 88 of the protein behaved similarly in HeLa and Jurkat T cells in that the mutations interfered with plasma membrane targeting and caused virus particles to assemble internally in CD63-positive LEs/MVBs, thereby reducing the amount of released virus. Interestingly, however, both wild-type Gag and the MA mutants assembled in such CD63-positive vesicles in macrophages. These results suggest that, although the altered MA residues retarget Gag from the plasma membrane to the LE/MVB in HeLa and Jurkat T cells, they are not determinants of LE/MVB localization in macrophages. These observations indicate that Gag trafficking to the LE/MVB compartment is regulated by different determinants in T cells and macrophages. Moreover, their finding that virus was released from macrophage and

HeLa cells with comparable efficiency provides compelling evidence that the LE/MVB is part of a productive pathway in the macrophage.

Macrophages might contain distinct classes of LE/MVB compartments and the wild-type Gag may target a subset of these. The nonproductive outcome that follows Gag targeting to the LE/MVB compartment in T and HeLa cells might be explained if the recruitment of cellular factors requires specific conditions existent specifically in the macrophage, where aspects of the LE/MVB machinery function constitutively. In this regard, adventitious expression of the human leukocyte antigen DR (HLA-DR), a subset of major histocompatibility class (MHC)-II molecules, induced relocation of infectious particle assembly to apparently intracellular compartments that bore LE/MVB markers (18). Thus, in cells such as macrophages where MHC-II molecules are normally expressed, the proteins may contribute to the utilization of MHC-II-containing compartments for virus assembly.

Recently, it was suggested by two independent groups that the LE/MVB vesicles with which HIV associates in macrophages are not endosomes, but an internally sequestered plasma membrane domain (**Figure 5**). Deneka et al. (12) observed that HIV-1 buds into an apparently internal membrane compartment that contains the tetraspanins CD81, CD9, and CD53, to which CD63 was recruited following HIV-1 infection. Though visually exhibiting an intracellular localization, these compartments were accessible to a membrane-impermeant tracer applied to the cells, indicating that they were connected to the cell surface. There was no overlap between these compartments and endocytic compartments labeled with endocytosed horse radish peroxidase-conjugated tracers. The authors concluded that the plasma membrane is the primary site of virus assembly and budding in macrophages. Moreover, that CD63 was recruited to sites of virion assembly on these morphologically internal plasma membrane compartments signifies that identification of LEs/MVBs based solely on the presence of the CD63 marker can be

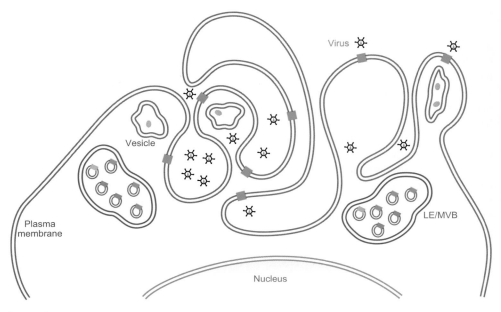

Figure 5

Depiction of internally sequestered plasma membrane domains on a convoluted macrophage surface. Budded virus particles are shown in apparently internal spaces formed by plasma membrane folds which are in proximity to late endosomes/multivesicular bodies (LEs/MVBs) and other vesicles.

misleading. Welsch et al. (98) utilized endocytosed bovine serum albumin–gold and a membrane-impermeant tracer to identify endosomes and the plasma membrane, respectively. Budding particles were found in seemingly intracellularly localized, virus-filled vacuolar structures that stained intensely with the tracer but lacked the endocytosed gold. Thus, Welsch et al. also concluded that the plasma membrane is the primary site of HIV-1 budding in infected macrophages. Presumably, this surface-accessible TEM is related to the membrane domain that serves as the HIV-1 exit site in HeLa and T cells, as noted above (57).

That some of the constituents at the budding site in macrophages are identical to those in T cells led to the expectation that the cellular protein composition of virions released from macrophages and T cells will exhibit significant overlap. This prediction was partially supported by proteomic analysis (**Figure 6**). These unexpected findings bring the concepts in the field full circle by introducing the notion that similar

(i.e., plasma membrane) organelles are targeted in both macrophages and T cells. With this new paradigm some apparent conundrums are resolved. For example, lipid rafts are recognized as sites of Gag assembly (39). Given their enrichment in the plasma membrane and dearth in LEs/MVBs (36), it would have been difficult to reconcile the role of lipid rafts if Gag assembly occurred in LEs/MVBs in macrophages and on the plasma membrane in T cells.

Cell-To-Cell Transmission of HIV in Living mDCs and Macrophages

In mature dendritic cells (mDCs), which are related to macrophages, Garcia et al. (19) found that HIV assembles in vesicles of endosomal origin that are nevertheless nonconventional and nonlysosomal. As was found for macrophages, the virus-containing compartment in mDCs also contained the tetraspanins CD81, CD82, and CD9, but little CD63 or LAMP-1 (lysosomal-associated membrane protein-1). Moreover, when allowed to contact T cells, the HIV-loaded mDCs redistributed CD81 and CD9, as well as the internalized HIV-1, to the infectious or viral synapse (46) formed at the mDC–T cell junction. The infectious synapse facilitates viral cell-to-cell transfer and propagation. In contrast, markers of the immunological synapse, MHC-II, and the T cell receptor were not redistributed. These findings provide evidence that HIV-1 selectively subverts components of the intracellular trafficking machinery in macrophage-related cell types to achieve efficient exit. Again, as was the case in macrophages, the virus compartment had several of the markers of the previously uncharacterized internally sequestered plasma membrane domain

Freed and collaborators (29) also report relocalization of Gag from an internal or apparently internal compartment to cell-cell junctions formed between infected macrophages and uninfected macrophages or T cells. Curiously, Freed et al. found that a MA mutant that targets LEs/MVBs does not relocalize to the viral synapse, suggesting that the internal compartment to which the wild-type Gag localizes in macrophages could be distinct from LEs/MVBs even though it bears LE/MVB markers, as Freed et al. and others have shown previously. Together, these observations suggest that aspects of trafficking and budding are differentially modulated by infection in T cells and macrophages even if they take place on the plasma membrane in both cell types.

Role of the L Domain

Efficient particle release from HeLa cells and macrophages, but not T cells, requires an intact L domain (11). However, targeting of Gag to the LE/MVB in either HeLa cells or macrophages does not require the p6 region in Gag that contains the Tsg101- and Alix/AIP-1-binding L domains. Thus, there appears to be no correlation between the L domain dependence of viral budding and the sites at which budding has been observed. From the perspec-

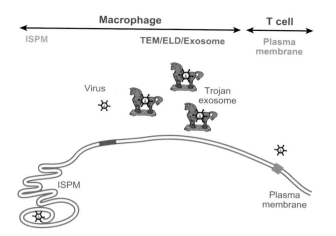

From MDMs	From exosomes	From T cells or nonhematopoietic origin
S100	Histone H4, 2B, 3.1	β2-microglobulin
Clathrin	Hsp70, 90	Tsg101
Arf1	CD2	Alix
Rab-related GTP-BP	CD9	CD42
Rho-related GTP-BP	CD18	CD43
>200 others	CD53	CD63
	CD81	CD81
	CD82	Hsp70
	Tsg101	Cofilin
	Alix	Actin
	Syntenin	Integrin α, β2
	Rab7, 11, GDI 1b	CyP A
	Gi2α	GPDH
	HLA-II-DR	
	Annexin A1, II, V	
	Actin	
	Cofilin	
	Tubulin	

Figure 6

Protein composition of virions released from monocyte-derived macrophages (MDMs). The virion-associated proteins identified to date by proteomic or other analyses include those found in particles derived from intracellular and surface membrane organelles. Although most of the constituents in virus particles purified from MDMs have not yet been detected in other sources, some of the proteins have been detected in virions released from T cells and cells of nonhematopoietic origin and some have been detected in exosomes. ISPM: intracellularly sequestered plasma membrane; TEM/ELD: tetraspanin-enriched microdomain/endosome-like domain; PM: plasma membrane. (From References 8, 89, and 97).

tive of the virus, this is not unexpected because the viral determinants of membrane targeting and viral particle untethering from the membrane are in different domains of the Gag precursor.

Role of ESCRT Proteins

Immuno-electron microscopy analysis of uninfected and HIV-1-infected primary human T cells and macrophages showed that the ESCRT proteins selected for analysis (Hrs, Tsg101, AIP1/Alix, and Vps4) were distributed throughout the endocytic pathway, including on the plasma membrane, with only a relatively small amount in the cytoplasm and on the LEs/MVBs defined by the CD63 marker (97). The ESCRT factors were twice as concentrated on components of the endocytic pathway (early and late endosomes/lysosomes) in macrophages compared with T cells. In contrast, the factors were twice as abundant on the plasma membrane of T cells compared with macrophages. Moreover, no significant relocalization of the factors was detected following HIV-1 infection. These intriguing findings suggest that the apparent differential budding site locations in T cells and macrophages are driven by the natural distribution of the cellular machinery required for budding.

Role of PI(4,5)P$_2$

To explain the original paradigm, i.e., cell-type-dependent budding locations, Ono & Freed (61) had proposed the existence of two different cellular binding partners for Gag (i.e., Gag receptors): one on the plasma membrane and the other on the LE/MVB. According to this model, the plasma membrane Gag receptor played a dominant role in HeLa cells and T cells, and the LE/MVB Gag receptor was proposed to be dominant in macrophages. However, if HIV-1 buds from the plasma membrane in both T cells and macrophages, as current evidence suggests, the same receptor should be operative in both cell types. An emerging candidate for a plasma membrane receptor is PI(4,5)P$_2$, a signaling lipid predominantly localized on the plasma membrane in most cell types (88). Ono et al. (60) discovered that depletion of PI(4,5)P$_2$ from the plasma membrane with 5-phosphatase IV caused Gag to be redirected to an LE/MVB compartment. In the same study, they demonstrated that promoting PI(4,5)P$_2$ formation on intracellular vesicles through expression of a constitutively active variant of ADP-ribosylation factor 6 (Arf-6) directed budding to the internal site. Viral release was inhibited significantly in both cases. Moreover, they also reported having preliminary data showing that in human MDMs PI(4,5)P$_2$ colocalizes with HIV-1 Gag on LEs and suggested the lipid plays a role in Gag targeting to LEs in this cell type. However, given that PI(4,5)P$_2$ is most likely abundant on the plasma membrane of the macrophage as it is in other cells, an additional factor(s) may be involved in regulating this selective Gag colocalization with intracellular PI(4,5)P$_2$ in this cell type. Alternatively, the vesicles bearing the colocalized Gag and PI(4,5)P$_2$ may not be LEs, but rather intracellular spaces delineated by CD63-positive plasma membrane extensions of the macrophage (12, 62, 98). In either case, the observation made by Ono & Freed (61) suggests that the distribution of PI(4,5)P$_2$ correlates to the apparent budding site location in T cells and macrophages.

Role of Cytosolic Ca^{2+}

Recent evidence also implicates free calcium (Ca^{2+}) in HIV release from both T cells and model cell systems (30, 67; L.S. Ehrlich, G.N. Medina, M.B. Khan, M.D. Powell, K. Mikoshiba & C.A. Carter, manuscript submitted). Induction of a transient rise in cytoplasmic Ca^{2+} by raising the extracellular Ca^{2+} concentration resulted in an increase in Gag assembly intermediates and VLPs in MVBs, leading to a significant rise in VLP release (67). Other studies showed that delivery of LE/MVB, or LE/MVB-derived transport vesicles, to the plasma membrane, fusion of docked transport vesicles with the plasma membrane, and reorganization of actin cytoskeleton that drives membrane remodeling require Ca^{2+} (16, 79, 80). The aforementioned targeting factor, PI(4,5)P$_2$, and the exocytosis factor, Ca^{2+}, are functionally linked, as the intracellular cytosolic Ca^{2+} concentration is coupled to PI(4,5)P$_2$ turnover (48). PI(4,5)P$_2$ hydrolysis by phospholipase C produces two signaling second messengers, inositol (1,4,5) triphosphate (IP3) and diacylglycerol. The diffusible product, IP3, binds and activates the IP3 receptor (IP3R) (49), which leads to release of Ca^{2+} from intracellular Ca^{2+} stores gated by the transmembrane channel formed by multimers of the IP3R protein. Emerging studies suggest that induction of host cell $PI(4,5)P_2$ turnover, activation of IP3R, and elevation of cytosolic Ca^{2+} are components of Gag function in facilitating virus release (L.S. Ehrlich, G.N. Medina, M.B. Khan, M.D. Powell, K. Mikoshiba & C.A. Carter, manuscript submitted).

Three isoforms of the IP3 receptor protein are expressed from three independent genes. T cells and macrophages express different IP3R isoforms, raising the possibility that this cellular machinery might also participate in budding in a cell-type-dependent manner (86, 87). According to Hattori et al. (33), IP3R types 1 and 3 play distinct roles in Ca^{2+} signaling and contribute to the pool of cytosolic Ca^{2+} in different manners. The type 1 IP3R is believed to set the threshold for the global Ca^{2+} response irrespective of its expression level. The type 3 IP3R is thought to control the activation state of types 1 and 2 by regulating the amount of Ca^{2+} in their surrounding area through the Ca^{2+} binding sites in the protein. Thus, the differential expression of the subtypes in T cells and macrophages might regulate the site of membrane docking or the efficiency of the budding process at that site.

Annexin 2, a Ca^{2+}-dependent protein that binds both Gag and $PI(4,5)P_2$, is believed to be important for infectious virus production in macrophages but not in T cells (73, 77). Through its interaction with $PI(4,5)P_2$, annexin 2 functions in membrane transport, exocytosis, actin polymerization, and clustering of lipid rafts. The annexin 2–Gag interaction appears to occur at the limiting membranes of LEs/MVBs. siRNA-targeted inhibition of gene expression did not result in a striking inhibition of virus release; however, proteolytic processing of Gag and incorporation of CD63 were inefficient and the released virions were not infectious. Thus, as an effector of Ca^{2+} signaling, annexin 2 may be another cellular factor that contributes to the cell type specificity of virus assembly events in the macrophage. Another calcium-dependent protein, ERK, which mediates p6 Gag phosphorylation required for release (34), is among several genes that are upregulated in HIV-infected macrophages (23).

Protein Composition of Virions

An antibody-based analysis of isolated virus showed a strong concordance between the protein composition of exosomes derived from dendritic cells (89) and virus released from macrophages (55). The recent proteomic and biochemical analysis of MDM-derived HIV-1 particles, purified using a technique optimized for removal of microvesicle and exosome contaminants, also supports the view that viral particles are released from macrophages through an exosomal pathway (**Figure 6**). Chertova et al. (8) found that the purified virions contained many common endosomal and MVB proteins, including 26 of 37 proteins previously found in exosomes. Hundreds of additional proteins not previously reported to be constituents of virus particles isolated from nonmacrophage cell types also were detected. Some of these, such as the S100 family of Ca^{2+} binding proteins and the Rab, Rac, and Rho GTPases, have known links to endosome- MVB-associated events. Annexin 2 and IP3R also were included among the proteins detected in the Chertova study. The multiplicity of incorporated host proteins and that many of the proteins are also detected in exosomes and in virions released from T cells confirm the complex nature of the virion assembly and release process.

FUTURE DIRECTIONS

Within the past decade, there has been a virtual explosion of new information pertaining to the mechanism by which HIV-1 infects cells and traffics to the site of assembly and release. The unique feature of infection that characterizes HIV-1 replication in the macrophage, i.e., the long-term persistence, appeared to agree well with the notion that the virus is sequestered within intracellular MVBs and released through an exosomal pathway. However, this hypothesis does not account for recent studies that provide compelling evidence that HIV-1 buds from the plasma membrane in both T cells and macrophages. If the budding site is not dependent on cell type, a new explanation is needed to reconcile the defining properties of the HIV-macrophage relationship. Future studies will also need to clarify the roles of the many viral and cellular determinants of productive infection that have been discovered, and efforts must be made to identify those determinants that make critical contributions to viral replication in macrophages and T cells. Some of these factors may be those that are incorporated into the mature virion but whose roles in establishment of infection in the target cell have yet to be elucidated. One intriguing candidate is Tsg101, a factor that binds Gag and plays a critical role in trafficking and release of assembled particles in the producer cell (20, 45, 93). Tsg101 is incorporated into virions released from macrophages and T cells (8); it localizes in the nucleus (100) and at the midbody of dividing cells (6, 50); it may modulate cell cycle progression (103); and it binds to and stabilizes cyclin-dependent kinase inhibitor 1A (CDKN1A/p21) (58), a known facilitator of macrophage infection (91). The postassembly role of this protein and others incorporated into the virion will most certainly serve as focal points for future investigations in the quest for discovery of novel antiviral strategies that may affect infection of the macrophage and its contribution to HIV-1 pathogenesis.

SUMMARY POINTS

1. In contrast to T cells, macrophages are more resistant to the cytopathic effects of HIV, and as a result, the macrophage produces and harbors the virus for a longer period, contributing significantly to dissemination throughout the host and person-to-person transmission.

2. Several aspects of macrophage infection are unique compared with those of T cells, including coreceptor selection upon entry, viral protein-mediated access to the nucleus, and preferential modulation of expression of cellular factors that function in transcriptional regulation and cell survival. During the productive phase of the infection, the major viral structural precursor polyprotein, Gag, may be targeted to different membrane compartments in macrophages and T cells through interplay of viral and cellular determinants.

3. A historically defined feature of HIV replication in the macrophage that is thought to be unique to this cell type is the accumulation of viral particles in LEs/MVBs. It has been widely assumed that this is the basis for the longevity of the HIV-infected macrophage despite continuous virus production. A model based on this view (the Trojan exosome hypothesis) holds that virus release occurs when LEs/MVBs traffic to the cell surface, fuse with the plasma membrane, and release their contents into the extracellular environment as exosomes.

4. Recent reports indicate that viral assembly can occur in LEs/MVBs of nonhematopoietic cells. Even more importantly, emerging studies suggest that budding and accumulation in LEs/MVBs may be rare in infected macrophages in vivo. Most confounding, new studies indicate that the plasma membrane may be the primary site of virus assembly and budding in both T cells and macrophages.

DISCLOSURE STATEMENT

The authors are not aware of any biases that might be perceived as affecting the objectivity of this review.

ACKNOWLEDGMENTS

We thank G. Medina for thoughtful comments and editing of the manuscript. LSE is supported by National Institutes of Health grant R01 AI 68463 (to CC).

LITERATURE CITED

1. Alkhatib G, Berger EA. 2007. HIV coreceptors: from discovery and designation to new paradigms and promise. *Eur. J. Med. Res.* 12:375–84
2. Bache KG, Brech A, Mehlum A, Stenmark H. 2003. Hrs regulates multivesicular body formation via ESCRT recruitment to endosomes. *J. Cell Biol.* 162:435–42
3. Booth AM, Fang Y, Fallon JK, Yang JM, Hildreth JE, Gould SJ. 2006. Exosomes and HIV Gag bud from endosome-like domains of the T cell plasma membrane. *J. Cell Biol.* 172:923–35
4. Bosch B, Clotet-Codina I, Blanco J, Pauls E, Coma G, et al. 2006. Inhibition of human immunodeficiency virus type 1 infection in macrophages by an alpha-v-integrin blocking antibody. *Antiviral Res.* 69:173–80
5. Brown A, Zhang H, Lopez P, Pardo CA, Gartner S. 2006. In vitro modeling of the HIV-macrophage reservoir. *J. Leukoc. Biol.* 80:1127–35
6. Carlton JG, Martin-Serrano J. 2007. Parallels between cytokinesis and retroviral budding: a role for the ESCRT machinery. *Science* 316:1908–12
7. Cassol E, Alfano M, Biswas P, Poli G. 2006. Monocyte-derived macrophages and myeloid cell lines as targets of HIV-1 replication and persistence. *J. Leukoc. Biol.* 80:1018–30
8. **Chertova E, Chertov O, Coren LV, Roser JD, Trubey CM, et al. 2006. Proteomic and biochemical analysis of purified human immunodeficiency virus type 1 produced from infected monocyte-derived macrophages. *J. Virol.* 80:9039–52**

8. Proteomic and biochemical characterization demonstrating that MDM-derived HIV contains ~70% of the proteins previously found in exosomes as well as unique proteins.

9. Coberley CR, Kohler JJ, Brown JN, Oshier JT, Baker HV, et al. 2004. Impact on genetic networks in human macrophages by a CCR5 strain of human immunodeficiency virus type 1. *J. Virol.* 78:11477–86

10. Demirov DG, Freed EO. 2004. Retrovirus budding. *Virus Res.* 106:87–102

11. Demirov DG, Orenstein JM, Freed EO. 2002. The late domain of human immunodeficiency virus type 1 p6 promotes virus release in a cell type-dependent manner. *J. Virol.* 76:105–17

12. Deneka M, Pelchen-Matthews A, Byland R, Ruiz-Mateos E, Marsh M. 2007. In macrophages, HIV-1 assembles into an intracellular plasma membrane domain containing the tetraspanins CD81, CD9, and CD53. *J. Cell Biol.* 177:329–41

13. Dettenhofer M, Yu XF. 1999. Proline residues in human immunodeficiency virus type 1 p6(Gag) exert a cell-type dependent effect on virus replication and virion incorporation of Pol proteins. *J. Virol.* 73:4696–704

14. Dianzani F, Castilletti C, Gentile M, Gelderblom H, Frezza F, Capobianchi MR. 1998. Effects of IFN-alpha on late stages of HIV-1 replication cycle. *Biochimie* 80:745–54

15. Dong X, Li H, Derdowski A, Ding L, Burnett A, et al. 2005. AP-3 directs the intracellular trafficking of HIV-1 Gag and plays a key role in particle assembly. *Cell* 120:663–74

16. Fader CM, Savina A, Sánchez D, Colombo MI. 2005. Exosome secretion and red cell maturation: exploring molecular components involved in the docking and fusion of multivesicular bodies in K562 cells. *Blood Cells Mol. Dis.* 35:153–57

17. Fassati A. 2006. HIV infection of nondividing cells: a divisive problem. *Retrovirology* 3:74–88

18. Finzi A, Brunet A, Xiao Y, Thibodeau J, Cohen EA. 2006. Major histocompatibility complex class II molecules promote human immunodeficiency virus type 1 assembly and budding to late endosomal/multivesicular body compartments. *J. Virol.* 80:9789–97

19. Garcia E, Pion M, Pelchen-Matthews A, Collinson L, Arrighi JF, et al. 2005. HIV-1 trafficking to the dendritic cell-T-cell infectious synapse uses a pathway of tetraspanin sorting to the immunological synapse. *Traffic* 6:488–501

20. Garrus JE, von Schwedler UK, Pornillos OW, Morham SG, Zavitz KH, et al. 2001. Tsg101 and the vacuolar protein sorting pathway are essential for HIV-1 budding. *Cell* 107:55–65

21. Gartner S, Markovits P, Markovitz DM, Kaplan MH, Gallo RC, Popovic M. 1986. The role of mononuclear phagocytes in HTLV-III/LAV infection. *Science* 233:215–19

22. Gendelman HE, Baca L, Turpin JA, Kalter DC, Hansen BD, et al. 1990. Restriction of HIV replication in infected T cells and monocytes by interferon-alpha. *AIDS Res. Hum. Retrovir.* 6:1045–49

23. Giri MS, Nebozhyn M, Showe L, Montaner LJ. 2006. Microarray data on gene modulation by HIV-1 in immune cells: 2000–2006. *J. Leukoc. Biol.* 80:1031–43

24. Goff A, Ehrlich LS, Cohen SN, Carter CA. 2003. Tsg101 control of human immunodeficiency virus type 1 Gag trafficking and release. *J. Virol.* 77:9173–82

25. Goff S. 2007. Host factors exploited by retroviruses. *Nat. Rev. Microbiol.* 5:253–63

26. Goodenow MM, Collman RG. 2006. HIV-1 coreceptor preference is distinct from target cell tropism: a dual-parameter nomenclature to define viral phenotypes. *J. Leukoc. Biol.* 80:965–72

27. Gottlinger HG, Dorfman T, Sodroski JG, Haseltine WA. 1991. Effect of mutations affecting the p6Gag protein on human immunodeficiency virus particle release. *Proc. Natl. Acad. Sci. USA* 88:3195–99

28. Gould SJ, Booth AM, Hildreth JE. 2003. The Trojan exosome hypothesis. *Proc. Natl. Acad. Sci. USA* 100:10592–97

29. Gousset K, Ablan SD, Coren LV, Ono A, Soheilian F, et al. 2008. Real-time visualization of HIV-1 GAG trafficking in infected macrophages. *PLoS Pathog.* 4:E1000015

30. Grigorov B, Arcanger F, Roingeard P, Darlix JL, Muriaux D. 2006. Assembly of infectious HIV-1 in human epithelial and T-lymphoblastic cell lines. *J. Mol. Biol.* 359:848–62

31. Harila K, Prior I, Sjöberg M, Salminen A, Hinkula J, Suomalainen M. 2006. Vpu and Tsg101 regulate intracellular targeting of the human immunodeficiency virus type 1 core protein precursor Pr55gag. *J. Virol.* 80:3765–72

32. Harris RS, Bishop KN, Sheehy AM, Craig HM, Petersen-Mahrt SK, et al. 2003. DNA deamination mediates innate immunity to retroviral infection. *Cell* 113:803–9

33. Hattori M, Suzuki AZ, Higo T, Miyauchi H, Michikawa T, et al. 2004. Distinct roles of inositol 1,4,5-trisphosphate receptor types 1 and 3 in Ca^{2+} signaling. *J. Biol. Chem.* 279:11967–75

12. Provides evidence that the plasma membrane is the primary site of HIV-1 budding in infected macrophages.

18. Suggests that MHC-II promotes HIV assembly and budding into LE/MVBs and that this occurs naturally in macrophages, which physiologically express MHC-II molecules.

28. Proposes the Trojan exosome hypothesis.

34. Hemonnot B, Cartier C, Gay B, Rebuffat S, Bardy M, et al. 2004. The host cell MAP kinase ERK2 regulates viral assembly and release by phosphorylating the p6 Gag protein of HIV-1. *J. Biol. Chem.* 279:32426–34

35. Hurley JH, Emr SD. 2006. The ESCRT complexes: structure and mechanism of a membrane-trafficking network. *Annu. Rev. Biophys. Biomol. Struct.* 35:277–98

36. Ikonen E. 2001. Roles of lipid rafts in membrane transport. *Curr. Opin. Cell Biol.* 13:470–77

37. Jacque JM, Stevenson M. 2006. The inner-nuclear-envelope protein emerin regulates HIV-1 infectivity. *Nature* 441:641–45

38. Jarrosson-Wuilleme L, Goujon C, Bernaud J, Rigal D, Darlix JL, et al. 2006. Transduction of nondividing human macrophages with gammaretrovirus-derived vectors. *J. Virol.* 80:1152–59

39. Joshi A, Freed EO. 2007. HIV-1 Gag trafficking. *Future HIV Ther.* 1:427–38

40. Jouvenet N, Neil SJ, Bess C, Johnson MC, Virgen CA, et al. 2006. Plasma membrane is the site of productive HIV-1 particle assembly. *PLoS Biol.* 4:e435

41. Katzmann DJ, Stefan CJ, Babst M, Emr SD. 2003. Vps27 recruits ESCRT machinery to endosomes during MVB sorting. *J. Cell Biol.* 162:413–23

42. Lindwasser OW, Resh MD. 2004. Human immunodeficiency virus type 1 Gag contains a dileucine-like motif that regulates association with multivesicular bodies. *J. Virol.* 78:6013–23

43. Lu Q, Hope LW, Brasch M, Reinhard C, Cohen SN. 2003. TSG101 interaction with HRS mediates endosomal trafficking and receptor down-regulation. *Proc. Natl. Acad. Sci. USA* 100:7626–31

44. Martin-Serrano J, Yarovoy A, Perez-Caballero D, Bieniasz PD. 2003. Divergent retroviral late-budding domains recruit vacuolar protein sorting factors by using alternative adaptor proteins. *Proc. Natl. Acad. Sci. USA* 100:12414–19

45. Martin-Serrano J, Zang T, Bieniasz PD. 2001. HIV-1 and Ebola virus encode small peptide motifs that recruit Tsg101 to sites of particle assembly to facilitate egress. *Nat. Med.* 7:1313–19

46. McDonald D, Wu L, Bohks SM, KewalRamani VN, Unutmaz D, Hope TJ. 2003. Recruitment of HIV and its receptors to dendritic cell-T cell junctions. *Science* 300:1295–97

47. Mervis RJ, Ahmad N, Lillehoj EP, Raum MG, Salazar FH, et al. 1988. The *gag* gene products of human immunodeficiency virus type 1: alignment within the *gag* open reading frame, identification of posttranslational modifications, and evidence for alternative gag precursors. *J. Virol.* 62:3993–4002

48. Mikoshiba K. 2006. Inositol 1,4,5-triphosphate (IP3) receptors and their role in neuronal cell function. *J. Neurochem.* 97:1627–33

49. Mikoshiba K. 2007. IP3 receptor/Ca^{2+} channel: from discovery to new signaling concepts. *J. Neurochem.* 102:1426–46

50. Morita E, Sandrin V, Chung HY, Morham SG, Gygi SP, et al. 2007. Human ESCRT and ALIX proteins interact with proteins of the midbody and function in cytokinesis. *EMBO J.* 26:4215–27

51. Morita E, Sundquist WI. 2004. Retrovirus budding. *Annu. Rev. Cell Dev. Biol.* 20:395–425

52. Mulky A, Cohen TV, Kozlov SV, Korbei B, Foisner R, et al. 2008. The LEM domain proteins emerin and LAP2α are dispensable for HIV-1 and MLV infection. *J. Virol.* 82:5860–68

53. Muller B, Patschinsky T, Krausslich HG. 2002. The late domain-containing p6 is the predominant phosphoprotein of human immunodeficiency virus type 1 particles. *J. Virol.* 76:1015–24

54. Neil SJ, Eastman SW, Jouvenet N, Bieniasz PD. 2006. HIV-1 Vpu promotes release and prevents endocytosis of nascent retrovirus particles from the plasma membrane. *PLoS Pathog.* 2:e39

55. Nguyen DG, Booth A, Gould SJ, Hildreth JE. 2003. Evidence that HIV budding in primary macrophages occurs through the exosome release pathway. *J. Biol. Chem.* 278:52347–54

56. Nydegger S, Foti M, Derdowski A, Spearman P, Thali M. 2003. HIV-1 egress is gated through late endosomal membranes. *Traffic* 4:902–10

57. Nydegger S, Khurana S, Krementsov DN, Foti M, Thali M. 2006. Mapping of tetraspanin-enriched microdomains that can function as gateways for HIV-1. *J. Cell Biol.* 173:795–807

58. Oh H, Mammucari C, Nenci A, Cabodi S, Cohen SN, Dotto GP. 2002. Negative regulation of cell growth and differentiation by TSG101 through association with p21(Cip1/WAF1). *Proc. Natl. Acad. Sci. USA* 99:5430–35

59. Okumura A, Lu G, Pitha-Rowe I, Pitha PM. 2006. Innate antiviral response targets HIV-1 release by the induction of ubiquitin-like protein ISG15. *Proc. Natl. Acad. Sci. USA* 103:1440–45

60. Ono A, Ablan SD, Lockett SJ, Nagashima K, Freed EO. 2004. Phosphatidylinositol (4,5) bisphosphate regulates HIV-1 Gag targeting to the plasma membrane. *Proc. Natl. Acad. Sci. USA* 101:14889–94

61. Ono A, Freed EO. 2004. Cell type-dependent targeting of human immunodeficiency virus type 1 assembly to plasma membrane and the multivesicular body. *J. Virol.* 78:1552–68

62. Orenstein JM. 2007. Replication of HIV-1 in vivo and in vitro. *Ultrastruct. Pathol.* 31:151–67

63. Orenstein JM, Meltzer MS, Phipps T, Gendelman HE. 1988. Cytoplasmic assembly and accumulation of human immunodeficiency virus types 1 and 2 in recombinant human colony-stimulating factor-1-treated human monocytes: an ultrastructural study. *J. Virol.* 62:2578–86

64. Ott DE, Coren LV, Copeland TD, Kane BP, Johnson DG, et al. 1998. Ubiquitin is covalently attached to the p6Gag proteins of human immunodeficiency virus type 1 and simian immunodeficiency virus and to the p12Gag protein of Moloney murine leukemia virus. *J. Virol.* 72:2962–68

65. Pelchen-Matthews A, Kramer B, Marsh M. 2003. Infectious HIV-1 assembles in late endosomes in primary macrophages. *J. Cell Biol.* 162:443–55

66. Peng G, Lei KJ, Jin W, Greenwell-Wild T, Wahl SM. 2006. Induction of APOBEC3 family proteins, a defensive maneuver underlying interferon-induced anti-HIV activity. *J. Exp. Med.* 203:41–46

67. Perlman M, Resh MD. 2006. Identification of an intracellular trafficking and assembly pathway for HIV-1 Gag. *Traffic* 7:731–45

68. Perno CF, Svicher V, Schols D, Pollicita M, Balzarini J, Stefano A. 2006. Therapeutic strategies towards HIV-1 infection in macrophages. *Antiviral Res.* 71:293–300

69. Poli G, Orenstein JM, Kinter A, Folks TM, Fauci AS. 1989. Interferon-alpha but not AZT suppresses HIV expression in chronically infected cell lines. *Science* 244:575–77

70. Popov S, Popova E, Inoue M, Göttlinger HG. 2008. Human immunodeficiency virus type 1 gag engages the Bro1 domain of ALIX/AIP1 through nucleocapsid. *J. Virol.* 82:1389–98

71. Poole E, Strappe P, Mok HP, Hicks R, Lever AM. 2005. HIV-1 Gag-RNA interaction occurs at a perinuclear/centrosomal site; analysis by confocal microscopy and FRET. *Traffic* 6:741–55

72. Raposo G, Moore M, Innes D, Leijendekker R, Leigh-Brown A, et al. 2002. Human macrophages accumulate HIV-1 particles in MHC II compartments. *Traffic* 3:718–29

73. Rescher U, Gerke V. 2004. Annexins—unique membrane binding proteins with diverse functions. *J. Cell Sci.* 117:2631–39

74. Resh MD. 2005. Intracellular trafficking of HIV-1 Gag: how Gag interacts with cell membranes and makes viral particles. *AIDS Rev.* 7:84–91

75. Rodriguez-Boulan E, Paskiet KT, Sabatini DD. 1983. Assembly of enveloped viruses in Madin-Darby canine kidney cells: polarized budding from single attached cells and from clusters of cells in suspension. *J. Cell Biol.* 96:866–74

76. Rudner L, Nydegger S, Coren LV, Nagashima K, Thali M, Ott DE. 2005. Dynamic fluorescent imaging of human immunodeficiency virus type 1 gag in live cells by biarsenical labeling. *J. Virol.* 79:4055–65

77. Ryzhova EV, Vos RM, Albright AV, Harrist AV, Harvey T, Gonzalez-Scarano F. 2006. Annexin 2: a novel human immunodeficiency virus type 1 Gag binding protein involved in replication in monocyte-derived macrophages. *J. Virol.* 80:2694–704

78. Saad JS, Miller J, Tai J, Kim A, Ghanam RH, Summers MF. 2006. Structural basis for targeting HIV-1 Gag proteins to the plasma membrane for virus assembly. *Proc. Natl. Acad. Sci. USA* 103:11364–69

79. Savina A, Fader CM, Damiani MT, Colombo MI. 2005. Rab11 promotes docking and fusion of multivesicular bodies in a calcium-dependent manner. *Traffic* 6:131–43

80. Savina A, Furlán M, Vidal M, Colombo MI. 2003. Exosome release is regulated by a calcium-dependent mechanism in K562 cells. *J. Biol. Chem.* 278:20083–90

81. Sharova N, Swingler C, Sharkey M, Stevenson M. 2005. Macrophages archive HIV-1 virions for dissemination in trans. *EMBO J.* 24:2481–89

82. Sherer NM, Lehmann MJ, Jimenez-Soto LF, Ingmundson A, Horner SM, et al. 2003. Visualization of retroviral replication in living cells reveals budding into multivesicular bodies. *Traffic* 4:785–801

83. Shun MC, Daigle JE, Vandegraaff N, Engelman A. 2007. Wild-type levels of human immunodeficiency virus type 1 infectivity in the absence of cellular emerin protein. *J. Virol.* 81:166–72

84. Stoorvogel W, Kleijmeer MJ, Geuze HJ, Raposo G. 2002. The biogenesis and functions of exosomes. *Traffic* 3:321–30

62. Ultrastructural examination of macrophages infected with HIV-1 in vitro and in vivo: results challenge the Trojan exosome hypothesis.

85. Strack B, Calistri A, Craig S, Popova E, Göttlinger HG. 2003. AIP1/ALIX is a binding partner for HIV-1 p6 and EIAV p9 functioning in virus budding. *Cell* 114:689–99

86. Sugiyama T, Furuya A, Monkawa T, Yamamoto-Hino M, Satoh S, et al. 1994. Monoclonal antibodies distinctively recognizing the subtypes of inositol 1,4,5-trisphosphate receptor: application to the studies on inflammatory cells. *FEBS Lett.* 354:149–54

87. Sugiyama T, Yamamoto-Hino M, Miyawaki A, Furuichi T, Mikoshiba K, Hasegawa M. 1994. Subtypes of inositol 1,4,5 trisphosphate receptor in human hematopoietic cell lines: dynamic aspects of their cell-type specific expression. *FEBS Lett.* 349:191–96

88. Takenawa T, Itoh T. 2001. Phosphoinositides, key molecules for regulation of actin cytoskeletal organization and membrane traffic from the plasma membrane. *Biochim. Biophys. Acta* 1533:190–206

89. Thery C, Boussac M, Veron P, Ricciardi-Castagnoli P, Raposo G, et al. 2001. Proteomic analysis of dendritic cell-derived exosomes: a secreted subcellular compartment distinct from apoptotic vesicles. *J. Immunol.* 166:7309–18

90. Valiathan RR, Resh MD. 2004. Expression of human immunodeficiency virus type 1 gag modulates ligand-induced downregulation of EGF receptor. *J. Virol.* 78:12386–94

91. Vazquez N, Greenwell-Wild T, Marinos NJ, Swaim WD, Nares S, et al. 2005. HIV-1 induced macrophage gene expression includes p21, a target for viral regulation. *J. Virol.* 79:4479–91

92. Verani A, Gras G, Pancino G. 2005. Macrophages and HIV-1: dangerous liaisons. *Mol. Immunol.* 42:195–212

93. VerPlank L, Bouamr F, LaGrassa TJ, Agresta B, Kikonyogo A, et al. 2001. Tsg101, a homologue of ubiquitin-conjugating (E2) enzymes, binds the L domain in HIV type 1 Pr55(Gag). *Proc. Natl. Acad. Sci. USA* 98:7724–29

94. von Schwedler UK, Stuchell M, Müller B, Ward DM, Chung HY, et al. 2003. The protein network of HIV budding. *Cell* 114:701–13

95. Wahl SM, Greenwell-Wild T, Peng G, Ma G, Orenstein JM, Vazquez N. 2003. Viral and host cofactors facilitate HIV-1 replication in macrophages. *J. Leukoc. Biol.* 74:726–35

96. Wahl SM, Greenwell-Wild T, Vázquez N. 2006. HIV accomplices and adversaries in macrophage infection. *J. Leukoc. Biol.* 80:973–83

97. Welsch S, Habermann A, Jager S, Muller B, Krijnse-Locker J, et al. 2006. Ultrastructural analysis of ESCRT proteins suggests a role for endosome-associated tubular-vesicular membranes in ESCRT function. *Traffic* 7:1551–66

98. Welsch S, Keppler OT, Habermann A, Allespach I, Krijnse-Locker J, Krausslich HG. 2007. HIV buds predominantly at the plasma membrane of primary human macrophages. *PLoS Pathog.* 3:e36

99. Wills JW, Craven RC. 1991. Form, function and use of retroviral gag proteins. *AIDS* 5:639–54

100. Xie W, Li L, Cohen SN. 1998. Cell cycle-dependent subcellular localization of the TSG101 protein and mitotic and nuclear abnormalities associated with TSG101 deficiency. *Proc. Natl. Acad. Sci. USA* 95:1595–600

101. Yamashita M, Emerman M. 2006. Retroviral infection of nondividing cells: old and new perspectives. *Virol.* 344:88–93

102. Yamashita M, Perez O, Hope TJ, Emerman M. 2007. Evidence for direct involvement of the capsid protein in HIV infection of nondividing cells. *PLoS Pathog.* 3:1502–10

103. Zhong Q, Chen Y, Jones D, Lee WH. 1998. Perturbation of TSG101 protein affects cell cycle progression. *Cancer Res.* 58:2699–702

104. Zou W, Kim JH, Handidu A, Li X, Kim KI, et al. 2007. Microarray analysis reveals that type 1 interferon strongly increases the expression of immune-response related genes in Ubp43 (Usp18) deficient macrophages. *Biochem. Biophys. Res. Commun.* 356:193–99

105. Shkriabai N, Datta SAK, Zhao Z, Rein A, et al. 2006. Interactions of HIV-1 Gag with assembly cofactors. *Biochemistry* 45:4077–83

97. A systematic ultrastructural analysis of ESCRT protein localization in T cells and macrophages that suggests cell-type-dependent localization of these proteins.

98. Provides evidence that the plasma membrane is the primary site of HIV-1 budding in infected macrophages.

102. Evidence that CA contributes to HIV's ability to infect nondividing cells like the macrophage and that the underlying mechanism is novel compared with that employed by the other viral proteins involved.

Antigenic Variation in *Plasmodium falciparum*

Artur Scherf, Jose Juan Lopez-Rubio, and Loïc Riviere

Biology of Host-Parasite Interactions Unit, CNRS URA2581, Institut Pasteur 75724 Paris, France; email: ascherf@pasteur.fr, jjlopez@pasteur.fr, lriviere@pasteur.fr

Annu. Rev. Microbiol. 2008. 62:445–70

The *Annual Review of Microbiology* is online at micro.annualreviews.org

This article's doi: 10.1146/annurev.micro.61.080706.093134

Key Words

var gene, epigenetic silencing and memory, telomere position effect, immune escape, merozoite phenotypic variation

Abstract

The persistence of the human malaria parasite *Plasmodium falciparum* during blood stage proliferation in its host depends on the successive expression of variant molecules at the surface of infected erythrocytes. This variation is mediated by the differential control of a family of surface molecules termed PfEMP1 encoded by approximately 60 *var* genes. Each individual parasite expresses a single *var* gene at a time, maintaining all other members of the family in a transcriptionally silent state. PfEMP1/*var* enables parasitized erythrocytes to adhere within the microvasculature, resulting in severe disease. This review highlights key regulatory mechanisms thought to be critical for monoallelic expression of *var* genes. Antigenic variation is orchestrated by epigenetic factors including monoallelic *var* transcription at separate spatial domains at the nuclear periphery, differential histone marks on otherwise identical *var* genes, and *var* silencing mediated by telomeric heterochromatin. In addition, controversies surrounding *var* genetic elements in antigenic variation are discussed.

Contents

INTRODUCTION

Phenotypic variation is a major survival strategy utilized by many pathogens that develop within a hostile environment such as the host's immune response (12). One of the most fascinating immune evasion strategies is the coordinated expression of variant surface molecules, which diminishes the immune clearance by the host and allows the establishment of prolonged chronic infections with successive waves of parasitemia (a process called antigenic variation). Recurring infection caused by a pathogen indicates the presence of sophisticated immune escape mechanisms.

Historic Overview

Plasmodium species infect and proliferate in the vertebrate host circulation system for extended periods, suggesting that they have developed common ways to escape immune destruction. The first indication of antigenic variation in a *Plasmodium* species was put forward by Cox (24), who observed relapse parasite populations in mice infected with *P. berghei*. Some years later, Brown & Brown (15) demonstrated the presence of antigenically distinct *P. knowlesi*–infected erythrocyte (IE) populations in chronically infected rhesus monkeys. Subsequently, antigenic variation was demonstrated for other nonhuman malaria species such as *P. chabaudi* in mice (87) and *P. fragile* in toque monkeys (57). In human malaria, relapse parasite populations were observed in primary *P. falciparum* infections inoculated for malaria therapy (92, 93). Experimental evidence for antigenic variation in *P. falciparum* was first shown in vivo using squirrel monkeys for experimental blood stage infection (59). The host's immune response to *P. falciparum* blood stage infection is determined largely by the immune system's recognition of the immunodominant and clonally variant surface molecule called PfEMP1 (*P. falciparum* erythrocyte membrane protein 1) (78). Genome sequence analysis of another human malaria species, *P. vivax*, revealed the existence of a large subtelomeric gene family that is expressed in a clonally variant manner at the surface of IEs (30). Rodent malaria models, despite their great potential for the study of antigenic variation in the natural host-parasite context, have lost some of their appeal as a model to study *P. falciparum* PfEMP1-mediated pathogenesis, because it became evident that the genes encoding this virulence factor (encoded by the *var* multigene family) are found only in *P. falciparum* (and in the closely related chimpanzee malaria

Phenotypic variation: the process by which a pathogen alters its surface proteins to adapt to changing host conditions

species *P. reichenowi*) (64). The complete life cycle of *P. falciparum* is shown in **Figure 1a**.

Since the discovery of the *var* gene family in 1995 (7, 112, 115), the literature on *var* genes has virtually exploded, leaving many researchers interested in malaria overwhelmed by the increasing complexity of this area. A number of excellent recent reviews have focused on the genomic organization and diversity of *var* genes (70, 74), how they traffic to the surface (119, 122), and their role in pathogenesis (50, 82, 91). Here, we focus mainly on the underlying molecular mechanism leading to variant surface antigen expression at the *P. falciparum* IE. We also review the latest data suggesting that merozoites may be involved in immune escape.

ANTIGENIC VARIATION AND PATHOGENESIS

Phenotypic variation can give rise to functionally different PfEMP1 adhesion ligands while changing the antigenic makeup of a molecule sufficiently to decrease or abolish recognition by the existing antibody response. Thus, functional diversity and immune escape are combined in the same surface IE molecules in *P. falciparum*. It is the functional diversity that may be a key factor in pathogenesis (for review see Reference 91). The type of adherence trait expressed by IEs may determine disease severity. A paradigm for adhesion-mediated pathogenesis is malaria during pregnancy (47), in which a single member of the *var* gene repertoire (called *var2CSA*) mediates adhesion to placental chondroitin sulfate A (33, 123). Expression of other adhesion phenotypes, such as Rosetting or intercellular adhesion molecule 1 (ICAM-1) binding (126) or particular *var* gene types (65), has also been linked to severe disease but the data are far less compelling owing to restricted access to patient organs for the investigation of the physiopathology process (95). These variant adhesion molecules are key players in the parasite's survival strategy of IEs, making them important targets for various intervention strategies.

Variant Surface Molecules and Immunity

It has been assumed that PfEMP1 is the major target of the protective antibody response (16, 49, 85, 113). With the discovery of novel variant gene families exported to the erythrocyte membrane, further studies are in progress to explore whether other surface molecules contribute in the protective immune response. Although PfEMP1 can induce a protective immune response in an experimental monkey model, switching expression to another variant yielded a new wave of parasitemia (6). Thus, PfEMP1 vaccines could prevent a particular form of the disease such as placental malaria, but they would not be suitable to protect, for example, a tourist from getting infected in malaria endemic regions. Even if the anti-severe-disease PfEMP1 vaccine concept appears feasible, obtaining strain-transcending immunity may present another major obstacle, given the high diversity between *var* gene repertoires in field isolates (see Generation of *var* Gene Repertoire Diversity, below).

Erythrocytes: The Perfect Hiding Place for Malaria Parasites?

The reason why intracellular malaria parasites have evolved to expose immunodominant antigens at the surface of the IE is not yet clear. In all malaria species analyzed, parasite-derived proteins are exposed at the IE surface, even in those species that survive without sequestration. It is generally accepted that *P. falciparum* mature blood stage forms drastically increase the erythrocyte rigidity (32), but apparently to a much lesser degree than other human malaria species such as *P. vivax*. Thus, it is assumed that these rigid IE would be trapped and destroyed in the spleen as abnormal or senescent cells. In vitro data emerged showing that parasite erythrocyte surface molecules can mediate interaction with different types of human immune cells, indicating that this interaction may lead to immune dysregulation (2, 3, 9, 31, 121). The use of transgenic parasites demonstrated that

Antigenic variation: the process by which a pathogen alters its surface proteins in order to evade a host immune response

IE: infected erythrocyte

PfEMP1: *P. falciparum* erythrocyte membrane protein 1

the expression of the immunodominant surface ligand PfEMP1 is linked to the suppression of host interferon-gamma (mainly by gamma delta T cells) in the early immune response against parasites in vitro (26). If this function operates in vivo, one may speculate that low interferon-gamma levels may improve parasite survival (38). Many malaria researchers, however, believe that the discussed functions may not be sufficient to explain the evolution of the IE to expose parasite molecules at its surface. An alternative hypothesis was presented by Saul (108), who proposed that a strong selective force could be due to erythrocytes, which

a **Life cycle of *Plasmodium falciparum***

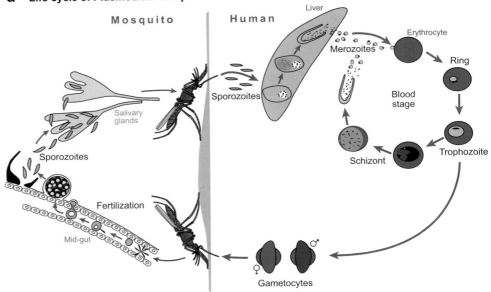

b **Variant parasite molecules trafficked to the erythrocyte membrane**

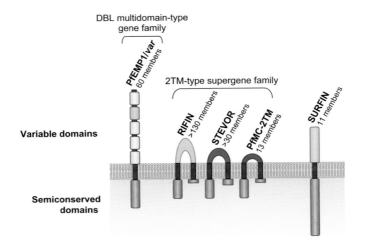

are not able to signal the presence of intracellular parasites to the host immune system (no major histocompatibility complex expression). The perfect hiding place within erythrocytes may have been an impasse for malaria parasites, because unchecked parasite proliferation may have killed the host before it could efficiently transmit itself to a mosquito vector. Rapid control of parasite proliferation may be achieved by exposing IEs to the immune system via the surface expression of a parasite antigen. However, to avoid elimination by the host immune response, sequential expression of related but variant surface proteins may solve the problem. Fine-tuned switch rates could enable prolonged infection and transmit the disease over long periods.

VARIANT MOLECULES EXPRESSED AT THE SURFACE OF INFECTED ERYTHROCYTES

Of the four malaria species that infect humans, only *P. falciparum* can be maintained in in vitro culture for prolonged periods (120). This method has opened the way for biochemical studies of variant antigens expressed at the surface of IEs. PfEMP1 was identified as a large polymorphic parasite-derived molecule at the surface of mature IEs using surface iodination experiments (78) and was linked in later studies to antigenic variation and cytoadhesion of mature IEs to endothelial cells (8, 105). Although the gene encoding PfEMP1 was on the list of "most wanted" virulence factors, its identity was unveiled ~10 years later and the gene family that codes for PfEMP1 was given the name *var* (7, 112, 115). The *P. falciparum* genome sequence analysis revealed that only a relatively small number of *var* genes (~60) per haploid genome serve as a reservoir in the immune evasion process (51). Characteristics of PfEMP1/*var* are shown in **Table 1**.

Do other surface-exposed non-PfEMP1 epitopes contribute to protective immunity and adhesion? In the search for novel surface antigens, genome sequence analysis and surface iodination led to the discovery of a second clonally variant gene family called *rif* (repetitive interspersed family) (39, 75) (**Figure 1b**). The trypsin-sensitive RIFIN antigen family of 30–45 kDa has a two-transmembrane (2TM) topology, which predicts that a short variant loop of ~170 amino acids is exposed at the erythrocyte surface. So far, no biological function can be associated to RIFIN and there is no

---

Figure 1

Life cycle of *Plasmodium falciparum* and variant gene families expressed during blood stage cycle. (*a*) When the female *Anopheles* mosquito takes a blood meal, sporozoites are injected into the human skin and infection starts with the invasion of liver cells. After several rounds of cell divisions, new parasite forms called merozoites are released into the bloodstream, where they invade erythrocytes. Merozoites subsequently develop into ring, trophozoite, and schizont stage parasites. Trophozoite-infected erythrocytes express parasite-derived adhesion surface molecules, resulting in sequestration of trophozoite and schizont stages in the deep vasculature. After completion of the blood stage cycle (48 h) new daughter cells (merozoites) are released that rapidly invade noninfected erythrocytes. Blood stage parasites establish chronic infections using antigenic variation and are responsible for all malaria symptoms. When certain forms of blood stage parasites (gametocytes) are picked up by a female *Anopheles* mosquito during a blood meal, they start another, different cycle of growth and multiplication in the mosquito. After 10–18 days, the parasites are found (as sporozoites) in the mosquito's salivary glands ready to infect another human host. (*b*) Schematic representation of gene families coding for variant molecules associated with the surface of *P. falciparum*–infected erythrocytes. All molecules shown are trafficked via vesicular structures (Maurer's clefts) to the erythrocyte membrane. The immunodominant PfEMP1/*var* molecule apparently is the first parasite antigen exposed at the erythrocyte surface (~18 h postinvasion) and is responsible for the adhesion of IE to microvasculature. Experimental data point to PfEMP1/*var* as the driving force for antigenic variation in *P. falciparum*. Members of the other gene families are apparently expressed at later moments of mature stages compared with the parasite adhesion to endothelial cells. Like *var* genes, *rif*, *stevor*, and *Pfmc-2TM* undergo clonal variation, but their biological role remains to be determined.

Table 1 *Plasmodium falciparum* variant gene families expressed at the erythrocyte membrane

	var	*rif*	*stevor*	*Pfmc-2TM*	*surf*
Function	Cytoadherence Immune escape Immunomodulation	Unknown	Unknown	Unknown	Unknown
Subcellular location	MC IE surface	MC/PV IE surface Merozoite	MC/PV IE surface Merozoite	MC/PV IE surface	MC/PV IE surface Merozoite
Molecular weight (kDa)	200–360 kDa	30–45 kDa	30–40 kDa	27 kDa	280–300 kDa
mRNA expression (hours postinvasion)	3–18	12–27	22–32	18–30	<6–40
Clonal variation	Yes	Yes	Yes	Yes	Not known
PRBC surface exposure:					
Antibody	18–48 h postinvasion	Not shown	Not shown	Not shown	Not shown
Surface iodination	Yes	Yes	Not shown	Not shown	No
Surface biotinylation	No	Not shown	Not shown	Not shown	Yes
Trypsin sensitive	Yes	Yes	Not shown	Not shown	Yes
Variable domain	≈150–320 kDa	≈17 kDa	≈12 kDa	≈1.7 kDa	≈100 kDa

Abbreviations: MC: Maurer's clefts; PV: parasitophorous vacuole; IE: infected erythrocyte.

published evidence that the variant loop is accessible to antibodies. The Malaria Genome Project has revealed a number of novel gene families that are preferentially located in different subtelomeric regions (51). Among these are two gene families that are exported to the erythrocyte membrane and belong to the 2TM superfamily. *stevor* has a structure similar to that of *rif* genes but is expressed much later in the blood stage cycle than PfEMP1 and RIFIN (for more details see **Table 1**) (66). The second family has been called *Pfmc-2TM* and has been associated with the erythrocyte membrane (76, 107). *stevor* and *Pfmc-2TM* are clonally variant, and their switch rates are similar to that of *var* genes (77). These data do support the idea that the hypervariable regions of the 2TM-type superfamily are under direct immune pressure and participate in antigenic variation. Both STEVOR and PfMC-2TM are released from Maurer's clefts, which are parasite-derived membranous structures in the IE cytoplasm, to the erythrocyte membrane (11, 76). However, there is no experimental evidence that these molecules contribute to the surface epitopes recognized by the variant-specific antibody immune response.

A proteomic approach of trypsin-released erythrocyte surface molecules identified a novel polymorphic gene family called *surf* (surface-associated interspersed genes) (127). One member of this small multigene family was co-transported with PfEMP1 and other variant molecules to the IE surface. It remains unknown whether the extracellular domain can be a target of the protective immune response and if switching in *surf* expression occurs.

A puzzling question arises from these recent observations: Once the IE sequestration in the deep microvasculature via PfEMP1 is established, why are other variant molecules expressed at the erythrocyte surface? We have much to learn of the intimate relationship between the parasite and the host cell.

Apart from PfEMP1, no compelling surface reactivity with antibodies has been shown for the other variant surface antigen families. Transcriptome and proteome analyses of clinical isolates, however, have revealed differences between the expression profile of exported gene products and that of laboratory parasites (27, 42, 48), suggesting that some variant molecules may be expressed higher in malaria patients. In addition, an intriguing idea is emerging that indicates that *rif* and *stevor* may contribute to immune escape of merozoites.

GENOME ORGANIZATION AND NUCLEAR ARCHITECTURE OF *var* GENES

Complete *P. falciparum* genome sequence analysis confirmed earlier observations locating *var* genes to the highly polymorphic chromosome end regions and a few central chromosome regions (51, 58, 106, 115). Approximately 60% of *var* members locate adjacent to a noncoding subtelomeric region composed of a mosaic of six distinct telomere-associated repeat elements (*TARE 1–6*) (**Figure 2a**). One to three *var* genes exist at telomeres pointing in different directions, either in tail-to-tail (most frequent), head-to-tail, or head-to-head orientation, and this location has been observed in different *P. falciparum* genomes (69). The central *var* genes are tandem arrayed in the head-to-tail orientation (51). The chromosomal location and orientation of *var* genes are associated with

a Organization of *Plasmodium falciparum var* genes

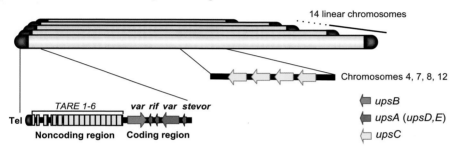

b *var* gene features

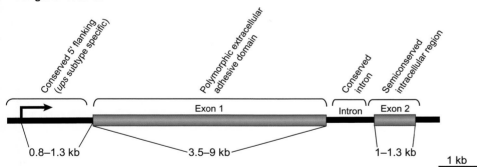

Figure 2

Genomic organization and nuclear position of *var* genes. (*a*) Subtelomeric regions of all 14 linear chromosomes have a common overall organization, with noncoding repeats of variable size called *TARE 1–6* (telomere-associated repeat elements) located next to telomere repeats. This region is followed by a member of the *var* gene family, frequently in combination with a second *var* gene transcribed in the opposite direction (toward the telomere). Several tandem-arrayed *var* gene clusters are found in central chromosome positions. *rif* genes are often interspersed with *var* genes. Chromosomal positions define one of the three major types of 5′-flanking regions (called *ups*) of *var* genes. (*b*) Common features between members of the *var* gene family are shown. All members contain two exons. The polymorphic extracellular domain (exon 1) contains variable numbers of Duffy-binding-like (DBL) adhesive domains, which can be categorized into five major types (alpha–epsilon) as well as cysteine-rich interdomain regions. The intracellular domain (exon 2) encodes a semiconserved amino acid terminal segment (70).

particular types of 5′-flanking regions called *ups*. The genes transcribed toward the centromere from intact chromosome ends generally have similar 5′-flanking gene sequences and are called the *upsB* type. Telomeric *var* genes that point toward the telomere are mostly of the *upsA* type. The *upsC* type is always associated with internal *var* clusters. In addition, some internal *var* genes are associated with *upsB*. Despite differences in the 5′-flanking region, all

var genes studied follow the rule of mutually exclusive expression. Characteristics of *var* genes are shown in **Figure 2b**. A comparative and exhaustive analysis of *var* genes from three different laboratory strains [3D7, HB3, and IT (FCR3)] showed that the overall organization of *var* genes is conserved in genetically distinct parasites (69). It remains unclear whether the orientation plays a role in the switching order and regulation of expression; however, certain telomeric *vars* (*upsA*- and *upsE*-type *var*) are linked to disease severity.

Can a small *var* repertoire provide enough genetic diversity for prolonged chronic infection observed in individual patients (92, 93)? Random sequence analysis of a *var* gene exon 1 region [Duffy-binding-like (DBL)-alpha domain] from genetically distinct parasites indicated that *var* repertoires may differ between parasites (73, 117). When a large set of *var* DBL-alpha probes from the HB3 parasites were hybridized with two other genetically distinct parasites (3D7 and Dd2), these parasites had only a limited overlap of *var* repertoire (**Figure 3a**) (45). This idea was

a   Minimal overlap between *var* gene repertoires

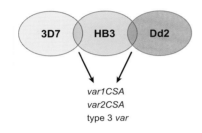

var1CSA
var2CSA
type 3 *var*

b   Nuclear architecture of telomeric and central *var* genes

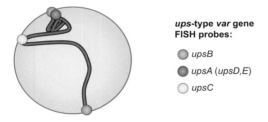

***ups*-type *var* gene FISH probes:**

- *upsB*
- *upsA* (*upsD,E*)
- *upsC*

c   Schematic view of telomere cluster

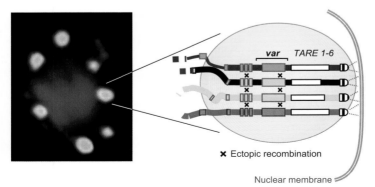

✗ Ectopic recombination

Nuclear membrane

Figure 3

var repertoires tend to show minimal overlap between different strains. (*a*) Hybridization of *var* gene probes from HB3 parasites revealed a low level of cross-hybridization among genotypically distinct laboratory-cultured parasites (3D7 and Dd2) (45). Only a limited set of semiconserved *var* genes (*var1CSA, var2CSA,* and type 3 *var*) was found when three sequenced *Plasmodium falciparum* genomes were compared (69). (*b*) Physical positioning of *var* genes in ring stage nucleus. All *var* genes analyzed localize to the nuclear periphery. Telomeric members form clusters with other telomeric *var* genes, whereas central *var* genes locate either outside of the cluster (103) or are linked to telomeric cluster. (*c*) Nuclear architecture of *var* genes and a model of enhanced recombination. FISH analysis shows that telomeric *var* genes are organized into foci in the nuclear periphery. The physical alignment of homologous subtelomeric regions involves the *TARE 1–6* elements (41) and most likely proteins bound to this region (86). Experimental evidence supports a role of these foci in *var* gene conversion events.

confirmed by comparing the complete set of sequenced *var* repertoires of 3D7, HB3, and IT (FCR3) (69) and large-sequence analysis of *var* DBL-alpha domain obtained from malaria patients from widespread geographic origins (1, 5). Taken together, these data suggest that the observed successive clinical episodes in malaria patients, which are caused by genetically distinct parasites (22), may depend on the *var* repertoire diversity of the parasite causing new infections.

Does the parasite's genomic organization of *var* gene or its nuclear architecture tell us something about the molecular mechanism that generates this enormous diversity? The overall structure of subtelomeric regions (including noncoding and coding regions) is conserved between the 28 chromosome ends, with the exception of subtelomeric chromosome truncations, which occasionally occur in laboratory-cultured parasites. Heterologous chromosomes carry large stretches of homologous DNA regions (50 to 80 kb) at their chromosome ends. *P. falciparum* subtelomeric DNA sequences are not found in malaria species other than *P. reichenowi* (64). Gene synteny between different malaria species is high for central chromosome regions but is low for subtelomeric regions (67), supporting the importance of this chromosome end compartment in species-specific host adaptations and the DNA diversification process.

Generation of *var* Gene Repertoire Diversity

FISH (fluorescent in situ hybridization) analysis showed that *P. falciparum* chromosomes do not randomly occupy the nucleus. Telomeric *var* genes localize to the nuclear periphery and form foci of four to seven chromosome ends (45). Surprisingly, internal *var* genes also locate to the perinuclear space (103) (see **Figure 3b**). One obvious result of physical tethering of homologous regions is the enhanced recombination between these sequences. Experimental data analyzing progeny clones from two genetic crosses supported the high rate of recombination in *var* genes leading to new *var* forms via

gene conversion (45, 117). Thus, the nuclear architecture may determine the hot spots of recombination in malaria parasites (**Figure 3c**). It is conceivable that *var* repertoires continually evolve during blood stage development, allowing an extended chronic phase in its host.

GENETIC ELEMENTS THAT CONTROL MONOALLELIC EXPRESSION

Although earlier work showed that mutually exclusive transcription of a single member of the *var* gene family is apparently under the control of epigenetic factors (110), one key question is which genetic elements nucleate the chromatin changes leading to monoallelic expression? A rather surprising observation implicating the *var* intron in the silencing process was made by Wellems and colleagues (28). A *luciferase* gene was constitutively transcribed in a transient transfection of *P. falciparum* blood stage parasites unless a *var* intron, placed at the 3′ end of the gene, led to complete repression of transcription in an S-phase-dependent manner (called pairing of *var* promoter and intron). Further analysis of the *var* intron-mediated silencing process showed that the intron has promoter activity (17, 37) (**Figure 4a**).

Work conducted in the Cowman laboratory (124, 125) suggested that a single noncoding region located upstream of a *var* gene (*upsB* and *upsC* types, of approximately 1.5 to 2 kb) detached from its native *var* gene (including the *var* intron) is sufficient to establish the *var* counting mechanism (124, 125). In this study, a *var* promoter activated by a drug-selectable marker, blasticidin deaminase (*bsd*), or human dihydrofolate reductase (*hdhfr*) can cooperate with the counting mechanism even when the *var* promoter is activated from a stable transfected episome. No coding *var* region was necessary to establish features linked to native endogenous *var* gene expression: Transcriptional activation occurs at the nuclear periphery, it silences all endogenous *var* genes, and it is stably inherited by the next generations in a manner similar to that of *var* genes

Counting mechanism: mutually exclusive expression of a single gene from within a multicopy gene family

hdhfr: human dihydrofolate reductase gene

a *var* silencing intron-mediated repression hypothesis

Transient transfection

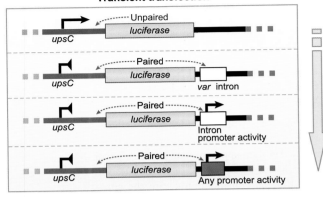

b Episomal activation of a *var* promoter

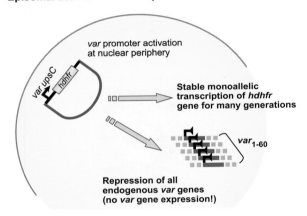

c *var* counting paired promoter hypothesis

Stable transfection

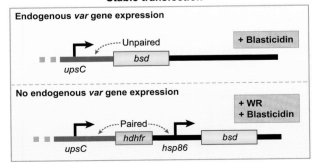

(**Figure 4***b*). The importance of the promoter region and the dispensable *var* antigen production for mutually exclusive *var* expression were confirmed independently by the Deitsch laboratory (36). The controversial data from the Cowman laboratory about the role of intron in silencing has stimulated new experiments, which have been published. It was shown that a *var* promoter repression similar to the intron-mediated silencing could be obtained by placing an active blood stage promoter at the 3′ end of the *bsd* or *luciferase* gene (**Figure 4***a*) (37). This could explain the intron-free silencing data observed in the constructs of the Cowman group (124, 125). Another appealing aspect of the work by Dzikowski et al. (37) is that any unpaired *var* promoter activity uncouples it from the parasite-counting machinery. This leads to an endogenous *var* gene (or second episomal paired *var* promoter) expressed simultaneously with an unpaired *var* promoter driving a selectable drug marker or *luciferase* gene (37, 125) (summarized in **Figure 4***c*). In summary, results obtained from constructs, which try to imitate the natural *var* gene transcription regulation, give support to the notion that *var* promoter pairing may play a role in silencing and the counting mechanism.

Figure 4

Genetic elements involved in the control of antigenic variation of *var* genes. (*a*) Constructs for transient *luciferase* reporter assays under the control of a *var* promoter (*upsC*). Pairing the promoter with a *var* intron (which has promoter activity) (17) or with another promoter (37) suggests a role of the intron in *var* gene silencing. (*b*) Stable transfection of a construct carrying a transcriptionally active *var* promoter (*upsC*) under the control of a drug-selectable marker (*hdhfr*) was used to analyze *var* promoter activity in the silent state (−drug) or active state (+ drug). The active *var* promoter on the plasmid is sufficient to promote monoallelic expression at the nuclear periphery and to nucleate chromatin-mediating stable inheritance of the active *var* promoter (124). (*c*) Different constructs of *var* promoters activated by a selectable drug marker gene (*bsd* and/or *hdhfr*), supporting the role of the intron in the *var* counting mechanism (37).

This *var* intron model has been challenged by studies investigating endogenous *var* genes. Two reports that investigated the transcription of a *var* gene in the native context do not support the intron model shown in **Figure 4a,c.** The first study did investigate the same *var* gene in the active and silent states (using tiled oligonucleotides along the *var2CSA* gene locus) (**Figure 5a**). An oligonucleotide specific for the *var2CSA* intron-adjacent exon 2 region did not show any significant transcription during the silent state of the gene at any moment during the life cycle stage. Thus, no intron promoter activity was detectable that could be linked to *var2CSA* gene repression. Analysis of other *var* exon 2 transcripts revealed that some but not all silent *var* genes transcribe exon 2 (102). Thus, the absence of sterile transcripts (unable to code for proteins) in some of the silent *var* genes indicates that continuous intron promoter activity is not an absolute necessity for silencing.

The second study showed that pairing a *var* promoter in its natural location with another promoter of a selectable drug marker gene approximately 4 kb downstream of the transcriptional start site leads to a constitutive transcription (truncated transcript of 7 kb) and uncoupling of the *var* promoter from mutually exclusive *var* transcription machinery (123) (**Figure 5b**). The opposite result might have been predicted on the basis of earlier described episomal *var* promoter studies. Thus, in the described context, the downstream promoter activity in *var2CSA* exon 1 could act as an antisilencer for the upstream *var* promoter, or alternatively the inserted selectable marker interrupts a distance-dependent *var*-silencing mechanism.

The latter study demonstrates that epigenetic *var* gene regulation is a fine-tuned mechanism that can easily be perturbed by changing the chromatin environment. Caution is needed when investigating epigenetic effects involved in *var* gene control. Ideally, an intron-less *var* gene in the chromosomal context may shed light on whether introns play a role in mutually exclusive expression.

a Promoter activity of a *var* intron in its native context

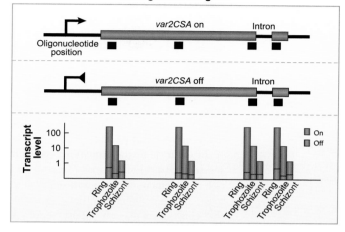

b Constitutive *var* promoter activity

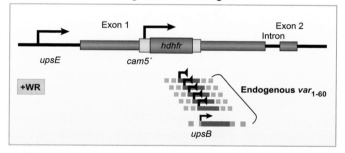

Figure 5

Analysis of an endogenous *var* gene does not support the postulated role of the *var* intron in silencing and counting. (*a*) Transcriptional analysis of tiled oligonucleotides along the *var2CSA* gene in an active and repressed state. Microarray data show the transcriptional upregulation of exon 1 and exon 2 of *var2CSA* in the active state but not in the silent state. No transcriptional activity was observed for exon 2 in the silent state, indicating that no constitutive intron promoter activity is needed for the silencing of this *var* gene (102). (*b*) A calmodulin promoter (*cam5′*) driving a selectable marker (*hdhfr*) inserted into the *var2CSA* gene leads to the constitutive activation of the *var2CSA* promoter, which results in the detachment from the *var* counting mechanism (123). The situation resembles the paired promoter described in **Figure 4c** but shows the opposite phenotype.

EPIGENETIC FACTORS ORCHESTRATE ANTIGENIC VARIATION

Although the molecular mechanisms that control mutually exclusive expression are far from understood, the central role for epigenetics in this process is becoming clear (110). *var* genes are transcribed by RNA polymerase II (72, 111), unlike the *vsg* genes in *Trypanosoma brucei*, and their expression is developmentally controlled throughout asexual blood stage. The highest level of *var* transcription occurred in early-ring-stage parasites (4–10 h postinvasion), whereas in later-stage parasites the transcriptional activity is silent (20, 72, 111). Regulation occurs at the transcription initiation level in a monoallelic manner (72, 110, 111).

In contrast to the paucity of recognizable specific transcription factors, *P. falciparum* has a rich complement of chromatin-modifying proteins, and this observation has fueled the idea that epigenetics play an important role in the control of gene expression. In this section we focus on the increasing number of studies showing the potential roles for chromatin structure and the effects of nuclear spatial organization and noncoding RNA (ncRNA) on monoallelic *var* gene expression.

Eukaryotic genomes are packaged into nucleosomes, whose position and chemical modification state can profoundly influence regulation of gene expression. *P. falciparum* possesses a rich but largely unexplored histone-modifying machinery (89, 109, 116, 118), and recent publications have pointed out its involvement in transcription regulation.

The first evidence linking histone modifications to *var* gene regulation comes from the work of Freitas-Junior et al. (46). Chromatin immunoprecipitation (ChIP) showed that the upstream regions of active *var* genes are associated with a reversible modification type (acetylation) of histone H4, whereas the histones associated with silenced *var* genes are hypoacetylated. Recent studies considerably refine the picture of chromatin modifications along the *var* gene in its various transcriptional states. For this analysis, a *var* gene with a unique 5′-flanking region (*var2CSA* gene) was chosen, allowing the investigation of chromatin changes in the active and silent states. ChIP analysis showed that the 5′-flanking region of the actively transcribed *var* gene during the ring stage is highly enriched for H3K4 di- and trimethylation (H3K4me2 and H3K4me3, respectively) (80). As mentioned above, the active *var* gene is transiently repressed at later stages in the cycle but remains in a state poised for transcription; the same *var* gene will be expressed in the next asexual life cycle. Lopez-Rubio et al. (80) showed that the poised *var* gene substantially loses the enrichment in H3K4me3, whereas H3K4me2 is maintained, presumably bookmarking this member for reactivation in the next cycle (**Figure 6a**).

In contrast, stably silenced *var* genes are associated with H3K9 trimethylation (H3K9me3) in the 5′ upstream (80) and coding regions (21, 80) during the entire asexual life cycle. H3K9me3 has been strongly correlated with gene silencing by heterochromatin in

Figure 6

Epigenetic factors implicated in *var* gene regulation. (*a*) Schematic of histone modifications associated with control of expression and epigenetic memory of *var* genes. Three different transcriptional states of *var* genes: Actively transcribed *var* gene in ring stages (*blue*) is associated with H3K4me2, H3K4me3, and H3K9ac in 5′-flanking regions. In the trophozoite/schizont stage the previously active *var* gene is transiently silenced (poised) (*blue*) but maintains the enrichment of H3K4me2. This association may transmit memory of the active *var* gene during cell division. Stably silenced *var* genes (5′-flanking region and exon 1) (*red*) are enriched in H3K9me3 throughout the asexual life cycle. (*b*) Protein domains present in *Plasmodium falciparum* that can recognize methylated lysines (chromo, tudor, and PHD) or acetylated lysines (bromo). Examples for methylation of H3K4 and acetylation of H3K9 are shown. (*c*) Schematic summary of epigenetic factors involved in antigenic variation. Nucleus-determined molecular events leading to monoallelic *var* infected erythrocyte surface expression in *P. falciparum*. Three major phenotypes and epigenetic factors involved in their control are shown. The role of the host immune system in the selection of variants and its potential participation in switching rate modulation are outlined. Other potential epigenetic factors contributing to antigenic variation are shown.

other eukaryotes, but this residue could also be acetylated H3K9ac, an epigenetic mark associated with active genomic regions. 5′-flanking regions of actively transcribed *var* genes are enriched in H3K9ac but not in H3K9me3, whereas in late coding regions of the same *var* gene there is enrichment in H3K9me3 but not in H3K9ac. This opposite distribution between marks is observed along the entire *var* gene in the repressed state (80). These results suggest a

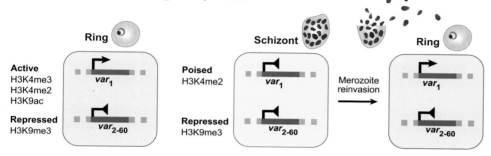

a Histone marks linked to *var* gene expression

Ring   Schizont   Ring

Active
H3K4me3
H3K4me2
H3K9ac   *var*$_1$

Repressed
H3K9me3   *var*$_{2-60}$

Poised
H3K4me2   *var*$_1$

Repressed
H3K9me3   *var*$_{2-60}$

Merozoite reinvasion

var$_1$

var$_{2-60}$

b Histone mark reading machinery in *Plasmodium falciparum*

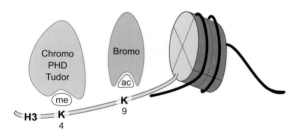

Chromo
PHD
Tudor

Bromo

(ac)

(me)

H3 — K$_4$ K$_9$

c Epigenetic factors contributing to antigenic variation

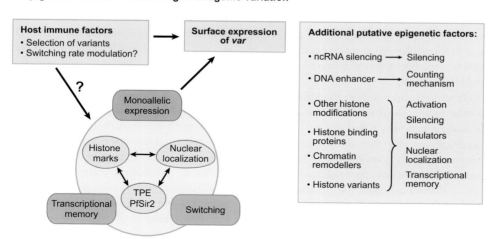

Host immune factors
• Selection of variants
• Switching rate modulation?

Surface expression of *var*

?

Monoallelic expression

Histone marks ⟷ Nuclear localization

TPE
PfSir2

Transcriptional memory   Switching

Additional putative epigenetic factors:

• ncRNA silencing ⟶ Silencing

• DNA enhancer ⟶ Counting mechanism

• Other histone modifications

• Histone binding proteins

• Chromatin remodellers

• Histone variants

Activation
Silencing
Insulators
Nuclear localization
Transcriptional memory

competition between acetylation and methylation at H3K9 at the 5′-flanking region of *var* genes to influence transcription activity and, together with the H3K4 methylation observations, underline the main role for this region (the 5′ upstream) in *var* gene regulation. It is of interest to explore whether all *ups var* subtypes behave in the same manner. Other histone modifications may possibly be implicated, providing a combination of marks dedicated to *var* monoallelic expression. Additional work is necessary to explore this hypothesis.

An active *var* gene locus is maintained over many life cycles. Inheritance of chromatin states (both on and off states) is still an active area of investigation. One hypothesis is that histone methylation at active genes could serve as a long-term signature for the propagation of the active state through cell division and maintain a state of specific cell differentiation (97). These marks could be faithfully replicated during DNA synthesis, enabling their epigenetic inheritance. Therefore, the enrichment of H3K4me2 and H3K9me3 observed in *var* loci throughout the life cycle could be a mark that stably transmits active and inactive *var* states for many generations.

Several studies link histone variants and ATP-dependent chromatin remodelers, both of which are present in *Plasmodium* species, with transcriptional memory (14, 71). It may be of interest to determine whether these factors are implicated in the maintenance of the virulence gene program of the parasite. Another mechanism of epigenetic inheritance in unicellular and multicellular organisms is methylation of DNA bases, but this modification is absent in malaria parasites.

Posttranslational histone modifications play a role in the regulation of gene expression in malaria parasites. Combination of covalent modifications may thus encode information, which most likely is interpreted by downstream modules that translate the modification pattern into a distinct outcome (68). These histone-binding factors (or reading machinery) harbor different protein domains that can recognize specific marks. Our understanding of how this histone code is written, modified, and read in *Plasmodium* species is poor. The *Plasmodium* genome database analysis points out the large number of proteins that may exert these functions (56, 63, 109, 116, 118), namely, methyltransferases, demethylases, acetyltransferases, deacetylases, and proteins containing domains that participate in the recognition of specific histone modifications (chromo, bromo, PHD, and tudor domains). One example of an effector protein with a potential homolog in the parasite is the heterochromatin protein 1 (HP1). HP1 possesses a chromo domain that can associate with methylated H3K9 and mediate gene silencing (**Figure 6b**). Because H3K9 methylation correlates with *var* gene silencing, PfHP1 could have a role in antigenic variation as suggested by work by the Hernandez-Rivas laboratory (R. Hernandez-Rivas, unpublished data).

In model systems, the nucleus seems to be divided into permissive and restrictive transcription compartments (52). In *Plasmodium*, several FISH studies indicate that differential subnuclear localization is implicated in *var* gene regulation. Silenced subtelomeric *var* genes associate with telomeric clusters at the nuclear periphery (35, 103, 124). Silenced chromosome central *var* genes also localize at the perinucleus (103) distinct from telomeric clusters, but another study localized a central *var* gene at a cluster (124). Activation of a *var* gene appears to require its relocation into a transcriptionally competent area, still in the nuclear periphery (35, 37, 86, 103). Whether the *var* expression site is outside the cluster (103) or linked to a telomeric cluster (86, 124) remains unclear. Nevertheless, the nuclear periphery is composed of distinct microenvironments with roles in either gene silencing or activation. Movements of *var* loci within the periphery would regulate their access to spatially restricted transcriptional machinery. Nuclear markers are needed that could help to better characterize this expression site.

Given that coactivation of neighboring *var* genes has not been observed, insulator elements that exist in *P. falciparum* most likely perform this function. It has been suggested that the intron could mediate this repression (44), but further work is necessary to investigate chromatin factors such as particular histone modifications or histone variants that may contribute to monoallelic expression.

Antisense transcription has been suggested as a general control mechanism for *Plasmodium* transcription (54, 90). Recent studies show the presence of ncRNA in blood stage parasites, supporting the hypothesis of a role for ncRNA in the assembly of heterochromatin and control of gene regulation (19, 79, 96). In *Schizosaccharomyces pombe* and higher eukaryotes, short double-stranded interfering RNAs have been implicated in transcriptional gene silencing, a process mediated by components of the RNA interference machinery (RNAi) and by H3K9 methylation (128). However, in *Saccharomyces cerevisiae* RNAi components and H3K9 methylation are absent, although gene repression can be mediated by the Sir complex and histone deacetylation. *P. falciparum* shares some characteristics with both models. PfSir2 (an ortholog to the *S. cerevisiae* protein, Sir2) is implicated in subtelomeric gene silencing, including subtelomeric *var* genes, and bioinformatic studies failed to identify RNAi gene candidates in *Plasmodium*, striking parallelisms with *S. cerevisiae*. Compared with *S. pombe*, H3K9 methylation is present in the parasite and is linked to *var* gene repression (21, 80). These particular features in *Plasmodium* emphasize the challenge of studying the epigenetic gene regulation in this organism, encouraging researchers to watch out for unique gene regulation mechanisms.

A novel silencing mechanism found in *S. cerevisiae* depends on antisense RNA stabilization and histone deacetylation (18). This process is distinct from the RNAi-mediated gene silencing described in *S. pombe* and other eukaryotes, although there are some parallelisms. Exploring whether this new silencing mechanism operates in *Plasmodium* would be of interest.

Finally, modulation of chromatin structure, nuclear architecture, and possibly ncRNA are working together to set up the precise transcriptional control of *var* genes (**Figure 6c**).

These processes could be specific to particular subclasses of *var* genes, for example, PfSir2-mediated silencing is restricted to the majority of subtelomeric *upsA*-type *var* genes, suggesting that activation of *var* genes silenced in a PfSir2-independent manner may occur with different switch rates. *Plasmodium* may also share some strategies for the regulation of antigenic variation with other protozoan parasites (81), such as the subnuclear expression site for *vsg* genes in African trypanosomes. A first genome-wide occupancy map of H3K9me3 and H3K9ac using ChIP-on-chip technology was published by Cui et al. (25). This study, although limited to coding regions, indicates that H3K9ac and H3K9m3 are regularly associated with active and silent *P. falciparum* blood stage genes, respectively. However, as shown for 5′-upstream regions of *var* genes (80), it will be crucial to include noncoding regions in subsequent arrays, a technical challenge in the case of the AT-rich *P. falciparum* genome. Epigenome analysis could reveal transcriptional activities of intergenic regions, which could have a role in differential gene regulation in *P. falciparum*.

Telomere Position Effect and Silencing of *var* Genes

Telomeres are not just the physical end of linear chromosomes, they exert essential functions, for example, in genome integrity and cell division (10, 40). One particular aspect relevant for this review is that telomeres recruit specific molecules that create a chromatin environment, which represses gene activity in a distance-dependent manner. This effect is also called the telomere position effect (TPE) and has been studied in detail in yeast (53, 94). The first indication of telomere-mediated repression in *P. falciparum* came from a study comparing transcription of one gene located 100 kb from telomere repeats and/or adjacent to telomere repeats (41). Quantitative RT-PCR analysis showed that transcription of the telomeric gene was about 50-fold lower than when the same gene was at the normal location. Telomere-associated *var* genes are distant

PfSir2: *P. falciparum* silent information regulator

from telomeres (20 to 35 kb), and *var* genes can be derepressed by the inactivation of the telomere-binding protein PfSir2 (35). In yeast, TPE is restricted to 3–5 kb adjacent to the telomere repeats. Apparently, PfSir2 spreads from telomere repeat DNA along *TARE 1–6* into the nearby coding regions including *var* gene promoters (46, 83). Although long considered nonfunctional junk DNA, *TARE 6* (also called rep20), a large repetitive element of 5 to 15 kb located adjacent to the first telomeric *var* gene, has a biological role in gene silencing. This DNA sequence may nucleate large amounts of PfSir2 and enhance the effect of TPE over a large distance to the coding regions. This idea is supported by two independent observations. First, *TARE 6* repeats are associated with PfSir2 in ChIP assays (46). Second, a drug-selectable marker integrated into the *TARE 6* region is repressed in a reversible manner (35). An alternative explanation of the effect of Pf-Sir2 on *var* gene repression could be a telomere fold-back mechanism that brings PfSir2 into the proximity of *var* genes (83). Given that Pf-Sir2 displays histone deacetylase activity (88), the enzymatic activity may modify histones at *var ups* sequences, leading to transcriptional repression.

Given that only a subset of telomeric *var* genes (mostly *upsA* type) was derepressed in PfSir2 mutant parasites, distinct mechanisms could control silencing of *upsB*- and *upsC*-type *var* genes. The nuclear periphery may recruit a number of different molecules required for the control of silencing of *var* gene subgroups. Further investigation of perinuclear chromatin is needed to gain a better insight into factors that control antigenic variation in malaria parasites.

SWITCHING OF *var* GENES

The *P. falciparum var* gene repertoire contains only 60 members, which is far less than the thousands of variant surface *vsg* genes of African trypanosomes. Consequently, a key question for chronic infection is how to keep antigen-switching at a rate that allows the emergence of a new variant without rapidly exhausting the entire arsenal. Several scenarios may be envisioned: The parasite has developed an intrinsic switch rate and order, which are adjusted to the time it takes for a protective antibody immune response to be mounted against the parasite by the host. Alternatively, the parasite may apply a more flexible strategy by means of a sensing machinery, which would change switch rates, depending on the presence of antibodies at the surface of the IE. Although some experimental data are compatible with a sensing mechanism in *P. knowlesi* (4), this has never been further investigated or demonstrated for other malaria species. A second key question is whether a switch order exists or is switching a random process? Experimentally, most studies are done on *P. falciparum* strains adapted to in vitro culture. Switch data on malaria patients are lacking and experimental models for *P. falciparum* in monkeys are expensive and mainly restricted to vaccine testing.

Molecular Mechanisms Underlying *var* Gene Switching

Malaria investigators have tried to find parallelisms with the existing paradigm on antigenic variation established in African trypanosomes. In fact, apparently all *var* genes are expressed by in situ activation from telomeric and central chromosome positions (110), whereas recombination of a silent copy into a telomeric expression site appears to be the major mechanism of switching in *T. brucei* (12, 84). In this organism, the switch order depends on the homology of the donor sequence and the expression site. Thus, switching of *var* genes in *P. falciparum* appears to rely only on epigenetic processes. Nevertheless, some reports established a relation between the switching and *var* gene deletion (29, 60, 62). Evidently, this type of switching is an accidental process due to the risk of rapidly exhausting the whole repertoire.

A rational approach to predict *var* switch orders is not possible because of our limited knowledge of the epigenetic processes that control antigenic variation. Experimental switch rates from laboratory parasites or, in very few cases, from infected patients have been obtained. Variant IE surface antigen switch rates in *P. falciparum* were first addressed in a study by Roberts et al. (105). The authors showed that parasites cultured in vitro in the absence of immune pressure switched spontaneously at a rate of 2%. This rate was in the range of that obtained for *P. chabaudi* switching in mice (approximately 1% to 0.04%) (13). These results illustrate that in vivo switch rates can vary drastically and that switch rates determined on cultured parasites reveal normally rapid switches.

Horrocks et al. (61) analyzed switch rate in a different manner. Parasites were selected for a particular variant adhesion phenotype and immediately cloned. These clones were used to measure both on rate (switching toward activation) and "off rate" (switching toward inactivation) switching of specific variants. The authors showed that each *var* gene possesses its own rate of switching. This implies the existence of a hierarchy of *var* gene expression, a kind of hard-wiring, contrary to the random switch rate model, which predicts that any variant can be expressed from a parental clone independent of the switch history (**Figure 7a**). The data from Horrocks et al. (61) indicated that no preferential switching between central or telomeric *var* genes occurs. Another study, however, suggested that central *var* genes switch more frequently than do subtelomeric *var* genes (43), indicating that gene location may be of importance. Their differing findings could be explained by the fact that both studies used distinct laboratory parasites known for their difference in phenotypic switching. It remains to be seen which type corresponds more to clinical isolates.

The first data of *var* switching in human malaria patients were obtained by using human volunteers infected with a cloned line (99). Several *var* transcripts dominated by a particular *var* type were observed. Switching measured by RT-PCR of transcribed *var* genes indicated rapid turnover of *var* transcripts (estimated switch rate around 16%), in contrast with the reported lower switching events in labora-

tory parasites (105). However, in the absence of quantitative RT-PCR these switch estimates remain open to discussion. The authors of this study proposed that the *var* switch process is reset, in other words, the same *var* gene is expressed as merozoites are released from the liver stage. Notably, this was also concluded in the *P. chabaudi* model (13). Would the reset to the same *var* gene in liver stage merozoites be advantageous in patients with a high

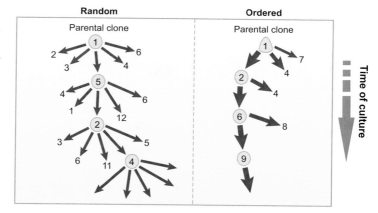

a **Model based on laboratory parasites**

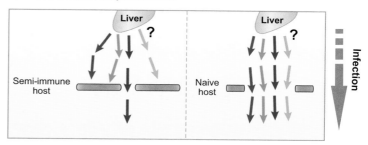

b **Model based on patient data**

Figure 7

Models for *var* gene switching in *Plasmodium falciparum*. (*a*) Schematic presentation of hypothetical switch modes. The random switching type predicts that *var* genes have similar switch rates. The ordered switching type postulates that each *var* gene possesses its own on and off rates of switching. Arrow thickness symbolizes the rate of switching, and numbers represent different *var* genes. (*b*) Model based on patient studies. Colored arrows represent different, expressed variants immediately after release of liver stage merozoites into the bloodstream. Either these merozoites are epigenetically programmed to express a number of different variants, or they are reset to one *var* member followed by high switch rates at the onset of blood stage infection. Both mechanisms may enhance infection in semi-immune persons.

level of immunity? In repeatedly infected hosts, liver stage merozoites expressing a distinct set of *var* genes may easily overcome existing immunity to a particular variant and establish a blood stage infection (**Figure 7b**). This hypothesis cannot be experimentally tested in *P. falciparum*, because hepatocyte cell lines do not produce viable merozoites, but this aspect may be addressed in mouse malaria species.

Mathematical Modeling

Finally, some mathematical calculations were designed to predict the switch rates that would be compatible with chronic infection. These models consider that variant IE surface antigens are the driving force for protective immunity. The modeling task is complicated by the difficulty of integrating different parameters of the host immune system (93, 98). The experimentally obtained switch rates for *P. falciparum* may result in the complete exhaustion of the *var* repertoire within a brief period (108). Most likely, high *var* switch rates apply only to some *var* genes, and the rate could be variable and probably much lower for many other *var* members.

A more recent mathematical model included a new parameter that could cope with the relatively high switch rates observed in laboratory parasites (104). The authors postulated that short-lived, partially cross-reactive immune responses to parasite-infected erythrocyte surface antigens could produce a cascade of sequentially dominant antigenic variants, each of which is immunologically distinct from its preceding types. This last point is striking because coordination of antigenic variation is provided by the host (55).

ANTIGENIC VARIATION IN MEROZOITES?

Phenotypic variation of merozoites was first discovered in the rodent malaria model *P. yoelii* (101). In this organism each merozoites originating from a single schizont expresses a distinct member of the Py235 multigene family. Py235 proteins are thought to determine the subset of erythrocytes that the parasites invade. In *P. falciparum*, invasion is determined by a related but small gene family called the reticulocyte-binding-like homolog (PfRh) genes (for details see **Table 2**). A second, small, invasion-related gene family is encoded by erythrocyte-binding antigen (EBA) genes. Members of both invasion-related families are expressed in a clonally variant manner (23, 34, 114). This may simply reflect the polymorphism on human erythrocytes and thus allow a broader spectrum of host cells to be infected. On the other hand, the capacity to switch on different members of the invasion pathway may also contribute to evading protective immune responses. A more puzzling observation is the mutually exclusive expression of two members of the merozoite rhoptry *clag* gene family, *clag3.1* and *clag3.2* (23). So far, these antigens have not been directly associated with the invasion process, but the recent data on *clag* suggest that phenotypic variation is used much more widely by the parasite than simply for immune evasion.

New observations raised the possibility that antigens encoded by variant multigene families such as *rif*, *surf*, and *stevor* have a role in merozoite biology. Members of each family associate with the apical end of merozoites, a region normally involved in the invasion process (11, 100, 127). Because at least two of these gene families are expressed in a clonally variant manner (**Table 2**), these variant antigens may camouflage conserved merozoite surface antigens required for the invasion process. This could protect the Achilles' heel of merozoites against immune attack by masking critical invasion molecules by variant proteins.

MALARIA *var*-IATIONS: FUTURE ORIENTATIONS

Thirteen years after the discovery of the antigenic variation genes we have gained considerable insight into the molecular process of

Table 2 Small variant gene families expressed in *Plasmodium falciparum* merozoites

	Rh	*Eba*	*RhopH1/clag*	*rif*	*stevor*	*surf*
Function in merozoites	Invasion pathway Binds to different erythrocyte receptors	Invasion pathway Some members bind glycophorin	Not shown	Not shown	Not shown	Not shown
Number of genes	6	4	5	>130	>30	10
Subcellular location of protein	Rhoptry neck	Microneme	Rhoptry protein	Apical region (RIFIN A-type) Merozoite cytosol (RIFIN B-type)	Apical region	Apical region
Phenotypic variation						
Altered invasion pathway	Yes	Yes	No	Not shown	Not shown	Not shown
Immune escape	Not shown	Not shown	No	Not shown	Not shown	Not shown
Differential expression of gene members	Yes (Rh2A, 2B, 4, 5)	Yes (eba140)	Yes (clag2, 3.1, 3.2)	Yes	Yes	Not shown
Antibody interferes with invasion	Yes	Yes	Not shown	Not shown	Not shown	Yes

Abbreviations: *Rh*: reticulocyte-binding homolog protein; *RhopH1*: high-molecular-mass rhoptry complex; *Eba*: erythrocyte-binding antigen.

immune evasion and pathogenesis in *P. falciparum*. At the individual parasite level (IE level) a strict molecular counting mechanism coordinates successive expression of different variants. The underlying molecular device, however, remains enigmatic not only in malaria parasites but also in other organisms that apply mutually exclusive expression of multigene families. Remembering expression of a particular *var* gene member for many parasite generations is maintained via epigenetic memory, which has been strongly linked to particular histone marks at 5′-flanking regions. A better insight into the writing and reading machinery may reveal how a single *var* gene is activated (by default, *var* genes are apparently repressed). Thus, the activation process from the default silent state is apparently the key event at the heart of understanding antigenic variation. We speculate

that a transenhancer could be the genetic element that controls the first step in the *var* activation procedure (81). A single copy element could solve the counting problem, because the postulated putative enhancer would associate only with a single *var* promoter at the same time.

The particular interest for antigenic variation is far beyond academic research activities given the molecule's key role in pathogenesis. Targeting the escape mechanism machinery appears to be a promising approach to tackle the parasite without having to deal with phenotypic variation of the surface molecule. The development of specific biological assays that target antigenic variation may uncover crucial genes required for *var* export to the IE surface, repression of the *var* gene family, or switching to new variants.

SUMMARY POINTS

1. *var* genes encode immunodominant clonally variant adhesion molecules (PfEMP1) expressed at the surface of IE. Adhesive interactions of IE with different host cells contribute to severe malaria and most likely to immune suppression.

2. Several distinct clonally variant molecules are expressed at the IE membrane, but their role in immune evasion remains unknown.

3. Mutually exclusive expression of a single *var* gene member is orchestrated by different epigenetic factors but does not require programmed DNA rearrangements.

4. The epigenetic memory of a monoallelic expressed *var* gene beyond mitosis and merozoite reinvasion is linked to a specific histone modification at the 5′-flanking region (H3K4 dimethylation), whereas stable repression is linked to H3K9 trimethylation along the 5′-noncoding and coding regions (exon 1).

5. A construct carrying a transcriptionally active *var* promoter (using a selectable marker gene) is transcribed at the nuclear periphery and promotes allelic exclusion of the parasite's endogenous *var* gene; no *var* product is required for asexual blood stage development.

6. The *var* intron-mediated silencing model has been challenged by several reports. Different experimental approaches have raised controversies, indicating the importance of the study of monoallelic expression in the chromosomal context.

7. *var* gene repertoires are diverse among genetically distinct parasites, and frequent gene conversion events between members of the family may be enhanced by the physical clustering of *var* genes in the nuclear periphery.

FUTURE ISSUES

1. Nuclear markers have yet to be developed to test the hypothesis that a specific *var* expression site exists and is required for monoallelic expression.

2. The reading machinery of the *P. falciparum* histone code that establishes heterochromatin regions and boundaries for the spreading of the transcribed *var* genes into neighboring heterochromatin should be explored.

3. Random mutagenesis transfection methods are in need of improvement to study the genes and genetic elements required for monoallelic expression and the silencing mechanism in endogenous *var* genes.

4. Researchers need to explore the role of ncRNA in the silencing process of clonally variant gene families.

5. The role of merozoite phenotypic variation in the parasite's immune escape needs further attention to understand better the biology of the non-*var* clonally variant gene families.

6. Analysis of the telomere-silencing mechanism is an important research avenue to explore the default *var*-silencing mechanism at the nuclear periphery.

7. The development of high-throughput screening assays may lead to the discovery of novel drugs that block the immune escape and/or the IE adhesion process.

DISCLOSURE STATEMENT

The authors are not aware of any biases that might be perceived as affecting the objectivity of this review.

ACKNOWLEDGMENTS

This work was supported by the European Commission (BioMalPar) contract No.: LSPH-CT-2004-503578 and by an ANR Microbiologie grant (No.: ANR-06-MIME-026-01) JJLR has financial support from the Human Frontier Science Program.

LITERATURE CITED

1. Albrecht L, Merino EF, Hoffmann EH, Ferreira MU, de Mattos Ferreira RG, et al. 2006. Extense variant gene family repertoire overlap in Western Amazon *Plasmodium falciparum* isolates. *Mol. Biochem. Parasitol.* 150:157–65

2. Baratin M, Roetynck S, Lepolard C, Falk C, Sawadogo S, et al. 2005. Natural killer cell and macrophage cooperation in MyD88-dependent innate responses to *Plasmodium falciparum*. *Proc. Natl. Acad. Sci. USA* 102:14747–52

3. Baratin M, Roetynck S, Pouvelle B, Lemmers C, Viebig NK, et al. 2007. Dissection of the role of PfEMP1 and ICAM-1 in the sensing of *Plasmodium falciparum*-infected erythrocytes by natural killer cells. *PLoS ONE* 2:e228

4. Barnwell JW, Howard RJ, Miller LH. 1983. Influence of the spleen on the expression of surface antigens on parasitized erythrocytes. *Ciba Found. Symp.* 94:117–36

5. Barry AE, Leliwa-Sytek A, Tavul L, Imrie H, Migot-Nabias F, et al. 2007. Population genomics of the immune evasion (var) genes of *Plasmodium falciparum*. *PLoS Pathog.* 3:e34

6. Baruch DI, Gamain B, Barnwell JW, Sullivan JS, Stowers A, et al. 2002. Immunization of *Aotus* monkeys with a functional domain of the *Plasmodium falciparum* variant antigen induces protection against a lethal parasite line. *Proc. Natl. Acad. Sci. USA* 99:3860–65

7. Baruch DI, Pasloske BL, Singh HB, Bi X, Ma XC, et al. 1995. Cloning of *P. falciparum* gene encoding PfEMP1, a malarial variant antigen and adherence receptor on the surface of parasitized human erythrocytes. *Cell* 82:77–87

8. Biggs BA, Anders RF, Dillon HE, Davern KM, Martin M, et al. 1992. Adherence of infected erythrocytes to venular endothelium selects for antigenic variants of *Plasmodium falciparum*. *J. Immunol.* 149:2047–54

9. Biswas AK, Hafiz A, Banerjee B, Kim KS, Datta K, Chitnis CE. 2007. *Plasmodium falciparum* uses gC1qR/HABP1/p32 as a receptor to bind to vascular endothelium and for platelet-mediated clumping. *PLoS Pathog.* 3:1271–80

10. Blasco MA, Gasser SM, Lingner J. 1999. Telomeres and telomerase. *Genes Dev.* 13:2353–59

11. Blythe JE, Yam XY, Kuss C, Bozdech Z, Holder AA, et al. 2008. *Plasmodium falciparum* STEVOR proteins are highly expressed in patient isolates and located in the surface membrane of infected red blood cells and the apical tip of merozoites. *Infect. Immun.* 76:3329–36

12. Borst P. 2003. Mechanisms of antigenic variation: an overview. In *Antigenic Variation*, ed. A Craig, A Scherf, pp. 1–16. London: Academic Press

13. Brannan LR, Turner CM, Phillips RS. 1994. Malaria parasites undergo antigenic variation at high rates in vivo. *Proc. Biol. Sci.* 256:71–75

14. Brickner DG, Cajigas I, Fondufe-Mittendorf Y, Ahmed S, Lee PC, et al. 2007. H2A.Z-mediated localization of genes at the nuclear periphery confers epigenetic memory of previous transcriptional state. *PLoS Biol.* 5:e81

15. Brown KN, Brown IN. 1965. Immunity to malaria antigenic variation in chronic infections of *Plasmodium knowlesi*. *Nature* 208:1286–88

16. Bull PC, Lowe BS, Kortok M, Molyneux CS, Newbold CI, Marsh K. 1998. Parasite antigens on the infected red cell surface are targets for naturally acquired immunity to malaria. *Nat. Med.* 4:358–60

17. Calderwood MS, Gannoun-Zaki L, Wellems TE, Deitsch KW. 2003. *Plasmodium falciparum var* genes are regulated by two regions with separate promoters, one upstream of the coding region and a second within the intron. *J. Biol. Chem.* 278:34125–32

18. Camblong J, Iglesias N, Fickentscher C, Dieppois G, Stutz F. 2007. Antisense RNA stabilization induces transcriptional gene silencing via histone deacetylation in *S. cerevisiae*. *Cell* 131:706–17

19. Chakrabarti K, Pearson M, Grate L, Sterne-Weiler T, Deans J, et al. 2007. Structural RNAs of known and unknown function identified in malaria parasites by comparative genomics and RNA analysis. *Rna* 13:1923–39

20. Chen Q, Fernandez V, Sundstrom A, Schlichtherle M, Datta S, et al. 1998. Developmental selection of *var* gene expression in *Plasmodium falciparum*. *Nature* 394:392–95

21. Chookajorn T, Dzikowski R, Frank M, Li F, Jiwani AZ, et al. 2007. Epigenetic memory at malaria virulence genes. *Proc. Natl. Acad. Sci. USA* 104:899–902

22. Contamin H, Fandeur T, Rogier C, Bonnefoy S, Konate L, et al. 1996. Different genetic characteristics of *Plasmodium falciparum* isolates collected during successive clinical malaria episodes in Senegalese children. *Am. J. Trop. Med. Hyg.* 54:632–43

23. Cortes A, Carret C, Kaneko O, Lim BY, Ivens A, Holder AA. 2007. Epigenetic silencing of *Plasmodium falciparum* genes linked to erythrocyte invasion. *PLoS Pathog.* 3:e107

24. Cox HW. 1959. A study of relapse *Plasmodium berghei* infections isolated from white mice. *J. Immunol.* 82:209–14

25. Cui L, Miao J, Furuya T, Li X, Su XZ, Cui L. 2007. PfGCN5-mediated histone H3 acetylation plays a key role in gene expression in *Plasmodium falciparum*. *Eukaryot. Cell* 6:1219–27

26. D'Ombrain MC, Voss TS, Maier AG, Pearce JA, Hansen DS, et al. 2007. *Plasmodium falciparum* erythrocyte membrane protein-1 specifically suppresses early production of host interferon-gamma. *Cell Host Microbe* 2:130–38

27. Daily JP, Scanfeld D, Pochet N, Le Roch K, Plouffe D, et al. 2007. Distinct physiological states of *Plasmodium falciparum* in malaria-infected patients. *Nature* 450:1091–95

28. Deitsch KW, Calderwood MS, Wellems TE. 2001. Malaria. Cooperative silencing elements in *var* genes. *Nature* 412:875–76

29. Deitsch KW, del Pinal A, Wellems TE. 1999. Intra-cluster recombination and *var* transcription switches in the antigenic variation of *Plasmodium falciparum*. *Mol. Biochem. Parasitol.* 101:107–16

30. del Portillo HA, Fernandez-Becerra C, Bowman S, Oliver K, Preuss M, et al. 2001. A superfamily of variant genes encoded in the subtelomeric region of *Plasmodium vivax*. *Nature* 12:839–42

31. Donati D, Zhang LP, Chene A, Chen Q, Flick K, et al. 2004. Identification of a polyclonal B-cell activator in *Plasmodium falciparum*. *Infect. Immun.* 72:5412–18

32. Dondorp AM, Kager PA, Vreeken J, White NJ. 2000. Abnormal blood flow and red blood cell deformability in severe malaria. *Parasitol. Today* 16:228–32

33. Duffy MF, Maier AG, Byrne TJ, Marty AJ, Elliott SR, et al. 2006. VAR2CSA is the principal ligand for chondroitin sulfate A in two allogeneic isolates of *Plasmodium falciparum*. *Mol. Biochem. Parasitol.* 148:117–24

34. Duraisingh MT, Triglia T, Ralph SA, Rayner JC, Barnwell JW, et al. 2003. Phenotypic variation of *Plasmodium falciparum* merozoite proteins directs receptor targeting for invasion of human erythrocytes. *EMBO J.* 22:1047–57

35. Duraisingh MT, Voss TS, Marty AJ, Duffy MF, Good RT, et al. 2005. Heterochromatin silencing and locus repositioning linked to regulation of virulence genes in *Plasmodium falciparum*. *Cell* 121:13–24

36. Dzikowski R, Frank M, Deitsch K. 2006. Mutually exclusive expression of virulence genes by malaria parasites is regulated independently of antigen production. *PLoS Pathog.* 2:e22

37. Dzikowski R, Li F, Amulic B, Eisberg A, Frank M, et al. 2007. Mechanisms underlying mutually exclusive expression of virulence genes by malaria parasites. *EMBO Rep.* 8:959–65

38. Fairhurst RM. 2007. Transgenic parasites: improving our understanding of innate immunity to malaria. *Cell Host Microbe* 2:75–76

39. Fernandez V, Hommel M, Chen QJ, Hagblom P, Wahlgren M. 1999. Small, clonally variant antigens expressed on the surface of the *Plasmodium falciparum*–infected erythrocyte are encoded by the *rif* gene family and are the target of human immune responses. *J. Exp. Med.* 190:1393–403

40. Figueiredo L, Scherf A. 2005. *Plasmodium* telomeres and telomerase: the usual actors in an unusual scenario. *Chromosome Res.* 13:517–24

41. Figueiredo LM, Freitas-Junior LH, Bottius E, Olivo-Marin JC, Scherf A. 2002. A central role for *Plasmodium falciparum* subtelomeric regions in spatial positioning and telomere length regulation. *EMBO J.* 21:815–24

42. Francis SE, Malkov VA, Oleinikov AV, Rossnagle E, Wendler JP, et al. 2007. Six genes are preferentially transcribed by the circulating and sequestered forms of *Plasmodium falciparum* parasites that infect pregnant women. *Infect. Immun.* 75:4838–50

43. Frank M, Dzikowski R, Amulic B, Deitsch K. 2007. Variable switching rates of malaria virulence genes are associated with chromosomal position. *Mol. Microbiol.* 64:1486–98

44. Frank M, Dzikowski R, Costantini D, Amulic B, Berdougo E, Deitsch K. 2006. Strict pairing of var promoters and introns is required for var gene silencing in the malaria parasite *Plasmodium falciparum*. *J. Biol. Chem.* 281:9942–52

45. Freitas-Junior LH, Bottius E, Pirrit LA, Deitsch KW, Scheidig C, et al. 2000. Frequent ectopic recombination of virulence factor genes in telomeric chromosome clusters of *P. falciparum*. *Nature* 407:1018–22

46. Freitas-Junior LH, Hernandez-Rivas R, Ralph SA, Montiel-Condado D, Ruvalcaba-Salazar OK, et al. 2005. Telomeric heterochromatin propagation and histone acetylation control mutually exclusive expression of antigenic variation genes in malaria parasites. *Cell* 121:25–36

47. Fried M, Duffy P. 1996. Adherence of *Plasmodium falciparum* to chondroitin sulfate A in the human placenta. *Science* 272:1502–4

48. Fried M, Hixson KK, Anderson L, Ogata Y, Mutabingwa TK, Duffy PE. 2007. The distinct proteome of placental malaria parasites. *Mol. Biochem. Parasitol.* 155:57–65

49. Fried M, Nosten F, Brockman A, Brabin BJ, Duffy PE. 1998. Maternal antibodies block malaria. *Nature* 395:851–52

50. Gamain B, Smith JD, Viebig NK, Gysin J, Scherf A. 2007. Pregnancy-associated malaria: parasite binding, natural immunity and vaccine development. *Int. J. Parasitol.* 37:273–83

51. Gardner MJ, Hall N, Fung E, White O, Berriman M, et al. 2002. Genome sequence of the human malaria parasite *Plasmodium falciparum*. *Nature* 419:498–511

52. Gasser S. 2001. Positions of potential: nuclear organization and gene expression. *Cell* 104:639–42

53. Gottschling DE, Aparicio OM, Billington BL, Zakian VA. 1990. Position effect at *S. cerevisiae* telomeres: reversible repression of Pol II transcription. *Cell* 63:751–62

54. Gunasekera AM, Patankar S, Schug J, Eisen G, Kissinger J, et al. 2004. Widespread distribution of antisense transcripts in the *Plasmodium falciparum* genome. *Mol. Biochem. Parasitol.* 136:35–42

55. Gupta S. 2005. Parasite immune escape: new views into host-parasite interactions. *Curr. Opin. Microbiol.* 8:428–33

56. Hakimi MA, Deitsch KW. 2007. Epigenetics in Apicomplexa: control of gene expression during cell cycle progression, differentiation and antigenic variation. *Curr. Opin. Microbiol.* 10:357–62

57. Handunnetti SM, Mendis KN, David PH. 1987. Antigenic variation of cloned *Plasmodium* fragile in its natural host *Macaca sinica*. Sequential appearance of successive variant antigenic types. *J. Exp. Med.* 165:1269–83

58. Hernandez-Rivas R, Mattei D, Sterkers Y, Peterson DS, Wellems TE, Scherf A. 1997. Expressed var genes are found in *Plasmodium falciparum* subtelomeric regions. *Mol. Cell. Biol.* 17:604–11

59. Hommel M, David PH, Oligino LD. 1983. Surface alterations of erythrocytes in *Plasmodium falciparum* malaria. *J. Exp. Med.* 157:1137–48

60. Horrocks P, Kyes S, Pinches R, Christodoulou Z, Newbold C. 2004. Transcription of subtelomerically located var gene variant in *Plasmodium falciparum* appears to require the truncation of an adjacent var gene. *Mol. Biochem. Parasitol.* 134:193–99

61. Horrocks P, Pinches R, Christodoulou Z, Kyes SA, Newbold CI. 2004. Variable var transition rates underlie antigenic variation in malaria. *Proc. Natl. Acad. Sci. USA* 101:11129–34

62. Horrocks P, Pinches R, Kyes S, Kriek N, Lee S, et al. 2002. Effect of var gene disruption on switching in *Plasmodium falciparum*. *Mol. Microbiol.* 45:1131–41

63. Iyer LM, Anantharaman V, Wolf MY, Aravind L. 2008. Comparative genomics of transcription factors and chromatin proteins in parasitic protists and other eukaryotes. *Int. J. Parasitol.* 38:1–31

64. Jeffares DC, Pain A, Berry A, Cox AV, Stalker J, et al. 2007. Genome variation and evolution of the malaria parasite *Plasmodium falciparum*. *Nat. Genet.* 39:120–25

65. Jensen AT, Magistrado P, Sharp S, Joergensen L, Lavstsen T, et al. 2004. *Plasmodium falciparum* associated with severe childhood malaria preferentially expresses PfEMP1 encoded by group A *var* genes. *J. Exp. Med.* 199:1179–90

66. Kaviratne M, Khan SM, Jarra W, Preiser PR. 2002. Small variant STEVOR antigen is uniquely located within Maurer's clefts in *Plasmodium falciparum*-infected red blood cells. *Eukaryot. Cell* 1:926–35

67. Kooij TW, Carlton JM, Bidwell SL, Hall N, Ramesar J, et al. 2005. A *Plasmodium* whole-genome synteny map: indels and synteny breakpoints as foci for species-specific genes. *PLoS Pathog.* 1:e44

68. Kouzarides T. 2007. Chromatin modifications and their function. *Cell* 128:693–705

69. Kraemer SM, Kyes SA, Aggarwal G, Springer AL, Nelson SO, et al. 2007. Patterns of gene recombination shape *var* gene repertoires in *Plasmodium falciparum*: comparisons of geographically diverse isolates. *BMC Genomics* 8:45

70. Kraemer SM, Smith JD. 2006. A family affair: var genes, PfEMP1 binding, and malaria disease. *Curr. Opin. Microbiol.* 9:374–80

71. Kundu S, Horn PJ, Peterson CL. 2007. SWI/SNF is required for transcriptional memory at the yeast GAL gene cluster. *Genes Dev.* 21:997–1004

72. Kyes S, Christodoulou Z, Pinches R, Kriek N, Horrocks P, Newbold C. 2007. *Plasmodium falciparum var* gene expression is developmentally controlled at the level of RNA polymerase II-mediated transcription initiation. *Mol. Microbiol.* 63:1237–47

73. Kyes S, Taylor H, Craig A, Marsh K, Newbold C. 1997. Genomic representation of *var* gene sequences in *Plasmodium falciparum* field isolates from different geographic regions. *Mol. Biochem. Parasitol.* 87:235–38

74. Kyes SA, Kraemer SM, Smith JD. 2007. Antigenic variation in *Plasmodium falciparum*: gene organization and regulation of the *var* multigene family. *Eukaryot. Cell* 6:1511–20

75. Kyes SA, Rowe JA, Kriek N, Newbold CI. 1999. Rifins: a second family of clonally variant proteins expressed on the surface of red cells infected with *Plasmodium falciparum*. *Proc. Natl. Acad. Sci. USA* 96:9333–38

76. Lavazec C, Sanyal S, Templeton TJ. 2006. Hypervariability within the *rifin*, *stevor* and *Pfmc*-2TM superfamilies in *Plasmodium falciparum*. *Nucleic Acids Res.* 34:6696–707

77. Lavazec C, Sanyal S, Templeton TJ. 2007. Expression switching in the *stevor* and *Pfmc*-2TM superfamilies in *Plasmodium falciparum*. *Mol. Microbiol.* 64:1621–34

78. Leech JH, Barnwell JW, Miller LH, Howard RJ. 1984. Identification of a strain-specific malarial antigen exposed on the surface of *Plasmodium falciparum*-infected erythrocytes. *J. Exp. Med.* 159:1567–75

79. Li F, Sonbuchner L, Kyes SA, Epp C, Deitsch KW. 2008. Nuclear noncoding RNAs are transcribed from the centromeres of *Plasmodium falciparum* and are associated with centromeric chromatin. *J. Biol. Chem.* 283:5692–98

80. Lopez-Rubio JJ, Gontijo AM, Nunes MC, Issar N, Hernandez Rivas R, Scherf A. 2007. 5′ flanking region of *var* genes nucleate histone modification patterns linked to phenotypic inheritance of virulence traits in malaria parasites. *Mol. Microbiol.* 66:1296–305

81. Lopez-Rubio JJ, Riviere L, Scherf A. 2007. Shared epigenetic mechanisms control virulence factors in protozoan parasites. *Curr. Opin. Microbiol.* 10:560–68

82. Mackintosh CL, Beeson JG, Marsh K. 2004. Clinical features and pathogenesis of severe malaria. *Trends Parasitol.* 20:597–603

83. Mancio-Silva L, Rojas-Meza A, Vargas M, Scherf A, Hernandez-Rivas R. 2008. Differential association of ORC and Sir proteins to telomeric domains in *Plasmodium falciparum*. *J. Cell Sci.* 121:2046–53

84. Marcello L, Barry JD. 2007. From silent genes to noisy populations-dialogue between the genotype and phenotypes of antigenic variation. *J. Eukaryot. Microbiol.* 54:14–17

85. Marsh K, Howard RJ. 1986. Antigens induced on erythrocytes by *P. falciparum*: expression of diverse and conserved determinants. *Science* 231:150–53

86. Marty AJ, Thompson JK, Duffy MF, Voss TS, Cowman AF, Crabb BS. 2006. Evidence that *Plasmodium falciparum* chromosome end clusters are cross-linked by protein and are the sites of both virulence gene silencing and activation. *Mol. Microbiol.* 62:72–83

87. McLean SA, Pearson CD, Phillips RS. 1982. *Plasmodium chabaudi*: antigenic variation during recrudescent parasitaemias in mice. *Exp. Parasitol.* 54:296–302

88. Merrick CJ, Duraisingh MT. 2007. *Plasmodium falciparum* Sir2: an unusual sirtuin with dual histone deacetylase and ADP-ribosyltransferase activity. *Eukaryot. Cell* 6:2081–91

89. Miao J, Fan Q, Cui L, Li J. 2006. The malaria parasite *Plasmodium falciparum* histones: organization, expression, and acetylation. *Gene* 369:53–65

90. Militello KT, Patel V, Chessler AD, Fisher JK, Kasper JM, et al. 2005. RNA polymerase II synthesizes antisense RNA in *Plasmodium falciparum*. *Rna* 11:365–70

91. Miller LH, Baruch DI, Marsh K, Doumbo OK. 2002. The pathogenic basis of malaria. *Nature* 415:673–79

92. Miller LH, Good MF, Milon G. 1994. Malaria pathogenesis. *Science* 264:1878–83

93. Molineaux L, Diebner HH, Eichner M, Collins WE, Jeffery GM, Dietz K. 2001. *Plasmodium falciparum* parasitaemia described by a new mathematical model. *Parasitology* 122:379–91

94. Mondoux MA, Zakian VA. 2006. Telomere position effect: silencing near the end. In *Telomeres*, ed. T de Lange, V Lundblad, EH Blackburn, pp. 261–316. Plainview, NY: Cold Spring Harbor Lab. 2nd ed.

95. Montgomery J, Mphande FA, Berriman M, Pain A, Rogerson SJ, et al. 2007. Differential *var* gene expression in the organs of patients dying of *Falciparum* malaria. *Mol. Microbiol.* 65:959–67

96. Mourier T, Carret C, Kyes S, Christodoulou Z, Gardner PP, et al. 2008. Genome-wide discovery and verification of novel structured RNAs in *Plasmodium falciparum*. *Genome Res.* 18:281–92

97. Ng HH, Robert F, Young RA, Struhl K. 2003. Targeted recruitment of Set1 histone methylase by elongating Pol II provides a localized mark and memory of recent transcriptional activity. *Mol. Cell* 11:709–19

98. Paget-McNicol S, Gatton M, Hastings I, Saul A. 2002. The *Plasmodium falciparum var* gene switching rate, switching mechanism and patterns of parasite recrudescence described by mathematical modelling. *Parasitology* 124:225–35

99. Peters J, Fowler E, Gatton M, Chen N, Saul A, Cheng Q. 2002. High diversity and rapid changeover of expressed *var* genes during the acute phase of *Plasmodium falciparum* infections in human volunteers. *Proc. Natl. Acad. Sci. USA* 99:10689–94

100. Petter M, Haeggstrom M, Khattab A, Fernandez V, Klinkert MQ, Wahlgren M. 2007. Variant proteins of the *Plasmodium falciparum* RIFIN family show distinct subcellular localization and developmental expression patterns. *Mol. Biochem. Parasitol.* 156:51–61

101. Preiser PR, Jarra W, Capiod T, Snounou G. 1999. A rhoptry-protein-associated mechanism of clonal phenotypic variation in rodent malaria. *Nature* 398:618–22

102. Ralph SA, Bischoff E, Mattei D, Sismeiro O, Dillies MA, et al. 2005. Transcriptome analysis of antigenic variation in *Plasmodium falciparum*—*var* silencing is not dependent on antisense RNA. *Genome Biol.* 6:R93

103. Ralph SA, Scheidig-Benatar C, Scherf A. 2005. Antigenic variation in *Plasmodium falciparum* is associated with movement of *var* loci between subnuclear locations. *Proc. Natl. Acad. Sci. USA* 102:5414–19

104. Recker M, Nee S, Bull PC, Kinyanjui S, Marsh K, et al. 2004. Transient cross-reactive immune responses can orchestrate antigenic variation in malaria. *Nature* 429:555–58

105. Roberts DJ, Craig AG, Berendt AR, Pinches R, Nash G, et al. 1992. Rapid switching to multiple antigenic and adhesive phenotypes in malaria. *Nature* 357:689–92

106. Rubio JP, Thompson JK, Cowman AF. 1996. The *var* genes of *Plasmodium falciparum* are located in the subtelomeric region of most chromosomes. *EMBO J.* 15:4069–77

107. Sam-Yellowe TY, Florens L, Johnson JR, Wang T, Drazba JA, et al. 2004. A *Plasmodium* gene family encoding Maurer's cleft membrane proteins: structural properties and expression profiling. *Genome Res.* 14:1052–59

108. Saul A. 1999. The role of variant surface antigens on malaria-infected red blood cells. *Parasitol. Today* 15:455–57

109. Sautel CF, Cannella D, Bastien O, Kieffer S, Aldebert D, et al. 2007. SET8-mediated methylations of histone H4 lysine 20 mark silent heterochromatic domains in apicomplexan genomes. *Mol. Cell Biol.* 27:5711–24

110. Scherf A, Hernandez-Rivas R, Buffet P, Bottius E, Benatar C, et al. 1998. Antigenic variation in malaria: in situ switching, relaxed and mutually exclusive transcription of *var* genes during intraerythrocytic development in *Plasmodium falciparum*. *EMBO J.* 17:5418–26

111. Schieck E, Pfahler JM, Sanchez CP, Lanzer M. 2007. Nuclear run-on analysis of *var* gene expression in *Plasmodium falciparum*. *Mol. Biochem. Parasitol.* 153:207–12

112. Smith JD, Chitnis CE, Craig AG, Roberts DJ, Hudson-Taylor DE, et al. 1995. Switches in expression of *Plasmodium falciparum var* genes correlate with changes in antigenic and cytoadherent phenotypes of infected erythrocytes. *Cell* 82:101–10

113. Staalsoe T, Shulman CE, Bulmer JN, Kawuondo K, Marsh K, Hviid L. 2004. Variant surface antigen-specific IgG and protection against clinical consequences of pregnancy-associated *Plasmodium falciparum* malaria. *Lancet* 363:283–89

114. Stubbs J, Simpson KM, Triglia T, Plouffe D, Tonkin CJ, et al. 2005. Molecular mechanism for switching of *P. falciparum* invasion pathways into human erythrocytes. *Science* 309:1384–87

115. Su XZ, Heatwole VM, Wertheimer SP, Guinet F, Herrfeldt JA, et al. 1995. The large diverse gene family *var* encodes proteins involved in cytoadherence and antigenic variation of *Plasmodium falciparum*–infected erythrocytes. *Cell* 82:89–100

116. Sullivan WJ Jr, Naguleswaran A, Angel SO. 2006. Histones and histone modifications in protozoan parasites. *Cell Microbiol.* 8:1850–61

117. Taylor H, Kyes SA, Newbold CI. 2000. *var* gene diversity in *P. falciparum* is generated by frequent recombination events. *Mol. Biochem. Parasitol.* 110:391–97

118. Templeton TJ, Iyer LM, Anantharaman V, Enomoto S, Abrahante JE, et al. 2004. Comparative analysis of Apicomplexa and genomic diversity in eukaryotes. *Genome Res.* 14:1686–95

119. Tonkin CJ, Pearce JA, McFadden GI, Cowman AF. 2006. Protein targeting to destinations of the secretory pathway in the malaria parasite *Plasmodium falciparum*. *Curr. Opin. Microbiol.* 9:381–87

120. Trager W, Jensen JB. 1976. Human malaria parasites in continuous culture. *Science* 193:673–75

121. Urban BC, Ferguson DJP, Pain A, Willcox N, Plebanski M, et al. 1999. *Plasmodium falciparum*–infected erythrocytes modulate the maturation of dendritic cells. *Nature* 400:73–77

122. van Ooij C, Haldar K. 2007. Protein export from *Plasmodium* parasites. *Cell Microbiol.* 9:573–82

123. Viebig NK, Gamain B, Scheidig C, Lepolard C, Przyborski J, et al. 2005. A single member of the *Plasmodium falciparum var* multigene family determines cytoadhesion to the placental receptor chondroitin sulphate A. *EMBO Rep.* 6:775–81

124. Voss TS, Healer J, Marty AJ, Duffy MF, Thompson JK, et al. 2006. A *var* gene promoter controls allelic exclusion of virulence genes in *Plasmodium falciparum* malaria. *Nature* 439:1004–8

125. Voss TS, Tonkin CJ, Marty AJ, Thompson JK, Healer J, et al. 2007. Alterations in local chromatin environment are involved in silencing and activation of subtelomeric *var* genes in *Plasmodium falciparum*. *Mol. Microbiol.* 66:139–50

126. Wahlgren M, Treutiger CJ, Gysin J. 1999. Cytoadherence and rosetting in the pathogenesis of severe malaria. In *Malaria: Molecular and Clinical Aspects*, ed. M Wahlgren, P Perlmann, pp. 289–328. The Netherlands: Harwood Acad.

127. Winter G, Kawai S, Haeggstrom M, Kaneko O, von Euler A, et al. 2005. SURFIN is a polymorphic antigen expressed on *Plasmodium falciparum* merozoites and infected erythrocytes. *J. Exp. Med.* 201:1853–63

128. Zaratiegui M, Irvine DV, Martienssen RA. 2007. Noncoding RNAs and gene silencing. *Cell* 128:763–76

Hijacking of Host Cellular Functions by the Apicomplexa

Fabienne Plattner and Dominique Soldati-Favre

Department of Microbiology and Molecular Medicine, Faculty of Medicine, University of Geneva CMU, 1211 Geneva 4, Switzerland; email: Fabienne.Plattner@medecine.unige.ch, dominique.soldati-favre@medecine.unige.ch

Annu. Rev. Microbiol. 2008. 62:471–87

The *Annual Review of Microbiology* is online at micro.annualreviews.org

This article's doi:
10.1146/annurev.micro.62.081307.162802

0066-4227/08/1013-0471$20.00

Key Words

Toxoplasma gondii, Plasmodium falciparum, Theileria parva, Cryptosporidium parvum, invasion, nutrients, immunity, apoptosis

Abstract

Intracellular pathogens such as viruses and bacteria subvert all the major cellular functions of their hosts. Targeted host processes include protein synthesis, membrane trafficking, modulation of gene expression, antigen presentation, and apoptosis. In recent years, it has become evident that protozoan pathogens, including members of the phylum Apicomplexa, also hijack their host cell's functions to access nutrients and to escape cellular defenses and immune responses. These obligate intracellular parasites provide superb illustrations of the subversion of host cell processes such as the recruitment and reorganization of host cell compartments without fusion around the parasitophorous vacuole of *Toxoplasma gondii*; the export of *Plasmodium falciparum* proteins on the surface of infected erythrocytes; and the induced transformation of the lymphocytes infected by *Theileria parva*, which leads to clonal extension.

Contents

INTRODUCTION

Viruses, which are completely dependent on the host cell molecular machinery for multiplication of their genomes as well as for assembly of new viral particles, are experts in manipulating the host metabolism. In contrast, most pathogenic bacteria are facultative intracellular organisms that enter the host cells primarily to avoid clearance by the immune system. These pathogens have coevolved with their host to establish an efficient infection (70). One of the main defense mechanisms that infected organisms use against pathogens (98) is the programmed cell death called apoptosis. Nevertheless, viruses (68) as well as bacteria (38) have learned over time to overcome this and other elaborate defense strategies.

Apicomplexa: a phylum of protozoan parasites that includes *Plasmodium* spp., the agent of malaria, and other pathogens of medical and veterinarian importance

Members of the phylum Apicomplexa are among the most successful, life-threatening pathogens in humans and animals. The completion of the genomes for several apicomplexans has greatly improved our knowledge, and two advancements have allowed researchers to study in depth the biology of some of these parasites. The first advancement is the establishment of a system to propagate the parasite in cell culture, and the second is the development of reliable tools for genetic manipulation. *Plasmodium* spp. and *Toxoplasma gondii* are convenient models for studying members of this phylum (51, 67).

As for bacteria and viruses, intracellular survival of apicomplexans critically depends on access to the host cell metabolites. Recent studies have highlighted novel roles of secreted parasite proteins in the modulation of host cellular functions, illustrating the enormous potential of apicomplexans to interfere with the signaling cascades, to modulate the metabolic pathways, to divert host metabolites and to exploit the lipid trafficking pathways of the host cells for their own purpose and benefit. In this review, we discuss four apicomplexan parasites (*Toxoplasma*, *Plasmodium*, *Cryptosporidium*, and *Theileria*) that drastically differ in their lifestyles and ecological niches. This review recapitulates and compares the broad variety of highly sophisticated strategies evolved by these parasites to subvert host processes to maximize access to nutrients, assure dissemination and transmission, neutralize host defenses, and avoid destruction.

DIVERSE LIFE CYCLES AND LIFESTYLES

The phylum Apicomplexa encompasses several thousand species that show a great level of diversity in their complex life cycles, including morphologically distinct life stages and one or more hosts (**Figure 1**). Some parasites need to be taken up by a definitive host to fulfill their sexual reproduction cycle and are transmitted to one or several intermediate hosts, in which only asexual reproduction occurs. For example, the female *Anopheles* mosquito serves both as definitive host and as vector for transmission for the *Plasmodium* spp., and the ixodid ticks fulfill the same tasks for *Theileria parva* and *Babesia bovis*. The sexual reproduction of *T. gondii* takes place exclusively in the guts of felines; however, the asexual form of this parasite can also be transmitted orally via the intermediate hosts (87). Other parasites such as *Cryptosporidium* and *Eimeria* spp. rely only on a single host for both sexual and asexual reproduction.

T. gondii is a ubiquitous parasite, infecting about one third of the world's population (91). In healthy humans, the infection by *T. gondii* is dampened by the immune system, turning toxoplasmosis into a mainly asymptomatic disease. It becomes life threatening only in immuno-compromised patients and as a congenital infection (52, 100). This parasite has the broadest host cell range, as it invades virtually all types of nucleated cells (**Figure 1a**). *Plasmodium falciparum*, the most notorious and deadly etiologic agent of malaria, is responsible for 500 million new infections each year, mainly in sub-Saharan Africa and South East Asia (83). Malaria is responsible for the deaths of 1–3 million people each year, mainly children and pregnant women (7). *Plasmodium* spp. infect hepatocytes and erythrocytes (**Figure 1b**) and most of the pathology they induce is associated with the drastic remodeling and lysis of the infected erythrocytes. *Cryptosporidium* spp., an agent of a diarrheal disease, infect mainly enterocytes in all type of vertebrates including humans (19) and cause severe losses in livestock (**Figure 1c**). *Theileria* spp. are responsible for major economic losses in the cattle industry, mainly in East and South Africa. These parasites invade lymphocytes (**Figure 1d**) and transform them, resulting in a lymphoproliferative disease in the host.

HOST CELL ENTRY AND ACCESS TO PRIVILEGED NICHES

In the case of *Toxoplasma* and *Cryptosporidium*, infection of the intermediate host is initiated by the ingestion of oocysts produced in the gut of the definitive hosts and shed in the feces, or in the case of *Plasmodium* and *Theileria*, infection is initiated by direct transmission of sporozoites during blood feeding by the vectors. Once inside the host organism, the parasites reach, invade, and replicate inside their respective target cells (epithelial cells, hepatocytes, leukocytes, erythrocytes and, for some species, specialized immune cells such as macrophages and dendritic cells). During this process, the parasites must overcome several biological barriers (e.g., mucus, epithelia, and blood-brain barrier) and drastically modify the infected cells such that they turn into suitable environments.

Host cell entry is an active process distinct from phagocytosis involving a substrate-dependent form of gliding motility. This active process considerably enlarges the repertoire of cells that can be infected by these parasites. For some life stages gliding motility allows migration across biological barriers and egress from host cells (48). The machinery powering gliding has been studied in great detail for *T. gondii* tachyzoites and *Plasmodium* sporozoites and merozoites (2). Gliding involves the concerted action of secretory adhesins, a myosin motor, and factors regulating actin dynamics. During invasion, proteins are discharged sequentially from microneme and rhoptry organelles (10). While micronemes primarily release adhesins involved in host cell recognition and attachment, rhoptry proteins contribute to the formation and modification of the parasitophorous vacuole membrane (PVM).

Host cell penetration is extremely rapid and completed within 20–30 s. During this process the host cell plasma membrane is invaginated and participates to the formation of the PVM. During this process, a ring-shaped structure called the moving junction (87) mediates the contact between the parasite and the host cell plasma membrane. This structure is pulled toward the posterior end of the parasite as the latter invades the cell and forms the parasitophorous vacuole (PV) (62) (**Figure 2a**). It is plausible that the small parasite rhoptry protein toxofilin transiently interacts with host cell cortical actin and depolymerizes it, thus facilitating parasite entry (46).

Gliding: a particular substrate-dependent type of motility conserved among apicomplexans that relies on a parasite acto-myosin system

Rhoptry: club-shaped secretory organelle with a bulbous part and a thinner neck part toward the apical end of the parasite

Micronemes: vesicular secretory organelles located at the apical end of the parasite

PVM: parasitophorous vacuole membrane

Parasitophorous vacuole (PV): a parasite-containing vacuole formed mainly from host cell lipids during invasion of the host cell

C. parvum also invades host cells by a mechanism dependent on parasite motility (97) and secretion from the apical organelles (13). However, in contrast to the mechanism described above, additional remodeling of host cell actin filaments is necessary for the establishment of a productive infection (33) (**Figure 2b**). During invasion, microvilli of the host enterocytes extend around the parasite and fuse together to form the PVM. Host actin, ezrin, and villin associate with the membrane surrounding the parasites (5). A band of electron-dense material of unknown composition is found just beneath the parasite. *Cryptosporidium*-induced activation of phosphoinositide 3-kinase (PI3K) leads to successive activation of Cdc42, N-WASP, and the Arp2/3 complex, resulting in actin filament polymerization just underneath the dense

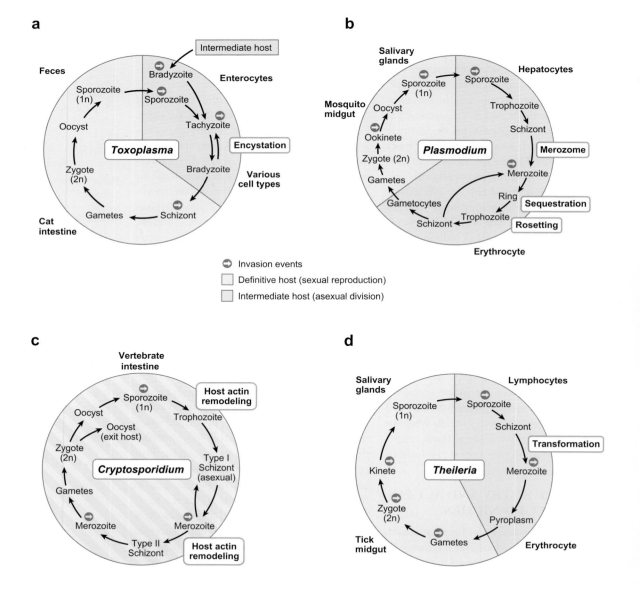

band and possibly also around the parasite (12, 35). Another pathway dependent on c-Src and cortactin also appears to contribute to actin remodeling during parasite entry (11).

Unlike most apicomplexans, host cell invasion by *T. parva* sporozoites is a passive internalization process (76) (**Figure 2c**). The sporozoite antigen p67 contributes to host cell recognition, and initial attachment to lymphocytes seems to involve MHC class I molecules (77). Subsequently, a zippering mechanism occurs concomitant with shedding of the surface proteins, leading to complete internalization of the parasite (96).

LIVING IN A NONFUSOGENIC PARASITOPHOROUS VACUOLE

Although all apicomplexans initially penetrate their host cell via the formation of a vacuole, they have evolved distinct strategies to escape or modify this vacuole and modulate their intracellular environments. *T. gondii*, *P. falciparum*, and *C. parvum* reside and replicate inside a nonfusogenic PV during their entire intracellular cycle. Such a safe niche confers resistance to some host cell defenses but significantly cuts the parasites off from host metabolites. These parasites have adopted different tactics to circumvent the problem, remodeling their vacuole to make it permissive to vital substances (3). In contrast parasites such as *T. parva* rapidly escape the vacuole and proliferate freely in the cytoplasm with direct access to host nutrients.

T. gondii: Intravacuolar Remodeling and Recruitment of Host Organelles

T. gondii infects biosynthetically active cells, including dendritic cells and macrophages, and

---

Figure 1

Simplified life cycles of four apicomplexans and their respective invasive stages (*red arrows*). (*a*) *Toxoplasma gondii* is transmitted to intermediate hosts by ingestion of oocysts from the environment or by ingestion of undercooked meat contaminated with bradyzoite-containing cysts. In the host's gastrointestinal tract, sporozoites differentiate into the rapidly dividing tachyzoite form, which can invade and multiply in all types of nucleated cells by endodyogeny. Under pressure of the immune system, tachyzoites differentiate into the slowly growing, cyst-forming bradyzoites and persist mainly in muscle and brain (encystation). Stage conversion to tachyzoites occurs and in the case of immunosuppression leads to reactivation of toxoplasmosis. Transmission to the definitive host (felids) is initiated by the invasion of the enterocytes and followed by sexual reproduction, leading to the production of infective oocysts excreted in cat feces. (*b*) *Plasmodium falciparum* sporozoites are transmitted into a vertebrate host via the mosquito bite during a blood meal. Salivary gland sporozoites migrate through the lymph or the bloodstream and invade hepatocytes. Thousands of merozoites are released into the bloodstream, invade erythrocytes and undergo multiple intraerythrocytic cycles. Some of the merozoites differentiate into gametocytes, which are taken up by the mosquito during a blood meal. Gametocytes differentiate into gametes, and sexual reproduction results in the formation of an ookinete, which traverses the mosquito midgut epithelium and develops into oocysts, leading to the release of sporozoites that migrate to the salivary glands. (*c*) The monoxenous life cycle of *Cryptosporidium parvum* begins when a vertebrate host ingests oocysts from the environment. Sporozoites are released from the oocysts in the gastrointestinal tract and establish infection in the enteric epithelium. Once internalized, the trophozoite undergoes asexual division by merogony to form a Type I schizont. After maturation, eight merozoites are released that invade neighboring enterocytes and either undergo further rounds of asexual division or develop into Type II schizonts. Four merozoites are released that invade neighboring cells and initiate gametogony. After fertilization, the zygote develops into an oocyst. Two types of oocysts are produced: those with thin walls, which stay in the host enterocyte and release infectious sporozoites that start a new cycle, and those with thick walls, which are excreted in the feces of the host. (*d*) *Theileria parva* sporozoites are transmitted by the tick during a blood meal. After invasion of lymphocytes, the parasite develops into intracytoplasmic multinucleated schizonts and induces proliferation of its host (transformation). The host cell and the schizont divide synchronously, leading to a clonal expansion of the infected lymphocytes. Some parasites form merozoites and are released by rupture of the host cell into the bloodstream, invade erythrocytes, and develop into piroplasms. Piroplasms are ingested by the tick during a blood meal and differentiate into gametes in the gut of the tick. Fertilization occurs, and the resulting zygotes invade epithelial cells from the gut and develop into kinetes. The kinetes migrate to and invade the salivary glands of the tick. Sporogony is initiated when the tick attaches to a host animal, resulting in the release of sporozoites into the saliva, ready for transmission to the host.

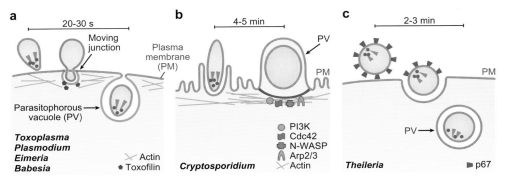

Figure 2

Host cell entry by apicomplexans and formation of the parasitophorous vacuole (PV). Rhoptries are depicted in green; micronemes are depicted in blue. (*a*) *Toxoplasma* and *Plasmodium*: active entry leading to the formation of a PV. Upon initial contact with the host cell, parasites reorient their apical end perpendicular to the host cell surface. Microneme and rhoptry proteins discharge their contents at the apical end and participate in the formation of the moving junction. Rhoptry-derived material is inserted into the nascent PV and into the host cell cytosol and nucleus. Microneme proteins are released onto parasite surface. (*b*) *Cryptosporidium*: active entry leading to the formation of an intracellular but extracytoplasmic vacuole. Microneme and rhoptry proteins are secreted during invasion, and the actin cytoskeleton of the host cell is remodeled underneath and around the invading parasite. The host cell PI3K is recruited below the attachment site of the parasite, and upon its activation, Cdc42, N-WASP, and finally Arp2/3 are also activated and mediate actin remodeling. The host enterocyte microvilli elongate, surround the parasite, and fuse to form the PV. (*c*) *Theileria*: passive uptake. Initial contact with a lymphocyte surface receptor followed by a zippering mechanism leads to parasite internalization. At the end of the process, the parasite escapes the vacuole and reaches the host cytosol by an unknown mechanism (not shown).

the parasite's major task is to remodel the PV primarily to avoid lysosomal fusion and to gain access to nutrients (**Figure 3***a*). During invasion, host transmembrane proteins are excluded from the forming PVM at the moving junction (63), leaving only host GPI-anchored proteins. The lipids forming the PV are derived mainly from the host cell plasma membrane (88, 95), although rhoptry-derived lipids also contribute to the process (43). The resulting nonphagosomal vacuole is a safe niche where the parasite can develop, as it does not fuse with the endosomal compartments and does not acidify over time (47, 64, 79).

The newly formed PVM of *T. gondii* is immediately modified to allow free diffusion of small molecules such as simple sugars, amino acids, nucleobases, and cofactors. Evidence for the presence of a pore in the PVM came from a study in which small fluorescent dyes and labeled peptides up to 1291 Da were injected into

T. gondii–infected host cells (74). This pore allows the passage of small molecules into the PV in the absence of transporters, whereas their subsequent uptake by the parasite is mediated via substrate-specific transporters or endocytosis. This molecular sieve plays a critical role in the context of the multiple parasite auxotrophies. *T. gondii* depends on glucose for energy and as a carbon source for fatty acid biosynthesis, and it is auxotrophic for amino acids such as arginine (37) and tryptophan (66). *T. gondii* salvages purines and pyrimidines (53, 75) despite its ability to synthesize pyrimidines de novo (36).

In the absence of biosynthetic machinery to produce sterols (18), *T. gondii* must scavenge this essential lipid from the host environment and store it. A plausible way for scavenging host cell lipids including cholesterol could be via the intimate association of host endoplasmic reticulum (ER) and mitochondria with the PVM (34,

a *Toxoplasma*

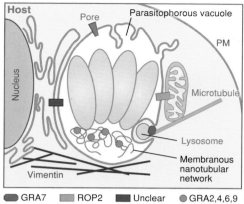

b *Plasmodium*

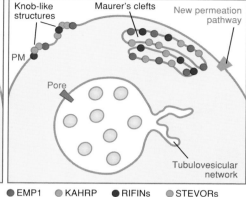

⬤ GRA7	▨ ROP2	■ Unclear	⬤ GRA2,4,6,9	⬤ EMP1	⬤ KAHRP	⬤ RIFINs	⬤ STEVORs

c *Cryptosporidium*

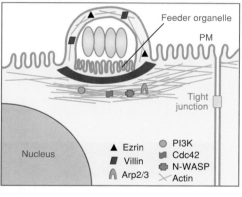

d *Theileria*

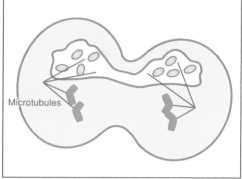

Figure 3

Parasite-induced host cell remodeling. (*a*) *Toxoplasma gondii*: recruitment of host endoplasmic reticulum and mitochondria. Vimentin surrounds the parasitophorous vacuole (PV) possibly to maintain it in close proximity to the host nucleus. A pore in the parasitophorous vacuole membrane (PVM) allows uptake of small nutrients. Subversion of the host cell microtubules is a possible path to internalize lysosomes into the PV, with GRA7 involved in the sealing of this vesicle. Several GRA proteins induce the formation of a membranous nanotubular network inside the PV, and some extensions are sent into the host cell cytoplasm. (*b*) Erythrocytes infected by *Plasmodium falciparum*: formation of Maurer's clefts in the host cytoplasm and knob-like structures at the surface of the host cell. A pore in the PVM allows uptake of small nutrients, and the new permeation pathway drastically increases the permeability of the erythrocyte to metabolites of the serum. A tubulovesicular network extends from the PV inside the cytoplasm.
(*c*) *Cryptosporidium parvum*: induction of massive host cytoskeletal actin remodeling. An actin platform at the host-parasite interface is formed and is possibly involved in vesicular trafficking to the PV. Ezrin and villin are found around the PV, whereas several host actin remodeling proteins are recruited to the interface by unknown parasite factors. (*d*) *Theileria parva:* association of the parasites with the host microtubules via the TaSE protein. Infected lymphocytes are transformed and undergo clonal expansion, and the multinucleated schizonts divide synchronously with the host cells. PM: plasma membrane.

41), which was confirmed by different experimental approaches (22, 81) (**Figure 4a**). The rhoptry protein TgROP2 is a transmembrane protein localized at the PVM that inserts into the host mitochondria membranes via its N-terminal domain exposed to the cytosol (80). A distinct role for TgROP2 and TgROP4 in iron uptake via binding to lactoferrin has also been reported recently (30). Because *T. gondii* lacks the mitochondrial genes required for de

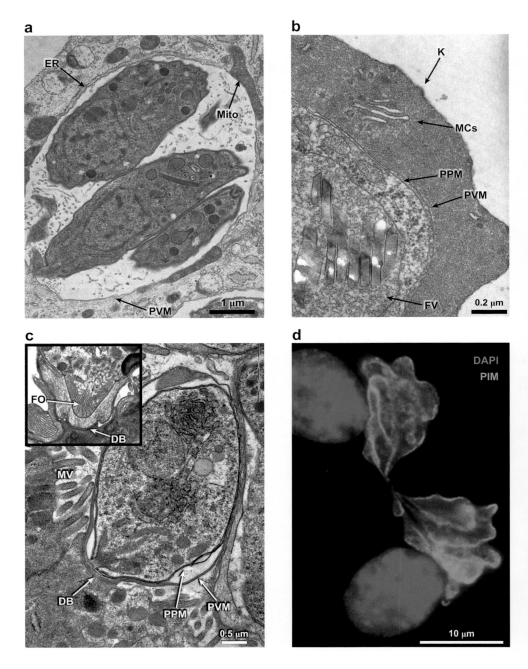

novo synthesis of lipoic acid, a cofactor for many biosynthetic pathways, the host mitochondria are likely the source of this lipid for the parasite (20). All these rhoptry proteins are originally found in the bulbous part of the organelle and do not have homologues in *P. falciparum* (6), suggesting that they play specific roles depending on the niche the parasite occupies.

In addition to host organelles, the host intermediate filament network also rearranges around the *Toxoplasma* parasitophorous vacuole. Host cell vimentin relocalizes to the vicinity of the parasite compartment and appears to dock it to the host cell nuclear surface (44).

Concomitant with the modifications due to PVM-host organelle association, the inside of the vacuole also undergoes major changes. A membranous nanotubular network (MNN) is built up in the lumen of the PV of *T. gondii*. It is first observed as a multilamellar structure in the posterior part of the parasite and then connects to the PVM during intracellular development (78). The MNN is presumed to increase the membranous surface for metabolic exchanges between the host cell cytoplasm and the parasite. Secretion by the dense granules has been implicated in the formation of this network. Several dense granule proteins (GRAs) are associated with the MNN, whereas others are detected at the PVM and in cytosolic extensions (61). Recently, the amphipathic α-helices of GRA2 have been involved in the formation

of the network (93). Because the GRAs do not show any homology to proteins from other eukaryotes, their precise function in the modification of the PV is still speculative.

A recent study has proposed a novel mechanism developed by the parasite to take up LDL-derived cholesterol from the host cell and deliver LDL and other endocytic vesicles to the PV. Structures called HOST (host organelle-sequestering tubulo-structures) are thought to be involved, in which the host microtubules are diverted and serve as a pathway to bring endosomal vesicles inside the PV (17). TgGRA7 has been implicated in the sealing mechanism during delivery of the vacuole.

Ultimately, the PVM of *T. gondii* undergoes a dramatic remodeling that ends with the formation of a cyst wall and coincides with the differentiation of the rapidly dividing tachyzoite stage into the slowly growing bradyzoite stage. This differentiation process leads to parasite encystation and the establishment of a persistent chronic infection.

P. falciparum: Extravacuolar Remodeling, Rosetting, and Sequestration

A molecular sieve at the PVM in the intraerythrocytic stages of the malaria parasite allows passage of molecules up to 1400 Da (27). *P. falciparum* like other apicomplexans is auxotrophic

MNN: membranous nanotubular network

GRA: dense granule protein

Figure 4

Illustration of the different types of host cell modifications induced by parasites. (*a*) Parasitophorous vacuole containing dividing *Toxoplasma gondii* parasites, and associated with host cell mitochondria (Mito) and endoplasmic reticulum (ER). Electron microscopy image courtesy of J.F. Dubremetz. (*b*) Erythrocyte infected with *Plasmodium falciparum* showing the parasite plasma membrane (PPM), the Maurer's clefts (MCs) and knobs (K) in the modified host cell. The food vacuole (FV) contains hemozoin crystals. Electron microscopy image courtesy of J.F. Dubremetz. (*c*) Attachment site of *Cryptosporidium muris* on the surface of rodent gastric cells. A maturing trophozoite with developing feeder organelle (FO) and dense band (DB) underneath the attachment site is shown. The inset shows a higher magnification of a fully developed FO. Electron microscopy image courtesy of A. Valigurová, with permission of Blackwell publishing group, from Valigurova et al. 2007. (6):495–510. The *Journal of Eukaryotic Microbiology*. (*d*) Concomitant division of a transformed lymphocyte and *Theileria annulata* schizont. Host and parasite nucleus are stained with DAPI and the parasite polymorphic immunodominant molecule (PIM) is in green. Immunofluorescence image courtesy of D. Dobbelaere. Abbreviations: MV, host cell microvilli; PVM, parasitophorous vacuole membrane.

for purines and synthesizes pyrimidines de novo. Nevertheless, during the intraerythrocytic stages, these parasites are confronted by one additional difficulty, as erythrocytes are not synthesizing new purines. As a result, the parasite heavily modifies the permeability of the erythrocyte plasma membrane for purine transport (39). The new permeation pathway, which results from parasite-induced modification of an erythrocyte plasma membrane channel (23), transports purines, sugars, anions, and inorganic cations across the erythrocyte plasma membrane; it is also called a plasmodial surface anion channel (26). *P. falciparum* also takes up and degrades hemoglobin in the food vacuole to access amino acids, except for isoleucine residues, which are missing from this multimeric protein complex (59).

In infected erythrocytes the development of the malaria parasite provokes several morphological changes, including the formation of the tubulovesicular network, which extends from the PVM to the periphery of the erythrocyte (**Figure 3b**). These structures have been implicated in the uptake of nucleosides and amino acids (57). The parasite induces the appearance of Maurer's clefts (MCs), which are plate-like structures with an electron-dense surface and a translucent lumen (84) (**Figure 4b**). The MCs can be considered a parasite-derived secretory compartment that contributes to the transport and targeting of parasite proteins to the surface of the infected erythrocytes (92).

Any alteration of an erythrocyte, including infection by a parasite, prones it to destruction by the spleen. To avoid splenic clearance, *P. falciparum* has evolved a strategy leading to cytoadherence of infected erythrocyte to the endothelium of blood vessels. This implies that parasite proteins capable of binding to host cell ligands such as intercellular adhesion molecule 1 (ICAM-1) are transported across the PVM and targeted to the erythrocyte plasma membrane. This process requires sophisticated trafficking machinery (16). The drawback of such a strategy is that parasite proteins are exposed to host circulation and become evident targets for the immune system. Malaria parasites have

evolved a solution to this problem by relying on antigenic variation, a mechanism that allows rapid turnover of the set of proteins sent to the infected erythrocyte surface. Several parasite proteins, including PfEMP1 encoded by the VAR gene family and which mediates cytoadherence, are sent to the erythrocyte surface, creating a visible alteration of the membrane called knobs (**Figure 4b**). KHARP and the RIFIN and STEVOR protein families also locate to the surface of the erythrocyte and some could play a role in immune evasion. All these proteins traffic through the MCs to reach their final destination on the erythrocyte plasma membrane (42).

A further strategy for immune evasion utilized by the malaria parasite consists of sending proteins to the surface of the infected erythrocyte that will bind to noninfected erythrocytes, a mechanism called rosetting. The advantage of rosetting in immune evasion is not completely understood. One report shows a positive correlation between rosetting and parasitemia (69), suggesting that merozoites leaving one erythrocyte could easily invade the surrounding cells, but an older study shows that this is not the case in vitro (14). Another possibility is that rosetting is a strategy to hide the infected cell from the immune system by docking noninfected erythrocytes around it, but this hypothesis needs further confirmation.

C. parvum: Intracellular Extracytoplasmic PV

C. parvum resides in an intracellular but extracytosolic niche at the apical side of enterocytes (**Figure 3c**). The parasite-induced modifications of the host actin cytoskeleton initiated during the invasion process are maintained and lead to the formation of an actin plaque at the host-parasite interface (32). Little information is available regarding the biological significance of the host actin remodeling process, but one hypothesis is that actin filaments open a track for vesicular trafficking that allows uptake of nutrients from the host cytosol. Furthermore, a feeder organelle made of extensively folded membrane structures develops above the dense band and is also thought to be important for nutrient uptake (21) (**Figure 4c**). Unfortunately, understanding the biology of *Cryptosporidium* has been hampered by the lack of methodologies for transfection and by the difficulties faced to reproduce the whole cycle in vitro despite recent progress (54). It is evident from genome analysis that the biosynthetic possibilities of this parasite are considerably reduced but likely compensated for by a large panoply of transporters (1, 85). Because of its particular localization at the apex of enterocytes, *C. parvum* might easily access nutrients directly from the lumen of the intestine. Two oxysterol binding proteins are found in *C. parvum*, and one of them localizes to the PVM, suggesting a role in lipid uptake directly from this compartment (102).

LIVING IN THE CYTOSOL

Theileria and *Babesia* reside directly in the host cell cytoplasm. Although both are internalized in a vacuole during invasion, the membrane surrounding *Babesia* is degraded shortly after parasite entry, leaving the parasites free access to host cell nutrients. The mechanism responsible for the exit from the vacuole is not understood.

T. parva: Transformation of the Host Cell

T. parva primarily invades metabolically fully active lymphocytes and has opted for a cytosolic niche that provides direct access to nutrients. Following dissolution of the PV, the parasite associates with host cell microtubules and induces a rapid and uncontrolled proliferation of the host cell, leading to clonal expansion of the transformed cell (29). This process is driven by the parasite, which creates a signaling platform on its surface called the IKK (IkappaB kinase) signalosome (45). The parasite assures that its own cell division is concomitant with host chromosome segregation and cell division to secure its proper transmission in the host daughter cells (**Figure 3d** and **Figure 4d**). Two parasite

proteins containing AT-hook motifs translocate to the host cell nucleus and could contribute to host cell transformations. TashATs might induce transcription changes that induce proliferation, whereas SuAT is involved in the modulation of the host cell cytoskeleton (89). Unlike other apicomplexans that rupture the host cells to spread to neighboring cells, *Theileria* spp. induce lymphocyte transformation and spread to daughter cells simply by association with the mitotic apparatus and even distribution in the daughter cells. The parasite protein TaSE appears to play a role in this process (73). The ability of this parasite to induce clonal expansion of its host cells is linked to the activation of NF-κB by the parasite (99). *Theileria* spp. are auxotrophs for all amino acids and cannot synthesize fatty acids; however, little is known about how the parasite acquires nutrients from its host (60, 65).

MODULATION OF THE IMMUNE RESPONSE AND APOPTOSIS PATHWAYS

A major challenge for intracellular pathogens is to manipulate the host signaling pathways to hide from or neutralize the effector cells of the immune system and control host cell apoptosis.

Dismiss the Defenses of Infected Nucleated Cells

In recent years, microarray technology has emerged as a powerful tool to investigate host-pathogen interactions. Several studies on apicomplexan infected cells showed that the parasite radically modifies host gene expression. Markedly, genes involved in MHC presentation, cytokines expression, and apoptotic pathways are modulated upon *C. parvum* (24) and *T. gondii* (4, 50) infections. Parasite effector molecules released into the host cell have recently been identified. Rhoptry proteins such as kinases, TgROP18 (31, 90), TgROP16, and phosphatase 2C access the host cytosol or nucleus and act as virulence factors in toxoplasmosis (71). TgROP16 interferes with the STAT3/6 pathway, leading to the downregulation of IL-12 (72). In *P. falciparum*, the circumsporozoite protein is exported to the hepatocyte cytoplasm, where it appears to outcompete NF-κB for entry into the nucleus, thus lowering the inflammatory response of these cells (82).

T. gondii has evolved multiple strategies to modulate and interfere with programmed cell death (9), including interference with mitochondrial pathways (8) as well as receptor-mediated (94) pathways. Other antiapoptotic strategies, such as the exploitation of Gi-protein-dependent PI3K signaling, have also been reported (49).

T. parva inhibits apoptosis by activation of the NF-κB pathway (58) and induces additional survival pathways, all of which converge in expression of the c-Myc oncogene (28). In hepatocytes infected with the malaria parasite, inhibition of apoptosis is mediated through the HGF/Met and PI3K signaling pathways (101) during early stages of infection. In later stages of infection, apoptosis is induced in the liver cells, but parasites block the activation of apoptotic signals, avoiding destruction of the infected cell and allowing release of parasite-filled vesicles (merosomes) into the bloodstream (86).

Subversion of Immune Effector Cells and Dissemination

T. gondii can survive and proliferate in macrophages by using several strategies, including inhibition of IL-12 and TNF-α, suppression of nitric oxide production, downregulation of MHC class II molecules, and activation of NF-κB (25). Furthermore, this parasite subverts the regulation of dendritic cell motility and uses this specific type of host cell as a Trojan horse to spread through the organism (55).

A characteristic of *T. gondii* infection is a fine-tuned activation/inhibition of the immune response designed to ensure host survival and chronic infection for transmission. These complex mechanisms of regulation have been recently reviewed for *T. gondii* (56) and for *P. falciparum* (15, 40).

SUMMARY POINTS

1. Many apicomplexans traverse biological barriers and actively penetrate host cells using a unique form of gliding motility. This active mode of penetration offers the opportunity to occupy unique privileged niches such as the erythrocytes in the case of malaria infection by *Plasmodium*. Active entry also leads to the formation of a unique nonphagosomal parasitophorous vacuole, which is segregated from the host cellular trafficking pathways, enabling *T. gondii* to survive and establish infection in virtually all cell types including macrophages. While the machinery governing host cell invasion has been well documented, less is known regarding the establishment of the vacuole, the host cell modifications leading to the acquisition of nutrients, and the host factors contributing to parasite replication and maintenance of infection.

2. Recent studies are only uncovering the tip of the iceberg, with the identification of exquisite strategies adopted by the apicomplexans to subvert the host cell processes to tailor their needs. *Theileria* dramatically interfere with the signaling cascades and induce transformation of the host cell, hence assuring their indefinite propagation. *P. falciparum* develops a complex tubular network to export proteins on the surface-infected erythrocyte, hence avoiding splenic clearance. *T. gondii* recruits host cell organelles and remodels lipid trafficking pathways to divert host preformed molecules, hence assuring intracellular survival and proliferation.

DISCLOSURE STATEMENT

The authors are not aware of any biases that might be perceived as affecting the objectivity of this review.

ACKNOWLEDGMENTS

FP is supported by the Swiss National Foundation FN 3100A0-102255/DS and currently 3100A0-116722/DS. It is part of the activities of the BioMalPar European Network of Excellence supported by a European grant (LSHP-CT-2004-503578) from the Priority 1 "Life Sciences, Genomics and Biotechnology for Health" in the sixth Framework Program and DS is an international scholar of the Howard Hughes Medical Institutes. We are grateful to Andrea Valigurová, Jean Francois Dubremetz, and Dirk Dobbelaere for kindly providing illustrations.

LITERATURE CITED

1. **Abrahamsen MS, Templeton TJ, Enomoto S, Abrahante JE, Zhu G, et al. 2004. Complete genome sequence of the apicomplexan, *Cryptosporidium parvum*. *Science* 304:441–45**

2. Baum J, Papenfuss AT, Baum B, Speed TP, Cowman AF. 2006. Regulation of apicomplexan actin-based motility. *Nat. Rev. Microbiol.* 4:621–28

3. Beyer TV, Svezhova NV, Radchenko AI, Sidorenko NV. 2002. Parasitophorous vacuole: morphofunctional diversity in different coccidian genera (a short insight into the problem). *Cell Biol. Int.* 26:861–71

4. Blader IJ, Manger ID, Boothroyd JC. 2001. Microarray analysis reveals previously unknown changes in *Toxoplasma gondii*-infected human cells. *J. Biol. Chem.* 276:24223–31

5. Bonnin A, Lapillonne A, Petrella T, Lopez J, Chaponnier C, et al. 1999. Immunodetection of the microvillous cytoskeleton molecules villin and ezrin in the parasitophorous vacuole wall of *Cryptosporidium parvum* (Protozoa: Apicomplexa). *Eur. J. Cell Biol.* 78:794–801

6. Bradley PJ, Sibley LD. 2007. Rhoptries: an arsenal of secreted virulence factors. *Curr. Opin. Microbiol.* 10:582–87

7. Breman JG. 2001. The ears of the hippopotamus: manifestations, determinants, and estimates of the malaria burden. *Am. J. Trop. Med. Hyg.* 64:1–11

8. Carmen JC, Hardi L, Sinai AP. 2006. *Toxoplasma gondii* inhibits UV light-induced apoptosis through multiple interactions with the mitochondrion-dependent programmed cell death pathway. *Cell Microbiol.* 8:301–15

9. **Carmen JC, Sinai AP. 2007. Suicide prevention: disruption of apoptotic pathways by protozoan parasites. *Mol. Microbiol.* 4:904–16**

10. Carruthers VB, Sibley LD. 1997. Sequential protein secretion from three distinct organelles of *Toxoplasma gondii* accompanies invasion of human fibroblasts. *Eur. J. Cell Biol.* 73:114–23

11. Chen XM, Huang BQ, Splinter PL, Cao H, Zhu G, et al. 2003. *Cryptosporidium parvum* invasion of biliary epithelia requires host cell tyrosine phosphorylation of cortactin via c-Src. *Gastroenterology* 125:216–28

12. Chen XM, Huang BQ, Splinter PL, Orth JD, Billadeau DD, et al. 2004. Cdc42 and the actin-related protein/neural Wiskott-Aldrich syndrome protein network mediate cellular invasion by *Cryptosporidium parvum*. *Infect. Immun.* 72:3011–21

13. Chen XM, O'Hara SP, Huang BQ, Nelson JB, Lin JJ, et al. 2004. Apical organelle discharge by *Cryptosporidium parvum* is temperature, cytoskeleton, and intracellular calcium dependent and required for host cell invasion. *Infect. Immun.* 72:6806–16

14. Clough B, Atilola FA, Pasvoi G. 1998. The role of rosetting in the multiplication of *Plasmodium falciparum*: rosette formation neither enhances nor targets parasite invasion into uninfected red cells. *Br. J. Haematol.* 100:99–104

15. Coban C, Ishii KJ, Horii T, Akira S. 2007. Manipulation of host innate immune responses by the malaria parasite. *Trends Microbiol.* 15:271–78

1. The autotrophies and auxotrophies, with highlights on the streamlined metabolic pathways.

9. A comprehensive review of the different steps in the intrinsic and extrinsic apoptotic pathways disrupted by as yet unidentified parasite effectors.

16. Cooke BM, Lingelbach K, Bannister LH, Tilley L. 2004. Protein trafficking in *Plasmodium falciparum*-infected red blood cells. *Trends Parasitol.* 20:581–89

17. Coppens I, Dunn JD, Romano JD, Pypaert M, Zhang H, et al. 2006. *Toxoplasma gondii* sequesters lysosomes from mammalian hosts in the vacuolar space. *Cell* 125:261–74

18. Coppens I, Sinai AP, Joiner KA. 2000. *Toxoplasma gondii* exploits host low-density lipoprotein receptor-mediated endocytosis for cholesterol acquisition. *J. Cell Biol.* 149:167–80

19. Corso PS, Kramer MH, Blair KA, Addiss DG, Davis JP, Haddix AC. 2003. Cost of illness in the 1993 waterborne *Cryptosporidium* outbreak, Milwaukee, Wisconsin. *Emerg. Infect. Dis.* 9:426–31

20. Crawford MJ, Thomsen-Zieger N, Ray M, Schachtner J, Roos DS, Seeber F. 2006. *Toxoplasma gondii* scavenges host-derived lipoic acid despite its de novo synthesis in the apicoplast. *EMBO J.* 25:3214–22

21. Current WL, Reese NC. 1986. A comparison of endogenous development of three isolates of *Cryptosporidium* in suckling mice. *J. Protozool.* 33:98–108

22. de Melo EJ, de Carvalho TU, de Souza W. 1992. Penetration of *Toxoplasma gondii* into host cells induces changes in the distribution of the mitochondria and the endoplasmic reticulum. *Cell Struct. Funct.* 17:311–17

23. Decherf G, Egee S, Staines HM, Ellory JC, Thomas SL. 2004. Anionic channels in malaria-infected human red blood cells. *Blood Cells Mol. Dis.* 32:366–71

24. Deng M, Lancto CA, Abrahamsen MS. 2004. *Cryptosporidium parvum* regulation of human epithelial cell gene expression. *Int. J. Parasitol.* 34:73–82

25. Denkers EY, Butcher BA. 2005. Sabotage and exploitation in macrophages parasitized by intracellular protozoans. *Trends Parasitol.* 21:35–41

26. Desai SA, Bezrukov SM, Zimmerberg J. 2000. A voltage-dependent channel involved in nutrient uptake by red blood cells infected with the malaria parasite. *Nature* 406:1001–5

27. Desai SA, Rosenberg RL. 1997. Pore size of the malaria parasite's nutrient channel. *Proc. Natl. Acad. Sci. USA* 94:2045–49

28. Dessauge F, Lizundia R, Baumgartner M, Chaussepied M, Langsley G. 2005. Taking the Myc is bad for *Theileria*. *Trends Parasitol.* 21:377–85

29. Dobbelaere DA, Kuenzi P. 2004. The strategies of the *Theileria* parasite: a new twist in host-pathogen interactions. *Curr. Opin. Immunol.* 16:524–30

30. Dziadek B, Dziadek J, Dlugonska H. 2007. Identification of *Toxoplasma gondii* proteins binding human lactoferrin: a new aspect of rhoptry proteins function. *Exp. Parasitol.* 115:277–82

31. El Hajj H, Lebrun M, Arold ST, Vial H, Labesse G, Dubremetz JF. 2007. ROP18 is a rhoptry kinase controlling the intracellular proliferation of *Toxoplasma gondii*. *PLoS Pathog.* 3:e14

32. Elliott DA, Clark DP. 2000. *Cryptosporidium parvum* induces host cell actin accumulation at the host-parasite interface. *Infect. Immun.* 68:2315–22

33. Elliott DA, Coleman DJ, Lane MA, May RC, Machesky LM, Clark DP. 2001. *Cryptosporidium parvum* infection requires host cell actin polymerization. *Infect. Immun.* 69:5940–42

34. Endo T, Pelster B, Piekarski G. 1981. Infection of murine peritoneal macrophages with *Toxoplasma gondii* exposed to UV light. *Z. Parasitenkd.* 65:121–29

35. Forney JR, DeWald DB, Yang S, Speer CA, Healey MC. 1999. A role for host phosphoinositide 3-kinase and cytoskeletal remodeling during *Cryptosporidium parvum* infection. *Infect. Immun.* 67:844–52

36. Fox BA, Bzik DJ. 2002. De novo pyrimidine biosynthesis is required for virulence of *Toxoplasma gondii*. *Nature* 415:926–29

37. Fox BA, Gigley JP, Bzik DJ. 2004. *Toxoplasma gondii* lacks the enzymes required for de novo arginine biosynthesis and arginine starvation triggers cyst formation. *Int. J. Parasitol.* 34:323–31

38. Gao LY, Kwaik YA. 2000. The modulation of host cell apoptosis by intracellular bacterial pathogens. *Trends Microbiol.* 8:306–13

39. Gero AM, O'Sullivan WJ. 1990. Purines and pyrimidines in malarial parasites. *Blood Cells* 16:467–84

40. Gowda DC. 2007. TLR-mediated cell signaling by malaria GPIs. *Trends Parasitol.* 23:596–604

41. Gupta N, Zahn MM, Coppens I, Joiner KA, Voelker DR. 2005. Selective disruption of phosphatidyl-choline metabolism of the intracellular parasite *Toxoplasma gondii* arrests its growth. *J. Biol. Chem.* 280:16345–53

42. Haeggstrom M, Kironde F, Berzins K, Chen Q, Wahlgren M, Fernandez V. 2004. Common trafficking pathway for variant antigens destined for the surface of the *Plasmodium falciparum*-infected erythrocyte. *Mol. Biochem. Parasitol.* 133:1–14

43. Hakansson S, Charron AJ, Sibley LD. 2001. *Toxoplasma* evacuoles: a two-step process of secretion and fusion forms the parasitophorous vacuole. *EMBO J.* 20:3132–44

44. Halonen SK, Weidner E. 1994. Overcoating of *Toxoplasma* parasitophorous vacuoles with host cell vimentin type intermediate filaments. *J. Eukaryot. Microbiol.* 1:65–71

45. Heussler VT, Rottenberg S, Schwab R, Kuenzi P, Fernandez PC, et al. 2002. Hijacking of host cell IKK signalosomes by the transforming parasite *Theileria*. *Science* 298:1033–36

46. Jan G, Delorme V, David V, Revenu C, Rebollo A, et al. 2007. The toxofilin-actin-PP2C complex of *Toxoplasma*: identification of interacting domains. *Biochem. J.* 401:711–19

47. Joiner KA, Fuhrman SA, Miettinen HM, Kasper LH, Mellman I. 1990. *Toxoplasma gondii*: fusion competence of parasitophorous vacuoles in Fc receptor-transfected fibroblasts. *Science* 249:641–46

48. Keeley A, Soldati D. 2004. The glideosome: a molecular machine powering motility and host-cell invasion by Apicomplexa. *Trends Cell Biol.* 14:528–32

49. Kim L, Denkers EY. 2006. *Toxoplasma gondii* triggers Gi-dependent PI 3-kinase signaling required for inhibition of host cell apoptosis. *J. Cell Sci.* 119:2119–26

50. Kim SK, Fouts AE, Boothroyd JC. 2007. *Toxoplasma gondii* dysregulates IFN-gamma-inducible gene expression in human fibroblasts: insights from a genome-wide transcriptional profiling. *J. Immunol.* 178:5154–65

51. Kooij TW, Janse CJ, Waters AP. 2006. *Plasmodium* postgenomics: better the bug you know? *Nat. Rev. Microbiol.* 4:344–57

52. Kravetz JD, Federman DG. 2005. Toxoplasmosis in pregnancy. *Am. J. Med.* 118:212–16

53. Krug EC, Marr JJ, Berens RL. 1989. Purine metabolism in *Toxoplasma gondii*. *J. Biol. Chem.* 264:10601–7

54. Lacharme L, Villar V, Rojo-Vazquez FA, Suarez S. 2004. Complete development of *Cryptosporidium parvum* in rabbit chondrocytes (VELI cells). *Microbes. Infect.* 6:566–71

55. Lambert H, Hitziger N, Dellacasa I, Svensson M, Barragan A. 2006. Induction of dendritic cell migration upon *Toxoplasma gondii* infection potentiates parasite dissemination. *Cell Microbiol.* 8:1611–23

56. Lang C, Gross U, Luder CG. 2007. Subversion of innate and adaptive immune responses by *Toxoplasma gondii*. *Parasitol. Res.* 100:191–203

57. Lauer SA, Rathod PK, Ghori N, Haldar K. 1997. A membrane network for nutrient import in red cells infected with the malaria parasite. *Science* 276:1122–25

58. Machado J Jr, Fernandez PC, Baumann I, Dobbelaere DA. 2000. Characterisation of NF-kappa B complexes in *Theileria parva*–transformed T cells. *Microbes Infect.* 2:1311–20

59. Martin RE, Kirk K. 2007. Transport of the essential nutrient isoleucine in human erythrocytes infected with the malaria parasite *Plasmodium falciparum*. *Blood* 109:2217–24

60. Mazumdar J, Striepen B. 2007. Make it or take it: fatty acid metabolism of apicomplexan parasites. *Eukaryot. Cell* 6:1727–35

61. Mercier C, Adjogble KD, Daubener W, Delauw MF. 2005. Dense granules: Are they key organelles to help understand the parasitophorous vacuole of all apicomplexa parasites? *Int. J. Parasitol.* 35:829–49

62. Michel R, Schupp K, Raether W, Bierther FW. 1980. Formation of a close junction during invasion of erythrocytes by *Toxoplasma gondii* in vitro. *Int. J. Parasitol.* 10:309–13

63. Mordue DG, Desai N, Dustin M, Sibley LD. 1999. Invasion by *Toxoplasma gondii* establishes a moving junction that selectively excludes host cell plasma membrane proteins on the basis of their membrane anchoring. *J. Exp. Med.* 190:1783–92

64. Mordue DG, Sibley LD. 1997. Intracellular fate of vacuoles containing *Toxoplasma gondii* is determined at the time of formation and depends on the mechanism of entry. *J. Immunol.* 159:4452–59

65. Pain A, Renauld H, Berriman M, Murphy L, Yeats CA, et al. 2005. Genome of the host-cell transforming parasite *Theileria annulata* compared with *T. parva*. *Science* 309:131–33

66. Pfefferkorn ER, Eckel M, Rebhun S. 1986. Interferon-gamma suppresses the growth of *Toxoplasma gondii* in human fibroblasts through starvation for tryptophan. *Mol. Biochem. Parasitol.* 20:215–24

45. NFκB is constitutively activated in *Theileria*-transformed cells and this activation is mediated by a unique recruitment of the IKK signalosomes on the parasite surface.

55. Shows that *T. gondii* subverts dendritic cell motility and likely exploits the host's natural pathways of cellular migration for parasite dissemination.

63. Demonstrates that host transmembrane proteins are excluded from the forming PVM during invasion, resulting in a nonfusogenic compartment.

67. Roos DS, Donald RG, Morrissette NS, Moulton AL. 1994. Molecular tools for genetic dissection of the protozoan parasite *Toxoplasma gondii*. *Methods Cell Biol.* 45:27–63

68. Roulston A, Marcellus RC, Branton PE. 1999. Viruses and apoptosis. *Annu. Rev. Microbiol.* 53:577–628

69. Rowe JA, Obiero J, Marsh K, Raza A. 2002. Short report: positive correlation between rosetting and parasitemia in *Plasmodium falciparum* clinical isolates. *Am. J. Trop. Med. Hyg.* 66:458–60

70. Roy CR, Mocarski ES. 2007. Pathogen subversion of cell-intrinsic innate immunity. *Nat. Immunol.* 8:1179–87

71. Saeij JP, Boyle JP, Coller S, Taylor S, Sibley LD, et al. 2006. Polymorphic secreted kinases are key virulence factors in toxoplasmosis. *Science* 314:1780–83

72. Saeij JP, Coller S, Boyle JP, Jerome ME, White MW, Boothroyd JC. 2007. *Toxoplasma* co-opts host gene expression by injection of a polymorphic kinase homologue. *Nature* 445:324–27

73. Schneider I, Haller D, Kullmann B, Beyer D, Ahmed JS, Seitzer U. 2007. Identification, molecular characterization and subcellular localization of a *Theileria annulata* parasite protein secreted into the host cell cytoplasm. *Parasitol. Res.* 101:1471–82

74. Schwab JC, Beckers CJ, Joiner KA. 1994. The parasitophorous vacuole membrane surrounding intracellular *Toxoplasma gondii* functions as a molecular sieve. *Proc. Natl. Acad. Sci. USA* 91:509–13

75. Schwartzman JD, Pfefferkorn ER. 1982. *Toxoplasma gondii*: purine synthesis and salvage in mutant host cells and parasites. *Exp. Parasitol.* 53:77–86

76. Shaw MK. 2003. Cell invasion by *Theileria* sporozoites. *Trends Parasitol.* 19:2–6

77. Shaw MK, Tilney LG, Musoke AJ, Teale AJ. 1995. MHC class I molecules are an essential cell surface component involved in *Theileria parva* sporozoite binding to bovine lymphocytes. *J. Cell Sci.* 108(Pt. 4):1587–96

78. Sibley LD, Niesman IR, Parmley SF, Cesbron-Delauw MF. 1995. Regulated secretion of multi-lamellar vesicles leads to formation of a tubulo-vesicular network in host-cell vacuoles occupied by *Toxoplasma gondii*. *J. Cell Sci.* 108(Pt. 4):1669–77

79. Sibley LD, Weidner E, Krahenbuhl JL. 1985. Phagosome acidification blocked by intracellular *Toxoplasma gondii*. *Nature* 315:416–19

80. Sinai AP, Joiner KA. 2001. The *Toxoplasma gondii* protein ROP2 mediates host organelle association with the parasitophorous vacuole membrane. *J. Cell Biol.* 154:95–108

81. Sinai AP, Webster P, Joiner KA. 1997. Association of host cell endoplasmic reticulum and mitochondria with the *Toxoplasma gondii* parasitophorous vacuole membrane: a high affinity interaction. *J. Cell Sci.* 110(Pt. 17):2117–28

82. Singh AP, Buscaglia CA, Wang Q, Levay A, Nussenzweig DR, et al. 2007. *Plasmodium* circumsporozoite protein promotes the development of the liver stages of the parasite. *Cell* 131:492–504

83. Snow RW, Guerra CA, Noor AM, Myint HY, Hay SI. 2005. The global distribution of clinical episodes of *Plasmodium falciparum* malaria. *Nature* 434:214–17

84. Spycher C, Rug M, Klonis N, Ferguson DJ, Cowman AF, et al. 2006. Genesis of and trafficking to the Maurer's clefts of *Plasmodium falciparum*-infected erythrocytes. *Mol. Cell Biol.* 26:4074–85

85. Striepen B, Pruijssers AJ, Huang J, Li C, Gubbels MJ, et al. 2004. Gene transfer in the evolution of parasite nucleotide biosynthesis. *Proc. Natl. Acad. Sci. USA* 101:3154–59

86. Sturm A, Amino R, van de Sand C, Regen T, Retzlaff S, et al. 2006. Manipulation of host hepatocytes by the malaria parasite for delivery into liver sinusoids. *Science* 313:1287–90

87. Su C, Evans D, Cole RH, Kissinger JC, Ajioka JW, Sibley LD. 2003. Recent expansion of *Toxoplasma* through enhanced oral transmission. *Science* 299:414–16

88. Suss-Toby E, Zimmerberg J, Ward GE. 1996. *Toxoplasma* invasion: The parasitophorous vacuole is formed from host cell plasma membrane and pinches off via a fission pore. *Proc. Natl. Acad. Sci. USA* 93:8413–18

89. Swan DG, Phillips K, Tait A, Shiels BR. 1999. Evidence for localisation of a *Theileria* parasite AT hook DNA-binding protein to the nucleus of immortalised bovine host cells. *Mol. Biochem. Parasitol.* 101:117–29

90. Taylor S, Barragan A, Su C, Fux B, Fentress SJ, et al. 2006. A secreted serine-threonine kinase determines virulence in the eukaryotic pathogen *Toxoplasma gondii*. *Science* 314:1776–80

72. The first identification of a parasite effector molecule that reprograms gene expression of the host cell.

91. Tenter AM, Heckeroth AR, Weiss LM. 2000. *Toxoplasma gondii*: from animals to humans. *Int. J. Parasitol.* 30:1217–58

92. Tilley L, Sougrat R, Lithgow T, Hanssen E. 2008. The twists and turns of Maurer's cleft trafficking in *P. falciparum*-infected erythrocytes. *Traffic* 9:187–97

93. Travier L, Mondragon R, Dubremetz JF, Musset K, Mondragon M, et al. 2008. Functional domains of the *Toxoplasma* GRA2 protein in the formation of the membranous nanotubular network of the parasitophorous vacuole. *Int. J. Parasitol.* 7:757–73

94. Vutova P, Wirth M, Hippe D, Gross U, Schulze-Osthoff K, et al. 2007. *Toxoplasma gondii* inhibits Fas/CD95-triggered cell death by inducing aberrant processing and degradation of caspase 8. *Cell Microbiol.* 9:1556–70

95. Ward GE, Miller LH, Dvorak JA. 1993. The origin of parasitophorous vacuole membrane lipids in malaria-infected erythrocytes. *J. Cell Sci.* 106(Pt. 1):237–48

96. Webster P, Dobbelaere DA, Fawcett DW. 1985. The entry of sporozoites of *Theileria parva* into bovine lymphocytes in vitro. Immunoelectron microscopic observations. *Eur. J. Cell Biol.* 36:157–62

97. Wetzel DM, Schmidt J, Kuhlenschmidt MS, Dubey JP, Sibley LD. 2005. Gliding motility leads to active cellular invasion by *Cryptosporidium parvum* sporozoites. *Infect. Immun.* 73:5379–87

98. Williams GT. 1994. Programmed cell death: a fundamental protective response to pathogens. *Trends Microbiol.* 2:463–64

99. Williams RO, Dobbelaere DA. 1993. The molecular basis of transformation of lymphocytes by *Theileria parva* infection. *Semin. Cell Biol.* 4:363–71

100. Wong SY, Remington JS. 1994. Toxoplasmosis in pregnancy. *Clin. Infect. Dis.* 18:853–61

101. Xiao GH, Jeffers M, Bellacosa A, Mitsuuchi Y, Van der Woude GF, Testa JR. 2001. Anti-apoptotic signaling by hepatocyte growth factor/Met via the phosphatidylinositol 3-kinase/Akt and mitogen-activated protein kinase pathways. *Proc. Natl. Acad. Sci. USA* 98:247–52

102. Zeng B, Zhu G. 2006. Two distinct oxysterol binding protein-related proteins in the parasitic protist *Cryptosporidium parvum* (Apicomplexa). *Biochem. Biophys. Res. Commun.* 346:591–99

Cumulative Indexes

Contributing Authors, Volumes 58–62

Eraso JM, 61:283–307
Ehrlich LS, 62:425–43

F

Falkovitz L, 58:143–59
Falkow S, 62:1–18
Filter JJ, 62:211–33
Fink DJ, 58:253–71
Finking R, 58:453–88
Fisher EJ, 60:425–49
Fry RC, 59:357–77
Fuerst JA, 59:299–328
Fuqua C, 61:401–21

G

Georgiou G, 60:373–95
Glorioso JC, 58:253–71
Golding BT, 60:27–49
Goldman WE, 60:281–303
Gottesman S, 58:303–28;
   59:379–405
Gralnick JA, 61:237–58
Grant SR, 60:425–49
Graumann PL, 61:589–618
Green R, 62:353–73

H

Haddad A, 59:91–111
Hand NJ, 59:91–111
Handelsman J, 61:23–34;
   62:375–401
Harris E, 62:71–92
Harrison MJ, 59:19–42
Harshey RM, 61:309–29
Hartmann E, 59:91–111
Hau HH, 61:237–58
Hecker M, 61:215–36
Heitman J, 60:69–105
Henderson IR, 62:153–69
Henriques AO, 61:555–88
Hinnebusch AG, 59:407–50
Holmes EC, 62:307–28
Horn M, 62:113–31
Horn MA, 61:169–89
Houben E, 59:329–355
Huynen MA, 59:191–209

I

Ito K, 59:211–31
Iyer LM, 61:453–75

J

Jacobs WR Jr, 61:35–50
Jakubowski S, 59:451–85
Janniere L, 61:309–29
Jetten MSM, 58:99–117
Jiménez-Sánchez A, 61:309–29
Jin DJ, 61:309–29
Johnson AD, 59:233–55

K

Kaguni JM, 60:351–71
Kaiser D, 58:75–98; 60:1–25
Kaplan S, 61:283–307
Kassen R, 58:207–31
Kawaoka Y, 59:553–86
Keiler KC, 62:133–51
Kew OM, 59:587–635
Klotz MG, 61:503–28
Kofoid E, 60:477–501
Kreuzer KN, 59:43–67
Krishnamoorthy V, 59:451–85
Krumholz LR, 60:149–66
Krystofova S, 61:423–52
Kugelberg E, 60:477–501
Kuipers OP, 62:193–210
Kumamoto CA, 59:113–33
Kyle JL, 62:71–92

L

Lakin-Thomas PL, 58:489–519
Lamed R, 58:521–54
Larrainzar E, 59:257–77
Law CJ, 62:289–305
Lee PA, 60:373–95
Leigh JA, 61:349–77
Levin PA, 61:309–29
Li F, 60:503–31
Li L, 61:423–52
Lin X, 60:69–105
Liti G, 59:135–53
Little AEF, 62:375–401
Loeb LA, 58:183–205
Lopez-Rubio JJ, 62:445–70

Louis EJ, 59:135–53
Ludden PW, 62:93–111
Luirink J, 59:329–55
Lynch M, 60:327–49

M

Mackenzie C, 61:283–307
Maguire ME, 60:187–209
Maiden MCJ, 60:561–88
Majdalani N, 59:379–405
Maloney PC, 62:289–305
Marahiel MA, 58:453–88
Martin B, 60:451–75
McDonald ME, 62:353–73
McLendon MK, 60:167–85
Metcalf WW, 61:379–400
Michiels T, 59:279–98
Mileykovskaya E, 61:309–29
Minsky A, 61:309–29
Misevic G, 61:309–29
Mole BM, 60:425–49
Moran CP Jr, 61:555–88
Morrissey JP, 59:257–77

N

Naffakh N, 62:403–24
Neidle EL, 58:119–42
Nohmi T, 60:231–53
Norris V, 61:309–29
Nyström T, 58:161–81

O

O'Gara F, 59:257–77
Ohlendorf DH, 58:555–85
Oremland RS, 60:107–30
Ornston LN, 59:519–51
Ortín J, 60:305–26

P

Pace NR, 61:331–47
Pallansch MA, 59:587–635
Pané-Farré J, 61:215–36
Papp-Wallace KM, 60:187–209
Park G, 61:423–52
Parke D, 59:519–51
Parra F, 60:305–26

Parrish CR, 59:553–86
Peñalva MA, 58:425–51
Peterson SB, 62:375–401
Pitcher RS, 61:259–82
Pizarro-Cerda J, 58:587–610
Plattner F, 62:471–87
Pohlschröder M, 59:91–111
Potrykus K, 62:35–51
Pratt JT, 61:131–48
Prudhomme M, 60:451–75
Puskás Á, 61:283–307

R

Raffa KF, 62:375–401
Rainey PB, 58:207–31
Rameix-Welti M-A, 62:403–24
Rappleye CA, 60:281–303
Reams AB, 58:119–42;
    60:477–501
Ripoll C, 61:309–29
Riviere L, 62:445–70
Robert V, 61:191–214
Roberts JW, 62:211–33
Robinson CJ, 62:375–401
Roh JH, 61:283–307
Rosenberg E, 58:143–59
Roth JR, 60:477–501
Rubio LM, 62:93–111

S

Sabatini R, 62:235–51
Saier M Jr, 61:309–29

Salas M, 61:1–22
Samson LD, 59:357–77
Santini JM, 60:107–30
Sariaslani FS, 61:51–69
Scherf A, 62:445–70
Schloss PD, 61:23–34
Schneemann A, 60:51–67
Scott JR, 60:397–423
Shankar S, 62:211–33
Shoham Y, 58:521–54
Shoukry NH, 58:391–424
Sibley LD, 62:329–51
Skarstad K, 61:309–29
Smidt H, 58:43–73
Smits WK, 62:193–210
Snel B, 59:191–209
Soldati-Favre D, 62:471–87
Sola I, 60:211–30
Stolz JF, 60:107–30
Strous M, 58:99–117
Sutter RW, 59:587–635

T

Tamayo R, 61:131–48
Thellier M, 61:309–29
Tommassen J, 61:191–214
Tomoiu A, 62:403–24
Tullman-Ercek D, 60:373–95

V

van der Donk WA, 61:477–501
van der Werf S, 62:403–24

van der Woude MW, 62:153–69
Veening J-W, 62:193–210
Vetting MW, 58:555–85
Viboud GI, 59:69–89
Vilchèze C, 61:35–50
Vinces MD, 59:113–33
Völker U, 61:215–36
von Heijne G, 59:329–55

W

Wagner-Döbler I, 60:255–80
Walker C, 58:391–424
Walker JJ, 61:331–47
Wall JD, 60:149–66
Wandersman C, 58:611–47
Wang D-N, 62:289–305
White AK, 61:379–400
Whiteway M, 61:529–53
Wileman T, 61:149–67
Willey JM, 61:477–501
Winkler WC, 59:487–517
Wright SJ, 61:423–52

Y

Yonath A, 58:233–51
Young DM, 59:519–51
Youngman EM, 62:353–73

Z

Zeng X, 61:283–307
Zuñiga S, 60:211–30

WITHDRAWN

ANNUAL REVIEWS
Intelligent Synthesis of the Scientific Literature

Annual Reviews – Your Starting Point for Research Online
http://arjournals.annualreviews.org

- Over 1150 Annual Reviews volumes—more than 26,000 critical, authoritative review articles in 35 disciplines spanning the Biomedical, Physical, and Social sciences—available online, including all Annual Reviews back volumes, dating to 1932

- Current individual subscriptions include seamless online access to full-text articles, PDFs, Reviews in Advance (as much as 6 months ahead of print publication), bibliographies, and other supplementary material in the current volume and the prior 4 years' volumes

- All articles are fully supplemented, searchable, and downloadable—see http://micro.annualreviews.org

- Access links to the reviewed references (when available online)

- Site features include customized alerting services, citation tracking, and saved searches

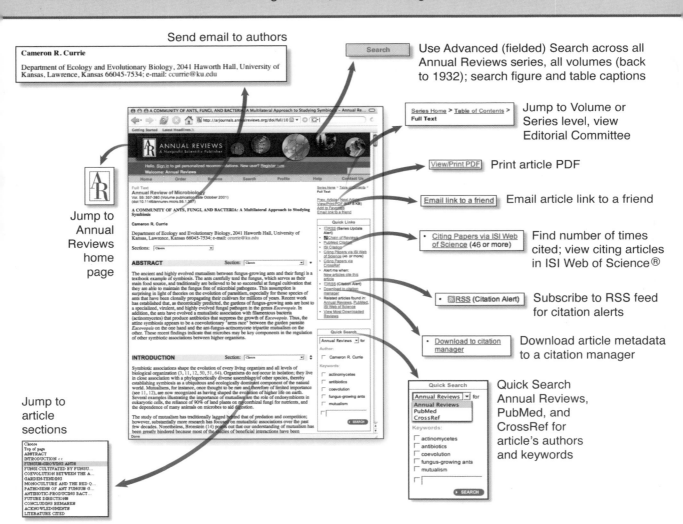

Send email to authors

Use Advanced (fielded) Search across all Annual Reviews series, all volumes (back to 1932); search figure and table captions

Jump to Volume or Series level, view Editorial Committee

Print article PDF

Email article link to a friend

Find number of times cited; view citing articles in ISI Web of Science®

Subscribe to RSS feed for citation alerts

Download article metadata to a citation manager

Quick Search Annual Reviews, PubMed, and CrossRef for article's authors and keywords

Jump to Annual Reviews home page

Jump to article sections